Volume I

Selected Statistical Papers of Sir David Cox

Design of Investigations, Statistical Methods and Applications

EDITED BY

D. J. Hand
Imperial College, London

A. M. Herzberg
Queen's University, Kingston

CAMBRIDGE
UNIVERSITY PRESS

CAMBRIDGE
UNIVERSITY PRESS

University Printing House, Cambridge CB2 8BS, United Kingdom

One Liberty Plaza, 20th Floor, New York, NY 10006, USA

477 Williamstown Road, Port Melbourne, VIC 3207, Australia

314-321, 3rd Floor, Plot 3, Splendor Forum, Jasola District Centre, New Delhi - 110025, India

79 Anson Road, #06-04/06, Singapore 079906

Cambridge University Press is part of the University of Cambridge.

It furthers the University's mission by disseminating knowledge in the pursuit of education, learning and research at the highest international levels of excellence.

www.cambridge.org
Information on this title: www.cambridge.org/9780521849395

First published 2005

A catalogue record for this publication is available from the British Library

ISBN 978-0-521-84939-5 Hardback
Volume 2 ISBN 9780521849401
2 volume set ISBN 9780521858168

Contents

Contents of Volume II

Foreword

Sir Ronald Fisher wrote in his preface to *Contributions to Mathematical Statistics*, a collection of his papers:

> In mathematics, more than in other disciplines, original sources often tend to be neglected, and sometimes are scarcely accessible to students. In more leisured times the best method of making original papers accessible has been the voluminous 'collected works' often published by learned societies in memory of their distinguished members. There is, I believe, a great deal to be said for the suggestion . . . that editorial notes made by the author in his lifetime might do much to aid the student, especially in understanding the original purpose in relation to the common knowledge, or common mis-apprehensions, current at the time the paper was written.

Although times have changed and some literature is more easily accessible than previously, such collections of papers where the author himself gives a critical assessment of his papers are rare. It was in this vein that we approached Professor Sir David Cox to make a self-assessment of his papers. He happily accepted our invitation.

Professor Cox is one of the seminal statistical thinkers of the twentieth and twenty-first centuries. It is important that his insights and recollections on the statistical literature and applications are known and that all this be recorded. The papers reprinted here were chosen by Professor Cox and the editors. A complete bibliography with annotations is given at the end of Volume II.

We take pleasure in presenting these volumes.

D. J. Hand
A. M. Herzberg

Preface

I am very grateful to my friends Agnes Herzberg and David Hand for suggesting this collection of papers and for all the work they have done in bringing the idea to fruition. They suggested that not only should there be a general preface but that each paper or group of papers should be preceded by a brief commentary. It was decided to include papers written up to and including 1993.

There is a temptation in writing such commentaries. I have tried to resist but doubt whether I have been fully successful. All my notes on and correspondence about earlier work were destroyed years ago in a wholly unsuccessful attempt to stem the flood of paper that, especially since the onset of the paperless office, threatens to engulf us. Thus my comments on what was in my mind when I wrote the papers are based on an increasingly frail memory. The temptation is to think that I knew more than I did, and to solidify vague ideas of the past beyond their real merit.

The papers are grouped into two volumes, the first on the whole more applied than the second.

A type of paper only briefly mentioned is that of specific applications although there are some such in Volume I. In one sense virtually all the papers are motivated by experience of applications, although this may be at second or third hand. In some selected cases notes have been added about relevant papers not reproduced in this collection.

Throughout my working life I have been exceptionally fortunate to have the chance of working with colleagues whom I respected and liked and from whom I learned much. A number of the papers are joint ones and without exception, I believe, these represent close collaboration. I am deeply grateful to all those with whom I have worked and especially to those involved in papers reproduced here.

I am grateful to Australian National University for their support and hospitality during the preparation of some of the commentaries.

It is a pleasure to thank David Tranah for his encouragement of the proposal and especially Diana Gillooly for her meticulous care and for much helpful advice.

D. R. Cox

1

Design of Investigations

IN 1945 I WENT TO WORK with H. E. Daniels at the Wool Industries Research Association, Leeds. Henry, who had worked there for almost 10 years, had accomplished much and in particular had convinced many of the scientists there of the key importance of design and appropriate statistical analysis. In essence the two wool industries, worsted and woollen, are concerned with producing uniform product from highly variable material, so that study of variability and its control are central issues. Henry left for Cambridge the next year but constantly returned fondly to recollections of his days in Leeds and the challenging problems he met there. I learned an enormous amount from Henry, not least the importance of meticulous sampling and measurement procedures. I also had the opportunity of working over the following few years with two very fine experimental scientists, Mr R. C. Palmer, a physical chemist, and Mr S. L. Anderson, a textile physicist. They might be described as old-style laboratory scientists who probed deeply by ingeniously devised experiments often using relatively simple apparatus built in the local workshops. Both were meticulous in using randomization when appropriate; I recall also Mr Anderson pointing out forcibly an occasion when my suggestion of randomizing for sample collection was really very silly and that a rigorously implemented scheme of systematic sampling would much better achieve the necessary objective of bias elimination. He also had clear views on the role and nature of causality in these contexts.

Design of experiments

While my published work on design largely dates from a later period it was strongly coloured by the experiences in Leeds. To be involved in study design is in some ways the most satisfying part of statistical work, yet as a component of the academic side of statistics design tends not to be well regarded. Perhaps more seriously the basic Fisherian ideas on design of experiments seem to have percolated very unevenly into laboratory practice. This may be because of the quite widespread notion that statistical design means complicated design. These surely have a place but they are not central to the key ideas. Further, to some extent the ideas on sampling have become immersed in arcane arguments about variance estimation in complex schemes. The increased emphasis on sampling problems in biology and in environmental investigations is particularly welcome from this perspective in leading to a concentration on essential issues.

D1   Some systematic experimental designs. *Biometrika* **38** (1951), 312–323.

Paper D1, the first of the papers on design, is based on a relatively complicated split-unit experiment, concentrating on the 'whole-unit' part. The simplified form involved a small number of experimental units, in fact nine, with three treatments to be compared and with a smooth time trend across the experiment corresponding to a slow oxidation process of intrinsic interest. Randomization seemed inefficient and inappropriate in such a small trial. There was a literature on systematic designs of the cyclic form $A, B, C, A, B, C, \ldots$, but these seemed also clearly inefficient for this particular problem.

Paper D1 proceeds from first principles to evaluate exactly or approximately the covariance matrix of the estimates resulting from an arbitrary design and fitting a linear model including treatment effects and a low-degree polynomial trend. A design is then chosen with good properties for the precision of key estimates, particular attention focuses on the average variance of simple contrasts of treatment levels, what is now called A-optimality. In the matrix calculations, analysis of covariance is useful although not essential; from one perspective, analysis of covariance is concerned with inverting partitioned matrices in a stage-by-stage way. While this is by no means the first paper to choose a design on the basis of a theoretical calculation, the approach was relatively unusual at the time. Incidentally, the experiment in question was successfully completed and led to clear conclusions.

D2   The design of an experiment in which certain treatment arrangements are inadmissible. *Biometrika* **41** (1954), 287–295.

Paper D2 is similar in spirit. Here several treatment applications are possible in sequence for the same batch of material but there are major constraints on the order in which they can be applied. Again a theoretical calculation of precision for an arbitrary design, made easier by the formalism of analysis of covariance, leads to the choice of an optimal design. There were various embellishments of the argument including the role of recovery of information. I am not sure whether the motivating metallurgical experiment was ever implemented; many years later Mr John Lewis, then of ICI Pharmaceuticals, pointed out a clinical application of the same idea.

Papers D1 and D2 predate the modern theory of optimal design, although not of course the qualitative notion of optimality which informally underpinned the classical work of Frank Yates and others in the 1930s. Presumably nowadays at a practical level appropriate designs would be found algorithmically on the computer, although the necessity to examine sensitivity to model assumptions remains. Both papers dispense with randomization as a key element of design.

This is for reasons that were compelling in the practical situations envisaged. Some of my more theoretical papers return to the issue of the interplay between randomization and statistical theory.

D3 The use of a concomitant variable in selecting an experimental design. *Biometrika* **44** (1957), 150–158. [Corrigenda, *Biometrika* **44** (1957), 534]

Paper D3 describes an investigation motivated by animal experiments and started at the suggestion of the late Dr Bernard Greenberg, then Head of Biostatistics at the University of North Carolina at Chapel Hill. A covariate is available on each experimental unit. Is it best that the covariate is used to form blocks, or as the basis of an analysis of covariance (least squares) adjustment, or should both methods be used in combination? This question recurs in various, usually more complex, forms. Several strategies were compared, the broad conclusion being that blocking is adequate so long as the correlation between response and covariate is not too high. A final sentence makes but does not develop the point that an important role for such covariates is the detection of interaction with the treatment effect. The whole issue of adjusting for baseline variables in addition to randomization resurfaces from time to time in the context of clinical trials.

D4 The use of control observations as an alternative to incomplete block designs. *J. R. Statist. Soc.* B **24** (1962), 464–471 (with M. Atiqullah).

Paper D4, in collaboration with the late M. Atiqullah, who continued to a distinguished career in Pakistan, explores a very traditional idea mentioned briefly by Frank Yates in his pioneering papers on incomplete block designs. This is to use whatever design is simplest and to achieve error control by interspersing a control treatment throughout the trial. The idea has, of course, particular appeal if comparisons with a baseline treatment are of special concern. The paper investigates various questions about the method, including the frequency of controls desirable. It is no surprise that the conclusions are entirely in line with Yates's comments on the matter. I have never seen the method used, although it does seem appealing in some contexts.

D5 Some systematic supersaturated designs. *Technometrics* **4** (1962), 489–495 (with K. H. V. Booth).

In the mid to late 1950s there was much interest in industrial statistical circles, especially in the United States, in a suggestion of F. E. Satterthwaite to use so-called random balance designs – in particular for problems with more factors than experimental units, when the term 'supersaturated design' is merited. In the simplest case of a 2^k system with an assumption of no interactions, the idea is that the factor levels should be randomized, possibly subject to marginal constraints on the numbers of upper and lower levels of each factor. It is then argued that in estimating the effect of any one factor the other factors can be treated as adding random error of zero mean.

The suggestion remained unpublished for some years and when finally a paper (Satterthwaite, 1959) did appear there were damaging criticisms, notably by George Box, in the discussion following the paper.

In one sense the issue is one of conditional inference. If it is clear that factor B, say, has a positive effect and, in the design used, the arrangement of its levels is very similar to that of another factor A whose effect is being studied, the randomization ceases to be relevant; the bias induced by B will be zero on the average, but its direction in the data under analysis is clear.

Nevertheless, it is apparent from analogy with broadly similar deterministic problems, such as the identification of an overweight coin in a minimum number of weighings, that in some situations meaningful supersaturated designs do exist. In Paper D5 Kathleen Booth and I formed a criterion assessing how close to orthogonality are the columns of an $n \times k$ matrix with elements 1 and -1 and with $n < k$ and then found by computer enumeration the arrangement or arrangements that minimize the lack of orthogonality. The paper represents one of the first attempts to find designs in this way. The work was done on the University of London Mercury computer and absorbed a very large amount of computer time. The statistical properties of the designs were then studied. If only one factor was effective, the others having zero effect, the anomalous factor was well identified, but as soon as more than a very small number of factors were effective, or when interaction was present, the individual factor effects became very unstable and potentially very confusing.

There remains the important issue of practical relevance. The question of what to do when there is a relatively large number of potential factors for screening is certainly a real one. Is use of supersaturated designs ever wise? I am not aware of detailed reports of specific applications, although Satterthwaite himself claimed a number of successes. I was told once that Bell Laboratories ran a substantial study of this form, but that the results were so confusing that even John Tukey's legendary skills at data analysis were defeated!

The limited results in Kathleen Booth's and my paper do indeed suggest that appreciable interaction or the presence of more than a very small number of effective factors will induce uninterpretable data. It seems likely that the judicious use of alternative approaches such as the grouping of factors into sets to produce a saturated design is preferable.

D6   Present position and potential developments: some personal views. Design of experiments and regression (with discussion). *J. R. Statist. Soc.* A **147** (1984), 306–315.

Paper D6 is a short review prepared in connection with the celebration of the 125th anniversary of the Royal Statistical Society. The list of supposedly open research issues may still be of interest, although most are by now resolved.

D7   A note on design when response has an exponential family distribution. *Biometrika* **75** (1988), 161–164.

While the formal theory of optimal design has been developed in some generality, the simplest results apply to linear regression models with simple error structure. Here the optimal designs do not depend on the regression parameters under study. Of course this by no means resolves the difficulties of application; sensitivity to the choice of model is always important and often specification of a design region requires an informal compromise between the requirements of precision and insensitivity to model failure at the edges of the region. In general, however, the optimal design depends on the unknown parameter under study.

The short Paper D7 shows the unsurprising result that, for the estimation of parameters in a canonical exponential model, standard linear model optimality arguments apply locally when all regression effects are small. It would probably be of some interest to provide guidance on when such approximate arguments are seriously misleading.

The paper was motivated by a rather different and in some ways more interesting issue connected with the relative advantages of matching and randomization in relatively simple comparisons involving exponential family distributions.

Randomization

A general theoretical issue connected with experimental design concerns the interplay between randomization and model-based analyses. Corresponding issues in sampling theory have been widely discussed. My understanding from the earlier literature, and to some extent from conversation, is that Frank Yates regarded randomization as of great value in establishing the structure of analysis required and the appropriate base for estimating error, but that the resulting analysis should typically be based on least squares estimates derived from an appropriate linear model. Oscar Kempthorne, in his very influential book and in subsequent papers, maintained that the randomization-based analysis is primary, least squares being a convenient approximation and, in those days, a necessary one in the light of computational limitations. R. A. Fisher, who after all invented both randomization and the randomization-based tests, was guarded on the subject. One reading is that, having been asked by E. S. Pearson about possible sensitivity to normality of the procedures Fisher was recommending, Fisher invented the permutation argument in effect for reassurance of the faint of heart, but not for normal use. Of course, from a pragmatic viewpoint, that the two approaches usually give very similar answers is reassuring and the matter can often be left there.

A different issue, raised implicitly in Paper D1, concerns establishing when randomization is inappropriate. Of course sometimes it may be so administratively difficult to implement that it is better to use some systematic procedure even if potentially biased. A more theoretically interesting point concerns the widely held notion that randomization is immaterial and possibly harmful in small experiments. Why is this? From the viewpoint of randomization-based tests, the issue is connected with some form of ancillarity. The randomization distribution and the associated standard errors are relevant if and only if the arrangement selected is one of a set that can be treated as indistinguishable in all material respects. But in small systems all individual arrangements have distinctive features. Put differently, any analysis-of-covariance-like procedure has, for a randomization justification, to consider the randomization conditionally on suitable functions of the covariates, and analysis-of-covariance-based adjustments are analysed from this perspective in Paper D8. There is a real difficulty here which is occasionally used as an argument against randomization in that it can always be argued that the design used is special. The larger the experiment the less force this consideration has.

A paper (Cox, 1956) not reproduced here also bears on these issues. It arose out of the comment by Oscar Kempthorne that standard analysis of covariance procedures of adjustment could not be justified by a randomization argument and therefore should not be used. The point of this short paper was that if weighted randomization is used, each arrangement having a probability of acceptance proportional to the corresponding residual sum of squares of the covariate, then some standard properties receive a randomization justification. I recall when the work was done being very uneasy, for a reason I did not quite understand, about the statistical relevance of this result. It was submitted to a rather theoretical journal with some expectation that it would be rejected; the referees' reports were in fact about the most unreservedly positive of any I have ever received. The idea is, however, surely flawed on the grounds that the different arrangements have recognizably different precisions making averaging over the randomization distribution for most, but perhaps not quite all, purposes irrelevant.

Another indirectly relevant discussion is that of Cox (1982), sketching properties of Efron's biased-coin randomization scheme. The ideas were developed in much more detail by my colleague Richard Smith (Smith, 1984). Insofar as the applications envisaged are to clinical trials, randomization-based estimation of error is much less important, the primary emphasis being on avoiding bias by the use of concealment.

D8   Randomization and concomitant variables in the design of experiments.
In *Statistics and Probability: Essays in Honor of C. R. Rao*, edited by
G. Kallianpur, P. R. Krishnaiah and J. K. Ghosh, pp. 197–202.
Amsterdam: North-Holland, 1982.

The first part of Paper D8 deals with the effect on the precision of estimates of imbalance in concomitant variables induced by randomization. It is remarked that this effect is random and assessment by expected values will be inadequate if the randomization shows after the event that a recognizably unbalanced design has been used. Then correction in analysis is required and may involve relatively large changes of precision. In a second part of the paper, the possibility is noted that in design one could re-randomize until a suitable quadratic form in the concomitant variables is sufficiently small.

Sampling

Formal design issues in sampling problems are very similar to those in experimental design, however different they may appear on the surface. The issues of bias avoidance and error control are after all the same. Of course this is in no way to blur the important interpretative distinction between experiments and observational investigations. Many of the issues are broadly similar, but the essential distinction in interpretation, especially over the power to establish causal effects, must not be undermined.

D9   Some sampling problems in technology. In *New Developments in Survey Sampling. A Symposium on the Foundations of Survey Sampling Held at the University of North Carolina, Chapel Hill, North Carolina*, edited by N. L. Johnson and H. Smith, Jr, pp. 506–527. New York: Wiley, 1969.

Paper D9 was given at a symposium on the foundations of sampling, centering largely around the work of V. P. Godambe. The first of the paper's two parts was an attempt to review sampling problems as they arose in technology, especially the issues involved in sampling various kinds of raw material. At that time, concern with the statistical problems of the traditional industries was higher than it is now, but this part of the paper may still be of some interest; the problems must still be around in a different guise, even if they are not often discussed. The second part of the paper addressed in more detail sampling problems originating in the textile industries, in particular length-biased sampling and what might be called recurrence-time sampling. The notion and terminology of length-biased sampling had been used for some time in the textile world, but this may have been the first paper to discuss the estimation issues involved in that context. Both parametric and nonparametric approaches are studied. Length-biased data are now recognized to arise quite widely. There are interesting connections also with the theory of stochastic point processes.

D10 On sampling and the estimation of rare errors. *Biometrika* **66** (1979), 125–132 (with E. J. Snell).

Paper D10 arose out of a question posed by an auditor asking for clarification of a sampling procedure for checking invoices known as 'monetary unit sampling'. The method was presented in the auditing world in a rather puzzling way according to which each invoice appeared to be divided into pounds (or even dollars) and each pound treated separately. My colleague Mrs Joyce Snell and I established the connection with probability-proportional-to-size sampling and essentially also to length-biased sampling. The problems of analysis were different from the usual ones, however, in that interest focused on those situations in which there were very few errors, or possibly none at all, and yet limits were required for the total error in the population. The recommended approach in the auditing literature, the so-called Stringer bounds, was complicated and remains rather obscure and also overconservative; we recommended a formal Bayesian approach based on the Poisson distribution and a simple prior. An interesting general point is that, when Joyce and I presented our results to a group of auditors, we tried to recommend an empirical Bayes approach in which a genuine prior distribution was used. This was emphatically rejected by the auditors on the grounds that the results in any particular case might be challenged in court and dependence on other experience would handicap acceptance of the conclusions. It is likely that the really critical issue is not so much how the data are analysed but rather how the accounts for study are actually chosen.

Summary

I imagine few statisticians involved in serious applications doubt the great importance, indeed almost the primary importance, of design in the broad sense that includes the design of measuring procedures and decisions about what and how to measure, as well as the design of experiments and of observational studies of various kinds. Nevertheless these topics do not have a high profile in the current statistical research literature. This may partly be because their fascination stems largely from involvement in the details of specific investigations in collaboration with sympathetic subject-matter workers. Moreover, as already noted, the statistical design of experiments has in some contexts been taken to be synonymous with the use of complicated designs; now ideas such as fractional replication are both elegant and also powerful in the right context, but in many fields the primary objectives of bias elimination and error control by relatively simple methods are more important.

Additional publications

Planning of Experiments. New York: Wiley, 1958.

Theory of the Design of Experiments. Boca Raton, Florida: Chapman and Hall/CRC Press, 2000 (with N. Reid).

The first book arose to a limited extent from lectures I gave in Cambridge, but initial plans to write a theoretical book were changed into the notion of something more suitable for subject-matter workers. Quite a few years elapsed before Nancy Reid and I put together lecture notes we had given in various places with new material to produce a more theoretical book.

Recent work on the design of experiments: a bibliography and a review. *J. R. Statist. Soc.* A **132** (1969), 29–67 (with A. M. Herzberg).

This review paper contained an annotated bibliography of publications in the design of experiments up to 1968, compiled with meticulous thoroughness largely by my co-author Agnes Herzberg. In addition Agnes contributed centrally to the discussion part of the paper. The bibliography may still be useful for tracing early work.

References

Cox, D. R. (1956). A note on weighted randomization. *Ann. Math. Statist.* **27**, 1144–1151.

Cox, D. R. (1982). A remark on randomization in clinical trials. *Utilitas Mathematica* **21A**, 245–252.

Satterthwaite, F. E. (1959). Random balance experimentation (with discussion). *Technometrics* **1**, 111–137 [Discussion 157–193].

Smith, R. L. (1984). Properties of biased coin designs in sequential clinical trials. *Ann. Statist.* **12**, 1018–1034.

[*From* Biometrika, *Vol.* 38. *Parts* 3 *and* 4, *December* 1951.]

SOME SYSTEMATIC EXPERIMENTAL DESIGNS

By D. R. COX

Statistical Laboratory, University of Cambridge

1. INTRODUCTION

Consider the following experiment. It is required to compare the effect in processing of a number of treatments applied to wool. The wool is divided into lots as alike as possible and the lots are numbered in random order. In week 1 lot 1 is processed with a certain treatment. In week 2 lot 2 is processed with, in general, a different treatment. And so on. The age of the wool affects its behaviour in processing so that, superimposed on any treatment differences, there will be a smooth trend due to aging. We want to estimate the treatment differences as simply and accurately as possible. In what order should the treatments be applied?

More generally, consider an experiment to compare a number of *treatments* on *plots* arranged in order in space or time, one treatment being applied to each plot. We assume that the measured quantity is

$$(\text{Treatment constant}) + (\text{Smooth trend}) + (\text{Random quantity}), \tag{1}$$

where the random quantity has zero mean, constant variance and is independent in different plots. We want to arrange the treatments to give simple and accurate estimates of the treatment differences.

One method is to arrange the experiment in randomized blocks, thus eliminating part of the trend from the error of the treatment comparisons.

A second method is to randomize completely the order of application of the treatments and to eliminate the trend by analysis of covariance on position. The calculations can be extensive if high-order trends are involved. Also when the above assumptions hold information is lost by the use of the method. The loss may be serious if the number of treatments and number of replicates are small (see §4·2).

A third and related method is to repeat the treatments always in the same order, e.g. $T_1 T_2 T_3 T_1 T_2 T_3 \ldots$, and to eliminate the trend by analysis of covariance on position (i.e. by fitting polynomials). Neyman (1929) discussed this in connexion with agricultural field trials and gave certain tables. The calculations are again lengthy, and to simplify them Hald (1948) has introduced generalized orthogonal polynomials the tabulation* of which would greatly reduce the labour of fitting the successive polynomials.

The discussion of a recent paper by Miss G. M. Cox (Cox, 1950) contains some comments on the use of polynomial fitting in the analysis of agricultural field trials. In particular, R. A. Fisher referred to his and Prof. van Uven's early work on the subject and criticized the method.

R. M. Williams (1949) has worked on a related problem in which there is no trend, but the errors in adjacent plots are correlated.

---

* Hald gives tables in some special cases.

2. OUTLINE OF METHOD

Let there be t treatments, n replicates of each treatment and $N = tn$ plots in all. Let the plots be equally spaced and suppose that the trend can be represented by a pth order polynomial

$$\alpha_1 \xi_1' + \ldots + \alpha_p \xi_p', \tag{2}$$

where ξ_μ' is the orthogonal polynomial of order μ for N equally spaced points given in Fisher & Yates's tables (1948). To begin with the value of p is assumed known *a priori*. The population mean of the measurement on the jth plot is

$$a_i + \sum_{\mu=1}^{p} \alpha_\mu \xi_\mu', \tag{3}$$

where a_i depends only on the treatment T_i applied to the jth plot.

The estimation of the a_i is easiest if the treatment differences are *completely orthogonal* to the trend. Now treatment differences are orthogonal to the linear part of the trend if and only if

$$s_{i1}' \equiv \Sigma_i \zeta_1' = 0 \quad (i = 1, \ldots, t), \tag{4}$$

where Σ_i denotes summation over all plots receiving treatment T_i. In general treatment differences are orthogonal to the complete trend if and only if

$$s_{i\mu}' \equiv \Sigma_i \xi_\mu' = 0 \quad (i = 1, \ldots, t; \ \mu = 1, \ldots, p). \tag{5}$$

If (5) is true we may estimate the treatment differences as if the trend were absent. The variance of a treatment comparison is $2\sigma^2/n$, where σ^2 is the error variance. It is shown in §4 that if (5) is not true the variance is in general greater than $2\sigma^2/n$.

Therefore the problem is solved if we can arrange the treatments to satisfy (5) or, if this is impossible, to satisfy (5) as nearly as possible.

We shall say that an arrangement of treatments such that (5) is true forms a design orthogonal to trends of order p.

3. SOME EXACT SOLUTIONS

When there are only two treatments and the trend can be represented by a cubic, some exact solutions are possible. The total number of plots, $N = 2n$, is even so that all the ξ_1' are odd integers. Now the sum of an odd number of odd integers cannot be zero, so that it is impossible to satisfy (4) for odd n. Therefore the number of replicates, n, must be even and equal to $2m$, say. Then ξ_1' takes the form

$$\ldots \quad -5 \quad -3 \quad -1 \quad \bigg| \quad 1 \quad 3 \quad 5 \quad \ldots.$$

There exist *symmetrical designs* in which if treatment T_i is applied in position $\xi_1' = r$ (called position r for short), the same treatment T_i is applied in position $-r$.

Example. $t = 2$, $n = 4$, $N = 8$:

$$
\begin{array}{cccc|cccc}
-7 & -5 & -3 & -1 & 1 & 3 & 5 & 7 \\
T_2 & T_2 & T_1 & T_1 & T_1 & T_1 & T_2 & T_2
\end{array}
$$

Since the odd-order orthogonal polynomials $\xi_{2\beta+1}'$ $(\beta = 0, 1, \ldots)$ are odd functions, all symmetrical designs have treatment differences orthogonal to $\xi_{2\beta+1}'$.

The problem is therefore solved if we can find m positions from $1, 3, 5, \ldots$ to take treatment T_1, such that

$$\Sigma_1' \xi_2' = 0, \tag{6}$$

where Σ_1' denotes summation over those positions in $1, 3, 5, \ldots$ which receive treatment T_1.

There are no solutions for odd m, but there are the following solutions for even m:

Eight replicates. $t = 2, m = 4, n = 8, N = 16, p = 3$:

ξ_1'	1	3	5	7	9	11	13	15
	T_1	T_2	T_2	T_1	T_2	T_1	T_1	T_2

(and the design formed by interchanging T_1 and T_2).

Twelve replicates. $t = 2, m = 6, n = 12, N = 24, p = 3$:

ξ_1'	1	3	5	7	9	11	13	15	17	19	21	23
	T_2	T_2	T_2	T_1	T_1	T_1	T_1	T_2	T_1	T_2	T_1	T_2
	T_2	T_2	T_1	T_2	T_1	T_1	T_1	T_1	T_2	T_2	T_2	T_1
	T_2	T_1	T_2	T_1	T_2	T_1	T_1	T_2	T_2	T_1	T_1	T_2
	T_2	T_2	T_1	T_1	T_2	T_1	T_2	T_1	T_1	T_2	T_1	T_2
	T_2	T_1	T_2	T_1	T_1	T_1	T_2	T_2	T_2	T_1	T_2	T_1
	T_2	T_2	T_1	T_2	T_1	T_2	T_1	T_1	T_1	T_1	T_2	T_2

(and the designs formed by interchanging T_1 and T_2).

Of these the second is recommended for practical use because it is most nearly orthogonal to ξ_4'.

There are probably solutions for all higher even m.

There are also non-symmetric designs orthogonal to ξ_1' and ξ_2'. Examples are:

Four replicates. $t = 2, m = 2, n = 4, N = 8, p = 2$:

ξ_1'	-7	-5	-3	-1	1	3	5	7
	T_2	T_1	T_1	T_2	T_1	T_2	T_2	T_1

Six replicates. $t = 2, m = 3, n = 6, N = 12, p = 2$:

ξ_1'	-11	-9	-7	-5	-3	-1	1	3	5	7	9	11
	T_1	T_2	T_1	T_2	T_2	T_2	T_1	T_1	T_1	T_2	T_1	T_2

Similar solutions exist for $m = 4$, etc., but are of no practical value, since it is better to use the above symmetric designs.

To summarize, there are exact solutions for two treatments as follows:

Cubic trend	Quadratic trend
$n = 8$	$n = 4$
$n = 12$	$n = 6$, etc.

probably $n = 4r$

A similar analysis for three and four treatments shows that there are no exact solutions for reasonably small n. Therefore it is necessary to consider designs which satisfy (5) only approximately.

4. GENERAL PROPERTIES OF APPROXIMATE SOLUTIONS*

4·1. *Solution by least squares*

When the treatment effects and the trend are not orthogonal, we set up least-squares equations for the estimates $\hat{a}_1, ..., \hat{a}_t, \hat{\alpha}_1, ..., \hat{\alpha}_p$. To do this it is convenient to introduce *normalized* orthogonal polynomials ξ_μ such that

$$\Sigma \xi_\mu = 0, \quad \Sigma \xi_\mu \xi_\nu = \delta_{\mu\nu} = \begin{cases} 1 & (\mu = \nu), \\ 0 & (\mu \neq \nu), \end{cases} \tag{7}$$

where the summation is over all plots. (The ξ_μ are simple multiples of the ξ'_μ tabulated in Fisher & Yates.) We write the trend as $\sum_{\mu=1}^{p} \alpha_\mu \xi_\mu$, and the sum of squares to be minimized is then

$$\sum_{i=1}^{t} \Sigma_i \left(x - a_i - \sum_{\mu=1}^{p} \alpha_\mu \xi_\mu \right)^2. \tag{8}$$

The least-squares equations are, after the use of (7),

$$Le = b, \tag{9}$$

where L is the $(t+p)$th order square matrix

$$\begin{pmatrix} n I_t & S \\ \hline S' & I_p \end{pmatrix}, \tag{10}$$

e is the column vector $\{\hat{a}_1, ..., \hat{a}_t, \hat{\alpha}_1, ..., \hat{\alpha}_p\}$ and b is the column vector

$$\{\Sigma_1 x, ..., \Sigma_t x, \Sigma \xi_1 x, ..., \Sigma \xi_p x\}.$$

In (10) I_j is the jth order unit matrix, S is the $t \times p$ matrix whose (i, μ)th element is $\Sigma_i \xi_\mu$, and S' is the transpose of S. (The general form of the equations can be seen from the special case (13) written out at length below.)

The inverse of the matrix (10) gives the variances and covariances of the $\hat{a}_i$ and hence the variances of the treatment comparisons $(\hat{a}_i - \hat{a}_j)$. Before considering the general case it is best to examine two important special cases.

4·2. *Two treatments*

When there are only two treatments the matrix (10) simplifies considerably. Since $\Sigma \xi_\mu = 0$, $\Sigma_1 \xi_\mu + \Sigma_2 \xi_\mu = 0$, so that $s_{1\mu} = -s_{2\mu} = s_\mu$, say. Thus

$$L = \begin{pmatrix} n & 0 & s_1 & \cdots & s_p \\ 0 & n & -s_1 & \cdots & -s_p \\ s_1 & -s_1 & 1 & & \\ \vdots & \vdots & & \ddots & \\ s_p & -s_p & & & \ddots \; 1 \end{pmatrix}.$$

By a Laplace development down the first two columns we get

$$|L| = n^2 - 2n \sum_{\mu=1}^{p} s_\mu^2.$$

* § 4·5 summarizes the main conclusions of § 4.

Similarly, the cofactors determining var $(\hat{a}_1)$, var $(\hat{a}_2)$, cov $(\hat{a}_1, \hat{a}_2)$ can be found and we get

$$\text{var}\,(\hat{a}_1 - \hat{a}_2) = \frac{2\sigma^2}{n}\left\{1 - \frac{2}{n}\sum_{\mu=1}^{p} s_{\mu}^2\right\}^{-1}. \tag{11}$$

Thus the minimum variance, achieved when all s_{μ} are zero, is the variance of the difference of two independent means. The efficiency E is

$$E = 1 - \frac{2}{n}\sum_{\mu=1}^{p} s_{\mu}^2. \tag{12}$$

Example. Consider three designs for two treatments, four replicates and cubic trend $(t = 2, n = 4, N = 8, p = 3)$.

ξ_1'	ξ_2'	ξ_3'	Order selected at random	Exact solution for quadratic trend	Best solution orthogonal to ξ_1' and ξ_3'
-7	7	-7	T_2	T_2	T_1
-5	1	5	T_1	T_1	T_2
-3	-3	7	T_1	T_1	T_2
-1	-5	3	T_2	T_2	T_1
1	-5	-3	T_2	T_1	T_1
3	-3	-7	T_2	T_2	T_2
5	1	-5	T_1	T_2	T_2
7	7	7	T_1	T_1	T_1
		$\Sigma_1 \xi_1'$	4	0	0
		$\Sigma_1 \xi_2'$	6	0	4
		$\Sigma_1 \xi_3'$	14	16	0
		s_1^2	$0\cdot0952$	0	0
		s_2^2	$0\cdot2143$	0	$0\cdot0952$
		s_3^2	$0\cdot7424$	$0\cdot9697$	0
		E	$0\cdot4740$	$0\cdot5152$	$0\cdot9524$

(For example $s_1 = \Sigma_1 \xi_1$, $s_1^2 = (\Sigma_1 \xi_1')^2/168$.)

From this example we conclude that
(i) an order selected at random may be very inefficient;
(ii) the exact solution for $p = 2$ is very inefficient for $p = 3$.

4·3. *Treatments orthogonal to all ξ_{μ} except one*

Another important special case is obtained when the treatments are orthogonal to all ξ_{μ} except one, which we call ξ. The matrix $\mathbf{S}$ consists of zeros except in one column which we denote $\{s_1, \ldots, s_t\}$. It is worth writing out the least-squares equations involving the $\hat{a}_i$ in non-matrix form. They are

$$\left.\begin{aligned}
n\hat{a}_1 \quad\quad\quad\quad\quad &+ s_1\hat{\alpha} = \Sigma_1 x, \\
n\hat{a}_2 \quad\quad\quad &+ s_2\hat{\alpha} = \Sigma_2 x, \\
&\vdots \\
n\hat{a}_t + s_t\hat{\alpha} &= \Sigma_t x, \\
s_1\hat{a}_1 + s_2\hat{a}_2 + \ldots + s_t\hat{a}_t + \quad \hat{\alpha} &= \Sigma\xi x.
\end{aligned}\right\} \tag{13}$$

The matrix L is a simple bordered matrix and

$$|L| = n^t - n^{t-1} \sum_{i=1}^{t} s_i^2. \tag{14}$$

Similarly, the cofactor of the first element is

$$n^{t-1} - n^{t-2} \sum_{i=2}^{t} s_i^2,$$

so that

$$\operatorname{var}(\hat{a}_1) = \frac{\sigma^2 \left\{ n^{t-1} - n^{t-2} \sum_{i=2}^{t} s_i^2 \right\}}{\left\{ n^t - n^{t-1} \sum_{i=1}^{t} s_i^2 \right\}}. \tag{15}$$

Similarly, we can find $\operatorname{var}(\hat{a}_2)$ and $\operatorname{cov}(\hat{a}_1, \hat{a}_2)$ and we get

$$\operatorname{var}(\hat{a}_1 - \hat{a}_2) = \sigma^2 \left\{ \frac{2n^{t-1} - n^{t-2} \left(s_1^2 + s_2^2 + 2 \sum_{3}^{t} s_i^2 \right) - 2n^{t-2} s_1 s_2}{n^t - n^{t-1} \sum_{1}^{t} s_i^2} \right\}. \tag{16}$$

One important consequence is that

$$\operatorname{var}(\hat{a}_1 - \hat{a}_2) \geqslant \frac{2\sigma^2}{n} \tag{17}$$

with equality if and only if $s_1 = s_2$.

We can find the variance of any treatment comparison in the same way. One natural measure of the efficiency of the design is the average variance $\bar{V}$ over all possible comparisons of pairs. Any particular $\hat{a}_i$ occurs in $(t-1)$ of the comparisons and does not occur in $(t-1)(t-2)/2$ of them. Therefore

$$\bar{V} = \frac{\sigma^2}{\left\{ n^t - n^{t-1} \sum_{1}^{t} s_i^2 \right\}} \left\{ 2n^{t-1} - \frac{2n^{t-2}}{t(t-1)} [(t-1) + (t-1)(t-2)] \sum_{1}^{t} s_i^2 - \frac{4n^{t-2}}{t(t-1)} \sum_{i>j} s_i s_j \right\}.$$

Now $\sum_{1}^{t} s_i = 0$, so that $2 \sum_{i>j} s_i s_j = - \sum s_i^2$. Thus

$$\bar{V} = \frac{2\sigma^2}{n} \left\{ \frac{n - (t-2)/(t-1) \sum s_i^2}{n - \sum s_i^2} \right\},$$

and the average efficiency E is

$$E = \{ 1 - \Sigma s_i^2 / n \} \left\{ 1 - \frac{(t-2)}{n(t-1)} \Sigma s_i^2 \right\}^{-1}. \tag{18}$$

This reduces to a special form of (12) in the special case $t = 2$.

4·4. General case

We now return to the general case $t \neq 2$ and $p \neq 1$ (equation (10)). By a Laplace development of (10) on the first t columns we get

$$|L| = n^t + \sum_{k=1}^{\{p, t\}} (-1)^k n^{t-k} [\Sigma |S^{(k)}|^2], \tag{19}$$

where $\{p, t\}$ is the smaller of p and t and $|S^{(k)}|$ is the determinant of a kth order minor of the matrix S, the summation being over all possible kth order minors, i.e. we can express $|L|$ in terms of the sums of squares of all minors that can be formed from S. The cofactor of the first element of L can be expressed in exactly the same way in terms of the minors of the matrix $\mathbf{S}_{(1)}$ formed by deleting the first row of $\mathbf{S}$. This leads directly to a formula for var $(\hat{a}_1)$. Similarly, cov $(\hat{a}_1, \hat{a}_2)$ can be obtained in terms of sums of products of certain minors formed from $\mathbf{S}$.

Now in practice we are interested in designs which are nearly orthogonal, i.e. for which the elements of $\mathbf{S}$ are small. In this case, if p and t are not too large, products of more than two of the elements of $\mathbf{S}$ can be neglected, and

$$
\left.
\begin{aligned}
|\mathbf{L}| &\simeq n^t - n^{t-1} \sum_{i,\,\mu} s_{i\mu}^2, \\[2mm]
\operatorname{var}(\hat{a}_1) &\simeq \left\{ \frac{n^{t-1} - n^{t-2} \sum\limits_{i \neq 1;\,\mu} s_{i\mu}^2}{|\mathbf{L}|} \right\} \sigma^2, \\[4mm]
\operatorname{var}(\hat{a}_1 - \hat{a}_2) &\simeq \left\{ \frac{2n^{t-1} - n^{t-2} \sum\limits_{\mu} (s_{1\mu}^2 + s_{2\mu}^2 + 2s_{1\mu} s_{2\mu} + 2 \sum\limits_{i \neq 1,\,2} s_{i\mu}^2)}{|\mathbf{L}|} \right\} \sigma^2.
\end{aligned}
\right\} \tag{20}
$$

Averaging over all comparisons of pairs, we get

$$
\overline{V} \simeq \frac{2\sigma^2}{n} \left\{ 1 - \frac{(t-2)}{n(t-1)} \sum_{i,\,\mu} s_{i\mu}^2 \right\} \left\{ 1 - \frac{1}{n} \sum_{i,\,\mu} s_{i\mu}^2 \right\}^{-1},
$$

so that the efficiency E is

$$
E \simeq \left\{ 1 - \frac{\sum s_{i\mu}^2}{n} \right\} \left\{ 1 - \frac{(t-2)}{n(t-1)} \sum s_{i\mu}^2 \right\}^{-1}. \tag{21}
$$

4·5. *Summary of* §4

The conclusions of §4 about non-orthogonal designs can be summarized as follows:

(i) Least squares equations (9) can be formed for the treatment and trend coefficients.

(ii) From these equations the variance of any treatment comparison can be obtained.

(iii) The solution is simple in the special case of two treatments (§4·2) or when the treatments are orthogonal to all relevant ξ'_μ except one (§4·3).

(iv) The efficiency of the design can be measured in terms of the variance of the difference between two treatments averaged over all comparisons of pairs of treatments. The efficiency is given by (12) and (18) in the above two special cases.

(v) The efficiency in the general case is given by a complicated expression which simplifies to (21) for designs that are nearly orthogonal.

(vi) The greatest accuracy in estimating treatment differences is obtained for orthogonal designs.

5. SOME APPROXIMATE SOLUTIONS

In this section the application of the results of §4 is illustrated. Consider experiments with three and four treatments and suppose the trend can be represented by a cubic curve ($p = 3$). If we restrict ourselves to symmetric designs automatically orthogonal to ξ'_1 and ξ'_3, the results of §4·3 apply.

In recording symmetric designs, marked by vertical dotted lines, only the treatments for positive ξ'_1 are shown. For example, the first design is in full:

ξ'_1	-11	-9	-7	-5	-3	-1	1	3	5	7	9	11
	T_1	T_2	T_3	T_3	T_2	T_1	T_1	T_2	T_3	T_3	T_2	T_1

In all cases further designs can be formed by interchanging $T_1, T_2, \ldots$. The best designs are:

Three treatments, four replicates $(t = 3, n = 4, N = 12, p = 3)$:

ξ_1'	1	3	5	7	9	11	
ξ_2'	-35	-29	-17	1	25	55	$\Sigma \xi_2'^2 = 12{,}012$
	T_1	T_2	T_3	T_3	T_2	T_1	

$$\Sigma_1 \xi_2' = 40, \quad \Sigma_2 \xi_2' = -8, \quad \Sigma_3 \xi_2' = -32,$$
$$\Sigma s_i^2 = \tfrac{2688}{12012} = 0 \cdot 2238,$$

Efficiency $= 0 \cdot 9712$.

Three treatments, six replicates $(t = 3, n = 6, N = 18, p = 3)$:

ξ_1'	1	3	5	7	9	11	13	15	17
ξ_2'	-40	-37	-31	-22	-10	5	23	44	68
(i)	T_1	T_2	T_1	T_0	T_2	T_1	T_1	T_0	T_1
(ii)	T_1	T_2	T_3	T_1	T_2	T_3	T_3	T_2	T_1

$$
\begin{array}{ccc}
& \text{(i)} & \text{(ii)} \\
\Sigma_1 \xi_2' = & -6 & 12, \\
\Sigma_2 \xi_2' = & -6 & -6, \\
\Sigma_3 \xi_2' = & 12 & -6.
\end{array}
$$

Efficiency $= 0 \cdot 9992$,

Design (ii) is recommended for practical use because it is the more nearly orthogonal to ξ_4'.

Four treatments, four replicates $(t = 4, n = 4, N = 16, p = 3)$:

ξ_1'	1	3	5	7	9	11	13	15
ξ_2'	-21	-19	-15	-9	-1	9	21	35
	T_1	T_2	T_3	T_4	T_4	T_3	T_2	T_1

$$\Sigma_1 \xi_2' = 28, \quad \Sigma_2 \xi_2' = 4, \quad \Sigma_3 \xi_2' = -12, \quad \Sigma_4 \xi_2' = -20, \quad \text{Efficiency} = 0 \cdot 9796.$$

Four treatments, six replicates $(t = 4, n = 6, N = 24, p = 3)$:

ξ_1'	1	3	5	7	9	11	13	15	17	19	21	23
ξ_2'	-143	-137	-125	-107	-83	-53	-17	25	73	127	187	253
	T_1	T_2	T_2	T_3	T_3	T_4	T_4	T_1	T_4	T_1	T_3	T_2

$$\Sigma_1 \xi_2' = 18, \quad \Sigma_2 \xi_2' = -18, \quad \Sigma_3 \xi_2' = -6, \quad \Sigma_4 \xi_2' = 6, \quad \text{Efficiency} = 0 \cdot 9999.$$

The following designs are also useful:

Three treatments, three replicates $(t = 3, n = 3, N = 9, p = 2)$:

ξ_1'	-4	-3	-2	-1	0	1	2	3	4
ξ_2'	28	7	-8	-17	-20	-17	-8	7	28
(a)	T_3	T_2	T_1	T_2	T_1	T_3	T_1	T_3	T_2
(b)	T_2	T_3	T_1	T_2	T_1	T_3	T_2	T_3	T_1

$$\text{(a)} \ \Sigma_1 \xi_2' = -36, \quad \Sigma_2 \xi_2' = 18, \quad \Sigma_3 \xi_2' = 18.$$

Efficiency $= 0 \cdot 8676$.

Exactly orthogonal to linear trend. Comparison of T_2 with T_3 orthogonal to quadratic trend.

$$(b)\ \Sigma_1 \xi_1' = \ \ \ 2, \ \ \Sigma_1 \xi_2' = 0,$$
$$\Sigma_2 \xi_1' = -3, \ \ \Sigma_2 \xi_2' = 3,$$
$$\Sigma_3 \xi_1' = \ \ \ 1, \ \ \Sigma_3 \xi_2' = -3.$$

Efficiency $\simeq 0.9584$.

It is hoped to give a fuller list of the useful solutions in a later paper. The method of construction is straightforward. For example, the design given for three treatments, six replicates and cubic trend is found as follows. Since the design is to be symmetric, treatment differences are automatically orthogonal to ξ_1' and ξ_3' so that we need only consider ξ_2'. For the 'positive' positions we have the following values of ξ_2':

ξ_1'	1	3	5	7	9	11	13	15	17
ξ_2'	-40	-37	-31	-22	-10	5	23	44	68
(a)	T_1	T_2	T_2	T_3	T_1	T_3	T_3	T_1	T_2
(b)	T_1	T_2	T_1	T_3	T_2	T_3	T_3	T_2	T_1
(c)	T_1	T_2	T_3	T_1	T_2	T_3	T_3	T_2	T_1

	(a)	(b)	(c)
$\Sigma_1 \xi_2'$	-12	-6	12,
$\Sigma_2 \xi_2'$	0	-6	-6,
$\Sigma_3 \xi_2'$	12	12	-6.

The problem is to select from these nine positions three for each treatment so that the sum of the values of ξ_2' over each treatment is as nearly zero as possible. A few minutes' trial shows that an exact solution is impossible and that the three designs above are the nearest solutions. The efficiency by (18) depends on the sum of squares of the $\Sigma_i \xi_2'$. Therefore (b) and (c) are the best solutions and by (18) their efficiency is 0.9992.

The designs given above for $p = 3$ are the best symmetric solutions. It is possible that more efficient solutions could be found by considering non-symmetric designs. The analysis of the non-symmetric designs would, however, be a little more complicated and the designs given above are, for most practical purposes, completely efficient.

A further useful property of some of the above designs is that certain treatment comparisons can be made with full accuracy even though the design as a whole is not fully efficient. For example, in design (a) for three treatments and three replicates, the comparison $(T_2 - T_3)$ is orthogonal to ξ_2 since $\Sigma_2 \xi_2' = \Sigma_3 \xi_2'$. Thus, for example, if T_1, T_2, T_3 are three levels of one treatment we can arrange that the linear part of 'between treatments' is fully accurate and orthogonal to the quadratic trend.

Some of the designs ($t = 3, n = 4$; $t = 4, n = 4$; $t = 3, n = 3$) turn out to be in blocks with one replicate of each treatment in each block, the arrangement of treatments within a block being systematic. Designs of this form were often advocated in the early literature of agricultural field trials. The present work shows that some of them are optimum provided that they are analysed correctly and that the assumptions on which the analysis is based are satisfied.

6. COMPARISON WITH OTHER DESIGNS

The example in §4·2 showed that the selection of a design at random and the elimination of the trend by least squares (analysis of covariance) may lead to considerable loss of efficiency in estimating the treatment differences. It is interesting to find the average efficiency over all possible designs. Suppose, first, that only linear trends are involved ($p = 1$), so that the analysis of §4·3 holds. We have from (18) when $(t-2)\,\Sigma s_i^2/n(t-1)$ is small compared with one,

$$E \simeq 1 - \frac{\Sigma s_i^2}{n(t-1)}, \tag{22}$$

where $s_i = \Sigma_i \xi_1$. To average (22) over all designs we need the following lemma:

LEMMA. *Let $N = tn$, where t, n are integers. Let $x_1, \ldots, x_N$ be N numbers such that $\Sigma x_i = 0$, $\Sigma x_i^2 = 1$. Divide the x_i at random into t blocks of n and let $s_1, \ldots, s_t$ be the sums of the x's in the separate blocks. Then the mean value of $s_1^2 + \ldots + s_t^2$ is $(N-n)/(N-1)$.*

The result follows on enumerating all arrangements, or can be deduced from formulae on sampling from finite populations.

Applying the Lemma to (22), we find that the average efficiency $\bar{E}$ is given by

$$\bar{E} \simeq 1 - \frac{1}{N-1}$$

$$= \frac{N-2}{N-1}. \tag{23}$$

In the same way the approximate result (21) gives in the general case

$$\bar{E} \simeq \left(\frac{N-p-1}{N-1} \right). \tag{24}$$

Equations (23) and (24) are exactly true for two treatments. (These are average efficiencies: in unfavourable cases the efficiency may be appreciably less than this.)

In the example given in §4·2, $t = 2, N = 8, p = 3$ and $\bar{E} = 4/7 = 0\cdot5714$.

In general the gain in efficiency due to the use of systematic designs is only large when the order of polynomial is not too small compared with the number of plots.

Another instructive comparison is with randomized blocks, t plots per block. If there is a linear trend of amount $b\sigma$ per plot, then in the absence of treatment effects the measured quantity for the t plots in any block can be written

$$\text{const.} + \epsilon_1 + b\sigma, \text{const.} + \epsilon_2 + 2b\sigma, \ldots, \text{const.} + \epsilon_t + tb\sigma,$$

where $\epsilon_1, \ldots, \epsilon_t$ are independently distributed with mean zero and variance σ^2. Two plots are selected at random to carry treatments T_1 and T_2 and the average variance of the comparison of T_1 with T_2 is, for this block,

$$\mathscr{E}\,[\epsilon_i + ib\sigma - \epsilon_j - jb\sigma]^2, \tag{25}$$

where (i,j) form a random sample without replacement from $1, \ldots, t$. After some algebra (25) becomes

$$2\sigma^2 \left[1 + \frac{t(t+1)\,b^2}{12} \right],$$

and for n blocks the variance is

$$\frac{2\sigma^2}{n} \left[1 + \frac{t(t+1)}{12} b^2 \right],$$

giving an efficiency

$$E = \left\{ 1 + \frac{t(t+1)}{12} b^2 \right\}^{-1}. \tag{26}$$

This shows, as would naturally be expected, that the steeper the trend and the larger the number of plots per block the less efficient is the randomized block design. These conclusions hold also for higher order trends.

To sum up, we have compared three ways of arranging the experiment:

(*a*) by a systematic design;

(*b*) by complete randomization and the use of covariance analysis to eliminate the trend; and

(*c*) by randomized blocks.

We can nearly always find a systematic design of efficiency very near unity, i.e. for which the variance of a treatment comparison is only slightly greater than $2\sigma^2/n$. The second method gives an average efficiency (24) independently of the amount of the trend. The third method gives an efficiency (26) decreasing as the amount of the trend increases.

For the analysis of the systematic designs to be valid we must be able to *assume* from our physical knowledge of the experiment that the measured quantity is formed by the addition of polynomial trend, treatment effect and completely random variation. This is not necessary when the design is randomized.

7. ANALYSIS OF RESULTS

In this section we show briefly how the results from one of these experiments should be analysed. The assumptions on which the analysis is based are stated at the end of §6. The method of analysis is a standard form of the method of least squares (analysis of covariance).

First, consider the exact designs for two treatments given in §3 of the paper. We can form an analysis of variance as follows. The treatment difference and sum of squares are obtained as if there was no trend. The trend coefficients and sums of squares are obtained as if there was no treatment difference. The residual sum of squares is obtained by subtraction:

	D.F.
Treatments	1
Trend	p
Residual	$2n-p-2$
Total	$2n-1$

The residual mean square s^2 estimates the error variance σ^2, and the estimated standard error of the treatment difference is $s(2/n)^{\frac{1}{2}}$. If the residual variation is normal the usual exact tests of significance hold.

In other cases the analysis is not as simple because trend and treatment effects are not exactly orthogonal. We can set the calculations out in the form of analysis of covariance or in the form of least squares theory. We can illustrate the latter method on the particular case of t treatments with cubic trend, the treatment differences being exactly orthogonal to ξ_1' and ξ_3'. We calculate sums of squares for ξ_1', ξ_2', and ξ_3' as if there were no treatment differences. We set up the least squares equations (13) for the joint estimation of the coefficient of ξ_2' and treatment effects. (The equations are in a form suitable for easy solution by iteration.) We thus get both the best estimates of the treatment effects and the sum of squares for fitting ξ_2' and treatment constants. We have the analysis of variance:

	S.S.	D.F.
Fitting ξ_1' and ξ_3'	U_1	2
Fitting treatments and ξ_2'	U_3	t
Fitting ξ_2'	U_2	1
Treatments	$U_3 - U_2$	$t-1$
Residual	$U_4 - U_1 - U_3$	$nt - t - 3$
Total	U_4	$nt - 1$

The residual mean square estimates the error variance σ^2, and the standard error of any treatment comparisons can be estimated directly from equation (16). If the distribution of the residual is normal, exact F and t tests can be made in the usual way.

8. Application of designs

It has been assumed that the order of trend p is given a priori. This is not normally true in practice, so that it is usually necessary to select a design with good properties for the largest p likely to occur. This value of p will depend on the nature of the experimental material and on the number of plots.

The solutions given above can be made the basis of more complicated designs. For example, the experiment may be repeated in a number of blocks, a trend being estimated from each block.

In the discussion it has been assumed that the main object of the experiment is to investigate the treatment differences. In some applications, such as that mentioned at the beginning of this paper, the trend itself is of direct physical interest. The same type of argument as used in §4 shows that the above systematic designs give optimum estimation of the trend.

Summary

Systematic experimental designs are given for use when a number of treatments are to be compared, one treatment being applied to each of a number of equally spaced plots. It is assumed that there is a smooth trend between plots and that the 'error' is independent in different plots. The designs enable the treatment effects (and the trend) to be estimated as simply and accurately as possible.

Sincere thanks are due to Mr R. C. Palmer, Mr F. J. Anscombe and Dr J. Wishart for helpful discussions.

REFERENCES

Cox, G. M. (1950). *Biometrics*, **6**, 317.
Fisher, R. A. & Yates, F. (1948). *Statistical Tables for Biological, Agricultural and Medical Research*. Oliver and Boyd.
Hald, A. (1948). *On the Decomposition of a Series of Observations*. Copenhagen: Gads Varlag.
Neyman, J. (1929). *The Theoretical Basis of Different Methods of Testing Cereals*. Part II. Warsaw: K. Buszczynski and Sons Ltd.
Williams, R. M. (1949). Unpublished Ph.D. Thesis, Cambridge University.

[*From* Biometrika, *Vol.* 41. *Parts* 3 *and* 4, *December* 1954.]

THE DESIGN OF AN EXPERIMENT IN WHICH CERTAIN TREATMENT ARRANGEMENTS ARE INADMISSIBLE

By D. R. COX

Statistical Laboratory, University of Cambridge

1. INTRODUCTION

Consider an experiment in which the experimental units are arranged in *sets* of k units, a set corresponding, for example, to a single production run of an industrial process. Suppose also that the k units in each set are arranged in order corresponding to the first, second, etc., *period* of the set. If there are both systematic differences between periods and an appreciable component of variance between sets, an experiment to compare a number of treatments, using these experimental units, would ordinarily be arranged in a Latin or Youden square, chosen to eliminate differences between periods and sets. Suppose, however, that for practical reasons there is a restriction on the order of treatments within each set, such as that the level of the treatment must not decrease from one period to the next in a set. Then Latin and Youden squares cannot be used. This paper is concerned with designs for such a situation; the method of construction is described and designs are listed for a few special cases. I have not hesitated to suggest designs that are nearly optimum whenever the determination of the strictly optimum design would be a very lengthy process.

The particular application for which this work was done concerned the properties of alloys cast from high-purity metals. Because of the high cost of the raw materials it was especially important to use an efficient experimental design. From a 20 lb. melt it was possible to cast four 5 lb. ingots, the smallest size convenient for further processing before testing. (In the language of the first paragraph the melt is a set, and the four ingots are the periods within a set, so that $k = 4$.) After an ingot had been cast, the composition of the remaining molten metal could be changed only by the *addition* of one of the alloying metals, no technique being available for selectively reducing the concentration of an alloying element. Therefore we have just the situation described above, in which the level of treatment cannot decrease from one period to the next in a set.

A further similar application is to certain large-scale experiments in steel-works, in which it is possible to tap several ladles from one cast of an open-hearth furnace. The effect of the addition of several elements could, with the present designs, be investigated by varying the chemical composition in successive ladles, rather than by the more costly and cumbersome procedure of varying the composition from cast to cast.

2. A SIMPLE CASE

Suppose that we wish to compare two treatments, A_{-1} and A_1, that $k = 4$ and that if A_1 occurs in one period of a set, all subsequent periods of that set must carry A_1. The only admissible sequences are:

Type	Period 1	Period 2	Period 3	Period 4
1	A_{-1}	A_{-1}	A_{-1}	A_{-1}
2	A_{-1}	A_{-1}	A_{-1}	A_1
3	A_{-1}	A_{-1}	A_1	A_1
4	A_{-1}	A_1	A_1	A_1
5	A_1	A_1	A_1	A_1

and the most general design is formed by taking n_i sets of type i, with $n_1 + \ldots + n_5 = N$ sets in all. Note that the treatment in period 1 is usually A_{-1} and in period 4 is usually A_1. Therefore some confounding of the treatment differences with periods is inevitable, unless only sequences 1 and 5 are used, in which case treatment differences are completely confounded with sets. Similarly some confounding with sets is inevitable, unless treatment differences are completely confounded with periods.

It will be assumed that the analysis of the results of such an experiment can be based on a linear set-up, in which any observation, y, is

(general mean) + (const. for set) + (const. for period)

+ (const. for treatment) + (random term),

where the random terms are uncorrelated and have zero mean and variance σ_r^2.

We shall first suppose that differences between sets and periods are to be eliminated completely. The best linear unbiased estimate, $\hat{\delta}^{(r)}$, of the treatment effect is obtained by the method of least squares and has variance $\lambda \sigma_r^2$, where λ depends only on the design, i.e. on $n_1, \ldots, n_5$. The best design is that which minimizes λ for given N.

To do the minimization, it is convenient to consider the problem as one of analysis of covariance. Introduce a concomitant variable, x, taking values -1 whenever A_{-1} is applied and 1 whenever A_1 is applied. The analysis of covariance of y and x, regarding the experiment as a two-way (set × period) arrangement, is

	s.s. of y	s.p. of x, y	s.s. of x
Between sets	$S(y, y)$	$S(x, y)$	$S(x, x)$
Between periods	$P(y, y)$	$P(x, y)$	$P(x, x)$
Residual	$R(y, y)$	$R(x, y)$	$R(x, x)$
Total	$T(y, y)$	$T(x, y)$	$T(x, x)$

By the theory of analysis of covariance, the residual regression coefficient $R(x, y)/R(x, x)$ is $\frac{1}{2}\hat{\delta}^{(r)}$, and by the formula for the variance of a regression coefficient

$$\text{var}\left(\tfrac{1}{2}\hat{\delta}^{(r)}\right) = \frac{\sigma_r^2}{R(x, x)}. \tag{1}$$

Thus we require to maximize $R(x, x)$ given N. First, it is clear by symmetry that for an optimum design $n_1 = n_5$, $n_2 = n_4$. Therefore

$$T(x, x) = 4N = 4(2n_1 + 2n_2 + n_3) \quad \text{and} \quad S(x, x) = \tfrac{1}{4}(2n_1 . 16 + 2n_2 . 4 + n_3 . 0),$$

since there are, for example, n_1 runs each with a total of -4 and n_1 each with a total of 4. We can find $P(x, x)$ similarly and thus get $R(x, x)$ by subtraction. The answer is

$$NR(x, x) = 12n_1 n_2 + 8n_1 n_3 + 6n_2 n_3 + 4n_2^2. \tag{2}$$

Now maximize $NR(x, x)$ for fixed N introducing a Lagrange multiplier μ. The solution of the resulting linear equations is $n_1 = \tfrac{1}{5}\mu$, $n_2 = n_3 = \tfrac{2}{5}\mu$.

The simplest case in which these equations have a solution in integers is $n_1 = 1, n_2 = n_3 = 2$ and hence the simplest design minimizing the variance per observation is

$$
\left.\begin{array}{cccc}
A_{-1} & A_{-1} & A_{-1} & A_{-1} \\
A_{-1} & A_{-1} & A_{-1} & A_1 \\
A_{-1} & A_{-1} & A_{-1} & A_1 \\
A_{-1} & A_{-1} & A_1 & A_1 \\
A_{-1} & A_{-1} & A_1 & A_1 \\
A_{-1} & A_1 & A_1 & A_1 \\
A_{-1} & A_1 & A_1 & A_1 \\
A_1 & A_1 & A_1 & A_1
\end{array}\right\} \tag{3}
$$

To use the design the order of the sets should be randomized.

Now $R(x, x) = 10$ so that
$$\operatorname{var}(\hat{a}^{(r)}) = \tfrac{2}{5}\sigma_r^2, \tag{4}$$

and σ_r^2 may be estimated from the residual mean square of y adjusting for regression on x, i.e. by the formula
$$\hat{\sigma}_r^2 = \frac{1}{20}\left[R(y, y) - \frac{R^2(x, y)}{R(x, x)}\right]. \tag{5}$$

There are a number of designs with which (4) may be compared:

(a) if the treatments are randomized over eight sets subject only to the requirement that each treatment occurs twice in each set and four times in each period, the variance of the estimate of the treatment effect is $\tfrac{1}{8}\sigma_r^2$; this design does not satisfy our basic requirement but may be considered for comparison;

(b) if the component of variance between sets is σ_1^2, the variance of the mean observation on a set is $\sigma_1^2 + \tfrac{1}{4}\sigma_r^2$, so that if four sets are done with constant treatment A_{-1} and four with constant treatment A_1, the variance of the estimate is $\tfrac{1}{2}\sigma_1^2 + \tfrac{1}{8}\sigma_r^2$.

This shows that if the design (3) and the estimate $\hat{a}^{(r)}$ are used when $\sigma_1^2 \ll \sigma_r^2$, there is an appreciable loss of accuracy compared with the simple design in which the treatment is kept constant in each set. This loss arises from the non-orthogonality with periods and sets; the period differences are systematic, so that information confounded with them is lost, but set differences may, after randomization, be regarded as additional random variation. Hence a second estimate $\hat{a}^{(s)}$, uncorrelated with $\hat{a}^{(r)}$, may be derived from the regression coefficient between sets in the analysis of covariance. Thus

$$\tfrac{1}{2}\hat{a}^{(s)} = \frac{S(x, y)}{S(x, x)}, \tag{6}$$

$$\operatorname{var}(\tfrac{1}{2}\hat{a}^{(s)}) = \frac{\sigma_s^2}{S(x, x)} = \frac{\sigma_s^2}{12}, \tag{7}$$

where $\sigma_s^2 = \sigma_r^2 + 4\sigma_1^2$ is the variance of the set totals on a unit basis and is estimated by

$$\hat{\sigma}_s^2 = \frac{1}{6}\left[S(y, y) - \frac{S^2(x, y)}{S(x, x)}\right]. \tag{8}$$

We combine $\hat{a}^{(r)}$ and $\hat{a}^{(s)}$ into a single estimate $\hat{a}$, where

$$\hat{a} = \left(\frac{R(x, x)}{\hat{\sigma}_r^2}\hat{a}^{(r)} + \frac{S(x, x)}{\hat{\sigma}_s^2}\hat{a}^{(s)}\right)\left(\frac{R(x, x)}{\hat{\sigma}_r^2} + \frac{S(x, x)}{\hat{\sigma}_s^2}\right)^{-1}, \tag{9}$$

$$\operatorname{var}(\hat{a}) = 4\left(\frac{R(x, x)}{\sigma_r^2} + \frac{S(x, x)}{\sigma_s^2}\right)^{-1} = \frac{2\sigma_r^2(\sigma_r^2 + 4\sigma_1^2)}{11\sigma_r^2 + 20\sigma_1^2}, \tag{10}$$

neglecting errors in weighting.

In particular when $\sigma_1^2 = 0$, var $(\hat{d}) = \frac{2}{11}\sigma_r^2$, showing a substantial reduction from (4), and indicating that recovery of information is a worth-while process in this case. Also var $(\hat{d})$ is less than $\frac{1}{8}\sigma_r^2 + \frac{1}{2}\sigma_1^2$, the variance in design (b) above, when $\sigma_1^2/\sigma_r^2 > \frac{1}{4}$.

To sum up, the design (3) should be used only when an appreciable component of variance between sets is expected and information contained in the set totals should be recovered unless $\sigma_r^2 \ll \sigma_1^2$.

If practical conditions demand the use of a number of sets other than 8, or a multiple of 8, (2) may be used to determine the best values of n_1, n_2 and n_3.

Optimum designs similar to (3) may be found for other values of k. Two examples are

$k=2$			$k=3$	
A_{-1}	A_{-1}	A_{-1}	A_{-1}	A_{-1}
A_{-1}	A_1	A_{-1}	A_{-1}	A_{-1}
A_{-1}	A_1	A_{-1}	A_{-1}	A_1
A_1	A_1	A_{-1}	A_1	A_1
		A_1	A_1	A_1
		A_1	A_1	A_1

In the preceding analysis it has been assumed that two fixed treatments are to be compared. If, however, the linear regression on a factor that may be varied continuously is required, the concomitant variable, x, is no longer confined to two values. If, for practical reasons, the level of the factor must lie between an upper and a lower limit, these may be made to correspond to $x = \pm 1$; any admissible set of k treatments is then defined by a non-decreasing sequence of k terms, all between ± 1. Although I have not obtained a formal proof, it is probably true that so long as only the linear effect is required, the designs in which only the lower and upper levels appear are still optimum.

3. A 2^3 FACTORIAL DESIGN

As an example of the design of a more complicated experiment, consider a 2^3 factorial experiment in sets with four periods per set, assuming that the three-factor interaction is unimportant and that the level of each factor must not decrease from period to period within a set. Introduce six concomitant variables x_1, x_2, x_3, x_4 ($= x_{23}$), x_5 ($= x_{31}$) and x_6 ($= x_{12}$) to represent the main effects A, B, C and the two-factor interactions BC, CA, AB. Thus x_2 takes values ± 1 according as B is at its upper or its lower level and x_5 equals 1 for the combinations $A_{-1}C_{-1}$, A_1C_1 and -1 for the combinations $A_{-1}C_1$, A_1C_{-1}. Twice the residual regression coefficient of y on x_i is thus $\hat{a}_i^{(r)}$, the least-squares estimate of the corresponding main effect or interaction.

Let R_{ij} denote the residual sum of products of x_i and x_j in the analysis of covariance and $R_i(y)$ the residual sum of products of x_i and y. Then the column $\hat{\mathbf{a}}^{(r)}$ of estimates, and their variance-covariance matrix, are given by

$$\mathbf{R}\hat{\mathbf{a}}^{(r)} = 2\mathbf{R}(y), \tag{11}$$

$$\operatorname{var}(\hat{\mathbf{a}}^{(r)}) = 4\mathbf{R}^{-1}\sigma_r^2. \tag{12}$$

We have now to choose the design to satisfy conditions on $\mathbf{R}^{-1}$; in particular we require the diagonal elements of $\mathbf{R}^{-1}$ to be small, especially those associated with the main effects. In a design in eight sets with one factor, (4) gives var $(\hat{a}^{(r)}) = \frac{2}{5}\sigma_r^2$. If several factors are investigated simultaneously the variance of a main effect cannot be less than the minimum variance obtained when that factor is investigated by itself. Therefore if we can find a design in eight sets for which the variance of a main effect is inappreciably greater than $\frac{2}{5}\sigma_r^2$, this design is effectively optimum for the estimation of main effects.

A common-sense procedure for eight runs is to arrange that each factor separately has the design (3) and that the factor combinations are distributed as symmetrically as possible. This gives the design

$$
\begin{pmatrix}
A_{-1}B_1C_{-1} & A_{-1}B_1C_{-1} & A_{-1}B_1C_1 & A_{-1}B_1C_1 \\
A_{-1}B_{-1}C_{-1} & A_{-1}B_1C_1 & A_{-1}B_1C_1 & A_1B_1C_1 \\
A_{-1}B_{-1}C_{-1} & A_{-1}B_1C_{-1} & A_{-1}B_1C_{-1} & A_1B_1C_1 \\
A_{-1}B_{-1}C_{-1} & A_{-1}B_{-1}C_{-1} & A_1B_1C_{-1} & A_1B_1C_{-1} \\
A_{-1}B_{-1}C_1 & A_{-1}B_{-1}C_1 & A_1B_1C_1 & A_1B_1C_1 \\
A_{-1}B_{-1}C_{-1} & A_1B_{-1}C_{-1} & A_1B_{-1}C_{-1} & A_1B_1C_1 \\
A_{-1}B_{-1}C_{-1} & A_1B_{-1}C_1 & A_1B_{-1}C_1 & A_1B_1C_1 \\
A_1B_{-1}C_{-1} & A_1B_{-1}C_{-1} & A_1B_{-1}C_1 & A_1B_{-1}C_1
\end{pmatrix} \tag{13}
$$

The matrix $\mathbf{R}$ of residual sums of products can be found easily. Main effects are orthogonal to interactions, i.e. the residual sum of products of any one of x_1, x_2, x_3 with any one of x_4, x_5, x_6 is zero. Thus $\mathbf{R}$ can be partitioned into two 3×3 matrices whose inverses may be found separately, i.e.

$$
\mathbf{R} = \begin{pmatrix} \mathbf{R}_1 & 0 \\ \hline 0 & \mathbf{R}_2 \end{pmatrix}, \qquad
\mathbf{R}^{-1} = \begin{pmatrix} \mathbf{R}_1^{-1} & 0 \\ \hline 0 & \mathbf{R}_2^{-1} \end{pmatrix},
$$

where
$$
\mathbf{R}_1 = \begin{pmatrix} 10 & 2 & -2 \\ 2 & 10 & -2 \\ -2 & -2 & 10 \end{pmatrix}, \quad
\mathbf{R}_1^{-1} = \frac{1}{56}\begin{pmatrix} 6 & -1 & 1 \\ -1 & 6 & 1 \\ 1 & 1 & 6 \end{pmatrix},
$$

$$
\mathbf{R}_2 = \begin{pmatrix} 22 & -2 & 4 \\ -2 & 22 & 4 \\ 4 & 4 & 8 \end{pmatrix}, \quad
\mathbf{R}_2^{-1} = \frac{1}{96}\begin{pmatrix} 5 & 1 & -3 \\ 1 & 5 & -3 \\ -3 & -3 & 15 \end{pmatrix}. \tag{14}
$$

Thus the variance of the estimate of a main effect is $\frac{24}{56}\sigma_r^2 = \frac{3}{7}\sigma_r^2$, which is inappreciably greater than $\frac{2}{5}\sigma_r^2$, the variance in a single-factor design. Hence (13) is effectively optimum for estimating main effects.

The recovery of information between sets will not be considered in detail, because with one replicate of the design only a single degree of freedom is available for estimating σ_s^2. If several replicates are done, recovery of information should be considered. From the set totals in (13) estimates may be formed of $a_1 - a_2$, a_3 but not of a_1, a_2 separately. Hence if (13) is repeated the names of the treatments should be changed.

With eight sets an alternative arrangement to (13) is with the treatments held constant in each set, each of the eight combinations in the 2^3 experiment occurring once. The variance of the estimate of a main effect is then $\frac{1}{2}(\sigma_1^2 + \frac{1}{4}\sigma_r^2)$. Therefore (13) is the more accurate only if

$$
\tfrac{3}{7}\sigma_r^2 < \tfrac{1}{2}\sigma_1^2 + \tfrac{1}{8}\sigma_r^2,
$$

i.e. if
$$
\sigma_1^2 > \tfrac{17}{28}\sigma_r^2. \tag{15}
$$

4. THE ESTIMATION OF CURVATURES

If it is required to estimate the curvature of the response to a factor in addition to the linear effect, we may introduce two concomitant variables for each main effect, the first, x', representing the linear component and the second, x'', the curvature. Thus with a factor at three equally spaced levels A_{-1}, A_0, A_1, x' takes values -1, 0, 1 and x'' values 1, -2, 1. Unless k and the number of factors are very small the number of possible designs is large and a formal mathematical analysis to determine the best design is rarely practicable.

However, when k is even we can write down a pair of sets from which the curvature may be estimated directly. Thus for $k = 4$

$$\left.\begin{array}{cccc} A_{-1} & A_{-1} & A_0 & A_0 \\ A_0 & A_0 & A_1 & A_1 \end{array}\right\} \tag{16}$$

gives an estimate of curvature from the difference of the observations on A_{-1}, A_1 and those on A_0. This estimate is clearly unaffected by period and set differences. Further, if a second factor B is superimposed on (16) according to the pattern

$$\left.\begin{array}{cccc} B_{-1} & B_1 & B_1 & B_1 \\ B_{-1} & B_{-1} & B_{-1} & B_1 \end{array}\right\} \tag{17}$$

the estimate of the curvature of A is unaffected by the main effect of B and by the linear-linear component of the interaction AB.

These results can be used in various ways by building up more complicated designs from the above pairs of sets. For example the following design in four sets is formed by interchanging A and B in the second pair:

$$\left.\begin{array}{cccc} A_{-1}B_{-1} & A_{-1}B_1 & A_0B_1 & A_0B_1 \\ A_0B_{-1} & A_0B_{-1} & A_1B_{-1} & A_1B_1 \\ A_{-1}B_{-1} & A_1B_{-1} & A_1B_0 & A_1B_0 \\ A_{-1}B_0 & A_{-1}B_0 & A_{-1}B_1 & A_1B_1 \end{array}\right\}. \tag{18}$$

If this design is analysed neglecting all interactions except the linear-linear one, the variance of a linear main effect is $\frac{1}{3}\sigma_r^2$, of the curvature of a main effect (defined as the regression coefficient of the response on x'') is $\frac{9}{160}\sigma_r^2$ and of the linear-linear interaction is $\frac{1}{4}\sigma_r^2$. Apart from the two curvatures, which are very slightly correlated, all estimates are independent. The most accurate estimate of the linear effect obtained in (3) with eight sets has a variance $\frac{1}{10}\sigma_r^2$ and hence the minimum variance in four sets is $\frac{1}{5}\sigma_r^2$. The increase in variance due to the inclusion of an estimate of curvature is not as large as might be expected. The accuracy of the linear effect relative to that of the curvature may be increased by the addition to (18) of sets derived from (3).

Another possibility is to introduce into the 2^3 experiment, (13), two new factors D and E whose curvature only is to be investigated. Hold D and E at their middle levels except in sets 2, 3, 6 and 7, and in these sets vary them in the following way:

$$\begin{array}{lcccc} \text{Set } 2 & D_{-1}E_0 & D_{-1}E_0 & D_0E_0 & D_0E_0 \\ 3 & D_0E_0 & D_0E_0 & D_0E_1 & D_0E_1 \\ 6 & D_0E_0 & D_0E_0 & D_1E_0 & D_1E_0 \\ 7 & D_0E_{-1} & D_0E_{-1} & D_0E_0 & D_0E_0 \end{array}\left.\right\}. \tag{19}$$

There is now a small amount of confounding with the interactions of A, B, C, but the loss of information is slight.

The number of possible designs is very large, and it is impossible to give a thorough account in a general paper such as this. The method is to decide how many factors are to be investigated, the number of sets available and the relative importance of linear effects and curvatures. Then build up from basic designs, such as (17), an arrangement with as much 'balance' as possible, and work out the matrix of the variances of the estimates. Next compare with simple designs such as (3) to see that too much accuracy has not been lost by the inclusion of extra factors. With practice this can be done fairly quickly, but clearly the work is only worth-while when economy in the design is of great importance.

5. ANALYSIS OF OBSERVATIONS

The analysis of the observations from one of the above designs is essentially a straight-forward process of multiple analysis of covariance, with the simplification that the matrix of sums of squares and products of the concomitant variables usually takes a simple form and can always be obtained and inverted beforehand. Also the concomitant variables are restricted to small integral values and sums of products with them can be formed rapidly.

Table 1 gives the analysis for the 2^3 design (13). Fictitious observations have been constructed by taking the main effects of A, B, C to be 2, -2, 1 and the interactions to be zero. The uncontrolled variation has been represented by period differences, by a random variation between sets of standard deviation 2, and by a random variation from unit to unit of standard deviation 1. A constant has been added to make all the observations positive.

Figures in bold type are constants for the design and may be entered before the observations are obtained. The procedure should be clear from the following notes:

(1) First form the analysis of variance of the observations, y, into correction, between sets (rows), between periods (columns) and residual.

(2) Then find the total sums of products of y with the six concomitant variables. For example x_{23} takes values 1 for $B_1 C_1$ and $B_{-1} C_{-1}$ and -1 otherwise. It is desirable to form the total sum of products separately by rows and by columns to obtain a check. Thus, by rows,
$$\Sigma y x_{23} = -3 \cdot 92 - 6 \cdot 44 + 7 \cdot 87 + \ldots - 6 \cdot 92 - 7 \cdot 35 = 48 \cdot 66.$$

(3) The sum of products for sets is obtained from the sum of products of the set totals of y and the concomitant variable. Thus for x_{23} it is
$$\tfrac{1}{4}(20 \cdot 36 \times 4 + 27 \cdot 50 \times 4) = 47 \cdot 86.$$

Similarly for periods. In this way the full analysis of covariance table is obtained, calculating the residual term as total + correction − rows − columns, since the individual terms have not be corrected.

(4) The estimates of the regression coefficients are formed by multiplying the residual sums of products by the appropriate coefficients, obtained from (14), given in bold type. Thus the regression coefficient on x_{31} is
$$\tfrac{1}{96}[1 \times (-1 \cdot 035) + 5 \times (-8 \cdot 605) + (-3) \times (-2 \cdot 06)] = -0 \cdot 3946.$$

The sum of squares for regression is obtained as the sum of products of the regression coefficients with the corresponding residual sum of products, and may, in this instance, be collected separately for main effects and for interactions. Thus, for main effects, the sum of squares is
$$1 \cdot 2895 \times 9 \cdot 705 + (-1 \cdot 4718) \times (-12 \cdot 385) + 0 \cdot 1230 \times 1 \cdot 595 = 30 \cdot 9390.$$

This leads to an analysis of variance from which the error variance is estimated by $1 \cdot 3845$ with 15 degrees of freedom (the population value is 1).

Table 1. *Analysis of 2^3 experiment*

$A_{-1}B_1C_{-1}$	3·92	$A_{-1}B_1C_{-1}$	6·44	$A_{-1}B_1C_1$	7·87	$A_{-1}B_1C_1$	9·81
$A_{-1}B_{-1}C_{-1}$	5·13	$A_{-1}B_1C_1$	2·99	$A_{-1}B_1C_1$	3·87	$A_1B_1C_1$	8·37
$A_{-1}B_{-1}C_{-1}$	4·10	$A_{-1}B_1C_{-1}$	1·89	$A_{-1}B_1C_{-1}$	5·27	$A_1B_1C_1$	8·27
$A_{-1}B_{-1}C_{-1}$	7·07	$A_{-1}B_{-1}C_{-1}$	5·81	$A_1B_1C_{-1}$	9·22	$A_1B_1C_{-1}$	8·80
$A_{-1}B_{-1}C_1$	10·78	$A_{-1}B_{-1}C_1$	9·01	$A_1B_1C_1$	9·01	$A_1B_1C_1$	12·13
$A_{-1}B_{-1}C_{-1}$	5·66	$A_1B_{-1}C_{-1}$	7·86	$A_1B_{-1}C_{-1}$	8·35	$A_1B_1C_1$	5·63
$A_{-1}B_{-1}C_{-1}$	2·97	$A_1B_{-1}C_1$	7·29	$A_1B_{-1}C_1$	8·68	$A_1B_1C_1$	7·68
$A_1B_{-1}C_{-1}$	5·90	$A_1B_{-1}C_{-1}$	5·75	$A_1B_{-1}C_1$	6·92	$A_1B_{-1}C_1$	7·35

Set (row) totals

Set	y	x_1	x_2	x_3	x_{23}	x_{31}	x_{12}
1	28·04	−4	4	0	0	0	−4
2	20·36	−2	2	2	4	0	0
3	19·53	−2	2	−2	0	4	0
4	30·90	0	0	−4	0	0	4
5	40·93	0	0	4	0	0	4
6	27·50	2	−2	−2	4	0	0
7	26·61	2	−2	2	0	4	0
8	25·92	1	−4	0	0	0	−4
	219·80	0	0	0	8	8	0

Period (column) totals

Period	y	x_1	x_2	x_3	x_{23}	x_{31}	x_{12}
1	45·53	−6	−6	−6	4	4	4
2	47·04	−2	−2	−2	0	0	−4
3	59·19	2	2	2	0	0	−4
4	68·04	6	6	6	4	4	4
	219·80	0	0	0	8	8	0

Analysis of covariance

	y^2	yx_1	yx_2	yx_3	yx_{23}	yx_{31}	yx_{12}	D.F.
Correction	1509·7512	0	0	0	54·95	54·95	0	1
Sets	1587·2470	4·995	−4·995	10·005	47·86	46·15	17·87	8
Periods	1552·3300	19·92	19·92	19·92	56·785	56·785	3·67	4
Residual	55·2266	9·705	−12·385	1·595	−1·035	−8·605	−2·06	21
Total*	1685·0524	34·62	2·54	31·52	48·66	39·38	19·48	32

* Checked by columns and by rows. Note that terms for sets and periods are uncorrected.

$$\frac{1}{56}\begin{pmatrix} 6 & -1 & 1 \\ -1 & 6 & 1 \\ 1 & 1 & 6 \end{pmatrix} \qquad \frac{1}{96}\begin{pmatrix} 5 & 1 & -3 \\ 1 & 5 & -3 \\ -3 & -3 & 15 \end{pmatrix}$$

Regr. coeff.	1·2895	−1·4718	0·1230	−0·0792	−0·3946	−0·0206
Contr. to s.s.	12·5146	18·2282	0·1962	0·0820	3·3955	0·0424
Variance	0·1483	0·1483	0·1483	0·0721	0·0721	0·2163
St. error	0·3851	0·3851	0·3851	0·2685	0·2685	0·4651

	D.F.	y^2	Mean square
Regr. on main effects (x_1, x_2, x_3)	3	30·9390	10·3130
interactions (x_{23}, x_{31}, x_{12})	3	3·5199	1·1733
Error	15	20·7677	1·3845
Residual (uncorrected for regr.)	21	55·2266	—

	Main effects			Interactions		
	A	B	C	BC	CA	AB
Estimate	2·5790	−2·9436	0·2460	−0·1584	−0·7892	−0·0412
St. error	0·7702	0·7702	0·7702	0·5370	0·5370	0·9302
True value	2	−?	1	0	0	0

(5) The variances of the regression coefficients are obtained by multiplying the error variance by the appropriate diagonal element of the matrices of coefficients. Confidence limits (or significance tests) for an individual regression coefficient follow from the t distribution in the usual way.

(6) Main effects and interactions are twice the corresponding regression coefficients.

6. Discussion

We have considered experimental design in a special situation where practical conditions severely limit the treatment arrangements that may be used. The limitation makes a substantial amount of non-orthogonality inevitable if certain portions of the uncontrolled variation (set and period differences) are to be eliminated from the error.

The following remarks are relevant to the application of the designs developed above and of other designs that may be worked out for similar situations:

(i) The design of even a simple experiment to compare two treatments requires care.

(ii) The number of factors that may be examined simultaneously is restricted.

(iii) It is better not to attempt to eliminate set differences unless there is an appreciable component of variance between sets (§§ 2, 3).

(iv) The correct method of analysis must be used, but this is usually straightforward once the sequence of calculations has been grasped.

(v) Finally, it must be repeated that the situation is a complicated one and that the use of special designs, while sometimes very well worth while, is usually only justified when economy in the experimental material is of great importance.

I am very grateful to Dr C. J. Anson, G. K. N. Group Research Laboratory, for suggesting the problem and for providing the description of the application, and to Dr J. Wishart for helpful comments on the draft of the paper.

[*From* Biometrika, *Vol.* 44. *Parts* 1 *and* 2, *June* 1957.]

THE USE OF A CONCOMITANT VARIABLE IN SELECTING AN EXPERIMENTAL DESIGN*

By D. R. COX

*Department of Biostatistics, School of Public Health, University of North Carolina
and Department of Statistics, University of California*†

1. INTRODUCTION

Suppose that we have t treatments for comparison using N experimental units, for example, animals, and that on each unit we have a preliminary observation, for example, the body weight, available before the treatments are allotted to the units. Suppose that it is expected that the final observation of interest, y, will, in the absence of treatment effects, be well correlated with the preliminary observation, x. Then a number of standard methods are available for exploiting this correlation in order to increase the precision of the estimated treatment effects.

The object of this paper is to compare the methods in some simple situations with a small number of experimental units. Attention is restricted to experiments in which each of the alternative treatments appears the same number, k, of items, so that $N = tk$.

2. SOME STANDARD METHODS

We list first the following methods for increasing the precision of the treatment comparisons.

Method I. An index of response is used, for example, y/x, y/x^α, $y - x$, etc. The index may be suggested by analysis of previous data or by general consideration of a plausible form for the regression relation between y and x.

Method II. The treatments are completely randomized over the experimental units and an adjustment made for regression on x by analysis of covariance. For simplicity only linear regression will be considered.

Method III. The experimental units are ranked in order of increasing x and then grouped into k blocks of t units each, the first block, for example, consisting of the t units with lowest values of x. A randomized block arrangement is then constructed, based on this grouping of the units.

Method IV. This is applicable when k is equal to, or is a multiple of, t. The units are again ranked in order of increasing x and a Latin square arrangement used to control not only differences between blocks but also variation associated with order within a block. Thus with $t = k = 4$, $N = 16$, the units are numbered from $1, ..., 16$, in order of increasing x and a Latin square set out, as in the following example:

| | Order within block | | | |
Block no.	1	2	3	4
1	$1:T_3$	$2:T_1$	$3:T_2$	$4:T_4$
2	$5:T_4$	$6:T_2$	$7:T_3$	$8:T_1$
3	$9:T_2$	$10:T_4$	$11:T_1$	$12:T_3$
4	$13:T_1$	$14:T_3$	$15:T_4$	$16:T_2$

* This paper was prepared with the partial support of the Air Research and Development Command, under contract with the U.S.A.F. School of Aviation Medicine.
† Present address, Birkbeck College, University of London.

A similar method, employing Youden squares, could be employed if k is not a multiple of t, but would require a lengthier analysis.

Method V. Methods II and III may be combined by using the randomized block design plus a covariance correction.

Method VI. It often happens that the preliminary observation x is not the only thing associated with the units that can be used to increase precision. One of a number of possibilities is that a grouping into randomized blocks should be made on the basis of the other criteria and then a covariance adjustment made for x.

Method VII. A systematic, or non-randomized, design may be used, chosen, for example, to give maximum precision under the hypothesis that the regression equation of y on x is a polynomial of low degree (Cox, 1951). The use of such a design has the disadvantages attendant on the lack of full randomization.

An eighth method is that of Papadakis (see, for example, Bartlett (1938)). This is useful in large experiments in which there are expected to be trends or serial correlation between experimental units adjacent in space or time, but will not be considered here.

3. BASIS FOR COMPARISON

Denote the t treatments by $T_1, ..., T_t$ and the corresponding estimated treatment means, after adjustment by covariance where appropriate, by $\hat{y}_1, ..., \hat{y}_t$. We measure the random errors associated with the design by the variance of the estimated difference between a pair of treatments, averaged over all pairs of treatments, i.e. by

$$V_t = \operatorname*{Ave}_{i,j,\, i \neq j} V(\hat{y}_i - \hat{y}_j). \tag{1}$$

This is a certain multiple of the population residual variance σ_0^2 and is not affected by errors in estimating σ_0^2. Call (1) the *true* (average) *imprecision* of the design; it is slightly different from the quantity suggested by Lucas and used by Greenberg (1953).

Often, however, we are interested in the *apparent imprecision*, V_a, making due allowance for the effective loss of information that arises in estimating the residual variance. To obtain V_a we use Fisher's factor $(f+3)/(f+1)$, where f is the number of degrees of freedom available to estimate the residual variance (Cochran & Cox, 1950, p. 26); that is, we put

$$V_a = V_t \left(\frac{f+3}{f+1} \right). \tag{2}$$

This will be supplemented by the rather arbitrary rule that if $f < 5$ no effective estimate of the residual variance is considered possible from the data alone.

From the point of view of the length of confidence intervals and the sensitivity of significance tests, a situation with $V_a = 1$ and with the variance estimated from the data is nearly equivalent to a situation in which the residual variance is known and $V_t = 1$.

The use of the average variance in (1) is a natural step but should be viewed critically, particularly in small experiments where there may be substantial variation in precision between different randomization patterns and between different comparisons within one randomization pattern. This is particularly so with Method II.

4. Calculation of imprecision

To compare the different methods it will be assumed that, in the absence of treatment effects, y and x have a bivariate normal frequency distribution with correlation coefficient, ρ, and with the variance of y for fixed x denoted by σ_0^2. If the effect of variation accounted for by x were completely removed, we should have $V_t = 2\sigma_0^2/k$. Therefore we write

$$V_t = \frac{2\sigma_0^2}{k} I_t, \tag{3}$$

$$V_a = \frac{2\sigma_0^2}{k} I_a, \tag{4}$$

and use I_t and I_a as indices of true and apparent imprecision. Clearly $I_a > I_t \geqslant 1$.
We now compute I_t and I_a for the various procedures set out in § 2.

Table 1. *Loss of precision from using wrong index of response*

ρ	Range of β_0/β within which $I_t \leqslant$			I_t if x ignored
	1·1	1·2	1·5	
0·2	$(-0{\cdot}55,\ 2{\cdot}55)$	$(-1{\cdot}19,\ 3{\cdot}19)$	$(-2{\cdot}46,\ 4{\cdot}46)$	1·04
0·4	$(0{\cdot}28,\ 1{\cdot}72)$	$(-0{\cdot}02,\ 2{\cdot}02)$	$(-0{\cdot}62,\ 2{\cdot}62)$	1·19
0·6	$(0{\cdot}68,\ 1{\cdot}32)$	$(0{\cdot}40,\ 1{\cdot}60)$	$(0{\cdot}06,\ 1{\cdot}94)$	1·56
0·8	$(0{\cdot}76,\ 1{\cdot}24)$	$(0{\cdot}67,\ 1{\cdot}33)$	$(0{\cdot}47,\ 1{\cdot}53)$	2·78
0·9	$(0{\cdot}82,\ 1{\cdot}18)$	$(0{\cdot}74,\ 1{\cdot}26)$	$(0{\cdot}59,\ 1{\cdot}41)$	5·26
0·95	$(0{\cdot}90,\ 1{\cdot}10)$	$(0{\cdot}85,\ 1{\cdot}15)$	$(0{\cdot}77,\ 1{\cdot}23)$	10·26

Method I. Suppose that the true regression coefficient of y on x is β and that the index of response is $y - \beta_0 x$. This situation has been considered by Gourlay (1953). It is easily shown that the index of true imprecision is

$$I_t^{(1)} = \left(1 - 2\rho^2 \frac{\beta_0}{\beta} + \rho^2 \frac{\beta_0^2}{\beta^2}\right)(1 - \rho^2)^{-1} \quad (\rho \neq 0), \\ 1 + \beta_0^2 \sigma_x^2/\sigma_y^2 \quad (\rho = 0). \tag{5}$$

If no attempt is made to use x, i.e. if $\beta_0 = 0$, $I_t = (1 - \rho^2)^{-1}$. Thus the attempted correction is an advantage whenever $0 < \beta_0/\beta < 2$. Table 1 shows the ranges of values of β_0/β for which $I_t \leqslant 1 \cdot 1$, $1 \cdot 2$, $1 \cdot 5$ and so tells us how near β and β_0 need to be to avoid losing a specified amount of information.

Fisher (1935, p. 163) criticized the use of indices of response based on inadequate assessments of the relation between y and x. It is no contradiction of these criticisms to conclude from Table 1 that, particularly if ρ is not near unity, β_0 does not need to be very near β to give a worth-while increase in precision. The use of non-linear indices such as y/x^α allows curvilinear regression to be accounted for. Note, however, that if the treatment effects are constant, independently of x, on the y scale, they will not be constant when the index is used; if the index and the original observation have equal physical significance there may well be no reason for expecting one rather than the other to be the appropriate scale for measuring treatment effects. However, if the object is the estimation of the effect on the long-run average of y of a change in treatments, there may occasionally be difficulties connected with a naïve use of average of y/x^α.

Method II. Denote the terms of the analysis of covariance as follows:

$$
\begin{array}{llll}
\text{Treatments} & T_{xx} & T_{xy} & T_{yy} \\
\text{Residual} & R_{xx} & R_{xy} & R_{yy} \\
\text{Total} & S_{xx} & S_{xy} & S_{yy}
\end{array}
$$

where, for example, R_{xy} denotes the residual sum of products of x and y. If $\bar{x}_i$ is the mean of x for the ith treatment and $\hat{y}_i$ the mean of y adjusted for regression on x,

$$
\left.
\begin{aligned}
V(\hat{y}_i - \hat{y}_j) &= \frac{2\sigma_0^2}{k}\left\{1 + \frac{(\bar{x}_i - \bar{x}_j)^2}{2R_{xx}}\right\}, \\
\operatorname*{Ave}_{i,j} V(\hat{y}_i - \hat{y}_j) &= \frac{2\sigma_0^2}{k}\left\{1 + \frac{T_{xx}}{(t-1)R_{xx}}\right\}
\end{aligned}
\right\}
\tag{6}
$$

This is conditional on fixed x's; but the x's are in fact a random sample from a normal population and the second term in brackets is therefore proportional to an F variate. Hence if we take expectations and remove the factors $2\sigma_0^2/k$, we obtain for the index of true imprecision

$$
\begin{aligned}
I_t^{(2)} &= 1 + \frac{1}{(k-1)t - 2} \\
&= \frac{N-t-1}{N-t-2}.
\end{aligned}
\tag{7}
$$

Also

$$
I_a^{(2)} = I_t^{(2)}\frac{N-t+2}{N-t},
\tag{8}
$$

since the residual degrees of freedom, after adjusting for regression, are $N - t - 1$.

Table 2 gives the values of (7) and (8) for various values of t, k with $N \leqslant 20$.

The increase in variance above that to be expected when the effect of x is completely eliminated can be described as due to errors in estimating the residual regression coefficient, or as arising from the non-orthogonality present in the linear set-up chosen to represent the populations sampled.

Method III. Consider first one block and a fixed set of x values $x_1, \ldots, x_t$. In the absence of treatment effects, the corresponding observations on y are $\beta x_1 + \epsilon_1, \ldots, \beta x_t + \epsilon_t$, where β is the regression coefficient of y on x and $\epsilon_1, \ldots, \epsilon_t$ are independent with constant mean and with variance σ_0^2, and determine the dispersion of y about its regression line on x. Hence if two positions are selected randomly to carry treatments T_i and T_j, the expected mean-square difference between the resulting observations is

$$
\sigma_0^2 + \frac{\beta^2}{t-1}\Sigma(x_i - \bar{x})^2 = \sigma_0^2\left\{1 + \frac{\beta^2}{(1-\rho^2)(t-1)}\frac{1}{\sigma_x^2}\frac{\Sigma(x_i - \bar{x})^2}{}\right\}.
$$

Hence, averaging over the k blocks, and over the distribution of the x's, we have that

$$
I_t^{(3)} = 1 + \frac{\rho^2}{1-\rho^2}W,
\tag{9}
$$

where W is the expected mean square of x within blocks, divided by the variance of x.

Table 2. *Measure of imprecision for various designs*

N	9	12	12	12	15	15	16	16	18	18	18	20	20	20
t	3	4	3	2	5	3	4	2	6	3	2	5	4	2
k	3	3	4	6	3	5	4	8	3	6	9	4	5	10
$I_t^{(2)}$	1·25	1·17	1·14	1·12	1·12	1·10	1·10	1·08	1·10	1·08	1·07	1·08	1·07	1·06
$I_a^{(2)}$	1·67	1·46	1·43	1·35	1·35	1·28	1·28	1·24	1·28	1·22	1·21	1·22	1·21	1·18
W	0·2498	0·2390	0·1715	0·1112	0·2325	0·1278	0·1633	0·0763	0·2282	0·1005	0·0654	0·1584	0·1214	0·0570
$I_t^{(3)}: \rho = 0$	1·00	1·00	1·00	1·00	1·00	1·00	1·00	1·00	1·00	1·00	1·00	1·00	1·00	1·00
0·4	1·05	1·05	1·03	1·02	1·04	1·02	1·03	1·01	1·04	1·02	1·01	1·03	1·02	1·01
0·6	1·14	1·13	1·10	1·06	1·13	1·07	1·09	1·04	1·13	1·06	1·04	1·09	1·07	1·03
0·8	1·44	1·42	1·30	1·20	1·41	1·23	1·29	1·14	1·41	1·18	1·12	1·28	1·22	1·10
0·9	2·07	2·02	1·73	1·47	1·99	1·54	1·70	1·33	1·97	1·43	1·28	1·68	1·52	1·24
0·95	3·31	3·21	2·59	2·03	3·15	2·18	2·51	1·71	3·11	1·93	1·61	2·47	2·12	1·53
$I_a^{(3)}/I_t^{(3)}$	*	1·29	1·29	1·33	1·22	1·22	1·20	1·25	1·18	1·18	1·22	1·15	1·15	1·20
W'	0·1015	—	—	—	—	—	0·0676	0·0436	—	0·0540	—	—	—	—
$I_t^{(4)}: \rho = 0$	1·00	—	—	—	—	—	1·00	1·00	—	1·00	—	—	—	—
0·4	1·02	—	—	—	—	—	1·01	1·01	—	1·01	—	—	—	—
0·6	1·06	—	—	—	—	—	1·04	1·02	—	1·03	—	—	—	—
0·8	1·18	—	—	—	—	—	1·12	1·08	—	1·10	—	—	—	—
0·9	1·43	—	—	—	—	—	1·29	1·19	—	1·23	—	—	—	—
0·95	1·94	—	—	—	—	—	1·63	1·40	—	1·50	—	—	—	—
$I_a^{(4)}/I_t^{(4)}$	*	—	—	—	—	—	1·29	*	—	1·29	—	—	—	—
$I_t^{(5)}$	1·25	1·17	1·17	1·20	1·12	1·12	1·11	1·14	1·10	1·10	1·12	1·08	1·08	1·11
$I_a^{(5)}$	*	1·56	1·56	*	1·41	1·41	1·36	1·47	1·32	1·32	1·41	1·26	1·26	1·36
$I_t^{(5)'}$	1·17	1·12	1·11	1·10	1·10	1·08	1·08	1·07	1·08	1·07	1·06	1·07	1·06	1·05
$I_a^{(5)'}$	*	1·50	1·48	*	1·38	1·35	1·32	1·38	1·30	1·28	1·33	1·24	1·24	1·29
$I_t^{(6)}$	1·50	1·25	1·25	1·33	1·17	1·17	1·14	1·20	1·12	1·12	1·17	1·10	1·10	1·14
$I_a^{(6)}$	*	1·67	1·67	1·87	1·46	1·46	1·40	1·50	1·35	1·35	1·46	1·28	1·28	1·40

The superfixes (2), (3), ... refer to the method of design. I_t denotes a measure of true variance, taking no account of errors in estimating the residual variance; I_a denotes a measure of apparent variance taking account of the residual degrees of freedom. $I = 1$ represents complete elimination of the variation associated with x. Cases where there are fewer than 5 residual degrees of freedom are marked by an asterisk. The values given for Method V are lower limits.

We need, in order to calculate W, to consider the following. Take an ordered sample $x_{(1)}, \ldots, x_{(n)}$ from a unit normal population. Divide this into blocks as described above and find the mean square within blocks. Then W is the expected value of this mean square and can be calculated for $N \leqslant 20$ from recently published tables of the second moments of order statistics in a normal sample (Teichroew, 1956); Table 2 gives some numerical values derived in this way.

The general conclusion from the values of $I_t^{(2)}$ and $I_t^{(3)}$ is that Method III is somewhat better than Method II if $\rho < 0.6$ and that Method II becomes appreciably better than Method III only when ρ is as large as 0.8 or more. It makes little difference if the comparison is based on I_a instead of on I_t. In larger experiments with moderate t and large k, both methods will be effective in reducing the value of I_t to near unity, except when ρ is very near unity, when the use of Method III will be inadvisable. However, Method III will remain reasonably efficient for any form of smooth regression between y and x, not just for linear regression. If the regression is linear, but the distribution of x is leptokurtic, the randomized block method is likely to be relatively less effective due to the end blocks having units with widely discrepant values of x.

Method IV. The argument is similar when a Latin square is used, except that W', equal to the expected residual mean square in the two-way array of x's formed from the rows and columns of the square, replaces W. Table 2 gives the value of W', and of $I_t^{(4)}$ in certain cases. If we have r squares, each of size $t \times t$, the residual degrees of freedom are $(t-1)(rt-r-1)$, when the residual within squares and the treatment $\times$ squares terms are combined. This number is small in the cases examined.

The additional precision gained by eliminating 'order within blocks' makes the critical value of ρ at which Methods II and IV are approximately equivalent equal to about 0.8.

Method V. This is the use of x simultaneously for blocking and for covariance correction. There are two possibilities. We may analyse the design as a randomized block with covariance, estimating the regression coefficient from the residual line of the analysis of covariance. Equation (6) applies with R_{xx} still defined as the residual sum of squares, this time in the randomized block analysis. We again require the expectation of T_{xx}/R_{xx}; over the randomization with the x's fixed

$$E(T_{xx}/R_{xx}) \geqslant E(T_{xx})/E(R_{xx}) = \frac{1}{k-1}, \tag{10}$$

with near equality in a large design. Hence

$$I_t^{(5)} \geqslant 1 + \frac{1}{(t-1)(k-1)}, \tag{11}$$

and the right-hand side of (11) will be used as an approximation to $I_t^{(5)}$; the error will be of the order of $[1/(t-1)(k-1)]^2$, as can be shown by the expansion methods of large-sample theory.

The second possibility is to analyse the design as if it were completely randomized. This would be in order if the arrangement into blocks has no effect on the y-values other than that due to correlation with the x's, and if the assumptions of the least squares model can be postulated, i.e. if we use more than pure randomization theory. In this case the argument parallel to (10) leads to

$$I_t^{(5)'} \geqslant 1 + \frac{1}{t(k-1)}, \tag{12}$$

again with near-equality in a large experiment.

The numerical values in Table 2 and a direct comparison of the formulae show that if $k > 3$ the lower limit for $I_l^{(5)}$ is greater than $I_l^{(2)}$. Hence, under the conditions postulated, Method V is inferior to Method II. Even with the second method of analysis, i.e. with (12), it seems unlikely that there is appreciable gain in *average* precision over Method II, under the conditions assumed.

Similar conclusions are reached from $I_a^{(5)}$ and $I_a^{(5)'}$.

Method VI. In this we consider a randomized block design with grouping based on a criterion separate from x. Quantitative investigation of this, based, for example, on the assumption that x, y and the property determining the grouping have a trivariate normal distribution, has not been attempted. We can, however, deal with two limiting cases. The system of blocking may be identical to that based on x. Equation (11) is then applicable. Or the criterion for grouping may be independent of x, in which case

$$I_l^{(6)} = \frac{(t-1)(k-1)-1}{(t-1)(k-1)-2}. \tag{13}$$

For the smaller values of $(t-1)(k-1)$ among the designs investigated, this is about $1 \cdot 15\, I_l^{(2)}$, showing that in these cases the additional system of blocking should be included only if there is a reasonable prospect of a reduction of 20 % or more in residual variance. For the larger values of $(t-1)(k-1)$, $I_l^{(6)}$ and $I_l^{(2)}$ are very nearly equal.

Method VII. This method is theoretically the most efficient one when the observations on y are built up of a polynomial trend on x plus treatment effect plus random error of constant mean and dispersion. The method is best illustrated by an example: suppose that $N = 9$, $t = k = 3$, that a second degree curve is considered adequate to represent the regression of y on x, and that the values of x are in order

$$-1 \cdot 3, \quad -0 \cdot 9, \quad -0 \cdot 8, \quad -0 \cdot 5, \quad -0 \cdot 2, \quad 0 \cdot 0, \quad 0 \cdot 4, \quad 0 \cdot 7, \quad 1 \cdot 1.$$

The most systematic procedure is to start by forming first and second degree orthogonal polynomials, ξ_1', ξ_2', from these observations. If $m_\alpha = \Sigma x_i^\alpha$, these are

$$\begin{aligned}
\xi_1' &= x_i - m_1/m_0, \\
\xi_2' &= x_i^2 - \left(\frac{m_0 m_3 - m_1 m_2}{m_0 m_2 - m_1^2}\right) x_i + \left(\frac{m_1 m_3 - m_2^2}{m_0 m_2 - m_1^2}\right).
\end{aligned} \tag{14}$$

The numerical values are

ξ_1': $-1 \cdot 13$, $-0 \cdot 73$, $-0 \cdot 63$, $-0 \cdot 33$, $-0 \cdot 03$, $0 \cdot 17$, $0 \cdot 57$, $0 \cdot 87$, $1 \cdot 27$,

ξ_2': $0 \cdot 89$, $0 \cdot 08$, $-0 \cdot 07$, $-0 \cdot 40$, $-0 \cdot 55$, $-0 \cdot 56$, $-0 \cdot 32$, $0 \cdot 07$, $0 \cdot 86$.

These are then normalized by dividing by $\sqrt{(\Sigma \xi_i'^2)}$ to give ξ_1 and ξ_2, namely,

ξ_1: $-0 \cdot 50$, $-0 \cdot 33$, $-0 \cdot 28$, $-0 \cdot 15$, $-0 \cdot 01$, $0 \cdot 08$, $0 \cdot 25$, $0 \cdot 39$, $0 \cdot 57$,

ξ_2: $0 \cdot 57$, $0 \cdot 05$, $-0 \cdot 04$, $-0 \cdot 26$, $-0 \cdot 35$, $-0 \cdot 36$, $-0 \cdot 21$, $0 \cdot 04$, $0 \cdot 55$.

We have to select from the nine units, three to receive T_1, etc. Let $\Sigma_1 \xi_1$, $\Sigma_1 \xi_2$, etc., denote the sum over those units receiving T_1 of ξ_1, ξ_2. This gives us six numbers. The treatment arrangement that minimizes the sum of squares of these six numbers is very nearly, or exactly, the most precise arrangement. Trial and error shows this arrangement to be

x	$-1 \cdot 3$	$-0 \cdot 9$	$-0 \cdot 8$	$-0 \cdot 5$	$-0 \cdot 2$	$0 \cdot 0$	$0 \cdot 4$	$0 \cdot 7$	$1 \cdot 1$
	T_1	T_3	T_2	T_2	T_3	T_1	T_1	T_3	T_2

(a non-randomized block design) with

	T_1	T_2	T_3
$\Sigma\xi_1$	$-0\cdot17$	$0\cdot14$	$0\cdot05$
$\Sigma\xi_2$	$0\cdot00$	$0\cdot25$	$-0\cdot26$

The sum of squares of these numbers is $S_{\min.} = 0\cdot1811$, and the value of I_t is, in general, approximately (Cox, 1951)

$$I_t^{(7)} \simeq \left\{1 - \frac{(t-2)}{k(t-1)} S_{\min.}\right\}\{1 - S_{\min.}/k\}^{-1}, \tag{15}$$

which in this case is $1\cdot03$. (The residual degrees of freedom would be 4 if both linear and quadratic trend were removed and 5 if only linear trend is removed.)

In general the systematic search for the optimum arrangement is tedious although it is usually possible to find quickly an arrangement that is nearly the best. If the degree of the polynomial is small compared with k, it will usually be possible to find an arrangement with $I_t^{(7)}$ negligibly greater than 1. The disadvantage of this method is the lack of randomization; the method is of most value in single small experiments.

5. DISCUSSION

In deciding what design to use in a particular case, we should consider

    (i) the values for imprecision given above;

    (ii) the extent to which departure from assumed conditions is likely to affect (i);

    (iii) the importance to be attached to simplicity of design, and analysis;

    (iv) the extent to which considerations other than precision are relevant.

The general conclusion from the calculations in §4 is that the methods based on covariance are preferable to the simpler methods based on blocking only if the correlation coefficient between y and x is at least $0\cdot6$, and that under the conditions postulated the systematic design, Method VII, is the most precise. For larger experiments all methods except the first are likely to have I near unity.

The main assumptions in the calculations are the linearity of the regression and the normality of the distribution of x. Non-normality of x should have little effect on the efficiency of covariance analysis, while I for the blocking methods will usually be an increasing function of the kurtosis of the distribution of x. If the regression is non-linear but smooth, blocking methods will remain effective, while covariance methods will not, unless the linear component accounts for most of the regression, or multiple covariance used.

The methods of design are all simple except Method VII. Details of analysis are, of course, simpler for methods not involving covariance.

There are two further considerations. The form of the relation between y and x may be of intrinsic interest, either in helping to understand the experimental material or in giving information useful in the design of further experiments. Also we may suspect that the treatment effects are not independent of x, i.e. that there is a treatment $\times x$ interaction. Such an interaction may give useful insight into the mechanism underlying the treatment effects and may also change any practical recommendations to be made from the experiment. If these considerations are relevant we shall normally prefer to use x quantitatively.

I am grateful to Dr B. G. Greenberg for very helpful discussion.

REFERENCES

BARTLETT, M. S. (1938). The approximate recovery of information from replicated experiments with large blocks. *J. Agric. Sci.* **28**, 418–20.

COCHRAN, W. G. & Cox, G. M. (1950). *Experimental Designs.* New York: John Wiley and Son.

Cox, D. R. (1951). Some systematic experimental designs. *Biometrika*, **38**, 312–23.

FISHER, R. A. (1935). *Design of Experiments.* Edinburgh: Oliver and Boyd.

GOURLAY, N. (1953). Covariance analysis and its applications in psychological research. *Brit. J. Statist. Psychol.* **6**, 25–34.

GREENBERG, B. G. (1953). Use of covariance and balancing in analytical surveys. *Amer. J. Publ. Hlth*, **43**, 692–9.

TEICHROEW, D. (1956). Tables of expected values of order statistics and products of order statistics for samples of size twenty and less from the normal distribution. *Ann. Math. Statist.* **27**, 410–26.

CORRIGENDA

Biometrika (1957), **44**, pp. 150–8

'The use of a concomitant variable in collecting an experimental design.' By D. R. Cox

Dr K. R. Nair has kindly pointed out that some of the results for Methods II and V of the above paper have been given by him in *Sankhya* (1942), **6**, 167–174. He has also noted that in formula (6) of my paper $(\bar{x}_i - \bar{x}_j)^2$ should read $k(\bar{x}_i - \bar{x}_j)^2$.　　　D.R.C.

[Biometrika, 44, 1957, p. 534]

The Use of Control Observations as an Alternative to Incomplete Block Designs

By M. Atiqullah and D. R. Cox

Birkbeck College, University of London

[Received February 1962. Revised June 1962]

Summary

The use of a control treatment as an alternative to incomplete block designs is considered, two possible methods of analysis being proposed. Theoretical efficiency factors are evaluated. These confirm conclusions of Yates (1936) that the efficiency is lower with a control than with an appropriate incomplete block design. The difference is not great for the larger values of k, the number of units per block, and the greater flexibility and simplicity of analysis of the methods with a control may lead to their being preferred, especially when a control treatment is necessary for other reasons, or when no simple incomplete block design is available. Extensions to the two-way elimination of error are considered.

1. Introduction

Consider an experiment in which the number of treatments is too large for a simple randomized block design to be used. There are available a large number of special types of incomplete block design, for example, balanced incomplete block designs, lattice designs, etc. (Cochran and Cox, 1957, Chapters 9–13). Nevertheless, occasionally it is considered essential to eliminate block differences, but either no convenient standard incomplete block design is available, or it is required to keep the analysis very simple.

A flexible method for dealing with such situations is to introduce a control treatment, C, on one or more units in each block and then to randomize the experimental treatments within replicates. If Y denotes a typical observation on a unit receiving one of the experimental treatments and X a typical observation on a unit receiving C, we may analyse the observations in a simple way by A: an analysis of covariance of the Y's using as the concomitant variable the mean of all X's in the corresponding block; or B: an analysis of variance, taking as the variate for analysis the difference between a Y and the mean of the X's in the corresponding block; or C: an analysis of variance, taking some other simple predetermined combination of Y and X, say Y minus β_0 times the mean of the X's in the corresponding block, where β_0 is a constant chosen perhaps on the basis of the analysis of preliminary data.

Method A is, in general, the most efficient of the three, in that the combination of Y and X for analysis is in effect determined from the data. Methods B and C may, however, sometimes be preferred for simplicity. We shall not consider method C further. In all the methods, the division of the units into blocks is ignored once the variables for analysis have been calculated, and the analysis, while not completely efficient, is standard.

The main previous work on the use of controls to increase precision is by Yates (1936); see also Cochran and Cox (1957, p. 387). Yates dealt briefly with a method equivalent to B of the present paper and then considered the use of systematically arranged controls. In this method a fertility index is constructed, using an appropriately weighted average of relevant control observations. Then analysis of covariance is used to adjust the observations on the treatments. Yates showed empirically that this gave lower efficiency than a lattice design. This use of systematically arranged controls is likely to be preferable to the methods of the present paper whenever, as in agricultural field trials, the variation of response is partly continuous. The present methods, on the other hand, are reasonable when the material falls naturally into distinct blocks, corresponding, for example, to batches of an industrial process.

2. NOTATION

Let there be t treatments $T_1, \ldots, T_t$ for comparison, which may in particular be the full set of treatments forming a complete factorial system. Suppose first that the experimental units are grouped into blocks of k units each, on c of which the control is applied.

Assume that a set of b blocks allows one replicate of the treatments, i.e. that $t = b(k-c)$. In general, when t is not an exact multiple of $k-c$, we suppose that certain treatments appear twice in the set of b blocks. Since we are interested in situations where $t \gg k - c$, this additional replication is unlikely to have an appreciable effect on efficiency. Finally, we suppose that the whole design consists of r replicates of the above. The treatments $T_1, \ldots, T_t$ are independently randomized within each replicate of the design.

We use Y as a generic symbol for an observation on one of the treatments $T_1, \ldots, T_t$ and X for an observation on C. Let θ_i denote the population treatment effect for T_i.

It will be assumed in assessing efficiency that the comparisons of interest are differences among the T's and that the efficiency of a design and method of analysis can be determined from

$$\text{ave var} \, (\hat{\theta}_i - \hat{\theta}_j); \tag{1}$$

the control C is assumed distinct from $T_1, \ldots, T_t$. The use of (1) is reasonable only when all comparisons $\hat{\theta}_i - \hat{\theta}_j$ have about the same variance. In practice, if one of the treatments is a standard treatment with which it is important to compare the other treatments, it will be natural to take this standard treatment for C. Comparisons with C will then be made with relatively high precision. Even if there is no such standard treatment, it will often be reasonable to take C to be one of the treatments under test, so that t is then one less than the total number of treatments. The quantity (1) will now give a conservative estimate of the efficiency of the design. If the treatments correspond to a factorial experiment, it will often be reasonable to take C as a "central" treatment.

If C is one of the treatments under test, it should be randomized along with $T_1, \ldots, T_t$. If C is not one of the treatments under test, it need not be randomized, and if there is a rational ordering of experimental units within the block a systematic arrangement of the controls will normally be better. However, in such a case, if $c > 1$, the use of a concomitant variable other than the mean of the control observations will be worth consideration (Yates, 1936).

3. THE MODELS FOR ANALYSIS AND FOR ASSESSING EFFICIENCY

Three methods of analysis were mentioned in section 1. In B and C, we have a straightforward analysis of variance into between replicates, between treatments and residual. A justification for the analysis can be based on the randomization of treatments within replicates, using the usual finite model of randomization theory (Kempthorne, 1955). A similar justification for method A, analysis of covariance, is more difficult because it is known that analysis of covariance is not exactly unbiased under the usual assumptions of randomization theory. However, it can be shown, by following the arguments of Cox (1956), that the analysis is unbiased in large samples.

To assess the efficiency of the methods, we make the much stronger assumption that an observation on the jth unit in the ith block of a particular replicate has, apart from replicate and treatment effects, the form

$$\mu + U_i + W_{ij}, \tag{2}$$

where $\{U_i\}, \{W_{ij}\}$ are sets of uncorrelated random variables of zero means and variances σ_u^2, σ_w^2. Thus σ_u^2 is the component of variance between blocks, whereas σ_w^2 is the component of variance within blocks. In particular (2) implies that the variances of control and treatment observations are the same. It is important, other things being equal, to take a control likely to lead to observations with small variance.

In calculating the efficiency of analysis of covariance, loss of efficiency arising from errors of estimation of the regression coefficient will be ignored. Note that under (2) none of the analyses A, B, C is fully efficient. The full maximum likelihood analysis will in general be very complicated.

4. Theoretical Efficiency

To calculate the efficiency of method A, analysis of covariance, we need the regression equation of Y on X. For the ith block of a particular replicate, let X_i be the mean observation on C and Y_{ij} an observation on a treatment. Ignoring treatment and replicate effects, we have that $X_i = \mu + U_i + W'_i$, $Y_{ij} = \mu + U_i + W_{ij}$, where W'_i is the mean of c values W_{ik}. Now the regression coefficient of Y_{ij} on X_i is

$$\beta = \frac{\text{cov}(X_i, Y_{ij})}{\text{var}(X_i)} = \frac{\sigma_u^2}{\sigma_u^2 + \sigma_w^2/c}.$$

We write
$$Y_{ij} = \mu(1-\beta) + \beta X_i + \{(1-\beta)U_i - \beta W'_i\} + W_{ij}$$
$$= \mu(1-\beta) + \beta X_i + R_i + W_{ij},$$

where R_i is a residual block effect common to all Y_{ij} in the block.

Consider the comparison by method A of two treatments within a replicate. With probability $(k-c-1)/(t-1)$, the two treatments lie in the same block, and with probability $(t-k+c)/(t-1)$, the treatments lie in different blocks. In the first case, the variance of the difference is $2\sigma_w^2$. In the second case, the variance, after adjusting for regression, is $2\{\sigma_w^2 + \text{var}(R_i)\} = 2\sigma_w^2(1+\beta/c)$.

Hence, in r replicates, the average variance (1) is

$$\frac{2\sigma_w^2}{r}\left\{\frac{(k-c-1)}{(t-1)} + \frac{(t-k+c)}{(t-1)}\left(1+\frac{\beta}{c}\right)\right\}. \tag{3}$$

Now a proportion $(k-c)/k$ of all units receive a treatment and hence the efficiency factor per unit for treatment comparisons is

$$E_A = \frac{(k-c)(t-1)}{k\{(k-c-1)+(t-k+c)(1+\beta/c)\}}. \tag{4}$$

This compares (3) with the variance that would be obtained if all experimental units were devoted to treatments and none to control and if the effects of inter-block variations were completely eliminated.

One situation in which the present method is likely to be useful is when a control is necessary for other reasons than to increase the precision of the treatment comparisons. It is then reasonable to consider an efficiency factor per treated unit. This factor is

$$E_A^* = \frac{(t-1)}{(k-c-1)+(t-k+c)(1+\beta/c)}.$$

Throughout the paper efficiency factor per treated unit will be denoted by an asterisk.

The efficiency factors for the analysis of differences, method B, are calculated similarly. We have that

$$E_B = \frac{c(t-1)(k-c)}{k(k+t-k)}, \quad E_B^* = \frac{c(t-1)}{(k+t-k)}. \tag{5}$$

For comparison, it is useful first to have the corresponding efficiency factor for a balanced incomplete block design, even though it is unlikely that the present methods would be used if a suitable balanced incomplete block design were available. For the intra-block and combined analyses, the efficiency factors are respectively

$$E_{bib(i)} = \frac{(k-1)t}{k(t-1)}, \quad E_{bib(c)} = \frac{\sigma_w^2 + \dfrac{t(k-1)}{(t-1)}\sigma_u^2}{\sigma_w^2 + k\sigma_u^2}. \tag{6}$$

A second basis for comparison is a design in which no attempt is made to eliminate differences between blocks. The efficiency factor is

$$E_{ign} = \frac{\sigma_w^2}{\sigma_w^2 + \dfrac{(t-k+c)}{(t-1)}\sigma_u^2}. \tag{7}$$

It is useful to examine first some special cases. When $k=2$, $c=1$, we cannot use analysis of covariance to estimate β. But, from (5), $E_B = \frac{1}{4}$. This is to be compared with, for large t, $E_{bib(c)} \sim (\sigma_w^2 + \sigma_u^2)/(\sigma_w^2 + 2\sigma_u^2)$, $E_{ign} \sim \sigma_w^2/(\sigma_w^2 + \sigma_u^2)$.

A good incomplete block design with two units per block will have an efficiency factor of about $\frac{1}{2}$. Thus, unless the use of C is desirable for other reasons, such an incomplete block design, if available, will be much preferable to method B. However, if $\sigma_u^2 > 3\sigma_w^2$, the use of method B would be preferable to ignoring the pairing of the experimental units.

We can consider for general k the limiting behaviour as $t \to \infty$. Then

$$E_B \sim \{c(k-c)\}/\{k(c+1)\}, \quad E_B^* \sim c/(c+1).$$

An approximately optimum value of c can now be obtained by treating c as a continuous variable; E_B is maximized when $c = -1 + \sqrt{(k+1)}$. With this value for c, $E_B \simeq 1 - 2\{\sqrt{(k+1)} - 1\}/k$. This tends slowly to unity as k increases. Thus with $k=8$, $E_B \simeq \frac{1}{2}$ and with $k=24$, $E_B \simeq \frac{2}{3}$. These results apply also to method A if $\sigma_u^2 \gg \sigma_w^2$.

If, on the other hand, $\sigma_u^2 = \sigma_w^2$, the approximate optimum of E_A is obtained when $c = -2 + \sqrt{(k+2)}$, with

$$E_A \simeq 1 - \frac{2\sqrt{(k+2)}}{k} + \frac{3}{k}.$$

This is rather greater than the value when $\sigma_u^2 \gg \sigma_w^2$.

In practice a reasonable compromise might often be to choose the value of c to be optimum if $\sigma_u^2 = \sigma_w^2$.

The corresponding value for a balanced incomplete block design, with recovery of information, is $\{\sigma_w^2 + (k-1)\,\sigma_u^2\}/\{\sigma_w^2 + k\sigma_u^2\}$ lying between 1 and $(k-1)/k$. If a design ignoring block differences is used the limiting efficiency factor is, from (7), $\sigma_w^2/(\sigma_u^2 + \sigma_w^2)$. Table 1 gives a few numerical results based on these formulae.

TABLE 1

Approximate efficiency factors when the number of treatments is large. Method A compared with balanced incomplete block designs

(i) E_A for $\sigma_u^2 \gg \sigma_w^2$, or, equivalently, E_B, when c is appropriately chosen.
(ii) E_A for $\sigma_u^2 = \sigma_w^2$, when c is appropriately chosen.

	E_A			$E_{bib(c)}$	
k	(i)	(ii)	$\sigma_u^2 = 0$	$\sigma_u^2 = \sigma_w^2$	$\sigma_u^2 \gg \sigma_w^2$
4	0·382	0·525	1·000	0·800	0·750
6	0·452	0·557	1·000	0·857	0·833
10	0·537	0·607	1·000	0·909	0·900
20	0·642	0·681	1·000	0·952	0·950

The general conclusions from Table 1 are first that the efficiency of A and B increases with k, the number of units per block. Secondly, for the values of k probably of most interest efficiencies are obtained about $\frac{1}{2}$ to $\frac{2}{3}$ of those for a balanced incomplete block design, if one exists. Thirdly, whenever $\sigma_u^2 > \sigma_w^2$, the use of method A or B will be preferable to ignoring the division into blocks.

Table 1 is concerned with large values of t. Rather larger efficiencies are obtained for moderate values of t. Thus with $k = 10$, $t = 20$, the values corresponding to (i) and (ii) in Table 1 are 0·608 and 0·760.

The efficiencies in the previous formulae and tables refer to the comparison of two treatments. However, comparisons of a treatment with the control will be made relatively precisely. For method A, the variance of the estimated differences between a treatment and C can be shown to be

$$\frac{\sigma_w^2}{r(c\sigma_u^2 + \sigma_w^2)} \left\{ \sigma_w^2 \frac{(tc+k-c)}{tc} + \sigma_u^2(c+1) \right\},$$

and by method B the variance is, directly, $\sigma_w^2(c+1)/(cr)$.

5. TWO-WAY ELIMINATION OF ERROR

In this section, we study briefly the use of control observations for the two-way elimination of error. We again calculate efficiency factors for two methods of analysis, A and B. In A, a double analysis of covariance is used taking as the two concomitant variables the mean of all control observations in the same row and the mean of all control observations in the same column. In B the variable for simple analysis of variance is the difference between the treatment observation and the sum of the two concomitant variables just mentioned.

Consider a design in which each replicate is a $p \times q$ rectangle, the control C being applied to the units in r completed rows and c complete columns; there is, however, no need to make observations on the rc units where control rows and columns intersect. The model (2) is generalized by supposing that, apart from replicate and treatment effects, an observation in the ith row and jth column has the form

$$\mu + U_i + V_j + W_{ij},$$

where $\{U_i\}$, $\{V_j\}$, $\{W_{ij}\}$ are sets of uncorrelated random variables of zero means and variances σ_u^2, σ_v^2, σ_w^2.

The concomitant variables associated with Y_{ij} are then $X_{i.} = \mu + U_i + V_i' + W_{i.}'$, $X_{.j} = \mu + U_j'' + V_j + W_{.j}''$, where the single prime denotes an average over c control columns and the double prime an average over r control rows. Now as i varies within a replicate, V_i' is constant, since it refers to a fixed set of columns, and similarly U_j'' is constant as j varies. Hence the quantities needed to find the regression equation of Y_{ij} on $X_{i.}$ and $X_{.j}$ are

$$\operatorname{cov}(Y_{ij}, X_{i.}) = \sigma_u^2, \quad \operatorname{cov}(Y_{ij}, X_{.j}) = \sigma_v^2, \quad \operatorname{cov}(X_{i.}, X_{.j}) = 0,$$

$$\operatorname{var}(X_{i.}) = \sigma_u^2 + \sigma_w^2/c, \quad \operatorname{var}(X_{.j}) = \sigma_v^2 + \sigma_w^2/r.$$

Thus, if we ignore treatment and replicate effects, the regression equation of Y_{ij} on $X_{i.}$ and $X_{.j}$ is

$$Y_{ij} = \mu(1 - \beta_u - \beta_v) + \beta_u X_{i.} + \beta_v X_{.j}$$
$$+ (1 - \beta_u) U_i + (1 - \beta_v) V_j + (W_{ij} - \beta_u W_{i.}' - \beta_v W_{.j}''),$$

where $\beta_u = c\sigma_u^2/(c\sigma_u^2 + \sigma_w^2)$, $\beta_v = r\sigma_v^2/(r\sigma_v^2 + \sigma_w^2)$.

Now the number of treatments t is equal to $(p-r)(q-c)$ and a proportion $t/(pq-rc)$ of all units receive a treatment. There are now three types of treatment comparison:

    (a) between units in the same row;
    (b) between units in the same column;
    (c) between units in different rows and columns.

The variances for these three types are easily calculated and we obtain the efficiency factor for analysis of covariance as

$$E_A = \frac{\left(1 - \dfrac{r}{p}\right)\left(1 - \dfrac{c}{q}\right)}{\left(1 - \dfrac{rc}{pq}\right)\left\{1 + \beta_u \dfrac{(t-q+c)}{c(t-1)} + \beta_v \dfrac{(t-p+r)}{r(t-1)}\right\}}. \tag{8}$$

The efficiency factor for method B can be calculated from first principles or by putting $\beta_u = \beta_v = 1$ in (8).

For given p, q, t, these formulae can be used to determine suitable values of r, c and the corresponding efficiency factors. A lower limit to the efficiency can again be obtained by letting $t \to \infty$. Two particular cases are of special interest:

    (a) a square design, with $p = q$, $r = c$, $t = (p-r)^2$;
    (b) a design with q fixed and small, so that $c = 1$, $t = (p-r)(q-1)$.

A detailed analysis will not be given, but Tables 2 and 3 give a few numerical results. They are based on optimum r determined by treating r as a continuous variable. For E_A it is assumed that $\sigma_u^2 = \sigma_v^2 = \sigma_w^2$. The efficiency factors E_A^*, E_B^* per treated unit are calculated using the same value of r.

For comparison with Tables 2 and 3 it can be shown that the efficiency factor if row and column variation is ignored and no controls are used is

$$E_{ign} = \cfrac{1}{1 + \cfrac{(t-q)\,\sigma_u^2}{(t-1)\,\sigma_w^2} + \cfrac{(t-p)\,\sigma_v^2}{(t-1)\,\sigma_w^2}}.$$

The numerical factors in the denominator will usually be fairly near 1, so that E_{ign} will be rather greater than $\sigma_w^2/(\sigma_u^2 + \sigma_v^2 + \sigma_w^2)$. Thus if $\sigma_u^2 + \sigma_v^2 > \sigma_w^2$, methods A and B will be preferable to ignoring row and column effects.

TABLE 2

Approximately optimum r and efficiency factors for square designs

t	36	64	100	144
r	1	1	2	2
E_A	0·404	0·424	0·444	0·464
E_A^*	0·538	0·530	0·623	0·620
r	2	3	3	3
E_B	0·300	0·355	0·389	0·413
E_B^*	0·538	0·628	0·623	0·620

TABLE 3

Approximately optimum r and efficiency factors for rectangular designs with c = 1

Values of r, E_A, E_A^* when $\sigma_u^2 = \sigma_v^2 = \sigma_w^2$

t	30	60	90	120
q	2	3	4	4
3	0·370	0·386	0·394	0·400
	0·604	0·618	0·627	0·626
		2	3	4
4	—	0·408	0·421	0·435
		0·585	0·603	0·614
		2	2	3
6	—	0·421	0·436	0·448
		0·575	0·572	0·593

One particular application of the methods is the arrangement of factorial experiments in rectangular arrays.

Another method of using control observations in a two-way lay-out is to use each control observation simultaneously as a row and column control. Thus, in a square design, we might put the control C on the units in the principal diagonal of the square. This method will be efficient if one, but not both, of the components σ_u^2, σ_v^2 are large. We shall not consider this further.

6. CONCLUSIONS

The following general conclusions have been obtained.

(i) Efficiency factors of around $\frac{1}{2}$ can be obtained for the one-way elimination of error, provided the number of units per block is not very small.

(ii) The use of a simple incomplete block design, if one is available, will be more efficient than the present methods, unless the use of a control is necessary for other reasons.

(iii) To find an appropriate c, the number of control units per block, and to estimate the precision that is likely to be attained, use the formulae of section 4. If analysis of covariance is to be used a prior estimate of σ_u^2/σ_w^2 is needed to choose c.

(iv) The efficiency factor will be lower when variation is to be controlled in two directions. Section 5 contains detailed formulae.

(v) If a control treatment is necessary for reasons other than the increase of precision, a more relevant efficiency factor is that per treated unit. The resulting values are appreciably greater than for the efficiency factor per unit.

REFERENCES

COCHRAN, W. G. and COX, G. M. (1957), *Experimental Designs*, 2nd ed. New York: Wiley.

COX, D. R. (1956), "A note on weighted randomization", *Ann. math. Statist.*, 27, 1144–1151.

KEMPTHORNE, O. (1955), "The randomization theory of experimental inference", *J. Amer. statist. Ass.*, 50, 946–967.

YATES, F. (1936), "A new method of arranging variety trials involving a large number of varieties", *J. agric. Sci.*, 26, 424–455.

Some Systematic Supersaturated Designs

KATHLEEN H. V. BOOTH AND D. R. COX

Birkbeck College, University of London

Supersaturated designs are factorial designs in which the number of factors exceeds the number of observations. Satterthwaite has suggested constructing such designs by a randomization procedure (random balance designs). In the present paper systematic designs are constructed which are, in a certain sense, as nearly orthogonal as possible. Examples of such designs are given constructed on an electronic computer and some properties of the designs are discussed.

1. INTRODUCTION

A supersaturated design is a factorial design in n observations in which the number of factors is more than $n - 1$. Satterthwaite (1959) suggested that such designs could be useful in a rapid preliminary investigation of a problem in which there are a large number of potentially relevant factors, but in which only a small proportion of these factors are thought likely to have important effects. He proposed to construct the design matrix at random, leading to so-called random balance designs. The idea that such randomization of the design really helps the interpretation of an experiment has been criticised; each element of the permutation reference set is a recognizably separate possibility and this seems to nullify the usefulness of this reference set for inference.

In this paper we propose systematic designs for tackling these problems and discuss briefly some of the properties of the designs. We have no experience of practical problems where such designs are likely to be useful; the conditions that interactions should be unimportant and that there should be a few dominant main effects seem very severe.

For further remarks on the problem, see the discussion following the papers of Satterthwaite (1959) and Budne. In particular, the contribution by Box is very relevant to the present paper, suggesting, among other things, the construction of systematic designs.

2. A CONDITION FOR NEAR ORTHOGONALITY

Consider two-level factorial experiments with f factors and n observations, n being even. Denote the two levels of a factor by $+1$ and -1. Then the design is determined by an $n \times f$ matrix of elements ± 1, the ith column r_i giving the sequence of factor levels for factor i in observations $1, 2, \cdots, n$. We consider only designs where all columns consist of $\frac{1}{2}n$ $+1$'s and $\frac{1}{2}n$ -1's. Now in an ordinary factorial experiment, assuming interactions can be ignored, we ensure the efficient and simple estimation of main effects by requiring the orthogonality of all design columns (Plackett & Burman, 1946). That is we require

$$r_i' r_j = 0 \qquad (i \neq j). \tag{1}$$

489

This condition cannot be satisfied for all i, j whenever $f > n - 1$; for otherwise the r_i, taken with a column of ones, would form a set of more than n orthogonal vectors in n dimensional space. Therefore we require to have (1) satisfied as nearly as possible. This requirement could be made precise in various ways. We shall use the following. First we require a minimum value for

$$\text{Max}_{i \neq j} |r_i' r_j| \,. \tag{2}$$

Then of two designs with the same value for (2), we prefer the one in which the number of pairs of columns attaining (2) is a minimum.

If n is divisible by 4, and $f > n - 1$, the value of (2) must be at least 4.

3. SOME DESIGNS

Table 1 gives seven designs for $n = 12, 18, 24$. Designs for intermediate values of f can be formed by dropping the final columns from the next largest design. Designs I and VII start with $n - 1$ factors arrayed in the regular Plackett and Burman design.

TABLE 1.

Supersaturated designs for f factors using n observations.
Each column corresponds to a factor, each row to an observation

DESIGN I. $f = 16, n = 12$

```
+ + + + + + + + + + + − − − − −
+ − + + + − − − + − − − − − − −
− + + + − − − + − − + + + − + +
+ + + − − − + − − + − − + + + +
+ + − − − + − − + − + + + − + −
+ − − − + − − + − + + + + + − +
− − − + − − + − + + + + + + − +
− − + − − + − + + + − + − + − +
− + − − + − + + + − − + + + + −
+ − − + − + + + − − − − + + + −
− − + − + + + − − − + − − − + +
− + − + + + − − − + − − − − − −
```

DESIGN II. $f = 20, n = 12$

```
+ − − + + + + − + + − − − + − − − − + −
− − − − + + − + − − + − + − − + − − − −
− − + + + − + + − + − − − − − − + + + +
− + − − − + + − + − − + − − − + − + + +
+ − + − − + + − + + − − + + + − + + − +
+ + + + + − + − − − + − + − + − − + − −
+ − − − − + − − − + + + − − + − + − + −
− + − + + − − − − + + + + + + − − − − +
+ + − − − − − + − + − − − + + + + + − −
+ − + − − − + + + − + − − − − − − − − +
− − + + − + + + + + + + + + + + + + + −
− + + + − − − − + − − + + + + + + + + +
```

TABLE 1 (Con't.)

DESIGN III. $f = 24, n = 12$

```
+ − − + + + + − + + − − − + − − − − − − − − + +
− − − + + − + − + − − + − + − − + − − − − + − − −
− − + + + − + + − + − − − − − − + + + − − + − −
− + − − − + + − + − + − − + − + + + + − − − − −
+ + + − + + − + + − − + + + − + − + − + − − − −
+ + + + + − + − − − + − + − + − + − + − + − − −
+ − − − − + − − − + + + − − + − + − − − + + − −
− + − + + − − − − + + + + + + − − − + + − + − +
+ + − − − − − + − + − − − + + + + + − − − − + +
+ − + − − − + + + − + − − − − − − + + + + + + +
− − + + − + + + + + + + + + + + + + − + + + + +
− + + + − − − − + − − + + + + + + + + + + + + +
```

DESIGN IV. $f = 24, n = 18$

```
+ − + − + − + − + − + − + − + + − − − − − − + +
− + − + − + − + − + − + − + − + − + − + − − − +
− − − − − − − − − − − − − − − + − − − − + − + −
+ + + − + − + + − − + − − − + + + + + + − + − +
+ + − + − + + − − + − − − − + + + + + + + + + −
+ − + − + + − − + − − − + + + + + + + + + + + +
− + − + + − − + − − − + + + + + + + + + + + + +
+ − + + − − + − − − + + + + − + + + + + + + + +
− + + − − + − − − + + + + − + − − + + + − − + −
+ + − − + − − − + + + + − + − + + − + − + − − −
+ − − + − − − + + + + − + − + − − + − − + − − −
− − + − − − + + + + − + − + + − + − − − − − + −
− + − − − + + + + − + − + + − − − − − + − + − −
+ − − − + + + + − + − + + − − − + − − + − − − +
− − − + + + + − + − + + − − + − + − − − + + − −
− − + + + + − + − + + − − + − − − + − + + + + +
− + + + + − + − + + − − + − − − − + + − − − − −
+ + + + − + − + + − − + − − − − + − − + − − − +
```

DESIGN V. $f = 30, n = 18$

```
− + + + − + − − − + − − + − − − − − + − − + + − + − + + + +
− + − + + − − − − + + − − − − − + − − + + − + − + + + + + −
− + + − − − − − + + + − + − − − − + − − + + − + − + + + + −
+ − − − − − + − + + − + − − − + − + − + − + − + + + + − − −
+ − + − − + − − − − + − − + − − + + − + − + + + + − − − − +
− − − + − + + − + − + − − − − − + + − + − + + + + − − − + −
− − − − + + − + − − − − + − + + − + − + + + + − − − + − − −
+ − − + + − − − + − − − + − − + − + − + + + + − − − + − − +
− − − + − − − − − + − + − + + − + − + + + + − − − + − − + +
− − − − − + + + − − − + + − − + − + + + + − − − + − − + + −
+ + + + + + + + + + + + + + + + − + + + + − − − + − − + + − +
+ + + + + + + + + + + + + + + + + + + + − − − + − − + + − + −
+ + + + + + + + + + + + + + + + + + + + − − − + − − + + − + +
− + + + + + + + + + + + + + + + + + − − − + − − + + − + − + +
+ + − − − − − − + − + + + + + + + − − − + − − + + − + − + +
+ + − − − − + + − + − − − − + − − − + − − + + − + − + + + +
+ + + − + − + − − − + − + − + − + − + − + − + − + − + − + −
− − + − + − − + − − − − − − + + − + − + − + − + − + − + − +
```

TABLE 1 (*Con't.*)

DESIGN VI. $f = 36, n = 18$

DESIGN VII. $f = 30, n = 24$

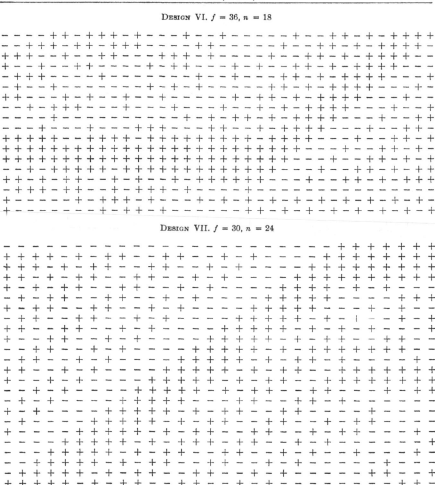

Table 2 gives for each design the frequency distribution of the sum of products $s = r'_i r_i$ over all pairs of columns of the design.

4. CONSTRUCTION OF DESIGNS

The designs were constructed on the University of London Mercury computer. An outline of the procedure is as follows.

A trial set of vectors was inserted into the machine, and the maximum cross-product, m, calculated. When, as usually occurred, several pairs of vectors had this maximum product, the final pair to be encountered was selected for improvement. The maximum cross-product and the number of pairs having this were then punched out.

A suitable trial vector, i.e. one having an equal number of $+1$'s and -1's, was then calculated and substituted for each vector of the pair in turn. If this

TABLE 2

Distribution of absolute value of sum of products, $|s| = |r'_i r_j|$

Design	$\|s\|$	Frequency					Number of pairs of rows
		0	2	4	6	8	
I	67			53			120
II	75			115			190
III	101			175			276
IV			193		83		276
V			281		154		435
VI			385		245		630
VII	295			81		59	435

substitution produced a reduction in the cross-product of the pair, the programme then checked that the cross-product of the new vector with all other members of the set was m, equality only being admitted if the original pair of vectors also had cross-product m. If these conditions were satisfied the trial vector was substituted permanently, and its value and location in the set punched out. The programme then returned to the starting point and recalculated the maximum cross-product, etc. If the conditions were not satisfied a new trial vector was calculated and the process repeated until an improvement was effected.

The programme could be terminated at any time, depending on the amount of computer time available, by throwing up a switch. The computer would then punch out the complete new set of vectors in a form suitable for re-input if more time were available later to try again.

Naturally, as the vector sets were improved, convergence became slower, and from considerations of machine time, each set was terminated when a run of $\frac{1}{2}$ hour had produced no improvement.

It may be mentioned here that the straightforward method of enumerating all possible sets and choosing the best is impracticable, since the computing time, even for the smallest set considered, is prohibitive.

A few practical details may be of interest. Since Mercury is a binary machine the vectors were represented by arrays of binary digits, 0 standing for $+1$, and 1 for -1. Using this method it was possible to effect a very rapid multiplication in calculating cross-products. The trial vectors were produced simply by counting, in binary, up to 2^n, n being the number of digits in the vector, and testing each number to see whether it had an equal number of 1's and 0's. In order to try the maximum range of possible numbers, the programme was made to punch out the point reached in this enumeration, and if a subsequent calculation were made this was taken as the starting point next time.

Three sizes of vector were considered, with 12, 18 and 24 digits. For the 12 and 24 digits vectors, orthogonal sets were available of 11 and 23 vectors, and in calculating, for example, the 12 × 16 set, these were used for the first 11 vectors and 5 of them repeated to complete the set. The *final* set of 12 × 16 was used, together with 4 repeats as a start for the 12 × 20 set, and so on. No

TABLE 3.

Variance of measure of non-orthogonality of two rows

Design	f	n	random balance	systematic	Design	f	n	random balance	systematic
I	16	12	13.1	7.07	IV	24	18	19.1	13.6
II	20	12	13.1	9.68	V	30	18	19.1	15.3
III	24	12	13.1	10.1	VI	36	18	19.1	17.4
					VI	30	24	25.0	11.4

orthogonal set is available for the 18 digit vectors, and here the orthogonal set of 16 × 15 was used as a basis and two digits added to each vector.

The programme is particularly convenient when, as on the Mercury at the University of London, it is much easier to obtain several runs of say, ½ hour, rather than one long run.

5. COMPARISON WITH RANDOM BALANCE DESIGNS

A simple way to compare the systematic designs with the random balance designs of Satterthwaite is to consider the variance of s, the sum of products $r_i' r_j$ of two columns chosen at random from the design. For the systematic designs, this variance is easily found from the information in Table 2. Thus, for design 1, with probability 67/120, s is zero, and with probability 53/120, $|s|$ is 4, the values $+4$, -4 being equally likely. Hence the variance of s is $E(s^2) = 53/120 \times 16 = 7.07$.

For the random balance designs, s is twice the total of a random sample of size $\frac{1}{2}n$ drawn without replacement from a finite population of $\frac{1}{2}n$ $+1$'s and $\frac{1}{2}n$ -1's. It follows from the formula for the variance of a mean in sampling a finite population that the variance of s is $n^2/(n-1)$. Table 3 summarizes the calculations. As would be expected, the systematic designs are, judged by this criterion, substantially better than random balance designs, especially when $f < 2n$. Now the distribution of s is nearly normal for random balance designs, whereas that for the systematic designs has a substantial negative kurtosis. Therefore a comparison based on the frequency of occurrence of large values of s would be even more in favour of the systematic designs.

Two more direct interpretations of s are as follows. Suppose that just two factors have effects, of magnitudes say $2\theta_1$ and $2\theta_2$, and that the error variance is σ^2. Suppose that the effects are large enough for us to be able to pick out these two factors as ones likely to have real effects and that we set up least squares equations for estimating θ_1 and θ_2, other factor effects being assumed zero. Then, the general mean having been eliminated, the matrix on the left-hand side of the least squares equations is

$$\begin{bmatrix} n & s \\ s & n \end{bmatrix}$$

and the variance-covariance matrix of the estimates is determined by the inverse matrix, and is

$$\sigma^2 \begin{bmatrix} n/(n^2 - s^2) & -s/(n^2 - s^2) \\ -s/(n^2 - s^2) & n/(n^2 - s^2) \end{bmatrix}. \tag{3}$$

For a given design and pair of factors, s will be a known constant. If we average over all positions for the two factors with effects, we have for the average variance of an estimated effect

$$\frac{\sigma^2}{n} E\left(1 - \frac{s^2}{n^2}\right)^{-1} \simeq \frac{\sigma^2}{n}\left\{1 + \frac{\operatorname{var}(s)}{n^2}\right\}. \tag{4}$$

It follows from (4) and the numerical values in Table 3 that, on the average, there is little increase in variance of estimation arising from non-orthogonality. The argument leading to (4) can be extended to show that if k effects are included for analysis, with design vectors selected at random from the design matrix, then the average variance of an estimated effect is approximately

$$\frac{\sigma^2}{n}\left\{1 + \frac{(k-1)\operatorname{var}(s)}{n^2}\right\}; \tag{5}$$

equation (4) is the special case $k = 2$.

These formulae are, however, likely to underestimate seriously the effect of non-orthogonality, unless the non-zero true effects are all very large. For one may well have to include in the least squares equations, together with each factor with a real effect, the factor or factors having highest s values with it. It is plausible that this would lead to about a fivefold increase in the coefficient of var(s) in (4) and (5).

There is a second interpretation of var(s). Consider the estimation of one factor effect, say the first. Suppose that all other factors have effects of magnitude 2Δ, the sign being chosen at random. Then to the ordinary estimate of the main effect of the first factor is added a random contribution from the other effects of variance

$$\frac{(f-1)}{4n^2} \Delta^2 \operatorname{var}(s), \tag{6}$$

to be added to the ordinary variance, σ^2/n.

Under these circumstances var(s) is a direct measure of the effect of non-orthogonality. More generally, the factor $(f-1)\Delta^2$ in the numerator of (6) can be replaced by the total mean square effect present in the other factors. Note that if $f \simeq 2n$, var(s) $\simeq n$, $n \gg 1$, (6) is approximately $\frac{1}{2}\Delta^2$. Thus if all factors except the first have effects about equal to the standard error, $\sigma/\sqrt{n}$, in an orthogonal experiment, the variance of the crude estimate of the effect of factor would in this case be increased, by non-orthogonality, by about 50 per cent.

REFERENCES

PLACKETT, R. L. and BURMAN, J. P. (1946), "The design of optimum multifactorial experiments," *Biometrika, 33*, 303–325.

SATTERTHWAITE, F. (1959), "Random balance experimentation," *Technometrics 1*, 111–137.

J. R. Statist. Soc. A (1984),
147, *Part* 2, *pp.* 306–315

Present Position and Potential Developments: Some Personal Views Design of Experiments and Regression

By D. R. COX

Fisher Memorial Lecture

[*The Chairman of the Fisher Memorial Committee*, Dr W. F. Bodmer, *in the Chair*]

SUMMARY

The first part of the paper gives a brief historical comment on the development of experimental design and some rather more extensive remarks on randomization. The second and longer part of the paper discusses various difficulties primarily of interpretation connected with regression analysis (in the broad sense). These include the selection of variables, the effect of complex error structure and the effect of errors in the explanatory variables. As an Appendix a few apparently open problems are outlined.

Keywords: BLOCKING; COMPLEX ERROR STRUCTURE; CORRELATION; ERRORS IN VARIABLES; GENERALIZED LINEAR MODEL; LINEAR MODEL; MODEL CHOICE; OBSERVATIONAL STUDY; OPTIMAL DESIGN; RANDOMIZATION; REGRESSION; STRATIFICATION

1. INTRODUCTION

I deeply appreciate being invited to give a Fisher Memorial Lecture.

The comprehensive discussion of design of experiments and of regression half-promised by the title is scarcely feasible. The first and shorter part of the paper makes some comments on experimental design and this is followed by a rather longer, although still highly selective, discussion of regression.

As an Appendix, some apparently unsolved problems are listed.

2. DESIGN OF EXPERIMENTS

For the applied statistician, design of experiments is one of the most interesting and satisfying aspects of our subject. In part this is because involvement in the planning phases of an investigation is usually a symptom of genuine "collaboration", as contrasted with "consultation" over difficulties of analysis at the late stages of a study.

One can distinguish three broad periods of innovation directly associated with applications.

Firstly, there is the period up to say 1950 with a primary motivation from agricultural and biometrical experimentation, the books of Fisher (1935) and Yates (1937) being outstanding accounts from that time. Recent work, much of it unpublished, by R. A. Bailey and T. P. Speed, sets many of the classical ideas in a rather "high-powered" mathematical framework. A key distinguishing feature of the "classical" approach is the notion that the analysis of variance table is a consequence of the treatment and unit structure and the randomization, rather than a deduction from an externally specified linear model. More broadly, while the very close links between analysis of variance and the theory of the linear model are of key importance, it would considerably undervalue analysis of variance to regard it as merely an appendage to the theory of the linear model.

Secondly, there is a period between about 1950 and 1970 with a primary motivation from

Present address: Department of Mathematics, Imperial College, London SW7 2BZ, UK

0035–9238/84/147306 $2.00

industrial experimentation. Here the emphasis has tended to be on treatment structure (response surface form and fractionation) and rather less on the detailed character of the error.

Finally, in recent years there has been much emphasis on clinical trials. Characterizing features are that, superficially at least, treatment structure is simple, experimental units (patients) enter serially, measurement of response may be delayed and that very often a large amount of prognostic information is initially available on each unit.

Of course this division into three phases does not imply lack of continuing interest, theoretical and applied, in the traditional areas. For an agricultural example, see Bartlett (1978) and Wilkinson *et al.* (1983) on the relation of field trials with spatial stochastic models. Here there are major unresolved issues as to both the practical fruitfulness of the approach, and the merits of rival superficially very similar methods of analysis. The role of cross-over designs attracts continuing attention and controversy in several applied fields. In industrial experimentation, in the UK at least, no doubt the major task is to try to get careful techniques of experimentation more widely adopted; so far as one can judge, basic ideas of experimental design are employed on an enormous scale in Japanese industrial research and development.

From a mathematical point of view, by far the major development in the past 20 years has been the flowering of a systematic theory of optimal experimental design; see Silvey (1980) for a concise account. This theory serves partly to consolidate and codify knowledge about traditional designs and, perhaps more importantly from the viewpoint of immediate application, offers a systematic approach to non-standard problems. For example, suppose that in an experiment with quantitative factors the design region is modified after an initial design has been partly implemented: how should the new design points be chosen? Packaged algorithms, hopefully merely mildly menacing to the user, are needed if such problems are to be tackled routinely under normal working conditions; see Welch (1982).

Limitations on the practical usefulness of designs derived via a formal optimality criterion arise mainly from the provisional character of aspects that have to be precisely specified in order to implement an algorithm. The situation is not essentially different from that with optimality considerations in other practical contexts; for the controversy in operational research between optimizing and so-called satisfycing, see Eilon (1972).

On the whole, the theory of optimal design makes rather strong *a priori* assumptions about the nature of error variability. It has not, for example, yet thrown special light on the role of randomization. That role has always been to some extent controversial; see, for example, the discussion between Fisher and Student in the 1930s.

The least controversial aspect of randomization is probably its role in eliminating systematic error, in particular by concealment. The empirical evidence for biases arising from the entry of personal choice in treatment assignment is so strong that failure to use some impersonally defined procedure can be damning; some element of randomization is often, although not always, the natural solution. The second-order theory of randomization, in conjunction with an assumption of unit-treatment additivity, yields, as noted above, a satisfying unity to the analysis of the classical designs. The relevance of the randomization distribution to the analysis of the design actually employed depends, however, on the absence of a "recognizable subset" in the randomization distribution. Here "recognizable" means with reference to some feature which there is a good scientific reason to judge relevant to the responses.

Put slightly differently, if there are "background ancillaries" the randomization distribution should be conditional on their observed values. Background ancillaries are features of the design for which there is reason, whether based on a formal model or not, to think that their value materially affects the precision of the treatment contrasts of interest. If such an ancillary is recognized at the design stage, then optimum or near-optimum values should be employed. If it is not recognized until the analysis stage, then any randomization analysis should be conditional on values equal or close to those realized in the design employed.

To the first order of asymptotic theory this is achieved via the asymptotic multivariate normal randomization distribution of the background ancillaries and the standard functions of the response variable. The approximate conditional distribution of the latter can thus be found, leading to a randomization version of the standard analysis of covariance adjustments.

For traditional medium-sized experiments with a limited number of prognostic features per experimental unit, blocking and randomization can be effective in controlling error. In clinical trials, however, quite apart from the special issue of serial entry, there may be difficulties of design associated with the availability of a large number of prognostic features, leading to a conflict between the desire for randomization and the need for balancing or conditioning. Peto, *et al.* (1976) recommended randomization with blocking on only a few key features, correcting by analysis of covariance for any serious imbalances that might arise. There has been some controversy (Simon, 1979) over this, especially for medium-sized trials where any need for substantial adjustments would both lower efficiency and reduce the intuitive appeal of the analysis.

Cox (1982a) suggested that the appropriate conditional randomization could be formally achieved by repeated rerandomization until a suitably defined measure of imbalance was sufficiently small; an approximate conditional randomization analysis would then in principle be possible.

An explicit notion of conditional randomization clarifies also the limited usefulness of randomization in small experiments, where any design in a potential randomization set is likely to be meaningfully different. It bears also on the desirability of rejecting "bad" designs thrown up by the randomization.

The penetration of statistical ideas on experimental design is very uneven as between different fields of application; particularly patchy areas are the physical sciences and biochemistry. It is possible that new developments or at least changes of emphasis are needed for these subjects.

A final general comment is that, while the distinction between experimental and observational studies is crucial and not to be blurred, there is much scope for parallel development and for improved techniques in observational studies imported partly or wholly from experimental design (Cochran, 1983). See, for example, Rosenbaum and Rubin's (1983) discussion of the possible effect of unobserved covariates on an observational study with binary response. Also, many of the issues connected with randomization in experimental design have parallels in sampling theory although the styles of discussion in the two literatures are surprisingly different.

For a thorough and lucid review of recent theoretical work on design of experiments, see Atkinson (1982).

3. REGRESSION ANALYSIS

3.1. *Introductory Comments*

Discussions of terminology and of refined distinctions between neighbouring subject areas are sometimes necessary and occasionally important, although possibly more often counterproductive and tedious. A distinction fairly widely accepted and surely important is that between regression and correlation. Suppose that on each individual there are two kinds of observation y and x, both possibly vectors:

(a) regression analysis is concerned with how the conditional distribution of y given x varies with x, in particular with how the "centre" of that distributions varies with x, and, in particular, with how the expected value varies with x;

(b) correlation analysis is concerned with departures from independence in the joint distribution of y and x, the two being treated on an equal footing.

Of course the product moment correlation coefficient as normally defined has a fairly direct interpretation in regression analysis. On the whole regression analysis is much simpler and more incisive than a comparable correlation analysis, essentially because the dimensionality of the distribution under analysis is reduced. Thus log linear analysis of contingency tables is a form of correlation analysis, whereas the singling out of one variable as a basis say for logistic regression analysis is an instance of regression analysis; it is inviting confusion to use the first when the second is appropriate.

Has the time come to abandon the term "regression" in favour of "dependence", i.e. to talk about dependence coefficients rather than regression coefficients, and so on? The older term "independent variable" for x has obvious drawbacks and seems to be passing out of use. To be attracted by regressand and regressor might be taken as indirect evidence of a biased education; response and explanatory variables seem the most evocative terms.

Linear regression analysis based on the method of least squares remains the most important and widely used and studied form of regression analysis. Many of the ideas, however, apply much more widely; see McCullagh and Nelder (1983) for an excellent account emphasizing so-called generalized linear models. Note, however, that many of the more complicated aspects such as hierarchical error and errors in the explanatory variables have not yet been studied in any generality.

It is trite that specification of the purpose of a statistical analysis is important. Several such purposes can be distinguished for regression analysis including:

(a) the derivation of a concise description of a body of data useful, for instance, in comparing different similar sets of data, and for characterizing individuals as high or low responses relatively to what might have been expected in the light of their explanatory variables;
(b) as a basis for prediction (or calibration) for new individuals;
(c) the estimation of regression coefficients including the identification of those explanatory variables with "important" effects.

Aspect (c) is closely allied to the prediction for a new individual of the consequences for the response of that individual to changing values of certain components of x, i.e. of intervention. It is the formulation implicit in comparative experiments and therefore also for observational studies performed in lieu of experiments. For that reason (c) will be given primary emphasis in this paper. We may call applications of that type analytical studies.

General issues connected with regression have been discussed by many authors. The following comments draw partly on Cox (1968) and Cox and Snell (1974).

3.2. *Choice of Representation*

Quite often, especially in observational studies, the dimension of x is fairly large and use of the "full" regression equation is for one reason or another inadvisable. How to deal with this situation is the aspect of regression that causes most concern in applications. To a large, although not entire, extent this problem is a product of the computer era, which has made large-scale fitting by least squares or maximum likelihood a fairly routine matter. Fisher (1938), however, fitted models containing seven explanatory variables (sea-level characteristics of members of an expedition) to explain as response, acclimatization at high altitude for nine individuals; a plausible estimate of error variance was available from the variation within individuals. See Draper and Joiner (1984) for a reanalysis.

The following miscellaneous comments are phrased in terms of least squares regression but are quite broadly applicable to analytical studies, i.e. to applications in which prediction or calibration is not an immediate aim.

(i) It will frequently be sensible to begin by dividing the explanatory variables into treatment variables, intrinsic variables (characterizing the individuals or exogenous, characterizing their environment) and non-specific variables identifying blocks, strata, litters and so on. In observational studies treatment variables identify aspects that would be introduced as treatments in a comparable experiment. Usually interest will focus on the effect of the treatment variables, and their interaction with intrinsic variables.

For example, in an observational study of the effects of smoking and alcohol in pregnancy on the infant's birth weight, measures of smoking and alcohol consumption would be treatment variables, age and sex intrinsic variables; social class would presumably best be considered as a surrogate treatment variable, hopefully incorporating some of the unmeasured variables that in an experimental study would have been controlled by randomization. In the teaching styles study (Aitkin *et al.*, 1981), variables characterizing teachers are treatment variables, initial test scores on children are intrinsic variables and schools or classes are nonspecific.

(ii) A key issue concerns the introduction of external information about the explanatory variables and their effect. While it would be a mistake to make sweeping statements about what is wise in all practical contexts, on the whole it seems to me best to use qualitative external information about the problem, of the kind set out below, especially in high-dimensional problems where the data alone are incapable of adequate resolution. Where assumptions are made, such as that a battery of related variables can be replaced by one of them, they should, where feasible, be tested for conformity with the data.

The development of a systematic approach to these matters of qualitative judgement is a suitable topic for a computerized "expert system" (IKBS).

A consequence of the inclusion of qualitative external information is that automatic variable selection rules have a role largely limited to the initial reduction of systems with so many explanatory variables that careful examination of all adequately fitting models is not feasible. There is a broad distinction between:

(a) methods in which some explanatory variables are omitted and some included in the final representation, those included being given "full weight" in a least squares or maximum likelihood procedure;
(b) methods in which all (or most) of the explanatory variables are included, but some have effects "shrunk" towards zero or some convenient reference level.

The occurrence in (b) of a continuous range of possibilities between complete inclusion and complete exclusion, rather than an all or nothing choice, is attractive from a commonsense point of view. It has, of course, an extensive literature under the name ridge regression and a fairly immediate and qualitatively appealing Bayesian interpretation is available. The barriers to practical use are, however, formidable unless the explanatory variables are all measured in comparable units.

We now discuss some of the qualitative considerations that can be useful.

(iii) Where there are several somewhat similar response variables, it will usually be wise to use the same explanatory variables for all. Thus one might take the union of the sets that seem appropriate for the response variables taken individually. For instance, in a study of the size of newborn infants, log weight, log length and log head circumference might be three response variables of interest. While somewhat different sets of explanatory variables might be included by separate analysis of the three variables, interpretation will probably be aided and the chance of overinterpretation reduced by using a common set of explanatory variables for the three analyses. (This is quite a separate issue from that of whether techniques of multivariate analysis are called for.)

(iv) If it appears that one or more suitable subsets of explanatory variables have been isolated as a basis for interpretation it will often be wise to add the omitted variables back one at a time and to examine the resulting estimates and their standard errors.

(v) When "main effects" have been isolated, it will be sensible to look for interactions, especially for treatment × treatment and treatment × specific variable interactions, where the corresponding main effects are appreciable (Cox, 1984).

(vi) When there are intermediate response variables included among the explanatory variables, a path model (involving conditional independencies) should if possible be formulated and tested. For contingency tables, see Edwards and Kreiner (1983).

(vii) Where an intrinsic explanatory variable with qualitative levels (e.g. sex) has an appreciable effect, it will be for consideration whether to split the data into sections for separate analysis.

It seems to me that there is some urgency for the incorporation into regression analysis packages of simple-minded guidelines, such as the above or improved versions thereof.

3.3 Effect of Intervention

A fairly common use of multiple regression is to predict not the outcome for a new individual, but rather the effect of an intervention, i.e. the imposition of a change in the system. Suppose that the explanatory variable x_1 is to be changed from x_1' to $x_1' + d_1$. A crucial question concerns the consequence for the other explanatory variables:

(a) any variables expected to change with x_1 in accordance with the same stochastic relations as hold in the baseline data are to be omitted from the regression equation;

(b) any variables to be held fixed should be included in the regression equation or be shown likely to have negligible effect;

(c) there remain variables that may change on intervention but not in the way holding in the data under analysis.

For (c) it is of course necessary to have some data about, or to make some assumptions concerning, the changes induced. It is important also, and often a severe limitation on empirical studies, that major changes in unmeasured explanatory variables may take place on intervention, other than those implicitly represented in the regression equation.

A much oversimplified example is where for a number of individuals a response y is measured and explanatory variables x_1 (a binary variable smoker, non-smoker); x_2, height; x_3, some aspect of diet. What is the effect for an individual of giving up smoking? Variable x_2 is of type (b), whereas diet is likely to change but not necessarily in the way indicated by the "between individual" variation. Plausible estimates must be made of the consequent change in x_3. Finally suppose that it is suspected that the unmeasured variable weight is important and that height appears in a direct regression equation solely as a surrogate for weight. Then, because weight may change after intervention, some attempt to estimate its effect is desirable.

Formally, in general, if β is the vector of parameters in the linear model, estimated by $\hat{\beta}$ with covariance matrix Ω_β, we are interested in $\alpha^T \beta$, where α is estimated by $\tilde{\alpha}$ with covariance matrix Ω_α. If random errors in $\tilde{\alpha}$ and $\hat{\beta}$ are uncorrelated

$$E(\tilde{\alpha}^T \hat{\beta}) = \alpha^T \beta,$$

$$\text{var } (\tilde{\alpha}^T \hat{\beta}) \simeq \tilde{\alpha}^T \Omega_\beta \tilde{\alpha} + \hat{\beta}^T \Omega_\alpha \hat{\beta},$$

so that approximate assessment of precision is possible. No doubt these formulae could be refined.

3.4. *Structure of Error*

In the context of second-order multiple regression there is a fairly well-established set of procedures for dealing with error variation in which the covariance matrix departs from the simple form $\sigma^2 I$ and in which it is possible to estimate the covariance matrix reasonably accurately. It is important to distinguish between the effect of the covariance matrix on the efficiency of the ordinary least squares estimate and its effect on estimated precision.

The extension of these results to generalized linear models is largely open. Sometimes, however, weighted least squares methods can be adapted after grouping of individuals into sets with approximately the same values of the explanatory variables. Empirical linearizing transformations, e.g. empirical logit transforms, can be found and analysed approximately, provided, of course, that the grouping adopted lends itself to a simple representation also of covariance structure. The widely appreciated problem of overdispersion, or less commonly underdispersion, relative to the binomial or Poisson distribution is a special case (Cox, 1982b).

If the grouping is not just a device employed for convenience but corresponds to grouping of the individuals with direct physical significance, there is the possibility that regression relations between and within groups are different; see, for instance, Cox and McCullagh (1982). Failure to recognize this can lead to major errors of interpretation.

3.5. *Concluding Remarks*

There are, of course, many further aspects of regression analysis. For instance, regression analysis of time series, "random effects" regression models and kernel regression methods have not even been mentioned. The usefulness of robust regression methods seems to me largely confined to situations where large bodies of data have to be summarized in a somewhat mechanical fashion. The current emphasis on regression diagnostics, the isolation of sensitive observations for individual study, is a formalization of more traditional attitudes to the analysis of data and much to be welcomed.

The topics discussed in this paper are very much a personal selection. It is a mark of the vigour and importance of the subject that, despite the very extensive literature developed over a long period, so much remains to be done.

ACKNOWLEDGEMENT

I am grateful to the Science and Engineering Research Council for their support.

APPENDIX

There follow outline statements of one or two open problems. The literature on both design of experiments and regression is so large that it is entirely possible that I have overlooked relevant work. Matters connected with very special designs have been excluded.

1. Is there a simple and useful non-null randomization theory of designs with binary (and other such) responses?

2. Is there a useful Bayesian theory of experimental design recognizing that the prior distribution for the interpreter of the results may be quite different from that of the planner of the investigation? (Equivalently obtaining the data may, of course quite legitimately, itself lead to a change in the prior distribution.)

3. While interesting work has been done on the design of experiments for estimating components of variance, there appears to be no systematic theory. Develop one.

4. What is the role in practice of cross-over designs with more periods than treatments?

5. Classify the kinds of residual effect that might arise, discuss the designs appropriate for each and examine the extent to which it is feasible to distinguish empirically between the different kinds of residual effect.

6. What is the theoretical asymptotic relative efficiency of the optimal randomized block design versus the optimal systematic design when spatial variation follows some (convenient) spatial stochastic model?

7. What is the asymptotic relative efficiency of both the types of design when spatial variation is a self-similar process?

8. Develop flexible families of designs for use when there are a number of factors with quantitative levels (response surface) combined with some factors with qualitative levels.

9. Develop sequential procedures for D- and D_S-optimal design based on maximum likelihood estimation in a fairly general model. What is the loss of efficiency if the design must be fixed in just two stages?

10. Examine, not solely in a clinical trial context, the so-called principle of analysis by intention to treat (ever randomized, always analysed), as opposed to analysis by treatment actually encountered.

11. In sequential experimentation, explicit changes in the objectives and protocol of the experiment are possible during the course of the investigation. Yet frequent major changes (a different "bright idea" each day) will usually be harmful. Can this usefully be formalized?

12. Develop a simple procedure for detecting whether an apparent dependence on say an explanatory variable x_q could be an artefact arising from measurement errors in one or more of the other explanatory variables.

13. Develop a procedure for binary logistic or probit regression taking account of regressions

"between" and "within" groups of individuals and more refined than the application of normal theory ideas to empirical logit or probit transforms.

14. Extend the discussion of 13 to other generalized linear models and to time series analysis.

15. Suppose that the regression of a response variable on a single explanatory variable is found to be non-linear. Show how to examine whether the non-linearity can be explained by errors of measurement in the explanatory variable. Generalize to multiple regression.

16. Examine in detail the effect in generalized linear models of errors in the explanatory variables.

17. Two (or more) sets of data are available in which the regression relations appear to be different. Suggest how to examine whether the sets are consistent with a single underlying relation, the errors of measurement of the explanatory variables being different in the different sets.

18. Obtain prediction limits, more accurate than those in the text, for the change in response induced by specified intervention on some of the explanatory variables. Extend to the generalized linear model.

19. Discuss carefully the effect on the choice of a multiple regression equation of specific objectives, in particular of the need to extrapolate to a particular point in the space of explanatory variables.

20. Develop versions of ridge regression for the generalized linear model; assume a subset of regression coefficients specified to which the "shrinkage" is not to be applied.

21. Develop a version of multiple regression in which the dependence on one or two explanatory variables of primary interest is assessed by "kernel" methods whereas adjustment for other explanatory variables is carried out by least squares.

22. Develop a systematic theory for the prediction (via intervals) of future values from generalized linear models.

REFERENCES

Aitkin, M., Anderson, D. and Hinde, J. (1981) Statistical modelling of data on teaching styles (with Discussion). *J. R. Statist. Soc.* A, **144**, 419–461.

Atkinson, A. C. (1982) Developments in the design of experiments. *Int. Statist. Rev.*, **50**, 161–177.

Bartlett, M. S. (1978) Nearest neighbour models in the analysis of field experiments (with Discussion). *J. R. Statist. Soc.* B, **40**, 147–174.

Cochran, W. G. (1983) *Planning and Analysis of Observational Studies*. New York: Wiley.

Cox, D. R. (1968) Notes on some aspects of regression analysis (with Discussion). *J. R. Statist. Soc.* A, **131**, 265–279.

———(1982a) Randomization and concomitant variables in the design of experiments. In *Statistics and Probability: Essays in Honor of C. R. Rao*, (G. Kallianpur, P. R. Krishnaiah, and J. K. Ghosh, eds), pp. 197–202. Amsterdam: North-Holland.

———(1982b) Some remarks on overdispersion. *Biometrika*, **70**, 269–274.

———(1984) Interaction (with Discussion). *Int. Statist. Rev.*, to appear.

Cox, D. R. and McCullagh, P. (1982) Some aspects of analysis of covariance (with Discussion). *Biometrics*, **38**, 541–561.

Cox, D. R. and Snell, E. J. (1974) The choice of variables in observational studies. *Appl. Statist.*, **23**, 51–59.

Draper, N. R. and Joiner, B. L. (1984) Residuals with one degree of freedom. *Amer. Statistn*, to appear.

Edwards, D. and Kreiner, S. (1983) The analysis of contingency tables by graphical models. *Biometrika*, **70**, 553–566.

Eilon, S. (1972) Goals and constraints in decision making. *Op. Res. Q.*, **23**, 3–15.

Fisher, R. A. (1935) *Design of Experiments*. Edinburgh: Oliver and Boyd (and subsequent editions).

———(1938) On the statistical treatment of the relation between sea-level characteristics and high-altitude acclimatization. *Proc. Roy. Soc.* B, **126**, 25–29.

McCullagh, P. and Nelder, J. A. (1983) *Generalized Linear Models*. London: Chapman and Hall.

Peto, R., Pike, M. C., Armitage, P., Cox, D. R., Howard, S. V., Mantel, M., McPherson, K., Peto, J. and Smith, P. G. (1976) Design and analysis of randomized clinical trials requiring prolonged observation of each patient. I. Introduction and design. *Brit. J. Cancer*, **34**, 585–612.

Rosenbaum, P. R. and Rubin, D. B. (1983) Assessing sensitivity to an unobserved binary covariate in an observational study with binary outcome. *J. R. Statist. Soc.* B, **45**, 212–218.

Silvey, S. D. (1980) *Optimal Design*. London: Chapman and Hall.

Simon, R. (1979) Restricted randomization designs in clinical trials. *Biometrics*, **35**, 503–512.

Welch, W. J. (1982) Branch-and-bound search for experimental designs based on *D*-optimality and other criteria. *Technometrics*, **24**, 41–48.

Wilkinson, G. N., Eckert, S. R., Hancock, T. W. and Mayo, O. (1983) Nearest neighbour (NN) analysis of field experiments (with Discussion). *J. R. Statist. Soc.* B, **45**, 151–211.

Yates, F. (1937) The design and analysis of factorial experiments. Techn. Commun. 35. Harpenden: Commonwealth Bureau of Soil Science.

Summary of discussion

S. C. Pearce gave an invited contribution to the discussion of the paper. He distinguishes sharply between the role of design in biological experiments and that in industrial experiments based essentially on the relative stability and reproducibility of physical and chemical systems. He considers also that there is a single-mindedness of approach in industrial experiments that makes formal consideration of optimality more appealing. He emphasizes also the difficulties in nonexperimental situations of approaching the study of the effect of notional interventions with their implications of causality and urges caution in terminology.

Biometrika (1988), **75**, 1, *pp.* 161–4
Printed in Great Britain

A note on design when response has an exponential family distribution

By D. R. COX

Department of Mathematics, Imperial College, London SW7 2BZ, U.K.

SUMMARY

Some simple design problems are considered when responses have an exponential family distribution with the difference between 'treatment' groups in the canonical parameter being the objective of study. It is shown that normal theory conclusions apply to a possibly surprising extent. The loss from randomizing followed by subsequent stratification is the equivalent of one observation per stratum. For binary data heterogeneity has to be quite large before failure to stratify induces serious loss, whereas unnecessary use of a matched pair design is fully efficient asymptotically.

Some key words: Binary data; Efficacy; Exponential family; Fisher information; Generalized linear model; Matched pair design; Optimal design; Randomization; Stratification.

1. INTRODUCTION

This note deals with some simple design problems when response has an exponential family distribution. One important special case concerns binary responses. The primary application is to randomized experiments, although the results are relevant also for observational studies. In §§2 and 3 we calculate respectively the loss of precision arising from imbalance within strata in the two 'treatment' groups under comparison and the loss arising from randomizing and ignoring strata.

Section 4 deals with a fairly general formulation in terms of explanatory variables.

We suppose that responses have a one-parameter exponential family distribution and that the parameter of interest is the difference between two groups in the canonical parameter. Thus for binary responses the logistic difference in the probabilities of success is considered.

The broad conclusion is that design considerations holding for normally distributed responses apply more widely.

2. STRATIFICATION

Consider the comparison of two conditions or treatments, C_0 and C_1, with individuals arranged in k strata. In the jth stratum let there be n_{j0}, n_{j1} individuals receiving C_0, C_1 with $n_j = n_{j0} + n_{j1}$, $n = \Sigma n_j$. We base assessments of precision on a model in which responses are independently distributed in an exponential family density with canonical parameter ϕ, that is with density

$$a(y) \exp \{\phi y - K(\phi)\}, \tag{1}$$

say. In the jth stratum, let ϕ take values α_j, $\alpha_j + \Delta$ in C_0, C_1, and let Δ be the parameter of interest. Let R_{j0}, R_{j1} be the total responses and $T_j = R_{j0} + R_{j1}$. Then the information about Δ contributed from the jth stratum is contained in the conditional density of, say, R_{j1} given $T_j = t_j$. This is obtained by first finding the joint distribution of R_{j0} and R_{j1} and thence that of T_j. On division the required conditional density has, for suitable functions $b(r_{1j}, t_j)$ and $L(\Delta, t_j)$, whose precise expression is irrelevant, the form

$$b(r_{1j}, t_j) \exp \{\Delta r_{1j} - L(\Delta, t_j)\}. \tag{2}$$

The conditional Fisher information about Δ from the jth stratum is thus $\mathrm{var}\,(R_{j1}|\,T_j = t_j, \Delta)$ and for many comparative purposes it will be enough to consider this at $\Delta = 0$, that is to write

$$I_j(t_j; n_{j0}, n_{j1}) = \mathrm{var}\,(R_{j1}|\,T_j = t_j, 0). \tag{3}$$

An alternative although closely related interpretation is that $\Sigma\,I_j$ is the Pitman efficacy of a test of $\Delta = 0$.

For instance, if the response is binary, R_{j0}, R_{j1} are the numbers of 'successes' in C_0, C_1, and (3) is the variance of the 'central' hypergeometric distribution; i.e.

$$I_j(t_j; n_{j0}, n_{j1}) = \frac{n_{j0}n_{j1}(n_j - t_j)t_j}{n_j^2(n_j - 1)}. \tag{4}$$

For most design purposes, we work with the expectation over T_j. In the binomial case with $\Delta = 0$, T_j has a binomial distribution with

$$E(T_j) = n_j\theta_j, \quad E\{(n_j - T_j)T_j\} = n_j(n_j - 1)\theta_j(1 - \theta_j),$$

say, so that the expected value of (4) is

$$I_j(n_{j0}, n_{j1}) = n_{j0}n_{j1}\theta_j(1 - \theta_j)/n_j = n_{j0}n_{j1}\nu_j/n_j, \tag{5}$$

where $\nu_j = K''(\alpha_j)$ is the variance of a single observation.

Now quite generally when $\Delta = 0$, $E(R_{j1}|\,T_j = t_j) = n_{j1}t_j/n_j$, and

$$E\,\mathrm{var}\,(R_{j1}|\,T_j = t_j) = \mathrm{var}\,(R_{j1}) - \mathrm{var}\,\{E(R_{j1}|\,T_j = t_j)\}.$$

Unconditionally when $\Delta = 0$, R_{j1} is the sum of independent and identically distributed random variables, as also is T_j; it follows that the second formula (5) holds in all cases.

Now a superficially easier route to this result is via the information matrix for the full likelihood based on (1), treating $(\alpha_1, \ldots, \alpha_k, \Delta)$ as the parameter vector. Inversion followed by isolation of the element corresponding to Δ leads to (5). For our purposes, however, this is unsatisfactory. Statistical interpretation would require the individual strata to be large, whereas the approach via conditional distributions applies in particular to the analysis of a large number of individually small strata.

If n_j is even and exact balance is achieved the information is

$$I_{Bj} = \tfrac{1}{2}n_j\nu_j. \tag{6}$$

On the other hand if the assignment is totally randomized, as would happen if the stratification were inserted retrospectively, then n_{j0}, n_{j1} are values of a binomial random variable of indices n_j and probability $\tfrac{1}{2}$, so that, on taking expectations in (5), the expected information is

$$I_{Rj} = \tfrac{1}{4}(n_j - 1)\nu_j. \tag{7}$$

Thus randomization rather than balancing induces an average loss equivalent to one observation per stratum. Note that for a stratum of size two, with probability $\tfrac{1}{2}$ both individuals receive the same treatment and two individuals are 'lost', whereas with probability $\tfrac{1}{2}$ exact balance is achieved, confirming that the expected loss is indeed one individual.

It is easily shown that over k strata, the expected loss of information is $\tfrac{1}{4}\Sigma\nu_j$ with variance $\tfrac{1}{8}\Sigma\,\nu_j^2(n_j - 1)/n_j$. If the α_j are nearly equal the expected loss corresponds to k individuals with a standard deviation $\sqrt{2}(k - \Sigma\,1/n_j)^{\frac{1}{2}}$. While expected loss of information is a natural design criterion, any design procedure that had an appreciable chance of leading to notable loss would be poor, even if the expected loss were small. For normally distributed responses of constant variance σ_0^2, we have $\nu_j = \sigma_0^2$ and these results are well known, on noting that the parameter of interest is the difference of means divided by σ_0^2.

3. Randomization ignoring stratification

We now consider the consequences of ignoring the stratification, randomizing the treatment allocation and basing the analysis on the comparison of the two group means. There is a corresponding interpretation for observational studies. Under a small difference Δ in the canonical parameter ϕ, the means are

$$\sum n_j K'(\alpha_j)/n = \sum n'_j \mu_j/n = \tilde{\mu}, \quad \tilde{\mu} + \Delta \sum n_j K''(\alpha_j)/n = \tilde{\mu} + \Delta \sum n_j \nu_j/n.$$

When $\Delta = 0$ the effective variance of response over the n individuals is

$$\tilde{\sigma}^2 = \sum n_j \nu_j/n + \sum n_j (\mu_j - \tilde{\mu})^2/(n-1). \tag{8}$$

Thus the local Fisher information about Δ, or equivalently the efficacy, is

$$(\Sigma \, n_j \nu_j/n)^2 \{\tilde{\sigma}^2 (1/m_0 + 1/m_1)\}^{-1},$$

where m_0 and m_1 are the numbers of individuals in the two groups.

Thus the information is

$$I_{\mathrm{CR}} = \rho(\sum n_j \nu_j/n) m_0 m_1/n, \tag{9}$$

where

$$\rho^{-1} = 1 + \frac{\sum n_j (\mu - \tilde{\mu})^2/(n-1)}{\sum n_j \nu_j/n} \tag{10}$$

is a dispersion index. In the balanced case $m_0 = m_1 = \frac{1}{2}n$, equation (9) becomes $\frac{1}{4}\rho \sum n_j \nu_j$, compared with $\frac{1}{4} \sum n_j \nu_j$ in the stratified balanced design, obtained by summing (6).

For the binomial situation with strata of equal size, $\mu_j = \theta_j$, $\nu_j = \theta_j(1 - \theta_j)$, say, with $\bar{\theta} = \Sigma \theta_j/k$; then, on replacing $n-1$ in (10) by n, we have that

$$\rho = 1 - \{\sum (\theta_j - \bar{\theta})^2/k\}\{\bar{\theta}(1 - \bar{\theta})\}^{-1}. \tag{11}$$

Now in the extreme situation when the individual θ_j are 0 and 1, $\rho = 0$, but in less extreme cases ρ can remain quite large even though the θ_j vary appreciably; thus $\rho = 0.84$ if $\bar{\theta} = \frac{1}{2}$ and the θ_j have standard deviation 0.2.

A particularly important special case is that of matched pairs. Comparison of (8) and (6) shows that, to the extent that efficacy is measured by Fisher information, or efficacy, there is no loss in unnecessary pairing, i.e. in stratifying when in fact $\rho = 1$. For binary data the above discussion shows that the loss from randomizing and failure to stratify is appreciable only in rather extreme cases. On the other hand the loss from randomizing in design and stratifying in analysis is 50%, as noted in § 2.

Note that these conclusions depend on the assumption that the difference in the canonical parameter is stable and is the focus of interest. For instance, for exponentially distributed random variables the canonical parameter is the rate, i.e. the reciprocal of the mean. Thus if the difference in rates is constant and the parameter of interest, there is asymptoically no loss of information in unnecessary pairing; it is, however, known (Lindsay, 1980) that, if the ratio of rates is constant and is the parameter of interest, the loss from unnecessary pairing is large.

4. Regression design

The previous discussion has been for simple comparisons of two groups or treatments, although the conclusions extend fairly directly to the comparison of a small number of unstructured treatments. Suppose now that for the ith individual there is a $p \times 1$ vector z_i of explanatory variables, and that we work with the model (1) with, for the ith individual, $\phi_i = g(\beta^{\mathrm{T}} z_i)$, where $g(x)$ is a known function and β is a vector of unknown parameters. Then the information matrix for estimating β has (r, s) element

$$\sum_i K''\{g(\rho^{\mathrm{T}} z_i)\}\{g'(\rho^{\mathrm{T}} z_i)\}^2 z_{ir} z_{is}. \tag{12}$$

Now if interest lies in estimation near to a null case in which response is an unknown constant, then the information matrix is proportional to $z^T z$, where z is the $n \times p$ matrix of explanatory variables, and the usual discussion tied to least-squares theory is directly applicable. The same is true if for some reason the first two factors in (12) vary only slowly with $\beta^T z_i$ over the range contemplated.

Therefore to this limited but nontrivial extent 'standard' results about regression design apply.

5. DISCUSSION

At first sight the comparison of alternative designs for a general exponential family distribution raises the usual difficulty of 'nonlinear' design, namely that the conclusions depend on the unknown parameter value under study, and that therefore any 'optimal' design must be sequential, or involve strong prior assumptions. This difficulty has been met here by concentrating on local arguments involving the estimation of small treatment differences. This is likely to be adequate, in particular whenever asymptotic maximum likelihood theory is applicable, and the treatment effect is within a few standard errors of zero. In other cases the familiar difficulties of nonlinear design will indeed arise.

Note also that, in the discussion of §§ 2 and 3, strata effects have been completely eliminated by conditioning. An alternative approach is to represent the strata effects as random variables and this would often be a reasonable formulation, especially if interest lay in a difference on a noncanonical parameter, when simple elimination by conditioning would not be available. The implications for design have not been explored.

ACKNOWLEDGEMENTS

I am grateful to Dr Alan Ebbutt, Glaxo Limited, whose questions led to this investigation, to Department of Statistics, Institute of Advanced Studies, Australian National University, Canberra for their hospitality while the work was done, and to the Science and Engineering Research Council for a Senior Fellowship.

REFERENCE

LINDSAY, B. G. (1980). Nuisance parameters, mixture models, and the efficiency of partial likelihood estimators. *Phil. Trans. R. Soc., Lond.* A **296**, 639–65.

[*Received March* 1987. *Revised July* 1987]

D8

G. Kallianpur, P.R. Krishnaiah, J.K. Ghosh, eds., *Statistics and Probability: Essays in Honor of C.R. Rao*
© North-Holland Publishing Company (1982) 197–202

RANDOMIZATION AND CONCOMITANT VARIABLES IN THE DESIGN OF EXPERIMENTS

D.R. COX

Department of Mathematics, Imperial College, London, U.K.

General comments are made on experimental design in the presence of concomitant observations. The circumstances under which appreciable non-orthogonality may arise are investigated. It is shown that under some circumstances repeated rerandomization, until a nearly balanced design emerges, is justifiable from the point of view of pure randomization theory.

1. Introduction

A central theme in statistical studies of experimental design is the control of error via information about the experimental units available before the assignment of treatments. Control is usually achieved by a judicious combination of balancing (blocking or stratification), adjustment (for example by analysis of covariance or by fitting some suitable special model) and randomization. Sometimes the initial information worth including is quite limited and then, at least for fairly simple treatment structures, a satisfactory design can usually be found by conventional methods.

It is useful to draw a non-rigid distinction between blocking and stratification as devices for the control of error. In stratification the levels correspond to factors classifying the experimental units (e.g., by age and sex), whereas in blocking the levels correspond to groups of supposedly similar individuals not necessarily meaningfully labelled. Further, there may well be a large number of replications of each treatment within a stratum, whereas normally the number of replications per block will be small.

We consider experiments comparing t treatments $T_1, \ldots, T_t$, equally replicated, interest attaching to

$$\operatorname*{ave\,var}_{u,v} (\hat{\tau}_u - \hat{\tau}_v), \tag{1}$$

where $\hat{\tau}_u$ is the estimated mean response for T_u. For simple experiments with one concomitant variable initially available per experimental unit, there will be a choice between using this variable to form blocks or as a basis for adjustment. Cox (1957) showed that with the criterion (1) it is only in very small experiments or when the relation between response and the concomitant variable is very strong that the choice is critical.

Consideration of a second use of concomitant variables, namely to detect interactions, tends to strengthen the case for a direct use of the concomitant variable, as against its indirect use to form blocks or strata.

197

Now if we use randomization followed by adjustment by analysis of covariance, there will be some inflation of variance arising from non-orthogonality, the amount depending on the lack of balance in the design realized and on the number of concomitant variables.

If there are initially available a considerable number of concomitant observations per experimental unit, it will often be a good idea to reduce the concomitant variables at the start to a small number of possibly derived variables. If, however, this cannot be done, stratification will typically be possible only for a few of the variables; the conventional approach is to randomize and to adjust for the remaining variables by analysis of covariance. We investigate this in more detail below.

The issues are of particular importance in connection with clinical trials where there has been some controversy over the best procedure when there are many prognostic (concomitant) variables. In clinical trials treatment assignment is nearly always sequential, a complication we largely ignore.

2. Theoretical development

Consider first an experiment with $n=2r$ experimental units to compare two treatments T_1 and T_2, each treatment being assigned to r units. Let a $p \times 1$ vector z of concomitant variables be available on each unit. Assume that linear methods of analysis based on the method of least squares are used.

Then if $\hat{\tau}_2 - \hat{\tau}_1$ is the adjusted difference between treatments after allowing for linear regression on z, we have under the usual second-order assumptions about error that

$$\mathrm{var}(\hat{\tau}_2 - \hat{\tau}_1) = \sigma^2 \left\{ \frac{2}{r} + (\bar{z}_2 - \bar{z}_1)^{\mathrm{T}} S_z^{-1} (\bar{z}_2 - \bar{z}_1) \right\}. \tag{2}$$

Here σ^2 is the residual variance per observation, $\bar{z}_1$ and $\bar{z}_2$ are the vector means of the concomitant variables for T_1 and T_2 in the design chosen and S_z is the matrix of sums of squares and products of z within treatments, eliminating block differences where appropriate.

More generally, if there are t treatments, each assigned to r experimental units,

$$\mathrm{ave\,var}(\hat{\tau}_u - \hat{\tau}_v) = \sigma^2 \left\{ \frac{2}{r} + \frac{2}{r(t-1)} \mathrm{tr}(B_z S_z^{-1}) \right\}, \tag{3}$$

where B_z is the matrix of sums of squares and products between treatments, i.e. of r times the sum of squares and products of deviations of means.

If treatment assignment is randomized

$$E_{\mathrm{R}}(S_z / d_w) = \Omega_z,$$

where E_{R} denotes expectation over the randomization, d_w is the degrees of freedom within treatments and Ω_z is the "finite population" covariance matrix, within blocks or strata where appropriate. Further, the randomization distribution of $\bar{z}_2 - \bar{z}_1$ is such that

$$E_{\mathrm{R}}(\bar{z}_2 - \bar{z}_1) = 0, \qquad E_{\mathrm{R}}\{(\bar{z}_2 - \bar{z}_1)(\bar{z}_2 - \bar{z}_1)^{\mathrm{T}}\} = 2\Omega_z / r.$$

Approximate multivariate normality will hold as $n \to \infty$ for fixed p, under weak conditions.

It follows that

$$\tfrac{1}{2} r(\bar{z}_2 - \bar{z}_1)^{\mathrm{T}} \Omega_z^{-1}(\bar{z}_2 - \bar{z}_1) = \tfrac{1}{2} r \| \bar{z}_2 - \bar{z}_1 \|_{\Omega_z}^2, \tag{4}$$

say, has expectation p and asymptotically a chi-squared distribution with p degrees of freedom. Hence also

$$\tfrac{1}{2} r \| \bar{z}_2 - \bar{z}_1 \|_{S_z/d_w}^2 \tag{5}$$

has approximately the same properties.

Thus, if we write W_p for a chi-squared random variable with p degrees of freedom, from (2) we have approximately

$$\mathrm{var}(\hat{\tau}_2 - \hat{\tau}_1) = \frac{2\sigma^2}{r} \left(1 + \frac{W_p}{d_w} \right); \tag{6}$$

the same formula follows from the more general form (3). The value of W_p depends on the particular design produced by the randomization.

If the numbers of blocks and treatments are small compared with n, d_w in (6) can be replaced by n. If we take expectations over the randomization, the inflation factor in (6) becomes

$$(1 + p/d_w) \simeq (1 + p/n), \tag{7}$$

in the special case just mentioned.

Despite the simplicity of these formulae, interpretation needs a little care.

3.   Preliminary discussion

An initial interpretation of (7) when $d_w \simeq n$ is that, because $1 + p/n \simeq n/(n-p)$, non-orthogonality is equivalent to a loss of p observations. This would suggest that so long, say, as $p < \tfrac{1}{10} n$ the effect of non-orthogonality is unimportant.

There are two reasons why this is oversimplified. First, in the final analysis it may well be that adjustment is carried out not for the full p concomitant variables but rather for a subset of p_0 variables which appear to show fairly clear connection with the response; some form of ridge regression is an alternative. The effect is likely to be to reduce the inflation factor $1 + W_p/d_w$, although presumably the factor will typically exceed $1 + W_{p_0}/d_w$.

The second effect works in the opposite direction. It is no defence of a particular experiment which is seriously unbalanced to show that the realized design is atypical. As a general guide in design, it is reasonable to require that most of the designs produced by the randomization are satisfactory. If $c_{p\varepsilon}^{*2}$ is the upper ε point of the chi-squared distribution with p degrees of freedom, we need not $p < \tfrac{1}{10} n$ but $c_{p\varepsilon}^{*2} < \tfrac{1}{10} n$ for perhaps $\varepsilon = 0.01$, an appreciably more stringent requirement. For $\varepsilon = 0.01$, $p = 1, 5, 10, 50, 100$ the minimum values of n are $67, 151, 232, 762, 1358$. Thus, quite large values of n are needed if the chance of appreciable imbalance is to be kept at 0.01.

The above discussion is for linear methods of analysis. It is likely that similar conclusions apply more broadly, for example to the maximum likelihood analysis of models for binary or survival data. There is, however, the technical difficulty that if the adjustments are relatively small their study requires greater refinement than the "first order" asymptotic theory of maximum likelihood estimation.

4. Randomization

So far we have assumed that a standard scheme of randomization is used, taking account of blocking or stratification as appropriate. There are two difficulties.

One is that adjustment by analysis of covariance does not have an exact second-order randomization theory. This can be overcome by weighted randomization (Cox 1956) or by reliance on asymptotic arguments.

A more serious point concerns the circumstances under which arguments based on randomization are convincing for the interpretation of the particular experimental arrangement actually used. For this it is necessary that the arrangement will not be distinguishable in a relevant respect from the reference set of arrangements used in the randomization calculations. A precise formulation will not be attempted.

This raises special difficulties in the present instance because the quantity W_p does classify the possible designs into sets of differing precisions, at least under the associated standard linear model. This reinforces the desirability of rejecting any arrangement for which W_p/d_w is appreciable; if the proportion of arrangements so rejected is not small, the whole relevance of the randomization scheme is suspect.

Of course, quite generally when randomization is used, the individual arrangements have distinguishing features. It is always possible that rational arguments will be produced for regarding the arrangement used as atypical, in which case an ad hoc analysis has to be considered. There is, however, an onus on those suggesting additional complications to show that substantial benefit results. The wide success of randomization depends on the empirical fact that often, especially when the randomization set contains many arrangements, there are indeed appreciable numbers of designs that can reasonably be regarded as equivalent. Even then, however, designs having some feature in an extreme form are best rejected, the effect on the randomization distribution being small provided that the proportion of rejected arrangements is small. In the present instance one would normally want to discard arrangements with $W_p/d_w > \frac{1}{10}$; see Section 3 for the conditions under which this will involve only a small proportion of arrangements.

There is, however, another possible procedure for randomization. Suppose that we randomize conditionally on $\|\bar{z}_2 - \bar{z}_1\|_{\Omega_z}$ being small. That is, we randomize, compute $\|\bar{z}_2 - \bar{z}_1\|_{\Omega_z}$ or, more generally $\mathrm{tr}(B_z\Omega_z^{-1})$, accept the design if $\|\bar{z}_2 - \bar{z}_1\|_{\Omega_z} < a$, for suitable a, and otherwise reject the design and rerandomize. In some special cases, notably when the z's identify a system of blocking, there may exist a large number of arrangements with $\|\bar{z}_2 - \bar{z}_1\|_{\Omega_z} = 0$ and an exact second moment randomization theory may hold. In general, however, we must argue approximately as follows.

Under the usual randomization model the response observed on unit i is $\xi_i + \tau_{[i]}$, where $T_{[i]}$ is the treatment applied to that unit; in addition, a vector of concomitant variables is available on each unit. Then under complete randomization the joint distribution of $(\bar{\xi}_2 - \bar{\xi}_1, \bar{z}_2 - \bar{z}_1)$, the differences between the means for T_2 and T_1, has zero expectation and covariance matrix

$$\frac{2}{r} \begin{pmatrix} \sigma_\xi^2 & \Omega_{\xi z} \\ \Omega_{\xi z}^{\mathrm{T}} & \Omega_z \end{pmatrix},$$

say, with elements defined by the finite population of ξ's and z's. To the extent that a multivariate normal approximation is adequate, the distribution of $\bar{\xi}_2 - \bar{\xi}_1$, when randomization is constrained so that $\|\bar{z}_2 - \bar{z}_1\|_{\Omega_z}$ is small, is normal with zero mean and variance

$$2\sigma_{\xi \cdot z}^2 / r,$$

in the usual notation. Finally, $\sigma_{\xi \cdot z}^2$ is consistently estimated from the empirical variance eliminating treatment effects and linear regression on z.

Now there are several requirements for the precise scheme of rerandomization:

(i) We want

$$\tfrac{1}{2} r \|\bar{z}_2 - \bar{z}_1\|_{\Omega_z}^2 / d_w \leqslant b \tag{8}$$

for some suitable b, e.g. $b \leqslant \tfrac{1}{20}$, so that negligible inflation of variance occurs.

(ii) We want the selected design to satisfy

$$\tfrac{1}{2} r \|\bar{z}_2 - \bar{z}_1\|_{\Omega_z}^2 \leqslant c_{p, 1-\alpha}^{*2} \tag{9}$$

for some fairly small, but not very small, value of α, so that $\|\bar{z}_2 - \bar{z}_1\|$ lies near the centre of its probability distribution under full randomization.

(iii) Except where special arguments can be employed, we want α defined by (9) not to be too small, so that the number of rerandomizations, geometrically distributed with mean $1/\alpha$, is not excessive.

(iv) If n_R is the number of distinct designs in the full randomization set, the mean in the constrained randomization set is approximately αn_R and this should be at least, say, 50–100 if an effective analysis based on the randomization distribution is to be in principle possible.

In very small experiments, having satisfied (8) and (9), condition (iv) will not be achieved. Then it seems necessary to abandon reliance on randomization theory, to assume a physical model for the random variation and to choose a systematic design; see Cox (1951) for the case where the z's determine a polynomial trend with "time".

Of course, if $p \ll n$ the much simpler procedure of Section 3, randomization with rejection of occasional extreme arrangements, will be satisfactory.

In clinical trials there is the further complication that entry is nearly always sequential. A scheme of "minimization" (Pocock 1979) could be used in which the assignment probabilities for a new individual with given z are, subject to concealed assignment, biased to the treatment leading to greater balance. The effect of such schemes on randomization analysis has not been fully discussed, although it is clear in a general way that any time trends in response should be eliminated from error.

References

Cox, D.R. (1951). Some systematic experimental designs. *Biometrika* **38**, 312–323.

Cox, D.R. (1956). A note on weighted randomization. *Ann. Math. Statist.* **27**, 1144–1151.

Cox, D.R. (1957). The use of a concomitant variable in selecting an experimental design. *Biometrika* **44**, 150–158.

Pocock, S.J. (1979). Allocation of patients to treatment in clinical trials. *Biometrics* **35**, 183–197.

SOME SAMPLING PROBLEMS IN TECHNOLOGY

D. R. Cox

Imperial College

SUMMARY

Some general aspects of technological sampling are
reviewed. Then some special procedures used in sampling
textile fibres and in work study are outlined. Finally
some theoretical problems raised by the last two topics
are analyzed. In particular, the estimation of frequency distributions from samples based in a known way
and from samples of 'recurrence-times' is developed.

1. INTRODUCTION

Most published statistical work on industrial sampling problems is concerned with schemes for acceptance
sampling. There the emphasis is not on the actual selection of the sample but rather on the intensity of
sampling that is appropriate and on the decision to be
taken on the basis of the sample. The relative neglect
of sample selection in books and journals on the industrial applications of statistics is probably explained
by sampling techniques being largely specific to particular applications. Much practical industrial sampling
may be rather haphazard; nevertheless to obtain reliable and reproducible test results, either for research
or control purposes, or in connexion with legal agree-
ments to supply material meeting a specification, carefully defined sampling procedures are essential. Thus
some British Standards, and comparable U.S. and Japanese
documents, contain careful discussion of sampling techniques; see, especially, the British Standards on sampling of coal and coke and of ores (British Standards
Institution, 1960, 1967), the former containing an excellent discussion of principles that should be widely
applicable. A bibliography on methods for sampling raw
materials is given by Bicking (1964); see, especially,
Duncan (1962) for some general discussion.

The plan of the present paper is as follows. In Section 2 some points that seem to arise quite commonly in industrial sampling are mentioned briefly. Then in Sections 3 and 4 the special problems of two fields are outlined, the sampling of textile fibres and work. sampling. Finally, in Section 5, some theoretical problems raised by the earlier discussion are analyzed.

2. SOME GENERAL POINTS

In this section some brief comments are given on a number of miscellaneous points that seem to arise fairly often in industrial sampling.

(i) Absence of Frame

A quite common difficulty is the initial absence of a good frame, i.e. of a clear specification of the position of each element of the population. Thus, suppose that material is stored say in sacks. It may be straight-forward to draw a random sample of sacks, but a procedure for sampling material from within a sack may be hard to specify unambiguously. One conclusion is that when practicable sampling should be done when material is, for example, moving on a conveyor belt, freely falling, etc., when each element of the population is identified by the time at which it passes a reference point. Then, provided that all material passing the sampling point at suitably chosen sampling points is taken, an unbiased method of sampling is obtained. Another possibility is to obtain a frame in terms of spatial coordinates by spreading the material uniformly over a circular or rectangular sampling area. If these procedures are not practicable, it will be necessary to specify that sample increments are taken from suitably spaced positions from within, say, the sack, but this will often leave open the possibility of appreciable biases.

(ii) Inaccessibility

In some situations, parts of the population may be completely or relatively inaccessible. It may be possible to stratify so that the inaccessible portions from a stratum sampled at low intensity. Another possibility, when sampling is to be repeated over time, is that occasional calibration observations can be used to establish a regression relation between values in the inaccessible and accessible portions of the population , or that there is a theoretical connexion. An example of the latter might arise in studying temperature distribution within a chemical reactor, when a theoretical temperature distribution may be available; measured temperatures at the sampling points can be used, in effect, to estimate an empirical correction to the theoretical temperature at an inaccessible point.

(iii) Distributional Aspects

Sometimes it may be required to estimate simply a mean for the population. In other cases, components of variance may also be required measuring say, important sources of variability in a complex process. In yet other situations, interest may lie in the extremes of the distribution over the population. Thus, a safety requirement may be formulated in terms of the maximum temperature achieved in a reactor, and in other situations the proportion of the population outside specification limits may be of special interest.

(iv) Adjustments

The possible availability of supplementary information and the need to apply adjustments or corrections to the sample observations raise problems on the whole more connected with analysis than with design. Two different aspects are involved. In one, allowance has to be made for differences between the observation obtained on a sampled unit and the "true" value. This may involve theoretical or empirical corrections, the calibration of a "quick" sampling procedure by occasional use of a definitive sampling technique, or, at its simplest, the reduction of the observed variance to allow for the variance explained by errors of measurement. The other possibllity is that supplementary information is available for the whole population; see(ii). The statistical analysis is straight-forward and well-known if only means and variances are of interest or if all distributions are normal, but estimation of extremes in non-normal populations may be more difficult.

(v) Components of Variance

The validation of sampling and testing methods for example by inter-laboratory trials, will often raise statistical problems of design and analysis connected with components of variance.

(vi) Formation and Division of Composite Samples

Sometimes the amount of material required for final analysis, e.g. for chemical analysis, is small, much smaller than can be handled without bias in the initial sampling. If separate analyses are not required for the individual sample portions (increments), it will then be common to form a composite sample by combining the increments. The composite sample is then reduced to the required size by thorough mixing, division into two and rejection of a randomly chosen half, the mixing, division and rejection being repeated until a final sample of the required size is obtained. If a direct measure of measurement error is required the process must be repeated independently on duplicate increments drawn at the same time. If estimates are required say of the variation between parallel machines, then the separate machines would normally be regarded as separate populations. More complex schemes of compositing may be useful when components of variance are to be estimated.

(vii) Time Series Aspects.

Quite often, there is a sequence of p pulations to be sampled and the sequence can be considered as a time series. For example, in a batch process, each batch may be regarded as a separate population. There are then typical times series problems in that, in effect, information about a particular population may be contributed by observations on neighbouring populations.

3. Sampling of Textile Fibres

This section is concerned with the problems of a very special field, namely, the sampling of textile fibres, for example for the estimation of fibre length distribution. Palmer and Daniels (1947) have given a general account of the problems and the following re-

marks are meant to explain the statistical points involved rather than to set out the experimental details.

Consider first an assembly of fibres all parallel
to an axis, the position of each fibre along the axis
being therefore defined by the coordinate of say the
left-hand end and the fibre length. The numbers of fibres in a typical population is usually extremely large;
the number crossing a particular cross-section may vary
from about 20 for a fine yarn to a thousand or more at
earlier stages of processing. The fibres are well mixed
in the early stages of processing and their left-ends
lie approximately in a Poisson process along the axis.

First there is a very important distribution between unbiased methods of sampling, in which each fibre
has an equal chance of selection, and length-biased
sampling in which the chance of selecting a particular
fibre is proportional to its length.

An unbiased sample is obtained in principle by
choosing very short sampling intervals along the axis
and taking all those fibres whose left-ends lie in the
sampling intervals. The experimental technique for doing this, cut-squaring, was developed by Daniels (1942).
Each fibre left-end is a point, and if the sampling intervals are randomly chosen, each fibre has an equal
chance of selection. In practice, because of the near
randomness of the fibre arrangement, the crucial step
is the isolation of the relevant fibres rather than the
positioning of the sampling intervals. Denote by $f(x)$
the density function of the fibre length, X, in the population and $f(x,y)$ the joint density of fibre length, X,
and another property Y, for example fibre diameter.

The second method of sampling is in principle as
follows. The assembly is gripped at a sampling point
and all fibres not gripped, i.e. not crossing the sampling point, gently combed out. The fibres remaining
constitute the sample, or an increment when a composite
sample is formed. Each fibre has a chance of selection
proportional to its length and so the density of length
in sampling by this method is

$$g(x) \;\; = \;\; \frac{x\,f(x)}{\mu} \;,$$
(3.1)

where μ is the mean of the unweighted density $f(x)$. The joint density of X and another variable Y is

$$g(x, y) = \frac{x\ f(x,\ y)}{\mu_X}, \qquad (3.2)$$

where μ_X is the mean of X in the unweighted density. The densities $g(x)$ and $g(x, y)$ are called length-biased. The marginal density of Y derived from (3.2) is the length-biased density of Y, etc.

Simple relations hold between the moments of $g(x)$ and those of $f(x)$ and, in particular,

$$E_g(X) = \mu(1 + \frac{\sigma^2}{\mu^2}), \qquad (3.3)$$

where σ^2, μ are the variance and mean of the unweighted distribution and E_g denotes an expectation with respect to the density $g(x)$. In general

$$E_g(X^r) = \frac{\mu_{r+1}}{\mu}, \qquad (3.4)$$

where μ_s is the s^u moment about the origin of the un-weighted distribution.

Practical difficulties are first that there may be breakage in combing out the unwanted fibres and, more interestingly from a statistical point of view, that non-parallelism of the fibres will mean that the sample is really extent-biased rather than length-biased, where extent is the length of the projection on to the axis of sampling.

In a variant of the method only the fibres to the right of the sampling point are removed, leaving a tuft of projecting fibres, called a combed tuft. Electrical or optical scanning, giving the thickness of this as a function of the distance from the sampling point, provides a very quick method of describing fibre length distribution. Imagine first that the length of each projecting fibre is measured. This would give obser-vations of a random variable Z whose distribution is found as follows. Given that a fibre selected is of length x, the sampling point is equally likely to lie anywhere along its length. Thus the conditional density of Z is rectangular over (0, x). Since (3.1) gives the density of the lengths of selected fibres, it follows that the density of Z is

$$k(z) = \int_z^\infty \frac{1}{x} \frac{x\, f(x)}{\mu}\, dx = \frac{1 - F(z)}{\mu}, (3.5)$$

say, where $F(z)$ is the cumulative distribution function of the unweighted distribution. This density is familiar in the theory of stochastic processes as the recurrence-time distribution.

It is easily shown, for example by evaluating the moment generating function, that

$$E_k(Z^r) = \frac{\mu_{r+1}}{(r+1)\mu}. \qquad (3.6)$$

In particular, $E_k(Z)$ is one-half the length-biased mean fibre length (3.3), as is physically obvious.

Of a tuft of N fibres, the expected number longer than z is

$$N \int_z^\infty k(u)\, du\ ,$$

so that the expected combed tuft curve is

$$c(z) = N \int_z^\infty k(u)\, du = \frac{N}{\mu} \int_z^\infty \{1 - F(u)\}\, du. (3.7)$$

In practice there are complications, because the scanning methods do not count fibres but measure approximately mass density and therefore depend on the correlation between cross-sectional area and length.

A useful property of (3.7) is that $c'(0)/c(0) = -1/\mu$, implying that the tangent to the combed tuft curve at $z = 0$ intersects the axis at μ. In the theoretical discussion of Section 5, it will be assumed that the measuring device in effect determines the individual 'recurrence-times'.

Palmer (1948) introduced an ingenious sampling procedure, dye sampling, for use with a plane web of non-parallel fibres. Imagine first that each fibre were to have one end, arbitrarily chosen, named as its left-end.

An unbiased method of sampling would take all fibres with
left-ends in randomly selected sampling areas. The procedure which in effect achieves this is to stamp a dyed
square on the web; fibres not having an end marked are rejected, those with one end marked receive weight $\frac{1}{2}$, and
those with both ends marked receive weight 1. The theoretical variance of the sample mean can be found, assuming a 'Poisson web'.

Size-biased samples arise in other contexts, for example, in work with assemblies of small particles, where
surface-area, volume- and mass-biased samples may arise
(Herdan, 1960). A rather different application is mentioned by Kriens (1963), in connexion with auditing. Items,
of which a running total of values is available, are to be
selected for checking. If sampling points are distributed
randomly over the total value of the items, a value-biased
sample will result.

4. WORK SAMPLING

As a second special field, consider some applications
to work study. Suppose that one or more individuals are
considered over a certain time period and that at any instant each individual is in one of a number of possible
states. For example, an individual may be a machine which
at any instant is either running or stopped, or may be an
operative who may be busy or idle. More realistically,
several types of stop may be distinguished or several
types of work by the operative, but for a general description it is enough to consider just two states.

One way of investigating such a system is by detailed
recording of its behaviour, for example by a continuous
recording device, or by observation with a stop watch.
Tippett (1935) discussed the estimation of the proportions
of times in the various states from observations merely
of the state occupied at discrete sampling points; under
suitable assumptions the observed proportion of observations say in the stopped state has a binomial sampling
variance. The method is widely applicable to systems for
which at any time point one of a number of clearly defined
and clearly identified states is occupied. In many situations the absence of quantitative observations is a
major advantage.

Practical aspects of the method, which is known

variously as the snap-reading method, the ratio-delay method and as activity sampling, are described for example by Barnes (1957).

In selecting the sampling instants, the usual principles of sample survey design will be relevant. Hill and Hill (1968) have given tables of random clock times for use when some form of random sampling is involved. In some applications systematic sampling will be in order and Davis (1955) and Cox (1961, pp. 86-90) have discussed its theoretical efficiency for particular types of stochastic process. Cox (1966) and Gaver and Mazunder (1967) have made a theoretical comparison of the snap-reading method with alternative schemes of sampling.

In practice, there are two rather different situations to be considered. In one there is a single individual, or at most a few individuals, and sampling is an occasional operation. In other applications, notably those originally considered by Tippett, there are a large number of individuals and a sampler is almost continuously patrolling the whole system. In many applications it will be important that individuals cannot predict when they will be sampled and this will preclude systematic sampling. Another widely important point is that a rule defining the sampling instants should be very precisely formulated; this is particularly important when there are stops that are very frequent but individually of short duration.

Note that the stopped times captured in the snap reading method form a length-biased sample and that the time measured forward from the sampling point to the completion of a stop has the recurrence-time distribution; the arguments of Section 3 are directly applicable.

5. SOME THEORETICAL RESULTS

5.1 Introduction

The techniques outlined in the previous two solutions raise theoretical problems, some of them common to the two situations. In connexion with sampling fibres there is a need to derive information about the unweighted

distribution from either the length-biased distribution
or from the recurrence-time distribution. Considerations
of the relative statistical efficiency of different methods of sampling are likely to be secondary in choosing
between methods. Nevertheless it is of interest to know
these efficiencies and they are obtained in Sections 5.2
and 5.3. Blumenthal (1967) has given an interesting discussion of these problems, although from a rather different viewpoint; in particular, most of his results assume that the coefficient of variation is known.

5.2 Estimation from Length-Biased Samples

In some contexts in which length-biased sampling is
used it may be quite reasonable to regard the length-
biased distribution as the object of study and in particular to consider the length-biased mean, variance,
etc., as parameters. In other situations, however, it
will be required to derive, from length-biased data,
estimates referring to the original distribution.

Suppose, therefore, that $X_1,\ldots,X_n$ form a random
sample of positive random variables having probability
density function (p.d.f.)

$$g(x) \;=\; \frac{x\,f(x)}{\mu} \quad (x>0), \qquad (5.1)$$

where μ is the mean of the p.d.f. $f(x)$. For the reasons
explained in Section 3, call $f(x)$ the unweighted p.d.f.
and $g(x)$ the weighted or length-biased p.d.f.

We examine methods that apply to general continuous
distributions and also those that defend on a paramative
form for $f(x)$. First consider the estimation of the unweighted mean μ, assumed finite. Now

$$E_g\left(\frac{1}{X}\right) \;=\; \int_0^\infty \frac{1}{x}\,\frac{x\,f(x)}{\mu}\,dx \;=\; \frac{1}{\mu}, \qquad (5.2)$$

where E_g denotes expectation with respect to the weighted
distribution. Thus

$$\frac{1}{\tilde{\mu}_g} \;=\; \frac{1}{n}\sum \frac{1}{X_i}$$

is an unbiased estimate of $1/\mu$. Also

$$E_g\left(\frac{1}{X^2}\right) = \frac{\mu_{-1}}{\mu} ,$$

where μ_r is the rth moment about the origin of the un-weighted distribution; we continue to write μ for μ_1. Thus

$$\text{var}_g\left(\frac{1}{\tilde{\mu}_g}\right) = \frac{1}{n}\text{var}_g\left(\frac{1}{X}\right) = \frac{\mu\mu_{-1}^{-1}}{\mu^2} . \qquad (5.3)$$

Provided that (5.3) is finite, $\tilde{\mu}_g$ is for large n asymptotically normally distributed with mean μ and variance

$$\frac{\mu^2(\mu\mu_{-1}-1)}{n} . \qquad (5.4)$$

On the other hand, the mean of an unweighted sample of size n estimates μ with variance $(\mu_2' - \mu^2)/n$. Therefore for estimating μ by these methods, the asymptotic efficiency of ordinary sampling relative to length-biased sampling is

$$\frac{\mu^2(\mu\mu_{-1}-1)}{(\mu_2 - \mu^2)} . \qquad (5.5)$$

Expression (5.5) can take any non-negative value. At one extreme there are unweighted distributions, for instance those with non-zero oridnate near the origin, for which μ_{-1} is infinite and μ, μ_2 finite. On the other hand, distributions with a suitably long upper tail will have μ_2 infinite and μ, μ_{-1} finite. Special unweighted distributions of interest are the log normal, for which (5.5) is unity, and the gamma distribution of index β, for which (5.5) is $\beta/(\beta - 1)$. Note especially that (5.5) refers to the estimation of a special parameter by methods that will not in general be efficient if the distribution has a given parametric form.

Next consider the estimation of the unweighted cumulative distribution function (c.d.f.), say for the purpose of graphical analysis of goodness of fit to a particular parametric form. To estimate the c.d.f. at say z, let

$$U_i(z) = \begin{cases} 1/X_i & \text{if } X_i \le z, \\ 0 & \text{if } X_i > z, \end{cases}$$

and

$$V_i = 1/X_i.$$

Then define

$$\tilde{F}_g(z) = \Sigma\, U_i(z)/\Sigma V_i. \tag{5.6}$$

We now show that for any fixed z, $F_g(z)$ is an estimate of $F(z)$, the unweighted c.d.f.

The asymptotic distribution of (5.6) is found by first examining the joint distribution of numerator and denominator. Write

$$\mu_r(z) = \int_0^z x^r f(x)\,dx$$

for the incomplete moments of the unweighted distribution; note that in particular $\mu_o(z) = F(z)$. Then it is easily shown that

$$E_g\{U_i(z)\} = \mu_o(z)/\mu, \quad E_g(V_i) = 1/\mu, \tag{5.7}$$

and that

$$\text{var}_g\{U_i(z)\} = \frac{\mu\mu_{-1}(z) - \{\mu_o(z)\}^2}{\mu^2}, \quad \text{var}_g(V_i) = \frac{\mu\mu_{-1} - 1}{\mu^2}$$

$$\text{cov}_g\{U_i(z), V_i\} = \frac{\mu\mu_{-1}(z) - \mu_o(z)}{\mu^2}$$

Provided that all terms are finite, the numerator and denominator of (5.6) have asymptotically a bivariate normal distribution and hence $\tilde{F}_g(z)$ is asymptotically normally distributed with mean

$$\frac{E_g\{U_i(z)\}}{E_g(V_i)} = \mu_0(z) = F(z)$$

and variance

$$\frac{\mu\mu_{-1}(z) - 2\mu\mu_{-1}(z)\mu_0(z) + \mu\mu_{-1}\{\mu_0(z)\}^2}{n} \qquad (5.8)$$

Now, for an unweighted sample, the proportion of ob-servations below z estimates F(z) with the binomial variance

$$\frac{F(z)\{1 - F(z)\}}{n} . \qquad (5.9)$$

Hence the efficiency of unweighted relative to weighted sampling for the estimation of F(z) by these methods is the ratio of (5.8) to (5.9).

Table 5.1 gives a few values for this when the unweighted distribution is log normal,

$$f(x) = \frac{1}{\sqrt{(2\pi)}\,\tau\,x} \quad \exp\left\{-\frac{(\log x - \lambda)^2}{2\,\tau^2}\right\}; \quad (5.10)$$

it is easy to express the incomplete moments of (5.10) in terms of the standardized normal integral. The tabulated values of F(z) correspond to simple points on a normal probability scale. The general conclusions are clear on general grounds. The upper tail is better estimated by length-biased sampling, the lower trial by unweighted sampling. More specifically, the distribution in the neighbourhood of the median is always better estimated by unbiased sampling, the relative efficiency there being $e^{\frac{1}{2}\tau^2}$. When τ is large and the distribution extremely skew, the region in which length-biased sampling is better is concentrated in the extreme upper tail; if we consider instead of a log normal distribution a long tailed distribution bounded away from zero the same effect would probably not occur.

Table 5.1. Efficiency of unweighted relative to weighted sampling for estimating F(z); log normal distribution, dispersion parameter τ

τ	0.2	0.5	1	2
F(z)				
.023	1.59	3.27	11.2	158
.159	1.30	2.00	4.53	33.2
.500	1.02	1.13	1.65	7.39
.841	0.793	0.601	0.503	1.44
.977	0.631	0.327	0.130	0.172

Suppose next that it is reasonable to base an analysis on a postulated functional form for the p.d.f. $f(x)$. Then the likelihood of a length-biased sample can be written down and routine methods of estimation applied. Note that sufficient statistics are the same for sampling $g(x)$ as for sampling $f(x)$. In particular, if $f(x)$ is a gamma density of mean μ and index β, then

$$g(x) = \frac{x\, f(x)}{\mu} = \frac{x\beta}{\mu^2} \left(\frac{\beta x}{\mu}\right)^{\beta-1} \frac{\exp(-\beta x/\mu)}{\Gamma(\beta)},$$

which is a gamma density of mean $\nu = \mu(1 + 1/\beta)$ and index $\gamma = \beta + 1$. It follows from the properties of maximum likelihood estimates that for a length-biased sample the maximum likelihood estimates $\hat{\nu}_g$ and $\hat{\gamma}_g$ are asymptotically independent with

$$\text{var}_g(\hat{\nu}_g) = \frac{\nu^2}{n\gamma}, \quad \text{var}_g(\hat{\gamma}_g) = \frac{1}{n\{\psi'(\gamma) - 1/\gamma\}}, \quad (5.11)$$

where $\psi(u) = d \log \Gamma(u)/du$. For large γ

$$\text{var}_g(\hat{\gamma}_g) \sim 2\gamma^2/n.$$

Thus, for estimating β length-biased sampling gives lower precision, corresponding to the change in argument in (5.11) from β to $\gamma = \beta + 1$. For estimating the mean μ, we have from the length-biased sample

$$\hat{\mu}_g = \hat{\nu}_g(1 - 1/\hat{\gamma}_g),$$

so that the asymptotic variance of $\hat{\mu}_g$ is

$$-\frac{\mu^2}{n(\beta+1)} \left[1 + \frac{1}{\beta^2(\beta+1)\{\psi'(\beta+1) - 1/(\beta+1)\}}\right]. \qquad (5.12)$$

This is to be compared with a variance of $\mu^2/(n\,\beta)$ for unweighted sampling. Some numerical values for the efficiency are given in Table 5.2.

Table 5.2. Efficiency of unweighted relative to weighted sampling for estimating by maximum likelihood the mean of a gamma distribution of index β

β	Rel. eff.	β	Rel.eff.
0.2	8.17	2	1.57
0.5	3.65	5	1.21
1	2.22	10	1.10

The distinction between this discussion and that at the beginning of the section is that in the earlier work an estimate $\tilde{\mu}_g$ of μ that is consistent for most distributions was used, whereas now an estimate based on the sufficient statistics for the gamma distribution is used. For small β the two estimates differ very appreciably.

Similar results can be obtained for the log normal distribution. A further case that is simple mathematically, even though rather unrealistic, arises when $f(x)$ is a normal density of mean μ and standard deviation σ, it being assumed that σ/μ is sufficiently small for negative values to arise with negligible probability. It can then be shown that in length-biased sampling

$$\text{var}_g(\hat{\mu}_g) = \frac{\sigma^2}{n(1 - 3\sigma^2/\mu^2)}.$$

The results of this subsection can be generalized in various ways. For example, the bias can be of a more complex form, or the estimation of other parameters can be considered. A particularly important extension arises

when the biased sampling of X affects the distribution
of a second variable Y correlated with X; see (3.2). A
simple extension of (5.6) leads to estimates of the unweighted properties of Y. For example, in textile applications one may be concerned with length-biased distribution of fibre diameter, fibre strength, etc. There
is a general connexion also with importance sampling in
Monte Carlo work.

5.3. Estimation from Recurrence Distributions

In the previous subsection we considered the estimation of quantities associated with the unweighted distribution from samples of the weighted or length-biased
distribution. We now consider the corresponding problems
when the recurrence-time distribution is sampled. That
is, we consider a random sample $Z_1, \ldots, Z_n$ of positive
random variables having the p.d.f.

$$k(z) = \frac{1 - F(z)}{\mu}, \qquad (5.13)$$

where $F(z)$ is the c.d.f. of the unweighted distribution
and it is assumed that $F(0) = 0$.

The first problem to be considered is the estimation
of μ for arbitrary densities $f(x)$. Since $k(0) = 1/\mu$, it
is in effect required to estimate the density at the
origin, a terminal of the distribution of the observations.
Previous work on the estimation of probability densities
seems to concentrate on estimation at points interior to
the range of variation. Here we sketch two methods for
the present problem.

The simplest method for large n is to choose a suitable grouping interval Δ and to consider histogram
heights near the origin. Let $p(a, b)$ be the proportion
of $Z_1, \ldots, Z_n$ in the interval (a, b). Then

$$E_k\left\{\frac{p(0, \Delta)}{\Delta}\right\} = \frac{1}{\Delta} \int_0^\Delta k(x)\,dx = k_0 + \tfrac{1}{2}k_0'\Delta + 0(\Delta^2),$$

$$(5.14)$$

$$E_k\left\{\frac{p(\Delta, 2\Delta)}{\Delta}\right\} = \frac{1}{\Delta} \int_\Delta^{2\Delta} k(x)\,dx = k_0 + \frac{3}{2}k_0'\Delta + 0(\Delta^2),$$

where $k_o = k(0)$ and $k_o' = k'(0)$.

Thus

$$\tilde{k}_o = \tfrac{1}{2}\{3p(0,\Delta) - p(\Delta,2\Delta)\} / \Delta \qquad (5.15)$$

is an approximately unbiased estimate of k_o. Its variance can be obtained using the multinomial distribution of $p(0,\Delta)$ and $p(\Delta,2\Delta)$; if both p's are small,

$$\frac{var_k(\tilde{k}_o)}{k_o^2} \sim \frac{5}{2n\,k_o\Delta} . \qquad (5.16)$$

In this n $k_o \Delta$ is approximately the expected number of observations in $[0,\Delta)$.

It is very easy to show that the somewhat more biased estimate $p(0,\Delta)/\Delta$ has smaller mean square error than $\tilde{k}_o$ if

$$\frac{k_o'^2}{k_o} < \frac{6}{n\Delta^3}$$

and to obtain from a higher order expansion a formula for the choice of Δ. These formulae are not directly helpful. If, however, it is thought that $k_o' = -f(0)/\mu = 0$, the use of the simpler estimate

$$p(0,\Delta)/\Delta \qquad (5.17)$$

would be sensible.

The major difficulty in estimating k_o is the choice of the amount of smoothing that is appropriate, i.e. the choice of Δ, or in more complex forms of estimate the selection of a weight function. In practice, several values of Δ will be tried, chosen after inspection of the data.

A closely related way of approaching the problem is by the use of the order statistics $Z_{(1)} \leq Z_{(2)} \leq \ldots \leq Z_{(n)}$. Roughly, $Z_{(1)}$ will estimate $(n\,k_o)^{-1}$ and so too will $Z_{(r)} - Z_{(r-1)}$, for r/n small. More precisely, expand the standard formula for the expected value of the order statistics about the origin. Then, for fixed r and large n, we have that

$$n\, E_k\{Z_{(r)}\} = \frac{r}{k_o}\,(1-\frac{1}{n}) - \frac{r(r+1)}{2n}\,\frac{k_o'}{k_o^3} + 0(\frac{1}{n^2})\,.$$

This suggests writing

$$U_r = \frac{n^2}{n-1}\,\{Z_{(r)} - Z_{(r-1)}\}\,,\qquad (5.18)$$

with $Z_{(0)} = 0$, when

$$E_k(U_r) = \frac{1}{k_o} - \frac{k_o'}{nk_o^3} + 0(\frac{1}{n^2})\,.\qquad (5.19)$$

The U_r's are approximately independently exponentially distribution. It is sensible to plot the U_r's against r and to estimate $1/k_o$ by extrapolating back a suitable block of observations to r = 0. The simple averaging of a block of observations, appropriate if $k_o'= 0$, is almost equivalent to the use of (5.17).

Now, having estimated $\mu = 1/k_o$ by say $\tilde{\mu}_k$, we can estimate the unweighted c.d.f. F(x) as follows. First, for a grouping interval Δ', define

$$\widetilde{F}_k(z) = 1 - \frac{\tilde{\mu}_k\, p(z - \tfrac{1}{2}\Delta',\ z+\tfrac{1}{2}\Delta')}{\Delta'}\,.\qquad (5.20)$$

This gives for $z = \tfrac{1}{2}\Delta'$, $\tfrac{3}{2}\Delta'$, ... estimates based on nonoverlapping intervals. These can then be amalgamated by a suitable algorithm (Brunk, 1955) to exploit the non-decreasing nature of F(z).

If Δ, Δ' are both small, if $z > 2\Delta + \tfrac{1}{2}\Delta'$ and if we ignore gains in precision from amalgamation, we have that asymptotically

$$var_k\{\widetilde{F}_k(z)\} = \frac{k(z)\{\Delta k_o + c\Delta' k(z)\}}{n\,\Delta\,\Delta'\,k_o^3}\,,\qquad (5.21)$$

where $c = \frac{5}{2}$ if the estimate (5.15) is used for k_o and c= 1 if (5.17) is used. Now $q(z) \leq q_o$ and it would often happen that $\Delta \gg \Delta'$. Then (5.21) becomes

$$\text{var}_k\{\tilde{F}_k(z)\} \simeq \frac{k(z)}{n\Delta'k_0^2} = \frac{\mu\{1-F(z)\}}{n\Delta'} . \qquad (5.22)$$

Thus by comparison with (5.9) for unweighted sampling, it follows that the efficiency of unweighted sampling relative to sampling $k(x)$ is at least

$$\frac{\mu}{\Delta'F(z)} . \qquad (5.23)$$

Since $F(z) \le 1$ and usually $\Delta' \ll \mu$, it follows that sampling the recurrence distribution will be very inefficient, at least when analyzed by the present method.

Suppose next that a parametric form can reasonably be taken for $f(x)$. The likelihood of $Z_1,\ldots,Z_n$ can be obtained; sufficient statistics for sampling $k(z)$ will not in general be the same as those for sampling $f(x)$. The alternative procedure of postulating a convenient non-increasing form for $k(z)$ and then deriving $f(x)$ or $F(x)$ will not be examined here.

If $f(x)$ is exponential on $(0,\infty)$, then $k(x) = f(x)$ and the two methods of sampling are equivalent. If $f(x)$ is normal with mean μ and standard deviation $\sigma = \eta^{-1}$, with σ/μ small, it can be shown that the expected second derivatives of the log likelihood L of the sample are given by

$$\frac{1}{n} E_k\left(-\frac{\partial^2 L}{\partial\mu^2}\right) = -\frac{1}{\mu^2} + \frac{\eta}{\mu} \int_{-\infty}^{\infty} \frac{\phi^2(v)}{\Phi(v)} \, dv = -\frac{1}{\mu^2} + \frac{0.903}{\mu\sigma} ,$$

$$\frac{1}{n} E_k\left(-\frac{\partial^2 L}{\partial\mu\partial\eta}\right) = \frac{1}{\eta\mu} \int_{-\infty}^{\infty} v \, \frac{\phi^2(v)}{\Phi(v)} \, dv = -\frac{\sigma}{\mu} 0.595, \qquad (5.24)$$

$$\frac{1}{n} E_k\left(-\frac{\partial^2 L}{\partial\eta^2}\right) = \frac{1}{\eta^3\mu} \int_{-\infty}^{\infty} v \, \frac{\phi^2(v)}{\Phi(v)} \, dv = \frac{\sigma^3}{\mu} 1.101$$

Thus for the maximum likelihood estimate of the mean μ, we have that asymptotically

$$\text{var}_k\,(\hat{\mu}_k) = \frac{\sigma^2}{n} \times \frac{1.72\,(\sigma/\mu)^{-1}}{1 - 1.72\,\sigma/\mu}. \qquad (5.25)$$

A more realistic case is when $f(x)$ has the log normal form (5.10), where it is convenient to write $\tau = 1/\kappa$. Then

$$\frac{1}{n}\,E_k\left(-\frac{\partial^2 L}{\partial \lambda^2}\right) = -\tau^{-1}\,e^{-\frac{1}{2}\tau^2}\int_{-\infty}^{\infty} \alpha'(u)\,\Phi(u)\,e^{-u\tau}\,du,$$

$$\frac{1}{n}\,E_k\left(-\frac{\partial^2 L}{\partial \lambda \partial}\right) = -\tau - \tau\,e^{-\frac{1}{2}\tau^2}\int_{-\infty}^{\infty} u\alpha'(u)\,\Phi(u)\,e^{-u\tau}\,du, \qquad (5.26)$$

$$\frac{1}{n}\,E_k\left(-\frac{\partial^2 L}{\partial \kappa^2}\right) = 3\tau^4 - \tau^3\,e^{-\frac{1}{2}\tau^2}\int_{-\infty}^{\infty} u^2\alpha'(u)\,\Phi(u)\,e^{-u\tau}\,du,$$

where $\alpha(u) = \phi(u)/\Phi(u)$, and $\alpha'(u) = -u\alpha(u) - \alpha^2(u)$. Now

$$\log\mu = \lambda + \tfrac{1}{2}\tau^2 = \lambda + \tfrac{1}{2}\kappa^{-2}, \text{ so that}$$

$$\frac{\text{var}_k(\hat{\mu}_k)}{\mu^2} = \text{var}_k(\hat{\lambda}) - \frac{2}{\kappa^3}\,\text{cov}_k(\hat{\lambda}, \hat{\kappa}) + \frac{1}{\kappa^6}\,\text{var}_k(\hat{\kappa}) \qquad (5.27)$$

can be obtained from (5.26). Table 5.3 gives some numerical values for $n\,\text{var}\,(\tilde{\mu})/\mu^2$, for various estimates $\tilde{\mu}$.

Table 5.3. Estimates of mean of log normal distribution. Squared coefficient of variation of estimate x n.

τ	Max.lik. from recurrence dist.	Mean of unweighted sample	Efficient estimat from unweight- ed sample
0.2	0.410	0.0408	0.0408
0.5	1.30	0.284	0.281
1	4.00	1.72	1.50
2	19.2	53.6	12.0

ACKNOWLEDGMENTS

I am grateful to Messrs. S. L. Anderson, T.W. Anderson, W.R. Buckland, R.D. Elston, W.A. Pridmore and V.J. Small for helpful comments and references, and to Mr. B.G.F. Springer, whose work was supported by the Science Research Council, for the numerical evaluation of some integrals.

REFERENCES

BARNES, R.M. (1957). Work sampling. New York: Wiley.

BICKING, C.A. (1964). Bibliography on sampling of raw materials and products in bulk. Tech. Assoc. of Pulp and Paper Industry, 47, 147-170.

BLUMENTHAL, S. (1967). Proportional sampling in life length studies, Technometrics, 9, 205-218.

BRITISH STANDARDS INSTITUTION (1960). The sampling of coal and coke, BS1017, Pt. 1 Coal, Pt. 2 Coke.

BRITISH STANDARDS INSTITUTION (1967). Sampling of iron and manganese ores, BS4103, Pt. I, Manual sampling.

BRUNK, H.D. (1955). Maximum likelihood estimates of monotone parameters, Ann. Math. Statist., 26, 607-616.

COX, D.R. (1955). Some statistical methods connected with series of events, J.R. Statist. Soc., B 17, 129-164.

COX, D.R. (1961). Renewal theory. London: Methuen.

COX, D.R. (1966). A note on the analysis of a type of reliability trial, J. Siam. Appl. Math., 14, 1133-1142.

DANIELS, H.E. (1942). A new technique for the analysis of fibre length distribution in wool, J. Text. Inst., 33, T 137- T 150.

DAVIS, H. (1955). A mathematical evaluation of a work sampling technique, Naval Res. Log. Q., 2, 111-117.

DUNCAN, A.J. (1962). Bulk sampling: problems and lines of attack. Technometrics 4, 319-344.

GAVER, D.P. and MAZUMDER, M. (1967). Statistical estimation in a problem of system reliability, Naval Res. Log. Q., 14, 473-488.

HERDAN, G. (1960). Small particle statistics. 2nd ed. London: Butterworth.

HILL, I.D. and HILL, P.A. (1968). Random times for activity sampling. Enfield, Middx: Inst. of Work Study Practitioners.

KRIENS, J. (1963). The procedures suggested by de Wolff and van Heerden for random sampling in auditing, Statist. Neerlandica, 17, 215-231.

PALMER, R.C. (1948). The dye sampling method of measuring fibre length destribution, J. Text. Inst., 39, T8-T22.

PALMER, R.C. and DANIELS, H.E. (1947). The sampling problem in single fibre testing, J. Text. Inst., 38, T94-T100.

TIPPETT, L.H.C. (1935). Statistical methods in textile research; uses of the binomial and Poisson distributions. A snap-reading method of making time studies of machines and operators in factory surveys, J. Text. Inst., 26, T51-75.

Biometrika (1979), **66**, 1, *pp.* 125–32
Printed in Great Britain

On sampling and the estimation of rare errors

BY D. R. COX AND E. J. SNELL

Department of Mathematics, Imperial College, London

SUMMARY

For each individual in a finite population a recorded value is available. A small proportion of the values is in error, all the errors being of the same sign and the maximum fractional error being known. It is required to obtain an upper confidence limit for the total error in the population. Various solutions, including a Bayesian one, are discussed for sampling with probability proportional to the recorded values, and the appropriateness of this form of sampling is examined.

Some key words: Auditing; Bayesian inference; Finite population; Length biased sampling; Optimal sampling; Poisson distribution; Probability proportional to size; Sampling; Superpopulation model.

1. INTRODUCTION

The work described in this paper was done in connexion with the auditing of accounts. Nevertheless we describe the problem in a general setting, partly to emphasize the breadth of potential applications and partly because it would be out of place to discuss in detail considerations specific to auditing.

Consider a finite population of N items with recorded values $x_1, ..., x_N$, assumed all positive. Suppose that the items are subject to errors $y_1, ..., y_N$ and hence to fractional errors $z_1, ..., z_N$, where $y_i = x_i z_i$. In more conventional language, x_i is the 'observed' value and $x_i - y_i$ the 'true' value of the ith item. Let

$$d_i = \begin{cases} 1 & (z_i \neq 0), \\ 0 & (z_i = 0). \end{cases}$$

Then $T_x = \Sigma x_i$ is the total recorded value of the population, $T_y = \Sigma y_i$ is the total error and $T_d = \Sigma d_i$ is the number of items in error. Note that $T_{xd} = \Sigma x_i d_i$ is the total recorded value of those items in error.

Suppose that it is required to estimate T_y by sampling. Now, T_y being a population total, the standard theoretical results of sample survey theory are available giving, on the basis of suitable sampling schemes, unbiased estimates of T_y and unbiased estimates of the sampling variance of these estimates. We consider here, however, situations in which most of the items are error-free. Indeed we may have populations and sample sizes such that with appreciable probability the whole sample is error-free. We may still wish to estimate T_y and in particular to give an upper confidence limit for its value. It is clear that estimated sampling variances are not much use for this, and are totally useless when no errors are observed.

Some assumption about the nature of the errors is needed and, with auditing partly in mind, we take initially situations in which $0 \leqslant z_i \leqslant 1$. This is relevant primarily when the recorded values are, if in error, too large and when true values and recorded values are non-negative. There is an immediate extension to the case when the errors all have the same sign and $0 \leqslant z_i \leqslant a$, with a known, and the results can, with a little more difficulty, be extended to $-b \leqslant z_i \leqslant a$, with $a, b > 0$ both known; that is, if errors of both signs can occur, bounds on the fractional error must be known or assumed, or some other corresponding assumption made.

A fairly typical situation in auditing is where a sample of perhaps 10^2–10^3 items gives either no items in error or perhaps one or a few items in error, with the corresponding z_i observed. We shall assume throughout that for the sampled items the z_i are observed without error, i.e. that the checking process is perfect. Of course in applications that would need critical examination if realistic upper levels are to be found for T_y and, under certain circumstances, some effort devoted to independent rechecking would be sensible.

A more general interpretation still of the investigation is that y_i is any further property of the ith item, most of the y_i being zero and the constraints on the z_i being satisfied.

2. A sampling scheme

The most appropriate form of sampling will depend, among other things, on the relative liability to error of items of various sizes, and on the relative costs of checking items of different sizes; see § 4. Often, however, the items with larger values will be relatively more important and sampling with probability proportional to size, that is x, will be sensible. To implement such sampling imagine the values cumulated, hence defining a series of points along a value-axis at $x_1, x_1 + x_2, ..., x_1 + ... + x_N = T_x$. A systematic sample at interval h is then taken, where h is determined according to the size of sample required. In auditing this is called monetary unit sampling (Arkin, 1963; Anderson & Teitlebaum, 1973; McRae, 1974, Chapter 17; Smith, 1976).

Obviously all items with value at least h are certain to be selected, indeed possibly more than once, i.e. they define a stratum with 100% sampling. Throughout the following discussion, we suppose that any such items have been removed for separate study. Then the chance of selecting an item is proportional to its x value. Throughout the mathematical argument we treat the sampling procedure as equivalent to random sampling with probability proportional to size. The approximation involved will be improved by randomizing the order of the items before sampling; sometimes it may be good to divide the population into subpopulations for separate sampling, possibly with different values of h. One of the attractions of the procedure is that it is easily implemented either by hand or on the computer.

The description of this procedure in the auditing literature is based on the notion of treating each unit, pound, dollar, etc., in a particular item, as an entity free from or subject to error. Because the proportion of units subject to error is small, various results based on the Poisson distribution are suggested. The different units forming an item are, however, not independent, and when one is sampled they are all sampled. Arguments based on treating the constituent 'pounds' as independent elements seem to us therefore suspect, although the conclusions are broadly correct.

3. Infinite population theory

There are various ways of approaching a theoretical treatment of this problem. The simplest is to treat the finite population as effectively infinite and the values x_i as random variables of density $p(x)$ and mean μ_x. Then

$$T_x/N = \mu_x. \tag{1}$$

Suppose that items of value x have small probability $\rho(x)$ of containing an error; conditional on a nonzero error, let the density of y, the magnitude of error, be $q(y|x)$. Then the total error and the total value of items subject to error are given respectively by

$$T_y/N = \int_0^\infty \int_0^\infty y q(y \mid x)\, p(x)\, \rho(x)\, dx\, dy, \tag{2}$$

$$T_{xd}/N = \int_0^\infty x p(x)\, \rho(x)\, dx. \tag{3}$$

Now, because sampling is with probability proportional to size, the density of x in the sample is $x p(x)/\mu_x$ and therefore the chance that an arbitrary item in the sample contains an error has the small value

$$\int_0^\infty x \rho(x)\, p(x)\, dx/\mu_x = T_{xd}/(N\mu_x) = T_{xd}/T_x, \tag{4}$$

and therefore the number, m, of errors observed in the sample has a Poisson distribution of mean nT_{xd}/T_x. Therefore if $\varpi^\alpha(m)$ is the upper $1 - \alpha$ confidence limit for the mean of a Poisson distribution, when m is observed, the upper confidence limit for T_{xd} is

$$T_x \varpi^\alpha(m)/n, \tag{5}$$

where m is the number of items in the sample with error, and a natural point estimate is mT_x/n.

In particular if the sample is error-free the upper 95% limit for T_{xd} is $3 \cdot 00 T_x/n$. This, interpreted as a conservative upper limit for T_y, is a central result in monetary unit sampling, as described in the auditing literature, and in particular can be used to fix a suitable n, and hence a suitable sampling interval $h \simeq T_x/n$.

In fact, however, the population property of interest is typically not T_{xd} but T_y and it follows from (2) and (3) that

$$\mu_z = E_w(Z \mid Z \neq 0) = \int_0^\infty \int_0^\infty \frac{y}{x} x p(x)\, \rho(x)\, q(y \mid x)\, dx\, dy \Big/ \int_0^\infty x p(x)\, \rho(x)\, dx,$$

so that

$$T_y = T_{xd}\, \mu_z, \tag{6}$$

where $E_w(.)$ denotes expectation under the weighted scheme of sampling used. Formula (6) illustrates the difficulty of the problem when no or very few errors are observed, for it is then impossible or difficult to estimate μ_z. An unbiased point estimate of T_y is

$$mT_x \bar{z}/n, \tag{7}$$

where $\bar{z}$ is the mean of the observed nonzero proportional errors, $\bar{z} = (z_1 + \ldots + z_m)/m$, say.

From another point of view (7) is just the Horvitz–Thompson estimator for a population total (Cochran, 1963, § 9·14); as already noted, when m is small and in particular zero, a sampling variance attached to (7) is useless for the analysis of the data. Possible ways of obtaining an upper limit for T_y include

(a) the typically very conservative assumption that $\mu_z = 1$, $T_y = T_{xd}$ leading to the use of (5);

(b) the use on the basis of previous experience of an *a priori* upper bound for μ_z, say μ_z^*, leading to the conservative upper limit $T_x \varpi^\alpha(m)\, \mu_z^*/n$;

(c) the use of an empirical Bayesian argument in which information is incorporated from previous similar experience, (b) being in a sense an extreme form;

5

(d) somewhat empirical modification of (b) such as the replacement of μ_z by the constant μ_z^* if $\bar{z} \leqslant \mu_z^*$ and by 1 if $\bar{z} > \mu_z^*$;

(e) application of a method recommended in the auditing literature based on the ordered nonzero proportional errors $z_{(1)} \geqslant \ldots \geqslant z_{(m)}$. In this the factor $\varpi^\alpha(m)$ in (5) is replaced by

$$\sum_{r=0}^{m} \{\varpi^\alpha(r) - \varpi^\alpha(r-1)\} z_{(r)}, \tag{8}$$

where $z_{(0)} = 1$, and $\varpi^\alpha(-1) = 0$. It was conjectured in an unpublished paper by A. D. Teitlebaum, that substitution of (8) into (5) leads to a conservative $1-\alpha$ upper confidence limit for T_y, whatever the distribution of errors. So far as we know, no proof of or counterexample to this is known. Numerical work at Imperial College by Mr D. Dutt not reported here has confirmed (8), the limits often being extremely conservative;

(f) the addition of a semiempirical constant to the Horvitz–Thompson estimated variance, suggested in an unpublished paper by H. L. Jones, the constant being chosen so that sensible results are obtained even when $m = 0$, when the Horvitz–Thompson estimated variance vanishes.

(g) calculation of a conservative upper confidence limit based on a set of multinomial probabilities, corresponding to probabilities that $0, 1, \ldots, 100p$ out of each pound examined are in error (Fienberg, Neter & Leitch, 1977; Neter, Leitch & Fienberg, 1978). The method of calculation is, however, quite complicated unless the number of items in error is very small and gives answers broadly similar to those of (e).

In §5 of the present paper we outline (c); a Bayesian theory is developed under special assumptions, taking into account the finite size of the population. For practical use, especially when m is likely to be very small, application of the simple methods (b) and (d) is, in our opinion, often likely to be sensible.

Note that in the above discussion no special assumptions are made concerning the relation between the occurrence of error and the value of the item.

4. Choice of sampling scheme

We now consider briefly the circumstances under which the sampling procedure of §2 is likely to be reasonably efficient; the discussion is similar to that used in studies of the efficiency of sampling with probability proportional to size. While in the present context estimated variances are not useful for the analysis of data, variances can be used as a reasonable guide in design.

Let the average cost of checking an item of recorded value x be $c(x)$ and consider for simplicity a family of sampling schemes in which the probability of selecting an item of value x is proportional to x^k; such schemes are easily implemented by a modification of the cumulative scheme described in §2. If $E_0(.)$ denotes expectation with respect to unweighted sampling, the expected cost of a sample of size n is

$$n E_0\{X^k c(X)\}/E_0(X^k). \tag{9}$$

We concentrate for the present discussion on the variance of the usual unbiased estimate of T_y, namely

$$\hat{T}_y^{(k)} = N E_0(X^k)\,(\Sigma\, y_i/x_i^k)/n. \tag{10}$$

Let

$$\phi(x) = \mathrm{pr}\,(Y \neq 0\,|\,X = x), \quad \mu(x) = E(Y\,|\,Y \neq 0, X = x), \quad \nu^2(x) = \mathrm{var}\,(Y\,|\,Y \neq 0, X = x)$$

describe the error distribution. It is then easy to show that for a given total cost C,

$$\mathrm{var}_k(\hat{T}_\nu^{(k)}) = \frac{N^2}{C}\left(E_0\{X^k c(X)\}\,E_0\left[\frac{\phi(X)\{\nu^2(X)+\mu^2(X)\}}{X^k}\right] - \{E_0(Y)\}^2\frac{E_0\{X^k c(X)\}}{E_0(X^k)}\right). \quad (11)$$

A simple special case that gives a general idea of the behaviour of (11) is obtained with

$$c(x) = c_0 x^{c_1}, \quad \mu(x) = \mu_0 x^{\mu_1}, \quad \nu(x) = \nu_0 x^{\nu_1}, \quad \phi(x) = \phi_0 x^{\phi_1} \quad (12)$$

and to take $\log X$ as normal, with mean ζ and variance τ^2; note that these are not consistent with the constraints on these quantities for all parameter values. Of course assumptions alternative to (12) may be called for in particular cases, for example $\phi(x) = \phi_0' + \phi_1' x$.

Introduction of (12) gives for the variance (11), on making the further approximation that terms in ϕ_0^2 are negligible,

$$\frac{N^2 c_0 \phi_0}{C}\exp\{(c_1+\phi_1)\zeta + \tfrac{1}{2}\tau^2(k+c_1)^2\}$$

$$\times\,[\nu_0^2\exp(2\nu_1\zeta)\exp\{\tfrac{1}{2}\tau^2(\phi_1+2\nu_1-k)^2\} + \mu_0^2\exp(2\mu_1\zeta)\exp\{\tfrac{1}{2}\tau^2(\phi_1+2\mu_1-k)^2\}]. \quad (13)$$

In particular the minimizing value of k can be obtained. One special case is when the cost and probability of error are independent of x, $c_1 = \phi_1 = 0$, and the magnitude of error proportional to x, $\mu_1 = \nu_1 = 1$. The optimal k minimizes $k^2 + (2-k)^2$ and so is $k = 1$. Use of unweighted sampling, $k = 0$, would increase (13) by a factor $e^{\tau^2} = 1 + \gamma_X^2$, where γ_X is the coefficient of variation of x in the population. With the same assumptions except that $c_1 = 1$, the cost being thus proportional to x, the function to be minimized is $(k+1)^2 + (2-k)^2$, leading to $k = \tfrac{1}{2}$. The values of (13) at $k = 0, \tfrac{1}{2}, 1$ are now in the ratio $e^{\tau^2/4} : 1 : e^{\tau^2/4}$. Thus the choice of sampling scheme is not critical unless the relative variation of x is extremely large. Note finally from (13) that simple discussion of the choice of k is possible whenever $\mu_1 = \nu_1$, that is whenever the coefficient of variation of the magnitude of error is independent of x. For then the optimizing value minimizes $(k+c_1)^2 + (\phi_1+2\mu_1-k)^2$ and the optimum is at $\tilde{k} = \tfrac{1}{2}(c_1+\phi_1+2\mu_1)$, showing a range of circumstances under which $k = 1$ is optimal.

5. PARAMETRIC FORMULATION: BAYESIAN APPROACH

In the discussion of §3 no parametric assumptions have been made. An alternative approach is to formulate the problem parametrically, either in infinite population form or in terms of a finite population drawn from an infinite superpopulation. Further a Bayesian analysis of the problem is natural whenever a series of populations of broadly similar properties is examined, the prior distribution being estimated from relevant data.

Of course the parametric formulation must depend on the context. Here we investigate the simplest form. We suppose that the probability ϕ that an item is defective is constant, independent of x, and small, so that Poisson approximations can be used. Given that an item is in error, let the proportional error have density $g(z; \mu)$, where μ is an unknown parameter, which can be taken when scalar to be the mean. The errors attaching to different items are assumed independent and identically distributed. We thus observe m, which has a Poisson distribution of mean $n\phi$, and values $z_1, \ldots, z_m$ in a random sample drawn from $g(z; \mu)$. In the infinite model formulation it is required to estimate $\phi\mu$, whereas in the finite population version it is required in effect to estimate the total error in the nonsampled part

5´

of the population. This is the random variable $T_y^{(u)}$ with moment generating function

$$E\{\exp(-sT_y^{(u)})\} = \prod_{i=1}^{R}\{1-\phi+\phi g^*(sx_i;\mu)\}, \tag{14}$$

where s is the argument of the moment generating function, $g^*(\,.\,;\mu)$ is the moment generating function corresponding to the density $g(\,.\,;\mu)$, and $R = N-n$ is the number of individuals in the population that are not sampled. Note that in the auditing problem errors detected will certainly be corrected, so that the population property of interest is precisely $T_y^{(u)}$.

The mean and variance from (14) are respectively

$$\phi\mu T_x^{(u)}, \quad \{\phi\sigma^2+\phi(1-\phi)\mu^2\}\{T_x^{(u)}\}^2(1+C_x^2)/R, \tag{15}$$

where C_x is the square of the coefficient of variation of x in the unsampled units and σ^2 is the variance of the density $g(\,.\,,.\,)$.

For the remainder of the discussion we suppose that the density $g(z;\mu)$ is exponential with mean μ. This may be reasonable under some circumstances. In others, as in the auditing problem, it is best regarded as a conservative approximation to a truncated exponential distribution over $(0,1)$. The theory for this, and for the truncated exponential distribution over $(0,1)$ plus an atom of probability at $z = 1$, can be developed, but are appreciably more complicated. Under the exponential model the minimal sufficient statistic is $(m,\bar{z})$, where $\bar{z} = (z_1+\ldots+z_m)/m$; of course $\bar{z}$ is not defined if $m = 0$. As noted previously, the occurrence of $m = 0$ with nonnegligible probability makes an 'exact' confidence interval solution of even the infinite model problem impossible.

For a Bayesian analysis, in the absence of specific information about a particular application, we investigate the simplest conjugate prior, giving ϕ and μ independent prior densities of the form

$$\frac{(a/\phi_0)^a\,\phi^{a-1}\exp(-a\phi/\phi_0)}{\Gamma(a)}\,\frac{\mu_0(b-1)}{\mu^2}\left(\frac{\mu_0(b-1)}{\mu}\right)^{b-1}\frac{\exp\{-(b-1)\mu_0/\mu\}}{\Gamma(b)}, \tag{16}$$

so that ϕ_0 and μ_0 are the prior means of ϕ and μ. The posterior density of (ϕ,μ) is now obtained by combining with the likelihood of $(m,\bar{z})$ and it follows on integration that $\psi = \phi\mu$ has the posterior form

$$\psi = \left\{\frac{m\bar{z}+(b-1)\mu_0)}{n+a\phi_0}\right\}\left(\frac{m+a}{m+b}\right)F_{2(m+a),2(m+b)}, \tag{17}$$

where F_{d_1,d_2} is a random variable with the standard F distribution with (d_1,d_2) degrees of freedom.

Now it is in principle desirable to use a homogeneous set of populations in defining the prior distributions, but this may often not be possible. Then from the interpretation of $1/\sqrt{a}$ and $1/\sqrt{b}$ as coefficients of variation of ϕ and $1/\mu$, it seems reasonable to consider a in the range say $\frac{1}{2}$ to 2 and b to be at least 3, the condition $b > 1$ being necessary to ensure that the prior mean is finite; if the exponential distribution for z is to be used to approximate to a truncated exponential distribution, then some combination such as $\mu_0 = \frac{1}{2}$, $b = 3$ will be conservative. Finally ϕ_0 will typically be very small, so that $a\phi_0 \ll n$, corresponding to the neglect of the exponential factor in the prior for ϕ. It would be quite wrong to suggest a Bayesian solution for universal use regardless of context, but the above choice, say $a = 1$, $b = 3$, $\mu_0 = \frac{1}{2}$, leads to

$$\psi = \frac{(m\bar{z}+1)(m+1)}{n(m+3)}F_{2(m+1),2(m+3)} \tag{18}$$

and hence to an upper confidence limit for the population total error remaining, which is $\psi T_x^{(u)}$, or to the order of approximation involved in the infinite population ψT_x.

Two other possibilities for brief consideration are that μ is assumed known and equal to μ_0; for example, it might be reasonable to assume that fractional errors are equally likely to be between 0 and 1, when $\mu_0 = \frac{1}{2}$. Then we let $b \to \infty$ and

$$\psi = \frac{\mu_0(m+a)}{n} F_{2(m+a),\infty}, \tag{19}$$

to be compared with the result from §3 that the upper confidence limit is $\mu_0 \varpi^\alpha(m)/n$; it is known that especially when $a \simeq \frac{1}{2}$ the agreement is close.

The other special possibility is to try for effective approximate confidence limits from (17) for $m \neq 0$ by passing to a prior corresponding to initial very vague knowledge, for which one reasonable choice is $a = 1$, $b = 1$. This leads for $m \neq 0$ to

$$\psi = \frac{m\bar{z}}{n} F_{2m+2,2m+2}. \tag{20}$$

The above discussion is for the infinite population version of the problem, where the quantity of interest is essentially $\psi = \phi\mu$. For the finite population version of the problem, we use the moment generating function (14), or alternatively the mean and variance derived from it, plus the posterior density (17), to show that conditionally on m and $\bar{z}$, the mean of $T_y^{(u)}$ is that derived from (17), whereas the variance is that from (17) multiplied by

$$1 + \frac{1}{R}(1 + C_x^2)\frac{2n(m+b-1)}{2m+a+b-1}, \tag{21}$$

and the simplest, if rather crude, way of allowing for this is to apply (17) with the factor $m+a$ and the degrees of freedom $2(m+a)$ both divided by (21).

It is natural to assess (20) by examining whether formal upper $1-\alpha$ limits for $\psi = \phi\mu$ calculated from the formula have approximately the desired confidence interval property There is a difficulty of principle in that if $m = 0$, that is if there are no observed errors, (20) is inapplicable; we cannot estimate $\phi\mu$ if there is no information about μ. The most satisfactory approach seems to be the following. Inference about ϕ, i.e. in effect T_{xd}, can be based on the Poisson distribution as in (5) and the required confidence interval property is assured. At least in principle we imagine limits computed for any m that arises. Inference about $\phi\mu$ can be attempted only when $m \neq 0$ and hence the question for consideration is whether an upper limit calculated from (20) has approximately the correct frequency property over a set of repetitions in which m and $\bar{z}$ vary, but taken conditionally on $m \neq 0$.

Some information may be gleaned from extreme cases. If all proportional errors, z, are the same, the limits (20) are too large by a factor very close to $\sqrt{2}$. If the numbers of non-zero errors are large, so that asymptotic theory is applicable, the limits are conservative, or not, according as the standard deviation of the distribution $g(z;\mu)$ is less than or exceeds the mean. Of the situations that might arise in an auditing context, the least favourable for the use of (20) is a mixture of a fairly high probability, p, of a very small value of $z = z_0$, say, with a complementary probability $(1-p)$ of unit z; this is broadly comparable with the asymptotic situation with a high coefficient of variation.

We are grateful to Mr J. Connolly, Mann Judd & Co., for presenting the problem which prompted this work.

References

Anderson, R. & Teitlebaum, A. D. (1973). Dollar-unit sampling. *Can. Chartered Accountant* **102**, No. 4, 30–9.

Arkin, H. (1963). *Handbook of Sampling for Auditing and Accounting.* New York: McGraw-Hill.

Cochran, W. G. (1963). *Sampling Techniques,* 2nd edition. New York: Wiley.

Fienberg, S., Neter, J. & Leitch, R. A. (1977). Estimating the total overstatement error in accounting populations. *J. Am. Statist. Assoc.* **72**, 295–302.

McRae, T. W. (1974). *Statistical Sampling for Audit and Control.* New York: Wiley.

Neter, J., Leitch, R. A. & Fienberg, S. E. (1978). Dollar unit sampling: multinomial bounds for total overstatement and understatement errors. *Accounting Rev.* **53**, 77–93.

Smith, T. M. F. (1976). *Statistical Sampling for Accountants.* London: Accountancy Age Books.

[*Received January* 1978. *Revised August* 1978]

2

Statistical Methods

THE PAPERS IN THIS SECTION concern, at least in part, new statistical methods for the analysis and interpretation of data. They have been divided into four main sections, each corresponding to a paper or papers published with formal discussion, with finally some miscellaneous topics. The groups are:

- point process data (series of events),
- binary data,
- survival data,
- multivariate analysis and graphical Markov models,
- miscellaneous.

Most, but not quite all, of the papers arose out of specific applications or groups of applications.

Point process data

SM1 Some statistical methods connected with series of events (with discussion). *J. R. Statist. Soc.* B **17** (1955), 129–164.

My first encounter with point process data was in connection with accidents to military aircraft in World War II. This was followed by more extensive experience in the wool textile industries; there, large amounts of data were collected on breakdowns especially in spinning and weaving. Paper SM1 arose out of these applications. It is a paper with too many fragmentary ideas, even though it is a substantially reduced version of an even more rambling paper. As was characteristic of papers of that period, there was emphasis on the avoidance of iterative calculations. It is an interesting question as to whether such a consideration is nowadays totally irrelevant; I suspect that the relative transparency of noniterative methods preserves some of their value.

Some of the aspects treated in the paper are as follows: graphical models for point process data, estimation of the variance–time function for a stationary process (i.e. estimation of $V(x)$, the variance of the number of points in an interval of length x), components of variance for counts, problems with several types of point and, in a long final section, sampling problems.

In many textile contexts, the Poisson process has too little variability to be a good representation of data, even when it is stationary in some sense. To cover this the notion of a Poisson process with an unobserved randomly varying rate was introduced and termed by M. S. Bartlett a 'doubly stochastic' Poisson process. The idea has many ramifications.

A final section of the paper, somewhat related to Paper D9, deals with techniques for sampling processes which alternate between two states, say a relatively rare failed state and a relatively frequent operating state. The methods are in a sense a development of what in management studies used to be called the snap-reading method, that is, point sampling of the process at widely separated time instants. I am not aware of developments or applications of the ideas in this part of the paper.

SM2 Multivariate point processes. In *Proceedings of the Sixth Berkeley Symposium on Mathematical Statistics and Probability Theory* **3**, edited by L. M. Le Cam, J. Neyman and E. L. Scott, pp. 401–448. Berkeley: University of California Press, 1972 (with P. A. W. Lewis).

In the early 1960s Peter A. W. Lewis came to London under an IBM Fellowship formally to work for a Ph.D. although he was already an established research worker. His work led on to a monograph (Cox and Lewis, 1966) incorporating some of his doctoral thesis (including the so-called Bartlett–Lewis cluster process), much of Paper SM1 and its subsequent development. A notable feature of the monograph was that Peter Lewis persuaded IBM to sponsor a suite of programmes covering all the techniques in the book, probably one of the first instances of systematic computer back-up for a book.

Paper SM2 is the culmination of a long collaboration with Peter. The first part of the paper is a systematic account of point processes of several types, developed in a mathematical style owing more to the massive contributions of M. S. Bartlett than to the more modern taste for mathematical abstractness. Both general theory and specific processes are discussed; see also Papers SP1, SP2, SP7–SP11 in Volume II of these papers.

The second part of the paper deals with statistical analysis concentrating on the analysis of long runs of a supposedly stationary series. The parallel discussion of numbers of related short series, panel data, is not attempted.

SM3 On the estimation of the intensity function of a stationary point process. *J. R. Statist. Soc.* B **27** (1965), 332–337.

The intensity function, or better the second-order intensity function, of a stationary point process is defined via the probability, given a point at time zero, that there is a point in $(x, x + \Delta)$ considered as a function of x; note that the point in question is not necessarily the first point after the origin and that the function will typically tend to $1/\rho$ as x increases, where ρ is the rate of the process. The function can be estimated from the histogram of intervals between all possible pairs of points. Properties of the estimates are derived and for the Poisson process related to the theory of U-statistics of Hoeffding.

An interesting general issue concerns the use in problems like this of smoothing techniques, in the present case the use of a kernel smoother rather than histogram smoothing. For many purposes, however, the objective is to use the estimated function for qualitative interpretation and comparison; this is aided by having essentially independent points of approximately known standard error and these are provided by a histogram-type estimate. The smoothness of the kernel estimate, some of it spurious, can hinder rather than aid interpretation.

Binary data

Up to the mid 1950s analysis of binary and similar response data was primarily in one of two forms. One was essentially some adaptation of the chi-squared test, departures from a simple null hypothesis, if found, being interpreted essentially qualitatively. A second theme was the fitting of dose–response curves especially in the context of bioassay, interest often being focused on the estimation of an ED50, the effective dose of, for example, a drug or an insecticide at which 50 per cent success is achieved. Most of the work used the integrated normal or so-called probit response curve; some employed the essentially equivalent logistic form. A prominent advocate of the latter was the influential medical statistician J. Berkson, although this aspect of his wide-ranging contributions tended to be focused on a largely academic dispute about the relative merits of fitting by maximum likelihood as compared with empirically weighted least squares applied to transformed responses, the method of minimum logit chi-squared. I am unclear who first used multiple logistic regression, although it may well have been J. Cornfield in the context of the Framingham study.

My own interest in binary data stemmed largely from contact with the experimental psychologists at Birkbeck College, London, who were primarily 'rat people', a line of investigation then much in favour. This interest led to the realization that the exponential family form of linear logistic regression provides a simple and unified treatment of many old and some new tests and of associated estimation problems. Essentially the linear logistic regression model provides a synthesis of many previous procedures and gave formal confidence intervals for many choices of parameter.

SM4 The regression analysis of binary sequences (with discussion).
  J. R. Statist. Soc. B **20** (1958), 215–242.

SM5 Two further applications of a model for binary regression. *Biometrika* **45** (1958), 562–565.

This short note covers two points that should have been in Paper SM4. One concerns the comparison of a set of personalistic probabilities as related to actual occurrence and nonoccurrence of the events in question. The other derives the analysis of matched-pair binary data from the logistic model, obtaining in particular McNemar's test for the null hypothesis of no effect together with simple confidence intervals for the logistic difference between the two conditions under comparison. These follow from a simple logistic model with individual pair effects and a condition (or treatment) effect. The striking feature of this test, namely that pairs in which both individuals respond in the same way play no role, shows both the strong consequences of model formulation and also the very demanding nature of what in Neyman–Pearson theory is called exact similarity.

SM6 The analysis of multivariate binary data. *J. R. Statist. Soc.* C **21** (1972), 113–120.

Paper SM6, a survey-like paper given to the Multivariate Study Group of the Royal Statistical Society, puts some emphasis on models of exponential family form. The role of permutation of names of categories as in some sense the analogue of linear transformations in standard multivariate analysis is mentioned. The table at the end of the paper sets out the similarities and differences between continuous and binary versions of some standard multivariate situations.

Additional publications

The Analysis of Binary Data. London: Methuen/Chapman and Hall, 1970.

Analysis of Binary Data, second edition. London: Chapman and Hall, 1989 (with E. J. Snell).

In retrospect I can see that I waited too long before writing the first version of this book. Most of the material was available quite a few years before. At the time it was still necessary to spend appreciable space on noniterative methods, notably the method of empirically weighted least squares applied to empirically transformed data. While such methods remain, I think, of value in some contexts, the amount of space devoted to them in the second edition, written with the very close involvement of Joyce Snell, was much reduced. Much more attention was paid to such issues as case-control studies, extensions to nonbinary data and so on.

Survival data

SM7 The analysis of exponentially distributed life-times with two types
of failure. *J. R. Statist. Soc.* B **21** (1959), 411–421.

In a sense my interest in survival (or failure) data dates from my first employment in the Department of Structural and Mechanical Engineering, Royal
Aircraft Establishment, Farnborough, where there was considerable attention to
the strength of materials and structures. The interest continued at the Wool Industries Research Association, where again strength testing played an important role.
It is also reflected less directly in Paper SM1 on series of point events. Paper SM7,
however, arose essentially as a comment on a paper by Mendenhall and Hader
(1958) written at a time of increasing interest in reliability issues, stemming in
part from the aerospace industry.

The specific data under discussion consist of a failure time and a type of failure
for each individual. The paper discusses the relation between two types of model
for such data. In one, each individual is 'assigned' a priori to a failure type with
a specified probability and associated with each is a characteristic distribution.
In the other, there are two independent failure times for each individual, one for
each type, and only the smaller is observed. Equivalently there are two hazard
functions, both defined conditionally on neither failure having yet occurred. It is
shown that the models cannot be distinguished empirically from this kind of data
and that in particular a model with independent component failure times is general
enough to fit any data of this sort. There are implications for the interpretation of
the method of competing risks.

The paper also gives some associated statistical techniques and explores briefly
the possibility that the component hazard functions are proportional. Some of the
ideas in the paper were recently relevant in work on the SARS epidemic of 2002–3.
The two types of model are essentially special cases of the pattern mixture models
and longitudinal models that appear in more recent discussions of missing value
problems.

SM8 Regression models and life-tables (with discussion). *J. R. Statist. Soc.*
B **34** (1972), 187–220.

Paper SM8 has been cited a fairly large number of times although no doubt
read rather less often. Much of its popularity stems from the implementation by
others of computerized procedures for its relatively painless use.

The paper came to be written because a number of friends identified a need
for methods for studying multiple dependence of survival time data that did not
make strong distributional assumptions. These friends certainly included Peter

Armitage, Ed Gehan and Marvin Zelen, but there may have been others. My first reaction was to suggest a Weibull-based formulation, essentially of accelerated life form, but this was thought too strong an assumption for general use. I think though this is an underused method.

Because of my interest in stochastic processes, an approach in terms of transition probabilities was natural, leading to a hazard-based formulation and thus fairly directly to the proportional hazards formulation. There was then the issue of how to analyse in the light of a semiparametric formulation in which baseline hazard is treated as arbitrary. It took some years of thought to see what to do; in fact the basic idea of partial likelihood occurred to me while I was feverish. In retrospect I am surprised that I persisted so long with the semiparametric approach in that by and large I am not very keen on nonparametric and semiparametric formulations. By allowing aspects to be arbitrary that will typically in some sense be smooth, such formulations tend to produce much unnecessary complication as compared with judicious parametric formulations, although of course these too have their hazards.

The paper itself then took little time to write with a well-chosen example with hidden depths supplied by Ed Gehan and some skilled programming help from Mrs Betty Chambers. The paper contains a good deal of supplementary work, for example: the use of time-dependent explanatory variables in particular to test the assumption of proportional hazards, the introduction of the product integral to synthesize discrete and continuous cases, multivariate failure times, and some pointers to estimation of the baseline hazard. Finally there was a critique of the proportional hazards assumption in the light of physical and physical-chemical models of failure and some discussion of the advantages of the accelerated life formulation.

The referees were reasonably sympathetic and the paper was accepted without too much revision. It is hard to be sure, but I suspect that nowadays it might have been rather difficult to get the paper published. It contains massive gaps and some lack of clarity in motivation of the method of analysis provided. I was certainly convinced, I am not quite sure why, that the estimates and procedures based on what came to be called partial likelihood were asymptotically correct, but I certainly could not have proved this when the paper was first submitted. The gap was largely filled by Paper Th8, reproduced in Volume II, page 175.

SM9 A note on multiple time scales in life testing. *J. R. Statist. Soc.* C **28** (1979), 73–75 (with V. T. Farewell).

Key elements in the study of survival data are the clear definition of both a time origin for each study individual and also an end-point. A third aspect concerns the measurement of the passage of time. Often this will just be clock time, but there are other possibilities; for example, in studying the life of car tyres one might use clock time, or kilometres driven, or kilometres weighting differently different kinds of road. Paper SM9 suggests a way of finding empirically the most appropriate time scale.

Additional publication

Analysis of Survival Data. London: Chapman and Hall, 1984 (with D. Oakes).

This book with David Oakes aimed to set out in reasonably accessible and self-contained form the main aspects of survival data. The field has become much more specialized since then, both in the elaboration of the mathematical methods used and in the variety of models studied, although the field has broadened considerably in the range of applications involved.

Multivariate analysis and graphical Markov models

SM10 Linear dependencies represented by chain graphs (with discussion). *Statistical Science* **8** (1993), 204–283 (with N. Wermuth).

Paper SM10 is one of the first of a series arising out of a collaboration which started in 1988 and is still continuing with Nanny Wermuth. This paper in a sense provides the empirical background for much of our subsequent work. In particular, once it is recognized that any pair of variables in a study may be such that one is a response to the other regarded as explanatory or may have to be treated on an equal footing, the need to distinguish a number of types of graphical representation of independencies is clear. Essentially three distinctions are needed. First, variables need to be blocked in such a way that all variables on an equal footing with one another are in the same block. Next, within a block there is a need to distinguish pairwise independence from fully conditional independence. Thirdly, in considering the dependence of one variable on those explanatory to it, there is a need to clarify the status of other variables on an equal footing with the putative response. Notation for doing this is introduced and, importantly, empirical examples are given for four component variables of the main possibilities.

SM11 On the calculation of derived variables in the analysis of multivariate responses. *J. Multivar. Anal.* **42** (1992), 162–170 (with N. Wermuth).

An important distinction in problems with a multivariate response is between situations in which invariance is desirable under a set of transformations of the component responses, often under arbitrary nonsingular linear transformations, and situations in which the component responses as recorded are individually meaningful. Paper SM11 deals with the former and proposes to transform the component responses so that, when the new variables are regressed on a set of explanatory variables, a seemingly unrelated regression structure is achieved. That is, each new component depends on just one of the explanatory variables. The calculations are related to, but rather different in spirit from, those of discriminant and canonical regression analysis. An example is given where appealing interpretable conclusions are reached.

An extension to time series analysis has been developed (Cox and Wermuth, 1999) and an application awaits publication.

SM12 Response models for mixed binary and quantitative variables. *Biometrika* **79** (1992), 441–461 (with N. Wermuth).

Paper SM12 develops a large number of possible models for the analysis of mixtures of discrete and quantitative variables and studies the relations between them in some detail.

Additional publication

Multivariate Dependencies. London: Chapman and Hall, 1996 (with N. Wermuth).

This book summarizes and extends the joint work described above and gives in some detail applications and further developments. The themes are the subject of continuing research.

Miscellaneous

SM13 An analysis of transformations (with discussion). *J. R. Statist. Soc.* B **26** (1964), 211–252 (with G. E. P. Box).

I first met George Box in the earlyish 1950s when he was at ICI Dyestuffs near Manchester in the days when ICI was one of the world's major enterprises, employing a large number of statisticians. George's highly original and practically relevant contributions had great influence far beyond Dyestuffs Division and beyond ICI itself. The suggestion and first outline of Paper SM13 arose one evening in about 1953 when I was staying with George in Sale and the basic idea of estimating a transformation by likelihood methods was certainly his; the need for a joint paper arose in both our minds as the only way to stop tedious suggestions that we write a paper together!

Progress after that slowed as we crossed and recrossed the Atlantic out of phase, and the nature of the paper changed as the rapid developments in computer power made the original expansion-based techniques that we tried obsolete. The final intention was to finish the paper on the building site that was becoming Carleton University, where we were kindly offered space in the few days before the International Statistical Institute met in Ottawa in 1963. The paper was indeed almost finished there. Then, as we prepared the final draft, we both realized virtually simultaneously that the Bayesian analysis that George by then favoured was fatally flawed; the flat priors took no account of the inevitable massive change in the magnitude of regression coefficients as one passed, say, from reciprocals to untransformed values. Two ways of dealing with this were hastily developed, one by a kind of approximate orthogonalization and the other by use of a data-dependent prior.

The paper was read to the Royal Statistical Society and unfortunately George was unable to be present. I confidently predicted an onslaught from Bayesian statisticians for the use of data-dependent priors but 40 years later the onslaught has still not come. Pericchi (1981) subsequently showed how the prior could arise by more standard arguments; in any case, in principle the notion that a prior distribution might depend on the data seems to me unexceptionable.

The most important aspect of the paper is probably the notion that estimating a transformation can be formalized even if in practice relatively informal methods are often adequate. The technique has, of course, the advantage that formal procedures are associated with the estimated transformation.

If one estimates within the family of power transformations, it will typically be best to use simple members of the family rather than, say, the transformation estimated by maximum likelihood.

The technique forces some compromise between linearity, constancy of variance and normality; for a quantitative discussion, see Paper Th21 in Volume II. It can be argued that separate representation of the mean and variance relation is preferable, although this must depend to some extent on whether the changes in variance are of intrinsic interest. When the variable under analysis is extensive (i.e. physically additive), the case for using averages on the original scale is, of course, particularly strong. In other cases, and certainly in the two real examples in the paper, transformation seems much the more insightful way to proceed. There is the additional point that in a sense constancy of variance is a form of absence of interaction; unit treatment additivity implies constant variance.

The method has been quite widely used.

Some subsequent discussion centred round the estimation of regression co-efficients whose interpretation changes, possibly rapidly, as a transformation is applied and as to what is the appropriate standard error in such cases. We contested (Box and Cox, 1982) some accounts which advocated inflated standard errors.

SM14 The role of statistical methods in science and technology. Inaugural Lecture, Birkbeck College, London, 1961.

The custom that new professors give a nontechnical inaugural lecture about their subject in principle and in actuality to the whole institution was still practiced at Birkbeck College in the late 1950s. The lecture aims to cover some of the main themes in statistics around that time. There are, in addition, some new points. There is an interesting botanical example of what is sometimes called the broken stick model, and there are remarks about optimal design applied to some counting problems in physics. The point that got me into some trouble, however, concerned the personalistic theory of probability as propounded by its charismatic exponent L. J. Savage. I thought this theory fascinating, but that its claim to take over the whole of probability was profoundly misguided. I therefore said in the lecture that the idea was a bold and imaginative step backwards. When the lecture was written up, should I include this remark? It was meant seriously and represented my view, if somewhat pungently, and so it had to stay. Some unknown friend drew Savage's attention to it. Our relations up to then had been very friendly. He was a powerful and eloquent advocate of his point of view and stern in his criticism of others, but just possibly not so keen on being answered back.

SM15 Combination of data. In *Encyclopedia of Statistical Sciences* **2**, edited by S. Kotz and N. L. Johnson, pp. 45–53. New York: Wiley, 1982.

'Combination of data' and 'combination of observations' are older terms for statistical methods, terms used especially in the physical sciences. In this short review the words are taken to mean combination of information from different sources. In the context of experiments, this used to be called analysis of series of experiments but, now in a broader context, tends to be named somewhat pretentiously meta-analysis or, less objectionably, overview. The basic ideas had been given a searching discussion by Yates and Cochran (1938), in the context of agricultural field trials, and developed in more detail in a series of papers by Cochran in the 1950s; see, for example, Cochran (1954). The main novelty in Paper SM15 is a fairly careful discussion of pooling in the presence of overdispersion with some attempt to address the possibility of systematic errors in some of the studies to be combined.

SM16 Interaction (with discussion). *Inter. Statist. Rev.* **52** (1984), 1–31.

Interaction, or perhaps more significantly absence of interaction, is one of the most important concepts in statistics. The formal literature is mostly about testing the null hypothesis of the absence of interaction. Paper SM16 concentrates on interpretation. Many of the points are essentially elementary, but even so often not given adequate emphasis.

Among the key points made are the following. Constancy of variance and absence of interaction are interrelated. The interpretation of interaction between two factors depends on whether one or both are treatments, intrinsic features of the study individuals or essentially levels of a random stratifying feature, as well as on whether the levels of the factors are quantitative or qualitative. (That is an implicit criticism of the high-jacking of the term 'factor' to refer only to the case with qualitative levels.) Then there is the important point that absence of interaction refers to a particular scale of measurement and is not invariant under nonlinear transformations of the response.

An issue not discussed in the paper is that interaction is one of the statistical ideas handicapped by an unsuitable name. Most scientists give the term a wholly or partially mechanistic interpretation and, while statistical interaction sometimes means that, more commonly it is concerned with the nonstability of some effect as a background variable changes. Absence of major interactions of that kind is an important basis for generalization of conclusions.

SM17 Notes on some aspects of regression analysis (with discussion).
J. R. Statist. Soc. A **131** (1968), 265–279 [Discussion 315–329].

This is a wide-ranging paper given with two other papers at a meeting of the Royal Statistical Society on regression. It deals with elementary but often overlooked issues, largely ones of interpretation, connected with regression analysis. Many of the points especially in Section 3 on multiple regression bear on the issue of interpreting regression coefficients causally, although that word is avoided. The extension to linear regression models in the canonical parameters of an exponential family model is indicated.

SM18 The choice of variables in observational studies. *J. R. Statist. Soc.* C **23** (1974), 51–59 (with E. J. Snell).

Paper SM18 is related to Paper SM17 and focuses more explicitly on observational studies. It is very broad in scope and correspondingly does not address details. Important matters such as trying to establish causality are mentioned; any value of the paper now is that it sets out many issues that remain important in designing and interpreting observational studies and so might be a useful starting point for detailed discussion. The paper was written at the request of the World Health Organization, but I have no idea what they did with it!

SM19 A general definition of residuals (with discussion). *J. R. Statist. Soc.* B **30** (1968), 248–275 (with E. J. Snell)

The notion of a residual as being what is left over after the systematic part of some variation has been removed or explained is obviously a very general one. Nevertheless, the formal definition is usually confined to linear models. In these, after estimating the systematic part, for example by the method of least squares, the difference between the observation and the corresponding fitted value is called the residual. Anscombe (1961) investigated the properties of such residuals very thoroughly. The probabilistic interpretation of the residuals is, in one sense, the estimation of the corresponding unobserved error term in an underlying linear model.

Joyce Snell's and my objective in Paper SM19 was to give a much more general definition of residuals and to find some of their properties. For this, what is sometimes called a functional model is used. That is, each observation is regarded as a given function of unknown parameters and an unobserved random variable and, of course, in general of explanatory variables regarded as fixed constants. That is, in a self-explanatory notation, $Y_i = g_i(\theta, \epsilon_i)$, where under the initial model the ϵ_i are independent and identically distributed random variables. Even in the continuous case there is considerable arbitrariness in this formulation in that, by choice of the functions g_i, the ϵ_i could be taken to have any distribution, for example the standard normal or the unit rectangular distribution. Once this choice is made, the residuals are defined by replacing θ by its maximum likelihood estimate $\hat{\theta}$ and then solving for R_i the equation $Y_i = g_i(\hat{\theta}, R_i)$. Study of the properties of R_i hinges on the higher-order properties including bias of the maximum likelihood estimate. A quite simple formula is obtained for that bias and used to define improved residuals that are intended to be closer in distributional form to the ϵ_i than are the 'crude' residuals R_i. Some rather simple illustrative examples are then given.

The ideas in the paper have occasionally been used in the analysis of binary data and of survival data.

There are different possible roles for residuals in the analysis of data. One is when a model is used to check data, for example for outliers. Another is when data, assumed provisionally to be reliable, are used to check models. Such procedures may be focused on specific kinds of departure or may be what M. S. Bartlett termed omnibus tests. Residuals may be used for all these purposes but it is probably for the last kind of application, the detection of unanticipated features, that they are especially suited.

SM20 Testing multivariate normality. *Biometrika* **65** (1978), 263–272 (with N. J. H. Small).

This work arose out of a dissatisfaction with the literature on examining consistency with the multivariate normal distributional form. Most of the methods only worked in a small number of dimensions and were formal analogues of familiar one-dimensional methods of looking at distributional shape. The most interesting departures from multivariate normal form, in addition to outliers, are likely to be systematic nonlinearities of structure of some sort.

The paper distinguishes between coordinate-dependent procedures which recognize the particular component structure in terms of which the data are expressed and procedures invariant under arbitrary nonsingular linear transformations of the components. The main emphasis is put on applying standard tests of nonlinearity by inclusion of quadratic terms. In the invariant case this is done for arbitrary linear combinations and the resulting test statistic maximized over choice of the combinations. I recall a firm prior belief, which Nicholas Small did not share, that the resulting distributions should have something to do with extreme value distributions; simulation showed he was right.

A practical example with eight variables on 90 individuals suggested that the methods are reasonably effective.

Related ideas were later used systematically in the analysis of data via graphical Markov models (Cox and Wermuth, 1994, 1996). Algorithms are available on psystat.sowi.uni-mainz.de.

SM21 Some remarks on the role in statistics of graphical methods. *J. R. Statist. Soc.* C **27** (1978), 4–9.

Paper SM21 arose from an invitation to give the opening talk at a conference on graphical methods organized at the University of Sheffield. Graphical methods have always been important but this was a time at which such methods were becoming both more accessible and more imaginative, in particular under the stimulus of John Tukey. His presence at the conference added much to it; I recall in particular that, there not being much time for discussion of my own paper, questions in John Tukey's unmistakable handwriting soon appeared on a notice board and it was made clear a number of times that answers of some sort or other were required.

As a preliminary to the paper, I had explored a little of the history of graphical methods and, while I benefited from colleagues and others, I was, no doubt naively, surprised at how much work was involved in getting the details even roughly right.

Most of the paper dealt with precepts for simple presentation, still often ignored. Recent developments have since made many new things possible, but they are not without problems. Indiscriminate use of colour can distract more than it helps and I do not think I am the only person to find perspective plots by and large distinctly unhelpful.

SM22 Analysis of variability with large numbers of small samples. *Biometrika* **73** (1986), 543–554 (with P. J. Solomon).

Paper SM22 arose out of Patty Solomon's involvement in a large primary prevention trial on hypertension. Nested or hierarchical arrangements are quite common in a number of fields and in their simplest form lead to sets each of a small number k of repeat observations, each set obtained under nominally the same conditions and in principle thus having its own mean value. Typical values of k are 2, 3 and 4. Replication is used primarily to improve precision of the estimated means but also to allow estimation of the within-set component of variance.

Occasionally, however, it may be fruitful to extract more information from the replicated observations and the paper sets out a number of simple methods, numerical and graphical, for doing this. Distinctions are possible between a common nonnormal distribution of errors and normal distributions with randomly varying or systematically changing variance. A further paper not reproduced here (Cox and Solomon, 1988) looked at corresponding problems with serial correlation.

SM23 Nonlinear component of variance models. *Biometrika* **79** (1992), 1–11 (with P. J. Solomon).

The title of Paper SM23 to some extent hides its content. The paper is an outcome of a continuing collaboration with Patty. A number of rather distinct ideas are explored. First, detection of departures from a standard normal-theory component of variance model is explored, aiming separately at the distributional forms of the upper and lower component errors and at the relation between the conditional variance of the lower component and the level of the upper component error. The relation between these three aspects is special when a transformation reduces the model to standard form.

The next part of the paper deals with Laplace expansion of log likelihood functions especially in models with random regression coefficient models. An application to a possible 'improvement' of the Mantel–Haenszel test is considered briefly.

SM24 On partitioning means into groups. *Scand. J. Statist.* **9** (1982), 147–152 (with E. Spjøtvoll).

Paper SM24 arose, I think, out of conversations with the Norwegian statistician the late Emil Spjøtvoll, when he came to London to give an invited series of special university lectures. We both were uneasy at some of the relatively complicated procedures that had been suggested for the interpretation of comparisons among a set of unstructured means. The paper records a very simple solution when it is adequate to arrange the means in groups, means in the same group not being clearly different but different groups having different overall means. The idea is simply to list all possible such groupings consistent with the data as judged by a standard test such as the F-test. Then a true grouping will not appear within the specified set in accordance with the standard F-distribution, no allowance for multiple testing being needed. The general point implicit in the discussion is that if the data are consistent with a number of different interpretations all should be shown in an initial assessment; of course, some or many possibilities may then be discarded on subject-matter grounds.

SM25 Some quick sign tests for trend in location and dispersion. *Biometrika* **42** (1955), 80–95 (with A. Stuart).

Paper SM25 is an excursion into the development of simple nonparametric tests and their assessment via Pitman's asymptotic relative efficiency. One minor detail, possibly still sometimes useful, concerns the advantage of showing power calculations plotted on a probability integral scale, thereby inducing approximate linearity in many cases.

Additional publication

Components of Variance. London and Boca Raton, Florida: Chapman and Hall/ CRC Press, 2002 (with P. J. Solomon).

This book covers both the rather traditional field of components of variance and its extension to multilevel modelling.

References

Anscombe, F. J. (1961). Examination of residuals. In *Proceedings of the Fourth Berkeley Symposium on Mathematical Statistics and Probability* **1**, edited by J. Neyman, pp. 1–36. Berkeley: University of California Press.

Box, G. E. P. and Cox, D. R. (1982). An analysis of transformations revisited, rebutted. *J. Amer. Statist. Assoc.* **77**, 209–210.

Cochran, W. G. (1954). The combination of estimates from different experiments. *Biometrics* **10**, 101–129.

Cox, D. R. and Lewis, P. A. W. (1966). *Statistical Analysis of Series of Events.* London: Methuen/New York: Wiley.

Cox, D. R. and Solomon, P. J. (1988). On testing for serial correlation in large numbers of small samples. *Biometrika* **75**, 145–148.

Cox, D. R. and Wermuth, N. (1994). Tests of linearity, multivariate normality and the adequacy of linear scores. *J. R. Statist. Soc.* C **43**, 347–355.

Cox, D. R. and Wermuth, N. (1996). *Multivariate Dependencies.* London: Chapman and Hall.

Cox, D. R. and Wermuth, N. (1999). Derived variables for longitudinal studies. *Proc. Nat. Acad. Sci.* USA **96**, 12273–12274.

Mendenhall, W. and Hader, R. J. (1958). Estimation of parameters of mixed exponentially distributed failure-time distributions from censored life test data. *Biometrika* **45**, 504–520.

Pericchi, L. R. (1981). A Bayesian approach to transformation to normality. *Biometrika* **68**, 35–43.

Yates, F. and Cochran, W. G. (1938). The analysis of groups of experiments. *J. Agric. Sci.* **28**, 556–580.

SM1

Journal of the Royal Statistical Society

SERIES B (METHODOLOGICAL)

Vol. XVII, No. 2, 1955

SOME STATISTICAL METHODS CONNECTED WITH SERIES OF EVENTS

By D. R. COX

Statistical Laboratory, University of Cambridge

[Read before the RESEARCH SECTION of the ROYAL STATISTICAL SOCIETY,
March 16th, 1955, Mr. E. C. FIELLER in the Chair]

SUMMARY.

THE paper deals with a number of problems of statistical analysis connected with events occurring haphazardly in space or time. The topics discussed include: tests of randomness, components of variance, the correlation between events of different types, and a modification of the snap-round method used in operational research.

1. INTRODUCTION

This paper is about the statistical analysis of events occurring haphazardly in space or time. There are many applications. The events may be electrical pulses in a nerve fibre, mutations, diseased plants in a field, customers arriving at a queueing point, fissions of bacteria, the emissions of radioactive particles, neps distributed over a thin web of textile sliver, slubs* distributed along the length of a yarn, the stops of a machine and so on. The requirements are a continuum, space or time, usually, but not necessarily, one-dimensional and a series of events which can be regarded as points in this continuum. Thus in the analysis of machine-stops the continuum is the running-time of the machine and each stop is characterized by the point in running-time at which it occurs. Often each event will have a numerical quantity attached to it, such as the duration of the stop, the magnitude of the electrical disturbance, the size of the slub and so on, but this we ignore except in §7.

The object of the paper is, quite deliberately, to touch on the solution of a number of problems rather than to give a very detailed discussion of any one. Consequently the treatment is often approximate and based on large-sample theory. Also I have in mind for the most part applications where simplicity of analysis is of some importance and, while some of the methods are not as simple as is desirable, things such as the iterative solution of complicated maximum likelihood equations have been avoided. Owing to the length of the paper it has unfortunately not been possible to include numerical examples of the various methods described.

The stimulus for much of this work came from the mill-surveys of the British Rayon Research Association, Manchester. I am indebted to the Director of Research, Mr. J. Wilson, for permission to quote examples and am very grateful to Mr. B. H. Wilsdon for suggesting a number of problems and for stimulating discussions.

2. GRAPHICAL METHODS

It is convenient to introduce some notation and to describe briefly some of the graphical methods that are useful for summarizing long series of events. As far as possible the symbol n

---

* A *nep* is a small knot of entangled fibres. A *slub* is a short abnormally thick place in a yarn or other fibrous strand.

is reserved for the number of events occurring in a certain time, or space, the symbol x for the interval between events, and the symbol t for a time (or space) co-ordinate.

Suppose that in a total time T, n events occur at times $t_1, \ldots, t_n$, in order. Let $x_{n,\,n+k} = t_{n+k} - t_n$ with $t_0 = 0$. Fix a time period τ and let $n_i(\tau)$ be the number of events in the interval $[(i-1)\tau, i\tau)$.

The main graphical methods are the following.

(i) Time is divided into equal intervals τ and $n_i(\tau)$ plotted against i. If there are large variations in the local rate of occurrence the restriction to a fixed τ is inconvenient.

(ii) A random-walk diagram can be plotted for the total number of events up to a time t as a function of t. From this the average rate of occurrence over any period can be read off. The disadvantages of this method are the difficulty of detecting sampling fluctuations in a cumulative diagram (see, however, §4.3) and the awkward shape of the curve making it difficult to plot extensive data concisely.

(iii) The last point can be dealt with by plotting $(n - a^{-1}t)$ against t, where a is a suitable constant approximately equal to the mean interval between events. If required, the value of n can be marked at suitable points along the graph. There is also the possibility of plotting $(t - b^{-1}n)$ against n but this is, on the whole, less useful.

(iv) The intervals $x_{1,\,2},\ldots,\ x_{n,\,n+1};\ \ldots$ can be plotted either against the serial number of the interval or against the time at the mid-point of the interval. This method deals with the objection to method (i) by, in effect, adjusting τ to the rate of occurrence. If it is required to examine the relation between the rate of occurrence and some smoothly varying external variable, the abscissa can be taken as the average of the external variable over the interval.

(v) A histogram can be formed for the interval between successive events and a scatter diagram can be obtained for successive intervals. These serve quite a different purpose from methods (i)–(iv).

3. Completely Random Series of Events

3.1 *General*

A fundamental role in the analysis is played by the concept of a completely random series of events, or Poisson process, defined formally by the conditions that the probability of an event occurring in an interval δt is $\lambda \delta t + o(\delta t)$ and that the probabilities associated with non-overlapping intervals are independent. It is very well known that the number of events in a fixed interval T is then a Poisson variate with mean λT and that the interval between a point and the kth succeeding event is distributed as $1/(2\lambda)$ times a χ^2 variate with $2k$ degrees of freedom. Statistical methods based on the analysis of intervals have been discussed by Maguire *et al.* (1952) and these and tests based on the Poisson distribution have been reviewed by Birnbaum (1954). These methods are basic for what follows.

In using the completely random series we do not usually suppose that events are caused by "chance", but we consider the random series either because its use simplifies the comparison of different series or because the detection of a departure from randomness may throw light on the mechanism underlying the series. One important case when the completely random series is likely to be a good representation of an observed series is when we observe a superposition of a large number of independent series. This was pointed out for telephone calls arriving at an exchange by Fry (1928) and special cases have been investigated in detail by Cox and Smith (1953, 1954). These considerations provide a "justification" for the completely random series analogous to the central limit "justification" of the normal distribution.

3.2 *Logarithmic Transformation*

We often have problems in which the parameter λ is not constant but is the product either of unknown parameters or of known constants with unknown parameters. This can happen in the following ways.

(i) If the interval of observation is not constant. For example in nep counting a section of sliver is weighed and the number of neps in it counted. This gives a series of observations n_i which are Poisson variates with means λW_i, where W_i is the weight of the ith section and is known. Or in observing numbers of stops of machines it will usually be impracticable to arrange that the observed running-time is the same for all machines.

(ii) If we have a regression problem in which λ varies as αY^β or as $\alpha e^{\beta Y}$, where Y is a known independent variable such as time.

(iii) It may be intuitively reasonable, or be suggested by inspection of the results, that, say, in a two-way arrangement, the row and column effects are multiplicative. Thus we may have observations on the stoppage rates of looms, the looms being classified by the sort of cloth being woven and by the weaver. If there are substantial differences between weavers and between sorts it is reasonable to see whether the data fit the hypothesis that the differences in stoppage rate between weavers are proportional for different sorts. In other words the λ appropriate to each observation is assumed to be the product of a weaver constant and a sort constant.

In all the problems just described, and particularly in the last two, it is natural, if a simple analysis is required, to make the problem linear by a logarithmic transformation and then to use estimates based on linear combinations of the observations.

Such a process is in general not fully efficient even if the particular linear estimates are derived by the method of least squares and the present use of the logarithmic transformation should be distinguished clearly from the transformations introduced by Fisher—see, for example, Fisher, (1954)—as an aid to the solution of maximum-likelihood equations. It is also, of course, quite different from transformations to stabilize variance or to induce normality. If it has been decided to make an approximate analysis, it is more important that the parameters estimated should be physically meaningful and easily interpreted than that conditions of normality and variance stability should be satisfied; hence it is usually reasonable to give precedence to the attainment of additivity.

If it is possible to obtain the intervals between successive sets of k events, the logarithmic transformation causes no difficulty. For if x is such an interval, throughout which λ is constant,

$$\log \frac{k}{x} = \log \lambda - \log \left(\frac{1}{2k} \chi^2_{2k} \right), \qquad . \qquad . \qquad . \qquad . \quad (3.1)$$

so that

$$E\left(\log \frac{k}{x}\right) = \log \lambda - \psi(k) \qquad . \qquad . \qquad . \qquad . \quad (3.2)$$

$$\simeq \log \lambda + \frac{1}{2k - \frac{1}{3} + \frac{1}{16k}}, \qquad . \qquad . \qquad . \qquad . \quad (3.3)$$

and

$$V\left(\log \frac{k}{x}\right) = \psi'(k) \qquad . \qquad . \qquad . \qquad . \quad (3.4)$$

$$\simeq \frac{1}{k - \frac{1}{2} + \frac{1}{10k}}. \qquad . \qquad . \qquad . \qquad . \quad (3.5)$$

In the formulae (3.2), (3.4) $\psi(k) = d \log \Gamma(k)/dk$ (Bartlett and Kendall, 1946). The approximate formulae (3.3), (3.5) give 1 per cent. accuracy at $k = 1$. In a regression problem the comparison of the residual variance with (3.5) tests the adequacy of the regression law.

If, however, we work with numbers of events in fixed-time intervals there is more difficulty. Cox (1953) has shown semi-empirically that if n is a Poisson variate with mean λT we can proceed as if $2\lambda T$ is distributed as χ^2_{2n+1}. This was shown to give accurate tests for the comparison of two rates of occurrence and for the determination of confidence limits for λ from a single observation n. In the paper just mentioned a small-sample logarithmic transformation was obtained from the above approximation but the argument for this was unsound. In effect one degree of freedom of the $2n + 1$ is a continuity correction and in combining a number of transformed values a continuity correction should only be applied once. This seems to make it difficult to construct a small-sample transformation for general use.

There is however one case that can be solved. Suppose that we make an analysis of the transformed observations using linear combinations with coefficients not depending on the observations, i.e. an "unweighted" analysis. Then if our linear estimates are to be unbiased and if we are to be able to find unbiased estimates of their sampling variance, we must try to find, corresponding to the Poisson variate n, a transformed value z_n and an estimated variance v_n such that ,

$$E(z_n) = \log \lambda, \qquad \qquad \qquad (3.6)$$

$$V(z_n) = E(v_n). \qquad \qquad \qquad (3.7)$$

It is obviously impossible to satisfy these exactly for all λ. For large n

$$z_n \sim \log \frac{n}{T} \text{ and } v_n \sim \frac{1}{n},$$

so that we try

$$z_n = \log \frac{n + \alpha}{T}, v_n = \frac{1}{n + \beta}, \qquad \qquad (3.8)$$

where α, β are to be determined.

Now if $\theta = \lambda T$ and $\delta n = n - \theta$,

$$z_n = \log \lambda + \log \left(1 + \frac{\delta n}{\theta} + \frac{\alpha}{\theta}\right)$$

and if this is expanded in series, taking δn, α to be small compared with θ, we get

$$E(z_n) = \log \lambda + \frac{\alpha - \frac{1}{2}}{\theta} + O\left(\frac{1}{\theta^2}\right). \qquad \qquad (3.9)$$

Therefore we take $\alpha = \frac{1}{2}$ when we have

$$E(z_n) = \log \lambda + O\left(\frac{1}{\theta^2}\right). \qquad \qquad (3.10)$$

With this value of α

$$V(z_n) = \frac{1}{\theta} + \frac{1}{2\theta^2} + O\left(\frac{1}{\theta^3}\right), \qquad \qquad (3.11)$$

while

$$E\left(\frac{1}{n + \beta}\right) = \frac{1}{\theta} + \frac{1 - \beta}{\theta^2} + O\left(\frac{1}{\theta^3}\right), \qquad \qquad (3.12)$$

so that we choose $\beta = \frac{1}{2}$ in order to satisfy (3.7).

Thus we have arrived at the transformation

$$z_n = \log \frac{n + \frac{1}{2}}{t} \text{ with } v_n = \frac{1}{n + \frac{1}{2}} \quad .$$

This has been derived by a series expansion in $1/\theta$ and hence we can reasonably expect (3.6), (3.7) to be satisfied down to say $\theta \simeq 5$. To check this $E(z_n)$, $V(z_n)$, $E(v_n)$ were calculated for $\theta = 1, 2, 3, 5$. The results are in the first part of Table 3.1.

TABLE 3.1

Properties of the Logarithmic Transformation for the Poisson Distribution

θ	$\log \theta$	$E(z_n)$	$V(z_n)$	Unmodified: $E(v_n)$	Modified: $E(v_n)$
1	0·000	0·169	0·505	1·076	0·616
2	0·693	0·719	0·473	0·640	0·471
3	1·099	1·100	0·372	0·420	0·358
5	1·690	1·607	0·228	0·231	0·223

Equation (3.6) for $E(z_n)$ is satisfied to a good approximation, even for θ as small as two, but the value of $E(v_n)$ (in the column headed "unmodified") is, for $\theta < 5$, appreciably greater than $V(z_n)$. This suggests changing the value of v_0 and some simple calculations showed that $v_0 = 3/4$ improved the agreement (see the column headed "modified"). Further small modifications say in the values of z_0, z_1, v_1 could no doubt be introduced, but there seems little to be gained by this.

To sum up we have the transformation

$$z_n = \log \frac{n + \frac{1}{2}}{t}, \qquad\qquad\qquad (3.13)$$

$$v_0 = 3/4, \quad v_n = 1/(n + \frac{1}{2}), \quad n \neq 0, \qquad\qquad (3.14)$$

which has the properties (3.6), (3.7), provided that the population mean is not less than about two. The analysis leading to (3.13), (3.14) is possibly of some general interest in indicating a method of constructing transformations to induce additivity.

Consider the efficiency of the analysis of the transformed variates. In the simplest case, for which the transformation is of course unnecessary, we have a large number m of quantities independently drawn from a Poisson distribution of mean θ. The variance of the best estimate of $\log \theta$ is $1/(m\theta)$ while the variance of the mean of the z's is $V(z_n)/m$. Hence the efficiency is $1/(\theta V(z_n))$ which, from (3.11), equals $1 - 1/(2\theta) + O(1/\theta^2)$; it is therefore high unless θ is very small. (The apparent efficiency computed from the values in Table 3.1 exceeds unity; this because the bias in the estimates from the transformed values upsets the formula for efficiency.) This is for a single unknown parameter, and the loss of efficiency is that arising from the use of *linear* combinations of the z's. In more complicated cases there will also be a loss of efficiency due to not giving a weight to each observation depending on its corresponding population value. This is the more important loss of efficiency; in some cases the major part of the loss could be avoided by a rough system of weighting not depending critically on the individual z's.

If an analysis is made of the transformed values weighting them in a way depending on the observations themselves, the requirements (3.6), (3.7) cease to be relevant. An extreme case is when the weights are determined entirely by the observations; thus if the n_i are drawn from a Poisson distribution of mean λT_i, $i = 1, \ldots, k$, $\log \lambda$ could be estimated by an expression of the type

$$\frac{\Sigma \sqrt{n_i}\, z'_{n_i}}{\Sigma \sqrt{n_i}}$$

and it can be shown that this would be very nearly unbiased if

$$z'_n = \log \left(\frac{n - \frac{1}{2} + \frac{1}{2k}}{T} \right). \qquad\qquad (3.15)$$

Note that $z_n > z'_n$ because the weighting in the second case tends to emphasize high values. A transformation is not necessary in such a simple situation but the argument serves to indicate the appropriate transformed variate.

In most practical cases in which weighted regression is used the coefficients of the z's will depend partly on the n's and partly on constants depending on the structure of the problem. The appropriate transformation is therefore likely to be intermediate between z_n, z'_n and this suggests the use of

$$z''_n = \log \frac{n}{T}, \quad v''_n = \frac{1}{n}, \qquad\qquad\qquad (3.16)$$

and this is in fact the form suggested in Cox (1953). Further investigation is needed to delimit the range of n for which (3.16) should be used, but it seems unlikely that serious error can arise if $n \gg \frac{1}{2}$, i.e. if few of the values of n are less than five.

To sum up we have derived the transformed variate (3.13) and the estimated variance (3.14) for use in an "unweighted" analysis and the transformation (3.16) for use in a "weighted" analysis. The first transformation has the properties required of it if no expectation is less than two and the second transformation is probably adequate if few observed frequencies are less than five.

4. Tests of Randomness

4.1 *General*

There are two main reasons for requiring a test of the randomness of an observed series of events:

  (i) the hope that the form of the non-randomness, if any, will throw light on the mechanism of the process;

  (ii) in order to justify the use of statistical methods based on the hypothesis of randomness, or to make the appropriate modification if non-randomness is present.

There are two consequences of these aims. First we usually require to measure, or at least to describe qualitatively, any non-randomnesss detected. Secondly we need to specify the type of non-randomness before we can discuss the relative merits of different tests. In applications we decide on the type of non-randomness, and hence on the appropriate test, either from prior knowledge or after preliminary inspection of the data. In the latter case there are the usual difficulties of "selection" which apply whenever the hypothesis for test, or the alternatives, are suggested by the data.

We first describe briefly a number of forms of non-random series and then consider the appropriate tests. To describe a non-random series we imagine that it is constructed by a probability mechanism in which the chance of an event between $(t, t + \delta t)$ is $\lambda(t) \delta t + o(\delta t)$, where $\lambda(t)$ is in general a function not only of t but of the whole history of the system up to t. We call $\lambda(t)$ the probability rate of occurrence; the series is random if and only if $\lambda(t)$ is constant.

4.2 *Some Types of Non-randomness*

Eight types of departure from randomness will be described briefly; there are no doubt other types of departure of importance in particular applications.

  (i) *Trend.*—If λ is a smooth function of the time t, we have a trend in the rate of occurrence. The easiest forms to deal with are probably $\lambda = \alpha e^{\beta t}$ and $\lambda = \alpha(t + t_0)^\beta$, where α, β are unknown and t_0 is known. The latter form would be appropriate if some process of growth has begun t_0 before the start of the period of observation.

  (ii) *Rate determined by a smooth external variable.*—In (i), instead of t being time it could be any smoothly varying quantity. Thus the stoppage rate of a loom might be a smooth function of relative humidity.

  (iii) *Independent intervals.*—If λ is a function only of the time elapsed since the preceding event, a series is obtained in which the intervals between successive events are independent random variables with the same distribution, exponential only when λ is constant. The main application is to renewal theory, where the events are the replacements of, say, a particular component of a machine and it is required to test whether the distribution of life-time is exponential. If $I(x)$ is defined as the ratio of the variance of the number of events in an interval of length x to the corresponding variance for a random series with the same mean interval, it is well-known (Feller, 1948), that as $x \to \infty$

$$I(x) \sim C^2, \qquad . \qquad . \qquad . \qquad . \qquad . \qquad (4.1)$$

where C is the coefficient of variation of life-time.

  (iv) *Autocorrelated intervals.*—Suppose that at time t, the two preceding events occurred at t', t'' and that λ is a function of $t - t'$ and $t - t''$. We then have a series analogous to a Markov stochastic process (Wold, 1948a and 1948b). An interesting special case is when, if n events have occurred before t, the probability of an event occurring between $(t, t + \delta t)$ is a function only of the interval x_n between the $(n - 1)^{th}$ and the nth events, i.e. of $t' - t''$ in the previous notation. It is rather difficult to find examples where this is directly relevant, but a rather similar situation occurs in bacterial growth (see, for example, Armitage (1952)).

It follows in this special case that conditional on u_n, u_{n+1} has the distribution $\lambda(u_n) e^{-\lambda(u_n)u_{n+1}}$. If the process is stationary, so that u_n, u_{n+1} have the same marginal distribution, $p(u)$ say, we have

$$p(u_{n+1}) = \int_0^\infty \lambda(u_n) e^{-\lambda(u_n)u_{n+1}} p(u_n) \, du_n.$$

(It can be shown that this equation has an exponential solution if and only if λ is constant.)   Wold considered in detail $\lambda(u) = au^{-\frac{1}{2}}$, but for the present purpose we require a form reducing to a constant for a special choice of parameter values.   The simplest thing to try is $\lambda(u) = \lambda_0 + \lambda_1 u$ where λ_1 is, if negative, sufficiently small for negative values of $\lambda(u)$ to be ignored.   If we substitute this into (4.1) the resulting integral equation, although it can be written in a fairly simple form in terms of the Laplace transform of $p(u)$, does not seem to have a simple exact solution.   If, however, $\lambda_1 << \lambda_0$ we have

$$p(u) = \lambda_0 e^{-\lambda_0 u} \left\{ 1 + \frac{\lambda_1}{\lambda_0} \left(\frac{1}{\lambda_0} - u \right) \right\} + O\left(\frac{\lambda_1^2}{\lambda_0^2}\right). \qquad \qquad (4.2)$$

A negative correlation between successive intervals corresponds to a positive λ_1 and hence, as would be expected, to an increase in the ordinate for small u.   Note that (4.2) can be written as an exponential expression to $O(\lambda_1^2/\lambda_0^2)$, so that the departure from randomness is represented only by a second order departure from the exponential distribution of intervals.

   (v) *Stochastic variations in* λ.—Suppose that there is a constant time S such that in $[nS, (n+1)S]$ λ has the constant value λ_n, where $\lambda_1, \lambda_2, \ \cdot \ \cdot \ \cdot$ are a random sample from some population. When the intervals within which λ is constant can be identified, this is in effect the situation known as heterogeneous Poisson sampling (Greenwood and Yule, 1920).   If, however, S, or the exact position of the intervals, are unknown, fresh problems arise.

   A generalization is that the probability rate of occurrence λ may vary in a stationary time-series, say with correlogram $\rho(u)$.   The preceding case has $\rho(u) = \text{Max} \ (1 - |u|/S, 0)$.   In most applications in which this model is used λ will change little over times comparable to the mean interval between events.

   For example it is plausible that, under constant conditions, with a homogeneous supply of weft yarn, the stops of a loom from weft breaks should form an approximately random series. Different consignments of weft would probably have different stoppage rates and hence the first model might be expected to apply with S equal to the running-time corresponding to a consignment of weft.   In practice there would probably be other causes of variation in stoppage rate so that the more general form, with λ a time-series, would be needed.

   Consider the general form described above and suppose that λ has mean $\bar{\lambda}$ and variance σ_λ^2. We again define $I(x)$ to be the ratio of the variance of the number of events occurring in an interval x to its value, $\bar{\lambda}x$, for a random series.   Since the process is stationary we can, without loss of generality, take the interval to be $[0, x]$.   Let $n(x)$ be the number of events in this interval.   Then

$$E[n(x) \mid \lambda(t), \ 0 \leqslant t \leqslant x] = \int_0^x \lambda(t) \, dt, \qquad \cdot \qquad \cdot \qquad (4.3)$$

$$V[n(x) \mid \lambda(t), \ 0 \leqslant t \leqslant x] = \int_0^x \lambda(t) \, dt. \qquad \cdot \qquad \cdot \qquad (4.4)$$

Now

$$I(x) \equiv V[n(x)]/(\bar{\lambda}x) = \{E_\lambda V[n(x) \mid \lambda(t)] + V_\lambda E[n(x) \mid \lambda(t)]\}/(\bar{\lambda}x)$$

$$= 1 + V_\lambda[\int_0^x \lambda(t) \, dt] \, (\lambda x)^{-1}$$

$$= 1 + \frac{2\sigma_\lambda^2}{\bar{\lambda}x} \int_0^x (x - u) \, \rho(u) \, du, \qquad \cdot \qquad \cdot \qquad (4.5)$$

the last term following from a well-known formula in the theory of stationary processes (Yule, 1945).

Consider the asymptotic form of (4.5) as x tends to infinity. For most of the correlograms considered in theoretical work, $\rho(u)$ tends to zero exponentially fast. Under the much weaker assumption that $\int_0^x u\rho(u)\,du$ is $o(x)$ and that $\int_0^\infty \rho(u)\,du$ is convergent to a non-zero value A_ρ,

$$\int_0^x (x-u)\,\rho(u)\,du \sim x \int_0^\infty \rho(u)\,du = xA_\rho. \qquad (4.6)$$

In this case

$$I(x) \sim 1 + \frac{2A_\rho\,\sigma_\lambda^2}{\bar{\lambda}}$$

$$= 1 + Q. \qquad (4.7)$$

say, where $Q = 2A_\rho\,\sigma_\lambda^2/\bar{\lambda}$ is an index of irregularity. In particular if $\rho(u) = \text{Max}\,(1 - |\,u\,|\,/S, 0)$

$$Q = S\sigma_\lambda^2/\bar{\lambda}. \qquad (4.8)$$

Note that while $I(x)$ has a limit when λ belongs to the class of stationary series corresponding to (4.6), there are important stationary series for which (4.6) does not hold. Thus there are series with a relatively large amount of very-long-term variation, for which $\rho(u) = O(u^{-\alpha})$, $0 < \alpha < 1$, and for these*

$$I(x) \sim kx^{1-\alpha}.$$

(vi) *Aggregated random series.*—Suppose that there are a series of "occurrence points" distributed completely randomly and that at any occurrence point there is a chance p_1 that there is one event, a chance p_2 of two events, and so on. This is relevant in certain queueing problems.

Consider the function $I(x)$ for this type of series. Let μ_n, C_n be the mean and coefficient of variation of the number of events per occurrence. Then, if r is the number of occurrences in an interval x,

$$E[n(x)\,|\,r] = r\mu_n, \qquad (4.9)$$

$$V[n(x)\,|\,r] = r\,\mu_n^2\,C_n^2 \qquad (4.10)$$

Since $E(r) = V(r) = \bar{\lambda}x$, say, it follows that

$$I(x) = \frac{V[n(x)]}{E[n(x)]} = \mu_n(1 + C_n^2). \qquad (4.11)$$

(vii) *Pooled output.*—This was referred to in §3.1. The series is formed by superimposing a number of independent series each of some simple form. Several practical examples are given in the papers referred to above.

(viii) *Simple series plus "noise".*—We may have a simple series, such as a completely regular series, with a completely random series superimposed on it. This is a special case of (vii). Thus the stops of a machine might be produced by two causes, one acting regularly, the other randomly.

It is convenient for the work in §5 to call series of events for which $I(x)$ tends to a limit as $x \to \infty$ *quasi-random*, and to denote the limit by $1 + Q$. Q is non-negative for types (v) and (vi) but can be negative for type (iii).

4.3. *Tests of Randomness*

Before discussing the more specialized tests it is worth mentioning the very neat and easily applied test due to Barnard (1953). This uses the result that if the series is random and given that n events have occurred in time T, then the instants of occurrence are a random sample of n from a rectangular distribution over $(0, T)$. Kolmogoroff's criterion is then applied to test the

---

* A full discussion of the limiting forms would be out of place.

agreement between the empirical and theoretical distribution functions. In terms of the graphical methods (ii) and (iii) this means measuring the maximum discrepancy between the random walk curve and the straight-line joining the ends of the curve. If large values are taken as significant the test is likely to be effective in detecting monotone trends.

In the remainder of this section we consider tests for detecting the types of non-randomness described in §4.2.

(i) *Trend.*—Consider first the trend $\lambda = \alpha e^{\beta t}$ and suppose that, in the notation of §2, n events occur at instants $t_1, \ldots, t_n$ in the interval $(0, T)$. The likelihood is

$$\alpha^n e^{\beta \Sigma t_i} \exp \left\{ - \int_0^T \alpha e^{\beta t} \, dt \right\},$$

where the last factor arises from the probabilities that no events occur in the open intervals $(0, t_1)$, $(t_1, t_2), \ldots, (t_n, T)$. Now n has a Poisson distribution with mean

$$\int_0^T \alpha e^{\beta t} \, dt$$

and hence the likelihood, conditional on n, is

$$n! \left(\frac{\beta}{e^{\beta T} - 1} \right)^n e^{\beta \Sigma t_i}. \qquad \qquad (4.12)$$

Thus $B = \Sigma t_i$ contains all the information about β. Equation (4.12) shows that on the null hypothesis $\beta = 0$ the t_i are, conditional on n, an ordered random sample of n from the rectangular population $(0, T)$. Hence

$$E(B) = \frac{nT}{2}, \quad V(B) = \frac{nT^2}{12} \qquad \qquad (4.13)$$

and B has the Irwin-Hall distribution which tends rapidly to normality as n increases. Thus we have a simple test of the significance of the trend.

If $\beta \neq 0$, B has the distribution of the sum of n values from the population

$$\frac{\beta}{e^{\beta T} - 1} e^{\beta t}, \quad 0 \leqslant t \leqslant T$$

and hence B is distributed with mean and variance given by

$$E(B) = n \frac{\beta T e^{\beta T} - e^{\beta T} + 1}{\beta (e^{\beta T} - 1)},$$

$$V(B) = n \frac{(e^{\beta T} - 1)^2 - \beta^2 T^2 e^{\beta T}}{\beta^2 (e^{\beta T} - 1)^2}.$$

If it is required to approximate to the distribution of B it would be natural to use the normal form for βT small, the χ^2 form for βT large and negative and the χ^2 form for $nT - B$ when βT is large and positive. When the normal approximation is used, the confidence distribution for β, assuming that the trend law is correct, is derived in principle by solving by iteration the equation

$$\hat{\beta}_\varepsilon = E(B) \Big|_{\beta = \hat{\beta}_\varepsilon} \pm g_\varepsilon \sqrt{V(B)} \Big|_{\beta = \hat{\beta}_\varepsilon},$$

where g_ε is the ε per cent. point of the unit normal distribution. The two roots of this equation for $\hat{\beta}_\varepsilon$ are the upper and lower points of the $100 - 2\varepsilon$ per cent. confidence interval for β. A simpler approximate regression method, which enables α, β to be estimated and the adequacy of the regression law to be tested, is given below.

If the trend is $\alpha e^{\beta_1 t + \beta_2 t^2}$, the quantities $\Sigma\, t_i$, $\Sigma\, t_i^2$ are jointly sufficient for β_1, β_2 and if the trend is $\alpha(t + t_0)^\beta$ the quantity $\Sigma \log (t_i + t_0)$ is sufficient. This last result can be proved directly or follows on noting that an exponential trend is converted into a power-law trend by a logarithmic stretching of the time scale.

Suppose now that the observations consist of the numbers of events in assigned time periods and let n_i be the number of events in the ith period of length T_i and let $\bar{t}_i$ denote the epoch at the centre of the ith period. Full estimation by maximum likelihood is too complicated; but if the variation of λ within the periods T_i is neglected, the appropriate statistic is the natural approximation to B, namely

$$B' = \Sigma\, n_i\, \bar{t}_i. \qquad . \qquad . \qquad . \qquad . \qquad . \qquad . \qquad (4.14)$$

Under the null hypothesis the n_i have a multinomial distribution, so that

$$E(B') = n \, \Sigma \, \bar{t}_i \, T_i / T$$

$$V(B') = n \, \Sigma \, \bar{t}_i^2 \, \frac{T_i(T - T_i)}{T^2}$$

$$- 2n \, \underset{i > j}{\Sigma} \, \bar{t}_i \bar{t}_j \, \frac{T_i T_j}{T^2}$$

$$\sim \frac{n}{T} \Sigma \, \bar{t}_i^2 \, T_i.$$

Note that B' (and B) is proportional to the linear regression coefficient of the n_i on time.

We now consider inefficient methods that are simpler, particularly in estimation problems. If the precise instants of occurrence are known, proceed as follows. Choose an integer k such that the probability rate of occurrence λ varies inappreciably over the intervals $x_{n,\,n+k}$ between k events. Calculate, in the notation of §2,

$$y_1 = x_{0,k} \quad \text{and} \quad \bar{t}_1 = \tfrac{1}{2}(t_0 + t_k)$$

$$y_2 = x_{k,2k} \quad \text{and} \quad \bar{t}_2 = \tfrac{1}{2}(t_k + t_{2k}),$$

and so on. For trends of the form $\lambda = \alpha(t + t_0)^\beta$ we consider the regression of $\log y_i$ on $\log (t_0 + \bar{t}_i)$ while for trends of the form $\lambda = \alpha e^{\beta t}$ we take the independent variable to be $\bar{t}_i$. The mean and residual variance of $\log y_i$ are given by (3.3) and (3.5).

Only straightforward linear regression calculations are involved. The adequacy of the form assumed for the regression curve would be tested by inserting a square term and testing it for significance. (Thus in the second case a regression $\lambda = \alpha e^{\beta t + \beta' t^2}$ would be taken.) A comparison of the observed residual mean square about the regression line with the theoretical value (3.5) tests whether the short term fluctuations can be accepted as completely random.

The asymptotic efficiency of these methods based on $\log y_i$ can be shown to be the same as the asymptotic efficiency of an estimate of the log of a normal population variance based on the mean of a number of log-variances, each with $2k$ degrees of freedom. This efficiency has been tabulated by Bartlett and Kendall (1946) from whose paper the following values are taken: $k = 1$, 61 per cent.; $k = 2$, 78 per cent.; $k = 3$, 84 per cent.; $k = 4$, 88 per cent.; $k = 5$, 90 per cent.

If the trend is of the form $\lambda = 1/(\alpha + \beta t)$, a simple and reasonably efficient test of the null hypothesis $\beta = 0$ can be obtained from the regression coefficient of y_i on $\bar{t}_i$. If, however, $\beta \neq 0$, the y_i are not of equal variance so that efficient estimation is difficult. Also this is not a convenient form for representing rapid increases in the rate of occurrence ($\beta < 0$).

The discussion of the approximate methods has been based so far on the intervals between events. Suppose now that only the numbers of events occurring in preassigned periods are known. For the ith such period let n_i be the number of events, T_i the length of the period, $\bar{t}_i$ the epoch at the centre of the period and λ_i the probability rate of occurrence, assumed effectively constant throughout the period and hence equal to its regression value at $\bar{t}_i$. To fit $\lambda = \alpha t^\beta$ we use the logarithmic transformation (3.13). To test the null hypothesis $\beta = 0$ we give the ith observation weight T_i; if the problem is primarily one of estimation, if λ_i varies appreciably, and if efficiency of estimation is important, smoothed values $\lambda_i{}^s$ can be obtained graphically and the ith observation given weight $T_i\lambda_i{}^s$. Alternatively the weighted transformation (3.16) can be used if none of the n_i is very small. If all the T_i are equal, a regression of the form $\alpha e^{\beta_1 t + \beta_2 t^2 + \cdots}$ can be fitted quickly, although not efficiently, by ordinary orthogonal polynomials. The above methods would be useful in any problem involving exponential or power-law regression of Poisson variates; such problems arise, for example, in some experimental studies of bacterial growth.

(ii) *Rate determined by a smooth external variable.*—The methods just described can be used.

(iii) *Independent intervals.*—Moran (1951) has examined the case when the alternative hypothesis is that the intervals are independently distributed in a Type III distribution with degrees of freedom different from two. The asymptotically optimum test is to apply Bartlett's homogeneity test to the intervals between successive events just as if these were estimates of variance each based on two degrees of freedom.

All the types of non-randomness described in §4.2 lead to a distribution of intervals that is not exactly exponential. However *any* test of randomness based on comparing the observed distribution of intervals with an exponential curve will be insensitive for types of departure from randomness other than (iii) and (vi). For example in case (v), in which λ is constant over stretches of length S, it can be shown that the asymptotic relative power of any test based solely on the distribution of intervals, tends to zero as $S\bar{\lambda}$ tends to infinity. In other words, as is qualitatively obvious, criteria depending on the overall distribution of intervals are useful only for detecting rapid short-term changes in λ.

(iv) *Autocorrelated intervals.*—A test for this should be based in some way on the serial correlation between successive intervals, but it is by no means clear that the ordinary product-moment correlation is appropriate, since the variates have a distribution departing widely from normality. Consider therefore the maximum-likelihood statistic. Let the intervals between successive events be $x_1, \ldots, x_{N+1}$. Suppose that the x's are derived from the model described in §4.2 (iv), so that the p.d.f. of x_{i+1} given x_i is

$$\lambda_0(1 + \lambda_1 x_i)\, e^{-\lambda_0(1 + \lambda_1 x_i) x_{i+1}}.$$

The log-likelihood of $x_1, \ldots, x_{N+1}$ involves a term (4.2) from the distribution of x_1; to avoid the complication that this introduces, we consider the log-likelihood of $x_2, \ldots, x_{N+1}$ conditional on x_1. This is

$$-\lambda_0 \sum_{i=1}^{N} x_{i+1} - \lambda_0\lambda_1 \sum_{i=1}^{N} x_i x_{i+1} + N \log \lambda_0 + \sum_{i=1}^{N} \log (1 + \lambda_1 x_i)$$

leading to the following maximum-likelihood equation for $\hat{\lambda}_1$:

$$\left(\sum_{i=1}^{N} x_{i+1} + \hat{\lambda}_1 \sum_{i=1}^{N} x_i x_{i+1} \right) \sum_{i=1}^{N} \left(\frac{x_i}{1 + \lambda_1 x_i} \right) = N \sum_{i=1}^{N} x_i x_{i+1}.$$

If we require to test the null hypothesis $\lambda_1 = 0$ from a large sample, we may suppose that $\hat{\lambda}_1 x_i \ll 1$ and then if

$$s_1 = \frac{1}{N} \sum_{i=1}^{N} x_{i+1} \simeq \frac{1}{N} \sum_{i=1}^{N} x_i, \quad s_2 = \frac{1}{N} \sum_{i=1}^{N} x_i^2, \quad s_{11} = \frac{1}{N} \sum_{i=1}^{N} x_i x_{i+1},$$

$$\hat{\lambda}_1 \simeq -\frac{s_{11} - s_1^2}{s_1(s_2 - s_{11})}. \qquad \qquad . \qquad . \qquad . \qquad . \qquad . \quad (4.15)$$

(In the conditions being considered $s_{11} \ll s_2$.) The numerator is proportional to the serial covariance and since, under the null hypothesis, s_1 and s_2 converge in probability to $1/\lambda_0$ and $2/\lambda_0^2$ respectively, the use of $\hat{\lambda}_1$ in a test of significance is asymptotically equivalent to the use of the serial product-moment correlation coefficient or to the serial covariance divided by the square of the mean.

From the usual asymptotic formulae of maximum-likelihood theory

$$V(\hat{\lambda}_1 \mid \lambda_1 = 0) = \frac{\lambda_0^2}{N}. \qquad . \qquad . \qquad . \qquad . \qquad (4.16)$$

(v) *Stochastic variations in* λ.—This is the most interesting case and there are several possibilities to be considered.

(a) Suppose that λ is constant within known periods of length S, varies randomly from period to period, and that $n_i = n_i(S)$ is the number of events in the ith period, $i = 1, \ldots, r, rS = T$. Then

$$\hat{I}_f(S) = \frac{\text{Corr. S.S. of } n_i}{(r-1) \times \text{mean of } n_i} \qquad . \qquad . \qquad . \qquad (4.17)$$

has expectation $1 + Q$, where Q is defined by (4.8). The ordinary χ^2_{r-1} criterion for testing excessive dispersion is $(r-1)\,\hat{I}_f(S)$.

(b) Suppose now that the same model holds but that S is unknown. Many of the following remarks apply also when λ is a general stationary time-series (case (c)). If the variation in λ from period to period is large, we can attempt to estimate the points at which λ changes. But if Q is small and it is required to estimate or test Q, a natural procedure is to estimate the curve of (4.5), i.e. to apply a test corresponding to (4.17) for a number of different time periods.

The theoretical curve for this type of variation of λ is

$$I(x) = 1 + \frac{Q}{S}\left(x - \frac{x^2}{3S}\right), \qquad x < S,$$

$$1 + Q\left(1 - \frac{S}{3x}\right), \qquad x \geqslant S. \qquad . \qquad . \qquad . \qquad (4.18)$$

Note that from (4.18)

$$I(S) = 1 + \tfrac{2}{3}Q, \qquad . \qquad . \qquad . \qquad . \qquad . \qquad (4.19)$$

while in (4.17) $E[\hat{I}_f(S)] = 1 + Q$. The difference arises because in (4.18) the intervals are randomly situated relative to the periods within which λ is constant, while in (4.17) the intervals coincide with these periods.

The estimation of the curve $I(x)$ has been discussed by Cox and Smith (1953). The general idea is to divide the series into successive intervals of length x_0 and then to count the numbers of events $n_i(x_0)$ in these intervals. The variance of the $n_i(x_0)$ leads to an estimate of $I(x_0)$; estimates of $I(mx_0)$ for integer m are found from the mth order moving averages of $n_i(x_0)$, i.e. from the numbers of events in $(0, mx_0), (x_0, \overline{m+1}\,x_0), \ldots$. A formula is given in the previous paper for the variance, under the null hypothesis, of an estimate derived in this way. From the formula the following large-sample results can be obtained:

If

$$m = 1, \quad V[\hat{I}(x)] \simeq \frac{2x}{T} + \frac{1}{\bar{\lambda}T}. \qquad . \qquad . \qquad . \qquad (4.20)$$

Here T is the total length of series for analysis; $1/\bar{\lambda}$ is the mean interval between events. The estimate is obtained from a series of non-overlapping intervals.

If

$$m = 2, \quad V[\hat{I}(x)] \simeq \frac{3x}{2T} + \frac{1}{\bar{\lambda}T}. \qquad . \qquad . \qquad . \qquad (4.21)$$

Here the estimate is obtained from the numbers of events in a series of intervals of length x, each interval overlapping the preceeding one by $\frac{1}{2}x$.

If

$$m = \infty, \quad V[\hat{I}(x)] \simeq \frac{4x}{3T} + \frac{1}{\bar{\lambda}T}. \qquad . \qquad . \qquad . \qquad (4.22)$$

This is the smallest variance obtainable by the present method; all possible intervals of length x are used.

It follows from these results that there is little to be gained by taking $m > 2$, and so the following modified computing scheme is recommended (Table 4.1).   With it the $I(x)$ curve can be estimated for a series of one to two hundred events in just over one hour.

(1) Denote the smallest x for which $I(x)$ is required by x_0. Write down in a column the numbers $n_1, n_2, \ldots$ of events in $(0, x_0)$, $(x_0, 2x_0)$, $\ldots$

(2) Form the sums of adjacent pairs of n_i to give the numbers of events in periods $2x_0$.

(3) Add the 1st, 3rd, 5th, $\ldots$ members of the last column in pairs to give the numbers of events in periods $4x_0$.

(4) Repeat this process for the subsequent columns.

(5) Work out the sum, $\ldots$, corrected sum of squares of each column. If M is the number of entries in a particular column find the variance by dividing by $M - 1$ for the first column and by $M - 2$ for the other columns.   Hence obtain $\hat{I}(x)$ by dividing by the mean for the column.

(6) The significance of the departure from unity is tested by taking $v\hat{I}(x)$ to be χ^2_v, where v is $M - 1$ for the first column and $\frac{2}{3}M/[1 + (\frac{3}{2} \times$ mean no.$)^{-1}]$ for the others.

The estimate of variance is unbiased under the null hypothesis but its properties under the alternative hypothesis have not been investigated.   Individual values of $\hat{I}(x)$ are tested for significant departures from unity as stated above: no overall test of the whole curve is available.

TABLE 4.1

Scheme for Computing $\hat{I}(x)$

Time Interval	x_0	$2x_0$	$4x_0$	$8x_0$ ...
	n_1	$n_1 + n_2$	$n_1 + \ldots + n_4$	$n_1 + \ldots + n_8$
	n_2	$n_2 + n_3$		
	n_3	$n_3 + n_4$	$n_3 + \ldots + n_6$	
	n_4	$n_4 + n_5$		
	n_5	$n_5 + n_6$	$n_5 + \ldots + n_8$	$n_5 + \ldots + n_{12}$
	n_6	$n_6 + n_7$		
	.	.	.	.
	.	.	.	.
	.	.	.	.

Under each column leave eleven rows for recording number of terms, sum, mean, uncorrected sum of squares, correction, corrected sum of squares, divisor, variance, $\hat{I}(x)$, degrees of freedom and significance attained.

The time period S can be estimated approximately by first estimating the limiting value $1 + \hat{Q}$ of the $\hat{I}(x)$ curve, and then finding the value of x, say $\hat{S}$, at which $\hat{I}(S) = 1 + \frac{2}{3}Q$. If the curve appears not to tend to a limit this is evidence of non-stationarity in the variation of λ, or at any rate of departure from the special type of stationary form considered in §4.2(v).

The analysis of series of events by the $I(x)$ curve is closely related to the use of analysis of variance in the investigation of ordinary time-series.   This was suggested by Tippett (1935b) and by Yule (1945).   The use (§5) of the limiting value $1 + Q$ to modify tests of significance based on the hypothesis of randomness is analogous to the important delta-infinity method of Jowett (1953) for ordinary time-series.

(c) If λ is a stationary time-series we apply the method just described.   This should give a reasonably sensitive test of randomness and a good estimate of Q.   The estimate $\hat{S}$ defined in

(b) will give a rough measure of the time-scale of any non-randomness detected, but a sensitive test of the form of the correlation is not practicable by this method. This is because the difference between $I(x)$ and unity is, from (4.5), proportional to

$$\int_0^x (x - u) \rho(u) \, du = \int_0^x dy \int_0^y \rho(v) \, dv \qquad . \qquad . \qquad . \qquad . \qquad (4.23)$$

Any periodicity, or similar property, of $\rho(u)$ is smoothed out in the double integration. If the primary purpose were to estimate the correlogram $\rho(u)$, or more specifically to test for periodicity, some method would have to be used depending directly on the correlation between numbers of events in intervals a certain distance apart. This will not be considered further: in many applications the estimation of Q will be the important thing, since it is this that enables us to modify tests based on the Poisson distribution to apply in more general cases (§5).

(d) Suppose finally that λ is constant within periods S, where S is unknown, and that it is decided to test for randomness by dividing the series into intervals of length x counting the number of events in each interval and applying the ordinary dispersion test, i.e. by estimating $I(x)$ for just one value of x. What value of x should we take?

If the series is of total length T there will be T/x periods and we shall test in the χ^2 distribution with this number of degrees of freedom less unity. The upper α per cent. point of this distribution is, asymptotically,

$$\frac{T}{x} + g_a \sqrt{\left(\frac{2T}{x}\right)} \qquad . \qquad . \qquad . \qquad . \qquad . \qquad (4.24)$$

if $x \ll T$, where g_a is the upper α per cent. point of the unit normal distribution. Hence, since the distribution of $\hat{I}(x)$ is asymptotically symmetrical, there is an approximately 50 per cent. chance of attaining significance if

$$\frac{T}{x} I(x) = \frac{T}{x} + g_a \sqrt{\left(\frac{2T}{x}\right)}, \qquad . \qquad . \qquad . \qquad . \qquad (4.25)$$

where $I(x)$ is defined by (4.18).

That is, with fixed x, S, T, α, there is 50 per cent. power if

$$Q = g_a \sqrt{\left(\frac{x}{T}\right)} \cdot \left(\frac{x}{S} - \frac{x^2}{3S^2}\right)^{-1}, \qquad x < S,$$

$$g_a \sqrt{\left(\frac{x}{T}\right)} \cdot \left(1 - \frac{S}{3x}\right)^{-1}, \qquad x \geqslant S. \qquad . \qquad . \qquad (4.26)$$

Maximum sensitivity is achieved when the Q to attain a given power is a minimum, i.e. we want to choose x to minimize (4.26). This gives $x = S$; if $x = 2S$ or $\frac{1}{2}S$ the value of Q is increased by only about 13 per cent. over its minimum value. Thus the loss of power is small even for quite large differences between x and S. This is all large-sample theory.

The broad conclusion is that if a single test is required, the series should be divided into sections of length the most plausible value of S. If, however, it is required to estimate Q the sections should be of length at least $2S$.

(vi) *Aggregated random series.*—There are several problems here depending on the form in which the observations become available. If the instants of occurrence are known one would count the number of times 2, 3, ... events occur within a time stretch Δ, where Δ is a suitable small time chosen to cover any dispersion of the events associated with one instant of occurrence. The parameters μ_n, σ_n^2 would then be estimated, or tested, from this distribution correcting for random coincidences. Details will not be given; a similar, but simpler, problem is treated in §6.2.

If only the numbers $n_i(\tau)$ of events in successive time intervals τ are known, $\tau \gg \Delta$, $\hat{I}(x)$ should, except for sampling fluctuations, have the constant value (4.11). Clearly it is not possible by this method to distinguish between the present type of non-randomness and the type in which λ is constant within each period τ and varies randomly from period to period.

(vii) *Pooled output.*—The statistical analysis of such series by a curve equivalent to $\hat{I}(x)$ has been considered in some detail in the two papers referred to in §4.2, and will not be discussed further here.

(viii) *Simple series plus "noise".*—This too will not be considered in detail. The $\hat{I}(x)$ curve would test randomness but is not likely to distinguish effectively between this and some other forms of pooled output. Mr. T. Gold, in unpublished work, has suggested that series of events should be examined for certain departures from randomness by finding the frequency distribution of the intervals between all pairs of events (not just successive events). He has constructed a very ingenious machine for obtaining this distribution. This method of analysis is clearly particularly appropriate when the observed series consists of a single, or small number, of regular series possibly obscured by a random series.

5. Testing and Measurement of Differences

5.1. *Comparison of Two Rates of Occurrence*

We now assume that we are dealing either with completely random series or with the quasi-random form considered in §4.2. Consider the comparison of the true rates of occurrence λ_i corresponding to two samples in which n_i events have occurred in time T_i, $i = 1, 2$. If the series are random a number of standard methods are available.

It would often be required not only to test the hypothesis that $\lambda_1 = \lambda_2$ but also to examine the magnitude of any difference that may exist; often the most comprehensible way of doing this is to obtain a confidence distribution for λ_1/λ_2. A simple approximate method for doing this can be based on the χ^2 approximation discussed in §3.2. With probability $(1 - 2\alpha)$.

$$\frac{t_2(n_1 + \tfrac{1}{2})}{t_1(n_2 + \tfrac{1}{2})} F_- < \frac{\lambda_1}{\lambda_2} < \frac{t_2(n_1 + \tfrac{1}{2})}{t_1(n_2 + \tfrac{1}{2})} F_+, \qquad . \qquad . \qquad . \qquad . \qquad (5.1)$$

where F_-, F_+ are the lower and upper α per cent. points of F with $(2n_1 + 1, 2n_2 + 1)$ degrees of freedom. The corresponding significance test is obtained by referring

$$F = \frac{t_1(n_2 + \tfrac{1}{2})}{t_2(n_1 + \tfrac{1}{2})} \qquad . \qquad . \qquad . \qquad . \qquad . \qquad (5.2)$$

to the same F distribution. These methods are closely connected with the exact results of Przyborowski and Wilenski (1940). If the times are measured up to the occurrence of the n_ith event, where n_i is preassigned, the $\tfrac{1}{2}$'s are to be omitted, the degrees of freedom are $(2n_1, 2n_2)$ and the test is exact.

If the series are quasi-random we must use large-sample approximations unless we are to make strong assumptions about the form of the series and to solve by maximum likelihood. In the large-sample approximation $\log_{10}(n_i/T_i)$ has mean $\log_{10}\lambda_i$ and variance

$$\omega_i = (1 + Q_i)(\log_{10} e)^2/n_i \qquad . \qquad . \qquad . \qquad . \qquad (5.3)$$

If the two series are independent $\log_{10}(n_1 T_2/n_2 T_1)$ therefore has mean $\log_{10}(\lambda_1/\lambda_2)$ and variance $\omega_1 + \omega_2$.

The following numerical example illustrates the methods.

Example.—Suppose that on loom 1 there are 30 stops in 75 running hours and on loom 2, 21 stops in 26 running hours. If the occurrence is random for both looms, the difference between looms is tested by $F = (72 \times 21\cdot5)/(26 \times 30\cdot5) = 2\cdot03$ with $(61, 43)$ degrees of freedom. The 1 and $\tfrac{1}{2}$ per cent. points are (Hald, 1952) $1\cdot98$, $2\cdot18$ so that the difference is significant at 2 per cent. For the 95 per cent. confidence interval for λ_1/λ_2 we need the lower and upper $2\tfrac{1}{2}$ per cent. points and these are $1/1\cdot73$ and $1\cdot77$ so that the confidence interval is $0\cdot279 < \lambda_1/\lambda_2 < 0\cdot872$.

Consider the example again using the log transformation. Assuming randomness, we have that $\log_{10}(30/75) = -0\cdot398$ and $\log_{10}(21/26) = -0\cdot0928$ are values from distributions of means

$\log_{10} \lambda_1$, $\log_{10} \lambda_2$ and standard deviations $0 \cdot 434\sqrt{(1/30)}$ and $0 \cdot 434\sqrt{(1/21)}$. Therefore $- 0 \cdot 398 - (- 0 \cdot 928) = - 0 \cdot 305$ is a value from a distribution of mean $\log_{10}(\lambda_1/\lambda_2)$ and standard deviation $0 \cdot 434\sqrt{(1/30 + 1/21)} = 0 \cdot 124$. We test the hypothesis $\lambda_1 = \lambda_2$ by referring $0 \cdot 305/0 \cdot 124 = 2 \cdot 47$ to the tables of the unit normal distribution. The 2 per cent. point is $2 \cdot 33$ and the 1 per cent. point is $2 \cdot 58$ so that we reach the same conclusion as before—significant at 2 per cent. The 95 per cent. confidence interval is on a log-scale $- 0 \cdot 305 \pm 1 \cdot 96 \times 0 \cdot 124 = (- 0 \cdot 548, - 0 \cdot 062)$, or in terms of λ_1/λ_2, $(0 \cdot 283, 0 \cdot 867)$.

Suppose now that it is known from previous experience of these stops that they occur not completely randomly, but in quasi-random series with $1 + Q \simeq 2$. The standard error in the above calculation is now multiplied by $\sqrt{2}$ and becomes $0 \cdot 175$. The normal deviate for testing the hypothesis $\lambda_1 = \lambda_2$ is now $0 \cdot 305/0 \cdot 175 = 1 \cdot 75$, corresponding to a probability of 8 per cent. The difference between significance at 2 per cent. and significance at 8 per cent. is probably big enough to be sometimes of practical importance, so that the example emphasizes the point that we should allow for non-randomness if significance tests of any accuracy are required.

The comparison of the rates of occurrence of several series where individual differences among the series are of interest raises no particularly new problems.

5.2. *Components of Variance: General Remarks*

Suppose that we have observations on k random or quasi-random series in the ith of which n_i events occur in time T_i, the corresponding population rates of occurrence being λ_i. Suppose also that we are not interested in individual comparisons among the λ_i but require to measure the variation of the λ_i. We do this by estimating a component of variance. There are four situations in which this may be a reasonable thing to do:

(i) in complicated sampling problems, components of variance determine the best scheme of sampling for estimating the overall mean rate of occurrence;

(ii) in some problems, a treatment or other systematic difference may be expected to affect not so much the average rate of occurrence of events as certain components of variance. Thus it may be of interest to determine whether on one type of loom the between-loom, or between-weaver, variations in stoppage rate are less than for another type of loom;

(iii) we may have a large amount of data on rates of occurrence classified in various ways. Thus in the weaving example we may have stoppage rates of several types for each of many looms, the looms being classified by types, by sort of cloth, by weavers, by mills, etc. And the object may be just to describe these data in an easily compared way. Components of variance seem the natural tool;

(iv) we may be directly interested in controlling the variation in λ and wish to determine the effect of eliminating certain sources of variation. This could be the case in investigations of congestion, where excessive variation in the rate of arrival of customers can appreciably increase the mean queueing-time (Cox, 1955).

All these situations are rather special and in many problems a detailed study of individual differences is likely to be more informative than the measurement by a single figure of the amount of variation of the λ_i.

5.3. *Estimation of a Single Component of Variance*

Consider the situation described at the beginning of §5.2 and suppose that the λ_i are randomly drawn from a population of mean $\bar{\lambda}$ and variance σ_λ^2. If the T_i are equal and the λ_i have a Pearson Type III distribution, the n_i follow a negative binomial distribution and the associated problems of estimation and testing have been considered extensively.

Suppose that the T_i vary appreciably. We can construct simple estimates of $\bar{\lambda}$ and σ_λ^2 by considering linear and quadratic forms in either the n_i or the logarithmic transforms, z_i. It is probably usually immaterial which method is used and as there is no point in applying a transformation without a positive gain, it is best to use the n_i, unless either

(a) the data form part of a larger set in which proportional variations in λ are expected, or

(b) the distribution of λ_i is expected to be such that the variance of log λ_i is more useful than the variance, or coefficient of variation, of λ_i itself. For example this would be the case if λ_i were log-normally distributed.

It is clearly impossible to make a general theoretical statement about whether or not the log-transformation should be applied, since the best thing to do depends in an essential way on the underlying distribution of the λ_i and on the magnitude of σ_λ^2.

We shall concentrate on methods based on the n_i. The difficulties are the usual ones in the analysis of observations of unequal weight (see, for example, Cochran, 1954). If σ_λ^2 were known, $\bar{\lambda}$ would be estimated by

$$ l = \Sigma \left(\sigma_\lambda^2 + \frac{1}{T_i} \right)^{-1} \frac{n_i}{T_i} \bigg/ \Sigma \left(\sigma_\lambda^2 + \frac{1}{T_i} \right)^{-1}, \qquad (5.4) $$

and the corresponding quadratic form on which to base an estimate of σ_λ^2 is

$$ S = \Sigma \left(\sigma_\lambda^2 + \frac{1}{T_i} \right)^{-1} \left(\frac{n_i}{T_i} - l \right)^2 \qquad (5.5) $$

This suggests a natural iterative procedure, but we shall confine ourselves to simple estimates. There are two important special cases of (5.4), (5.5).

First if $\sigma_\lambda^2 \ll 1/T_i$, we take

$$ l_1 = \Sigma n_i / \Sigma T_i \qquad (5.6) $$

and

$$ S_1 = \Sigma T_i \left(\frac{n_i}{T_i} - l_1 \right)^2 = \Sigma \frac{n_i^2}{T_i} - \frac{(\Sigma n_i)^2}{\Sigma T_i} \qquad (5.7) $$

The corresponding estimate of σ_λ^2 is

$$ s^2_{\lambda(1)} = (S_1 - (k-1) l_1) \bigg/ \left(\Sigma T_i - \frac{\Sigma T_i^2}{\Sigma T_i} \right) \qquad (5.8) $$

and

$$ V(l_1) = \frac{\bar{\lambda}}{\Sigma T_i} + \sigma_\lambda^2 \frac{\Sigma T_i^2}{(\Sigma T_i)^2}, \qquad (5.9) $$

so that the estimated variance of l_1 is

$$ V_e(l_1) = \frac{l_1}{\Sigma T_i} + s^2_{\lambda(1)} \frac{\Sigma T_i^2}{(\Sigma T_i)^2} \qquad (5.10) $$

Secondly if the superimposed variation is predominant, so that $\sigma_\lambda^2 \gg 1/T_i$, we take

$$ l_2 = \frac{1}{k} \Sigma \left(\frac{n_i}{T_i} \right) \qquad (5.11) $$

and

$$ S_2 = \Sigma \left(\frac{n_i}{T_i} - l_2 \right)^2 = \Sigma \frac{n_i^2}{T_i^2} - \frac{[\Sigma (n_i/T_i)]^2}{k}. \qquad (5.12) $$

The corresponding estimate of σ_λ^2 is

$$ s^2_{\lambda(2)} = \frac{S_2}{k-1} - \frac{l_2}{k} \Sigma \frac{1}{T_i} \qquad (5.13) $$

and

$$ V(l_2) = \frac{\bar{\lambda}}{k^2} \Sigma \frac{1}{T_i} + \frac{\sigma_\lambda^2}{k} \qquad (5.14) $$

so that the estimated variance of l_2 is

$$V_e(l_2) = \frac{S_2}{k(k-1)}. \qquad . \qquad . \qquad . \qquad . \qquad (5.15)$$

The estimate of σ_λ^2 would often be expressed as a coefficient of variation.
To decide which method to use, the following considerations are relevant.

(i) If the testing of the hypothesis $\sigma_\lambda^2 = 0$ is the most important thing, method one should be used. S_1/l_1 is the usual χ^2_{k-1} criterion for dispersion. If a large value of $s^2_{\lambda(1)}$ is obtained and it is important to estimate $\bar{\lambda}$ accurately, a revised estimate

$$\Sigma\left(s^2_{\lambda(1)} + \frac{1}{T_i}\right)^{-1}\frac{n_i}{T_i}\bigg/\Sigma\left(s^2_{\lambda(1)} + \frac{1}{T_i}\right)^{-1}. \qquad . \qquad . \qquad (5.16)$$

could be found fairly quickly.

(ii) Suppose that the estimation of $\bar{\lambda}$ is of prime importance, that the T_i have a coefficient of variation of not more than 20–30 per cent. with no very extreme values, that we cannot rely on σ_λ^2 being small and that it is desirable to fix on a method for routine use.* Then it is almost immaterial which method is used and a decision should be reached from whatever prior evidence is available about σ_λ^2. For if $\sigma_\lambda^2 \gg 1/T_i$,

$$\frac{V(l_1)}{V(l_2)} = \frac{k\,\Sigma\,T_i^2}{(\Sigma\,T_i)^2} = 1 + C^2_T, \qquad . \qquad . \qquad (5.17)$$

where C_T is the coefficient of variation of T_i. On the other hand if $\sigma_\lambda^2 = 0$

$$\frac{V(l_2)}{V(l_1)} = \frac{(\Sigma\,T_i^{-1})(\Sigma\,T_i)}{k^2} \simeq 1 + C^2_T, \qquad . \qquad . \qquad (5.18)$$

since, if $\bar{T}$ is the mean of the T_i,

$$\frac{1}{T_i} \simeq \frac{1}{\bar{T}}\left\{1 - \frac{T_i - \bar{T}}{\bar{T}} + \frac{(T_i - \bar{T})^2}{\bar{T}^2}\right\}, \qquad . \qquad . \qquad (5.19)$$

so that

$$\frac{1}{k}\Sigma\,T_i^{-1} \simeq \frac{1}{\bar{T}}(1 + C^2_T). \qquad . \qquad . \qquad (5.20)$$

Hence the loss of information due to using l_1 when l_2 should be used, and due to using l_2 when l_1 should be used, are the same and are small if C_T is small.

(iii) If it is required to decide quickly which method to use from a rapid inspection of the results, the following procedure can be used.

(a) Measure the T_i in units such that they are of order one.

(b) Find $u_i = n_i/T_i$ and by the use of the range, estimate the standard deviation s_u of u_i. (In doing this take no account of observations with very small T_i.)

(c) If $s^2_u > 3\bar{u}/\bar{T}$ use method two, if $s^2_u < \bar{u}/\bar{T}$ use method one. In other cases a more detailed examination is necessary if much rests on the choice of method.

(d) In estimating $\bar{\lambda}$ inspect the data to see that with method one there is no discrepant observation with large T_i, and with method two no discrepant observation with small T_i.

If the individual series are quasi-random instead of random, factors $(1 + Q)$ are to be inserted in an obvious way.

Many of the above remarks are, with obvious modifications, relevant to the problem of estimating a component of variance from normally distributed variates arranged in a one-way classification with unequal numbers.

---

* These conditions hold in the application to nep counting, where T_i is the weight of the ith section. The variation in the T_i arises because it is not practicable to select sections of exactly constant weight.

Finally it should be noted that there are corresponding formulae for use with the logarithmic transforms of §3.3. Also there are a number of problems connected with the estimation of several components in more complicated situations, but these will not be discussed here.

6. Correlation Between Different Types of Event

6.1. *General*

Up to now we have considered events of only one type. A number of problems arise in examining the relation between the occurrences of events of different types A, B and three such problems will be considered briefly. These arise when there is

(i) a correlation between the precise instants of occurrence of A and B;

(ii) a correlation between the superimposed variations of the λ values of A and B;

(iii) an effect on the rate of occurrence of A persisting for some time after the occurrence of each B.

We could construct a general set-up to cover all three cases, and many others, by allowing the probability that an A event occurs in $(t, t + \delta t)$ to be a function of the instants of occurrence of all preceding A and B events and possibly also of external parameters. In a brief treatment it is, however, convenient to separate (i)–(iii).

6.2. *Correlation between Instants of Occurrence*

As an example suppose that events A are faults in a piece and events B stops of the loom. Some, but not all, stops cause faults and some faults are not caused by stops. Suppose first that the points of occurrence of events are known and that n_A and n_B events of the two types have occurred in a total time T.

Suppose that we can fix a small time-interval Δ such that if an A and a B event are separated by less than Δ we can regard this as suggesting a "true" coincidence. We can obtain a simple significance test for the existence of coincidences as follows. Mark off on the time axis, intervals of length 2Δ centred at the events B. This will determine a set $S_B(\Delta)$ of points of total length $T_B(\Delta)$ equal to $2n_B(\Delta)$ if no two B events occur within Δ of each other, and less than $2n_B\Delta$ otherwise. Given the total number n_A of A events and given that the A events are distributed randomly, the number n_{AB} of A events falling in $S_B(\Delta)$ follows a binomial distribution with parameter $T_B(\Delta)/T$ and index n_A. In the special case when $n_B\Delta \ll T$ and when a negligible number of pairs of B events fall within Δ, this means that we test n_{AB} in the Poisson distribution with mean $2n_An_B\Delta/T$. Notice that the assumptions underlying this test are not symmetrical with respect to A and B; no assumption is made about the distribution of the events B and these play the role of an independent variable in a regression problem.

For the remainder of the analysis we consider the approximate case in which $n_B\Delta \ll T$ and in which the occurrence of pairs of B events less than Δ apart can be neglected. Assume that there is a probability p of a B event producing an A event coincident with it, i.e. occurring within Δ of the B event, and that the remaining A events are distributed completely randomly, and suppose that it is required to estimate p. The number n_{AB} of coincidences is, conditional on n_A, n_B, of the form $x + y$, where x is a binomial variate with parameter p and index n_B, and y is, conditional on x, a Poisson variate with expectation $2(n_B - x)(n_A - x) \Delta/T$.

Hence

$$E(n_{AB}) = pn_B + \frac{2\Delta n_B(1 - p)(n_A - pn_B + p)}{T}, \qquad . \qquad . \qquad . \quad (6.1)$$

and, since $n_B\Delta/T$ is small, p can be estimated by

$$\hat{p} = \frac{n_{AB}}{n_B} - \frac{2\Delta(n_B - n_{AB})(n_An_B - n_{AB}n_B + n_{AB})}{Tn_B}. \qquad . \qquad . \quad (6.2)$$

If further $n_B \gg 1$, this takes the form

$$\hat{p} \simeq \frac{n_{AB}}{n_B} - \frac{2\Delta(n_B - n_{AB})(n_A - n_{AB})}{T}. \qquad . \qquad . \qquad . \quad (6.3)$$

We have, from (6.3), neglecting terms in Δ^2,

$$V(\hat{p}) \simeq \frac{p(1-p)}{n_B} + \frac{4\Delta}{T} p(1-p)(n_A + n_B - 2pn_B). \qquad (6.4)$$

The number of coincidences would tend to be high if there were a common trend in the rates of occurrence, even if there were no detailed correspondence in the instants of occurrence. If such trends are suspected, the series should be broken into sections within each of which the trend is negligible, and the expectation of n_{AB} found separately for each section.

A further refinement would be to attach to each pair of A, B events a score depending on their distances apart. The best system of scoring to adopt would be determined by the distribution of the recording errors in the positions of individual points. For simplicity one would take zero score for two events separated by more than say Δ_0.

The above methods require a knowledge of the instants of occurrence of the events. If only the numbers of events occurring in different intervals are available, indirect methods for estimating p are in principle possible. For the ith period denote the numbers of events and the length of the period by $(n_A{}^i, n_B{}^i, T_i)$. Then if there is a constant probability p that a B event will produce an A, and if the number of A's that are not caused by B's has mean aT_i and variance bT_i and is uncorrelated with n_B, and with the number of A's caused by B's, then

$$E(n_A{}^i \mid n_B{}^i, T_i) = pn_B{}^i + aT_i, \qquad (6.5)$$

$$V(n_A{}^i \mid n_B{}^i, T_i) = p(1-p) n_B{}^i + b T_i. \qquad (6.6)$$

The last expression will often be approximately proportional to T_i. Then $n_A{}^i/\sqrt{T_i}$ has approximately constant variance and

$$E(n_A{}^i/\sqrt{T_i} \mid n_B{}^i, T_i) = pn_B{}^i/\sqrt{T_i} + a\sqrt{T_i}, \qquad (6.7)$$

so that the regression, through the origin of $n_A{}^i/\sqrt{T_i}$ on $n_B{}^i/\sqrt{T_i}$ and $\sqrt{T_i}$ leads to the required estimate.

Although the assumptions on which (6.7) is based seem mild from a mathematical point of view, they could easily be false in a practical application. Indeed in a number of cases in which I have applied the method to problems in which p could be estimated directly, the regression method has completely failed, due primarily to variation from period to period in the quantity a.

6.3 *Long-term Correlation*

Suppose that the series is divided into sections, in the ith of which the two types of event occur randomly and independently at rates $\lambda_A{}^i$, $\lambda_B{}^i$, where $\lambda_A{}^i$, $\lambda_B{}^i$ are random variables with correlation coefficient ρ_{AB}. For example A and B may be stops of two different types occurring in weaving; on any one loom the stops may occur independently but the variation between looms in the population stoppage rates may be correlated.

This is a typical problem in components of covariance to be tackled by methods similar to those discussed in §5.3 for components of variance. There would be two types of estimate, an unweighted one appropriate for the estimation of components known to be appreciable, and a weighted one appropriate for testing components for significance. Also each method could be used with or without a logarithmic transformation of the numbers of events.

The main criterion for deciding whether or not to use the logarithmic transformation is whether the correlation between λ_A and λ_B is most naturally expressed on a linear or a log scale. This is to be decided partly by inspecting the data and partly by prior considerations; for example if the individual λ_A's and λ_B's vary over an appreciable range it would sometimes be natural to expect some rough dependence of the form $\lambda_A \propto \lambda_B{}^\alpha$ and in such a case the log transformation is natural.

It may sometimes happen that such a component of covariance, or the corresponding regression or correlation coefficient, can be given a fundamental interpretation and in this case the use of efficient methods of estimation would be important. But, at any rate in the applications to machine-stops, the main use is descriptive, in the reduction to simple comparable form of large amounts of data based on very different periods of observation. Efficiency of estimation is then not important.

6.4 *Persistent Effects*

Suppose that B is a relatively rarely occurring event in the nature of a "treatment" applied with the object of modifying the rate of occurrence of A. An example in weaving is when A is a particular type of stop called banging-off. When banging-off occurs frequently, the weaver sometimes calls for the overlooker to adjust the loom. Such adjustments constitute the events B.

There are a number of statistical problems that can arise in this type of situation. If the system is expected to behave in a very reproducible way, a reasonable hypothesis to consider is that the distribution of A's following a particular B is described by a steady trend in the probability rate of occurrence, e.g. by $\lambda = \alpha e^{\beta t}$, the rate λ returning to the value α after each event B. Estimates of α, β can then be obtained from each section between two successive B's, either graphically or by the regression methods of §4.3 (i). The last few observations of each set should be discarded if there is a strong correlation between the occurrence of B and the immediately preceding rate of occurrence of A. If the different estimates of α (and of β) differ only by sampling errors, the data have been summarized in a neat and concise way. This method was applied to the bangings-off on several looms, but the estimates of α and of β were found to vary appreciably, even for one loom.

Suppose now that a more restricted problem is considered, namely that of estimating and testing the local change in the rate of occurrence of A produced by the events B. Measure the local rate of occurrence of A immediately following a B event by y_a, defined as the logarithm of the interval from B up to the kth following A event, where k is a suitable small integer. If λ_a is the corresponding probability rate of occurrence, we have the mean and variance of from (3.3) and (3.5), replacing λ in these formulae by λ_a. A quantity y_b can be defined similarly as the logarithm of the interval from B to the kth proceeding A event, but its expectation and variance would be given by (3.3) and (3.5), with $\lambda = \lambda_b$, only if the events B were independent of local fluctuations in the occurrences of A. This would not in general be so. Thus suppose that in the weaving example stops occur completely randomly independently of B. Then the overlooker will on the whole attend to the loom when there has just been a run of stops close together and a naïve comparison of y_a with y_b would indicate, incorrectly, that the overlooker had caused an improvement.

Essentially B is a treatment and to assess its effect we must have a "control", i.e. a series of instants at which B does not occur but which are otherwise similar to the instants at which B does occur. If the occurrence of B is entirely under the experimenter's control a number of interesting problems of experimental design arise. We shall, however, consider only the case where the events B are not under experimental control; this is certainly the situation with loom investigations under production conditions.

A natural procedure is to take a number of stretches of $2k + 1$ events containing no B and to calculate y_a, y_b as if the centre event were B. Plot y_a against y_b, distinguishing between the true B events and the dummies just described. The general qualitative interpretation of such a graph is as follows; detailed analysis, possibly using the above results about the mean and variance of y_a, may be necessary to decide the significance of any conclusions suggested by the graph.

(1) If the curves for B and for control are parallel, this suggests that B has a constant proportional effect on the rate of occurrence of A, of amount measured by the separation of the curves. If the curves are coincident there is no evidence that B affects the rate of occurrence of A.

(2) If the curves for B and for control are not parallel, the proportional effect of B is not constant. If the curve for B is horizontal and if the variation of y_a is in agreement with the theoretical value (3.5), the data are consistent with the probability rate of occurrence always returning to the same value after each event. In other cases the interpretation is more complicated. Thus attention by the overlooker may have a slight effect when λ_b is low and a large effect when λ_b is high; this would be shown by a convergence of the curves for large y_b.

The success of the method depends in part on there being no strict rule linking B with the immediately preceding occurrences of A. Thus if the overlooker attended to the loom whenever k bangings-off occurred within an interval T_0 there would be no control and B points with the same y_b, although we might get some information from whether there appeared to be a jump between the two curves at $y_b = \log T_0$. In general we must consider carefully whether or not points with the same y_b are genuinely comparable and some element of personal judgement will be involved. Such an analysis is not so convincing as that of a properly designed experiment, for in the latter we know from the randomization that comparable sections with and without an event B differ in no systematic way other than that due to B itself. In the analysis of the banging-off data we can obtain a series of "control" instants which appear to be comparable with the treated instants, but we cannot be sure that the overlooker did not decide to attend to the loom for some reason not discernible from the data, making the B instants systematically different from the "control". This sort of difficulty is of course inherent in any system not involving randomization; its importance will depend on the particular application.

A similar method of constructing an artificial "control" would apply if we wished to assess the effect of the event B on some quantitative variable. For example we may have daily routine test data over a long period and wish to find the local effect of certain action taken rarely and irregularly.

7. SAMPLING

7.1. *General*

There are many problems connected with sampling series of events that raise no particularly new points. For example if neps are distributed in the form considered in §4.2 (v), the variance of the number of neps in sections of weight w is, for large w, $w\,(A + B/w)$, and if it takes a time C to select a section for counting and a time $D + En$ to count n neps, a familiar calculation determines the optimum sampling weight.

One problem that does raise more special considerations is to decide under what circumstances it advisable to record the precise times of occurrence of events. This, whether done by continuous observation with a stop-watch or by automatic recording, is likely in many cases to be more expensive than counting numbers of events falling in assigned intervals. It seems from the discussion in the preceding sections that

(i) if the $I(t)$ tests of randomness of §4.3 is to be used, either the numbers of events in intervals of *constant* length should be recorded or the precise instants of occurrence;

(ii) if the tests of §4.3 (iii), (iv), (vi) are to be used or if the correlation problems of §§6.2, 6.4 are involved a knowledge of the instants of occurrence is advisable;

(iii) in most other cases it would be enough to know the numbers of events in convenient fairly small time intervals.

The remainder of this section on sampling is concerned with a very special, but important, problem arising frequently in operational research.

7.2. *The Determination of Proportional Times*

To be explicit suppose that the events are stops of a machine. With each stop is associated a time, the repair-time, so that clock-time is divided into alternating periods running and stopped. In many applications each repair-time is divided into a period awaiting attention and a servicing-time and there are several types of stop; these complications will be ignored to begin with.

There are a number of quantities about such a system that we may require to estimate, the main ones being

(i) the average rate of occurrence of stops;

(ii) the mean repair-time per stop;

(iii) the proportion of time for which the machine is running.

There are two main methods used, continuous time-study with a stop-watch and the snap-round method (Tippett, 1935a). The first gives all three quantities—and of course distributions as well as averages. The snap-round method consists simply in observing at a large number of instants whether the machine is running or stopped, leading to an estimate of (iii). If a direct count is then made to determine the stoppage-rate, the mean repair-time can be deduced.

A method similar to the snap-round method is used in other fields, for example by plant ecologists who take a systematic sample of points along a line in order to determine proportional areas of different types of vegetation.

Mr. P. D. Vincent of the Shirley Institute has recently investigated in detail the relative accuracies of the two approaches, considered as methods of estimating the mean repair-times of various operations in cotton weaving. Continuous observation is much the more accurate but its disadvantages as a general method seem to be

(a) that the repair-times obtained in the presence of an observer may not be representative;
(b) that continuous observation of repair-times may be unacceptable.

It is therefore at least of theoretical interest to examine a method of determining simultaneously the quantities (i)–(iii) without the direct timing of operations.

7.3 *Description and Theory of the Modified Snap-round Method*

The basic idea is that instead of recording just the state of the system at an isolated time-instant, the behaviour is followed qualitatively over a fixed time-period h; the observer is supposed to have a method of measuring out, sufficiently accurately, just this one length of time.*

In each period of observation one of the following outcomes is obtained: the machine is

(i) running throughout the period of observation, denoted by R;
(ii) stopped at the beginning but running at the end; denoted by SR;
(iii) running at the beginning but stopped at the end, denoted by RS;
(iv) stopped at the beginning and has not been restarted by the end, denoted by SS;
(v) running at the beginning, stops, is restarted and is running at the end, denoted by RsR.

We shall ignore in the theory the possibility that two or more stops occur in the period of observation; if this happens it should be counted as (v).

Let a large number N of independent observations be made and let the sample proportions of observations of the five types be $\hat{p}_1, \ldots, \hat{p}_5$ and the corresponding probabilities be $p_1, \ldots, p_5$. (The independence means that observations must not be made close together on one machine. In many applications there would be a number of similar machines and these could be sampled in turn.) Since the process is reversible in time $p_2 = p_3 = \frac{1}{2}p_{23}$, say. Conditional on a stop having occurred, i.e. conditional on the observation being of types (ii)–(v), let p_m, p_s, p_r be the probabilities of types (ii) or (iii), type (iv) and type (v). Clearly these are the only quantities that give information about the distribution of repair-times and it is the estimation of this that we consider first.

Assume that the instants of observation are randomly dispersed with respect to the repair-times and that the frequency curve of repair-time is $f(l)$ with mean m. Consider first those repair-times of length l. Given that stop has been recorded, the period of observation must *begin* at some instant in the interval of length $(h + l)$ shown in Fig. 7.1. By the hypothesis of the random-

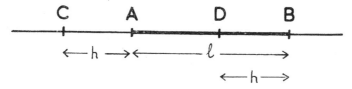

FIG. 7.1.—A period of observation and a repair-time.

ness of the points of observation, the initial point is rectangularly distributed over this interval. We can now calculate the probabilities $p_m(l)$, $p_s(l)$, $p_r(l)$ of observations of type m, s, r, given l; the values are

$$p_m(l) = \frac{2 \operatorname{Min}(h, l)}{h + l}, \qquad \qquad (7.1)$$

* I am very grateful to Mr. N. L. Webb for discussions of this method.

$$p_s(l) = \frac{\text{Max } (l - h, 0)}{h + l}, \qquad \qquad \text{(7.2)}$$

$$p_r(l) = \frac{\text{Max } (h - l, 0)}{h + l}. \qquad \qquad \text{(7.3)}$$

For example to prove the formula for $p_s(l)$, note that no observation SS is possible if $h > l$, while if $h < l$ the observation is SS if the initial point falls in the section AD of Fig. 7.1 of length $l - h$. Hence the required probability is the change that a rectangular variate of range $h + l$ falls in a section of length $l - h$. Similarly for the other probabilities.

Now let l vary in the frequency distribution $f(l)$. Each interval of length l is exposed to the "risk" of being observed for a period $l + h$ and hence the frequency distribution of intervals in the sample is $(l + h) f(l)/(m + h)$. Therefore the final probabilities p_m, p_s, p_r are

$$p_m = \frac{2}{m + h} \int_0^\infty \text{Min } (h, l) f(l) \, dl, \qquad \qquad \text{(7.4)}$$

$$p_s = \frac{1}{m + h} \int_h^\infty (l - h) f(l) \, dl, \qquad \qquad \text{(7.5)}$$

$$p_r = \frac{1}{m + h} \int_0^h (h - l) f(l) \, dl. \qquad \qquad \text{(7.6)}$$

Since these three probabilities add up to unity we can, even in theory, estimate only two aspects of the distribution $f(l)$. It turns out that we can make a simple estimate of the mean repair-time m. For

$$p_s - p_r = \frac{1}{m + h} \int_0^\infty (l - h) f(l) \, dl$$

$$= \frac{m - h}{m + h}.$$

Therefore we take

$$\hat{m} = h \cdot \frac{1 - \hat{p}_r + \hat{p}_s}{1 + \hat{p}_r - \hat{p}_s} \qquad \qquad \text{(7.7)}$$

as our estimate of m.

To estimate the coefficient of variation, C, of repair-time, we must assume a functional form for $f(l)$ and the most convenient seems to be the log-normal. That is we assume that

$$f(l) = \frac{1}{\sqrt{(2\pi)} \, \sigma l} \exp \left\{ - \frac{(\log l - \mu)^2}{2\sigma^2} \right\}, \qquad \qquad \text{(7.8)}$$

so that

$$m = e^{\mu + \frac{1}{2}\sigma^2}, \qquad \qquad \text{(7.9)}$$

$$C^2 = e^{\sigma^2} - 1. \qquad \qquad \text{(7.10)}$$

It then follows from (7.4) and (7.8) that

$$p_m = \frac{2m}{m + h} G\left(\frac{\log h/m}{\sigma} - \tfrac{1}{2}\sigma \right) + \frac{2h}{m + h} G\left(-\frac{\log h/m}{\sigma} - \tfrac{1}{2}\sigma \right), \qquad \text{(7.11)}$$

where

$$G(x) = \frac{1}{\sqrt{(2\pi)}} \int_{-\infty}^x e^{-\frac{1}{2}t^2} \, dt.$$

Hence an estimate $\hat{\sigma}$, and therefore also an estimate $\hat{C}$, is obtained by substituting $\hat{m}$ for m and $\hat{p}_m$ for p_m in (7.11) and solving the resulting equation for σ. This procedure will be discussed in more detail below.

If there are n observations in which stops are recorded, $\hat{p}_m$, $\hat{p}_s$, $\hat{p}_r$ follow a multinomial distribution so that

$$V(\hat{p}_m) = p_m(1 - p_m)/n, \quad C(\hat{p}_s, \hat{p}_r) = -p_s p_r/n, \text{ etc.} \qquad . \qquad . \qquad . \qquad (7.12)$$

Therefore we find, by the usual method for calculating approximate standard errors of functions, that

$$V\left(\frac{\hat{m}}{m}\right) \simeq \frac{4[1 - p_m - (p_r - p_s)^2]}{[1 - (p_r - p_s)^2]^2 \, n} . \qquad . \qquad . \qquad . \qquad (7.13)$$

Now consider the estimation of the proportion ω of time for which the machine is stopped; this is the quantity estimated in Tippett's original method. We now consider all the observations, including those, R, in which no stop is observed, and so work with the proportions $\hat{p}_1, \ldots, \hat{p}_5$. If we were to consider just the initial points of observation as forming a simple snap-round, we would estimate ω by $\hat{p}_2 + \hat{p}_4$. If we were to consider just the final points of observation we would take $\hat{p}_3 + \hat{p}_4$. Therefore we put

$$\hat{\omega} = \tfrac{1}{2}\hat{p}_{23} + \hat{p}_4. \qquad . \qquad . \qquad . \qquad . \qquad (7.14)$$

If there are N observations in all

$$V(\hat{\omega}) = \frac{1}{4N} [p_{23} + 4p_4 - (p_{23} + 2p_4)^2]. \qquad . \qquad . \qquad . \qquad (7.15)$$

If $\omega \ll 1$ so that $p_{23}, p_4 \ll 1$,

$$V\left(\frac{\hat{\omega}}{\omega}\right) \simeq \frac{p_4 + \tfrac{1}{4} p_{23}}{N \omega^2} \qquad . \qquad . \qquad . \qquad . \qquad (7.16)$$

$$= \frac{(p_s + \tfrac{1}{4} p_m)}{N \omega (p_s + \tfrac{1}{2} p_m)}, \qquad . \qquad . \qquad . \qquad (7.17)$$

since $(1 - p_1) p_s = p_4$, etc. and $\omega = p_4 + \tfrac{1}{2} p_{23} = (p_s + \tfrac{1}{2} p_m)(1 - p_1)$.

In N simple snap-rounds

$$V\left(\frac{\hat{\omega}}{\omega}\right) = \frac{1 - \omega}{N\omega} \simeq \frac{1}{N\omega}. \qquad . \qquad . \qquad . \qquad (7.18)$$

Next we consider the estimation of the mean stoppage-rate μ, expressed in stops per unit of running time. Clearly

$$\mu = \omega/[m(1 - \omega)] \qquad . \qquad . \qquad . \qquad . \qquad (7.19)$$

so that we put

$$\hat{\mu} = \hat{\omega}/[\hat{m}(1 - \hat{\omega})]$$

or

$$h\hat{\mu} = \frac{(\tfrac{1}{2}\hat{p}_{23} + \hat{p}_4)}{(1 - \tfrac{1}{2}\hat{p}_{23} - \hat{p}_4)} \cdot \frac{(1 + \hat{p}_r - \hat{p}_s)}{(1 - \hat{p}_r + \hat{p}_s)}$$

$$= \frac{\tfrac{1}{2}\hat{p}_{23} + \hat{p}_5}{1 - \tfrac{1}{2}\hat{p}_{23} - \hat{p}_4}. \qquad . \qquad . \qquad . \qquad (7.20)$$

If we assume that $\omega \ll 1$, we get

$$V\left(\frac{\hat{\mu}}{\mu}\right) \simeq \frac{\tfrac{1}{4} p_{23} + p_5}{N(\tfrac{1}{2} p_{23} + p_5)^2} \qquad . \qquad . \qquad . \qquad (7.21)$$

$$= \frac{(\tfrac{1}{4} p_m + p_r)(\tfrac{1}{2} p_m + p_s)}{(\tfrac{1}{2} p_m + p_r)^2 \, N \, \omega}. \qquad . \qquad . \qquad (7.22)$$

Finally it is useful to rewrite $V(\hat{m}/m)$ as given by (7.13) in terms of the total number N of observations. The result is

$$V\left(\frac{\hat{m}}{m}\right) \simeq \frac{4[1 - p_m - (p_r - p_s)^2](\frac{1}{2} p_m + p_s)}{[1 - (p_r - p_s)^2]^2 N \omega}. \qquad . \qquad . \qquad . \quad (7.23)$$

7.4 *Numerical Results and Discussion*

Since there are essentially three independent quantities among $p_1, \ldots, p_5$ and two simple relations connecting them with m, ω, it is only necessary to give one further relation in order to

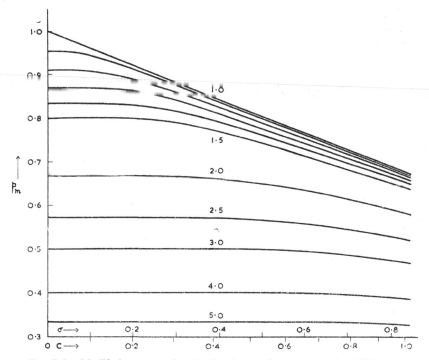

FIG. 7.2.—Modified snap-round method.   Curves of p_m against σ (and C) for the following values of h/m: $1 \cdot 0 \ (0 \cdot 1),\ 1 \cdot 5 \ (0 \cdot 5),\ 3 \cdot 0,\ 4 \cdot 0,\ 5 \cdot 0$.

determine the theoretical values of the proportions in any situation.   This additional relation is most conveniently given as a series of graphs of p_m against σ for different values of m/h.   It would not be necessary to use these graphs in a practical application, except to estimate σ, but we do need them to make theoretical estimates of the accuracy of the method.   It is clear from (7.7) and (7.11) that p_m is unchanged, and p_r and p_s are interchanged, by changing m to m', where $m'/h = h/m$. Hence it is only necessary to give curves for $m/h \geqslant 1$.

Fig. 7.2 gives the required curves.   They will now be used to determine an appropriate value for the period of observation h and to investigate the accuracies of the estimates $\hat{m}, \hat{\omega}, \hat{\mu}$ and $\hat{\sigma}$ (or $\hat{C}$).   The formulae (7.23), (7.17) and (7.22) give the coefficients of variation squared of $\hat{m}, \hat{\omega}, \hat{\mu}$ in the form of a multiplier A_m, A_ω, A_μ times $1/(N\omega)$.   Table 7.1 gives A_m, A_ω, A_μ for various values of m/h and for $\sigma = 0, 0 \cdot 2, 0 \cdot 4$.

A precise general statement about an optimum value for h is not possible, both because a compromise is necessary in deciding which variance, or combination of variances, to minimize

TABLE 7.1

Coefficients for Determining the Accuracy of the Estimates $\hat{m}$, $\hat{\omega}$ $\hat{\mu}$.

	A_m			A_ω			A_μ		
m/h	$\sigma=0$	$\sigma=0\cdot2$	$\sigma=0\cdot4$	$\sigma=0$	$\sigma=0\cdot2$	$\sigma=0\cdot4$	$\sigma=0$	$\sigma=0\cdot2$	$\sigma=0\cdot4$
1/3	·444	·444	·445	·500	·500	·500	·278	·278	·278
1/2	·375	·375	·386	·500	·500	·505	·375	·375	·376
1/1·5	·278	·281	·333	·500	·501	·520	·444	·445	·453
1/1·4	·245	·251	·321	·500	·502	·526	·459	·460	·472
1/1·2	·150	·189	·308	·500	·511	·546	·486	·494	·518
1	0	·159	·317	·500	·540	·579	·500	·540	·579
1·2	·184	·227	·369	·583	·593	·622	·600	·613	·655
1·4	·342	·352	·450	·643	·644	·661	·700	·703	·736
1·5	·417	·421	·499	·667	·667	·680	·750	·752	·780
2	·750	·751	·772	·750	·750	·752	1·000	1·000	1·010
3	1·333	1·333	1·336	·833	·833	·834	1·500	1·500	1·501

$$V(\hat{m}/m) \simeq A_m/(N\omega), \quad V(\hat{\omega}/\omega) \simeq A_\omega/(N\omega), \quad V(\hat{\mu}/\mu) \simeq A_\mu/(N\omega).$$

and because Table 7.1 gives a comparison for constant number of observations, not for a constant total amount of observer's time. Clearly if the period h is reduced, more observations can be taken in a given total time; in a particular application a fairly precise allowance for this could be introduced, but the most that can be said in general is that as h is decreased, N could be increased less rapidly than in an inverse proportionality to h.

The following conclusions can be drawn.

(i) The optimum value of h for estimating m is near to m; this was to be expected on common-sense grounds. The coefficients in Table 7.1 suggest that slightly larger values of h are to be preferred, but this is likely to be at least balanced by the increase in N that is possible for smaller h.

(ii) The choice of h is critical for $\sigma = 0$, but not for the larger values of σ. Hence if σ is known to be small, it is worth-while either making a moderately accurate preliminary estimate of m or else changing the value of h during the study to make it near to m. If $\sigma \geqslant 0\cdot2$, however, the choice of h is not critical and any value within say ± 30 per cent. of m will do.

(iii) In the range near $m = h$, A_m is smaller than A_ω and A_μ, so that m is estimated with smaller fractional error than are ω and μ.

(iv) The quantities ω and μ are usually estimated more accurately from low values of h; for, although A_ω, A_μ increase as h decreases, this increase will usually be more than offset by the increase in N that is possible for smaller h. Hence if it is important to estimate ω and μ accurately and rather less important to estimate m, a value of m/h less than unity would usually be appropriate.

(v) Accurate estimation of σ (or C) is not possible, even in principle, without a very large number of observations. To take the most favourable case, suppose that h has been taken very nearly equal to m and that errors of estimation of m can be neglected. Then the relation between p_m and σ is, from (7.8),

$$p_m = 2\,G(-\tfrac{1}{2}\sigma), \quad . \qquad . \qquad . \qquad . \qquad . \qquad (7.24)$$

which in the relevant range is linear with slope $-1/\sqrt{(2\pi)}$. Hence in this case

$$V(\hat{\sigma}) \simeq \frac{1}{2\pi}\,V(\hat{p}_m)$$

$$= \frac{1}{2\pi}\,\frac{p_m(1-p_m)}{n}$$

$$= \frac{p_m(1-p_m)(p_s + \tfrac{1}{2}p_m)}{2\pi\,N\,\omega}. \qquad . \qquad . \qquad . \qquad (7.25)$$

Also

$$\sigma \simeq \frac{1}{\sqrt{(2\pi)}} (1 - p_m) \qquad \qquad \qquad (7.26)$$

and in the situation we are considering $p_s = \frac{1}{2}(1 - p_m)$, so that

$$V\left(\frac{\hat{\sigma}}{\sigma}\right) \simeq \frac{p_m}{2(1 - p_m) N \omega}. \qquad \qquad (7.27)$$

For the particular cases $\sigma = 0$, $0\cdot2$, $0\cdot4$, this becomes ∞, $5\cdot78/(N\omega)$, $2\cdot66/(N\omega)$. These are very much larger than the corresponding values of the coefficient of variation squared of $\hat{m}$, $\hat{\omega}$, $\hat{\mu}$ given in Table 7.1 for $m/h = 1$. To sum up, estimation of σ is likely to be inaccurate and if it is to be attempted it is important that h should be near to m.

We now make a comparison with continuous observation using a stop-watch to measure individual running and stopped times. To take a particular case let $\sigma = 0\cdot2$ and let $h = m = 1$ unit. Then $V(\hat{m}/m) = 0\cdot16/(N\omega)$, etc. Now if the machine were timed continuously for a period $Nh = N$, $N\omega$ stops would on the average be observed, and with $\sigma \simeq U = 0\cdot2$,

$$V(\hat{m}/m) \simeq 0\cdot04/(N\omega),$$

when the individual repair-times are measured. Hence as a method of estimating m, the modified snap-round method is about 25 per cent. efficient. For estimating ω and μ, however, the modified snap-round method gives $V(\hat{\omega}/\omega) \simeq V(\hat{\mu}/\mu) \simeq 0\cdot540/(N\omega)$, while continuous observation over the same period gives for random stops $1/(N\omega)$. Hence the modified snap-round method is the more efficient in this case, because in effect each isolated period of observation h counts stops occurring over a longer period.

7.5 *Summing-up of the Method*

The practical usefulness of the method depends to a considerable extent on non-statistical considerations and cannot be examined here. It does, however, follow from the calculations in the previous section that the method is capable in theory of estimating ω and μ rather more accurately than a comparable amount of continuous observation, and of estimating m with lower accuracy than attained by continuous observation. In many cases, however, if sufficient observations are taken to ensure accurate estimates of ω and μ, the resulting variance of $\hat{m}$ will be small enough for practical purposes. It is desirable that h should be chosen to be near to m, but the choice is not critical unless either σ is small, or estimation of σ is to be attempted. It is necessary, therefore, that there should be some prior knowledge of m but this would usually be so.

The theory has been worked out assuming there to be only one type of repair-time. If there were several, and this would usually be the case, the only difficulty is connected with the choice of h. If no two repair-times differed by more than say a factor of two, a compromise h could be found that would permit reasonably accurate estimation of the individual m's, ω's and μ's. But if say one type of repair-time is much longer than the others special methods would have to be used to estimate it.

Various modifications can be made to the method by introducing complications, such as the following.

(i) If the estimation of ω and μ were of particular importance, it might be economical to use a mixture of ordinary snap-rounds ($h = 0$) and the present method ($h \simeq m$);

(ii) More than two times of observation could be introduced in various ways. Thus if the machine is stopped at the end of the period h, observation could be continued say for further $\frac{1}{2}h$. Or different times could be used for different types of stop. Or a method analogous to the "up and down" method of biological assay (Dixon & Mood, 1948) could be used in which h is increased every time SS is observed and decreased every time RsR is observed. And so on.

A final point is that, as in the ordinary snap-round method, it is essential that the choice of periods for observation should not be affected by the state of the machine.

References

ARMITAGE, P. (1952), "The statistical theory of bacterial populations subject to mutation", *J.R. Statist. Soc.*, B, **14**, 1–33.

BARNARD, G. A. (1953), "Time intervals between accidents", *Biometrika*, **40**, 212–213.

BARTLETT, M. S. & KENDALL, D. G. (1946), "The statistical analysis of variance-heterogeneity and the logarithmic transformation", *Suppl. J.R. Statist. Soc.*, **8**, 128–138.

BIRNBAUM, A. (1954), "Statistical methods for Poisson processes and exponential populations", *J. Amer. Statist. Ass.*, **49**, 254–266.

COCHRAN, W. G. (1954), "The combination of estimates from different experiments", *Biometrics*, **10**, 101–129.

COX, D. R. (1953), "Some simple approximate tests for Poisson variates", *Biometrika*, **40**, 354–360.

—— (1955), "The statistical analysis of congestion", *J.R. Statist. Soc.*, A, (to appear).

—— & SMITH, W. L. (1953), "The superposition of several strictly periodic sequences of events", *Biometrika*, **40**, 1–11.

—— —— (1954), "On the superposition of renewal processes", *Biometrika*, **41**, 91–99.

DIXON, W. J. & MOOD, A. M. (1948), "A method for obtaining and analyzing sensitivity data", *J. Amer. Statist. Ass.*, **43**, 109–126.

FELLER, W. (1948), "On probability problems in the theory of counters", *Courant Anniversary Volume*, 105–115. New York: Interscience Publishers.

FISHER, R. A. (1954), "The analysis of variance with various binomial transformations", *Biometrics*, **10**, 130–139.

FRY, T. C. (1928), *Probability and its Engineering Uses*. New York: van Nostrand.

GREENWOOD, M. & YULE, G. U. (1920). "An inquiry into the nature of frequency distributions representative of multiple happenings with particular reference to the occurrence of multiple attacks of disease or of repeated accidents", *J.R. Statist. Soc.*, **83**, 255–279.

HALD, A. (1952), *Statistical Tables and Formulae*. New York: J. Wiley.

JOWETT, G. H. (1955), "The comparison of means of industrial time-series", *Applied Statistics*, **4**, 32–46.

KEMPTHORNE, O. (1952), *The design and analysis of experiments*. New York: J. Wiley.

MAGUIRE, B. A., PEARSON, E. S. & WYNN, A. H. A. (1952), "The time intervals between industrial accidents", *Biometrika*, **39**, 168–180.

MORAN, P. A. P. (1951), "The random division of an interval—Part II," *J.R. Statist. Soc.*, B, **13**, 147–150.

PRZYBOROWSKI, J. & WILENSKI, H. (1940), "Homogeneity of results in testing samples from Poisson series", *Biometrika*, **31**, 313–323.

TIPPETT, L. H. C. (1935a), "Statistical methods in textile research. Part 3: a snap-reading method of making time-studies of machines and operatives in factory surveys", *J. Text. Inst.* **26**, T13–33.

—— (1935b), "Some applications of statistical methods to the study of variation of quality in the production of cotton yarn," *Suppl. J. R. Statist. Soc.*, **2**, 27–55.

WOLD, H. (1948a), "Sur les processus stationnaires ponctuels", *Colloques Internationaux du C.N.R.S.*, **13**, 75–86.

—— (1948b), "On stationary point processes and Markov chains", *Skand. Aktuar.*, **31**, 229–240.

YULE, G. U. (1945), "On a method of studying time-series based on their internal correlations", *J.R. Statist. Soc.*, **108**, 208–225.

Summary of discussion

When the paper was read to the Royal Statistical Society a vote of thanks was proposed by C. A. B. Smith, seconded by K. D. Tocher and the discussion continued by E. S. Pearson, M. S. Bartlett, T. Lewis, G. H. Jowett, E. S. Page, D. J. Bartholomew and R. A. Fairthorne.

Smith points out various contexts in which time-varying Poisson processes might arise. Pearson gently suggests that the methods are overcomplicated for the industrial contexts from which they primarily arose. Bartlett makes connections with the general theory of point processes. Lewis discusses problems connected with the testing of prototypes. Jowett makes connections with other work on time series subject to erratic fluctuations of unknown form. Page relates the discussion to that of cusum charts. Bartholomew reports on the properties of various tests of randomness. Finally, Fairthorne objects to the term 'snap-round methods' used in the paper for cross-sectional sampling.

MULTIVARIATE POINT PROCESSES

D. R. COX

IMPERIAL COLLEGE, UNIVERSITY OF LONDON

and

P. A. W. LEWIS*

IMPERIAL COLLEGE, UNIVERSITY OF LONDON

and

IBM RESEARCH CENTER

1. Introduction

We consider in this paper events of two or more types occurring in a one dimensional continuum, usually time. The classification of the events may be by a qualitative variable attached to each event, or by their arising in a common time scale but in different physical locations. Such multivariate point processes, or multitype series of events, are to be distinguished from events of one type occurring in an n dimensional continuum and considered, for example, by Bartlett [2]. It is of course possible to have multivariate point processes in, say, two dimensions, for example, the locations of accidents labelled by day of occurrence, but we do not consider this extension here.

Multivariate series of events arise in many contexts; the following are a few examples.

EXAMPLE 1.1. Queues are a well-known situation in which bivariate point processes arise as the input and output, although interest in the joint properties of the input and output processes is fairly recent (for example, Daley [16] and Brown [7]). The two processes occur simultaneously in time. Many of the variants on simple queueing situations which have been considered give rise to more than two point processes.

EXAMPLE 1.2. An important and rich source of multivariate point processes is neurophysiology (Perkel, Gerstein, and Moore [41]). Information is carried along nerve bundles by spikes which occur randomly in time. (The spikes are extremely narrow and, at least in many situations, their shape and height do not appear to vary or carry information.) The neuronal spike trains of different types may be observations at different locations with no definite knowledge of physical connection, or may be the inputs and outputs to nerve connections (neurons).

EXAMPLE 1.3. When the events are crossings of a given level by a real valued stochastic process in continuous time, the up crossings and down crossings of the

*Support under a National Institutes of Health Special Fellowship (2–FO3–GM 38922–02) is gratefully acknowledged. Present address: Naval Postgraduate School, Monterey, California.

401

level constitute a very special bivariate point process in which the two types of events alternate (Leadbetter [27]). However, up crossings of two different levels produce a general type of bivariate point process which is of interest, for example, in reliability investigations.

EXAMPLE 1.4.   In reliability studies over time, of continuously operating machines such as computers, the failures are more often than not labelled according to the part of the system in which they occurred, or according to some other qualitative characterization of the failure, for example, mechanical or electrical. One might also be interested in studying interactions between preventive maintenance points and failures occurring during normal operation. Again a comparison between failure patterns in separately located computers (Lewis [29]) might be of interest in determining whether some unknown common variable, such as temperature and/or humidity, influences reliability.

EXAMPLE 1.5.   Cox [11] has considered the problem of analyzing events of two types in textile production. The two types of event may be breakdowns in the loom and faults in the cloth, or different types of breakdown of the loom. The continuum is length of thread rather than time.

EXAMPLE 1.6.   In the analysis of electrocardiograms the trace is continuous, but both regular heart beats and various types of ectopic heart beats occur. It is therefore of interest to analyze electrocardiograms as bivariate event processes, even though defining the precise time of occurrence of the event (heartbeat) may present some problems.

EXAMPLE 1.7.   Traffic studies are a rich source of multivariate point processes. Just two possibilities are that the events may be the passage of cars by a point on a road when the type of event is differentiated by direction of travel, or we may consider passage of cars past two different positions.

EXAMPLE 1.8.   Finally, physical phenomena such as volcanoes or earthquakes (Vere-Jones [45], [46], [47]) may have distinguishing features of many kinds—generally highly compacted attributes of the process, for example, the general location of the origin of the earthquake.

Multivariate point processes can be regarded as very special cases of univariate point processes in which a real valued quantity is associated with each point event, that is, special cases of what Bartlett [3] has called, rather generally, line processes. In particular, if the real valued quantity takes only two possible values, we have in effect a bivariate process of events of two types.

Three broad types of problems arise for multivariate point processes. The first are general theoretical and structural problems of which the most outstanding is the problem of characterizing the dependence and interaction between a number of processes. This is the only general theoretical question we will consider in any detail; it is intimately connected with the statistical analysis of bivariate point processes.

The second type of problem is the calculation of the properties of special stochastic models suggested, for example, by physical considerations. This in general is a formidable task even for quite simple models.

Thirdly, there are problems of statistical analysis. These include:

(a) comparing rates in independent processes (Cox and Lewis [14], Chapter 9) from finite samples;

(b) assessing possible dependence between two processes from finite samples;

(c) determining, again from finite samples, the probabilistic structure of a mechanism which transforms one process into a second quite clearly dependent process.

The range of the problems will become clear in the main body of the paper. The topics considered are briefly as follows.

In Section 2, we give some notation and define various types of interevent sequences and counting processes which occur in bivariate point processes. Concepts such as independence of the marginal processes and stationarity and regularity of the complete, .bivariate process are defined. The ideas of this section are illustrated by considering two independent renewal processes and also the semi-Markov process (Markov renewal process).

In Section 3, we study dependence and correlation in bivariate point processes, defining complete intensity functions and second order cross intensity functions and cross spectra, giving their relationship to covariance time surfaces. Doubly stochastic bivariate point processes are defined and their cross intensity function is given. Other simple models of bivariate point processes are defined through the complete intensity and cross intensity functions. In this way, various degrees of interaction between events in the bivariate process can be specified. A class of bivariate Markov interval processes is defined.

In Section 4, a simple delay model with marginal Poisson processes is considered in some detail. Other special physical models are considered briefly in Section 5.

General comments on bivariate Poisson processes are given at the end of Section 5; a bivariate Poisson process is defined simply as a bivariate point process whose marginal processes are Poisson processes.

Statistical procedures are considered in Section 6, including the estimation of second order cross intensity functions and cross spectra, as well as covariance time surfaces. Tests for dependence in general and particular situations are considered, and statistical procedures for some special processes are given.

Throughout, emphasis is placed on concepts rather than on mathematical details and a number of open questions are indicated. For the most part we deal with bivariate processes, that is, with events of two types; the generalization of more than two types of events is on the whole straightforward.

2. General definitions and ideas

2.1. *Regularity*. Throughout Section 2, we deal with bivariate processes, that is, processes of events of two types called type a and type b. The process of, say, type a events alone is called a *marginal process*.

In a univariate point process such as a marginal process in a bivariate point process, regularity is defined by requiring that in any interval of length Δt

(2.1) $Pr\{(\text{number events in } \Delta t) > 1\} = o(\Delta t).$

Regularity is intuitively the nonoccurrence of multiple events.

For bivariate processes, we say the process is *marginally regular* if its marginal processes, considered as univariate point processes, are both regular. The bivariate process is said to be *regular* if the process of superposed marginal events is regular, that is, if the process of events regardless of type is regular. This type of regularity, of course, implies *marginal regularity*.

A simple, rather degenerate, bivariate process is obtained by taking three Poisson processes, say I, II, and III, and superposing processes I and II to obtain the events of type a and superposing II and III to obtain the events of type b (Marshall and Olkin [33]). Clearly, the bivariate process is marginally regular but not regular. However, if the events of type b are made up of process III events superposed with process II events delayed by a fixed amount, the resulting bivariate process is regular. A commonly used alternative to the word *regular* is *orderly*.

2.2. *Independence and stationarity.* Independence of the marginal processes in a bivariate process is intuitively defined as independence of the number of events (counts) in any two sets of intervals in the marginal processes. The more difficult problem of specifying dependence (and correlation) in the bivariate process is central to this paper and will be taken up in the next section.

In the sequel, we will be primarily concerned with *transient* or *stationary* bivariate point processes, as opposed to *nonhomogeneous* processes. The latter type of process is defined roughly as one with either an evolutionary or cyclic trend, whereas a *transient* process is roughly one whose probabilistic structure eventually becomes stationary (time invariant). There are a number of types of stationarity which need to be defined more carefully.

DEFINITION 2.1 (Simple stationarity). *Let $N^{(a)}(t_1^{(1)}, t_1^{(1)} + \tau_1^{(1)})$ be the number of events of type a in the interval $(t_1^{(1)}, t_1^{(1)} + \tau_1^{(1)}]$ and $N^{(b)}(t_2^{(1)}, t_2^{(1)} + \tau_2^{(1)})$ be the number of events of type b in the interval $(t_2^{(1)}, t_2^{(1)} + \tau_2^{(1)}]$. The bivariate point process is said to have* simple stationarity *if*

(2.2) $Pr\{N^{(a)}(t_1^{(1)}, t_1^{(1)} + \tau_1^{(1)}) = n^{(a)}; N^{(b)}(t_2^{(1)}, t_2^{(1)} + \tau_2^{(1)}) = n^{(b)}\}$
$$= Pr\{N^{(a)}(t_1^{(1)} + y, t_1^{(1)} + \tau_1^{(1)} + y) = n^{(a)};$$
$$N^{(b)}(t_2^{(1)} + y, t_2^{(1)} + \tau_2^{(1)} + y) = n^{(b)}\},$$

for all $t_1, t_2, \tau_1^{(1)}, \tau_2^{(1)}, y > 0$.

In other words, the joint distribution of the number of type a events in a fixed interval and the number of type b events in another fixed interval is invariant under translation.

Simple stationarity of the bivariate process implies an analogous property for the individual marginal processes and for the superposed process.

In the sequel, we assume that for the marginal processes considered individually the probabilities of more than one type a event in τ_1 and more than one type b event in τ_2 are, respectively, $\rho_a \tau_1 + o(\tau_1)$ as $\tau_1 \to 0$ and $\rho_b \tau_2 + o(\tau_2)$ as $\tau_2 \to 0$, where ρ_a and ρ_b are finite.

Simple stationarity and these finiteness conditions imply that the univariate forward recurrence time relationships in the marginal processes and the pooled processes hold (Lawrance [25]).

If in addition the process is regular, Korolyuk's theorem implies that ρ_a, ρ_b, and $\rho_a + \rho_b$ are, respectively, the rates of events of types a, events of types b, and events regardless of type.

DEFINITION 2.2 (Second order stationarity).   *By extension, we say that the bivariate point process has* second order stationarity *(weak stationarity) if the joint distribution of the number of type a events in two fixed intervals and the number of type b events in another two fixed intervals is invariant under translation.*

This type of stationarity is necessary in the sequel for the definition of a time invariant cross intensity function. Clearly, it implies second order stationarity for the marginal processes considered individually and for the superposed marginal processes.

DEFINITION 2.3 (Complete stationarity).   *By extension,* complete stationarity *for a bivariate point process is invariant under translation for the joint distribution of counts in arbitrary numbers of intervals in each process.*

2.3. *Asynchronous counts and intervals.*   In specifying stationarity, we did not mention the time origin or the method of starting the process. There are three main possibilities.

(i) The process is started at time $t = 0$ with initial conditions which produce stationarity, referred to as *stationary initial conditions*.

(ii) The process is transient and is considered beyond $t = 0$ as its start moves off to the left. The process then becomes stationary as the start moves to minus infinity. There is generally a specification of the state of the process at $t = 0$ known as the *stationary equilibrium conditions*.

Note that in both (i) and (ii) stationarity is defined by invariance under shifts to the right.

(iii) In a stationary point process, a time is specified without knowledge of the events and is taken to be the origin, $t = 0$. The time $t = 0$ is said to be an *arbitrary time* in the (stationary) process, selected by an *asynchronous sampling* of the process.

Now there is associated with the stationary bivariate point process a counting process $\mathbf{N}(t_1, t_2) = \{N^{(a)}(t_1), N^{(b)}(t_2)\}$, where

(2.3) $N^{(a)}(t_1)$ is the number of type a events in $(0, t_1]$,
            $N^{(b)}(t_2)$ is the number of type b events in $(0, t_2]$,

and a bivariate sequence of intervals $\{X^{(a)}(i), X^{(b)}(j)\}$, where, assuming regularity of the process, $X^{(a)}(1)$ is the forward recurrence time in the process of type a events (that is, the time from $t = 0$ to the first type a event), $X^{(a)}(2)$ is the time between the first and second type a events, and so forth; and the $\{X^{(b)}(j)\}$ sequence is defined similarly.

Note that for asynchronous sampling of a stationary process the indices i and j can take negative values; in particular, $\{X^{(a)}(-1), X^{(b)}(-1)\}$ are the bivariate backward recurrence times.

There is a fundamental relationship connecting the bivariate counting processes with the bivariate interval processes; this is a direct generalization of the relationship for the univariate case:

$$(2.4) \qquad N^{(a)}(t_1) < n^{(a)}, \qquad N^{(b)}(t_2) < n^{(b)},$$

if and only if

$$(2.5) \qquad \begin{aligned} S^{(a)}(n^{(a)}) &= X^{(a)}(1) + \cdots + X^{(a)}(n^{(a)}) > t_1, \\ S^{(b)}(n^{(b)}) &= X^{(b)}(1) + \cdots + X^{(b)}(n^{(b)}) > t_2 \end{aligned}$$

Probability relationships are written down directly from these identities connecting the bivariate distribution of counts with the bivariate distribution of the sums of intervals $S^{(a)}(n^{(a)})$ and $S^{(b)}(n^{(b)})$.

Equations (2.4) and (2.5) can be used, for example, to prove the asymptotic bivariate normality of $\{N^{(a)}(t_1), N^{(b)}(t_2)\}$ for a broad class of bivariate point processes.

2.4. *Semisynchronous sampling.* In a univariate point process, *synchronous* sampling of the stationary process refers to the placement of the time origin at an arbitrary event and the examination of the counts and intervals following this arbitrary event (Cox and Lewis [14], Chapter 4, and McFadden [35]). In more precise terms (Leadbetter [27], [28], and Lawrance [24]), the notion of an arbitrary event in a stationary point process is the event $\{N(0, \tau) \geq 1\}$ as τ tends to zero, and the distribution function $F(t)$ of the interval between the arbitrary event and the following event is defined to be

$$(2.6) \qquad 1 - F(t) = \lim_{\tau \to 0+} Pr\{N(\tau, \tau + t) = 0 \,|\, N(0, \tau) \geq 1\}.$$

In bivariate point processes, the situation is more complex. Synchronous sampling of the marginal process of type a events produces *semisynchronous sampling* of the process of b events from an arbitrary a event, and *vice versa*.

The bivariate counting processes and intervals following these two types of sampling are denoted as follows:

(a) for semisynchronous sampling of b by a,

$N_a^{(a)}(t_1)$ is the number of type a events following an origin at a type a event;

$N_a^{(b)}(t_1)$ is the number of type b events following an origin at a type a event;

$\{X_a^{(a)}(i)\}$ is, for $i = 1$, the time from the origin at a type a event to the next event of type a, and for $i = 2, 3, \cdots$, the intervals between subsequent type a events;

$\{X_a^{(b)}(j)\}$ is, for $j = 1$, the time from the origin at a type a event to the first subsequent type b event, and for $j = 2, 3, \cdots$, the intervals between subsequent type b events;

(b) for semisynchronous sampling of a by b, the subscript becomes b instead of a, in the above expressions, indicating the nature of the origin.

Note that in general (Slivnyak [44]) the sequences $\{X_a^{(a)}(i)\}$ and $\{X_b^{(b)}(j)\}$, being the synchronous interval processes in the marginal point processes, are stationary, whereas $\{X_a^{(b)}(i)\}$ and $\{X_b^{(a)}(j)\}$, the semisynchronous intervals, are in general not stationary. Also for independent processes, the semisynchronous sequences are identical with the asynchronous sequences $\{X^{(a)}(i)\}$ and $\{X^{(b)}(j)\}$.

2.5. *Pooling and superposition of processes.* In discussing regularity, we referred to the superposition of the two marginal processes in the bivariate process. This is the univariate process of events of both types considered without specification of the event type and is referred to simply as the *superposed process*. Study of the superposed process of rate $\rho_a + \rho_b$ is an intimate part of the analysis of the bivariate process. Asynchronous sampling of the superimposed process gives counts and intervals denoted by $N^{(\cdot)}(t_1)$ and $\{X^{(\cdot)}(i)\}$, whereas synchronous sampling, that is, the process considered conditionally (in the Khinchin sense) on the existence of an event of an unspecified type at the origin, gives $N_{\cdot}^{(\cdot)}(t_1)$ and $\{X_{\cdot}^{(\cdot)}(i)\}$.

Semisynchronous sampling of the superposed process by events of type a or type b is also possible and the notation should be clear.

We call the superposed process with specification of the event type the *pooled* process. The original bivariate process can then be respecified in terms of the process

$$(2.7) \qquad \{X_{\cdot}^{(\cdot)}(i), \qquad T_{\cdot}(i)\},$$

where $T_{\cdot}(i)$ is a binary valued process indicating the type of the ith event after the origin in the superposed process with synchronous sampling. Clearly, the marginal processes of event types, that is, $\{T(i)\}$, $\{T_{\cdot}(i)\}$, $\{T_a(i)\}$ and $\{T_b(i)\}$ are themselves of interest. Note that they are in general not stationary processes for all types of sampling and are related to the processes defined in Sections 2.3 and 2.4. Thus, for example,

$$(2.8) \qquad \{X_a^{(\cdot)}(1) \leqq x; T_a(1) = a\} \Leftrightarrow \{X_a^{(a)} < X_a^{(b)}; X_a^{(a)} \leqq x\},$$

with much more complicated statements relating events of higher index i. The binary sequence of event types has no counterpart in univariate point process.

Thus, there are many possible representations of a bivariate point process. Which is the most fruitful is likely to depend on the particular application.

As a very simple practical example of these representations, consider a generalization of the alternating renewal process. We have a sequence of positive random variables $W(1), Z(1), W(2), Z(2), \cdots$, representing operating and repair intervals in a machine. It is natural to assume that $W(i)$ and $Z(i)$ are mutually correlated but independent of other pairs of operating and repair times. Type a events, occurring at the end of the $W(i)$ variables, are machine failures. Type b events occur at the end of $Z(i)$ variables and represent times at which the machine goes back into service.

Specification of the process is straightforward and simple in terms of the pooled process variables $\{X_{\cdot}^{(\cdot)}(i), T_{\cdot}(i)\}$, $\{X_{a}^{(\cdot)}(i), T_{a}(i)\}$, and so forth. However, marginally the type b events are a renewal process, whereas the type a events are a nonrenewal, non-Markovian point process and the dependency structure expressed through the intervals in the marginals is complex.

2.6. *Successive semisynchronous sampling.* Finally, we mention the possibility of successive semisynchronous samples of the marginal process of type b events by a events. The origin is at an a event, as in ordinary semisynchronous sampling and connected with this a event is the time forward (or backward) to the next b event. Subsequent a events are associated with the times forward (or backward) to the next b event. It is not clear how generally useful this procedure is in studying bivariate point processes. It has been used, however, by Brown [7] in studying identifiability problems in $M/G/\infty$ queues; see also Section 6.4.

2.7. *Palm-Khinchin formulae.* In the theory of univariate point processes, there are relations connecting the distributions of sums of synchronous intervals and sums of asynchronous intervals. Similar relationships connect the synchronous and asynchronous counting processes. These relationships are sometimes called the Palm-Khinchin formulae and are given, for example, by Cox and Lewis ([14], Chapter 4).

The best known of these relations connects the distributions of the synchronous and asynchronous forward recurrence times in a stationary point process with finite rate ρ (Lawrance [25]). In the context of the marginal process of type a events,

$$(2.9) \qquad \rho_a\{1 - F_{X^{(a)}}(t)\} = D_t^+ F_{X^{(a)}}(t),$$

where D_t^+ denotes a right derivative. For moments when the relevant moments exist we have

$$(2.10) \qquad E\{(X^{(a)})^r\} = \frac{\rho_a E\{(X_a^{(a)})^{r+1}\}}{r+1}.$$

Palm-Khinchin type formulae for bivariate point processes have been developed by Wisniewski [50], [51]. They are far more complex than those for univariate processes, both in terms of the number of relationships involved and in the analytical problems encountered. Thus, on the first point there are not only interval relationships, but also relationships between the probabilistic structures of the binary sequences $\{T(i)\}$ and $\{T.(i)\}$ and between the probabilistic structures of the binary sequence $\{T(i)\}$ and the binary sequences $\{T_a(i)\}$ and $\{T_b(i)\}$.

On the second point, the analytical problems are illustrated by the following argument. It is easily shown that an arbitrarily selected point in the (univariate) superposed process is of type a with probability $\rho_a/(\rho_a + \rho_b)$. Thus, any probabilistic statement about the variables $\{X^{(\cdot)}_\cdot(i), T_\cdot(i)\}$, say $g(X^{(\cdot)}_\cdot(1), T_\cdot(1), \cdots)$, is expressible in terms of the same probabilistic statement for $\{X^{(\cdot)}_a(i), T_a(i)\}$ and $\{X^{(\cdot)}_b(i), T_b(i)\}$,

$$(2.11) \qquad g(X^{(\cdot)}_\cdot(1), T_\cdot(1), \cdots) = \frac{\rho_a}{\rho_a + \rho_b} g(X^{(\cdot)}_a(1), \cdots) + \frac{\rho_b}{\rho_a + \rho_b} g(X^{(\cdot)}_b(1), \cdots).$$

Now if a relationship between $g(X^{(\cdot)}(1), T(1), \cdots)$, and $g(X^{(\cdot)}_\cdot(1), T_\cdot(1), \cdots)$ exists, we can relate the asynchronous sequence to the two semisynchronous sequences through (2.11). But the usual univariate Palm-Khinchin formulae relate univariate distributions of *sums* of asynchronous intervals to univariate distributions of sums of synchronous intervals. Clearly, formulae relating *joint* properties of asynchronous intervals and types to joint properties of synchronous intervals $X_\cdot(i)$ and types $T_\cdot(i)$ are needed if one is, for example, to relate, through (2.11) and generalizations of (2.8), bivariate distributions of asynchronous forward recurrence times $\{X^{(a)}(1), X^{(b)}(1)\}$ to the bivariate distributions of the semisynchronous forward recurrence times $\{X^{(a)}_a(1), X^{(b)}_a(1)\}$ and $\{X^{(a)}_b(1), X^{(b)}_b(1)\}$,

Lawrance [26] has noted this need for extended Palm-Khinchin formulae and conjectured results in the univariate case.

Of Wisniewski's results [50], [51], we cite here only two moment formulae. These relate the moments of the joint asynchronous forward recurrence times $\{X^{(a)}(1), X^{(b)}(1)\}$ with the moments of both of the semisynchronous forward recurrence times, $\{X^{(a)}_a(1), X^{(b)}_a(1)\}$ and $\{X^{(a)}_b(1), X^{(b)}_b(1)\}$. The feature that probabilistic properties of both semisynchronous sequences are needed to determine probabilistic properties of the asynchronous sequence is characteristic of all these relationships, and follows from (2.11).

We have for the bivariate analogues to (2.10), for $r = 1$,

$$(2.12) \qquad \tfrac{1}{2}\rho_a E[\{X^{(a)}_a(1)\}^2] + \tfrac{1}{2}\rho_b E[\{X^{(b)}_b(1)\}^2]$$
$$= E\{X^{(a)}\} + E\{X^{(b)}\}$$
$$= \rho_a E\{X^{(a)}_a(1)\, X^{(b)}_a(1)\} + \rho_b E\{X^{(a)}_b(1)\, X^{(b)}_b(1)\}$$

and

$$(2.13) \qquad 12 E\{X^{(a)}(1)\, X^{(b)}(1)\}$$
$$= \rho_a E[3X^{(a)}_a(1)\, \{X^{(b)}_a(1)\}^2 + 3\{X^{(a)}_a(1)\}^2\, X^{(b)}_a(1) - \{X^{(a)}_a(1)\}^3]$$
$$+ \rho_b E[3X^{(b)}_b(1)\, \{X^{(a)}_b(1)\}^2 + 3\{X^{(b)}_b(1)\}^2\, X^{(a)}_b(1) - \{X^{(b)}_b(1)\}^3].$$

The interesting feature of (2.12) is that the correlation between semisynchronous forward recurrence times is a function only of the properties of the marginal processes and not of the dependency structure of the bivariate point process. Moreover, (2.13) shows that if we use correlation between the asynchronous forward recurrence times as a measure of dependence in the bivariate point process, this dependence only affects the third order joint moments of the semisynchronous forward recurrence times.

2.8. *Examples*.   To illustrate the definitions and concepts introduced above, we consider two very simple bivariate point processes. The analytical details developed here will be used in Section 6 in considering the statistical analysis of bivariate point processes.

EXAMPLE 2.1. *Independent renewal processes*.   Consider two *delayed* renewal processes $\{X^{(a)}(1); X_a^{(a)}(i), i = 2, 3, \cdots\}$ and $\{X^{(b)}(1); X_b^{(b)}(j), j = 2, 3, \cdots\}$, where using a shortened notation,

$$(2.14) \qquad G^{(a)}(x) = Pr\{X^{(a)}(1) \leqq x\} = \frac{\int_0^x \{1 - F^{(a)}(u)\} \, du}{E_a(X)},$$

$$(2.15) \qquad Pr\{X_a^{(a)}(i) \leqq x\} = F^{(a)}(x), \quad E_a(X) = \int_0^\infty x \, dF^{(a)}(x),$$

with similar definitions for the process of type b events. The distribution of the variable $X^{(a)}(1)$ in (2.14) and the analogous distribution for $X^{(b)}(1)$ are the stationary initial conditions for the marginal renewal processes, and clearly the independence of the processes implies that these distributions (jointly) give stationarity to the bivariate process. Because of independence there is no difference between semisynchronous and asynchronous sampling; the process is defined completely in terms of the properties of the asynchronous and synchronous intervals.

Properties of intervals in the superposed process, and properties of successive intervals and event types in the pooled process, that is, $\{X^{(\cdot)}(i), T(i)\}$ are very difficult to obtain explicitly. The sequences $X^{(\cdot)}(i)$ and $T(i)$ are neither stationary nor independent, but contain transient effects. We have for example

$$(2.16) \qquad Pr\{X^{(\cdot)}(1) > x; T(1) = a\} = Pr\{X^{(b)}(1) > X^{(a)}(1) > x\}$$

$$= \int_x^\infty \{1 - G^{(b)}(y)\} \, dG^{(a)}(y)$$

and, marginally,

$$(2.17) \qquad Pr\{T(1) = a\} = \int_0^\infty \{1 - G^{(b)}(y)\} \, dG^{(a)}(y).$$

The only simple case is where the two renewal processes are Poisson processes with parameters ρ_a and ρ_b. Then, of course, $\{X^{(\cdot)}(i)\}$ is a Poisson process of rate $\rho_a + \rho_b$ and $T(i)$ is an independent binomial sequence

$$(2.18) \qquad Pr\{T(1) = a\} = Pr\{T(1) = a | X^{(\cdot)} > x\} = \frac{\rho_a}{\rho_a + \rho_b}.$$

EXAMPLE 2.2. *Semi-Markov processes (Markov renewal processes)*.   The two state semi-Markov process is the simplest bivariate process with dependent structure and plays, in bivariate process theory, a role similar to that played in univariate process theory by the renewal process. It is, in a sense, the closest one gets in bivariate processes to a regenerative process. The process is defined in terms of the sequences $\{X_a^{(\cdot)}(i), T_a(i)\}$, and $\{X_b^{(\cdot)}(i), T_b(i)\}$, the type processes $\{T_a(i)\}$ and $\{T_b(i)\}$ being Markov chains with transition matrix

$$(2.19) \qquad \mathbf{P} = \begin{pmatrix} p_{aa} & p_{ab} \\ p_{ba} & p_{bb} \end{pmatrix} = \begin{pmatrix} \alpha_1 & 1 - \alpha_1 \\ 1 - \alpha_2 & \alpha_2 \end{pmatrix},$$

while the distributions of the random variables $X_a^{(\cdot)}(i)$, $X^{(\cdot)}(i)$, and $X_b^{(\cdot)}(i)$ depend only on the type of events at i and $(i - 1)$. Thus, illustrating the regenerative nature of the process, we define $F_{aa}(x)$ to be, for $i \geq 2$,

$$
\begin{aligned}
(2.20) \qquad F_{aa}(x) &= Pr\{X_a^{(\cdot)}(i) \leq x \,|\, T_a(i) = a, T_a(i-1) = a\} \\
&= Pr\{X^{(\cdot)}(i) \leq x \,|\, T.(i) = a, T.(i-1) = a\} \\
&= Pr\{X^{(\cdot)}(i) \leq x \,|\, T(i) = a, T(i-1) = a\}
\end{aligned}
$$

with equivalent definitions for $F_{ab}(x)$, $F_{ba}(x)$, $F_{bb}(x)$. Thus, the effect of the initial sampling disappears when the type of the first subsequent events is known.

The joint distributions of the time from the origin to the first event and the type of the first event $(i = 1)$ are either quite arbitrary initial conditions, or initial conditions established by the kind of sampling involved at the origin and denoted by the subscript on the interval random variable. Thus for asynchronous sampling, we get stationary initial conditions which are specified by the joint distribution of $X^{(\cdot)}(i)$ and $T(1)$.

These stationary equilibrium conditions (Pyke and Schaufele [43]) are that $T(1) = a$ and $T(1) = b$ have probabilities p_a and p_b, where

$$(2.21) \qquad \{p_a, p_b\} = \{p_a, p_b\}\, \mathbf{P} = \left\{ \frac{1 - \alpha_2}{2 - \alpha_1 - \alpha_2}, \frac{1 - \alpha_1}{2 - \alpha_1 - \alpha_2} \right\},$$

the equilibrium probabilities of the Markov chain, and the time from the origin to an event of type a has distribution function

$$(2.22) \qquad \frac{p_{ba} \int_0^x R_{ba}(u)\, du}{E\big(X_b^{(b)}(1)\big)} + \frac{p_{aa} \int_0^x R_{aa}(u)\, du}{E\big(X_a^{(a)}(1)\big)},$$

with a similar definition for the time to an event of type b.

Cinlar [9] has reviewed the properties of semi-Markov processes. Our view of these processes, being related to statistical problems arising in the analysis of bivariate point processes, will be somewhat different from the usual one. Thus, note that in the marginal processes the regenerative property of the semi-Markov process implies that the times between events of type a, $X_a^{(a)}(i), i = 1, 2, \cdots$, are independent and identically distributed, as are the $X_b^{(b)}(j), j = 1, 2, \cdots$. Therefore, the marginal processes are renewal processes and we say that the semi-Markov process is a *bivariate renewal process*. Since the types of successive renewals (events) form a Markov chain, the process is also called a Markov renewal process ([9], p. 130). However, the two marginal renewal processes together with the Markov chain of event types do not determine the process.

The dependency structure of this bivariate renewal process can also be examined through joint properties of forward recurrence times in the process. The joint forward recurrence times $\{X_a^{(a)}(1), X_a^{(b)}(1)\}$ for semisynchronous sampling of b by a are, in the terminology of semi-Markov process theory, the first passage times from state a to state a and from state a to state b, with similar definitions for $\{X_b^{(a)}(1), X_b^{(b)}(1)\}$. Denoting the marginal distributions of these random variables by $F_a^{(a)}(x)$ and so forth, we have the equations

$$(2.23) \qquad F_a^{(a)}(x) = p_{aa}F_{aa}(x) + p_{ab}F_{ab}(x) * F_b^{(a)}(x),$$

$$(2.24) \qquad F_a^{(b)}(x) = p_{ab}F_{ab}(x) + p_{aa}F_{aa}(x) * F_a^{(b)}(x),$$

$$(2.25) \qquad F_b^{(a)}(x) = p_{ba}F_{ba}(x) + p_{bb}F_{bb}(x) * F_b^{(a)}(x),$$

$$(2.26) \qquad F_b^{(b)}(x) = p_{bb}F_{bb}(x) + p_{ba}F_{ba}(x) * F_a^{(b)}(x),$$

where $*$ denotes Stieltjes convolution and $F_{aa}(x)$ and so forth, are defined in (2.20).

These equations can be solved using Laplace–Stieltjes transforms. Thus, if $\mathscr{F}_a^{(a)}(s)$ is the Laplace–Stieltjes transform of $F_a^{(a)}(x)$, and so forth, we get

$$(2.27) \qquad \mathscr{F}_a^{(a)}(s) = \alpha_1 \mathscr{F}_{aa}(s) + \frac{(1 - \alpha_1)(1 - \alpha_2)\mathscr{F}_{ab}(s)\mathscr{F}_{ba}(s)}{1 - \alpha_2 \mathscr{F}_{bb}(s)},$$

$$(2.28) \qquad \mathscr{F}_b^{(b)}(s) = \alpha_1 \mathscr{F}_{bb}(s) + \frac{(1 - \alpha_1)(1 - \alpha_2)\mathscr{F}_{ab}(s)\mathscr{F}_{ba}(s)}{1 - \alpha_1 \mathscr{F}_{aa}(s)}.$$

From these results, we can write down joint forward recurrence time distributions using the regenerative properties of the process. For example,

$$(2.29) \quad R_{X_a^{(a)}(1), X_a^{(b)}}(x_1, x_2)$$

$$= Pr\{X_a^{(a)}(1) > x_1, X_a^{(b)} > x_2\}$$

$$= \begin{cases} \alpha_1 R_{aa}(x_1) + (1 - \alpha_1)\left\{R_{ab}(x_1) + \displaystyle\int_{x_2}^{x_1}[1 - F_b^{(a)}(x_1 - u)]\,dF_{ab}(u)\right\} \\ \qquad\qquad\qquad\qquad\qquad\qquad\qquad \text{if } x_1 \geqq x_2, \\[2ex] (1 - \alpha_1)R_{ab}(x_2) + \alpha_1\left\{R_{aa}(x_2) + \displaystyle\int_{x_1}^{x_2}[1 - F_a^{(b)}(x_2 - u)]\,dF_{aa}(u)\right\} \\ \qquad\qquad\qquad\qquad\qquad\qquad\qquad \text{if } x_2 \geqq x_1. \end{cases}$$

It is actually much simpler, because of the regenerative nature of the process, to express results in terms of the order statistics and order types associated with $R(x_1, x_2)$. These aspects of the process are worked out in greater detail by Wisniewski [50].

Note that the process derived previously as two independent Poisson processes is a very particular form of semi-Markov process. The question then arises whether there are any other semi-Markov processes with Poisson marginals and the answer is clearly yes. For example, when $\alpha_1 = \alpha_2 = 0$ we have the special case of an alternating renewal process and choosing $F_{ab}(x)$ and $F_{ba}(x)$ to be distributions of random variables proportional to chi square variables with one degree of freedom gives Poisson marginals. The example shows in fact that one can produce any desired marginal renewal processes in a semi-Markov process, as is also clear from (2.27) and (2.28).

From equations such as (2.14) and (2.15), it can be shown that no bivariate process of independent renewal marginals is a semi-Markov process unless the marginals are also Poisson processes. The dependency structure in a semi-Markov process is actually better characterized by the second order cross intensity function, which we introduce in the next section, rather than by joint moments of forward recurrence times. This cross intensity function together with the two distributions of intervals in the marginal renewal processes (or equivalently the intensity functions of the marginal renewal processes) completely specifies the semi-Markov process.

3. Dependence and correlation in bivariate point processes

3.1. *Specification*. We now consider in more detail the specification of the structure of bivariate point processes. It is common in the study of particular stochastic processes to find that physically the same process can be specified in several equivalent but superficially different ways. A simple and familiar example is the stationary univariate Poisson process which can be specified as:

(a) a process in which the numbers of events in disjoint sets have independent Poisson distributions with means proportional to the measures of the sets;

(b) a renewal process with exponentially distributed intervals;

(c) a process in which the probability of an event in $(t, t + \Delta t]$ has an especially simple form, as $\Delta t \to 0$.

We call these three specifications, respectively, the counting, the interval, and the intensity specifications. Univariate point processes can in general be specified in these three ways, if the initial conditions are properly chosen.

While the counting specification (a) for bivariate point processes is in principle fundamental, it is often too complicated to be very fruitful. If the joint characteristic functional of the process, defined by an obvious generalization of the univariate case, can be obtained in a useful form, this does give a concise representation of all the joint distributions of counts; even then, such a characteristic functional would usually give little insight into the physical mechanism generating the process.

Often, special processes are most conveniently handled through some kind of interval specification, especially when this corresponds rather closely to the physical origin of the process. In particular, the two state semi-Markov process

is most simply specified in this way, as shown in Section 2.8. The two main types
of interval specifications discussed in Section 2 were the specifications in terms
of the intervals in the marginal processes, or the intervals and event types in the
pooled process. The latter is the basic specification for the semi-Markov process.
Other processes, such as various kinds of inhibited processes and the bivariate
Poisson process of Section 4.1 are specified rather less directly in terms of
relations between intervals and event types.

However, in some ways the most convenient general specification is through
the intensity. Denote by $\mathcal{H}_t$ the history of the process at time t, that is, a complete specification of the occurrences in $(-\infty, t]$ measured backwards from an
origin at t, then two time points t', t'' have the same history if and only if the
observed sequences $\{x^{(a)}(-1), x^{(a)}(-2), \cdots\}$, $\{x^{(b)}(-1), x^{(b)}(-2), \cdots\}$ are identical if measured from origins at t' and at t''.

Then a marginally regular process is specified by

$$(3.1) \qquad \lambda^{(a)}(t; \mathcal{H}_t) = \lim_{\Delta t \to 0+} \frac{Pr\{N^{(a)}(t, t + \Delta t) \geq 1 | \mathcal{H}_t\}}{\Delta t},$$

$$(3.2) \qquad \lambda^{(b)}(t; \mathcal{H}_t) = \lim_{\Delta t \to 0+} \frac{Pr\{N^{(b)}(t, t + \Delta t) \geq 1 | \mathcal{H}_t\}}{\Delta t},$$

$$(3.3) \qquad \lambda^{(ab)}(t; \mathcal{H}_t) = \lim_{\Delta t \to 0+} \frac{Pr\{N^{(a)}(t, t + \Delta t) N^{(b)}(t, t + \Delta t) \geq 1 | \mathcal{H}_t\}}{\Delta t}.$$

We call these functions the *complete intensity functions* of the process. If the
process is regular $\lambda^{(ab)}(t; \mathcal{H}_t) = 0$. Given the functions (3.1) through (3.3) and
some initial conditions, we can construct a discretized realization of the process,
although this is, of course, a clumsy method of simulation if the interval
specification is at all simple.

One advantage of the complete intensity specification is that one can generate
families of models of increasing complexity by allowing more and more complex
dependency on $\mathcal{H}_t$. This may be useful, for instance, in testing consistency of
data with a given type of model, for example, a semi-Markov process. Further,
if the main features of $\mathcal{H}_t$ that determine the intensity functions can be found,
an appropriate type of model may be indicated.

As an example of a complete intensity specification, consider the two state
semi-Markov process. Here the only aspects of $\mathcal{H}_t$ that are relevant, if at least
one event has occurred before t, are the backward recurrence time to the previous event and the type of that event $\{x^{(\cdot)}(-1), t(-1)\}$. Any initial conditions
disappear once one event has occurred. For convenience, we write the partial
history as (u, a), if the preceding event is of type a and (u, b) if it is of type b. Then
assuming that the process is regular, that is, that none of the interval distributions
has an atom at zero, we have

$$(3.4) \qquad \lambda^{(a)}\{t; (u, a)\} \equiv \lambda_a^{(a)}(u) = \frac{p_{aa} f_{aa}(u)}{p_{aa} R_{aa}(u) + p_{ab} R_{ab}(u)},$$

$$(3.5) \qquad \lambda^{(b)}\{t; (u, a)\} \equiv \lambda_a^{(b)}(u) = \frac{p_{ab} f_{ab}(u)}{p_{aa} R_{aa}(u) + p_{ab} R_{ab}(u)},$$

with similar expressions defining $\lambda_b^{(a)}(u)$ and $\lambda_b^{(b)}(u)$ when the partial history is (u, b). These complete intensities are analogues of the hazard function, or age specific failure rate, which can be used to specify a univariate renewal process.

The semi-Markov process is characterized by the dependence on $\mathcal{H}_t$ being only on u and the type of the preceding event. We can generalize the semi-Markov process in many ways, for instance by allowing a dependence on both of the backward recurrence times $x^{(a)}(-1)$ and $x^{(b)}(-1)$; see Section 3.8.

3.2. *Properties of complete intensity functions.* We now consider briefly some properties of complete intensity functions. It is supposed for simplicity that the process is regular and that it is observed for a time long enough to allow initial conditions to be disregarded. For the semi-Markov process "long enough" is the occurrence of at least one event.

(i) If the process is completely stationary, as defined in Section 2.3, the intensity functions depend only on $\mathcal{H}_t$ and not on t.

(ii) Nonstationary generalizations of a given stationary process can be produced by inserting into the intensity a function either of t, for example, $e^{\gamma t}$, $\exp\{\gamma \cos(\omega_0 t + \phi)\}$, or of the numbers of events that have occurred since the start of the process.

(iii) The intensity specification of a stationary process is unique in the sense that if we have two different intensity specifications and can find a set of histories of nonzero probability such that, say, the first intensity specification gives greater intensity of events of type a than the second, then the two processes are distinguishable from suitable data. Note that this is not the same question as whether two different specifications containing unknown parameters are distinguishable.

(iv) The events of different types are independent if and only if $\lambda^{(a)}$ and $\lambda^{(b)}$ involve $\mathcal{H}_t$ only through the histories of the separate processes of events of type a and type b, denoted, respectively, by $\mathcal{H}_t^{(a)}$ and $\mathcal{H}_t^{(b)}$.

(v) We can call the process purely a dependent if both $\lambda^{(a)}$ and $\lambda^{(b)}$ depend on $\mathcal{H}_t$ only through $\mathcal{H}_t^{(a)}$. In many ways the simplest example of such a process is obtained when both intensities depend only on the backward recurrence time in the process of events of type a, that is, on the time $u^{(a)}$ measured back to the previous event of type a. Denote the intensities by $\lambda^{(a)}(u^{(a)})$ and $\lambda^{(b)}(u^{(a)})$. Then the events of type a form a renewal process; if in particular $\lambda^{(a)}(\cdot)$ is constant, the events of type a form a Poisson process. If simple functional forms are assumed for the intensities, the likelihood of data can be obtained in a fairly simple form and hence an efficient statistical analysis derived.

(vi) A different kind of purely a dependent process is derived from a shot noise process based on the a events, that is, by considering a stochastic process $Z^{(a)}(t)$ defined by

$$(3.6) \qquad Z^{(a)}(t) = \int_0^\infty g(u) \, dN^{(a)}(t - u).$$

We then take $\lambda^{(b)}(t)$ and possibly also $\lambda^{(a)}(t)$ to depend only on $Z^{(a)}(t)$. In particular if $g(u) = 1(u < \Delta)$ and $g(u) = 0(u \geq \Delta)$, the intensities depend only on the number of events of type a in $(t - \Delta, t)$. Hawkes [22] has considered some processes of this type.

(vii) The intensity functions look in one direction in time. This approach is therefore rather less suitable for processes in a spatial continuum, where there may be no reason for picking out one spatial direction rather than another.

(viii) Some simple processes, for example, the bivariate Poisson process of Section 4.1, have intensity specifications that appear quite difficult to obtain.

3.3. *Second order cross intensity functions.* In Section 3.2, we considered the complete intensity functions which specify probabilities of occurrence given the entire history $\mathscr{H}_t$. For some purposes, it is useful with stationary (second order) processes to be less ambitious and to consider probabilities of occurrence conditionally on much less information than the entire history $\mathscr{H}_t$. We then call the functions corresponding to (3.1) through (3.6) incomplete intensity functions. For example, using the notation of (3.4) and (3.5), both

$$(3.7) \qquad \lim_{\Delta t \to 0+} Pr \frac{\{N^{(a)}(t, t + \Delta t) \geq 1 | (u, a)\}}{\Delta t}$$

and the corresponding function when the last event is of type b are defined for any regular stationary process, even though they specify the process completely only for semi-Markov processes.

A particularly important incomplete intensity function is obtained when one conditions on the information that an event of specified type occurs at the time origin. Again for simplicity, we consider stationary regular processes. Write

$$(3.8) \qquad h_a^{(a)}(t) = \lim_{\Delta t \to 0+} Pr \frac{\{N^{(a)}(t, t + \Delta t) \geq 1 | \text{type } a \text{ event at } 0\}}{\Delta t},$$

$$(3.9) \qquad h_a^{(b)}(t) = \lim_{\Delta t \to 0+} Pr \frac{\{N^{(b)}(t, t + \Delta t) \geq 1 | \text{type } a \text{ event at } 0\}}{\Delta t},$$

with similar definitions for $h_b^{(a)}(t)$, $h_b^{(b)}(t)$ if an event of type b occurs at 0. We call the function (3.9) a second order cross intensity function. For nonregular processes, it may be helpful to introduce intensities conditionally on events of both types occurring at 0.

Note that the cross intensity functions $h_a^{(b)}(t)$ and $h_b^{(a)}(t)$ will contain Dirac delta functions as components if, for example, there is a nonzero probability that an event of type b will occur exactly τ away from a type a event. For the process to be regular, rather than merely marginally regular, the cross intensity functions must not contain delta functions at the origin.

Note too that $h_a^{(b)}(t)$ is well defined near $t = 0$ and will typically be continuous there.

If following an event of type a at the origin, subsequent events of type a are at times $S_a^{(a)}(1)$, $S_a^{(a)}(2)$, $\cdots$ and those of type b at $S_a^{(b)}(1)$, $S_a^{(b)}(2)$, $\cdots$, we can write

$$(3.10) \qquad h_a^{(a)}(t) = \sum_{r=1}^{\infty} f_{S_a^{(a)}(r)}(t),$$

$$(3.11) \qquad h_a^{(b)}(t) = \sum_{r=1}^{\infty} f_{S_a^{(b)}(r)}(t),$$

where $f_U(\cdot)$ is the probability density function of the random variable U.

If the process of type a events is a renewal process, (3.10) is a function familiar as the renewal density. For small t the contribution from $r = 1$ is likely to be dominant in all these functions.

The intensities are defined for all t. However, $h_a^{(a)}(t)$ and $h_b^{(b)}(t)$ are even functions of t. Further, it follows from the definition of conditional probability that

$$(3.12) \qquad h_b^{(a)}(t)\rho_b = h_a^{(b)}(-t)\rho_a,$$

where ρ_a and ρ_b are the rates of the two processes. For processes without long term effects, we have that as $t \to \infty$ or $t \to -\infty$

$$(3.13) \qquad h_a^{(a)}(t) \to \rho_a, \qquad h_b^{(a)}(t) \to \rho_a, \qquad h_b^{(b)}(t) \to \rho_b.$$

Sometimes it may be required to calculate the intensity function of the superposed process, that is, the process in which the type of event is disregarded. Given an event at the time origin, it has probability $\rho_a/(\rho_a + \rho_b)$ of being a type a, and hence the intensity of the superposed process is

$$(3.14) \qquad h_{\cdot}^{(\cdot)}(t) = \frac{\rho_a}{\rho_a + \rho_b}\{h_a^{(a)}(t) + h_a^{(b)}(t)\} + \frac{\rho_b}{\rho_a + \rho_b}\{h_b^{(a)}(t) + h_b^{(b)}(t)\}.$$

This is a general formula for the intensity function of the superposition of two, possibly dependent, processes.

3.4. *Covariance densities.* For some purposes, it is slightly more convenient to work with covariance densities rather than with the second order intensity functions; see, for example, Bartlett [4]. To define the *cross covariance density*, we consider the random variables $N^{(a)}(0, \Delta't)$ and $N^{(b)}(t, t + \Delta''t)$ and define

$$(3.15) \qquad \gamma_a^{(b)}(t) = \lim_{\Delta't, \Delta''t \to 0+} \frac{\text{Cov}\{N^{(a)}(0, \Delta't), N^{(b)}(t, t + \Delta''t)\}}{\Delta't\,\Delta''t}$$

$$= \rho_a h_a^{(b)}(t) - \rho_a\rho_b.$$

It follows directly from (3.14) or from (3.12) and (3.15), that $\gamma_a^{(b)}(t) = \lambda_b^{(a)}(-t)$. Note that an *autocovariance density* such as $\gamma_a^{(a)}(t)$ can be written

$$(3.16) \qquad \gamma_a^{(a)}(t) = \rho_a\delta(t) + \gamma_{a,\text{cont}}^{(a)}(t) = \rho_a\delta(t) + \rho_a\{h_b^{(a)}(t) - \rho_a\},$$

where the second terms are continuous at $t = 0$ and $\delta(t)$ denotes the Dirac delta function.

We denote by $V^{(ab)}(t_1, t_2)$ the covariance between $N^{(a)}(t_1)$ and $N^{(b)}(t_2)$ in the stationary bivariate process. This is called the *covariance time surface*. Then

$$(3.17) \qquad V^{(ab)}(t_1, t_2) = \text{Cov }\{N^{(a)}(t_1), N^{(b)}(t_2)\}$$

$$= \text{Cov }\left\{\int_0^{t_1} dN^{(a)}(u), \int_0^{t_2} dN^{(b)}(v)\right\}$$

$$= \int_0^{t_1} \int_0^{t_2} \gamma_b^{(a)}(u - v) \, du \, dv.$$

In the special case $t_1 = t_2 = t$, we write $V^{(ab)}(t, t) = V^{(ab)}(t)$. It follows from (3.17) that

$$(3.18) \qquad V^{(ab)}(t) = \int_0^t (t - v)\{\gamma_b^{(a)}(v) + \gamma_a^{(b)}(v)\} \, dv.$$

Note that in (3.18) a delta function component at the origin can enter one but not both of the cross covariance densities. If, in (3.18), we take the special highly degenerate case when the type b and type a processes coincide point for point, we obtain the well-known variance time formula

$$(3.19) \qquad V^{(aa)}(t) = \text{Var }\{N^{(a)}(t)\}$$

$$= \rho_a t + 2 \int_{0+}^t (t - v)\gamma_a^{(a)}(v) \, dv.$$

An interesting question concerns the conditions under which a set of functions $\{\gamma_a^{(a)}(t), \gamma_b^{(a)}(t), \gamma_a^{(b)}(t), \gamma_b^{(b)}(t)\}$ can be the covariance densities of a bivariate point process. Now for all α and β

$$(3.20) \qquad \text{Var }\{\alpha N^{(a)}(t_1) + \beta N^{(b)}(t_2)\}$$

$$= \alpha^2 V^{(aa)}(t_1) + 2\alpha\beta V^{(ab)}(t_1, t_2) + \beta^2 V^{(bb)}(t_2)$$

$$\geq 0.$$

Thus for all t_1 and t_2,

$$(3.21) \qquad V^{(aa)}(t_1) \geq 0, \qquad \{V^{(ab)}(t_1, t_2)\}^2 \leq V^{(aa)}(t_1) V^{(bb)}(t_2), \qquad V^{(bb)}(t_2) \geq 0.$$

The conditions (3.21) can be used to show that certain proposed functions cannot be covariance densities. It would be interesting to know whether corresponding to any functions satisfying (3.21) there always exists a corresponding stationary bivariate point process.

Nothing special is learned by letting t_1 and $t_2 \to 0$ in (3.21). If, however, we let $t_1 = t_2 \to \infty$, we have under weak conditions that

$$(3.22) \qquad V^{(aa)}(t) \sim t\left\{\rho_a + 2 \int_{0+}^\infty \gamma_a^{(a)}(v) \, dv\right\} = \rho_a t I^{(aa)},$$

$$(3.23) \qquad V^{(ab)}(t) \sim t \int_0^\infty \{\gamma_b^{(a)}(v) + \gamma_a^{(b)}(v)\} \, dv = (\rho_a \rho_b)^{1/2} t I^{(ab)},$$

$$(3.24) \qquad V^{(bb)}(t) \sim t \left\{ \rho_b + 2 \int_0^\infty \gamma_b^{(b)}(v)\, dv \right\} = \rho_b t I^{(bb)},$$

where the right sides of these equations define three asymptotic measures of dispersion I. The conditions

$$(3.25) \qquad I^{(aa)} \geqq 0, \qquad \{I^{(ab)}\}^2 \leqq I^{(aa)} I^{(bb)}, \qquad I^{(bb)} \geqq 0$$

must, of course, be satisfied, in virtue of (3.21).

3.5. *Some special processes.* The second order intensity functions, or equivalently the covariance densities, are not the most natural means of representing the dependencies in a point process, if these dependencies take special account of the nearest events of either or both types. Thus for the semi-Markov process, the second order intensity functions satisfy integral equations; see, for example, Cox and Miller [15], pp. 352–356. The relation with the defining functions of the process is therefore indirect. Thus, while in principle the distributions defining the process could be estimated from data via the second order intensity functions, this would be a roundabout approach, and probably very inefficient.

We now discuss briefly two processes for which the second order intensity functions are more directly related to the underlying mechanism of the process.

Consider an arbitrary regular stationary process of events of type a. Let each event of type a be displaced by a random amount to form a corresponding event of type b; the displacements of different points are independent and identically distributed random variables with probability density function $p(\cdot)$. Denote the probability density function of the difference between two such random variables by $q(\cdot)$. Then a direct probability calculation for the limiting, stationary, process shows that $\big($Cox [12]$\big)$

$$(3.26) \qquad h_a^{(b)}(t) = p(t) + \int_{-\infty}^\infty h_a^{(b)}(v) p(t-v)\, dv,$$

$$(3.27) \qquad h_b^{(b)}(t) = \int_{-\infty}^\infty h_a^{(a)}(v) q(t-v)\, dv.$$

In particular, if the type a events form a Poisson process, $h_a^{(a)}(v) = \rho_a$, so that

$$(3.28) \qquad h_a^{(b)}(t) = p(t) + \rho_a, \qquad h_b^{(b)}(t) = \rho_a.$$

The constancy of $h_b^{(b)}(t)$ is an immediate consequence of the easily proved fact that the type b events on their own form a Poisson process. The results (3.28) lead to quite direct methods of estimating $p(\cdot)$ from data and to tests of the adequacy of the model. For positive displacements the type a events could be the inputs to an $M/G/\infty$ queue, the type b events being the outputs. Generalizations of this delay process are considered in Sections 4 and 5.

As a second example, consider a *bivariate doubly stochastic Poisson process.* That is, we have an unobservable real valued (nonnegative) stationary bivariate process $\{\mathbf{\Lambda}(t)\} = \{\Lambda_a(t), \Lambda_b(t)\}$. Conditionally on the realized value of this process, we observe two independent nonstationary Poisson processes with rates, respectively, $\Lambda_a(t)$ for the type a events and $\Lambda_b(t)$ for the type b events. Then, by first arguing conditionally on the realized value of $\{\Lambda(t)\}$, we have a stationary bivariate point process with

(3.29)
$$\gamma_a^{(a)}(t) = \rho_a \delta(t) + c_\Lambda^{(aa)}(t), \qquad \gamma_b^{(a)}(t) = c_\Lambda^{(ab)}(t),$$
$$\gamma_b^{(b)}(t) = \rho_b \delta(t) + c_\Lambda^{(bb)}(t),$$

where $E\{\Lambda(t)\} = (\rho_a, \rho_b)$ and the c_Λ are the auto and cross covariance functions of $\{\Lambda(t)\}$.

Again there is a quite direct connection between the covariance densities and the underlying mechanism of the process. Two special cases are of interest. One is when $\Lambda_a(t) = \Lambda_b(t) \rho_a/\rho_b$, leading to some simplification of (3.29). Another special case is

(3.30)
$$\Lambda_a(t) = \rho_a + R_a \cos(\omega_0 t + \theta + \Phi),$$
$$\Lambda_b(t) = \rho_b + R_b \cos(\omega_0 t + \Phi).$$

In this

(3.31)
$$E(R_a) = E(R_b) = 0,$$
$$E(R_a^2) = \sigma_{aa}, \qquad E(R_{ab}) = \sigma_{ab}, \qquad E(R_{bb}) = \sigma_{bb}.$$

and the random variable Φ is uniformly distributed over $(0, 2\pi)$ independently of R_a and R_b. Further, ρ_a, ρ_b, ω_0, and θ are constants and, to keep the Λ nonnegative, $|R_a| \leqq \rho_a, |R_b| \leqq \rho_b$. This defines a stationary although nonergodic process $\{\Lambda(t)\}$.

Specifications (3.30) and (3.31) yield

(3.32)
$$\gamma_a^{(a)}(t) = \rho_a \delta(t) + \sigma_{aa} \cos(\omega_0 t), \qquad \gamma_b^{(a)}(t) = \sigma_{ab} \cos\{\omega_0(t + \theta)\},$$
$$\gamma_b^{(b)}(t) = \rho_b \delta(t) + \sigma_{bb} \cos(\omega_0 t).$$

Of course this process is extremely special. Note, however, that fairly general processes with a sinusoidal component in the intensity can be produced by starting from the complete intensity functions of a stationary process and either adding a sinusoidal component or multiplying by the exponential of a sinusoidal component; the latter has the advantage of ensuring automatically a nonnegative complete intensity function.

3.6. *Spectral analysis of the counting process.* For Gaussian stationary stochastic processes, study of spectral properties is useful for three rather different reasons:

(a) the spectral representation of the process itself may be helpful;

(b) the spectral representation of the covariance matrix may be helpful;

(c) the effect on the process of a stationary linear operator is neatly expressed.

For point processes a general representation analogous to (a) has been discussed by Brillinger [6]. Bartlett [1] has given some interesting second order theory and applications in the univariate case. For doubly stochastic Poisson processes, which are of course very special, we can often use a full spectral representation for the defining $\{\Lambda(t)\}$ process; indeed the $\{\Lambda(t)\}$ process may be nearly Gaussian.

If we are content with a spectral analysis of the covariance density, we can write, in particular for the complex valued *cross spectral density function*,

$$(3.33) \qquad g_b^{(a)}(\omega) = \frac{1}{2\pi} \int_{-\infty}^{\infty} e^{-i\omega t} \gamma_b^{(a)}(t)\, dt = g_a^{(b)}(-\omega).$$

Because of the mathematical equivalence between the covariance density and the spectral density, the previous general and particular results for covariance densities can all be expressed in terms of the spectral properties. While these will not be given in full here, note first that the measures of dispersion in (3.22) through (3.24) are given by

$$(3.34) \qquad \begin{aligned} \rho_a I^{(aa)} &= 2\pi g_a^{(a)}(0), \qquad (\rho_a \rho_b)^{1/2} I^{(ab)} = 2\pi g_b^{(a)}(0), \\ \rho_b I^{(bb)} &= 2\pi g_b^{(b)}(0). \end{aligned}$$

All the results (3.26) through (3.32) can be expressed simply in terms of the spectral properties. Thus, from (3.29), the spectral analysis of the bivariate doubly stochastic Poisson process leads directly to the spectral properties of the process $\{\Lambda(t)\}$ on subtracting the "white" Poisson spectra. Thus, spectral analysis of a doubly stochastic Poisson process is likely to be useful whenever the process $\{\Lambda(t)\}$ has an enlightening spectral form. In the special case (3.32) where $\{\Lambda(t)\}$ is sinusoidal,

$$(3.35) \qquad g_a^{(a)}(\omega) = \frac{\rho_a}{2\pi} + \frac{\sigma_{aa}}{2\pi} \delta(\omega - \omega_0),$$

$$(3.36) \qquad g_b^{(a)}(\omega) = \frac{\sigma_{ab}}{2\pi} e^{i\omega\theta} \delta(\omega - \omega_0).$$

The complex valued cross spectral density can be split in the usual way into real and imaginary components, which indicate the relative phases of the fluctuations in the processes of events of type a and type b. We can also define the coherency as $|g_b^{(a)}(\omega)|^2/\{g_a^{(a)}(\omega)g_b^{(b)}(\omega)\}$; for the doubly stochastic process driven by proportional intensities, $\Lambda_b(t) \propto \Lambda_a(t)$, and the coherency is one for all ω, provided that the "white" Poisson component is removed from the denominator.

The natural analogue for point processes of the stationary linear operators on a real valued process is random translation, summarized in (3.26) and (3.27). It follows directly from these equations and from the relation between covariance densities and intensity functions that

$$(3.37) \qquad g_a^{(b)}(\omega) = \rho_a p^\dagger(\omega) + g_a^{(a)}(\omega)p^\dagger(\omega),$$

$$(3.38) \qquad \left\{g_b^{(b)}(\omega) - \frac{\rho_b}{2\pi}\right\} = q^\dagger(\omega)\left\{g_a^{(a)}(\omega) - \frac{\rho_a}{2\pi}\right\},$$

where $p^\dagger(\omega)$, $q^\dagger(\omega)$ are the Fourier transforms of $p(t)$ and $q(t)$. A more general type of random translation for bivariate processes is discussed in Section 5.

3.7. *Variance and covariance time functions.* For univariate point processes the covariance density or spectral functions are mathematically equivalent to the variance time function $V^{(aa)}(t)$ of (3.19), which gives as a function of t the variance of the number of events in an interval of length t. This function is useful for some kinds of statistical analysis; examination of its behavior for large t is equivalent analytically to looking at the low frequency part of the spectrum.

For bivariate point processes, it might be thought that the variance time function $V^{(aa)}(t)$, $V^{(bb)}(t)$, and the covariance time function $V^{(ab)}(t)$ of (3.18) are equivalent to the other second order specifications. This is not the case, however, because it is clear from (3.18) that only the combinations $\gamma_b^{(a)}\omega) + \gamma_a^{(b)}(\omega)$ can be found from $V^{(ab)}(t)$ and this is not enough to fix the cross covariance function of the process.

The cross covariance density can, however, be found from the covariance time surface, $V^{(ab)}(t_1, t_2)$ of (3.17).

The covariance time function and surface are useful for some rather special statistical purposes.

The variance time function of the superposed process is

$$(3.39) \qquad V^{(\cdot\cdot)}(t) = V^{(aa)}(t) + 2V^{(ab)}(t) + V^{(bb)}(t);$$

this is equivalent to the relation (3.14) for intensity functions.

3.8. *Bivariate interval specifications; bivariate Markov interval processes.* As has been mentioned several times in this section, the second order intensity functions and their equivalents are most likely to be useful when the dependencies in the underlying mechanism do not specifically involve nearest neighbors, or other features of the process that are most naturally expressed serially, that is, through event number either in the pooled process or in the marginal processes rather than through real time.

For processes in which an interval specification is more appropriate, there are many ways of introducing functions wholly or partially specifying the dependency structure of the process. For a stationary univariate process we can consider the sequence of intervals between successive events as a stochastic process, indexed by serial number, that is, as a real valued process in discrete time. The second order properties are described by an autocovariance sequence which, say for events of type a, is

$$(3.40) \qquad \gamma_x^{(aa)}(j) = \mathrm{Cov}\left\{X_a^{(a)}(k),\, X_a^{(a)}(k+j)\right\}, \qquad j = 0, \pm 1, \cdots.$$

McFadden [35] has shown that the autocovariance sequence is related to the distribution of counts by the simple, although indirect, formula

$$(3.41) \qquad \gamma_x^{(aa)}(j) = \frac{1}{\rho_a} \left[\int_0^\infty Pr\{N^{(a)}(t) = j\} \, dt - \frac{1}{\rho_a} \right].$$

Stationarity of the $X_a^{(a)}(i)$ sequence is discussed in Slivnyak [44].

If the distribution of $N^{(a)}(t)$ is given by the probability generating function

$$(3.42) \qquad \phi^{(a)}(z, t) = \sum_{j=0}^\infty z^j Pr\{N^{(a)}(t) = j\}$$

with Laplace transform $\phi^{(a)*}(z, s)$, it will be convenient to substitute in (3.41) the result

$$(3.43) \qquad \int_0^\infty Pr\{N^{(a)}(t) = j\} \, dt = \left[\frac{1}{j!} \frac{\partial^j \phi^{(a)*}(z, s)}{\partial z^j} \right]_{z=0, s=0+}.$$

The sequence (3.40) and the analogous one for events of type b summarize the second order marginal properties. To study the joint properties of various kinds of intervals between events, the following are some of the possibilities.

(i) The two sets of intervals $\{X_a^{(a)}(r), X_b^{(b)}(r)\}$ may be considered as a bivariate process in discrete time, that is, we may use serial number in each process as a common discrete time scale. Cross covariances and cross spectra can then be defined in the usual way. While this may occasionally be fruitful, it is not a useful general approach, because for almost all physical models events in the two processes with a common serial number will be far apart in real time. Another problem is that if the process is sampled semisynchronously, say on a type a event, the sequence $X_a^{(b)}(r)$ is not a stationary sequence, although it will generally "converge" to the sequence $X_b^{(b)}(r)$. Again, sufficiently far out from the sampling point, events in the two processes with common serial number will be far apart in real time.

(ii) We may consider the intervals between successive events in the process taken regardless of type, that is, the superposed process. This gives a third covariance sequence, namely, $\gamma_x^{(\cdot\cdot)}(j)$. For particular processes this can be calculated from (3.41) applied to the pooled process, particularly if the joint distribution of the count $N^{(a)}(t)$ and $N^{(b)}(t)$ are available.

In fact, if the joint distribution of $\{N^{(a)}(t), N^{(b)}(t)\}$ is specified by the joint probability generating function $\phi^{(ab)}(z_a, z_b, t)$ with Laplace transform $\phi^{(ab)*}(z_a, z_b, s)$, we have from (3.41) and (3.43) that

$$(3.44) \qquad \gamma_x^{(\cdot\cdot)}(j) = \frac{1}{\rho_a + \rho_b} \left\{ \frac{1}{j!} \frac{\partial^j \phi^{(ab)*}(z, z, s)}{\partial z^j} - \frac{1}{\rho_a + \rho_b} \right\},$$

the derivative being evaluated at $z = 0, s = 0+$.

A limitation of this approach is, however, that independence of the type a and type b events is not reflected in any simple general relation between the three covariance sequences; this is clear from (3.44). Consider also the process of two independent renewal processes of Section 2.3; the covariance sequence for the superposed sequence is complex and not directly informative.

(iii) The discussion of (i) and (ii) suggests that we consider some properties of the intervals in the pooled sequence, that is, the superposed process with the type of each event being distinguished. Possibilities and questions that arise include the following.

(a) The sequence of event types can be considered as a binary time series. In particular, it might be useful to construct a simple test of dependence of the two series based on the nonrandomness of the sequence of event types. Such a test would, however, at least in its simplest form, require the assumption that the marginal processes are Poisson processes.

(b) We can examine the distributions and in particular the means of the backward recurrence times from events of one type to those of the opposite type, that is, $X_b^{(a)}(-1)$ and $X_a^{(b)}(-1)$. If the two types of events are independently distributed, the two "mixed" recurrence times should have marginal distributions corresponding to the equilibrium recurrence time distributions in the marginal process of events of types a and b.

(c) A more symmetrical possibility similar in spirit to (b) is to examine the joint distribution of the two backward recurrence times measured from an arbitrary time origin, that is, of $X^{(a)}(-1)$ and $X^{(b)}(-1)$; the marginal distributions are, of course, the usual ones from univariate theory. If the events of the two types are independent, the two recurrence times are independently distributed with the distribution of the equilibrium recurrence times. Note, however, the discussion following (2.13). It would be possible to adapt (b) and (c) to take account of forward as well as of backward recurrence times.

(d) Probably the most useful general procedure for examining dependence in a bivariate process through intervals is to consider intensities conditional on the two separate asynchronous backward recurrence times. This is not quite analogous to the use of second order intensities of Section 3.3. Denote the realized backward recurrence times from an arbitrary time origin in the stationary process by u_a, u_b, that is, $u_a = x^{(a)}(-1)$, $u_b = x^{(b)}(-1)$. We then define the *serial intensity functions* for a stationary regular process by

$$(3.45) \qquad \lambda^{(a)}(u_a, u_b) = \lim_{\Delta t \to 0+} \frac{Pr\{N^{(a)}(\Delta t) \geq 1 | U_a = u_a, U_b = u_b\}}{\Delta t},$$

with an analogous definition for $\lambda^{(b)}(u_a, u_b)$. These are, in a sense, third order rather than second order functions, since they involve occurrences at three points.

Now these two serial intensities are defined for all regular stationary processes, but they are complete intensity functions in the sense of Section 3.1 only for a very special class of process that we shall call *bivariate Markov interval processes*. These processes include semi-Markov processes and independent renewal processes; note, however, that in general the marginal processes associated with a bivariate Markov interval process are not univariate renewal processes (as with semi-Markov processes) and this is why we have not called the processes bivariate renewal processes.

As an example of this type of process, consider the alternating process of Section 2.5 with disjoint pairwise dependence. Denote the marginal distribution of the $W(i)$ by $G(x)$, and the conditional distribution of the $Z(i)$, given $W(i) = w$, by $F(z|w)$. If $\bar{G}(x) = 1 - G(x)$, $\bar{F}(z|w) = 1 - F(z|w)$, and the probability densities $g(x)$ and $f(z|w)$ exist, then

$$(3.46) \qquad \gamma^{(a)}(u_a, u_b) = \frac{g(u_a)}{\bar{G}(u_a)}, \qquad \gamma^{(b)}(u_a, u_b) = \frac{f(u_a|u_b - u_a)}{\bar{F}(u_a|u_b - u_a)}.$$

These are essentially hazard (failure rate) functions.

Thorough study of bivariate Markov interval processes would be of interest. The main properties can be obtained in principle because of the fairly simple Markov structure of the process. In particular if $p(u, v)$ denotes the bivariate probability density function of the backward recurrence times from an arbitrary time, that is, of (U_a, U_b) or $(X^{(a)}(-1), X^{(b)}(-1))$, then

$$(3.47) \qquad \frac{\partial p(u, v)}{\partial u} + \frac{\partial p(u, v)}{\partial v} = -\{\lambda^{(a)}(u, v) + \lambda^{(b)}(u, v)\}p(u, v),$$

and

$$(3.48) \qquad p(0, v) = \int_0^\infty p(u, v)\lambda^{(a)}(u, v)\, du, \qquad p(u, 0) = \int_0^\infty p(u, v)\lambda^{(b)}(u, v)\, dv.$$

From the normalized solution of these equations, some of the simpler properties of the process can be deduced.

More generally, (3.45) may be a useful semiqualitative summary of the local serial properties of a bivariate point process. It does not seem possible to deduce the properties of the marginal processes given just $\lambda^{(a)}(u_a, u_b)$ and $\lambda^{(b)}(u_a, u_b)$, except for very particular processes such as the Markov interval process. For the alternating process with pairwise disjoint dependence, we indicated this difficulty in Section 2.5.

4. A bivariate delayed Poisson process model with Poisson noise

In previous sections, we defined bivariate Poisson processes to be those processes whose marginal processes (processes of type a events and type b events) are Poisson processes. Bivariate Poisson processes with a dependency structure which is completely specified by the second order intensity function arise from semi-Markov (Markov renewal) processes. The complete intensity function is also particularly simple.

Other bivariate Poisson processes can be constructed and in the present section we examine in some detail one such process. Its physical specification is very simple, although the specification of its dependency structure via a complete intensity function is difficult. The details of the model also illustrate the definitions introduced in Section 2.

General considerations on bivariate Poisson processes will be given in the next section.

4.1. *Construction of the model.* Suppose we have an unobservable main or generating Poisson process of rate μ. Events from the main process are delayed (independently) by random amounts Y_a with common distribution $F_a(t)$ and superposed on a "noise" process which is Poisson with rate λ_a. The resulting process is the observed marginal process of type a events. Similarly, the events in the main process are delayed (independently) by random amounts with common distribution $F_b(t)$ and superposed with another independent noise process which is Poisson with rate λ_b. The resulting process is then the marginal process of type b events. It is not observed which type a and which type b events originate from common main events.

In what follows, we assume for simplicity that the two delays associated with each main point are independent and positive random variables. The process has a number of possible interpretations. One is as an immigration death process with immigration consisting of couples "arriving" and type a events being deaths of men and type b events being deaths of women. Other queueing or service situations should be evident. The Poisson noise processes are added for generality and because they lead to interesting complications in inference procedures. In particular applications, it might be known that one or both noise processes are absent.

Various special cases are of interest. Thus, if delays of both types are equal with probability one, we have the Marshall–Olkin process [34] mentioned in Section 2. Without the added noise and if delays on one side (say, the a event side) are zero with probability one, we have the delay process of Section 3.5 or, equivalently, an $M/G/\infty$ queue, where type a events are arrivals and type b events are departures. The noise process on the a event side would correspond to independent balking in the arrival process.

4.2. *Some simple properties of the model.* If we consider the transient process from its initiation, it is well known (for example, Cox and Lewis [14], p. 209) that the processes are nonhomogeneous Poisson processes with rates that are, respectively,

$$(4.1) \qquad \rho_a(t_1) = \lambda_a + \mu F_a(t_1),$$

$$(4.2) \qquad \rho_b(t_2) = \lambda_b + \mu F_b(t_2).$$

Furthermore, the superposed process is a generalized branching Poisson process whose properties are given by Lewis [30] and Vere-Jones [47]. Thus, at each point in the main or generating process there are, with probability $(\lambda_a + \lambda_b)/$

$(\lambda_a + \lambda_b + \mu)$, no subsidiary events, and, with probability $\mu/(\lambda_a + \lambda_b + \mu)$, two subsidiary events. In the second case, the two subsidiary events are independently displaced from the main or parent event by amounts having distributions $F_a(\cdot)$ and $F_b(\cdot)$.

It is also known (Doob [17]) that as $t \to \infty$, or the origin moves off to the right, the marginal processes become simple stationary Poisson processes of rates

$$(4.3) \qquad \rho_a = \lambda_a + \mu, \qquad \rho_b = \lambda_b + \mu,$$

respectively, for any distributions $F_a(u)$ and $F_b(u)$. The superposed process is then a stationary generalized branching Poisson process of rate $\lambda_a + \lambda_b + 2\mu$. The bivariate process is unusual in this respect, since there are very few dependent point processes whose superposition has a simple structure. The properties of the process of event types $\{T(i)\}$ or $\{T.(i)\}$ are, however, by no means simple to obtain, as will be evident when we consider bivariate properties below. Note too that stationarity of the marginal and superposed process does not imply stationarity of the bivariate process. A counterexample will be given later when initial conditions are discussed.

Asymptotic results for the bivariate counting process $\{N^{(a)}(t_1), N^{(b)}(t_2)\}$ can be obtained by a simple generalization of the methods of Lewis [30]. If, for simplicity, $t_1 = t_2 = t$, the intuitive basis of the method is that when t is very large, the proportion of events that are delayed from the generating process until after t goes (in some sense) to zero and the process behaves as though all events are concentrated at their generating event, that is, like the Marshall–Olkin process. Thus,

$$(4.4) \qquad E\{N^{(a)}(t)\} = \mathrm{Var}\,\{N^{(a)}(t)\} = V^{(aa)}(t) \sim \rho_a t,$$

$$(4.5) \qquad E\{N^{(b)}(t)\} = \mathrm{Var}\,\{N^{(b)}(t)\} = V^{(bb)}(t) \sim \rho_b t,$$

$$(4.6) \qquad \mathrm{Var}\,\{N^{(\cdot)}(t)\} \sim (\rho_a + \rho_b + 2\mu)t,$$

and therefore

$$(4.7) \qquad \mathrm{Cov}\,\{N^{(a)}(t),\ N^{(b)}(t)\} = V^{(ab)}(t) \sim \mu t.$$

The asymptotic measures of dispersion $I^{(aa)}$, $I^{(ab)}$, and $I^{(bb)}$ defined in equations (3.22) to (3.24) are therefore 1, $\mu/(\rho_a\rho_b)^{1/2}$, and 1. Result (4.7) will be useful in a statistical analysis of the process. By similar methods (Lewis, [30]), one can establish the joint asymptotic normality of the bivariate counting process.

Another property of the process which is simple to derive is the second order cross intensity function (3.8) or the covariance density function (3.14). In fact because of the Poisson nature of the main process and the independence of the noise processes from the main process, there is a contribution to the covariance density only if the type b event is a delayed event and the event at τ is the same event appearing in the type a event process with its delay of Y_a. Thus using (3.10), we get the cross covariance function

$$(4.8) \qquad \gamma_b^{(a)}(\tau) = \mu f_{Y_a - Y_b}(\tau) = \gamma_a^{(b)}(-\tau)$$

if $F_{Y_a}(\cdot)$ and $F_{Y_b}(\cdot)$ are absolutely continuous.

If $F_{Y_a}(\cdot)$ and $F_{Y_b}(\cdot)$ have jumps, there will be delta function components in the cross intensity. In particular, when Y_a and Y_b are zero with probability one, there is a delta function component at zero and the process is marginally regular but not regular.

Result (4.8) will be verified from the more detailed results we derive next for the asynchronously sampled, stationary bivariate process. For this we must first consider detailed results for the transient process.

4.3. *The transient counting process.*   The number of events of type a in an interval $(0, t_1]$ following the start of the process is denoted by $N_0^{(a)}(t_1)$ and the number of events of type b in $(0, t_2]$ by $N_0^{(b)}(t_2)$

Assume first that $t_2 \geq t_1 > 0$.

Now if a main event occurs at time v in the interval $(0, t_1]$, then it contributes either one or no events to the type a event process in $(0, t_1]$ and one or no events to the type b event process in $(0, t_2]$. This bivariate binomial random variable has generating function

$$(4.9) \qquad 1 + (1 - z_1)(1 - z_2)F_a(t_1 - v)F_b(t_2 - v)$$
$$+ (z_2 - 1)F_b(t_2 - v) + (z_1 - 1)F_a(t_1 - v).$$

Since we will be using the conditional properties of Poisson processes in our derivation, we require the time v to be uniformly distributed over $(0, t_1]$ and the resulting generating function for the contribution of each main point is obtained by integrating (4.9) with respect to v from 0 to t_1 and dividing by t_1. After some manipulation, this gives

$$(4.10) \qquad 1 + \frac{(1 - z_1 - z_2 + z_1 z_2)}{t_1} \int_0^{t_1} F_a(v)F_b(t_2 - t_1 + v) \, dv$$
$$+ \frac{(z_2 - 1)}{t_1} \int_0^{t_1} F_b(t_2 - t_1 + v) \, dv + \frac{(z_1 - 1)}{t_1} \int_0^{t_1} F_a(v) \, dv$$
$$= Q(z_1, z_2, t_2, t_1).$$

Now assume that there are k_1 events from the main Poisson process of rate μ in $(0, t_1]$, and k_2 main events in $(t_1, t_2]$. Then using the conditional properties of the Poisson process and the independence of the number of main events in $(0, t_1]$ and $(t_1, t_2]$, we get for the conditional generating function of $N_0^{(a)}(t_1)$ and $N_0^{(b)}(t_2)$

$$(4.11) \qquad \exp\{\lambda_a t_1(z_1 - 1) + \lambda_b t_2(z_2 - 1)\} \{Q(z_1, z_2, t_2, t_1)\}^{k_1}$$
$$\cdot \left\{ 1 + \frac{(z_2 - 1)}{(t_2 - t_1)} \int_0^{t_2 - t_1} F_b(u) \, du \right\}^{k_2}.$$

Removing the conditioning on the independently Poisson distributed number of events k_1 and k_2, we have for the logarithm of the joint generating function of $N_0^{(a)}(t_1)$ and $N_0^{(b)}(t_2)$

$$(4.12) \quad \psi_0(z_1, z_2; t_1, t_2)$$
$$= \log \phi(z_1, z_2; t_1, t_2)$$
$$= \rho_a t_1(z_1 - 1) + \rho_b t_2(z_2 - 1)$$
$$- \mu(z_2 - 1) \int_0^{t_2} R_b(u) \, du - \mu(z_1 - 1) \int_0^{t_1} R_a(u) \, du$$
$$+ \mu(1 - z_1 - z_2 + z_1 z_2) \int_0^{t_1} F_a(u) F_b(t_2 - t_1 + u) \, du,$$

where $\rho_b = \lambda_a + \mu$, $\rho_b = \lambda_b + \mu$, $R_b(u) = 1 - F_b(u)$, and we still have $t_2 \geq t_1$.

A similar derivation gives the result for $t_1 \geq t_2$ and we can write for the general case

$$(4.13) \quad \psi_0(z_1, z_2; t_1, t_2)$$
$$= \rho_a t_1(z_1 - 1) + \rho_b t_2(z_2 - 1)$$
$$- \mu(z_2 - 1) \int_0^{t_2} R_b(v) \, dv - \mu(z_1 - 1) \int_0^{t_1} R_a(v) \, dv$$
$$+ \mu(1 - z_1)(1 - z_2) \int_0^{\min(t_1, t_2)} F_a(t_1 - v) F_b(t_2 - v) \, dv.$$

The expected numbers of events (4.1) and (4.2) in the marginal processes also come out of (4.13), as do the properties of the transient, generalized branching Poisson process obtained by superposing events of type a and type b. Moreover, when the random variables Y_a and Y_b have fixed values, $\psi_0(0, 0; t_1, t_2)$ gives the logarithm of the survivor function of the bivariate exponential distribution of Marshall and Olkin [33], [34].

Note that $\psi_0(z_1, z_2; t_1, t_2)$ is the generating function of a bivariate Poisson variate, that is, a bivariate distribution with Poisson marginals. It is, in fact, the bivariate form of the multivariate distribution which Dwass and Teicher [18] showed to be the only infinitely divisible Poisson distribution:

$$(4.14) \quad \phi(\mathbf{z}) = \exp\left\{ \sum_{i=1}^n a_i(z_i - 1) + \sum_{i<j} a_{ij}(z_i - 1)(z_j - 1) + \cdots \right.$$
$$\left. + a_{1,2,\cdots,n} \prod_{i=1}^n (z_i - 1) \right\}.$$

However, since the coefficients in (4.13) depend on t_1 and t_2, the joint distribution of events of type a in two disjoint intervals and events of type b in another two disjoint intervals will not have the form (4.14). This is clearly only true for the highly degenerate Marshall–Olkin process of Section 2.

4.4. *The stationary asynchronous counting process.* To derive the properties of the generating function of counts in the stationary limiting process, or equivalently the asynchronously sampled stationary process, we consider first the number of events of type a in $(t, t_1]$ and of type b in $(t, t_2]$. Because of the independent interval properties of the main and noise Poisson processes of rates λ_a, λ_b, and μ, respectively, this is made up independently from noise and main events occurring in $(t, \max(t_1, t_2)]$, whose generating function is given by (4.13), and by main events occurring in $(0, t]$ and delayed into $(t, t_1]$ or $(t, t_2]$.

Consider, therefore, the generating function of the latter type of events. A main event at v in $(0, t]$ generates either one or no type a events in $(t, t_1]$ and either one or no events of type b in $(t, t_2]$. The generating function of this bivariate binomial random variable is

$$(4.15) \qquad 1 + (z_1 - 1)p_a + (z_2 - 1)p_b + (z_1 - 1)(z_2 - 1)p_a p_b,$$

where

$$(4.16) \qquad p_a = p_a(1; t_1; t; v) = R_a(t - v) - R_a(t + t_1 - v)$$

and

$$(4.17) \qquad p_b = p_b(1; t_2; t; v) = R_b(t - v) - R_b(t + t_2 - v).$$

If the start time v is assumed to be uniformly distributed over $(0, t]$, then the generating function becomes

$$(4.18) \qquad 1 + (z_1 - 1)\frac{\bar{p}_a}{t} + (z_2 - 1)\frac{\bar{p}_b}{t} + (z_1 - 1)(z_2 - 1)\frac{\overline{p_a p_b}}{t},$$

where

$$(4.19) \qquad \bar{p}_a = \int_0^t \{R_a(v) - R_a(v + t_1)\}\, dv,$$

$$(4.20) \qquad \bar{p}_b = \int_0^t \{R_b(v) - R_b(v + t_2)\}\, dv,$$

$$(4.21) \qquad \overline{p_a p_b} = \int_0^t \{R_a(v) - R_a(v + t_1)\}\{R_b(v) - R_b(v + t_2)\}\, dv.$$

It follows from (4.19) and (4.20) that if t_1 and t_2 are finite, we have, even if $E(Y_a)$ and $E(Y_b)$ are infinite,

$$(4.22) \qquad \lim_{t \to \infty} \bar{p}_a = \int_0^{t_1} R_a(v)\, dv, \qquad \lim_{t \to \infty} \bar{p}_b = \int_0^{t_2} R_b(v)\, dv;$$

and since $\overline{p_a p_b} \leq \bar{p}_b$ for all t, t_1, t_2, we have that

$$(4.23) \qquad \lim_{t \to \infty} \overline{p_a p_b} = \int_0^\infty \{R_a(v) - R_a(v + t_1)\}\{R_b(v) - R_b(v + t_2)\}\, dv$$

exists for finite t_1 and t_2.

The results (4.19) through (4.23) are used, as in the derivation of (4.12), to obtain the cumulant generating function of the contribution of delayed events of type a and type b to $(t, t_1]$ and $(t, t_2]$ when $t \to \infty$. This is

$$(4.24) \quad \psi^+(z_1, z_2; t_1, t_2)$$

$$= \mu(z_1 - 1) \int_0^{t_1} R_a(v) \, dv + \mu(z_2 - 1) \int_0^{t_2} R_b(v) \, dv$$

$$+ \mu(z_1 - 1)(z_2 - 1) \int_0^\infty \{R_a(v) - R_a(v + t_1)\} \{R_b(v) - R_b(v + t_2)\} \, dv.$$

Combined with (4.13), we have for the stationary bivariate process the result

$$(4.25) \quad \psi(z_1, z_2; t_1, t_2)$$

$$= \rho_a t_1 (z_1 - 1) + \rho_b t_2 (z_2 - 1) + \mu(z_1 - 1)(z_2 - 1)$$

$$\cdot \left[\int_0^{\min(t_1, t_2)} F_a(t_1 - v) F_b(t_2 - v) \, dv \right.$$

$$\left. + \int_0^\infty \{R_a(v) - R_a(v + t_1)\} \{R_b(v) - R_b(v + t_2)\} \, dv \right].$$

Note that this is the cumulant generating function of a bivariate Poisson distribution and that the covariance time function (3.17) is the term in (4.25) multiplying $(z_1 - 1)(z_2 - 1)$;

$$(4.26) \quad V^{(ab)}(t_1, t_2) = \mu \int_0^{\min(t_1, t_2)} R_a(t - v_1) R_b(t_2 - v) \, dv$$

$$+ \mu \int_0^\infty \{R_a(v) - R_a(v + t_1)\} \{R_b(v) - R_b(v + t_2)\} \, dv.$$

Differentiation of this expression with respect to t_1 and t_2 gives, after some manipulation, the covariance density (4.8), as predicted by the general formula (3.17).

Thus, if the densities associated with $R_a(\cdot)$ and $R_b(\cdot)$ exist, we can express (4.25) as

$$(4.27) \quad \psi(z_1, z_2; t_1, t_2)$$

$$= \rho_a t_1 (z_1 - 1) + \rho_b t_2 (z_2 - 1)$$

$$+ (z_1 - 1)(z_2 - 1) \mu \int_0^{t_1} \int_0^{t_2} f_{Y_a - Y_b}(u - v) \, du \, dv.$$

There are a number of alternative forms for and derivations of this distribution.

The behavior of $V^{(ab)}(t_1, t_2)$, although it is clearly a monotone nondecreasing function of both t_1 and t_2, is complex and will not be studied further here. In (4.7), we saw that along the line $t_1 = t_2 = t$ it is asymptotically μt.

We have also not established the complete stationarity of the limiting bivariate process; this follows from the fact that the delay depends only on the distance from the Poisson generating event, and can be established rigorously using bivariate characteristic functionals.

The complete intensity functions for this process (3.1) cannot be written down and although the second order intensity function is simple it does not specify the dependency structure of the process completely, as it does for the bivariate semi-Markov process. Note too that the cross covariance function (4.8) is always positive, so that there is in effect no inhibition of type a events by b events. In fact from the construction of the process, it is clear that just the opposite effect takes place. We examine the dependency structure of the delay process in more detail here by looking at the joint asynchronous forward recurrence time distribution. This distribution is of some interest in itself.

4.5. *The joint asynchronous forward recurrence times.* In the asynchronous process of the previous section, the time to the kth event of type a, $S^{(a)}(k)$, has a gamma distribution with parameter k and $S^{(b)}(h)$ has a gamma distribution with parameter h. Thus, the joint distribution of these random variables is a bivariate gamma distribution of mixed marginal parameters k and h which is obtained from the generating function (4.25) via the fundamental relationship (2.5). We consider only the joint forward recurrence times $S^{(a)}(1) = X^{(a)}(1)$ and $S^{(b)}(1) = X^{(b)}(1)$ which have a bivariate exponential distribution:

$$(4.28) \qquad R_{ab}(t_1, t_2) = Pr\{X^{(a)}(1) > t_1, X^{(b)}(1) > t_2\}$$
$$= \exp\{\psi(0, 0; t_1, t_2)\}$$
$$= \exp\{-\rho_a t_1 - \rho_b t_2 + V^{(ab)}(t_1, t_2)\}.$$

Clearly, this bivariate exponential distribution reduces to the distribution discussed by Marshall and Olkin [33] in the degenerate case when there are no delays (or fixed delays [34]). For no delays

$$(4.29) \qquad R_{ab}(t_1, t_2) = -\rho_a t_1 - \rho_b t_2 + \mu \min(t_1, t_2).$$

The bivariate exponential distribution (4.27) is not the same as the infinitely divisible exponential distribution discussed by Gaver [20], Moran and Vere-Jones [38], and others. Whenever the delay distributions $R_a(\cdot)$ and $R_b(\cdot)$ have jumps, $R_{ab}(t_1, t_2)$ will have singularities.

For the correlation coefficient, we have

$$(4.30) \qquad \rho_a \rho_b \, \text{Corr}\,\{X^{(a)}(1), X^{(b)}(1)\} = \int_0^\infty \int_0^\infty R_{ab}(t_1, t_2)\, dt_1\, dt_2 - \frac{1}{\rho_a \rho_b}.$$

It is not possible to integrate this expression explicitly except in special cases. However, since $V^{(ab)}(t_1, t_2) \geq 0$, we clearly have that the correlation coefficient is greater than zero.

For the special case (4.29) the correlation (4.30) is $1/\{(1 + (\lambda_a + \lambda_b)/\mu)\}$.

We do not pursue further here the properties of the process obtainable from the joint distribution of counts (4.25) of the synchronous counting process $\{N^{(a)}(t_1), N^{(b)}(t_2)\}$. However, it is useful to summarize what useful properties can be derived for this or any other bivariate process such as those given in the next section, from this bivariate distribution.

(i) The marginal generating functions ($z_1 = 0$ or $z_2 = 0$) give the correlation structure of the marginal interval process through equation (3.43). This is trivial for the delay process.

(ii) The generating function with $z_1 = z_2$ gives the correlation structure of the intervals in the superposed process through (3.44). For the delay process this is the interval correlation structure of a clustering (branching) Poisson process.

(iii) The covariance time surface and cross intensity and marginal intensity functions can be obtained. Again for the delay process this is trivial.

(iv) The joint distribution of the asynchronous forward recurrence times $\{X^{(a)}(1), X^{(b)}(1)\}$ can be calculated. Other functions of interest are the smaller and larger of $X^{(a)}(1)$ and $X^{(b)}(1)$, and the conditional distributions and expectations, for example, $E\{X^{(a)}(1)|X^{(b)}(1) = x\}$. The latter is difficult to obtain for the delay process, the regression being highly nonlinear.

(v) In principle, one can obtain not only the distributions of the smaller and larger of $X^{(a)}(1)$ and $X^{(b)}(1)$, but also the order type (jointly or marginally) since

$$(4.31) \qquad Pr\{T(1) = a\} = Pr\{X^{(a)}(1) < X^{(b)}(1)\}.$$

(vi) It is not possible to obtain the complete distributions of types, for example, $Pr\{T(1) = a; T(2) = a\}$ from the bivariate distribution of asynchronous counts, since these counts are related to the sums of intervals by (2.5). For this information, we need more complete probability relationships, that is, for the pooled process $\{X^{(\cdot)}(1), T(1); X^{(\cdot)}(2), \cdots\}$. Note too that $\{T(i)\}$ is not a stationary binary sequence.

It is possible to obtain distributions of semisynchronous counting processes for the delay processes although we do not do this here. One reason for doing this is to obtain information on the distribution of the stationary sequences $T.(i)$. Thus,

$$(4.32) \qquad Pr\{T.(0) = a, T.(1) = b\} = \frac{\rho_a}{\rho_a + \rho_b} Pr\{X_a^{(b)}(1) < X_a^{(a)}(1)\},$$

and so forth, from which the correlation coefficient of lag one is obtained. It is not possible to carry the argument to lags of greater than one solely with joint distributions of sums of semisynchronous intervals.

4.6. *Stationary initial conditions.* We discuss here briefly the problem of obtaining stationary initial conditions for the delay process, since this has some bearing on the problems considered in this paper.

Note that for the marginal processes in the delayed Poisson process the numbers of events generated before t which are delayed beyond t have, if $E(Y_a) < \infty$ and $E(Y_b) < \infty$, Poisson distributions with parameters $\mu E(Y_a)$ and $\mu E(Y_b)$, respectively, when $t \to \infty$. Denote these random variables by $Z^{(a)}$ and $Z^{(b)}$. If the transient process of Section 4.3 is started with an additional number

Z_a of type a events which occur independently at distances $\bar{Y}_a(1), \cdots, \bar{Y}_a(Z_a)$ from the origin, where

$$(4.33) \qquad Pr\{\bar{Y}_a(i) \leqq t\} = \int_0^t \frac{R_a(u)\, du}{\{E(Y_a)\}},$$

and with an additional number Z_b of type b events which occur independently at distances $\bar{Y}_b(1), \cdots, \bar{Y}_b(z_b)$ from the origin, where the common distribution of the $\bar{Y}_b(j)$ is directly analogous to that of the $\bar{Y}_a(i)$, then the marginal processes are stationary Poisson processes (Lewis [30]). However, the bivariate process is not stationary. This can be verified, for instance, by obtaining the covariance density from the resulting generating function and noting that it depends on t_1 and t_2 separately, and not just on their difference.

In obtaining stationary initial conditions, the joint distribution of Z_a and Z_b is needed. Without going into the details of the limiting process, the generating function for these random variables is clearly (4.24) when $t_1 \to \infty$ and $t_2 \to \infty$. Thus,

$$(4.34) \quad \psi_{Z_a, Z_b}(z_1, z_2)$$
$$= (z_1 - 1)\mu E(Y_a) + (z_2 - 1)\mu E(Y_b) + (z_1 - 1)(z_2 - 1)\mu$$
$$\cdot E\{\min(Y_a, Y_b)\},$$

where $E\{\min(Y_a, Y_b)\} = \int_0^\infty R_a(v) R_b(v)\, dv$. This is the generating function (4.14) of a bivariate Poisson distribution.

Further details of this model, including the complete stationary initial conditions, will be given in another paper.

5. Some other special processes

We discuss here briefly several important models for bivariate point processes. The specification of the models is through the structure of intervals and is based on direct physical considerations, unlike, say, the bivariate Markov process with its specification of degree of dependence through the complete intensity functions. At the end of the section we consider the general problem of specifying the form of bivariate Poisson processes.

5.1. *Single process subject to bivariate delays.* The bivariate delayed Poisson process of the previous section can be generalized in several ways. First, the delays Y_a and Y_b might be correlated since, for instance, in the example of a man and wife in a bivariate immigration death process, their residual lifetimes would be correlated. Again Y_a and Y_b may take both positive and negative values. The stationary analysis of the previous section goes through essentially unchanged although specifying initial conditions is difficult. The covariance function (4.8) is the same except that $f_{Y_a - Y_b}(t)$ is, of course, no longer a simple convolution.

Another extension is to consider main processes which are, say, regular stationary point processes with rate μ and intensity function $h_\mu(t)$. Then the cross intensity function for the bivariate process, $h_a^{(b)}(t)$, becomes

$$(5.1) \qquad \frac{\lambda_a}{\lambda_a + \mu}\, \rho_b + \frac{\mu}{\lambda_a + \mu}\left\{\lambda_b + f_{Y_a - Y_b}(t) + \int_{-\infty}^{\infty} h_\mu(u) f_{Y_a - Y_b}(t - u)\, du\right\},$$

with a similar expression for $h_b^{(a)}(t)$. These should be compared with (3.26). Except when the main process is a renewal process, explicit results beyond the intensity function are difficult to obtain. For the renewal case an integral equation can be written down, as also for branching renewal processes (Lewis, [32]); from the integral equation higher moments of the bivariate counting process can be derived.

5.2. *Bivariate point process subject to delays.* Instead of having a univariate point process in which each point (say the ith) is delayed by two different amounts $Y_a(i)$ and $Y_b(i)$ to form the bivariate process, one can have a main bivariate point process in which the ith type a event is delayed by $Y_a(i)$ and the jth type b event is delayed by $Y_b(j)$, thus forming a new bivariate point process. This does not reduce to the bivariate delay process of Section 5.1 although it is conceptually similar.

The simplest illustration is where there is error (jitter) in recording the positions of the points. Usually the errors are taken to be independently distributed, although Y_a and Y_b may have different distributions. Another situation is an immigration death process with two different types of immigrants.

If the main process has cross intensities $\bar{h}_a^{(b)}(t)$ and $\bar{h}_b^{(a)}(t)$, then the delayed bivariate process (with no added Poisson noise) has cross intensity

$$(5.2) \qquad h_a^{(b)}(t) = \int_{-\infty}^{\infty} \bar{h}_a^{(b)}(t - v) f_{Y_a - Y_b}(v)\, dv.$$

It will not be possible from data to separate properties of the jitter process from those of the underlying main process, unless strong special assumptions are made.

An interesting situation occurs when the main process is a semi-Markov process with marginal processes which are Poisson processes as, for example, in Section 2. Then the delayed bivariate point process is, in equilibrium, a bivariate Poisson process.

5.3. *Clustering processes.* Univariate clustering processes (Neyman and Scott [39], Vere-Jones [47], and Lewis [30]) are important. Each main event generates one subsidiary sequence of events and the subsidiary sequences have a finite number of points with probability one. The subsidiary processes are independent of one another but can be of quite general structure. When the subsidiary processes are finite renewal processes, the clustering process is known as a Bartlett–Lewis process; when the events are generated by independent delays from the initiating main event, the process is known as a Neyman–Scott cluster process.

The bivariate delay process and delayed bivariate process described in the previous two subsections are special cases of bivariate cluster processes and clearly both types of main process are possible for these cluster processes. As an example of a bivariate main process generating two different types of subsidiary process, Lewis [29] considered computer failure patterns and discussed the possibility of two types of subsidiary sequences, one generated by permanent component failures and the other by intermittent component failures.

There are many possibilities that will not be discussed here. Some general points of interest are, however, the following.

(i) When the main process is a univariate Poisson process, producing a bivariate clustering Poisson process, bivariate superposition of such processes again produces a bivariate clustering Poisson process. The process is thus infinitely divisible.

(ii) Both the marginal processes and the superposed processes are (generalized) cluster processes. Thus, we can use known results' for those processes and expressions such as (3.39) and (3.14) to find variance time curves and cross intensities for the bivariate process. When the main process is a semi-Markov process, the marginal processes are clustering (or branching) renewal processes (Lewis, [32]).

(iii) The analysis in Section 4 can be used for these processes when the main process is a univariate Poisson process. Bivariate characteristic functionals are probably also useful.

5.4. *Selective inhibition.* A simple, realistic and analytically interesting model arises in neurophysiological contexts. We have two series of events, the first called the inhibitory series of events and the second the excitatory series of events, occurring on a common time scale. Each event in the inhibitory series blocks only the next excitatory event (and blocks it only if no following inhibitory event occurs before the excitatory event). This is the simplest of many possibilities.

Although only the sequence of noninhibited excitatory events (the responses) is usually studied, Lawrance has pointed out that there are a number of bivariate processes generated by this mechanism, in particular the inhibitory events and the responses [24], [25]. These may constitute the input and output to a neuron, and are the only pair we consider here. In particular, we take the excitatory process to be a Poisson process with rate ρ_a and the inhibiting process to be a renewal process with interevent probability distribution function $F_b(x)$. The response process has dependent intervals unless the inhibitory process also is a Poisson process.

When the excitatory process is a renewal process with interevent probability distribution function $F_a(x)$ and the inhibitory process is Poisson with rate ρ_b, the process of responses is a renewal process. This follows because the original renewal process is in effect being thinned at a rate depending only on the time since the last recorded response and such an operation preserves the renewal property. This bivariate renewal process is not a semi-Markov process, as can be seen by attempting to write down the complete intensity functions (3.4) and (3.5). The complete intensity functions become simple only for the trivariate process of inhibitory events, responses, and nonresponses.

Coleman and Gastwirth [10] have shown that it is possible to pick $F_a(x)$ so that the responses also form a Poisson process. The covariance density of this bivariate Poisson process can be obtained; it is always negative (personal communication, T. K. M. Wisniewski).

Other forms of selective inhibition can be postulated; some have been discussed by Coleman and Gastwirth [10]. Another possibility is the simultaneous inhibition, as above, of two excitatory processes by a single, unobservable inhibitory process. When the inhibitory process is Poisson and the excitatory processes are renewal processes, the two response processes are a bivariate renewal process.

There are, of course, many other neurophysiological models, generally more complicated than the selective inhibition models and many times involving the doubly stochastic mechanism discussed in Section 3.5. An interesting example is given by Walloe, Jansen, and Nygaard [48].

5.5. *General remarks on bivariate Poisson processes.* In this and previous sections, we have encountered several examples of bivariate Poisson processes, defined as bivariate point processes in which the marginal processes are Poisson processes.

(i) The degenerate Poisson process of Marshall and Olkin was discussed in Section 2.

(ii) The process in (i) is a special case of a broad family of bivariate Poisson processes generated by bivariate delays on univariate Poisson processes. Several other examples arise in considering delays on bivariate Poisson processes.

(iii) Semi-Markov processes have renewal marginals and a broad class of bivariate Poisson processes is obtained by choosing the marginal processes to be Poisson processes. Delays added to these particular semi-Markov processes again produce bivariate Poisson processes.

(iv) A rather special case arises when a Poisson process inhibits a renewal process.

Another example is mentioned because it illustrates the problem considered in Section 3.8 of specifying dependency structure in terms of the bivariate, discrete time sequence of marginal intervals. Thus, we can start the process and require that the intervals in the marginals with the same serial index be bivariate exponentials. Any bivariate exponential distribution may be used, such as (4.28) or those of Gaver [20], Plackett [42], Freund [19], and Griffiths [21]. The interval structure is stationary, as is the counting process of the marginals, which are Poisson processes. The bivariate counting process is, however, not stationary. It is not clear whether one gets the counting process to be stationary, as defined in Section 2, when moving away from the origin, but since the time lag between the dependent intervals increases indefinitely as $n \to \infty$ the process is degenerate and tends to almost independent Poisson processes a long time from the origin.

No general structure is known for bivariate Poisson processes. There follow some general comments and some open questions.

(i) The bivariate Poisson process as defined is infinitely divisible in that bivariate superposition produces bivariate Poisson processes. However, of the above models of bivariate Poisson processes, only the bivariate delayed Poisson process keeps the same dependency structure under bivariate superposition.

(ii) Does unlimited bivariate superposition produce two independent Poisson processes? The answer is, generally, yes (Cinlar [8]).

(iii) It can be shown that successive independent delays on bivariate Poisson processes (and most bivariate processes) produces in the limit a process of independent Poisson processes. This can be seen from (5.2) and (5.3), but needs bivariate characteristic functionals for a complete proof.

(iv) The numbers of events of the two types in an interval $(0, t]$ in a bivariate Poisson process have a bivariate Poisson distribution. Some general properties of such distributions are known (Dwass and Teicher [18]), the bivariate Poisson distribution (4.33) is the only infinitely divisible bivariate Poisson distribution. An open question of interest in investigating bivariate Poisson processes is whether, when Z_1, Z_2, and $Z_1 + Z_2$ have marginally Poisson distributions of means μ_1, μ_2, and $\mu_1 + \mu_2$, Z_1 and Z_2 are independent. If this is so, a bivariate Poisson process in which the superposed marginal process is a Poisson process must have the events of two types independent.

(v) The broad class of stationary bivariate Poisson processes arising from delay mechanisms have positive cross covariance densities, that is, no "inhibitory effect." For the semi-Markov process with Poisson marginals, it is an open question as to whether cross covariance densities which take on negative values exist. In particular, for the alternating renewal process with identical gamma distributions of index one for up and down times, the cross covariance is strictly positive. The only model which is known to produce a bivariate Poisson process with strictly negative covariance density is the Poisson inhibited renewal process described earlier in this section.

6. Statistical analysis

6.1. *General discussion.* We now consider in outline some of the statistical problems that arise in analyzing data from a bivariate point process. If a particular type of model is suggested by physical considerations, it will be required to estimate the parameters and test goodness of fit. In some applications, a fairly simple test of dependence between events of different types will be the primary requirement. In yet other cases, the estimation of such functions as the covariance densities will be required to give a general indication of the nature of the process, possibly leading to the suggestion of a more specific model. In all cases, the detection and elimination of nonstationarity may be required.

There is one important general distinction to be drawn, parallel to that between correlation and regression in the analysis of quantitative data. It may be that both types of event are to be treated symmetrically, and that in particular the stochastic character of both types needs analysis. This is broadly the attitude implicit in the previous sections. Alternatively one type of event, say b, may be causally dependent on previous a events, or it may be required to predict events of type b given information about the previous occurrences of events of both types. Then it will be sensible to examine the occurrence of the b's conditionally on the observed sequence of a's and not to consider the stochastic mechanism generating the a's; this is analogous to the treatment of the independent variable in regression analysis as "fixed." Note in particular that the pattern of the a's might be very nonstationary and yet if the mechanism generating the b's is stable, simple "stationary" analyses may be available.

In the rest of this section, we sketch a few of the statistical ideas required in analyzing this sort of data.

6.2. *Likelihood analyses.* If a particular probability model is indicated as the basis of the analysis when the model is specified except for unknown parameters, in principle it will be a good thing to obtain the likelihood of the data from which exactly or asymptotically optimum procedures of analysis can be derived, for example, by the method of maximum likelihood; of course, this presupposes that the usual theorems of maximum likelihood theory can be extended to cover such applications. Unfortunately, even for univariate point processes, there are relatively few models for which the likelihood can be obtained in a useful form. Thus, one is often driven to rather *ad hoc* procedures.

Here we note a few very particular processes for which the likelihood can be calculated.

In a semi-Markov process, the likelihood can be obtained as a product of a factor associated with the two state Markov chain and factors associated with the four distributions of duration; if sampling is for a fixed time there will be one "censored" duration. Moore and Pyke [36] have examined this in detail with particular reference to the asymptotic distributions obtained when sampling is for a fixed time, so that the numbers of intervals of various types are random variables.

A rather similar analysis can be applied to the bivariate Markov process of intervals of Section 3.8, although a more complex notation is necessary. Let

$$(6.1) \qquad L^{(a)}(x; v, w) = \exp\left\{ -\int_0^x \lambda^{(a)}(z + v, z + w)\, dz \right\},$$

$$(6.2) \qquad L^{(b)}(x; v, w) = \exp\left\{ -\int_0^x \lambda^{(b)}(z + v, z + w)\, dz \right\}.$$

We can summarize the observations as a sequence of intervals between successive events in the pooled process, where the intervals are of type aa, ab, ba, or bb. We characterize each interval by its length x and by the backward recurrence time at the start of the interval measured to the event of opposite type. Denote this by v if measured to a type a event and by w if measured to a type b event.

Then the contribution to the likelihood of the length of the interval and the type of the event at the end of the interval is

(6.3) $\lambda^{(a)}(x, w + x)\, L^{(a)}(x; 0, w)\, L^{(b)}(x; 0, w)$ for an aa interval,

(6.4) $\lambda^{(b)}(x, w + x)\, L^{(a)}(x; 0, w)\, L^{(b)}(x; 0, w)$ for an ab interval,

(6.5) $\lambda^{(a)}(v + x, x)\, L^{(a)}(x; v, 0)\, L^{(b)}(x; v, 0)$ for a ba interval,

(6.6) $\lambda^{(b)}(v + x, x)\, L^{(a)}(x; v, 0)\, L^{(b)}(x; v, 0)$ for a bb interval.

Thus, once the intensities are specified parametrically the likelihood can be written down and, for example, maximized numerically.

Now the above discussion is for the "correlational" approach in which the two types of event are treated symmetrically. If, however, we treat the events of type b as the dependent process and argue conditionally on the observed sequence of events of type a, the analysis is simplified, in effect by replacing $\lambda^{(a)}(\cdot, \cdot)$ and $L^{(a)}(\cdot, \cdot, \cdot)$ by unity.

A particular case of interest is when the intensities are linear functions of their arguments. This, of course, precludes having the semi-Markov process as a special case.

A further example of when a likelihood analysis is feasible is provided by the bivariate sinusoidal Poisson process of (3.30) with R_a, R_b, and Φ regarded as unknown parameters. An analysis in terms of exponential family likelihoods is obtained by taking (3.20) to refer to the log intensity; for the univariate analysis see Lewis [31].

In the bivariate case, we reparametrize and have

(6.7) $$\lambda_a(t) = \frac{\rho_a \exp\{R_a \cos(\omega_0 t + \theta + \Phi)\}}{I_0(R_a)}$$

and

(6.8) $$\lambda_b(t) = \frac{\rho_b \exp\{R_b \cos(\omega_0 t + \Phi)\}}{I_0(R_b)},$$

where $I_0(R_b)$ is a zero order modified Bessel function of the first kind. It is convenient to assume that observation on both processes is for a common fixed period t_0, where $\omega_0 t_0$ is an integral multiple of 2π, say $2\pi p$. Then $\int_0^{t_0} \lambda_a(u)\, du = \rho_a t_0$ and $\int_0^{t_0} \lambda_b(u)\, du = \rho_b t_0$.

If $n^{(a)}$ type a events are observed in $(0, t_0]$ at times $t_1^{(a)}, \cdots, t_{n^{(a)}}^{(a)}$ and $n^{(b)}$ type b events at times $t_1^{(b)}, \cdots, t_{n^{(b)}}^{(b)}$, then, using the likelihood for the nonhomogeneous bivariate Poisson process

(6.9) $$\prod_{i=1}^{n^{(a)}} \lambda_a(t_i^{(a)}) \prod_{j=1}^{n^{(b)}} \lambda_b(t_j^{(b)}) \exp\{-\rho_a t_0 - \rho_b t_0\},$$

we find that the set of sufficient statistics for $\{\rho_a, \rho_b, R_a \cos(\theta + \Phi), R_a \sin(\theta + \Phi),$ $R_b \cos \Phi, R_b \sin \Phi\}$ are $\{n^{(a)}, n^{(b)}, \mathscr{A}_a(\omega_0), \mathscr{B}_a(\omega_0), \mathscr{A}_b(\omega_0), \mathscr{B}_b(\omega_0)\}$, where

$$(6.10) \qquad \mathscr{A}_a(\omega_0) = \sum_{i=1}^{n^{(a)}} \cos(\omega_0 t_i^{(a)}), \qquad \mathscr{B}_a(\omega_0) = \sum_{j=1}^{n^{(b)}} \sin(\omega_0 t_j^{(a)}),$$

with similar definitions for $\mathscr{A}_b(\omega_0)$ and $\mathscr{B}_b(\omega_0)$.

Typically, if $R_a = R_b = R$, maximum likelihood estimates of R and tests of $R = 0$ are based on monotone functions of $\mathscr{A}_a(\omega_0)$, $\mathscr{B}_a(\omega_0)$, $\mathscr{A}_b(\omega_0)$, and $\mathscr{B}_b(\omega_0)$. The estimation and testing procedures are formally equivalent to tests for directionality on a circle from two independent samples when the direction vector has a von Mises distribution (Watson and Williams, [49]).

Other trend analyses can be carried out with a similar type of likelihood analysis if the model is a nonhomogeneous bivariate Poisson process.

For most other special models, including quite simple ones such as the delayed Poisson process of Section 4, it does not seem possible to obtain the likelihood in usable form; it would be helpful to have ways of obtaining useful pseudo-likelihoods for such processes.

For testing goodness of fit, it may sometimes be possible to imbed the model under test in some richer family; for instance, agreement with a parametric semi-Markov model could be tested by fitting some more general bivariate Markov interval process and comparing the maximum likelihoods achieved. More usually, however, it will be a case of finding relatively *ad hoc* test statistics to examine various aspects of the model.

In situations in which the model of independent renewal processes or the semi-Markov model may be relevant, the following procedures are likely to be useful. To test consistency with an independent renewal process model, we may:

(a) examine for possible nonstationarity,

(b) test the marginal processes for consistency with a univariate renewal model (Cox and Lewis [14], Chapter 6),

(c) test for dependence using the estimates of the cross intensity given in the next section, or test that the event types do not have the first order Markov property.

If dependence is present, it may be natural to see whether the data are consistent with a semi-Markov process. (Note, however, that the family of independent renewal models is not contained in the family of semi-Markov models.) To test for the adequacy of an assumed parametric semi-Markov model, we may, for example, proceed as follows:

(a) examine for possible nonstationarity,

(b) test the sequence of event types for the first order Markov property (Billingsley [5]),

(c) examine the distributional form of the four separate types of interval,

(d) examine the dependence of intervals on the preceding interval and the preceding event type.

6.3. *Estimation of intensities and associated functions.*   If a likelihood based analysis is not feasible, we must use the more empirical approach of choosing aspects of the process thought to be particularly indicative of its structure and estimating these aspects from the data. In this way we may be able to obtain estimates of unknown parameters and tests of the adequacy of a proposed model.

In the following discussion, we assume that the process is stationary. With extensive data, it will be wise first to analyze the data in separate sections, pooling the results only if the sections are reasonably consistent.

The main aspects of the process likely to be useful as a basis for such procedures are the frequency distributions of intervals of various kinds, the second order functions of Section 3.3 through 3.7 and, the bivariate interval properties, in particular the serial intensity functions (3.42). As stressed in Section 3, it will often happen that one or other of the above aspects is directly related to the underlying mechanism of the process and hence is suitable for statistical analysis.

Estimation of the univariate second order functions does not need special discussion here. We therefore merely comment briefly on the estimation of the serial intensity functions and the cross properties; for the latter the procedures closely parallel the corresponding univariate estimation procedures.

6.3.1. *Cross intensity function.*   To obtain a smoothed estimate of the cross intensity function $h_a^{(b)}(t)$, choose a grouping interval Δ and count the total number of times a type b event occurs a distance between t and $t + \Delta$ to the right of a type a event; let the random variable corresponding to this number be $R_a^{(b)}(t, t + \Delta)$. In practice, we form a histogram from all possible intervals between events of type a and events of type b. We now follow closely the argument of Cox and Lewis ([14], p. 122) writing, for observations over $(0, t_0)$,

$$(6.11) \quad R_a^{(b)}(t, t + \Delta) = \left\{ \int_{u=0}^{t_0-t-\Delta} \int_{x=t}^{t+\Delta} + \int_{u=t_0-t-\Delta}^{t_0-t} \int_{x=t}^{t_0-u} \right\} dN^{(a)}(u) \, dN^{(b)}(u + x).$$

Now for a stationary process

$$(6.12) \qquad E\{dN^{(a)}(u) \, dN^{(b)}(u + x)\} = \rho_a h_a^{(b)}(x) \, du \, dx,$$

and a direct calculation, plus the assumption that $h_a^{(b)}(x)$ varies little over $(t, t + \Delta)$, gives

$$(6.13) \qquad E\{R_a^{(b)}(t, t + \Delta)\} = (t_0 - t - \tfrac{1}{2}\Delta)\rho_a \int_t^{t+\Delta} h_a^{(b)}(x) \, dx,$$

thus leading to a nearly unbiased estimate of the integral of the cross intensity over $(t, t + \Delta)$.

If the type b events are distributed in a Poisson process independently of the type a events, we can find the exact moments of $R_a^{(b)}(t, t + \Delta)$, by arguing conditionally both on the number of type b events and on the whole observed process of type a events (see Section 6.4). To a first approximation, $R_a^{(b)}(t, t + \Delta)$

has (conditionally) a Poisson distribution of mean $n^{(a)}n^{(b)}\Delta/t_0$ provided that Δ is small and, in particular, that few type a events occur within Δ of one another. This provides the basis for a test of the strong null hypothesis that the type b events follow an independent Poisson process; it would be interesting to study the extent to which the test is distorted if the type b events are distributed independently of the type a events, although not in a Poisson process.

6.3.2. *Cross spectrum.* Estimation of the cross spectrum is based on the cross periodogram, defined as follows. For each marginal process, we define the finite Fourier–Stieltjes transforms of $N^{(a)}(t)$ and $N^{(b)}(t)$ (Cox and Lewis [14], p. 124) to be

$$(6.14) \quad H_{t_0}^{(a)}(\omega) = (2\pi t_0)^{-1/2} \sum_{\ell=1}^{n^{(a)}} \exp\{i\omega t_\ell^{(a)}\} = (2\pi t_0)^{-1/2}\{\mathscr{A}_{t_0}^{(a)}(\omega) + i\mathscr{B}_{t_0}^{(a)}(\omega)\},$$

$$(6.15) \quad H_{t_0}^{(b)}(\omega) = (2\pi t_0)^{-1/2} \sum_{j=1}^{n^{(b)}} \exp\{i\omega t_j^{(b)}\} = (2\pi t_0)^{-1/2}\{\mathscr{A}_{t_0}^{(b)}(\omega) + i\mathscr{B}_{t_0}^{(b)}(\omega)\}.$$

The cross periodogram is then

$$(6.16) \qquad\qquad \mathscr{I}_{t_0}^{(ab)}(\omega) = H_{t_0}^{(a)}(\omega)\, \bar{H}_{t_0}^{(b)}(\omega)$$

(Jenkins [23]). Thus, the estimates of the amplitude and phase of harmonic components of fixed frequencies in a nonhomogeneous bivariate Poisson model considered in the previous section are functions of the empirical spectral components. It can also be shown, as for the univariate case (Lewis [31]), that $\mathscr{I}_{t_0}^{(ab)}(\omega)$ is the Fourier transform of the unsmoothed estimator of the cross intensity function obtained from all possible intervals between events of type a and events of type b.

The distribution theory of $\mathscr{I}_{t_0}^{(ab)}(\omega)$ for independent Poisson processes follows simply from the conditional properties of the Poisson processes. Thus, we find that $A_{t_0}^{(a)}(\omega)$ and $B_{t_0}^{(a)}(\omega)$ have the (conditional) joint generating function

$$(6.17) \qquad\qquad [I_0\{(\xi_a^2 + \xi_b^2)^{1/2}\}]^{n^{(a)}}$$

if $\omega t_0 = 2\pi p$, from which it can be shown, for example, that $\mathscr{A}_{t_0}^{(a)}(\omega)$ and $\mathscr{B}_{t_0}^{(b)}(\omega)$ go rapidly to independent normal random variables with means 0 and standard deviations $\frac{1}{2}t_0\rho_a$ as $n^{(a)}$ becomes large. Consequently, the real and imaginary components of the cross periodogram have double exponential distributions centered at zero with a variance which does not decrease as t_0 increases.

At two frequencies ω_1 and ω_2 such that $\omega_1 t_0 = 2\pi p_1$ and $\omega_2 t_0 = 2\pi p_2$, the real components of $\mathscr{I}_{t_0}^{(ab)}(\omega_1)$ and $\mathscr{I}_{t_0}^{(ab)}(\omega_2)$ are asymptotically uncorrelated, as are the imaginary components. Consequently, smoothing of the periodogram is required to get consistent estimates of the in phase and out of phase components of the cross spectrum. The problems of bias, smoothing, and computation of the spectral estimates are similar to those for the univariate case discussed in detail by Lewis [31].

Note that the smoothed intensity function or the smoothed spectral estimates can be used to estimate the delay probability density function in the one sided Poisson delay model (see equations (3.26) and (3.27)) and the difference of the delays in the two sided (bivariate) Poisson delay model. In the first case, the estimation procedure is probably much more efficient, in some sense, than the procedure discussed by Brown [7] unless the mean delay is much shorter than the mean time between events in the main Poisson process.

6.3.3. *Covariance time function.*   Another problem that arises with the bivariate Poisson delay process is to test for the presence of the Poisson noise and to estimate the rate μ of the unobservable main process. Since the covariance time curve $V^{(ab)}(t) \sim \mu t$, we can estimate μ by estimating $V^{(ab)}(t)$ and also test for Poisson noise by comparing the estimated measures of dispersion $I^{(aa)}$, $I^{(bb)}$, and $I^{(ab)}$, defined in (3.22), (3.23), and (3.24). Care will be needed over possible nonstationarity.

The simplest method for estimating $V^{(ab)}(t)$ is to estimate the variance time curves $V^{(aa)}(t)$, $V^{(bb)}(t)$, and $V^{(\cdot\cdot)}(t)$ with the procedures given by Cox and Lewis ([14], Chapter 5) and to use (3.39) to give an estimate of $V^{(ab)}(t)$.

There is no evident reason for estimating the covariance time surface $C(t_1 t_2)$ along any line except $t_1 = t_2$.

6.3.4. *Serial intensity function.*   Estimation of the serial intensity functions raises new problems, somewhat analogous to the analysis of life tables. Consider the estimation of $\lambda^{(a)}(u_a, u_b)$ of (3.42). One approach is to pass to discrete time, dividing the time axis into small intervals of length Δ. Each such interval is characterized by the values of (u_a, u_b) measured from the center of the interval if no type a event occurs within the interval, and by the values of (u_a, u_b) at the type a event in question if one such event occurs; we assume for simplicity of exposition that the occurrence of multiple type a events can be ignored. Thus, each time interval contributes a binary response plus the values of two explanatory variables (u_a, u_b); the procedure extends to the case of more than two explanatory variables, and to the situation in which multiple type a events occur within the intervals Δ.

We can now do one or both of the following:

(a) assume a simple functional form for the dependence on (u_a, u_b) of the probability $\lambda^{(a)}(u_a, u_b)\Delta$ of a type a event and fit by weighted least squares or maximum likelihood (Cox [13]);

(b) group into fairly coarse "cells" in the (u_a, u_b) plane and find the proportion of "successes" in each cell.

It is likely that standard methods based on an assumption of independent binomial trials are approximately applicable to such data and, if so, specific assumptions about the form of the serial intensities can be tested. In particular, we can test the hypothesis that the process is, say, purely a dependent, making the further assumption to begin with that the dependence is only on u_a.

By extensions of this method, that is, by bringing in dependencies on more aspects of the history at time t than merely u_a and u_b, it may be possible to build up empirically a fairly simple model for the process.

6.4. *Simple tests for dependence.* As noted previously, it may sometimes be required to construct simple tests of the null hypothesis that the type a and type b events are independent, as defined in Section 2.2. This may be done in various ways. Much the simplest situation arises when we consider the dependence of, say, the type b events on the type a events, argue conditionally on the observed type a process, and consider the strong null hypothesis that the type b events form an independent Poisson process. Then, conditionally on the total number of events of type b, the positions of the type b events, $t_1^{(b)}, \cdots, t_{n(a)}^{(b)}$ are independently and uniformly distributed over the period of observation. Thus in principle, the exact distribution of any test statistic can be obtained free of nuisance parameters.

The two simplest of the many possible test statistics are probably:

(a) particular ordinates of the cross intensity function, usually that near the origin; equivalently we can use the statistic $R_a^{(b)}(0, \Delta)$ of Section 6.3, directly;

(b) the sample mean recurrence time backwards from a type b event to the nearest preceding type a event.

The null distribution of $R_a^{(b)}(0, \Delta)$ can be found as follows. Place an interval of length Δ to the right of each type a event. (It is assumed for convenience that either there is a type a event at the origin, or that the position of the last type a event before the origin is available.) Let $\pi_0, \pi_1, \pi_2, \cdots, \pi_{n(a)}$ be the proportion of the observed interval $(0, t_0)$ covered jointly by $0, 1, 2, \cdots, n^{(a)}$ of these intervals Δ. Then, if there are $n^{(b)}$ events of type b in all, the null distribution of $R_a^{(b)}(0, \Delta)$ is that of the sum of $n^{(b)}$ independent random variables each taking the value i with probability $\pi_i, i = 1, \cdots, n^{(a)}$.

Similarly, for the second test statistic, we can find the null distribution as follows. Regard the sequence of intervals between successive type a events as a finite population $x = \{x_1, \cdots, x_N\}$, say. This includes the intervals from 0 to the first type a event and from the last type a event to t_0. If t_0 is preassigned, $N = n^{(a)} + 1$. Note that $\Sigma x_i = t_0$. Then the null distribution of the test statistic is that of the mean of n_b independent and identically distributed random variables each with probability density function

$$(6.18) \qquad \frac{1}{t_0} \sum_{i=1}^{N} U(x; x_i),$$

where

$$(6.19) \qquad U(x; x_i) = \begin{cases} 1, & 0 \le x \le x_i, \\ 0, & \text{otherwise.} \end{cases}$$

Thus, in particular, the null mean and variance of the test statistic are

$$(6.20) \qquad \frac{\Sigma x_i^2}{2t_0}, \qquad \frac{1}{n^{(b)}} \left\{ \frac{1}{3t_0} \cdot \Sigma x_i^3 - \frac{(\Sigma x_i^2)^2}{4t_0^2} \right\}.$$

A strong central limit effect may be expected.

The tests derived here may be compared with similar ones in which the null distribution is derived by computer simulation, permuting at random the observed sequences of intervals (Perkel [40]; Moore, Perkel, and Segundo [37]; Perkel, Gerstein, and Moore [41]). In both types of procedure, it is not clear how satisfactory the tests are in practice as general tests of independence, when the type b process is not marginally Poisson. Note, however, that in order to obtain a null distribution for (a) and (b) above it is necessary to assume only that one of the marginal processes is a Poisson process.

If it is required to treat the two processes symmetrically, taking the null hypothesis that there are two mutually independent Poisson processes, there are many possibilities, including the use of the estimated cross spectral or cross intensity functions or of a two sample test based on the idea that, conditionally on $n^{(a)}$ and $n^{(b)}$, the times to events in the two processes are the order statistics from two independent populations of uniformly distributed random variables.  Again, in the symmetrical case when both marginal processes are clearly not Poisson processes, tests of independence based on the cross spectrum are probably the best broad tests. For this purpose, investigation of the robustness of the distribution theory given in Section 6.3 would be worthwhile.

$$\diamond \quad \diamond \quad \diamond \quad \diamond \quad \diamond$$

We are indebted to Mr. T. K. M. Wisniewski, Dr. A. J. Lawrance, and Professor D. P. Gaver for helpful discussions during the growth of this paper.

REFERENCES

[1] M. S. BARTLETT, "The spectral analysis of point processes," *J. Roy. Statist. Soc. Ser. B*, Vol. 25 (1963), pp. 264–296.

[2] ———, "The spectral analysis of two-dimensional point processes," *Biometrika*, Vol. 51 (1964), pp. 299–311.

[3] ———, "Line processes and their spectral analysis," *Proceedings of the Fifth Berkeley Symposium on Mathematical Statistics and Probability*, Berkeley and Los Angeles, University of California Press, 1967, Vol. 3, pp. 135–154.

[4] ———, *An Introduction to Stochastic Processes*, Cambridge, Cambridge University Press, 1966 (2nd ed.).

[5] P. BILLINGSLEY, "Statistical methods in Markov chains," *Ann. Math. Statist.*, Vol. 32 (1961), pp. 12–40.

[6] D. R. BRILLINGER, "The spectral analysis of stationary interval functions," *Proccedings of the Sixth Berkeley Symposium on Mathematical Statistics and Probability*, Berkeley and Los Angeles, University of California Press, 1972, Vol. 1, pp. 483–513.

[7] M. BROWN, "An $M/G/\infty$ estimation problem," *Ann. Math. Statist.*, Vol. 41 (1970), pp. 651–654.

[8] E. CINLAR, "On the superposition of m-dimensional point processes," *J. Appl. Prob.*, Vol. 5 (1968), pp. 169–176.

[9] ———, "Markov renewal theory," *Adv. Appl. Prob.*, Vol. 1 (1969), pp. 123–187.

[10] R. COLEMAN and J. L. GASTWIRTH, "Some models for interaction of renewal processes related to neuron firing," *J. Appl. Prob.*, Vol. 6 (1969), pp. 38–58.

[11] D. R. Cox, "Some statistical methods connected with series of events," *J. Roy. Statist. Soc. Ser. B*, Vol. 17 (1955), pp. 129–164.

[12] ———, "Some models for series of events," *Bull. Inst. Internat. Statist.*, Vol. 40 (1963), 737–746.

[13] ———, *Analysis of Binary Data*, London, Methuen; New York, Barnes and Noble, 1970.

[14] D. R. Cox and P. A. W. Lewis, *The Statistical Analysis of Series of Events*, London, Methuen; New York, Barnes and Noble, 1966.

[15] D. R. Cox and H. D. Miller, *The Theory of Stochastic Processes*, London, Methuen, 1966.

[16] D. J. Daley, "The correlation structure of the output process of some single server queueing systems," *Ann. Math. Statist.*, Vol. 39 (1968), pp. 1007–1019.

[17] J. L. Doob, *Stochastic Processes*, New York, Wiley, 1953.

[18] M. Dwass and H. Teicher, "On infinitely divisible random vectors," *Ann. Math. Statist.*, Vol. 28 (1957), pp. 461–470.

[19] J. E. Freund, "A bivariate extension of the exponential distribution," *J. Amer. Statist. Assoc.*, Vol. 56 (1961), pp. 971–977.

[20] D. P. Gaver, "Multivariate gamma distributions generated by mixture," *Sankhyā Ser. A*, Vol. 32 (1970), pp. 123–126.

[21] R. C. Griffiths, "The canonical correlation coefficients of bivariate gamma distributions," *Ann. Math. Statist.*, Vol. 40 (1969), pp. 1401–1408.

[22] A. G. Hawkes, "Spectra of some self-exciting and mutually exciting point processes," *Biometrika*, Vol. 58 (1971), pp. 83–90.

[23] G. M. Jenkins, "Contribution to a discussion of paper by M. S. Bartlett," *J. Roy. Statist. Soc. Ser. B*, Vol. 25 (1963), pp. 290–292.

[24] A. J. Lawrance, "Selective interaction of a Poisson and renewal process: First-order stationary point results," *J. Appl. Prob.*, Vol. 7 (1970), pp. 359–372.

[25] ———, "Selective interaction of a stationary point process and a renewal process," *J. Appl. Prob.*, Vol. 7 (1970), pp. 483–489.

[26] ———, "Selective interaction of a Poisson and renewal process: The dependency structure of the intervals between responses," *J. Appl. Prob.*, Vol. 8 (1971), pp. 170–184.

[27] M. R. Leadbetter, "On streams of events and mixtures of streams," *J. Roy. Statist. Soc. Ser. B*, Vol. 28 (1966), pp. 218–227.

[28] ———, "On the distribution of times between events in a stationary stream of events," *J. Roy. Statist. Soc. Ser. B*, Vol. 31 (1969), pp. 295–302.

[29] P. A. W. Lewis, "A branching Poisson process model for the analysis of computer failure patterns," *J. Roy. Statist. Soc. Ser. B*, Vol. 26 (1964), pp. 398–456.

[30] ———, "Asymptotic properties and equilibrium conditions for branching Poisson processes," *J. Appl. Prob.*, Vol. 6 (1969), pp. 355–371.

[31] ———, "Remarks on the theory, computation and application of the spectral analysis of series of events," *J. Sound Vib.*, Vol. 12 (1970), pp. 353–375.

[32] ———, "Asymptotic properties of branching renewal processes," *J. Appl. Prob.*, to appear.

[33] A. W. Marshall and I. Olkin, "A multivariate exponential distribution," *J. Amer. Statist. Assoc.*, Vol. 62 (1967), pp. 30–44.

[34] ———, "A generalized bivariate exponential distribution," *J. Appl. Prob.*, Vol. 4 (1967), pp. 291–302.

[35] J. A. McFadden, "On the lengths of intervals in a stationary point process," *J. Roy. Statist. Soc. Ser. B*, Vol. 24 (1962), pp. 364–382.

[36] E. H. Moore and R. Pyke, "Estimation of the transition distributions of a Markov renewal process," *Ann. Inst. Statist. Math.*, Vol. 20 (1968), pp. 411–424.

[37] G. P. Moore, D. H. Perkel, and J. P. Segundo, "Statistical analysis and functional interpretation of neuronal spike data," *Ann. Rev. Psychology*, Vol. 28 (1966), pp. 493–522

[38] P. A. P. Moran and D. Vere-Jones, "The infinite divisibility of multivariate gamma distributions," *Sankhyā Ser. A*, Vol. 31 (1969), pp. 191–194.

[39] J. NEYMAN and E. L. SCOTT, "Statistical approach to problems of cosmology," *J. Roy. Statist. Soc. Ser. B*, Vol. 20 (1958), pp. 1–29.

[40] D. H. PERKEL, "Statistical techniques for detecting and classifying neuronal interactions," *Symposium on Information Processing in Sight Sensory Systems*, Pasadena, California Institute of Technology, 1965, pp. 216–238.

[41] D. H. PERKEL, G. L. GERSTEIN, and G. P. MOORE, "Neuronal spike trains and stochastic point processes II. Simultaneous spike trains," *Biophys. J.*, Vol. 7 (1967), pp. 419–440.

[42] R. L. PLACKETT, "A class of bivariate distributions," *J. Amer. Statist. Assoc.*, Vol. 60 (1965), pp. 516–522.

[43] R. PYKE and R. A. SCHAUFELE, "The existence and uniqueness of stationary measures for Markov renewal processes," *Ann. Math. Statist.*, Vol. 37 (1966), pp. 1439–1462.

[44] I. M. SLIVNYAK, "Some properties of stationary flows of homogeneous random events," *Theor. Probability Appl.*, Vol. 7 (1962), pp. 336–341.

[45] D. VERE-JONES, S. TURNOVSKY, and G. A. EIBY, "A statistical survey of earthquakes in the main seismic region of New Zealand. Part I. Time trends in the pattern of recorded activity," *New Zealand J. Geol. Geophys.*, Vol. 7 (1964), pp. 722–744.

[46] D. VERE-JONES and R. D. DAVIES, "A statistical survey of earthquakes in the main seismic region of New Zealand, Part II. Time series analysis," *New Zealand J. Geol. Geophys.*, Vol. 9 (1966), pp. 251–284.

[47] D. VERE-JONES, "Stochastic models for earthquake occurrence," *J. Roy. Statist. Soc. Ser. B*, Vol. 32 (1970), pp. 1–62.

[48] L. WALLOE, J. K. S. JANSEN, and K. NYGAARD, "A computer simulated model of a second order sensory neuron," *Kybernetik*, Vol. 6 (1969), pp. 130–140.

[49] G. S. WATSON and E. J. WILLIAMS, "On the construction of significance tests on the circle and the sphere," *Biometrika*, Vol. 43 (1956), pp. 344–352.

[50] T. K. M. WISNIEWSKI, "Forward recurrence time relations in bivariate point processes," *J. Appl. Prob.*, Vol. 9 (1972).

[51] ———, "Extended recurrence time relations in bivariate point processes," to appear.

On the Estimation of the Intensity Function of a Stationary Point Process

By D. R. Cox

Birkbeck College, University of London

[Received December 1964. Revised February 1965]

SUMMARY

The intensity function of a stationary point process gives as a function of x the conditional probability, given an event at time t, of an event in the interval $(t+x, t+x+\Delta x)$. A very nearly unbiased estimate of the grouped intensity function is constructed. Its sampling properties are investigated when the data are generated by a Poisson process.

1. INTRODUCTION

CONSIDER a stationary process of point events in which there is zero probability that two or more events occur simultaneously. Let ρ be the mean rate of occurrence, so that as $\Delta t \to 0$

$$\text{prob}\{\text{one event in } (t, t+\Delta t)\} = \rho \Delta t + o(\Delta t), \qquad (1)$$

for all t. The second-order properties of the process in continuous time are determined by the intensity function $m(x)$, defined for $x > 0$, by

$$m(x) = \lim_{\Delta x \to 0} \frac{\text{prob}\{\text{event in } (t+x, t+x+\Delta x) | \text{event at } t\}}{\Delta x}. \qquad (2)$$

When the stationary point process is an equilibrium renewal process, the function $m(x)$ is the renewal density associated with the corresponding ordinary renewal process; see, for instance, Cox (1962) for the definitions of these terms. For a general stationary point process without multiple occurrences, the covariance density (Bartlett, 1955, p. 166) is defined for $x > 0$ by

$$\gamma(x) = \lim_{\Delta_1, \Delta_2 \to 0} \frac{\text{cov}\{\text{no. events in } (t, t+\Delta_1), \text{ no. events in } (t+x, t+x+\Delta_2)\}}{\Delta_1 \Delta_2}.$$

This is related to the intensity function $m(x)$ by

$$\gamma(x) = \rho\{m(x) - \rho\}, \qquad (3)$$

from which a spectral density can be defined (Bartlett, 1963). If the definitions are extended to $x = 0$, contributions proportional to Dirac delta functions have to be added.

For a Poisson process, $m(x) = \rho$. For a renewal process it is known that under very weak conditions on the distribution of intervals $m(x) \to \rho$ as $x \to \infty$, where ρ is the reciprocal of the mean interval between events. In general, it follows easily from (1) and the stationarity that the expected number of events in an interval of length t

starting from an arbitrary time point is ρt, whereas, from (2), the corresponding expectation in an interval starting from an arbitrary event is

$$\int_{0+}^{t} m(x)\,dx.$$

It can be seen from this correspondence that, for a broad class of processes without long-term after-effects and in which $m(x)$ tends to a limit as $x \to \infty$, that limit is ρ.

In the present note we are concerned, however, not with the general probabilistic properties of $m(x)$, but with its estimation from a realization. A point estimate is obtained. Then some properties of the estimate are given when the data come from a Poisson process. We do not consider here the distribution of the estimates under other models, nor do we examine extensions to point processes in two or more dimensions, or to processes with several types of event.

2. A POINT ESTIMATE

Suppose that the process is observed over an interval $(0, t_0)$, starting from an arbitrary time. Let $N(t)$ be the sample counting function, having $N(0) = 0$ and jumping by one at each event; denote by n the total number of events observed. Now the estimation of $m(x)$ raises the same problems as does the estimation of other density functions, for example probability or spectral density functions. Some smoothing is required for graphical presentation and for the derivation of estimates with reasonably small variances; see Parzen (1962) for a formal procedure for the estimation of a probability density function and for references to earlier work. Here, however, we consider only a simple procedure directly analogous to the formation of a histogram. For a grouping interval τ, we estimate the grouped intensities

$$m_g(r\tau + \tfrac{1}{2}\tau) = \frac{1}{\tau}\int_{r\tau+0}^{r\tau+\tau} m(x)\,dx \quad (r = 0, 1, \ldots). \tag{4}$$

In practical work it is usually advisable to examine a number of values of τ; for formal study of limiting properties it would be reasonable to allow τ to tend slowly to zero as $t_0 \to \infty$. Here, however, we regard τ as fixed.

It is natural to base an estimate of (4) on the total number S_r of pairs of events separated by an interval between $r\tau$ and $r\tau + \tau$. That is we take, in principle, the $\tfrac{1}{2}n(n-1)$ (positive) intervals between all possible ordered pairs of events and form them into a histogram with groups of width τ. The formal expression for the number S_r is

$$S_r = \int_{u=0}^{t_0-r\tau} \int_{x=r\tau}^{r\tau+\tau} dN_u\,dN_{u+x}$$

$$= \int_{u=0}^{t_0-r\tau-\tau} \int_{x=r\tau}^{r\tau+\tau} dN_u\,dN_{u+x} + \int_{u=t_0-r\tau-\tau}^{t_0-r\tau} \int_{x=r\tau}^{t_0-u} dN_u\,dN_{u+x}. \tag{5}$$

To evaluate the expectation of (5), we write (1) as $E(dN_u) = \rho\,du$ and (2) as

$$E(dN_u\,dN_{u+x}) = \rho\,du\,E(dN_{u+x}|\text{event at } u)$$

$$= \rho m(x)\,du\,dx. \tag{6}$$

It follows easily from (6) that the expectation of the first double integral in (5) is exactly

$$\rho\tau(t_0 - r\tau - \tau)\,m_g(r\tau + \tfrac{1}{2}\tau).$$

When $\tau/(t_0-r\tau)$ is small, as would always be the case in practice, the second integral in (5) is only a small correction term and can be approximated by assuming $m(x)$ to be constant over the relevant range and equal to $m_g(r\tau+\frac{1}{2}\tau)$. Then the expectation of the second integral is

$$\tfrac{1}{2}\rho\tau^2 m_g(r\tau+\tfrac{1}{2}\tau).$$

With this approximation,

$$E(S_r) = \rho\tau(t_0-r\tau-\tfrac{1}{2}\tau)\, m_g(r\tau+\tfrac{1}{2}\tau). \tag{7}$$

Since $\hat{\rho} = n/t_0$ has expectation ρ, we are led to define

$$\hat{m}_g(r\tau+\tfrac{1}{2}\tau) = \frac{S_r t_0}{n\tau(t_0-r\tau-\tfrac{1}{2}\tau)}. \tag{8}$$

The qualitative reason for the factor $t_0/(t_0-r\tau-\frac{1}{2}\tau)$ is that the effective length of the series decreases with increasing r.

It is of interest to note a sample result corresponding to the probabilistic property, noted in Section 1, that $m(x)\to\rho$ as $x\to\infty$. Since the total number of pairs of intervals is $\frac{1}{2}n(n-1)$, we have that

$$\sum_{r=0}^{r_0} S_r = \tfrac{1}{2}n(n-1), \tag{9}$$

where $r_0 = (t_0/\tau)-1$ and to avoid unimportant end effects we suppose that t_0/τ is an integer. On combining (8) and (9), we have that

$$\frac{\Sigma w_r\hat{m}_g(r\tau+\tfrac{1}{2}\tau)}{\Sigma w_r} = \frac{n-1}{t_0-2\tau+\tau^2/t_0} \simeq \hat{\rho} = \frac{n}{t_0}, \tag{10}$$

where $w_r = (t_0-r\tau-\frac{1}{2}\tau)/t_0$. Equation (10) implies that if the sample function is fluctuating around an average independent of r, that average must be nearly $\hat{\rho}$.

3. PROPERTIES FOR A POISSON PROCESS

When we use the estimates (8) to interpret possible systematic departures from a Poisson process, we are interested in the sampling properties of (8) under the null hypothesis of a Poisson process. Under this null hypothesis, the estimates (8) all have expectations exactly ρ, but rather than examine the sampling deviations from ρ, it is more useful to examine the deviations from $\hat{\rho}$. Alternatively, we argue that to obtain sampling properties of the Poisson process independent of the nuisance parameter ρ, we should consider the distribution of S_r conditionally on the observed value of n, the total number of events. In this conditional distribution, we can regard the positions of events as independently rectangularly distributed over $(0, t_0)$.

We therefore consider the following problem. Let $U_1, ..., U_n$ be independently rectangularly distributed over $(0, t_0)$ and let

$$V_{ij}^{(r)} = \begin{cases} 1 & \text{if } r\tau\leqslant|U_i-U_j|<r\tau+\tau, \\ 0 & \text{otherwise.} \end{cases} \tag{11}$$

Then, in the notation of Section 2,

$$S_r = \sum_{j>i} V_{ij}^{(r)}. \tag{12}$$

The mean and variance of (12) are best obtained from the conditional mean of $V_{ij}^{(r)}$ given U_i, namely that if $2r\tau+\tau<t_0$,

$$E(V_{ij}^{(r)}\,|\,0\leqslant U_i<r\tau)=\tau/t_0, \tag{13}$$

$$E(V_{ij}^{(r)}\,|\,r\tau\leqslant U_i=u_i<r\tau+\tau)=\frac{\tau}{t_0}+\frac{u_i-r\tau}{t_0}, \tag{14}$$

$$E(V_{ij}^{(r)}\,|\,r\tau+\tau\leqslant U_i<t_0-r\tau-\tau)=2\tau/t_0, \tag{15}$$

with expressions analogous to (14) and (13) in the ranges $t_0-r\tau-\tau<U_i<t_0-r\tau$ and $t_0-r\tau<U_i<t_0$. These follow directly from (11), since U_j is distributed independently of U_i. The unconditional mean of $V_{ij}^{(r)}$ is obtained from (13)–(15) by integrating with respect to the uniform distribution of U_i and is

$$E(V_{ij}^{(r)})=2\frac{\tau}{t_0}\frac{r\tau}{t_0}+\frac{2}{t_0}\int_{r\tau}^{r\tau+\tau}\left(\frac{\tau}{t_0}+\frac{u-r\tau}{t_0}\right)du+\frac{2\tau}{t_0}\frac{(t_0-2r\tau-2\tau)}{t_0}$$

$$=\frac{2\tau(t_0-r\tau-\tfrac{1}{2}\tau)}{t_0^2}. \tag{16}$$

Thus

$$E(S_r)=\frac{n(n-1)\,\tau(t_0-r\tau-\tfrac{1}{2}\tau)}{t_0^2} \tag{17}$$

and, from (8),

$$E\{\hat{m}_g(r\tau+\tfrac{1}{2}\tau)\}=\frac{n-1}{t_0}. \tag{18}$$

To find var (S_r), we have first that since $V_{ij}^{(r)}$ is a $(0,1)$ variable, its expected square is given by (16). Further, conditionally on U_i, the random variables $V_{ij}^{(r)}$, $V_{il}^{(r)}$ $(j\neq l)$ are independent. Hence $E(V_{ij}^{(r)}\,V_{il}^{(r)}\,|\,U_i=u_i)$ can be obtained directly from (13)–(15), and the unconditional expectation is then found to be

$$E(V_{ij}^{(r)}\,V_{il}^{(r)})=\frac{4\tau^2}{t_0^2}-\frac{6r\tau^3}{t_0^3}-\frac{10\tau^3}{3t_0^3}. \tag{19}$$

Finally $V_{ij}^{(r)}$ and $V_{ln}^{(r)}$ are independent when i,j,l,n are all different. Now enumeration shows that

$$E(S_r^2)=\tfrac{1}{2}n(n-1)E(V_{ij}^{(r)})$$

$$+n(n-1)(n-2)E(V_{ij}^{(r)}\,V_{il}^{(r)})$$

$$+\tfrac{1}{4}n(n-1)(n-2)(n-3)\{E(V_{ij}^{(r)})\}^2, \tag{20}$$

whence, on using (16), (17) and (19), we have that

$$\text{var}\,(S_r)=\frac{n(n-1)\,\tau(t_0-r\tau-\tfrac{5}{2}\tau)}{t_0^2}+n(n-1)(2rn+\tfrac{2}{3}n+\tfrac{2}{3})\frac{\tau^3}{t_0^3}$$

$$-n(n-1)(4n-6)(r+\tfrac{1}{2})^2\frac{\tau^4}{t_0^4}; \tag{21}$$

the terms have been collected in powers of τ/t_0, which would normally be very small. Finally

$$\text{var}\{\hat{m}_g(r\tau+\tfrac{1}{2}\tau)\} = \frac{t_0^2\,\text{var}\,(S_r)}{n^2\,\tau^2(t_0-r\tau-\tfrac{1}{2}\tau)^2}.$$ (22)

For many purposes an adequate approximation to (22) is

$$\frac{n-1}{n\tau(t_0-r\tau+\tfrac{1}{2}\tau)}.$$ (23)

A rough approach to the distribution theory, likely to be valid when $r\tau/t_0$ is small, is to neglect the end effects in (13)–(15), i.e. to suppose that (15) is always applicable. To this order of approximation all $V_{ij}^{(r)}$ are independent and hence S_r has a binomial distribution corresponding to $\tfrac{1}{2}n(n-1)$ trials each with probability of "success" $2\tau/t_0$. That is,

$$E(S_r) = \frac{n(n-1)\,\tau}{t_0}, \quad \text{var}\,(S_r) \sim \frac{n(n-1)\,\tau}{t_0}.$$ (24)

These are indeed the first terms of an expansion of (17) and (21). This argument, and more generally the approximate equality of (17) and (21), suggest that S_r should have nearly a Poisson distribution. We shall not examine this approximation further. As $n \to \infty$, the random variables S_r are asymptotically normally distributed (Hoeffding, 1948), because they are so-called U-statistics.

Thus it will help in interpreting a graph of $\hat{m}_g(r\tau+\tfrac{1}{2}\tau)$ against $r\tau+\tfrac{1}{2}\tau$ to put approximate "control limits" at

$$\frac{n-1}{t_0} \pm k\sqrt{\text{var}\,(\hat{m}_g)},$$ (25)

taking say $k=2$.

4. Covariance of Two Estimates

We can apply the arguments of Section 3 to examine the covariance of S_{r_1} and S_{r_2} for $r_1 < r_2$. We have then that

$$E(V_{ij}^{(r_1)} V_{ij}^{(r_2)}) = 0,$$ (26)

since it is certain that at least one of $V_{ij}^{(r_1)}$ and $V_{ij}^{(r_2)}$ is zero. When i, j, l, m are all different, $V_{ij}^{(r_1)}$ and $V_{lm}^{(r_2)}$ are independent, so that the only essentially new calculation needed is of

$$E(V_{ij}^{(r_1)} V_{il}^{(r_2)}) \quad (j \neq l).$$ (27)

The derivation is an extension of (13)–(15). In fact if $2r_2\tau+\tau<t_0$, we have that

$$E(V_{ij}^{(r_1)} V_{il}^{(r_2)}|0<U_i<r_1\tau) = \tau^2/t_0^2,$$

$$E(V_{ij}^{(r_1)} V_{il}^{(r_2)}|r_1\tau \leqslant U_i = u_i<r_1\tau+\tau) = \left(\frac{\tau}{t_0}+\frac{u_i-r_1\tau}{t_0}\right)\frac{\tau}{t_0},$$ (28)

$$E(V_{ij}^{(r_1)} V_{il}^{(r_2)}|r_1\tau+\tau \leqslant U_i<r_2\tau) = 2\tau^2/t_0^2, \quad \text{etc.}$$

We have that

$$\mathrm{cov}\,(S_{r_1}, S_{r_2}) = -n(n-1)\frac{2\tau^2}{t_0^2} + n(n-1)(2r_1 n + n - 2r_1 + 2r_2)\frac{\tau^3}{t_0^3}$$

$$-n(n-1)(n-\tfrac{3}{2})(2r_1+1)(2r_2+1)\frac{\tau^4}{t_0^4}. \tag{29}$$

Note that, in accordance with the discussion of (24), the leading term of (29) is the covariance in a multinomial distribution with index $\tfrac{1}{2}n(n-1)$ between the numbers of occurrences in two cells, each with probability $2\tau/t_0$. To this order of approximation the correlation coefficient between any two of the estimates $\hat{m}_g(r\tau+\tfrac{1}{2}\tau)$ is $-2\tau/t_0$.

A useful partial check on the correctness of (21) and (29) is obtained by taking r_1 even. Then $\mathrm{var}\,(S_{r_1} + S_{r_1+1})$ can be calculated from (21) and (29) and also directly from (21) with τ replaced by 2τ and r by $\tfrac{1}{2}r_1$.

We shall not examine in detail the construction of tests based on combinations of estimates $\hat{m}_g(r\tau+\tfrac{1}{2}\tau)$. To a first approximation

$$\sum_{r=0}^{s-1}\left\{\hat{m}_g(r\tau+\tfrac{1}{2}\tau) - \frac{n-1}{t_0}\right\}^2 \Big/ \mathrm{var}\,\{\hat{m}_g(r\tau+\tfrac{1}{2}\tau)\} \tag{30}$$

is distributed as chi-squared with s degrees of freedom, provided that $s\tau \ll t_0$. Similarly, the standard error of linear combinations of the estimates (8) can be calculated.

5. DISCUSSION

A number of estimated intensity functions for empirical series are given by Cox and Lewis (1966). In a general way, the analysis of the intensity function is equivalent to that of the spectrum of the point process (Bartlett, 1963). Substantial practical experience would be desirable before reaching a firm conclusion about the relative merits of the two approaches.

The referee has made the very interesting suggestion of estimating the intensity function by first converting the point process into a real-valued process, by summing impulse functions attached to each event. Then standard techniques for estimating autocorrelation functions and spectral density functions are available and yield estimates of certain weighted averages of the intensity function or spectral density function of the point process. A comparison with the method of the present paper will not be attempted; however, the estimate (8) is very simply related to the original data and this must be an advantage in interpretation.

REFERENCES

BARTLETT, M. S. (1955), *Introduction to the Theory of Stochastic Processes*. Cambridge University Press.
—— (1963), "The spectral analysis of point processes", *J. R. statist. Soc.* B, **25**, 264–281.
Cox, D. R. (1962), *Renewal Theory*. London: Methuen.
—— and LEWIS, P. A. W. (1966), *Statistical Analysis of Series of Events*. London: Methuen. To appear.
HOEFFDING, W. (1948), "A class of statistics with asymptotically normal distributions", *Ann. math. Statist.*, **19**, 293–325.
PARZEN, E. (1962), "On estimation of a probability density function and mode", *Ann. math. Statist.*, **33**, 1065–1076.

Journal of the Royal Statistical Society

SERIES B (METHODOLOGICAL)

Vol. XX, No. 2, 1958

THE REGRESSION ANALYSIS OF BINARY SEQUENCES

By D. R. Cox

Birkbeck College, University of London

[Read before the RESEARCH SECTION of the ROYAL STATISTICAL SOCIETY, March 5th, 1958, Professor G. A. BARNARD in the Chair]

SUMMARY

A SEQUENCE of 0's and 1's is observed and it is suspected that the chance that a particular trial is a 1 depends on the value of one or more independent variables. Tests and estimates for such situations are considered, dealing first with problems in which the independent variable is preassigned and then with independent variables that are functions of the sequence. There is a considerable amount of earlier work, which is reviewed.

1. INTRODUCTION

Suppose that there are for analysis one or more series of trials, the observation on any one trial taking one of two forms, such as "success" or "failure", "defective" or "not-defective", and so on. Denote the possible observations by 0 and 1, each series of trials therefore giving a sequence of 0's and 1's. Suppose further that corresponding to each trial there are one or more independent variables and that we suspect that the probability that a particular trial gives say the outcome 1 depends on the corresponding values of the independent variables. In this paper we consider methods for estimating and testing such dependencies. If the observations were continuous instead of being (0, 1) variates, the problems would be treated by the standard methods of simple and multiple regression.

The following are some examples. Haldane and Smith (1948) have considered a test for a birth-order effect in the occurrence of hereditary abnormalities. The data here consist of a series for each family such as $N A A N$, meaning that the first child is normal, the second has the particular abnormality under study, and so on. The independent variable is the serial number of the birth. Alternatively, the independent variable may be maternal age and, in an extension of the problem, both maternal age and serial number are available and it is required to examine whether there is a dependence on birth order when any effect of maternal age is eliminated.

In certain learning studies in experimental psychology a task is attempted a number of times in succession and the primary observation is success or failure at each trial. We may wish to examine the dependence of the probability of success on such variables as the number of preceding trials, the number of previously rewarded successes, and so on. The last of these variables is rather difficult to deal with because it is not preassigned, but is a function of the outcomes of the preceding trials in the sequence.

VOL. XX. NO. 2.

I

Some standard problems appear as special cases of the situation considered here. Thus, if the independent variable takes on just two levels and if different trials are statistically independent, the numbers of 0's and 1's at the two levels can be written in a 2×2 contingency table and the usual theory is applicable. If the independent variable takes on k values at each of which there are a number of trials, the data can be written as a $2 \times k$ table, and the procedure that we shall obtain is to a first approximation equivalent to the standard method of isolating from χ^2 a single degree of freedom for linear regression on the column variable.

If the independent variable is at several levels corresponding to different doses of a drug and if we are interested in the "position" and "slope" of the dependence, we have the usual situation in assays with quantal response. In this paper, however, we shall be concerned with what, in continuous problems, would be the slope of the regression line, regarding the position of the line as a nuisance parameter. Hence the present paper is not of much direct relevance to assay work.

2. General Formulation

Let $Y_1, \ldots, Y_n$ be mutually independent random variables each taking the values 0 and 1 and let $x_1, \ldots, x_n$ be a set of fixed numbers. In the simplest form of our problem there is suspected to be a relation between $\theta_i = \mathrm{pr}\,(Y_i = 1)$ and x_i. In a more complicated form there are two sets of fixed numbers $\{x_i\}$ and $\{u_i\}$ and we are concerned with the joint dependence between θ_i and x_i, u_i. In further forms of the problem we consider x_i which are functions of $Y_1, \ldots, Y_{i-1}$, such as Y_{i-1} and $Y_1 + \ldots + Y_{i-1}$, the Y_i therefore no longer being independent.

Consider first the case with a single preassigned independent variable. We need a parametric representation of the relation between θ_i and x_i that will have a simple intuitive meaning, that will be a reasonable approximation to relations likely to occur in practice and that will be manageable mathematically. A linear relation is unsuitable, except over a narrow range, because of the restriction that θ_i must lie in [0, 1] and, in the absence of special considerations for a particular problem, the best form seems to be the logistic law

$$\mathrm{logit}\ \theta_i \equiv \log\{\theta_i/(1 - \theta_i)\} = \alpha + \beta x_i, \tag{1}$$

i.e.

$$\theta_i \equiv \mathrm{pr}\,(Y_i = 1) = e^{\alpha + \beta x_i}/(1 + e^{\alpha + \beta x_i}), \tag{2}$$

and

$$1 - \theta_i \equiv \mathrm{pr}\,(Y_i = 0) = 1/(1 + e^{\alpha + \beta x_i}). \tag{3}$$

This form has been extensively used in work on bioassays, notably by Berkson.

As stated above, we shall be concerned with inference about β, regarding α as a nuisance parameter. The interpretation of β is that if θ_i is small β is the fractional increase in θ_i per unit increase in x_i, whereas if $1 - \theta_i$ is small, β is the fractional decrease in $1 - \theta_i$ per unit increase in x_i. If all the values of θ_i considered lie near θ^0, (2) represents a linear relation between θ_i and x_i with slope $\beta\theta^0(1 - \theta^0)$.

In more complicated problems, for example with several independent variables, we shall use natural generalizations of (1).

All the significance tests to be developed below for null hypotheses expressing the absence of regression, are non-parametric, in that the logistic law enters only into the derivation of the test criterion and not into the development of its sampling distribution under the null hypothesis.

3. Remarks on Sampling Without Replacement

It will appear in §4 that the distributional problems to be solved in connection with the present problems depend on sampling without replacement from a finite population. It is therefore convenient first to recall some results about such sampling.

Let $x_1, \ldots, x_n$ denote the finite population and let $X_{0,y}$ be a random variable defined as the sum of y numbers selected randomly without replacement.* The moments of $X_{0,y}$ are

$$m_{0,y} = E(X_{0,y}) = ym_1, \tag{4}$$

$$\sigma^2_{0,y} = V(X_{0,y}) = y(n-y) m_2/(n-1), \tag{5}$$

$$E[(X_{0,y} - m_{0,y})^3] = y(n-y)(n-2y) m_3/[(n-1)(n-2)], \tag{6}$$

$$E[(X_{0,y} - m_{0,y})^4] = \frac{y(n-y)}{(n-1)(n-2)(n-3)} \{(n^2 - 6ny + n + 6y_2) m_4$$
$$+ 3(y-1) n(n-y-1) m_2^2\}, \quad . \tag{7}$$

where

$$m_1 = \Sigma x_i/n, \tag{8}$$

$$m_r = \Sigma (x_i - m_1)^r/n, \qquad r > 1. \tag{9}$$

These formulae are given, for example, by Irwin and Kendall (1944).

In numerical work it is preferable to calculate the m_r from the power sums about the origin:

$$m_1 = s_1/n, \tag{10}$$

$$m_2 = (ns_2 - s_1^2)/n^2, \tag{11}$$

$$m_3 = (n^2s_3 - 3ns_2s_1 + 2s_1^3)/n^3, \tag{12}$$

$$m_4 = m(n^3s_4 - 4n^2s_1s_3 + 6ns_2s_1^2 - 3s_1^4)/n^4, \tag{13}$$

where

$$s_r = \Sigma x_i^r.$$

For our theoretical purposes it is more useful, however, to write $f = y/n$ and to make an expansion with f fixed and n large. This gives

$$m_{0,y} = ym_1, \tag{14}$$

$$\sigma^2_{0,y} \sim ym_2(1 - f + 1/n), \tag{15}$$

$$\gamma_{1(0,y)} \sim \frac{(1 - 2f) \gamma_1}{\sqrt{\{f(1-f) n\}}}, \tag{16}$$

$$\gamma_{2(0,y)} \sim -\frac{6}{n} + \frac{[1 - 6f(1-f)] \gamma_2}{nf(1-f)}, \tag{17}$$

* The notation is chosen to fit in with the application considered in §4.

where γ_1 and γ_2 refer to cumulant ratios of the finite population. The term independent of γ_2 in (17) could be removed by redefinition of the γ_2's in terms of the K's of finite sampling theory instead of the m's.

Equations (16) and (17) suggest that in samples for which y and $n - y$ are both appreciable, $X_{0,y}$ will be nearly normally distributed and that the "central limit effect" will be relatively more rapid than in samples of size Min $(y, n - y)$ drawn with replacement. Rigorous proofs that as $n \to \infty$ with f fixed, $(X_{0,y} - m_{0,y})/\sigma_{0,y}$ converges in distribution function, under suitable restrictions, to the unit normal law are given by Madow (1948), Hoeffding (1951) and Motoo (1957); the first two authors, but not the last, require all moments of the population to behave suitably as $n \to \infty$.

For special forms of finite population, explicit expressions may be attained. Thus, if r of the x_i are equal to one and the remaining $(n - r)$ values are zero, $X_{0,y}$ has the hypergeometric distribution with probability generating function

$$\sum_{s} \frac{\binom{r}{s}\binom{n-r}{y-s}}{\binom{n}{y}} \zeta^s. \tag{18}$$

If $x_i = i$ for $i = 1, \ldots, n$, Haldane and Smith (1948) have shown that the probability generating function is

$$\frac{1}{\binom{n}{y}} \prod_{r=1}^{y} \left(\frac{\zeta^r - \zeta^{n+1}}{1 - \zeta^r} \right) \tag{19}$$

and have expressed the cumulants of $X_{0,y}$ in terms of Bernouilli's numbers.

A generalization, that will be needed in §5, concerns sampling from a bivariate finite population. We suppose that there are n pairs $(x_1, u_1), \ldots, (x_n, u_n)$, that a random sample of y pairs is drawn without replacement and that we consider the joint distribution of the random variables defined as before. The joint sampling cumulants of $X_{0,y}$ and $U_{0,y}$ do not seem to have been considered in the literature. The cumulants up to the third order are given by

$$\kappa_{11(0,y)} = E[(X_{0,y} - m_{0,y}^{(x)})(U_{0,y} - m_{0,y}^{(u)})] = y(n - y) m_{11}/(n - 1), \tag{20}$$

$$\kappa_{21(0,y)} = E[(X_{0,y} - m_{0,y}^{(x)})^2(U_{0,y} - m_{0,y}^{(u)})] = \frac{y(n - y)(n - 2y) m_{21}}{(n - 1)(n - 2)}, \tag{21}$$

where

$$m_{\alpha\beta} = \frac{1}{n} \sum (x_i - m_1^{(x)})^\alpha (u_i - m_1^{(u)})^\beta. \tag{22}$$

The simplest way to prove say (21) is to note first that the expectation must be symmetric in i, of degree $(2, 1)$ and invariant under separate translations of the x_i and the u_i and hence must be of the form $A(y, n) m_{21}$. To determine A, suppose that $x_i = u_i$, when m_{21} becomes m_{30}, and the expectation is the third moment of $X_{0,y}$. The form of A follows immediately from (6). Mixed fourth moments can be calculated similarly.

4. SINGLE PREDETERMINED INDEPENDENT VARIABLE

4.1. *General theory.*—Let $x_1, \ldots, x_n$ be, as in §2, fixed quantities. The likelihood of observations $Y_1, \ldots, Y_n$ is

$$\exp\{\Sigma(\alpha + \beta x_i) \, Y_i\}/\Pi(1 + e^{\alpha + \beta x_i}) = e^{\alpha Y + \beta X}/\Pi(1 + e^{\alpha + \beta x_i}), \qquad (23)$$

where $Y = \Sigma \, Y_i$, $X = \Sigma \, Y_i x_i$. Thus X, Y are jointly sufficient for the parameters α, β, as was noted by Garwood (1941).

Further, it is clear that Y, which is the total number of 1's, can tell us nothing about β, which is concerned with the way the distribution of 1's changes with x_i. Hence, to make an inference about β, we argue conditionally on the observed value of Y. The arguments for doing this as a matter of principle raise interesting general issues analogous to the corresponding ones for the 2×2 contingency table. The point will not be discussed here.

Consider the distribution of X conditionally on the observed value of Y, which will be denoted by y. Now

$$\text{pr}(X = x, \, Y = y) = \frac{N_{x,y} \, e^{\alpha y + \beta x}}{\Pi(1 + e^{\alpha + \beta x_i})} \qquad (24)$$

where $N_{x,y}$ is the number of distinct ordered sets of y numbers, taken from $x_1, \ldots, x_n$, whose sum is x. Therefore

$$\text{pr}(Y = y) = \frac{\Sigma_x N_{x,y} \, e^{\alpha y + \beta x}}{\Pi(1 + e^{\alpha + \beta x_i})}, \qquad (25)$$

where Σ_x denotes summation over all values taken by the random variable X. Finally, if we take the ratio of (24) to (25), we have that

$$\text{pr}(X = x \mid Y = y) = \frac{N_{x,y} \, e^{\beta x}}{\Sigma_x N_{x,y} \, e^{\beta x}}, \qquad (26)$$

and the nuisance parameter α has been eliminated. When $\beta = 0$, (26) is the distribution of the random variable $X_{0,y}$ of §3. If $X_{\beta,y}$ denotes the random variable X considered conditionally on y when β is the true parameter value, we have by (26) that

$$\text{pr}(X_{\beta,y} = x) = \frac{N_{x,y}}{\binom{n}{y}} \frac{e^{\beta x}}{M_{0,y}(\beta)}, \qquad (27)$$

where $M_{0,y}(s)$ is the moment generating function of $X_{0,y}$.

Thus the moment generating function of $X_{\beta,y}$ is given by

$$M_{\beta,y}(s) = \frac{M_{0,y}(\beta + s)}{M_{0,y}(\beta)}, \qquad (28)$$

and the corresponding relation for cumulant generating functions is

$$K_{\beta,y}(s) = K_{0,y}(\beta + s) - K_{0,y}(\beta). \qquad (29)$$

In particular it follows from (29) that the cumulants of $X_{\beta,y}$ for general values of β are given by the derivatives of the cumulant generating function for $X_{0,y}$ evaluated at β instead of at the origin.

If a manageable explicit expression were available for the distribution of $X_{0,y}$ or for its moment generating function, direct calculation of, or approximation to, the properties of $X_{\beta,y}$ would be possible. But in general there seems little that can be done except to obtain from (29) a series expansion for the cumulants of $X_{\beta,y}$ in ascending powers of β; thus

$$\kappa_{\beta,y}{}^{(1)} = E(X_{\beta,y}) = m_{0,y} + \beta\sigma^2{}_{0,y} + \frac{1}{2}\beta^2\sigma^3{}_{0,y}\,\gamma_{1(0,y)} + \cdots , \tag{30}$$

$$\kappa_{\beta,y}{}^{(2)} = V(X_{\beta,y}) = \sigma^2{}_{0,y} + \beta\sigma^3{}_{0,y}\,\gamma_{0(0,y)} + \cdots , \tag{31}$$

etc., the general result being that

$$\kappa_{\beta,y}{}^{(r)} = \sum_{t=0}^{\infty} \kappa_{0,y}{}^{(t+r)}\,\frac{\beta^t}{t!} . \tag{32}$$

Bennett (1956) has obtained a special case of (29) and the resulting expansions. His choice of criterion was intuitive.

Another approximation to (29), appropriate when $y/n \ll 1$, is to regard $X_{0,y}$ as the sum of a random sample drawn with replacement (Haldane and Smith, 1948). This can be useful when the cumulant generating function of the population can be expressed simply.

It follows from the asymptotic normality of $X_{0,y}$ that a sufficient condition that $X_{\beta,y}$ is nearly normal when y and $n - y$ are both large is that $\beta\sqrt{(m_2 y)} = o(1)$. In fact the simplest approximation based on (30) and (31) is to take $X_{\beta,y}$ as normally distributed with mean and variance

$$m_{0,y} + \beta\,\sigma^2{}_{0,y} \text{ and } \sigma^2{}_{0,y}, \tag{33}$$

respectively. This can also be derived directly from (27), by using the normal approximation to

$$N_{x,y} \Big/ \binom{n}{y}$$

and at the same time approximating to $M_{0,y}(\beta)$ by the moment generating function of a normal random variable. Further terms of the expansions (30), etc., can be recovered by an Edgeworth expansion of (27). A normal approximation involved in calculating the power function of the exact test in a 2×2 contingency table was obtained by Patnaik (1948) using a similar argument; Stevens (1951) pointed out that if the parameter corresponding to β is not small the approximation can be badly wrong.

4.2. *The* 2×2 *contingency table.*—If the x_i take on two values, say 0 and 1, the statistic X is the entry in the lower right hand corner of the 2×2 contingency table :

$$x_i$$

	0	1
0		
Y_i		
1		

and the conditional distribution (27) reduces to the distribution given by Fisher (1935) as appropriate for inference about the parameter $e^\beta = \psi$, where

$$\psi = \frac{\text{pr}\,(Y_i = 1 \mid x_i = 1)}{\text{pr}\,(Y_i = 0 \mid x_i = 1)} \times \frac{\text{pr}\,(Y_i = 0 \mid x_i = 0)}{\text{pr}\,(Y_i = 1 \mid x_i = 0)} . \tag{34}$$

The significance test of the null hypothesis $\beta = 0$ based on the hypergeometric distribution (18) of X is, of course, the familiar exact test of association in the 2×2 table.

As noted in §4.1, the normal approximation to the non-null distribution of X based on the first terms of (30) and (31) has been given by Patnaik. An alternative normal approximation, based on an expansion around the mode of the distribution of X, has been obtained by Cornfield (1956), and this should be much the more satisfactory method when $|\beta|$ is large.

To compare the different approximations, consider the example analysed by Cornfield.

Example 1.—The data in Table 1 were obtained in a survey of physicians.

TABLE 1

Distribution of Physicians

	Controls	Lung Cancer Patients
Smokers	32	60
Non-smokers	11	3

Cornfield (1956) has obtained as 95 per cent. confidence limits for ψ the values $0 \cdot 030$ and $0 \cdot 623$; he shows that the corresponding exact tail area probabilities are $3 \cdot 8$ per cent. and $2 \cdot 4$ per cent. respectively, instead of the required $2 \cdot 5$ per cent. His procedure involves the iterative solution of a quartic equation.

The simplest approximate solution by the methods of §4.1 uses the first terms of (30) and (31), that is regards X as normally distributed with mean $m_{0,y} + \beta \sigma^2_{0,y}$ and variance $\sigma^2_{0,y}$, where X is observed to be 3, $y = 14$ and the finite population sampled consists of 43 values with $x_i = 0$ and 63 with $x_i = 1$, so that $n = 106$. Hence by (10) and (11)

$$m_1 = 63/106 = 0 \cdot 5943, \quad m_2 = m_1(1 - m_1) = 0 \cdot 2411,$$

so that by (4) and (5)

$$m_{0,y} = 8 \cdot 321, \quad \sigma^2_{0,y} = 2 \cdot 957, \quad \sigma_{0,y} = 1 \cdot 720.$$

With an observed value of 3, 95 per cent. confidence limits for the true mean based on a normal approximation with a continuity correction are $(- 0 \cdot 871, 6 \cdot 871)$ and since the true mean is, by (33), approximately $m_{0,y} + \beta \sigma^2_{0,y}$, it follows that the limits for β are $(- 3 \cdot 109, - 0 \cdot 490)$. Thus the limits for $\psi = e^\beta$ are $(0 \cdot 045, 0 \cdot 613)$, compared with Cornfield's values of $(0 \cdot 030, 0 \cdot 623)$.

Although the limits calculated by this simple approximation appear to agree remarkably well with Cornfield's values, there is a serious difficulty connected with the lower value. For the random variable X is non-negative, so that the approximation which gives $- 0 \cdot 871$ as a possible value for its expectation cannot be sensible (see Stevens, 1951). This particular difficulty is unlikely to arise if none of the entries in the contingency table is small.

The effect of taking additional terms in the expansion for the cumulants of $X_{\beta,y}$ is discouraging. We find directly from (12) and (16) that $\gamma_{1(0,y)} = -0.08112$ and is so small that the inclusion of terms of an Edgeworth expansion to replace the normal multiplier 1.96 is hardly worth considering. (The true skewness of $X_{\beta,y}$ for the smaller values of β in the confidence range is large and positive.) If we assume $X_{\beta,y}$ to be normal with mean

$$m_{0,y} + \beta\sigma^2_{0,y} + \frac{1}{2}\beta^2\sigma^3_{0,y}\,\gamma_{1(0,y)}$$

and variance

$$\sigma^2_{0,y} + \beta\gamma_{1(0,y)}\,\sigma^3_{0,y},$$

we obtain as new confidence limits $(0.0624, 0.643)$, and the left-hand limit is worse than that given by the first approximation. The reason is that, as already noted, the first approximation to the value of γ_1 has the wrong sign. I have not investigated the effect of including further terms in the series expansions.

Example 1 has been dealt with in some detail, partly because the problem of getting a simple method of calculating limits for the odds-ratio in a 2×2 table is of intrinsic interest, and partly in order to illustrate the type of difficulty to be expected in general in applying the formulae of §4.1. The general conclusion is that the first approximation has worked well up to a point and that the refinement of the approximation by the inclusion of further terms is likely to be unsuccessful when one of the observed frequencies in the table is small. It seems reasonable to expect that the inclusion of further terms will be useful if none of the frequencies is small and β is not large, but the extensive numerical investigation that is desirable to specify at all precisely just when the method works has not been undertaken.

4.3. *The $2 \times k$ contingency table.*—A direct generalization of the situation just considered, where the independent variable x took on two values, arises when the independent variable takes on a comparatively small number, k say, of values, there being several observations of Y_t at each value of x. That is, we have a $2 \times k$ contingency table, in which the k-fold classification carries a numerical score.

It is well known that in such cases a test of significance of trend can be obtained by extracting from χ^2_{k-1} for the whole table, a single degree of freedom for linear regression on x (Yates, 1948; Armitage, 1955; Cochran, 1955). This procedure is almost equivalent to the test based on (33); see especially Armitage's paper. In fact, Armitage obtains his test by a formal linear regression analysis of the $(0, 1)$ variables Y_t on x_i, using the usual formulae of least-squares theory. It is clear, however, that with the x's and Y, the total number of 1's, regarded as fixed, X is equivalent to the linear regression coefficient.

Now as the number of observations Y_t recorded at each x value increases, the total at each x value becomes normally distributed so that the usual results about least-squares theory for normal random variables imply the asymptotic efficiency of the χ_1 test against linear alternatives. A new result from §4.1 is that the test statistic is uniquely appropriate for testing against all logistic alternatives, even in small samples, in that it uses the sufficient statistic X. A further point is that by regarding X in the way we have as the sum of a random sample obtained without replacement, a routine procedure is available for refining the normal approximation, or, if special precision is desired in measuring significance, the exact probability may be obtained by enumeration; see also Haldane (1940).

As noted in §2, the procedure is non-parametric in that the logistic assumption enters only into the derivation of the test statistic and not into the specification of the null hypothesis.

Example 2.—Suppose that there are seven equally spaced values of x, -3, $\ldots$, 3, and that eight trials are made at each value. The finite population of x values is thus discrete rectangular and formulae (10)–(13) give $m_1 = m_3 = 0$, $m_2 = 4$, $m_4 = 28$. Suppose further that of the 56 observed Y_t, there are 23 observations equal to 1 and 33 equal to 0. Then $n = 56$, $y = 23$ in the formulae (4)–(7) [or (14)–(17)] and we find that $m_{0,y} = 0$, $\sigma^2_{0,y} = 55 \cdot 2$ and that the fourth cumulant of $X_{0,y}$ is $-200 \cdot 4$, giving $\gamma_{2(0,y)} = -0 \cdot 066$. Of course $\gamma_{1(0,y)} = 0$.

Therefore significance is tested by seeing whether X, equal to $\sum_{-3}^{3} i y_i$, where y_i is the number of 1's when $x = i$, is significantly extreme in a distribution of mean 0, variance $55 \cdot 2$, $\gamma_1 = 0$ and $\gamma_2 = -0 \cdot 066$. The tables of Pearson and Merrington (1951) can be used to assess the change, due to non-normality, in the usual significance limits, or alternatively the Fisher-Cornish inversion of the Edgeworth series can be employed.

If we are to use the present method to estimate β, the expansions

$$E(X_{\beta,y}) = 55.2\,\beta - 33.4\,\beta^3 + \ \ldots \ ,$$
$$V(X_{\beta,y}) = 55.2 - 100.2\,\beta^2 + \ \ldots$$

must be applied. The second terms are small relative to the first if $|\beta| \leqslant \frac{1}{4}$ so that we can reasonably expect good results from the terms just displayed only in this range. Otherwise other methods must be employed.

In general, at least four methods are available for interval estimation of the parameter β in this and similar situations. First it would be possible to work from the distribution (26) which is appropriate for inference about β and which is free of nuisance parameters. Unless the value of y is such that the possible values of X more extreme than the observed value are fairly small in number and are such that the $N_{x,y}$ can be found reasonably simply, this method is impracticable.

The second method is to use formulae based on (29), i.e. the expansions (30), etc. As has already been remarked, this approach gives linear regression formulae as a first approximation and these may be expected to work reasonably well when the slope is small and particularly when the theoretical probabilities lie between $\frac{1}{5}$ and $\frac{4}{5}$, over which range the logistic curve is substantially linear. In the examples discussed above the inclusion of a small number of further terms is not successful. As noted in §4.2, further investigation is desirable to find just when additional terms in these expansions can be put in with advantage.

The third and fourth methods do not use the conditional distribution (26). Maximum likelihood may be applied to (23) and tables for this, and a simple iterative calculating scheme, are given by Berkson (1957). The asymptotic standard error can be found in the usual way. The fourth method, which is not iterative, is that of minimum logit χ^2 (Berkson,

1955), i.e. the method of least squares with empirical weights applied to suitably transformed observations; again an asymptotic standard error can be found by the usual regression formulae. The last two methods are asymptotically equivalent; Berkson (1955) has shown for a range of special cases that from the point of view of minimizing the mean square error of a point estimate of β, the latter method is to be preferred. The relevance or otherwise of this for general practical use raises very interesting issues: see Berkson's papers and Silverstone (1957).

It is recommended that where the significance test is of primary importance the method of the present paper should be used and that if estimation is of primary importance and the sample size is not too small, one of the last two methods should be applied.

4.4. *The serial order test.*—If $x_i = i$ and if x is interpreted to correspond to serial order, the test statistic X is the sum of the rank numbers of those observations giving the outcome 1. The use of this as a test of randomness has been proposed by Haldane and Smith (1948), who gave tables for testing the statistic in small samples, as well as the large sample approximation to its distribution. They dealt also with the case when observations corresponding to some serial numbers are missing.

It follows from §4.1 that this test is optimum against logistic alternatives. It is also of interest to remark that the test amounts to Wilcoxon's two-sample test (see, for instance, Siegel, 1957, p. 72 and Birnbaum, 1956) applied to the ranks corresponding to (*a*) the 0's and (*b*) the 1's. This then is a situation where Wilcoxon's test is optimum, although the problem is different from that for which Wilcoxon's test is most commonly applied, namely the comparison of the locations of two samples. It would be interesting to know if there are any location problems for which Wilcoxon's test is optimum.

The approximate procedures analogous to those described in §4.3 lead to confidence intervals for β. Another approach, from which I have not succeeded in getting a useful working method, is to use (19) and (29) to obtain the cumulant generating function of $X_{\beta, y}$, to approximate to the resulting sums by integrals and hence to obtain an approximation to the distribution when Min $(y, n - y)$ is large and β is not necessarily small.

If the probability that any particular trial has outcome 1 is always small, so that $e^{\alpha + \beta t}/(1 + e^{\alpha + \beta t}) \sim e^{\alpha + \beta t} << 1$ and if x is thought of as time, we have very nearly a point process in continuous time. The test statistic is the sum of the times at which a 1 is observed and the procedure and set-up become those considered by Cox (1955).

4.5. *The combination of data from several series.*—In some applications, including that of the previous subsection, there may be several series for analysis, all assumed to have the same value of β. Thus in the illustrative example discussed by Haldane and Smith, the children of 51 families are classified as normal or phenylketonuric. The independent variable is the serial number of the birth, so that there are 51 series for combination.

If it is assumed that the probability that a child is normal depends only on the serial number of the birth, and is the same for all the families involved, we consider the total rank sum and test it conditionally on the grand total number of normals; this is in effect the procedure of §4.3 using a $2 \times k$ contingency table with the column classification determined by birth order.

The procedure used by Haldane and Smith depends on the total rank sum considered conditionally on the separate total numbers of normals, family by family. This is easily seen to be the theoretically optimum procedure when the nuisance parameters α are different for each family and the logistic regression coefficient β is constant. In fact if the affix j

refers to the jth series (family), the likelihood is from (23)

$$\exp\{\Sigma \alpha^{(j)} Y^{(j)} + \beta \Sigma X^{(j)}\}/ \Pi_i (1 + e^{\alpha^{(j)}+\beta x_{ij}^{(j)}}),\qquad (35)$$

so that $Y^{(1)}, \ldots, Y^{(k)}, X = \Sigma X^{(j)}$ are the jointly sufficient statistics, and the nuisance parameters $\alpha^{(1)}, \ldots, \alpha^{(k)}$ are eliminated by considering the distribution of X conditionally on $Y^{(1)}, \ldots, Y^{(k)}$.

4.6. *Test with α known.*—So far we have supposed that α is an unknown nuisance parameter and this has led us to consider the distribution of X conditionally on the observed value of Y, the total number of 1's. Suppose now the null hypothesis is that 1's occur independently at constant known probability $\theta_0 = e^{\alpha_0}/(1 + e^{\alpha_0})$ and it is required for the test to be sensitive against alternatives in which the probability at the ith trial depends on x_i. In particular if $x_i = i$, we require a test for trend from the value θ_0.

It follows from (13) that the simple sufficient statistic for β is again $X = \Sigma x_i Y_i$, but this time we work with its unconditional distribution, so that X is now the sum of n independent random variables. Under the null hypothesis

$$E(X) = \theta_0 \Sigma x_i,\qquad (36)$$

$$V(X) = \theta_0 (1 - \theta_0) \Sigma x_i^2,\qquad (37)$$

and the distribution is asymptotically normal as $n \to \infty$ provided, for example, that for some $r > 2$

$$\lim_{n\to\infty} \sum_{i=1}^{n} |x_i|^r /\{ \sum_{i=1}^{n} x_i^2\}^{r/2} = 0.\qquad (38)$$

This condition, which follows immediately from Lyapunov's form of the central limit theorem, covers in particular the case $x_j = j$. For small n, an exact test can be made quite easily.

We can examine the non-null distribution for small enough β either by replacing the logistic law by the asymptotically equivalent linear regression or by a direct argument based on approximating by the central limit theorem to the terms in the exact distribution of X. The result is that if as $n \to \infty$ the X's and β are such that $\beta^2 \sum_1^n x_i^2 = o(1)$, X is normal with mean $\theta_0 \Sigma x_i + \beta\theta_0(1 - \theta_0) \Sigma x_i^2$ and variance $\theta_0(1 - \theta_0) \Sigma x_i^2$. Thus, as would be expected on general grounds, we get equivalent results in testing β in large samples by (a) regarding α as unknown and using §4.2 or by (b) estimating θ_0, i.e. α_0, from the average proportion of 1's and then treating this as a known true value, using the present method.

Example 3.—Suppose that an experimental task is attempted 9 times and that the correct response at each trial, which may be L or R, is determined by randomization. Success or failure is recorded at each trial and it is required to test the null hypothesis that successes occur randomly with probability $\frac{1}{2}$ against the alternative that there is a trend in the success rate. Suppose that the results of one run are $F S F S S F S S S$.

If success S is denoted by 1 and if $x_i = i$, i.e. we examine for regression on serial order, the test statistic is the sum of the ranks of the S's, i.e. 35. From (36) and (37) the theoretical mean is

$$\frac{\theta_0\, n(n+1)}{2} = 22\tfrac{1}{2}$$

and the theoretical variance

$$\theta_0(1-\theta_0)\,\frac{n(n+1)(2n+1)}{6} = \frac{285}{4}$$

giving a standard error of 8·44. Hence, after correcting for continuity, the deviation from expectation is 1·422 times the standard error, corresponding to significance at the 15½ per cent. level in a two-sided test, using the normal approximation. A simple enumeration shows the exact level to be $41/256 = 0\cdot160$.

5 Two Predetermined Independent Variables

Suppose now that the Y_i are mutually statistically independent and that there are two independent regression variables for each trial, x_i and u_i say. We generalize the logistic law (2) into

$$\text{pr}\,(Y_i = 1) = e^{\alpha+\beta x_i+\gamma u_i}/(1 + e^{\alpha+\beta x_i+\gamma u_i}), \tag{39}$$

where β and γ are partial regression coefficients. A direct generalization of (23) leads to the consideration of the joint distribution of $X = \Sigma\, x_i Y_i$ and $U = \Sigma\, u_i Y_i$, conditionally on the observed value of $Y = \Sigma\, Y_i$. Further the relation for the cumulant generating function corresponding to (29) is that

$$K_{\beta,\gamma,y}\,(s_1, s_2) = K_{0,y}\,(\beta + s_1,\, \gamma + s_2) - K_{0,y}\,(\beta,\,\gamma). \tag{40}$$

If we simplify the notation by denoting the cumulants of (X, U) when $\beta = \gamma = 0$ by κ_{ij}, instead of by $\kappa_{ij(0,y)}$ (section 3) we have

$$E(X - \kappa_{10}) = \kappa_{20}\beta + \kappa_{11}\gamma + \tfrac{1}{2}\kappa_{30}\beta^2 + \kappa_{21}\beta\gamma + \tfrac{1}{2}\kappa_{12}\gamma^2 + \;\cdots\;, \tag{41}$$

$$E(U - \kappa_{01}) = \kappa_{11}\beta + \kappa_{02}\gamma + \tfrac{1}{2}\kappa_{21}\beta^2 + \kappa_{12}\beta\gamma + \tfrac{1}{2}\kappa_{03}\gamma^2 + \;\cdots\;, \tag{42}$$

etc. If the second degree terms are dropped, we can construct unbiased estimates of β and γ and these are the estimates given by a formal linear regression analysis of the (0, 1) observations on the x's and the u's. Improved estimates may for small β and γ be found by equating the right hand sides of (41) and (42) to $X - \kappa_{10}$ and $U - \kappa_{01}$ respectively.

We shall consider the special problem of testing the null hypothesis $\beta = 0$ with γ an arbitrary nuisance parameter.

The method is to construct a statistic having expectation β, as far as the terms retained in (41) and (42). Let $X' = X - \kappa_{10}$, $U' = U - \kappa_{01}$ and write

$$\begin{pmatrix} \kappa_{20} & \kappa_{11} \\ \kappa_{11} & \kappa_{02} \end{pmatrix}^{-1} = \begin{pmatrix} \kappa^{20} & \kappa^{11} \\ \kappa^{11} & \kappa^{02} \end{pmatrix},$$

denoting by Δ the determinant of the maxtrix on the right. As noted above, the first approximations to unbiased estimates of β, γ are respectively

$$B_1 = \kappa^{20} X' + \kappa^{11} U', \quad C_1 = \kappa^{11} X' + \kappa^{02} U'. \tag{43}$$

If we evaluate the expectation of B_1 using (41) and (42) and replace the second order terms by their sample estimates with sign reversed, we get for the required estimate of β

$$B_1 - \tfrac{1}{2}\Delta[(\kappa_{30}\kappa_{02} - \kappa_{21}\kappa_{11})(B_1^2 - \kappa^{20}) + 2(\kappa_{21}\kappa_{02} - \kappa_{12}\kappa_{11})(B_1 C_1 - \kappa^{11})$$
$$+ (\kappa_{12}\kappa_{02} - \kappa_{03}\kappa_{11})(C_1^2 - \kappa^{02})]. \tag{44}$$

The first approximation to the variance of (43) is κ^{20}; the second approximation is very complicated.

Alternative procedures analogous to those of §4.3 are to apply maximum likelihood or minimum logit χ^2; these result in standard multiple regression calculations, iterative in the first method, non-iterative in the second.

6. Regression on the Result of the Preceding Trial

In this section we deal briefly with the situation in which the probability of a 1 at the ith trial may depend on the observation at the preceding trial. Suppose that the probability of a 1 following a 1 is

$$e^{\alpha+\beta}/(1 + e^{\alpha+\beta}) \tag{45}$$

and following a 0 is

$$e^{\alpha}/(1 + e^{\alpha}); \tag{46}$$

this is, of course, merely a special way of writing the transition probabilities in a simple Markov chain.

The likelihood of $Y_2, \ldots, Y_n$ given Y_1 is

$$\frac{e^{\alpha \sum\limits_{2}^{n} Y_i + \beta \sum\limits_{2}^{n} Y_i Y_{i-1}}}{(1 + e^{\alpha})^{\sum\limits_{2}^{n}(1-Y_{i-1})}(1 + e^{\alpha+\beta})^{\sum\limits_{2}^{n} Y_{i-1}}} \tag{47}$$

the likelihood having to be taken conditionally on Y_1, since the distribution of Y_1 is undefined. The expression (47) is complicated by end effects and the jointly sufficient set of statistics consists of Y_1 and Y_n in addition to $\sum Y_i$ and $\sum Y_i Y_{i-1}$. A formal simplification results from making the set-up circular, by defining $Y_0 = Y_n$. Then the likelihood is

$$\frac{e^{\alpha \sum Y_i + \beta \sum Y_i Y_{i-1}}}{(1 + e^{\alpha})^{n - \sum Y_i}(1 + e^{\alpha+\beta})^{\sum Y_i}}, \tag{48}$$

and $\sum Y_i = Y$ and $\sum Y_i Y_{i-1} = X$ are jointly sufficient. The distribution of X conditionally on $Y = y$ is

$$\mathrm{pr}\,(X = x \mid Y = y) = \frac{M_{x,y}\, e^{\beta x}}{\sum M_{x,y}\, e^{\beta x}}$$
$$= \frac{M_{x,y}\, e^{-\frac{1}{2}\beta w}}{\sum M_{x,y}\, e^{-\frac{1}{2}\beta w}}, \tag{49}$$

say, where w is the total number of runs in the circular array of 0's and 1's, equal to $2(y - x)$, and where $M_{x,y}$ is the number of distinct samples with y 1's for which $X = x$. The difference from (26) is that the null distribution is that of the number of runs in random circular permutation of y 1's and $n - y$ 0's.

There is now the familiar difficulty, that arises also in the study of ordinary serial correlation coefficients, of deciding whether to deal in practice with a quantity defined artificially in order to simplify the mathematics. Unless there is some physical justification of he circular definition, it seems best to take the test statistic, W, equal to the number of runs of 0's and 1's in the observed non-circular sequence Y_i and to take its distribution conditionally on $Y = y$ as given approximately by the analogue of (49), namely

$$\text{pr}(W = w \mid Y = y) = \frac{M'_{w,y}\, e^{-\frac{1}{2}\beta w}}{\Sigma\, M'_{w,y}\, e^{-\frac{1}{2}\beta w}}, \tag{50}$$

where $M'_{w,y}$ is the number of different samples of y 1's and $n - y$ 0's having the total number of runs equal to w.

Now W is formally identical with Wald and Wolfowitz's run statistic for testing for the difference between two populations on the basis of a random sample drawn from each. In that application (Wald and Wolfowitz, 1942), with samples of size $n - y$ and y, the observations are ranked and then those from the first sample are scored 0 and those from the second sample scored 1. The total number of runs is used as a test statistic: the authors show that this test is consistent for all possible alternatives, although the procedure is clearly very inefficient for many particular types of alternative.

The null distribution is determined by the numbers $M'_{x,y}$ and the exact distribution has been given by Wald and Wolfowitz and earlier by Stevens (1939), who considered both the circular and non-circular cases.* Wald and Wolfowitz show also that the total number of runs W is asymptotically normal as $n \to \infty$ with y/n fixed, with

$$E(W) \sim \frac{2y(n - y)}{n}, \tag{51}$$

$$V(W) \sim \frac{4y^2(n - y)^2}{n^3}. \tag{52}$$

It follows from (5), using a relation for cumulant generating functions similar to (29), that if β is small, W is asymptotically normal with mean

$$\frac{2y(n - y)}{n} - \frac{1}{2}\beta V(W) \tag{53}$$

and variance (52), so that approximate estimation of β is possible by the methods used before.

A small sample test can be made from the tables of Swed and Eisenhart (1943) or alternatively by reduction to a 2×2 contingency table in the way very clearly set out by Stevens. This can be done exactly in the circular case. In the non-circular case, it is usually probably best to work not with the total number of runs but with say W_1, the number of runs of 1's, which may be equal to or be one more or one less than the number

---

* The problem had been considered before this as a combinatorial one.

of runs of 0's. The quantity W_1 may be tested by the usual exact test applied to the 2×2 contingency table

W_1	$y - W_1$
$n - y + 1 - W_1$	$W_1 - 1$

The more general problems of inference in Markov chains with more than two states have been considered by a number of authors (Bartlett, 1951; Hoel, 1954; Good, 1955; Anderson and Goodman, 1957).

7. Cumulative Score as an Independent Varibale

As was explained in §i, it is in some applications right to assume that the probability of a 1 on the ith trial is a function of the number of 1's occurring in the first $(i - 1)$ trials, i.e. of $Y_1 + \ldots + Y_{i-1}$. A significance test for such dependence could be obtained by using the serial order test of §4.5, although these would presumably be some loss of power in doing this. For estimation purposes, however, it is necessary to treat the problem afresh.

Suppose that the probability that a trial has outcome 1, given that there have been v 1's in the previous trials, is

$$e^{\alpha + \beta v}/(1 + e^{\alpha + \beta v}). \tag{54}$$

Consider a sequence $Y_1, \ldots, Y_n$ in which there are y 1's and let r_0 be the serial number of the first 1, $r_0 + r_1$ the serial number of the second 1 and so on, $r_0 + \ldots + r_{y-1}$ being the serial number of the last 1, and r_y being defined as $n - r_0 - \ldots - r_{y-1}$. Then the likelihood is

$$\frac{e^{y\alpha + \frac{1}{2}y(y-1)\beta}}{\prod\limits_{k=0}^{y} (1 + e^{\alpha + k\beta})^{r_k}}, \tag{55}$$

so that the full sequence $r_0, \ldots, r_y$ is required for sufficiency.

A more intuitive approach is to consider the sum of products analogous to those that have occurred in the previous sections, namely $\Sigma Y_i(Y_1 + \ldots + Y_{i-1})$. This, however, is equal to $\frac{1}{2}y(y - 1)$, so that this approach leads nowhere. Another possibility is to note that if β is small, or large, the likelihood (55) involves the r's only through Σkr_k and $\Sigma r_k = n$. This suggests that the use of $R = \Sigma kr_k$ as a test statistic will be locally optimum, large values of R corresponding to negative values of β.

Under the null hypothesis $\beta = 0$ each ordering of the r_i has equal probability, with the exception that if one of the r_i is 0 it must occur at a fixed position, namely at the end. Let $y' = y + 1$ if all r_i are non-zero, i.e. if the series ends with a 0, and let $y' = y$ if $r_y = 0$. Thus y' is the number of non-zero r's, and the $y'!$ permutations of the r's are equally likely. If y' is not too small a satisfactory test should be obtained from the distribution of R in the universe of permutations of the non-zero r's observed. In this

$$E_p(R) = \frac{1}{2}(y' - 1)n, \tag{56}$$

$$V_p(R) = \frac{y'(y' + 1)}{12} \Sigma (r_i - \bar{r})^2, \tag{57}$$

where E_p, V_p refer to expectations over the permutations just mentioned and the sum of squares in (57) is over the non-zero r's. The last formula is most neatly proved by noting that for arbitrary $\{r_0, \ldots, r_y\}$ we have by symmetry and invariance considerations that

$$V_p(R) \equiv E_p\{\Sigma\, k(r_k - \bar{r})\}^2 = B\,\Sigma\,(r_i - \bar{r})^2, \qquad (58)$$

where B depends only on y. The expectations of both sides are easily calculated when $r_0, \ldots, r_y$ are uncorrelated random variables of mean zero and unit variance, and B is then determined. Asymptotic normality follows from Hoeffding's theorem (Hoeffding, 1951).

Example 4.—In a learning experiment the following sequence is obtained in 20 trials, a correct response being denoted by 1 and an incorrect one by 0:

$$0, 0, 0, 1, 0, 0, 0, 1, 0, 1, 1, 1, 0, 1, 1, 1, 0, 1, 0, 1,$$

Here $r_0 = r_1 = 4$, $r_2 = r_5 = r_8 = r_9 = 2$, $r_3 = r_4 = r_6 = r_7 = 1$, $r_{10} = 0$, with $n = 20$, $y' = y = 10$, and $R = 72$. From (56) and (57), the theoretical mean on the null hypothesis is 90 and the variance

$$\frac{10 \times 11}{12} \left\{ 52 - \frac{20^2}{10} \right\} = 110.$$

Hence the standardized deviation from expectation, after correction for continuity, is $17\cdot5/10\cdot49 = 1\cdot67$, significant at about the $9\frac{1}{2}$ per cent. level in the normal tables.

If information from several series is to be combined, the values of R are added.

The statistic R does not seem to lend itself to a simple approximate estimation procedure. If an estimate of β is required, as would usually be the case, the simplest procedure is to consider the formal maximum likelihood or minimum χ^2 procedures from (55).

One industrial application in which this sort of situation might arise is in the study of failure rates for mechanisms under modification. At each trial the mechanism may not fail (0) or may fail (1). After a failure, but not after a non-failure, the mechanism is modified. The parameter $-\beta$ measures the rate of decrease, due to modification, in the probability of failure.

A more frequently occurring application, however, is probably the one suggested above to the analysis of learning experiments. There has been much interesting work recently on stochastic models for the representation of learning in simple situations (Bush and Mosteller, 1955; Cane, 1956; Audley, 1956, 1957; Audley and Jonckheere, 1956) and some of these studies have led to formidable statistical problems of fitting and testing. When these studies aim at linking the observations to a neuro-physiological mechanism, it is reasonable to take the best model practicable and to wrestle as vigorously as possible with the resulting statistical complications. If, however, the object is primarily the reduction of data to a manageable and revealing form, it seems fair to take for the probability of a success, 1, as simple an expression as possible that seems to the right general shape and which is flexible enough to represent the various possible dependencies that one wants to examine. For this the logistic seems a good thing to consider.

The simplest case is the one considered above where the probability of a correct response is assumed to depend only on the number of previous correct responses, or in a slightly more complicated case, only on the number of previous rewarded correct responses. The equation (54) may be compared with that of Audley (1956), which in the present notation is that the probability that a trial has outcome 1 is

$$(\alpha + \beta v)/(1 + \beta v). \tag{59}$$

A difficulty that might arise with some applications of (59), although not with the present one, is that β must be non-negative. The maximum likelihood solution for (59) has been used by Audley.

A fairly general model that should cover a variety of cases is to set up a logistic dependence between the probability of a 1 and such quantities as (i) the number of preceding correct responses (or rewarded responses), (ii) the number of (penalized) incorrect responses, and (iii) the result of the preceding trial. Much of the interest of such an analysis would presumably lie in finding whether, under a range of circumstances, the dependencies can be represented with a small number of such independent variables. Maximum likelihood fitting, using the tables of Berkson (1957), results in a straightforward iterative multiple regression calculation. The number of linear equations to be solved at each step is one more than the number of independent variables.

An alternative, very simple non-iterative, method of analysis is to use the method of minimum logit χ^2; grouping of the independent variables would be required to ensure that there are few combinations for which the responses are all 0 or all 1. This method is recommended for finding a first approximation for use in the iterative solution of the maximum likelihood equations.

Some of this work was done at the I.M.S. Summer Institute, 1957, and support from the National Science Foundation is gratefully acknowledged. I wish also to thank Dr. A. Birnbaum for some helpful discussions.

REFERENCES

ANDERSON, T. W. & GOODMAN, L. A. (1957), "Statistical inference about Markov chains", *Ann. Math. Statist.*, **28**, 89–110.

ARMITAGE, P. (1955), "Tests for linear trends in proportions and frequencies", *Biometrics*, **11**, 375–386.

AUDLEY, R. J. (1956), *Stochastic Processes and the Description of Learning Behaviour in Choice Situations*. Ph.D. thesis, Univ. of London.

——— (1957), "A stochastic description of the learning behaivour of an individual subject", *Q. J. Exptl. Psychol.*, **9**, 12–20.

——— & JONCKHEERE, A. R. (1956), "Stochastic processes and learning behaviour", *Brit. J. Statist. Psychol.*, **9**, 87–94.

BARTLETT, M. S. (1951), "The frequency goodness of fit for probability chains", *Proc. Camb. Phil. Soc.*, **47**, 86–95.

BENNETT, B. M. (1956), "On a rank order test for the equality of probability of an event", *Skand. Akt.* **20**, 11–18.

BERKSON, J. (1955), "Maximum likelihood and minimum χ^2 estimates of the logistic function", *J. Amer. Statist. Assoc.*, **50**, 130–162.

——— (1957), "Tables for the maximum likelihood estimate of the logistic function", *Biometrics*, **13**, 28–34.

BIRNBAUM, Z. W. (1956), "On the use of the Mann-Whitney statistic", *Proc. 3rd. Berkeley Symp. on Math. Statist. and Prob..*, **1**, 13–17.

BUSH, R. R. & MOSTELLER, F. (1955), *Stochastic Models for Learning*. New York: J. Wiley.

CANE, V. R. (1956), "Some statistical problems in experimental psychology", *J. R. Statist. Soc. B*, **18**, 177–201.

COCHRAN, W. G. (1955), "Some methods for strengthening the common χ^2 tests", *Biometrics*, **10**, 417–451.

CORNFIELD, J. (1956), "Statistical problem arising from retrospective studies", *Proc. 3rd. Berkeley Symp. on Math. Statist. and Prob.*, **3**, 135–148.

COX, D. R. (1955), "Some statistical methods connected with series of events", *J. R. Statist. Soc.*, B, **17**, 129–164.

FISHER, R. A. (1935), "The logic of inductive inference", *J. R. Statist. Soc.*, **98**, 39–54.

GARWOOD, F. (1941), "The application of maximum likelihood to dosage-mortality curves", *Biometrika*, **32**, 46–58.

GOOD, I. J. (1955), "The likelihood ratio test for Markoff chains", *Biometrika*, **42**, 531–533.

HALDANE, J. B. S. (1940), "The mean and variance of χ^2 when used as a test of homogeneity when the expectations are small", *Biometrika*, **31**, 346–355.

—— & SMITH, C. A. B. (1948), "A simple exact test for birth-order effect", *Ann. Eugenics*, **14**, 117–124.

HOEFFDING, W. (1951), "A combinatorial central limit theorem", *Ann. Math. Statist.*, **22**, 558–566.

HOEL, P. G. (1954), "A test for Markoff chains", *Biometrika*, **41**, 430–433.

IRWIN, J. O. & KENDALL, M. G. (1944), "Sampling moments of moments for a finite population", *Ann. Eugenics*, **12**, 138–142.

MADOW, W. G. (1948), "On the limiting distributions of estimates based on samples from finite universes", *Ann. Math. Statist.*, **19**, 535–545.

MOTOO, M. (1957), "On Hoeffding's combinatorial central limit theorem", *Ann. Inst. Statist. Math.*, **8**, 145–154.

PATNAIK, P. B. (1948), "The power function of the test for the difference between two proportions in a 2×2 table", *Biometrika*, **35**, 157–175.

PEARSON, E. S. & MERRINGTON, M. (1951), "Tables of the 5% and 0·5% points of Pearson curves expressed in standard measure", *Biometrika*, **38**, 4–10.

SIEGEL, S. (1957), *Non-parametric Methods of Statistics*. New York: McGraw-Hill.

SILVERSTONE, H. (1957), "Estimating the logistic curve", *J. Amer. Statist. Assoc.*, **52**, 567–577.

STEVENS, W. L. (1939), "Distribution of groups in a sequence of alternatives", *Ann. Eugenics*, **9**, 10–17.

—— (1951), "Mean and variance of an entry in a contingency table", *Biometrika*, **38**, 468–470.

SWED, F. S. & EISENHART, C. (1943), "Tables for testing randomness of grouping in a sequence of alternatives", *Ann. Math. Statist.*, **14**, 66–87.

WALD, A. & WOLFOWITZ, J. (1940), "On a test of whether two samples are from the same population", *Ann. Math. Statist.*, **11**, 147–162. Reprinted in Selected Papers in Statistics and Probability, A. Wald, McGraw Hill.

YATES, F. (1948), "The analysis of contingency tables with groupings based on quantitative characters", *Biometrika*, **35**, 176–181.

Summary of discussion

When the paper was read to the Royal Statistical Society a vote of thanks was proposed by C. A. B. Smith, seconded by G. H. Jowett and the discussion continued by V. R. Cane, D. V. Lindley, D. E. Barton, H. E. Daniels, G. M. Jenkins and G. A. Barnard. A written contribution was received from A. Stuart.

The first three speakers give interesting examples where the methods in the paper or some extension of them might be helpful. Smith's examples are in phonetics and genetics. Jowett raises industrial and ecological issues and Cane outlines two applications in experimental psychology. Lindley points out that the argument used in the paper to justify conditional inference can be replaced by one due to Lehmann and Scheffé on the structure of exactly similar critical regions; Barnard raises difficulties arising from the Neyman–Pearson notion of exact similarity and in a second contribution Lindley comments. Barton gives some valuable historical references and also points out relevant results by F. N. David. Daniels, in an especially prescient contribution, connects the conditional distributions to saddle-point approximations and to likelihood ratio statistics. This comment connects with themes in higher-order asymptotics taken up by many workers 30 years later. Jenkins develops the analysis of a two-state Markov chain in correlational terms.

[*From* Biometrika, *Vol. 45. Parts* 3 *and* 4, *December* 1958.]

Two further applications of a model for binary regression

By D. R. COX

Birkbeck College, University of London

1. *Introduction.* In a recent paper (Cox, 1958), I have discussed some aspects of a logistic model for analysing regression when the dependent variable can take only two values, say 0 and 1. In the present note two further applications are presented of what is essentially the same model. The first is to the analysis of 2×2 contingency tables based on matched pairs, and the second is to the testing of the agreement between an observed binary sequence and a corresponding sequence of probabilities.

2. *The 2×2 contingency table with matched pairs.* Consider the form taken by a simple comparison of matched pairs when the observations are $(0, 1)$ variables. Let there be n pairs of individuals, the pairing usually being such that the two individuals in any one pair tend to be alike. Let one member of each pair belong to group A, the other to group B, the assignment being randomized if a comparative experiment is involved. An observation, taking one of two values 0 and 1, is made on each individual. For the ith pair, let these be represented by random variables Y_{ia}, Y_{ib}. The possible observations on a pair, writing that on A first, are $(0, 0)$, $(0, 1)$, $(1, 0)$ and $(1, 1)$.

It is possible to form a 2×2 contingency table from the data

	Group A	Group B
0		
1		
	n	n

McNemar (1947) seems to have been the first to point out that the usual χ^2 significance test for such a table is invalid, because it ignores the correlation induced by pairing. He recommended that the significance of the difference between A and B should be tested by rejecting the pairs $(0, 0)$ and $(1, 1)$, and by examining whether the proportion of $(1, 0)$'s among the remaining 'mixed' observations $(0, 1)$ and $(1, 0)$ is consistent with binomial variation with chance $\frac{1}{2}$. Mosteller (1952) and Cochran (1950) have given further accounts of this test and Cochran has discussed extensions to the comparison of more than two groups. Stuart (1957) has recently obtained a test equivalent to McNemar's by arguments based on the theory of stratified sampling.

This work raises two problems. Are there circumstances under which the test is optimum, and is there a corresponding estimation procedure? To deal with these questions we must set up a parametric model covering the non-null case. The simplest such model seems to be the following. Let all random variables be mutually independent and let there be a parameter λ_i characteristic of the ith pair and a parameter ψ describing the true difference between A and B, such that

$$\Pr(Y_{ia} = 1)/\Pr(Y_{ia} = 0) = \lambda_i, \tag{1}$$

$$\Pr(Y_{ib} = 1)/\Pr(Y_{ib} = 0) = \psi\lambda_i. \tag{2}$$

If we write $\lambda_i = e^{\alpha_i}$, $\psi = e^{\beta}$, we have the logistic model of the earlier paper.

It follows by the arguments of that paper, in particular of §4·5, that the jointly sufficient set of statistics consists of (i) $\sum_{i=1}^{n} Y_{ib}$, (ii) the pair totals $(Y_{ia} + Y_{ib})$, $i = 1, ..., n$. Further, optimum inference about ψ, regarding $\lambda_1, ..., \lambda_n$ as unknown nuisance parameters, is based on the distribution of (i) conditionally on the set (ii). Now whenever $Y_{ia} + Y_{ib} \neq 1$, the contribution of the ith pair to (i) is fixed. Hence, the conditional distribution just mentioned is equivalent to that of R = number of pairs $(0, 1)$ conditionally on the observed value of M = number of pairs $(0, 1)$ or $(1, 0)$.

Now a simple calculation from (1) and (2) shows that

$$\Pr(Y_{ia} = 0,\ Y_{ib} = 1 \mid Y_{ia} + Y_{ib} = 1) = \psi/(1+\psi) = \theta,\ \text{say}. \tag{3}$$

Therefore R, conditionally on the observed value of M, has a binomial distribution

$$\Pr(R = r \mid M = m) = \binom{m}{r} \theta^r (1-\theta)^{m-r}. \tag{4}$$

In particular the optimum test of the null hypothesis $\psi = 1$, $\theta = \frac{1}{2}$ is McNemar's test, and confidence intervals for θ and hence for ψ are obtained in the usual way for a binomial parameter. The significance test can be looked on as the very special case of Haldane & Smith's (1948) test for a serial order effect obtained when each series contains just two items.

Example. Mosteller (1952) illustrated the test on an experiment in which each of 100 subjects used both of two drugs A and B, the response being a dichotomy 'not-nausea', 'nausea' (0 and 1, say). 81 subjects never had nausea, i.e. gave the observation $(0, 0)$, 9 subjects gave $(1, 0)$, i.e. had nausea with A but not with B, 1 subject gave $(0, 1)$ and 9 gave $(1, 1)$. The significance test of the null hypothesis that the drugs are equally liable to induce nausea amounts to testing whether a division of 10 trials into $(0, 1)$ in significantly extreme in a binomial distribution with chance $\frac{1}{2}$. The exact significance level in a two-sided test is $11/512 \simeq 0{\cdot}021$; as an approximation to this, we get from a χ^2 test, corrected for continuity, that significance is attained at very nearly the $0{\cdot}025$ level. A table of 95 % confidence limits for the binomial probability (Hald, 1952) gives $(0{\cdot}003,\ 0{\cdot}445)$ as the limits for θ and hence the odds factor ψ is between $1/300$ and $4/5$.

Tests and interval estimates comparing the values of ψ in different experiments can be done by familiar techniques for binomial variates.

3. *Test of agreement between a sequence and a set of probabilities.* Let $Y_1, \ldots, Y_n$ be mutually independent random variables each taking the values $(0, 1)$ and let $p_1, \ldots, p_n$ be a given set of numbers, $0 \leqslant p_i \leqslant 1$. Suppose that it is required to use observations on $Y_1, \ldots, Y_n$ to test the hypothesis that

$$\Pr(Y_i = 1) = p_i, \quad (i = 1, \ldots, n). \tag{5}$$

For example, a weather forecaster might put forward each day a number purporting to be the probability that it will rain the following day. It might then be required to test whether the observed occurrences of rain are consistent with these probabilities.

If n is large, we may group the trials into sets each with nearly constant p_i; then the observed proportion of 1's in each set can be compared with the corresponding p_i. Let n be too small for this test to be used.

One method of deriving a small sample test, when special alternatives to (5) are not available, is to consider a family of probabilities derived from (5). This family is characterized by a continuous parameter β and

$$\log\{\Pr_\beta(Y_i = 1)/\Pr_\beta(Y_i = 0)\} = \beta \log\{p_i/(1-p_i)\}. \tag{6}$$

The null hypothesis (5) corresponds to $\beta = 1$. If $\beta > 1$, the suggested probabilities p_i show the right general pattern of variation, but do not vary enough. If $0 < \beta < 1$, the suggested probabilities vary too much. If $\beta < 0$, the p_i vary in the wrong direction and if $\beta = -1$, the p_i are the complements of the true probabilities.

The log likelihood under (6) of an observed series $y_1, \ldots, y_n$ is

$$\beta \Sigma y_i \log p_i + \beta \Sigma (1-y_i) \log(1-p_i) - \Sigma \log\{p_i^\beta + (1-p_i)^\beta\}. \tag{7}$$

Hence, the sufficient statistic is obtained by scoring

$$X_i = \begin{cases} \log(2p_i) & \text{when} \quad Y_i = 1; \\ \log[2(1-p_i)] & \text{when} \quad Y_i = 0, \end{cases} \tag{8}$$

and by considering a total score $X = \Sigma X_i$. The factor 2 is included to make the expected score positive and to arrange that an event of probability $\frac{1}{2}$ scores 0.

Under the null hypothesis $\beta = 1$,

$$E_1(X) = n \log 2 + \Sigma p_i \log p_i + \Sigma (1-p_i) \log(1-p_i), \tag{9}$$

$$V_1(X) = \Sigma p_i(1-p_i)\{\log[p_i/(1-p_i)]\}^2. \tag{10}$$

Provided that n is not very small and that none of the p_i is near 0 or 1, the distribution of X is nearly normal.

In principle it would be possible to calculate confidence intervals for β from an observed value $X = x$. If x significantly exceeds (9), this is evidence that $\beta > 1$.

Example. Suppose that there are 16 trials, 8 of which have outcome 1 and 8 have outcome 0. Let the p_i corresponding to the zero observations be 0·1, 0·1, 0·2, 0·2, 0·4, 0·5, 0·6, 0·7, and corresponding to the unit observations 0·3, 0·3, 0·5, 0·6, 0·6, 0·8, 0·9, 0·9.

Thus the score for the first observation recorded as 0 is $\log [2(1-p_i)] = \log 1·8 = 0·255$, and the score for the first observation recorded as 1 is $\log (2p_i) = \log 0·6 = -0·222$. We find that the total observed score $x = 1·106$ and that under the hypothesis $\beta = 1$, equations (9) and (10) give

$$E_1(X) = 1·030, \quad V_1(X) = 0·785,$$

so that there is excellent agreement with expectation. Under the hypothesis $\beta = 0$, i.e. that 1's occur randomly with constant chance $\tfrac{1}{2}$, we find

$$E_0(X) = n \log 2 + \tfrac{1}{2}\Sigma \log [p_i(1-p_i)] = -1·329,$$
$$V_0(X) = \tfrac{1}{4}\Sigma \{\log [p_i/(1-p_i)]\}^2 = 1·314.$$

The observed value differs significantly from $E_0(X)$ at the 5 % level. Thus the data support the idea that 1's do not occur with constant chance $\tfrac{1}{2}$ and are in excellent agreement with the suggested probabilities.

The family (6), on which the test just described is based, is especially appropriate when the sequence $\{p_i\}$ is known to be correct at and near $p = \tfrac{1}{2}$ but possibly incorrectly spread around $p = \tfrac{1}{2}$. Thus we may call the test based on (9) and (10) a test for spread. A natural generalization is to replace (6) by

$$\log \{\Pr_{\beta,\alpha}(Y_i = 1)/\Pr_{\beta,\alpha}(Y_i = 0)\} = \beta \log \{p_i/(1-p_i)\} + \alpha, \tag{11}$$

the null hypothesis being that $\beta = 1$, $\alpha = 0$. The pair of sufficient statistics are X, as defined previously, and $Y = \Sigma Y_i$. Under the null hypothesis, X, Y are nearly jointly normally distributed with the mean and variance of X given by (9) and (10) and with

$$E_1(Y) = \Sigma p_i, \quad V_1(Y) = \Sigma p_i(1-p_i), \tag{12}$$
$$C_1(X, Y) = \Sigma p_i(1-p_i) \log [p_i/(1-p_i)]. \tag{13}$$

Note that if the p_i are symmetrically arranged about $\tfrac{1}{2}$, X and Y are uncorrelated.

A test for bias ignoring spread will be based on Y alone, i.e. solely on the observed total number of 1's. If both bias and spread are of interest, it is necessary to specify the relative importance to be attached to each, if an optimum small-sample procedure is to be found. Since it is rarely possible to do this, a sensible practical approach is to find the observed values x and y and to see whether

$$(x - E_1(X), y - E_1(Y)) \begin{pmatrix} V_1(X) & C_1(X, Y) \\ C_1(X, Y) & V_1(Y) \end{pmatrix}^{-1} \begin{pmatrix} x - E_1(X) \\ y - E_1(Y) \end{pmatrix} \tag{14}$$

is significantly large in the χ^2 distribution with 2 degrees of freedom. The expression (14) is, except for a factor $\tfrac{1}{2}$, the exponent in the bivariate normal distribution of X and Y; it is the likelihood ratio statistic for testing the hypothesis that X, Y have the bivariate normal distribution (9), (10), (12) and (13), against the hypothesis that X, Y have arbitrary means, but the same covariance matrix as under the null hypothesis. This, of course, does not allow for the fact that the covariance matrix varies in a determined way with the parameters α and β. However, the determination of the correct likelihood ratio criterion requires the maximum likelihood estimation of α and β, which is tedious.

Example. Consider the data that were analysed previously. We have that the observed value of Y is $y = 8$ and that $E_1(Y) = 7·7$, $V_1(Y) = 2·930$, $C_1(X, Y) = -0·090$. Therefore, the observed value of Y, as well as that of X, agrees well with its expectation under the suggested scheme of probabilities and the need for a combined test hardly arises. The formal details of such a test are that

$$(1·106-1·030, 8-7·7) \begin{pmatrix} 0·785 & -0·090 \\ -0·090 & 2·930 \end{pmatrix}^{-1} \begin{pmatrix} 1·106-1·030 \\ 8-7·7 \end{pmatrix} \tag{15}$$

is to be tested as χ^2 with 2 degrees of freedom. The value of expression (15) is 0·01: a value smaller than this would arise by chance only about 1 in 100 times.

There are further problems connected with the general situation discussed here. First, the same set of observations can be consistent with several alternative sequences of probabilities and it may be required to consider which sequence is preferable. It seems reasonable to prefer that sequence of probabilities for which the information in Shannon's sense is a minimum, for this implies minimum uncertainty concerning the outcome of the realized sequence. According to (9), this amounts to preferring the probabilities for which $E_1(X)$ is a minimum. Secondly, it happens in some applications that the probabilities p_i are not given, but have to be estimated from data by fitting a particular type of model, often to the same data with which goodness of fit is to be tested. In such cases, the most satisfactory test of goodness of fit is likely to be obtained by fitting a model containing additional parameters and testing estimates of the additional parameters for significance from zero. The approach of the present section is relevant only when there are available no special forms of alternative specific to the problem.

REFERENCES

COCHRAN, W. G. (1950). The comparison of percentages in matched samples. *Biometrika*, **37**, 256–66.

COX, D. R. (1958). The regression analysis of binary sequences. *J.R. Statist. Soc.* B, **20**, to appear.

HALD, A. (1952). *Statistical Tables and Formulas.* New York: Wiley and Sons.

HALDANE, J. B. S. & SMITH, C. A. D. (1948). A simple exact test for birth-order effect. *Ann. Eugen., Lond.*, **14**, 117–24.

MCNEMAR, Q. (1947). Note on the sampling error of the difference between correlated proportions or percentages. *Psychometrika*, **12**, 153–7.

MOSTELLER, F. (1952). Some statistical problems in measuring the subjective response to drugs. *Biometrics*, **8**, 220–6.

STUART, A. (1957). Comparison of frequencies in matched samples. *Brit. J. Statist. Psychol.* **10**, 29–32.

The Analysis of Multivariate Binary Data[†]

By D. R. Cox

Imperial College, London

SUMMARY

A brief review is given of the main methods and models for the analysis of multivariate binary data. The relation with standard second-order techniques is discussed.

Keywords: BINARY DATA; QUANTAL RESPONSE; MULTIVARIATE ANALYSIS; MULTI-DIMENSIONAL CONTINGENCY TABLE; CHI-SQUARED; LATENT STRUCTURE MODEL; PRINCIPAL COMPONENTS; CLUSTERING; DISCRIMINANT ANALYSIS; LOGISTIC MODEL; PROBIT ANALYSIS; PERMUTATIONS; MODES; BAHADUR REPRESENTATION

1. INTRODUCTION

IT is fairly common to have multivariate data in which the individual variates are binary, i.e. take one of just two possible values which can be coded as 0 and 1. While there has been appreciable work on the analysis of such data there is no thoroughly developed body of methods and theory to correspond to so-called second order or normal theory methods. The object of the present paper is to review methods that have been proposed and to outline some new proposals, which do, however, need much further development.

The two main methods in common use are probably the following:

(a) to apply second-order methods just as if the 0's and 1's are quantitative observations;

(b) to use the theory of multidimensional contingency tables leading in one approach to a series of chi-squared tests and to the partition of a total chi-squared (Lancaster, 1969; Plackett, 1969).

Method (a) has the major advantage of simplicity and is effective when the dependencies are of a simple form; it can, however, take no account of effects which depend essentially on the interrelationships of variables taken three or more at the time. The use of chi-squared will not be considered further, mainly because of its rather strong emphasis on significance testing.

Except for a few remarks at the end on data with mixed binary and quantitative variates, we deal only with binary variates. It is, however, likely that most of the discussion can be extended to variates with, say, three or four levels of response.

2. STUDIES OF DEPENDENCE AND OF ASSOCIATION

A first important distinction is between binary variables representing responses and those representing explanatory variables or factors. In fact, if there is just one response variable, we have an essentially univariate situation analogous to analysis

[†] Based on a talk given to the Multivariate Study Group, Royal Statistical Society, April 1971.

113

of variance or multiple regression. Cox (1970) has given a connected account and little more will be said here.

In fact, if the single binary dependent variable is denoted by Y, we are essentially concerned with the dependence of $E(Y) = P(Y = 1) = \theta$ on other variables in the problem. If we assume a linear representation for $E(Y)$ and apply the ordinary least squares formulae, we get simple and reasonable procedures provided that the probability lies between about 0·2 and 0·8; outside that range however there are difficulties arising partly from the need for weighting and more seriously from fitted values outside the range $(0, 1)$. In that case the use of a linear logistic model will usually be best, i.e. we assume a linear model for $\log\{\theta/(1-\theta)\}$ and fit, normally by maximum likelihood. Although computationally more complicated, the situation is in principle then exactly comparable to multiple regression.

While the above may well be the most frequently arising situation, from now on we shall concentrate on the genuinely multivariate situation in which there are several, and indeed possibly many, binary response variates. We then have to study the association between these variables and not just the dependence of one variate on others.

The central problem is thus to describe the joint distribution of a set of binary variables. Once this has been done we can deal with such problems as

   (a) concise comparison of two or more samples and the construction of discriminant functions;
   (b) reductions of dimensionality analogous to principal components;
   (c) clustering.

Second-order multivariate analysis owes some of its relative theoretical simplicity to the remarkable fact that *all* aspects of the multivariate normal distribution are determined by the means and covariance matrix in all $p(p+3)/2$ parameters. It is, however, an open empirical question how frequently the scientifically useful information is contained in means and covariances, even when the normal distribution is superficially a tolerable fit.

The oldest approach to multivariate binary data is to define indices of association following essentially Yule. Goodman and Kruskal have reviewed and extended this work (1954, 1959, 1963). We shall not consider it further, aiming to work more directly with the distribution itself.

3. MULTIVARIATE BINARY DISTRIBUTIONS

We now discuss the main ways of describing such distributions. This is a desirable preliminary to the study of methods of analysis. The data of Solomon (1961), reproduced as Table 1, serve as a fairly typical example, although the problems are of course different in emphasis if the number p of variates is appreciably greater than the value four involved here. (Note, however, that there is some doubt as to how the illustrative data are best regarded. Following others we shall treat the data as samples from two four-variate populations. It may, however, be better considered as a single population sampled "retrospectively".)

There follow brief notes on eight kinds of model.

(i) *Independent variables.* The component binary random variables $Y_1, ..., Y_p$ may be treated as independent. Of course this greatly simplifies such things as the comparison of samples and the construction of discriminant functions. A central question is then how large the departures from independence have to be to make the procedures based on independence misleading. Independence gives a model with p parameters.

(ii) *Arbitrary multinomial distributions.* Another simple model, in a sense complementary to (i), is to treat the sample as a multinomial one with $2^p - 1$ independent parameters corresponding to the 2^p distinct observations that can be obtained. The disadvantages of this are that it is only applicable when n, the number of observations, is fairly large and p is fairly small, there being a reasonable number of observations in each cell, and also that it gives little insight into the structure of the data.

TABLE 1

Distribution of four binary variates in two groups

	Low I.Q. group	High I.Q. group
1111	62	122
1110	70	66
1101	31	33
1100	41	25
1011	283	329
1010	253	247
1001	200	172
1000	305	217
0111	14	20
0110	11	10
0101	11	11
0100	14	9
0011	31	56
0010	46	55
0001	37	64
0000	82	53
Total	1,491	1,491

Source: Solomon (1961).

(iii) *Logistic models.* The remaining models are intermediate between (i) and (ii), and allow the presence of special kinds of dependence. The simplest, most flexible, and in many ways the most important models are probably the logistic representations of the probabilities. Write $Z_i = 2Y_i - 1$, so that the Z's take values ± 1. Suppose that

$$\log P(Z_1 = z_1, ..., Z_p = z_p) = \alpha_1 z_1 + ... + \alpha_p z_p + \alpha_{12} z_1 z_2 + ... + \alpha_{p-1,} z_{p-1} z_p,$$

$$+ ... - \Lambda, \quad (1)$$

where Λ is a normalizing constant; e^Λ is a sum of exponentials chosen to make the probabilities sum to unity. If only the first degree terms are included, we have the independence model (i), whereas if all terms up to $z_1 ... z_p$ are taken we have in effect the general multinomial model (ii). What we hope for is that only a fairly limited number of terms need to be included, usually selected in the light of the data. There are of course many special cases; for instance the analogue of the "equal correlation" case of normal theory is to take $\alpha_{ij} = \alpha$, $\alpha_{ijk} = \dots = 0$. Note that if we put in all first and second degree terms there are $\frac{1}{2}p(p+1)$ parameters, nearly as many as in normal theory. This is rather disconcerting in that one might expect that binary data would support substantially fewer parameters than quantitative data and one suspects that multivariate normal theory models are often over-parameterized.

The interpretation of the parameters is best seen by considering conditional distributions. For example, conditionally on $Z_2 = z_2, \dots, Z_p = z_p$,

$$\frac{1}{2}\log\left\{\frac{P_c(Z_1 = 1)}{P_c(Z_1 = -1)}\right\} = \alpha_1 + \alpha_{12} z_2 + \dots + \alpha_{1p} z_p + \alpha_{123} z_2 z_3 + \dots,$$

where the subscript c indicates a conditional probability. Unfortunately, although such conditional probabilities have a simple form, marginal probabilities do not. In particular, $\log\{P(Z_1 = 1)/P(Z_1 = -1)\}$ and for $p > 2$

$$\log\{P(Z_1 = 1 \mid Z_2 = z_2)/P(Z_1 = -1 \mid Z_2 = z_2)\}$$

are not in general simply related to the α's.

The fitting and testing of models like (i) for example, by maximum likelihood are fairly well developed. However, the main emphasis should be placed on the exploitation of such a model, for example to facilitate the comparison of sets of data, etc.

If all 2^p cells are occupied, the models can be analysed in terms of the log frequencies (Plackett, 1969) but otherwise maximum likelihood methods will normally be the best to use. A simple first step, when all or nearly all cells are occupied, is to compute ranked factorial contrasts from the log frequencies and to plot on a semi-normal scale, indicating on the plot the theoretical standard error $\Sigma\sqrt{(1/n_{ijk})}$.

The formal fitting of any but the simplest of these models is likely to be an effective approach only for fairly small values of p, because of the large number of parameters involved. Incidentally note that the likelihood ratio discriminant between two groups is a simple function of the variates one at a time if and only if the coefficients of all second and higher order terms are identical in the two groups. A similar result holds when more than two groups are compared. This suggests that if second-order techniques are applied to binary data, "product" variates should be included if major departures from the above homogeneity conditions are likely.

The logistic model is implicit or explicit in a good deal of work on multivariate binary data; for some historical comments and general discussion see Mantel (1966).

(iv) *Additive model.* It would be possible to set out a representation analogously to that of (iii) but directly in terms of probabilities rather than log probabilities, or more generally in terms of some other function than the logarithm. Which is better is really to be settled empirically, but the additive models have two disadvantages. There can be difficulties with values outside the range $(0, 1)$ and "independence" is not achieved as the simplest special case. The second difficulty, but not the first, is overcome by the Bahadur (1961) representation. According to this any p-variate binary distribution can be written in a series as follows. First let $\theta_i = P(Y_i = 1)$ and introduce the standardized variables $U_i = (Y_i - \theta_i)/\sqrt{\{\theta_i(1 - \theta)\}}$. Call

$$\rho_{12\dots k} = E(U_1 \dots U_k)$$

the kth order correlation between $Y_1, \dots, Y_k$, with, of course, an analogous definition for any other subset of variates. Then the joint distribution of the Y's can be written in the form

$$P(\mathbf{Y} = \mathbf{y}) = \prod_{i=1}^{p} P(Y_i = y_i)$$

$$\times \left\{ 1 + \sum_{i>j} \rho_{ij} u_i u_j + \sum_{i>j>k} \rho_{ijk} u_i u_j u_k + \ldots + \rho_{12\ldots p} u_1 \ldots u_p \right\}.$$

The second factor gives the effect of departures from independence. This representation is similar in spirit to but probably less useful than (iii).

(v) *Modal clustering model.* An entirely different kind of model has been discussed in an unpublished thesis by A. F. Ebbutt. In its simplest form this is developed as follows:

(a) There is first a subset of variables C which take constant values.
(b) Variables not in C are independently distributed.
(c) The variables in C are, next, subject to independent fairly small changes of misclassification.
(d) Finally there may be two or more sets C or within a given C there may be two or more "modes".

In a large number of dimensions it is not easy to disentangle this situation, but Ebbutt has produced methods that will do it.

(vi) *Latent class analysis.* This is a special representation introduced by Lazarsfeld (1950). It amounts to assuming a mixture of say k classes within each of which the variates are independently distributed. For a review of estimation of parameters, etc. see, for example, Madansky (1969). This is a very special model likely to be useful when there is strong prior expectation that the classes have a clear physical existence.

(vii) *Transformations by permutation.* A central role is played in second-order multivariate methods by linear transformations and especially by orthogonal transformations. If we apply the same techniques to binary data we are in effect assuming that linear functions of binary variates have a useful interpretation; it is not clear that this is always so, although with the inclusion of product variates the linear functions cover a much wider range of aspects. In some cases, however, a different kind of transformation may be more relevant, namely permutation of the defining component variates.

A simple example will clarify the idea. Consider two variates

$$Y_1 = \begin{cases} 1 & \text{if husband votes Labour,} \\ 0 & \text{otherwise,} \end{cases} \qquad Y_2 = \begin{cases} 1 & \text{if wife votes Labour,} \\ 0 & \text{otherwise.} \end{cases}$$

For some purposes we might get a simpler representation of the data in terms of two new binary variates Y'_1 and Y'_2 defined as follows:

$$Y'_1 = \begin{cases} 1 & \text{if husband and wife discordant,} \\ 0 & \text{otherwise,} \end{cases} \qquad Y'_2 = Y_1.$$

A third possibility is to take $Y''_1 = Y'_1$, $Y''_2 = Y_2$. These three are the only essentially distinct ways of using the 2^2 distinct responses to define two binary variables. The relationships are as follows:

Y_1	Y_2	Y'_1	Y'_2	Y''_1	Y''_2
0	0	0	0	0	0
0	1	1	0	1	1
1	0	1	1	1	0
1	1	0	1	0	1

The most obvious criterion for choosing between representations is to aim at independence of the defining variates.

In general, a given set of 2^p cells can be used to define p binary variates in many different ways. In fact there are $(2^p - 1)!/p!$ essentially different sets, i.e. ones that cannot be obtained from one another by interchanging 0's and 1's and permuting variables. Dr P. Bloomfield has pointed out that by restricting attention to transformations directly related to the original set of variates, this number may be reduced by a substantial factor. The large number of possibilities for $p > 3$ is an advantage in that it means we have almost as rich a choice of possible transformations as in the continuous case; on the other hand computationally it will be an embarrassment! The best approach almost certainly depends on the magnitude of p. For small values of p, the recognition of simple structure in a full logistic fit may be the best approach, but for larger values of p an iterative approach working on pairs of variates at a time is probably better. This remains to be explored.

There are now many things that might be done. Transform if possible to independence. The marginal probabilities of the new variates correspond to the eigenvalues in principal component analysis. For two samples transform so that discrimination is achieved with just a few of the new variates; procede similarly with more than two sets. It may be necessary to restrict the permutations to some meaningful subgroup. Until algorithms have been developed for implementing these ideas it is hardly possible to assess their usefulness.

(viii) *Relation with underlying continuous distribution.* One historically important way of obtaining binary distributions is to start with one or more continuous, possibly normal, distributions of unobserved variates $W_1, ..., W_p$ and to suppose that $Y_i = 1$ if and only if, say, $W_i > 0$. This is quite often a useful heuristic device, but seems unnecessary otherwise, unless the W's are of intrinsic interest.

In this section a number of ways of describing multivariate binary distributions have been outlined. This is very probably not a complete list. While most of the models suggested above have associated with them schemes for formal estimation and significance testing, it must be put strongly that this is not the aspect of prime importance. The usefulness of the models lies in their application to describe complex data concisely, to facilitate comparisons between sets of data, etc.

Probably the most flexible model in general is the logistic one of (iii). If it is required to include dependence on a further set of variables, $\mathbf{x}$ or other forms of structure, suitable terms can be added to the right-hand side. If there are $q \mathbf{x}$ variables the simplest thing would be to add pq linear terms representing the effects $x_i z_j$; this leads also to a representation of mixed quantitative and binary variates. For we can combine say a multivariate normal distribution of the quantitative variates with a logistic representation for the conditional distribution of the binary variates given the quantitative ones. More generally every term in the model (1) may be allowed to depend on **x**.

4. DISCUSSION

The previous section has concentrated on the description of multivariate binary distributions, this seeming a necessary preliminary to a rational discussion of methods of analysis. In most cases the form of the analysis for a given model is fairly obvious; it may, however, help to draw some parallels with familiar second-order techniques and Table 2 sets out to do this concisely. A full discussion would require a much longer paper.

For a short bibliography of papers on these topics, see Cox (1970, Appendix B).

TABLE 2

Outline of some broad problems and appropriate second order and binary techniques

	Second order *(normal theory)*	Binary
Internal problems		
Description of single sample	Calculation of means and covariance matrix Search for special structure Transformations	Calculation of marginal proportions and pairwise logistic differences. Plots of these Search for special structure Fitting of more elaborate logistic or other model
Reduction in dimensions	Principal components	Permutational principal components. Recognition of meaningful sets of independent variables in logistic representation
Clustering	Various	Modal clustering
Search for hypothesized underlying structure	Factor analysis	Latent class analysis
Univariate external problems		
Dependence of univariate response on complex explanatory variables	Multiple regression and analysis of variance	Multiple regression often logistic
Multivariate external problems		
Comparison of two or more sample means	Hotelling's T^2, etc.	Adaptations of T^2 to examine marginal proportions
Full comparison of two or more samples	Analysis also of covariance matrices	Fitting and comparison of logistic or other models
Discriminant analysis	Linear discriminant function	Fitting of logistic or other models and estimation of likelihood ratio
Relation between set of variates and fixed explanatory vectors	Canonical regression	Study of fitted logistic model with added **x** dependence Permutational analysis
Relation between two or more sets of variates	Canonical correlation	If feasible reduce to the previous case

Note added in proof. Since the talk on which this paper was based was given and the paper itself accepted for publication, there have been a number of papers on this general topic, especially on the log linear model; see in particular the special multivariate issue of *Biometrics*, March 1972, Vol. 28, No. 1.

REFERENCES

BAHADUR, R. R. (1961). A representation of the joint distribution of responses to *n* dichotomous items. In *Studies in Item Analysis and Prediction* (H. Solomon, ed.), pp. 158–176. Stanford, Calif.: Stanford University Press.

Cox, D. R. (1970). *The Analysis of Binary Data.* London: Methuen.

GOODMAN, L. A. and KRUSKAL, W. H. (1954, 1959, 1963). Measures of association for cross classifications. *J. Amer. Statist. Ass.*, **49**, 732–764; **54**, 123–163; **58**, 310–364.

LANCASTER, H. O. (1969), Contingency tables of higher dimensions. *Bull. Int. Statist. Inst.*, **43**, I, 143–151.

LAZARSFELD, P. F. (1950). Logical and mathematical foundation of latent structure analysis. In *Measurement and Prediction* (S. A. Stouffer *et al.*, eds.), pp. 362–412. Princeton, N. J.: Princeton University Press.

MADANSKY, A. (1968). Latent structure. In *Int. Encl. of Social Sciences*, Vol. 9, pp. 33–38. New York: Macmillan and Free Press

MANTEL, N. (1966). Models for complex contingency tables and polychotomous dosage response curves. *Biometrics*, **22**, 83–95.

PLACKETT, R. L. (1969). Multidimensional contingency tables. *Bull. Int. Statist. Inst.*, **43**, I, 133–142.

SOLOMON, H. (1961). Classification procedures based on dichotomous response vectors. In *Studies in Item Analysis and Prediction* (H. Solomon, ed.), pp. 177–186. Stanford, Calif.: Stanford University Press.

The Analysis of Exponentially Distributed Life-Times with Two Types of Failure

By D. R. Cox

Birkbeck College, University of London

[Received April, 1959]

SUMMARY

A NUMBER of alternative probability models are considered for the interpretation of failure data when there are two or more types of failure. Some of the statistical techniques that can be used for such data are illustrated on an example discussed recently by Mendenhall and Hader.

1. Introduction

Mendenhall and Hader (1958) have recently given an interesting account of a model for the analysis of failure-time distributions when there are two, or more, types of failure. They illustrate their theory by analysing some data on the failure-times of radio transmitter receivers; the failures were classed into two types, those confirmed on arrival at the maintenance centre and those unconfirmed. In the present paper their example is used to illustrate and distinguish between a number of models that can be used for this type of data.

The essential feature of the problem is that we have independent individuals exposed to risk, and that on failure an individual is withdrawn from risk. We observe, for example, that individual number one fails after life-time t, and that the failure is say of the first type: this means that we know the time at which failure of the first type occurs, but only that failure of the second and other types had not occurred by time t. In many applications, including Mendenhall and Hader's, the sample contains individuals that have not failed at the end of the period of the observation. Thus in their example no receivers were operated after 630 hr.

Data like this arise in several fields in addition to industrial life-testing. For example, in medical and actuarial work the estimation and comparison of death rates from a particular cause requires corrections for deaths from other causes. In particular Seal (1954) and Elveback (1958) have discussed the more theoretical aspects of this in connection with actuarial work and given numerous references. Sampford (1954) has dealt with similar problems in bioassays. In tensile strength testing there may be two or more types of failure, for example jaw breaks and fractures in the centre of the test specimen; here the observation is load on failure, not life on failure. A further interesting application is in experimental psychology. Audley (1957) has interpreted latency measurements in learning experiments by postulating independent Poisson processes of A-responses and B-responses; the first process for which an event occurs is considered to determine the nature (A or B) of the response and the time at which it occurs.

We shall consider in the present paper a number of probability models that can be used for these problems; the models become identical in special cases. The choice between the models is a question partly of which fits the data best, and partly of which set of assumptions is the most reasonable physical representation of the process. If one type of model fits much better than another, information may be obtained about the underlying mechanism.

For the most part we consider here systems in which the frequency distributions involved are exponential. This is a serious restriction, even though it is known that in some applications distributions close to the exponential are obtained. There are two situations in which an exponential distribution would be expected on general grounds. First failure may be due to an external point occurrence, for example an accident, arising randomly in an age-independent way. Secondly there may be many more or less independent causes of failure, when the observed failure time is the smallest of a number of independent random variables; under some rather special conditions the resulting distribution will be exponential. On the other hand, if, for example, there is a single process of wear going on at a fairly steady rate, the distribution of failure time may be far from exponential.

2. Some Probability Models

We consider first two models that involve exponential distributions for the component life-times. We suppose that there are two types of failure; the generalization when there are more than two is straightforward.

Model A. Independent Poisson risks.—Imagine that failures of Types I and II occur independently in Poisson processes with parameters λ_1 and λ_2; that is we have random variables T_1, T_2 independently distributed with p.d.f.'s $\lambda_1 e^{-\lambda_1 t}$, $\lambda_2 e^{-\lambda_2 t}$. The random variable T_i can be interpreted as the time of failure from cause i, if the other cause of failure were inoperative. The observed type and time of failure is determined by the smaller of T_1 and T_2.

The main properties of this model are that:

(i) the probability that a failure is of type I is $\lambda_1/(\lambda_1 + \lambda_2)$, independently of the time at which failure occurs;

(ii) the p.d.f. of failure time of all individuals, and of those whose type of failure is given, is $(\lambda_1 + \lambda_2) e^{-(\lambda_1+\lambda_2)t}$.

Model B. Single risk.—Suppose that individuals are of two types, the chance that an individual is of the first type being θ. An individual of type I is subject only to the risk of failure of the first type and the p.d.f. of failure-time is $h_1(t) = a_1 e^{-a_1 t}$; similarly the p.d.f. of failure-time of the second type is $h_2(t) = a_2 e^{-a_2 t}$. The probability that failure is of the first type given that it occurred at time t is

$$\frac{\theta h_1(t)}{\theta h_1(t) + (1 - \theta) h_2(t)} \tag{1}$$

and is independent of t if and only if $a_1 = a_2$, when the process is completely equivalent to Model A with $a_1 = a_2 = \lambda_1 + \lambda_2$, $\theta = \lambda_1/(\lambda_1 + \lambda_2)$.

Model B is the one fitted by Mendenhall and Hader (1958). Non-standard statistical problems arise only if there are individuals who have not failed at the end of the period of observation. Bartlett (1953) has considered a similar model in connection with the estimation from cloud chamber tracks of the mean life-times of unstable particles, when two types of particle are present.

Model B generalizes in a straightforward way when the distributions $h_i(t)$ are not exponential.

We consider now a very general model involving arbitrary distributions.

Model C. General independent risks.—Suppose that the random variables T_1, T_2 of Model A are independently distributed with continuous distribution functions $F_1(t)$, $F_2(t)$. Then the probability, say $g_1(t)\,\delta t + o(\delta t)$, that failure occurs between $(t, t+\delta t)$ and is of the first type is given by

$$g_1(t) = F'_1(t)[1 - F_2(t)] \tag{2}$$

and similarly

$$g_2(t) = F'_2(t)[1 - F_1(t)]. \tag{3}$$

Now in this model the probability that a failure is of type I, given that it occurs at t, is equal to $g_1(t)/[g_1(t) + g_2(t)] = \pi_1(t)$, say. This probability does not involve t if and only if

$$\frac{F'_1(t)}{1 - F_1(t)} = \psi \frac{F'_2(t)}{1 - F_2(t)} \tag{4}$$

for some constant ψ; that is, if and only if

$$1 - F_1(t) = [1 - F_2(t)]^\psi. \tag{5}$$

The probability that a failure is of type I is $\psi/(1 + \psi)$. Armitage (1959) has used this condition in a study of the comparison of different survivor curves. Equation (4) shows that the conditional failure rates (forces of mortality) of the two types of failure are in constant ratio for all t. Our final model involves this condition.

Model D. Independent proportional risks.—Here we assume the conditions of Model C, plus the requirement that (4) is satisfied.

Thus in the model time of failure and type of failure are statistically independent, and the p.d.f. of time of failure among all individuals, and among individuals whose type of failure is given, is the same, namely $g_1(t) + g_2(t)$.

3. Comparison and Interpretation of Models

Obviously the form of model B in which the $h_i(t)$ are arbitrary, can be used to fit any data of the type considered here, namely data in which only one failure-time is observed for each individual. It is convenient to begin this section by showing that model C also is all-embracing in the same way. For let

$$G(x) = \int_0^x [g_1(t) + g_2(t)]\, dt \tag{6}$$

be the probability of failure before x. Then it follows from (2) and (3), or can be seen directly, that

$$1 - G(x) = [1 - F_1(x)][1 - F_2(x)]. \tag{7}$$

If we substitute in (2) we get that

$$\frac{F'_1(x)}{1 - F_1(x)} = \frac{g_1(x)}{1 - G(x)} \tag{8}$$

from which

$$1 - F_1(x) = \exp\left\{-\int_0^x \frac{dG_1(u)}{1 - G(u)}\right\}, \tag{9}$$

where $G_1(x)$ is the integral of $g_1(x)$. There is a similar equation for $F_2(x)$. The distribution functions $F_i(x)$ determine and are determined by the functions $g_i(t)$.

The equality of the two sides of (8) is seen alternatively by noting that both are equal to the conditional failure rate from failure of type I at age t. In fact (9) is the actuarial life-table analysis of the system. For $dG_1(u)/[1 - G(u)]$ is the conditional failure rate for Type I failures, and equation (9) is that expressing a distribution function in terms of the conditional failure rate.

It follows that no data of the present type can be inconsistent with model C. In order to test the applicability of the model it would be essential to have a different type of data; for example it may be possible to get information about the values of both T_1 and T_2 for the same individual, or to divide the individuals into rational subgroups.

In some applications the single-risk model B can be ruled out on general grounds, it being clear that all individuals are subject to both types of risk. In other cases, as in the application to cloud chamber tracks, there are two distinct types of individual and the single-risk model is the appropriate one.

Model A, with independent Poisson risks, is the simplest, and is of course implied in Markovian birth-death and allied processes. Harris *et al.* (1950), Moran (1953) and Meier (1955) have studied some of the associated estimation problems. The generalization with independent proportional risks, model B, is likely to arise in two main circumstances. First there may be similar processes of wear and ageing underlying the two types of failure. A second, quite different, possibility is that there is "really" only one type of failure, and that when failure occurs, a random event with constant probability $\psi/(1 + \psi)$ determines the type to which the failure is assigned.

Now in Mendenhall and Hader's example, the classification of failures is based on whether or not the failure is confirmed on arrival at the maintenance centre. Various probability models may be worth considering in situations like this, and the choice between them must depend on detailed knowledge not available to me in the present case. The following points may, however, be made:

(a) It seems reasonable to regard each individual as subject to both risks, and hence not to use model B.

(b) The two types of failure may not correspond to physically different processes of wear. A possible model is then the one mentioned in the preceding paragraph, i.e. model D, with $\psi/(1 + \psi)$ interpreted as the probability that when a failure has occurred it is classed as confirmed.

(c) In some circumstances, failures of only one type may arise completely randomly. For example, the unconfirmed failures may have been due to temporary misuse of the equipment. This suggests the use of model C with $F_2(t)$ exponential, and $F_1(t)$ general.

4. Some Statistical Procedures

A first step in the analysis of this sort of data is to examine whether the proportion of failures of Type I varies in time; if it does vary models A and D are excluded. A simple test is to group the times of failure and to test by χ^2 the hypothesis that the probability $\pi_1(t)$ that a failure is of Type I is constant.

A test more sensitive against alternatives that $\pi_1(t)$ changes monotonically in time is as follows; let $t_{11}, \ldots, t_{1r_1}$ be the times of failure of the first type, $t_{21}, \ldots, t_{2r_2}$ those of the second type, individuals that do not fail being ignored. Then under the null hypothesis H_0 that $\pi_1(t)$ is independent of t, $\bar{t}_{1.} = \Sigma t_{1t}/r_1$ is the observed value of a random variable T_1, having the distribution of the mean of a sample of size r_1 drawn randomly without replacement from the finite population $\{t_{11}, \ldots, t_{2r_2}\}$ of size $r_1 + r_2$. Thus under H_0

$$E(T_1) = \frac{\sum\limits_{i,j} t_{ij}}{r_1 + r_2} = \bar{t}_{..}, \tag{10}$$

$$V(T_1) = \frac{r_2}{r_1(r_1 + r_2)(r_1 + r_2 - 1)} \sum\limits_{i,j} (t_{ij} - \bar{t}_{..})^2. \tag{11}$$

The random variable T_1 is nearly normally distributed when r_1 and r_2 are large; an exact test can in principle always be made. The test is equivalent to the permutation t test for comparing two sample means and is optimum (Cox, 1958) when the alternative hypothesis is that $\pi_1(t)/[1 - \pi_1(t)] = g_1(t)/g_2(t) = \alpha e^{\beta t}$.

If this test is used it needs to be supplemented by inspection of the data to check that no large changes, not contributing to $T_1 - E(T_1)$, get overlooked.

Table 1

Frequency Distributions of Times of Failure of Radio Transmitters (after Mendenhall and Hader)

(hr.)	Type I (confirmed)	Type II (unconfirmed)	Total	Fitted frequency (exponential)
	(Observed frequency)			
0–	26	15	41	55·9
50–	29	15	44	47·7
100–	28	22	50	40·0
150–	35	13	48	34·3
200–	17	11	28	29·0
250–	21	8	29	24·6
300–	11	7	18	20·9
350–	11	5	16	17·7
400–	12	3	15	15·0
450–	7	4	11	12·7
500–	6	1	7	10·8
550–	9	2	11	9·2
600–629	6	1	7	4·8
Not failed at 630 hr.	—	—	44	46·4
Total	218	107	369	369·0

Example.—Table 1 shows the data of Mendenhall and Hader's example, given in full in their paper, grouped by time of failure. The first test of the constancy of $\pi_1(t)$ is the standard χ^2 test applied to the 12×2 contingency table formed from the first two columns of Table 1. We get $\chi^2_{12} = 9 \cdot 37$. For the more sensitive test based on (10) and (11), we have from the grouped data $\bar{t}_{1.} = 229 \cdot 7$. The mean and standard deviation of this under the null hypothesis are $218 \cdot 5$ and $6 \cdot 18$, so that the departure from expectation is $1 \cdot 81$ times the standard error; the corresponding multiple using the ungrouped data is $2 \cdot 05$. There is thus some evidence, by no means conclusive, that $\pi_1(t)$ increases with time. In fact the proportion of failures of Type I is reasonably constant at about 2/3 up to about 400 hr and then increases to about 4/5.

Since there is some doubt about the reality of this effect it is worth continuing the analysis in two parts, one taking $\pi_1(t)$ constant, the other not.

If we assume $\pi_1(t)$ to be constant, the next step is to test whether the Poisson form, model A, applies, or whether the more general model D with non-exponential distributions must be fitted. Suppose that the failure-times, without regard to type of failure, are $t_1, \ldots, t_r$ and that there are s individuals that have not failed at the end of their periods of observation $t'_1, \ldots, t'_s$. If these are sampled from a Poisson process with parameter λ, the maximum likelihood estimate of λ is (Epstein and Sobel, 1953)

$$\hat{\lambda} = \frac{r}{\Sigma t_i + \Sigma t'_j}. \tag{12}$$

If the frequency distribution is grouped, and censored at or near a fixed value, fitted frequencies can be calculated and a χ^2 goodness of fit test applied.

Example.—The estimate of $\hat{\lambda}$ from the total column of Table 1 is given by $50\hat{\lambda} = 0 \cdot 1646$ and the fitted frequencies are given in the last column. We get $\chi^2_{12} = 16 \cdot 66$, which is well short of the tabulated 10 per cent. point. Note however that, whereas there is excellent agreement with the fitted frequencies above 200 hr, there is a systematic departure up to 200 hr., the departure being of similar form for type I and for Type II failures. The adoption of a finer grouping for the small failure-times does not throw more light on this.

If $s = 0$, so that we have an uncensored sample of failure-times, a test of the exponential form more sensitive against smooth departures can be obtained in a number of ways. One of the simplest, and one that does not depend unduly on the small observations, which are often subject to severe errors of recording, etc., is that based on Sherman's statistic S, the ratio of the mean deviation to the mean. Sherman (1957) has provided tables of its distribution for $r \leqslant 21$; in Sherman's notation $S = 2\tilde{\omega}_n$, $r = n + 1$. In larger samples

$$S = \frac{\sum\limits_{i=1}^{r} |t_i - \bar{t}|}{r\bar{t}} \tag{13}$$

can be taken approximately normally distributed with mean $e^{-1}[1 - (2r)^{-1}]$ and variance $0 \cdot 05908/r - 0 \cdot 01237/r^2$ (Bartholomew, 1954).

This test, at any rate in its simple form, cannot be used on the example because of the censoring. One way of getting a "smooth test" for this is by postulating a functional form reducing to the exponential as a special case, to estimate parameters by maximum likelihood and to test for departure from exponential form by comparison against a large-sample standard error. The most suitable functional form seems to be the Weibull type, with distribution function

$$1 - \exp\{-(\gamma t)^{1+\beta}\}, \qquad (14)$$

the null hypothesis being that $\beta = 0$.

However the test obtained like this is quite cumbersome, even when the approximation that β is small is used. Moreover it depends rather critically on the small observations. We shall therefore not go into details here.

Instead, in the special case when censoring is at a fixed point, $t_j' = t_c$, $t_i < t_c$, we can find a simple test by the method of Cochran (1955). In this a weighted sum of deviations between observed and expected frequencies is compared with its large-sample standard error. Let f_i be the observed frequency and m_i the fitted frequency in the ith group. We need to choose weights g_i determining a test statistic

$$L_g = \Sigma\, g_i(f_i - m_i). \qquad (15)$$

Cochran gives the large-sample variance $V(L_g)$ when the g_i are preassigned and also proves that if the g_i are chosen as functions of the sample to maximize $L_g^2/V(L_g)$, then the maximized ratio is equal to the χ^2 goodness-of-fit criterion. Here, however, we want to choose "smooth" weights, and a reasonable procedure is to take a family of weights with a small number of adjustable parameters, chosen to maximize the above ratio; the maximized value will, under the null hypothesis, be distributed as χ_k^2, where k is the dimensionality of the family of weights.

Example.—Number the 14 groups of Table 1 from 1, . . . , 14, counting the group of censored observations as number 14. How shall the deviations be weighted? One might first consider linear weighting $g_i = i$; however the estimating equation for the unknown parameter is such that this weighting amounts very nearly to testing the difference between fitted and observed frequencies for the censored observations. This may indeed be one right thing to test, if, for example, it is required to examine the adequacy of the exponential curve for extrapolation, but it would best be done directly by giving all groups, except the last, zero score. In the example it is clear that good agreement with the null hypothesis would be obtained.

One way of selecting a family of "smooth" weights is to take a general linear combination of the second and third degree orthogonal polynomials for 14 equally-spaced points, $\xi_2' + \lambda\xi_3'$ say. The resulting ratio $L_g^2/V(L_g)$, maximized with respect to λ, can be tested as χ_2^2. It is clear from Table 1 that $\xi_2' + \lambda\xi_3'$ does not correlate well with the deviations and that a significant value of χ^2 will not result. Indeed if one had no prior knowledge of the pattern of deviations expected, a many-parameter family of weights would be needed to produce a large $L_g^2/V(L_g)$, and it is unlikely that very high significance would result. From a practical point of view, it would hardly be justifiable to pursue the point far.

An extremely rough argument is that if we were to fit a k parameter family of weights, which accounted for all the real variation present, the remaining degrees of freedom in the total χ^2 would on the average contribute unity each. Thus with a total $\chi_{12}^2 = 16\cdot66$, we can expect roughly that if fitting a k parameter family of weights leaves only random variation, the remaining $12 - k$ degree of freedom contribute $12 - k$, thus giving for testing significance $\chi_k^2 = 16\cdot66 - (12 - k) = 4\cdot66 + k$. Now at the 5 per cent. level this value of χ_k^2 is significant only if $k = 1$, i.e. only if the weights are preassigned. In general, if we found that $k \leqslant k_0$ for significance, we would have to consider whether the observed pattern of deviations is such as would, on prior grounds, have been included in a k_0 parameter pattern of weights for picking out systematic deviations.

In the present case the observed departures from the fitted distribution have little effect on the mean value but would be important if, for example, one wished to test whether a log-normal distribution is a better fit than an exponential. Bias in recording small values also could produce such a departure from the exponential distribution.

To sum up, there is an interesting suggestion of a departure from the exponential form, in the lower part of the distribution, but this cannot be regarded as firmly established.

Thus Model A cannot definitely be ruled out, even though there is evidence against it on two counts.

In the present example there is no point in trying to fit a more elaborate distribution, but such fitting would be necessary if it were required to estimate the mean of the complete distribution of failure-times. Distributions that might be fitted are

(a) a Weibull distribution, (14);

(b) a Γ distribution;

(c) a log-normal distribution.

The first two, but not the third, reduce to the exponential as a special case, and it is known (Irwin, 1942) that the log-normal and the exponential are hard to distinguish from data. Moreover, a Weibull or Γ distribution with coefficient of variation less than one is very difficult to distinguish from a log-normal distribution. The main differences between the families occur when there is a high proportion of very short failure-times, as in a Weibull distribution, (14), with $\beta < 0$, or in a Γ distribution with infinite ordinate at the origin. Such a distribution is appreciably different from a log-normal distribution, both near the origin and in having a relatively shorter tail. A consequence of the long tail of the log-normal distribution is that the conditional failure-rate, after increasing to a maximum, tends to zero for very long failure-times (Gumbel, 1958). This is often implausible on general grounds. In the example the departure is in the direction of the log-normal. Note that the region of small values which provides much information for discriminating between distributional types is the region of least importance in determining the mean failure-time.

There are sometimes theoretical reasons for expecting (a), which is an extreme-value distribution, but if the object of the analysis is not to discriminate between different stochastic models, but is simply to estimate mean values, it is reasonable to choose, among distributions that fit the data equally well, that for which the parameters are most easily estimated from censored data. Very often we shall then be free to choose between (a), (b), and (c). However, if the object is to examine distributional form, and especially if the conditional failure rate is of prime concern, the log-normal may be excluded on account of the property noted in the last paragraph.

The Weibull distribution and the log-normal are easily fitted graphically to censored data, the Γ distribution less easily. Efficient numerical estimation is much the easiest for the log-normal, since methods developed for the normal distribution (Gupta, 1952;

Sarhan and Greenberg, 1956, 1958; Blom, 1958) can be applied to the log-failure-times. It may be advisable to censor extreme observations in the lower part of the distribution.

Finally, we discuss briefly several things that can be done if the first test shows that $\pi_1(t)$ varies with t.

(a) *Test of the form of one distribution*

It may be suspected that one distribution function, say $F_2(t)$, is exponential, the other, $F_1(t)$, being of unknown form. An exact small-sample test of the hypothesis that $F_2(t)$ is exponential would be of interest; the following approximate procedure is based on a life-table analysis.

Divide the range of failure-times into intervals of width h so short that in any one interval few failures occur and both conditional failure rates are effectively constant. At the start of the ith interval let n_i individuals be in service, and let N_{i0}, N_{i1}, N_{i2} be the numbers surviving the ith interval, and failing from Type I and II failures respectively. Let $\lambda_{i1}, \lambda_{i2}$ be the corresponding conditional failure rates.

Then

$$E(N_{i0}) = n_i e^{-(\lambda_{i1}+\lambda_{i2})h} \tag{16}$$

$$E(N_{i2}) = \frac{n_i \lambda_{i2}}{\lambda_{i1} + \lambda_{i2}} [1 - e^{-(\lambda_{i1}+\lambda_{i2})h}]. \tag{17}$$

Thus

$$h\lambda_{i2} = \frac{E(N_{i2})}{n_i} \left[1 - \frac{E(N_{i0})}{n_i}\right]^{-1} \left\{-\log \frac{E(N_{i0})}{n_i}\right\} \tag{18}$$

$$\simeq \frac{E(N_{i2})}{n_i} + \frac{E(N_{i2})(n_i - E(N_{i0}))}{2 n_i^2}.$$

This suggests putting

$$h\hat{\lambda}_{i2} = \frac{N_{i2}}{n_i} + \frac{N_{i2}(N_{i1} + N_{i2})}{2 n_i^2}. \tag{19}$$

For some purposes it is more useful to consider

$$\theta_{i2} = 1 - e^{-\lambda_{i2}h} \simeq \lambda_{i2}h - \tfrac{1}{2}\lambda_{i2}^2 h^2,$$

the conditional probability of not surviving the ith interval when only Type II failures operate; we take

$$\hat{\theta}_{i2} = \frac{N_{i2}}{n_i} + \frac{N_{i1}N_{i2}}{2 n_i^2}. \tag{20}$$

To the order of approximation considered here

$$\hat{\theta}_{i2} = \frac{N_{i2}}{n_i - \tfrac{1}{2} N_{i1}}. \tag{21}$$

Approximately

$$V(\hat{\theta}_{i2}) = \theta_{i2}/n_i. \tag{22}$$

Formulae of this type are widely used in actuarial work; Elveback (1958) has made a thorough theoretical analysis.

We can test whether θ_{i2} is constant by a χ^2 test applied to the estimates $\hat{\theta}_{i2}$, or to the estimates $\hat{\lambda}_{i2}$, if for some reason unequal grouping intervals are used; the single degrees of freedom in χ^2 for linear, quadratic, etc. increase in θ_{i2} can be picked out in the usual way. This procedure applied to the pooled failure-times provides another solution to the problem of testing a single censored sample for agreement with the exponential form. An approximation in the procedure, additional to that in (20) and (22), is that the n_i are regarded as constant, whereas in fact they are random variables. This does not affect the approximations (20) and (22), as is easily seen by considering expectations, first conditionally on n_i, and then as n_i varies.

Example.—There is some doubt whether this analysis is completely appropriate for Mendenhall and Hader's example, because it is not clear that the two types of failure are physically distinct. Hence it may not be right to represent them by independent random variables. However, if it is correct to treat the failures as genuinely distinct, it would, in some applications, be reasonable to see whether the unconfirmed failures occur randomly; the calculations for this are shown in Table 2. The λ_{i2} are estimated from (19), the interval estimate for the last group being multiplied by 50/30.

Under the null hypothesis that the $\hat{\lambda}_{i2}$ are constant an estimate $\hat{\lambda}_2$ is obtained by weighting the $\hat{\lambda}_{i2}$ and then the estimated standard errors $D(\hat{\lambda}_{i2})$ of the individual $\hat{\lambda}_{i2}$'s are obtained from (22); a special calculation is needed for the last group. The pattern of differences $\hat{\lambda}_{i2} - \hat{\lambda}_2$ is quite systematic, although the total χ^2_{12} is only 9·35. Components of χ^2 for linear and quadratic regression can, if required, be isolated in the usual way by least squares fitting.

TABLE 2

Life-table Analysis to Test whether the Underlying Distribution of Unconfirmed Failure-times is Exponential

Hr.	Number at risk n_i	Type I failures N_{i1}	Type II failures N_{i2}	N_{i2}/n_i	$50\,\hat{\lambda}_{i2}$	$D(50\,\hat{\lambda}_{i2})$	$\hat{\theta}_{i1}$	$\hat{\theta}_{i2}$
0–	369	26	15	0·0406	0·0429	0·0121	0·0719	0·0421
50–	328	29	15	0·0457	0·0488	0·0129	0·0904	0·0478
100–	284	28	22	0·0775	0·0843	0·0138	0·1024	0·0813
150–	234	35	13	0·0556	0·0613	0·0152	0·1537	0·0597
200–	186	17	11	0·0591	0·0636	0·0171	0·0941	0·0618
250–	158	21	8	0·0506	0·0553	0·0185	0·1363	0·0540
300–	129	11	7	0·0543	0·0850	0·0205	0·0876	0·0566
350–	111	11	5	0·0450	0·0483	0·0221	0·1013	0·0473
400–	95	12	3	0·0316	0·0341	0·0239	0·1283	0·0336
450–	80	7	4	0·0500	0·0534	0·0261	0·0897	0·0522
500–	69	6	1	0·0145	0·0162	0·0281	0·0876	0·0151
550–	62	9	2	0·0323	0·0398	0·0296	0·1475	0·0346
600–629	51	6	1	0·0196	0·0378	0·0421		

Weighted mean 0·0543

(b) *Life-table for estimating $F_1(t)$ and $F_2(t)$*

We may take the most general model C, with arbitrary distributions $F_i(t)$ and estimate $F_i(t)$ from (9) by replacing $G_i(t)$ and $G(t)$ by the corresponding sample step functions, possibly after grouping. As noted in section 2, this is a life-table method in which a conditional failure rate is calculated for each type of failure and then converted into a distribution function. With ungrouped data this is very nearly the product limit method

(Kaplan and Meier, 1958); with grouped data we may estimate conditional failure probabilities $\hat{\theta}_{i1}$, $\hat{\theta}_{i2}$ using (20), then converting them into survival probabilities by finding, for example,

$$\prod_{j=1}^{i} (1 - \hat{\theta}_{j1}) \tag{23}$$

which is the probability of surviving to the end of the ith interval, when only failures of Type I operate.

(c) *Parametric form for the underlying distributions*

We may assume parametric forms for $F_i(t)$ and then estimate the parameters say by maximum likelihood. Mr. P. Chandler, in some as yet unpublished work, has considered this in connection with an application to strength testing. He assumed that the distributions $F_i(t)$ are logistic and showed that a close approximation to the maximum likelihood estimates can be obtained simply.

(d) *Parametric form for the observed distributions*

A fourth possibility is to fit a simple parametric form to the observed distributions $g_i(t)$ and then to substitute the fitted functions $\hat{g}_i(t)$ into (9) in order to estimate $F_i(t)$.

Comparison of methods (b), (c) and (d) will not be attempted here.

REFERENCES

ARMITAGE, P. (1959), "The comparison of survival curves", *J. R. Statist. Soc.* A, **122**, 279–300.
AUDLEY, R. J. (1957), "A stochastic description of the learning behaviour of an individual subject", *Q. J. Exptl. Psychol.*, **9**, 12–20.
BARTHOLOMEW, D. J. (1954), "Note on the use of Sherman's statistic as a test of randomness", *Biometrika*, **41**, 556–558.
BARTLETT, M. S. (1953), "On the statistical estimation of mean life-times", *Phil. Mag.*, **44**, 249–262.
BLOM, G. (1958), *Statistical Estimates and Transformed Beta-Variables*. New York: Wiley.
COCHRAN, W. G. (1955), "A test of a linear function of the deviations between observed and expected numbers", *J. Amer. Statist. Ass.*, **50**, 377–397.
COX, D. R. (1958), "The regression analysis of binary sequences", *J. R. Statist. Soc.* B, **20**, 215–242.
ELVEBACK, L. (1958), "Estimation of survivorship in chronic disease; the actuarial method", *J. Amer. Statist. Ass.*, **53**, 420–440.
EPSTEIN, B. & SOBEL, M. (1953), "Life testing", *J. Amer. Statist. Ass.*, **48**, 486–502.
GUPTA, S. (1952), "Estimation of the mean and standard deviation of a normal population from a censored sample", *Biometrika*, **39**, 260–273.
IRWIN, J. O. (1942), "The distribution of the logarithm of survival times when the true law is exponential", *J. Hyg., Camb.*, **42**, 328–333.
GUMBEL, E. J. (1958), *Statistics of Extreme Values*. New York: Columbia University Press.
HARRIS, T. E., MEIER, P. & TUKEY, J. W. (1950), "Timing of the distribution of events between observations", *Human Biology*, **22**, 249–270.
KAPLAN, E. L. & MEIER, P. (1958), "Non-parametric estimation from incomplete observations", *J. Amer. Statist. Ass.*, **53**, 457–481.
MEIER, P. (1955), "Note on estimation in a Markov process with constant transition rates", *Human Biology*, **27**, 121–124.
MENDENHALL, W. & HADER, R. J. (1958), "Estimation of parameters of mixed exponentially distributed failure time distributions from censored life test data", *Biometrika*, **45**, 504–520.
MORAN, P. A. P. (1951), "Simple estimation methods for evolutive processes", *J. R. Statist. Soc.* B, **13**, 147–150.
SAMPFORD, M. R. (1954), "The estimation of response-time distributions, III", *Biometrics*, **10**, 531–561.
SARHAN, A. E. & GREENBERG, B. G. (1956), "Estimation of location and scale parameters by order statistics from singly and doubly censored samples, I", *Ann. Math. Statist.*, **27**, 427–451.
—— —— (1958), "Estimation of location and scale parameters by order statistics from singly and doubly censored samples, II", *Ann. Math. Statist.*, **29**, 79–105.
SEAL, H. L. (1954), "The estimation of mortality and other decremental probabilities", *Skand. Akt.*, **37**, 137–162.
SHERMAN, B. (1957), "Percentiles of the ϖ_n statistic", *Ann. Math. Statist.*, **28**, 259–261.

Regression Models and Life-Tables

By D. R. Cox

Imperial College, London

[Read before the ROYAL STATISTICAL SOCIETY, at a meeting organized by the
Research Section, on Wednesday, March 8th, 1972, Mr M. J. R. HEALY in the Chair]

SUMMARY

The analysis of censored failure times is considered. It is assumed that on
each individual are available values of one or more explanatory variables.
The hazard function (age-specific failure rate) is taken to be a function of
the explanatory variables and unknown regression coefficients multiplied
by an arbitrary and unknown function of time. A conditional likelihood is
obtained, leading to inferences about the unknown regression coefficients.
Some generalizations are outlined.

Keywords: LIFE TABLE; HAZARD FUNCTION; AGE-SPECIFIC FAILURE RATE; PRODUCT
LIMIT ESTIMATE; REGRESSION; CONDITIONAL INFERENCE; ASYMPTOTIC THEORY;
CENSORED DATA; TWO-SAMPLE RANK TESTS; MEDICAL APPLICATIONS; RELIABILITY
THEORY; ACCELERATED LIFE TESTS.

1. INTRODUCTION

LIFE tables are one of the oldest statistical techniques and are extensively used by
medical statisticians and by actuaries. Yet relatively little has been written about
their more formal statistical theory. Kaplan and Meier (1958) gave a comprehensive
review of earlier work and many new results. Chiang in a series of papers has, in
particular, explored the connection with birth–death processes; see, for example,
Chiang (1968). The present paper is largely concerned with the extension of the
results of Kaplan and Meier to the comparison of life tables and more generally to
the incorporation of regression-like arguments into life-table analysis. The arguments
are asymptotic but are relevant to situations where the sampling fluctuations are
large enough to be of practical importance. In other words, the applications are
more likely to be in industrial reliability studies and in medical statistics than in
actuarial science. The procedures proposed are, especially for the two-sample
problem, closely related to procedures for combining contingency tables; see Mantel
and Haenzel (1959), Mantel (1963) and, especially for the application to life tables,
Mantel (1966). There is also a strong connection with a paper read recently to the
Society by R. and J. Peto (1972).

We consider a population of individuals; for each individual we observe either
the time to "failure" or the time to "loss" or censoring. That is, for the censored
individuals we know only that the time to failure is greater than the censoring time.

Denote by T a random variable representing failure time; it may be discrete or
continuous. Let $\mathscr{F}(t)$ be the survivor function,

$$\mathscr{F}(t) = \mathrm{pr}\,(T \geqslant t)$$

and let $\lambda(t)$ be the hazard or age-specific failure rate. That is,

$$\lambda(t) = \lim_{\Delta t \to 0+} \frac{\mathrm{pr}\,(t \leqslant T < t + \Delta t \mid t \leqslant T)}{\Delta t}. \tag{1}$$

Note that if T is discrete, then

$$\lambda(t) = \sum \lambda_{u_j} \delta(t - u_j), \tag{2}$$

where $\delta(t)$ denotes the Dirac delta function and $\lambda_t = \text{pr}(T = t \,|\, T \geqslant t)$. By the product law of probability $\mathscr{F}(t)$ is given by the product integral

$$\mathscr{F}(t) = \overset{t-0}{\underset{u=0}{\mathscr{P}}} \{1 - \lambda(u) \, du\} = \lim \prod_{k=0}^{r-1} \{1 - \lambda(\tau_k)(\tau_{k+1} - \tau_k)\}, \tag{3}$$

the limit being taken as all $\tau_{k+1} - \tau_k$ tend to zero with $0 = \tau_0 < \tau_1 < ... < \tau_{r-1} < \tau_r = t$. If $\lambda(t)$ is integrable this is

$$\exp\left\{ -\int_0^t \lambda(u) \, du \right\}, \tag{4}$$

whereas if $\lambda(t)$ is given by (2), the product integral is

$$\prod_{u_j < t} (1 - \lambda_{u_j}), \tag{5}$$

If the distribution has both discrete and continuous components the product integral is a product of factors (4) and (5).

2. THE PRODUCT-LIMIT METHOD

Suppose observations are available on n_0 independent individuals and, to begin with, that the failure times are identically distributed in the form specified in Section 1. Let n individuals be observed to failure and the rest be censored. The rather strong assumption will be made throughout that the only information available about the failure time of a censored individual is that it exceeds the censoring time. This assumption is testable only if suitable supplementary information is available. Denote the distinct failure times by

$$t_{(1)} < t_{(2)} < ... < t_{(k)}. \tag{6}$$

Further let $m_{(i)}$ be the number of failure times equal to $t_{(i)}$, the multiplicity of $t_{(i)}$; of course $\sum m_{(i)} = n$, and in the continuous case $k = n$, $m_{(i)} = 1$.

The set of individuals at risk at time $t - 0$ is called the *risk set* at time t and denoted $\mathscr{R}(t)$; this consists of those individuals whose failure or censoring time is at least t. Let $r_{(i)}$ be the number of such individuals for $t = t_{(i)}$. The product-limit estimate of the underlying distribution is obtained by taking estimated conditional probabilities that agree exactly with the observed conditional frequencies. That is,

$$\hat{\lambda}(t) = \sum_{i=1}^{k} \frac{m_{(i)}}{r_{(i)}} \delta(t - t_{(i)}). \tag{7}$$

Correspondingly,

$$\hat{\mathscr{F}}(t) = \overset{t-0}{\underset{u=0}{\mathscr{P}}} \{1 - \hat{\lambda}(u) \, du\} = \prod_{t_{(i)} < t} \left\{ 1 - \frac{m_{(i)}}{r_{(i)}} \right\}. \tag{8}$$

For uncensored data this is the usual sample survivor function; some of the asymptotic properties of (8) are given by Kaplan and Meier (1958) and by Efron (1967) and can be used to adapt to the censored case tests based on sample cumulative distribution function.

The functions (7) and (8) are maximum-likelihood estimates in the family of all possible distributions (Kaplan and Meier, 1958). However, as in the uncensored case, this property is of limited importance and the best justification is essentially (7). The estimates probably also have a Bayesian interpretation involving a very "irregular" prior.

If the class of distributions is restricted, either parametrically or by some such condition as requiring $\lambda(t)$ to be monotonic or smooth, the maximum-likelihood estimates will be changed. For the monotone hazard case with uncensored data, see Grenander (1956). The smoothing of estimated hazard functions has been considered by Watson and Leadbetter (1964a, b) for the uncensored case.

3. Regression Models

Suppose now that on each individual one or more further measurements are available, say on variables $z_1, ..., z_p$. We deal first with the notationally simpler case when the failure-times are continuously distributed and the possibility of ties can be ignored. For the jth individual let the values of $\mathbf{z}$ be $\mathbf{z}_j = (z_{1j}, ..., z_{pj})$. The z's may be functions of time. The main problem considered in this paper is that of assessing the relation between the distribution of failure time and $\mathbf{z}$. This will be done in terms of a model in which the hazard is

$$\lambda(t; \mathbf{z}) = \exp(\mathbf{z}\boldsymbol{\beta}) \lambda_0(t), \tag{9}$$

where $\boldsymbol{\beta}$ is a $p \times 1$ vector of unknown parameters and $\lambda_0(t)$ is an unknown function giving the hazard function for the standard set of conditions $\mathbf{z} = \mathbf{0}$. In fact $(\mathbf{z}\boldsymbol{\beta})$ can be replaced by any known function $h(\mathbf{z}, \boldsymbol{\beta})$, but this extra generality is not needed at this stage. The following examples illustrate just a few possibilities.

Example 1. Two-sample problem. Suppose that there is just one z variable, $p = 1$, and that this takes values 0 and 1, being an indicator variable for the two samples. Then according to (9) the hazards in samples 0 and 1 are respectively $\lambda_0(t)$ and $\psi\lambda_0(t)$, where $\psi = e^\beta$. In the continuous case the survivor functions are related (Lehmann, 1953) by $\mathscr{F}_1(t) = \{\mathscr{F}_0(t)\}^\psi$. There is an obvious extension for the k sample problem.

Example 2. The two-sample problem; extended treatment. We can deal with more complicated relationships between the two samples than are contemplated in Example 1 by introducing additional time-dependent components into $\mathbf{z}$. Thus if $z_2 = tz_1$, where z_1 is the binary variable of Example 1, the hazard in the second sample is

$$\psi e^{\beta_2 t} \lambda_0(t). \tag{10}$$

Of course in defining z_2, t could be replaced by any known function of t; further, several new variables could be introduced involving different functions of t. This provides one way of examining consistency with a simple model of proportional hazards. In fitting the model and often also in interpretation it is convenient to reparametrize (10) in the form

$$\rho \exp\{\beta_2(t - t^*)\}, \tag{11}$$

where t^* is any convenient constant time somewhere near the overall mean. This will avoid the more extreme non-orthogonalities of fitting. All the points connected with this example extend to the comparison of several samples.

Example 3. Two-sample problem with covariate. By introducing into the models of Examples 1 and 2 one or more further z variables representing concomitant variables, it is possible to examine the relation between two samples adjusting for the presence of concomitant variables.

Example 4. *Regression*. The connection between failure-time and regressor variables can be explored in an obvious way. Note especially that by introducing functions of t, effects other than constant multiplication of the hazard can be included.

4. ANALYSIS OF REGRESSION MODELS

There are several approaches to the analysis of the above models. The simplest is to assume $\lambda_0(t)$ constant, i.e. to assume an underlying exponential distribution; see, for example, Chernoff (1962) for some models of this type in the context of accelerated life tests. The next simplest is to take a two-parameter family of hazard functions, such as the power law associated with the Weibull distribution or the exponential of a linear function of t. Then standard methods such as maximum likelihood can be used; to be rigorous extension of the usual conditions for maximum-likelihood formulae and theory would be involved to cover censoring, but there is little doubt that some such justification could be given. This is in many ways the most natural approach but will not be explored further in the present paper. In this approach a computationally desirable feature is that both probability density and survivor function are fairly easily found. A simple form for the hazard is not by itself particularly advantageous, and models other than (9) may be more natural. For a normal theory maximum-likelihood analysis of factorial experiments with censored observations, see Sampford and Taylor (1959), and for the parametric analysis of response times in bioassay, see, Sampford (1954).

Alternatively we may restrict $\lambda_0(t)$ qualitatively, for example by assuming it to be monotonic or to be a step function (a suggestion of Professor J. W. Tukey). The latter possibility is related to a simple spline approximation to the log survivor function.

In the present paper we shall, however, concentrate on exploring the consequence of allowing $\lambda_0(t)$ to be arbitrary, main interest being in the regression parameters. That is, we require our method of analysis to have sensible properties whatever the form of the nuisance function $\lambda_0(t)$. Now this is a severe requirement and unnecessary in the sense that an assumption of some smoothness in the distribution $\mathscr{F}_0(t)$ would be reasonable. The situation is parallel to that arising in simpler problems when a nuisance parameter is regarded as completely unknown. It seems plausible in the present case that the loss of information about β arising from leaving $\lambda_0(t)$ arbitrary is usually slight; if this is indeed so the procedure discussed here is justifiable as a reasonably cautious approach to the study of β. A major outstanding problem is the analysis of the relative efficiency of inferences about β under various assumptions about $\lambda_0(t)$.

The general attitude taken is that parametrization of the dependence on z is required so that our conclusions about that dependence are expressed concisely; of course any form taken is provisional and needs examination in the light of the data. So far as the secondary features of the system are concerned, however, it is sensible to make a minimum of assumptions leading to a convenient analysis, provided that no major loss of efficiency is involved.

5. A CONDITIONAL LIKELIHOOD

Suppose then that $\lambda_0(t)$ is arbitrary. No information can be contributed about β by time intervals in which no failures occur because the component $\lambda_0(t)$ might conceivably be identically zero in such intervals. We therefore argue conditionally on the set $\{t_{(i)}\}$ of instants at which failures occur; in discrete time we shall condition also on the observed multiplicities $\{m_{(i)}\}$. Once we require a method of analysis holding for all $\lambda_0(t)$, consideration of this conditional distribution seems inevitable.

For the particular failure at time $t_{(i)}$, conditionally on the risk set $\mathscr{R}(t_{(i)})$, the probability that the failure is on the individual as observed is

$$\exp\{\mathbf{z}_{(i)}\,\boldsymbol{\beta}\}\bigg/ \sum_{l\in\mathscr{R}(t_{(i)})} \exp\{\mathbf{z}_{(l)}\,\boldsymbol{\beta}\}. \tag{12}$$

Each failure contributes a factor of this nature and hence the required conditional log likelihood is

$$L(\boldsymbol{\beta}) = \sum_{i=1}^{k} \mathbf{z}_{(i)}\,\boldsymbol{\beta} - \sum_{i=1}^{k} \log\left[\sum_{l\in\mathscr{R}(t_{(i)})} \exp\{\mathbf{z}_{(l)}\,\boldsymbol{\beta}\}\right]. \tag{13}$$

Direct calculation from (13) gives for $\xi, \eta = 1, ..., p$

$$U_\xi(\boldsymbol{\beta}) = \frac{\partial L(\boldsymbol{\beta})}{\partial \beta_\xi} = \sum_{i=1}^{k}\{z_{(\xi i)} - A_{(\xi i)}(\boldsymbol{\beta})\}, \tag{14}$$

where

$$A_{(\xi i)}(\boldsymbol{\beta}) = \frac{\sum z_{\xi l}\exp(z_l\,\boldsymbol{\beta})}{\sum \exp(z_l\,\boldsymbol{\beta})}, \tag{15}$$

the sum being over $l\in\mathscr{R}(t_{(i)})$. That is, $A_{(\xi i)}(\boldsymbol{\beta})$ is the average of z_ξ over the finite population $\mathscr{R}(t_{(i)})$, using an "exponentially weighted" form of sampling. Similarly

$$\mathscr{I}_{\xi\eta}(\boldsymbol{\beta}) = -\frac{\partial^2 L(\boldsymbol{\beta})}{\partial \beta_\xi \partial \beta_\eta} = \sum_{i=1}^{k} C_{(\xi\eta i)}(\boldsymbol{\beta}), \tag{16}$$

where

$$C_{(\xi\eta i)}(\boldsymbol{\beta}) = \{\sum z_{\xi l} z_{\eta l}\exp(z_l\,\boldsymbol{\beta})/\sum \exp(z_l\,\boldsymbol{\beta})\} - A_{(\xi i)}(\boldsymbol{\beta})\,A_{(\eta i)}(\boldsymbol{\beta}) \tag{17}$$

is the covariance of z_ξ and z_η in this form of weighted sampling.

To calculate the expected value of (16) it would be necessary to know the times at which individuals who failed would have been censored had they not failed. This information would often not be available and in any case might well be thought irrelevant; this point is connected with difficulties of conditionality at the basis of a sampling theory approach to statistics (Pratt, 1962). Here we shall use asymptotic arguments in which (16) can be used directly for the estimation of variances, $\boldsymbol{\beta}$ being replaced by a suitable estimate. For a rigorous justification, assumptions about the censoring times generalizing those of Breslow (1970) would be required. It would not be satisfactory to assume that the censoring times are random variables distributed independently of the z's. For instance in the two-sample problem censoring might be much more severe in one sample than in the other.

Maximum-likelihood estimates of $\boldsymbol{\beta}$ can be obtained by iterative use of (14) and (16) in the usual way. Significance tests about subsets of parameters can be derived in various ways, for example by comparison of the maximum log likelihoods achieved. Relatively simple results can, however, be obtained for testing the global null hypothesis, $\boldsymbol{\beta} = 0$. For this we treat $\mathbf{U}(0)$ as asymptotically normal with zero mean vector and with covariance matrix $\mathscr{I}(0)$. That is, the statistic

$$\{\mathbf{U}(0)\}^{\mathrm{T}}\{\mathscr{I}(0)\}^{-1}\{\mathbf{U}(0)\} \tag{18}$$

has, under the null hypothesis, an asymptotic chi-squared distribution with p degrees of freedom.

We have from (14) and (15) that

$$\mathbf{U}_\xi(0) = \sum_{i=1}^{k} (z_{(\xi i)} - A_{(\xi i)}), \tag{19}$$

where $A_{(\xi i)} = A_{(\xi i)}(0)$ is the mean of z_ξ over $\mathcal{R}(t_{(i)})$. Further, from (16),

$$\mathscr{I}_{\xi \eta}(0) = \sum_{i=1}^{k} C_{(\xi \eta i)}, \tag{20}$$

where $C_{(\xi \eta i)} = C_{(\xi \eta i)}(0)$ is the covariance of z_ξ and z_η in the finite population $\mathcal{R}(t_{(i)})$. The form of weighted sampling associated with general β has reduced to random sampling without replacement.

6. ANALYSIS IN DISCRETE TIME

Unfortunately it is quite likely in applications that the data will be recorded in a form involving ties. If these are small in number a relatively *ad hoc* modification of the above procedures will be satisfactory. To cover the possibility of an appreciable number of ties, we generalize (9) formally to discrete time by

$$\frac{\lambda(t; \mathbf{z})\, dt}{1 - \lambda(t; \mathbf{z})\, dt} = \exp(\mathbf{z}\boldsymbol{\beta}) \frac{\lambda_0(t)\, dt}{1 - \lambda_0(t)\, dt}. \tag{21}$$

In the continuous case this reduces to (9); in discrete time $\lambda(t; \mathbf{z})\, dt$ is a non-zero probability and (21) is a logistic model.

The typical contribution (12) to the likelihood now becomes

$$\exp\{\mathbf{s}_{(i)}\, \boldsymbol{\beta}\} \Bigg/ \sum_{l \in \mathcal{R}(t_{(i)}; m_{(i)})} \exp\{\mathbf{s}_{(l)}\, \boldsymbol{\beta}\}, \tag{22}$$

where $\mathbf{s}_{(i)}$ is the sum of $\mathbf{z}$ over the individuals failing at $t_{(i)}$ and the notation in the denominator means that the sum is taken over all distinct sets of $m_{(i)}$ individuals drawn from $\mathcal{R}(t_{(i)})$.

Thus the full conditional log likelihood is

$$\sum_{i=1}^{k} \mathbf{s}_{(i)}\, \boldsymbol{\beta} - \sum_{i=1}^{k} \log \left[\sum_{l \in \mathcal{R}(t_{(i)}; m_{(i)})} \exp\{\mathbf{s}_{(l)}\, \boldsymbol{\beta}\} \right].$$

The derivatives can be calculated as before. In particular,

$$U_\xi(0) = \sum_{i=1}^{k} \{s_{(\xi i)} - m_{(i)}\, A_{(\xi i)}\}, \tag{23}$$

$$\mathscr{I}_{\xi \eta}(0) = \sum_{i=1}^{k} \frac{m_{(i)}\{r_{(i)} - m_{(i)}\}}{\{r_{(i)} - 1\}} C_{(\xi \eta i)}. \tag{24}$$

Note that (24) gives the exact covariance matrix when the observations $z_{(\xi i)}$ and the totals $s_{(\xi i)}$ are drawn randomly without replacement from the *fixed* finite populations $\mathcal{R}(t_{(1)}), \ldots, \mathcal{R}(t_{(k)})$. In fact, however, the population at one time is influenced by the outcomes of the "trials" at previous times.

7. THE TWO-SAMPLE PROBLEM

As an illustration, consider the two-sample problem with the proportional hazard model of Section 3, Example 1. Here $p = 1$ and we omit the first suffix on the indicator variable. Then

$$U(0) = n_1 - \sum_{i=1}^{k} m_{(i)} A_{(i)}, \tag{25}$$

$$\mathscr{I}(0) = \sum_{i=1}^{k} \frac{m_{(i)}\{r_{(i)} - m_{(i)}\}}{\{r_{(i)} - 1\}} A_{(i)}\{1 - A_{(i)}\}, \tag{26}$$

where $A_{(i)}$ is the proportion of the risk population $\mathscr{R}(t_{(i)})$ that have $z = 1$, i.e. belong to sample 1, and n_1 is the total number of failures in sample 1. An asymptotic two-sample test is thus obtained by treating

$$U(0)/\sqrt{\mathscr{I}(0)} \tag{27}$$

as having a standard normal distribution under the null hypothesis. This is different from the procedure of Gehan who adapted the Wilcoxon test to censored data (Gehan, 1965; Efron, 1967; Breslow, 1970). The test has been considered in some detail by Peto and Peto (1972).

The test (27) is formally identical with that obtained by setting up at each failure point a 2×2 contingency table (sample 1, sample 2) (failed, survived). To test for the presence of a difference between the two samples the information from the separate tables can then be combined (Cochran, 1954; Mantel and Haenzel, 1959; Mantel, 1963). The application of this to life tables is discussed especially by Mantel (1966). Note, however, that whereas the test in the contingency table situation is, at least in principle, exact, the test here is only asymptotic, because of the difficulties associated with specification of the stopping rule. Formally the same test was given by Cox (1959) for a different life-table problem where there is a single sample with two types of failure and the hypothesis under test concerns the proportionality of the hazard function for the two types.

When there is a non-zero value of β, the "weighted" average of a single observation from the risk population $\mathscr{R}(t_{(i)})$ is

$$A_{(i)}(\beta) = \frac{e^{\beta} A_{(i)}}{1 - A_{(i)} + e^{\beta} A_{(i)}} \tag{28}$$

and the maximum-likelihood equation $U(\beta) = 0$ gives, when all failure times are distinct,

$$\sum_{i=1}^{k} \frac{e^{\hat{\beta}} A_{(i)}}{1 - A_{(i)} + e^{\hat{\beta}} A_{(i)}} = n_1. \tag{29}$$

If $\hat{\beta}$ is thought to be close to some known constant, it may be useful to linearize (29). In particular, if $\hat{\beta}$ is small, we have as an approximation to the maximum-likelihood estimate

$$\hat{\beta}_0 = (n_1 - \sum A_{(i)}) / \sum A_{(i)}\{1 - A_{(i)}\}.$$

The procedures of this section involve only the ranked data, i.e. are unaffected by an arbitrary monotonic transformation of the time scale. Indeed the same is true for any of the results in Section 4 provided that the z's are not functions of time. While

8

the connection with the theory of rank tests will not be explored in detail, it is worth examining the form of the test (27) for uncensored data with all failure times distinct. For this, let the failure times in sample 1 have ranks $c_1 < c_2 < ... < c_{n_1}$ in the ranking of the full data. At the ith largest observed failure time, individuals with ranks $n, n-1, ..., i$ are at risk, so that

$$A_{(i)} = \frac{1}{n-i+1} \sum_{l=1}^{n_1} H(c_l - i),$$ (30)

where $H(x)$ is the unit Heaviside function,

$$H(x) = \begin{cases} 1 & (x \geqslant 0), \\ 0 & (x < 0). \end{cases}$$ (31)

Thus, by (25),

$$U(0) = n_1 - \sum_{l=1}^{n_1} \sum_{i=1}^{c_l} \frac{1}{n-i+1}$$

$$= n_1 - \sum_{l=1}^{n} e_{nc_l},$$ (32)

where e's are the expected values of the order statistics in a random sample of size n from a unit exponential distribution. The test based on (32) is asymptotically fully efficient for the comparison of two exponential distributions (Savage, 1956; Cox, 1964). Further, by (26),

$$\mathscr{I}(0) = \sum_{l=1}^{n_1} e_{nc_l} - \sum_{l=1}^{n_1} (1 + 2n_1 - 2l) v_{nc_l},$$ (33)

where

$$v_{nc_l} = \sum_{i=1}^{c_l} \frac{1}{(n-i+1)^2}$$ (34)

is the variance of an exponential order statistic.

Here the test statistic is, under the null hypothesis, a constant minus the total of a random sample of size n_1 drawn without replacement from the finite population $\{e_{n1}, ..., e_{nn}\}$. The exact distribution can in principle be obtained and in particular it can be shown that

$$E\{U(0)\} = 0, \quad \text{var}\{U(0)\} = \frac{n_1(n-n_1)(n-e_{nn})}{n(n-1)}.$$ (35)

There is not much point in this case in using the more complicated asymptotic formula (33), especially as fairly simple more refined approximations to the distribution of the test statistic are available (Cox, 1964). It can easily be verified that

$$E\{\mathscr{I}(0)\} \sim \text{var}\{U(0)\}.$$ (36)

8. Estimation of Distribution of Failure-time

Once we have obtained the maximum-likelihood estimate of β, we can consider the estimation of the distribution associated with the hazard (10) either for $z = 0$, or for some other given value of z. Thus to estimate $\lambda_0(t)$ we need to generalize (7).

To do this we take $\lambda_0(t)$ to be identically zero, except at the points where failures have occurred, and carry out a separate maximum-likelihood estimation at each such failure point. For the latter it is convenient to write the contribution to $\lambda_0(t)$ at $t_{(i)}$ in the form

$$\frac{\pi_{(i)} \exp(-\boldsymbol{\beta}\bar{\mathbf{z}}_{(i)})}{1-\pi_{(i)}+\pi_{(i)} \exp(-\boldsymbol{\beta}\bar{\mathbf{z}}_{(i)})}\, \delta(t-t_{(i)}),$$

where $\bar{\mathbf{z}}_{(i)}$ is an arbitrary constant to be chosen; it is useful to take $\bar{\mathbf{z}}_{(i)}$ as approximately the mean in the relevant risk set. The maximum-likelihood estimate of $\pi_{(i)}$ can then be shown to satisfy

$$\hat{\pi}_{(i)} = \frac{m_{(i)}}{r_{(i)}} - \frac{\hat{\pi}_{(i)}(1-\hat{\pi}_{(i)})}{r_{(i)}} \sum_{j \in R(t_{(i)})} \frac{\exp\{\hat{\boldsymbol{\beta}}(\mathbf{z}_j - \bar{\mathbf{z}}_{(i)})\}-1}{1-\hat{\pi}_{(i)}+\hat{\pi}_{(i)} \exp\{\hat{\boldsymbol{\beta}}(\mathbf{z}_j - \bar{\mathbf{z}}_{(i)})\}}, \tag{37}$$

which can be solved by iteration. The suggested choice of $\bar{\mathbf{z}}_{(i)}$ is designed to make the second term in (37) small. Note that in the single-sample case, the second term is identically zero. Once (37) is solved for all i, we have by the product integral formula

$$\hat{\mathscr{F}}_0(t) = \prod_{t_{(i)}<t} \left\{1-\frac{\hat{\pi}_{(i)} \exp(-\hat{\boldsymbol{\beta}}\bar{\mathbf{z}}_{(i)})}{1-\hat{\pi}_{(i)}+\hat{\pi}_{(i)} \exp(-\hat{\boldsymbol{\beta}}\bar{\mathbf{z}}_{(i)})}\right\}. \tag{38}$$

For an estimate at a given non-zero $\mathbf{z}$, replace $\exp(-\hat{\boldsymbol{\beta}}\bar{\mathbf{z}}_{(i)})$ by $\exp\{\hat{\boldsymbol{\beta}}(\mathbf{z}-\bar{\mathbf{z}}_{(i)})\}$. Alternative simpler procedures would be worth having (Mantel, 1966).

9. BIVARIATE LIFE TABLES

We now consider briefly the extension of life-table arguments to multivariate data. Suppose for simplicity that there are two types of failure time for each individual represented by random variables T_1 and T_2. For instance, these might be the failure-times of two different but associated components; observations may be censored on neither, one or both components. For analogous problems in bioassay, see Sampford (1952).

The joint distribution can be described in terms of hazard functions $\lambda_{10}(t), \lambda_{20}(t), \lambda_{21}(t|u), \lambda_{12}(t|u)$, where

$$\left.\begin{aligned}
\lambda_{p0}(t) &= \lim_{\Delta t \to 0+} \frac{\mathrm{pr}(t \leqslant T_p < t+\Delta t \,|\, t \leqslant T_1, t \leqslant T_2)}{\Delta t} \quad (p=1,2), \\[2mm]
\lambda_{21}(t|u) &= \lim_{\Delta t \to 0+} \frac{\mathrm{pr}(t \leqslant T_2 < t+\Delta t \,|\, t \leqslant T_2, T_1=u)}{\Delta t} \quad (u<t),
\end{aligned}\right\} \tag{39}$$

with a similar definition for $\lambda_{12}(t|u)$. It is easily shown that the bivariate probability density function $f(t_1,t_2)$ is given by

$$f(t_1,t_2) = \exp\left[-\int_0^{t_1-0}\{\lambda_{10}(u)+\lambda_{20}(u)\}\,du - \int_{t_1+0}^{t_2-0}\lambda_{21}(u|t_1)\,du\right]\lambda_{10}(t_1)\,\lambda_{21}(t_2|t_1), \tag{40}$$

for $t_2 \geqslant t_1$, with again an analogous expression for $t_2 \leqslant t_1$. It is fairly easy to show formally that a necessary and sufficient condition for the independence of T_1 and T_2 is

$$\lambda_{12}(t|u) = \lambda_{10}(t), \quad \lambda_{21}(t|u) = \lambda_{20}(t), \tag{41}$$

as is obvious on general grounds. Note also that if $\mathscr{F}(t_1, t_2)$ is the joint survivor function

$$\lambda_{10}(t) = -\frac{1}{\mathscr{F}(t,t)}\left[\frac{\partial \mathscr{F}(t,u)}{\partial t}\right]_{u=t}, \quad \lambda_{12}(t|u) = -\frac{\partial^2 \mathscr{F}(t,u)}{\partial t\,\partial u}\bigg/\frac{\partial \mathscr{F}(t,u)}{\partial u}. \tag{42}$$

Dependence on further variables z can be indicated in the same way as for (11). The simplest model would have the same function of z multiplying all four hazard functions, although this restriction is not essential.

Estimation and testing would in principle proceed as before, although grouping of the conditioning u variable seems necessary in the parts of the analysis concerning the function $\lambda_{12}(t|u)$ and $\lambda_{21}(t|u)$.

Further generalizations which will not, however, be explored here are to problems in multidimensional time and to problems connected with point processes (Cox and Lewis, 1972; Cox, 1972).

10. An Example

To illustrate some of the above results, it is convenient to take data of Freireich *et al.* used by Gehan (1965) and several subsequent authors. Table 1 gives the ordered times for two samples of individuals; censored values are denoted with asterisks. Table 2 outlines the calculation of the simple test statistic $U(0)$ and its asymptotic variance. The failure instants and their multiplicities $m_{(i)}$ are listed; $A_{(i)}$ is the proportion of the relevant risk population in sample 1.

TABLE 1

Times of remission (weeks) of leukemia patients
(Gehan, 1965, from Freireich *et al.*)

Sample 0 (drug 6-MP)	6*, 6, 6, 6, 7, 9*, 10*, 10, 11*, 13, 16, 17*, 19*, 20*, 22, 23, 25*, 32*, 32*, 34*, 35*
Sample 1 (control)	1, 1, 2, 2, 3, 4, 4, 5, 5, 8, 8, 8, 8, 11, 11, 12, 12, 15, 17, 22, 23

* Censored.

The value of $U(0) = n_1 - \sum m_{(i)} A_{(i)}$ is 10·25 with an asymptotic standard error $\sqrt{\mathscr{I}(0)}$ of 2·50. The critical ratio of just over 4 compares with about 3·6 for the generalized Wilcoxon test of Gehan (1965). The overwhelming significance of the difference is in line with one's qualitative impression of the data.

The technique used to find $\hat{\beta}$ was direct computation of the log likelihood as a function of β and of a further parameter γ to be described in a moment. This, while not the best way of getting maximum-likelihood estimates on their own, is useful in enabling various approximate tests and confidence regions to be found in a unified manner.

To examine possible departures from the simple model of proportional hazards, the procedure of Example 2 of Section 3 was followed, taking as in (11) the hazard in sample 1 to be a time-dependent multiple of that in sample 0 of the form

$$\exp\{\beta + \gamma(t-10)\}\,\lambda_0(t); \tag{43}$$

the arbitrary constant 10 is inserted to achieve approximate orthogonality of estimation of the two parameters, being chosen as a convenient value in the centre of the range.

A test of the global null hypothesis $\beta = \gamma = 0$ could be done via the test statistic (20) but is not very relevant here. Instead the log likelihood (15) was computed

TABLE 2

Main quantities for the test of the null hypothesis for the data of Table 1

"Failure" time		Risk population			Multiplicity	
Sample 0	Sample 1	No. in sample 0	No. in sample 1	$r_{(i)}$	$A_{(i)}$	$m_{(i)}$
23	23	6	1	7	0·1429	2
22	22	7	2	9	0·2222	2
	17	10	3	13	0·2308	1
16		11	3	14	0·2143	1
	15	11	4	15	0·2667	1
13		12	4	16	0·2500	1
	12, 12	12	6	18	0·3333	2
	11, 11	13	8	21	0·3810	2
10		15	8	23	0·3478	1
	8, 8, 8, 8	16	12	28	0·4286	4
7		17	12	29	0·4138	1
6, 6, 6		21	12	33	0·3636	3
	5, 5	21	14	35	0·4000	2
	4, 4	21	16	37	0·4324	2
	3	21	17	38	0·4474	1
	2, 2	21	19	40	0·4750	2
	1, 1	21	21	42	0·5000	2

$$U(0) = n_1 - \sum m_{(i)} A_{(i)} = 10\cdot25;$$

$$\mathscr{I}(0) = \sum \frac{m_{(i)}\{r_{(i)} - m_{(i)}\}}{r_{(i)} - 1} A_{(i)}\{1 - A_{(i)}\} = 6.2570.$$

directly for a grid of points in the (β, γ) plane. Note that in (15) the first term is $21\beta - 28\gamma$; for instance, the coefficient -28 is the sum of the values $(t - 10)$ over the individuals in sample 1. The logarithmic second term is simple for those time points at which there is a single completed time, $m_{(i)} = 1$; for example corresponding to the time 7 there is a term in the log likelihood

$$-\log(17 + 12e^{\beta - 3\gamma}),$$

the risk set at this time consisting of 17 individuals from sample 0 and 12 from sample 1. For points of higher multiplicity, the situation is more complicated, because all possible samples of size $m_{(i)}$ from the risk population have to be considered; fortunately all the samples have the same totals of the two relevant variables. For example, for the point 6, of multiplicity 3, we have to consider the total of all samples of size 3 drawn from the relevant risk population and this leads to a term

$$-\log\left\{\binom{21}{3} + \binom{21}{2}\binom{12}{1}e^{\beta - 4\gamma} + \binom{21}{1}\binom{12}{2}e^{2\beta - 8\gamma} + \binom{12}{3}e^{3\beta - 12\gamma}\right\}. \qquad (44)$$

To avoid unduly large numbers, it might often be convenient to divide each term in the logarithm by a suitable constant, but this was not done in the present case.

The maximum-likelihood estimate of β when $\gamma = 0$ is $\hat{\beta} = 1.65$. Thus the ratio of the hazards is estimated as $e^{\hat{\beta}} = 5.21$; if the distributions were exponential, this would be the ratio of means. Confidence limits for β, subject to $\gamma = 0$, can be obtained either by computing the second derivative $\mathscr{I}(\hat{\beta})$ or directly from the log likelihood. With the latter method, approximate 95 per cent confidence limits for β of $(0.78, 2.60)$ are obtained from those values for which the log likelihood is within $\frac{1}{2} \times 1.96^2 = 1.92$ of its maximum value. An alternative test of the null hypothesis $\beta = 0$ is obtained by comparing the log likelihood at $\beta = 0$ and $\beta = \hat{\beta}$; the difference of 7.43 corresponds to chi-squared of 14.9 and hence to a standardized deviate of 3.86, in reasonable agreement with test based on $U(0)$.

The inclusion of the extra parameter γ provides a test of the adequacy of the assumption of simply related hazards. In fact the additional log likelihood achieved by the extra parameter, about 0.01, is small, even suspiciously small. Confidence limits for γ are, at the 95 per cent level, approximately 0.12 and 0.14. Thus any marked departure from the proportional hazard model is not likely to be a smooth monotonic change with t. Further details of the likelihood function will not be given here. It is, however, quadratic to a close approximation and the particular parametrization chosen achieved almost exact orthogonality.

Finally, we consider graphical techniques, which are likely to be particularly useful for data more extensive than the present set. A first step is to obtain unconditional estimates of the separate survivor functions by (8). For sample 1 this gives the ordinary sample survivor function, there being no censoring. For sample 0, we get the product limit estimate. Now consider estimation of the survivor functions under the model of proportional hazards; the constrained maximum-likelihood estimates of the survivor functions in the two samples are given by (37) and (38). Iterative solution of the 17 equations of the form (37) took in all $\frac{1}{2}$ sec. on the CDC 6600; $\bar{z}$ was chosen separately for each risk set so that $e^{\hat{\beta}\bar{z}}$ equalled the mean of $e^{\hat{\beta}z}$ over the risk set in question.

Fig. 1 shows the four estimated functions. Discrepancy with the model of proportional hazards would be shown by clear departures of the conditional from the unconstrained survivor curves. More elaborate versions of this analysis are certainly possible, in which, for instance, plots are made on a non-linear scale, or in which residuals from the constrained fit are formed, or in which the analysis is presented in tabulated rather than graphical form. The graphical analysis confirms the consistency of the data with a model of proportional hazards.

Only a very brief note will be added here about alternative approaches to the analysis. If exponential distributions are assumed the relevant statistics are the total periods at risk, namely 359 weeks and 182 weeks, and the total numbers of failures 9 and 22 respectively. Approximate 95 per cent confidence limits for the log ratio of means can be obtained via the F distribution with $(18, 44)$ degrees of freedom. They are 0.83 and 2.43, as compared with 0.78 and 2.60 from the earlier analysis.

An analysis with a step function for $\lambda_0(.)$ is barely feasible with the limited amount of data available. The procedure is to divide the time scale into cells, for instance 0–10 weeks and 11–20 weeks. Numbers of failures and periods at risk are calculated for each cell and hence ratios of rates derived. Provided they are consistent for the different cells the ratios can then be combined into a single summary statistic with confidence limits. In the present example this approach does not lead to essentially different conclusions.

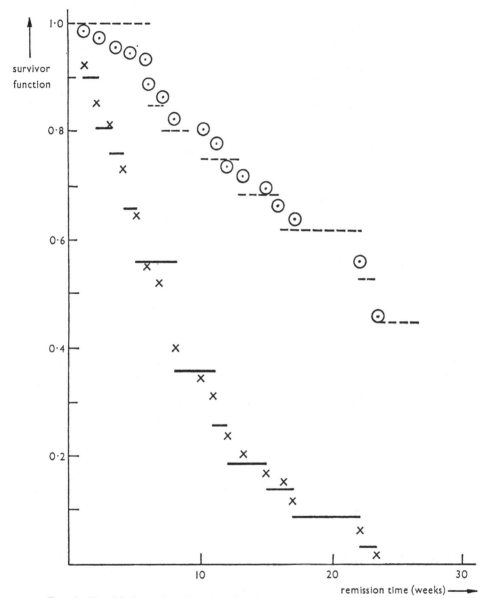

Fig. 1. Empirical survivor functions for data of Table 1. Product limit estimate, – – – – – –, sample 0 (6-MP); ———, sample 1 (control). Estimate constrained by proportionality: ⊙, sample 0; ×, sample 1. For clarity, the constrained estimates are indicated by the left ends of the defining horizontal lines.

A third possibility is the use of the Weibull distribution. If we assume a common index in the two samples we may fit by maximum-likelihood distribution functions in the form

$$1-\exp\{-(\rho x/\kappa)^{\nu}\}, \quad 1-\exp\{-(\kappa\rho x)^{\nu}\}.$$

The maximum-likelihood estimate of the index is $\hat{\nu} = 1\cdot3$ and the maximized log-likelihoods show that this is just significantly different from $\nu = 1\cdot0$ at the 5 per cent level. The explanation of the departure probably lies largely in the deficiency of small failure times in sample 0. Fitting of different indexes for the two samples has not been attempted. Approximate 95 per cent confidence limits for the log ratio of means can be derived in the usual way from the maximized log likelihoods and are $0\cdot71$ and $2\cdot10$; the maximum-likelihood estimate is $\log(\hat{\kappa}^2) = 1\cdot31$.

The data have been analysed in some detail to illustrate a number of relevant points. Many applications are likely to be more complicated partly because of larger sample sizes and partly because of the presence of a number of explanatory variables.

11. Physical Interpretation of Model

The model (9), which is the basis of this paper, is intended as a representation of the behaviour of failure-time that is convenient, flexible and yet entirely empirical. One of the referees has, however, suggested adding some discussion of the physical meaning of the model and in particular of its possible relevance to accelerated life testing. Suppose in fact that there is a variable s, called "stress", and that life tests are carried out at various levels of s. For simplicity we suppose that s is one-dimensional and that each individual is tested at a fixed level of s. The usual idea is that we are really interested in some standard stress, say $s = 1$, and which to use other values of s to get quick laboratory results as a substitute for a predictor of the expensive results of user trials.

Now in order that the distribution of failure-time at one level of stress should be related to that at some other level, the relationship being stable under a wide range of conditions, it seems necessary that the basic physical process of failure should be common at the different stress levels; and this is likely to happen only when there is a single predominant mode of failure. One difficulty of the problem is that of knowing enough about the physical process to be able to define a stress variable, i.e. a set of test conditions, with the right properties.

One of the simplest models proposed for the effect of stress on the distribution of failure-time is to assume that the mechanism of failure is identical at the various levels of s but takes place on a time-scale that depends on s. Thus if $\mathscr{F}(t; s)$ denotes the survivor function at stress s, this model implies that

$$\mathscr{F}(t; s) = \mathscr{F}\{g(s)\,t; 1\}, \tag{45}$$

where $g(s)$ is some function of s with $g(1) = 1$. Thus the hazard function at stress s is

$$g(s)\,\lambda_0\{g(s)\,t\}, \tag{46}$$

where $\lambda_0(.)$ is the hazard at $s = 1$. In particular if $g(s) = s^{\beta}$ and if $z = \log s$ this gives

$$e^{\beta z}\,\lambda_0(e^{\beta z}\,t). \tag{47}$$

This is similar to but different from the model (9) of this paper. A special set of conditions where (47) applies is where the individual is subject to a stream of shocks of randomly varying magnitudes until the cumulative shock exceeds some time-independent tolerance. If, for instance, all aspects of the process except the rate of incidence of shocks are independent of s, then (45) will apply.

If, however, the shocks are non-cumulative and failure occurs when a rather high threshold is first exceeded, failures occur in a Poisson process with a rate depending on s. A special model of this kind often used for thermal stress is to suppose that failure corresponds to the excedence of the activation energy of some process; then by the theory of rate processes (47) can be used with $\lambda_0(.) = 1$ and z equal to the reciprocal of absolute temperature.

As a quite different model suppose that some process of ageing goes on independently of stress. Suppose further that the conditional probability of failure at any time is the product of an instantaneous time-dependent term arising from the ageing process and a stress-dependent term; the model is non-cumulative. Then the hazard is

$$h(s)\,\lambda_0(t), \tag{48}$$

where $h(s)$ is some function of stress. Again if $h(s) = s^\beta$, the model becomes

$$e^{\beta z}\,\lambda_0(t) \tag{49}$$

exactly that of (9), where again $\lambda_0(t)$ is the hazard function at $s = 1$, $z = 0$. One special example of this model is rather similar to that suggested for (46), except that the critical tolerance varies in a fixed way with time and the shocks are non-cumulative, the rate of incidence of shocks depending on s. For another possibility, see Shooman (1968).

If hazard or survivor functions are available at various levels of s we might attempt an empirical discrimination between (46) and (48). Note, however, that if we have a Weibull distribution at $s = 1$, $\lambda_0(.)$ is a power function and (46) and (48) are identical. Then the models cannot be discriminated from failure-time distributions alone. That is, if we did want to make such a discrimination we must look for situations in which the distributions are far from the Weibull form. Of course the models outlined here can be made much more specific by introducing explicit stochastic processes or physical models. The wide variety of possibilities serves to emphasize the difficulty of inferring an underlying mechanism indirectly from failure times alone rather than from direct study of the controlling physical processes.

As a basis for rather empirical data reduction (9), possibly with time-dependent exponent, seems flexible and satisfactory.

ACKNOWLEDGEMENTS

I am grateful to the referees for helpful comments and to Professor P. Armitage, Mr P. Fisk, Dr N. Mantel, Professors J. W. Tukey and M. Zelen for references and constructive suggestions.

REFERENCES

Breslow, N. (1970). A generalized Kruskal–Wallis test for comparing samples subject to unequal patterns of censoring. *Biometrika*, **57**, 579–594.
Chernoff, H. (1962). Optimal accelerated life designs for estimation. *Technometrics*, **4**, 381–408.
Chiang, C. L. (1968). *Introduction to Stochastic Processes in Biostatistics.* New York: Wiley.

COCHRAN, W. G. (1954). Some methods for strengthening the common χ^2 tests. *Biometrics*, **10**, 417–451.

COX, D. R. (1959). The analysis of exponentially distributed life-times with two types of failure. *J. R. Statist. Soc. B*, **21**, 411–421.

—— (1964). Some applications of exponential ordered scores. *J. R. Statist. Soc. B*, **26**, 103–110.

—— (1972). The statistical analysis of dependencies in point processes. In *Symposium on Point Processes* (P. A. W. Lewis, ed.). New York: Wiley (to appear).

COX, D. R. and LEWIS, P. A. W. (1972). Multivariate point processes. *Proc. 6th Berkeley Symp.* (to appear).

EFRON, B. (1967). The two sample problem with censored data. *Proc. 5th Berkeley Symp.*, **4**, 831–853.

GEHAN, E. A. (1965). A generalized Wilcoxon test for comparing arbitrarily single-censored samples. *Biometrika*, **52**, 203–224.

GRENANDER, U. (1956). On the theory of mortality measurement, I and II. *Skand. Akt.*, **39**, 90–96, 125–153.

KAPLAN, E. L. and MEIER, P. (1958). Nonparametric estimation from incomplete observations. *J. Am. Statist. Assoc.*, **53**, 457–481.

LEHMANN, E. L. (1953). The power of rank tests. *Ann. Math. Statist.*, **24**, 23–43.

MANTEL, N. (1963). Chi-square tests with one degree of freedom: extensions of the Mantel–Haenszel procedure. *J. Am. Statist. Assoc.*, **58**, 690–700.

—— (1966). Evaluation of survival data and two new rank order statistics arising in its consideration. *Cancer Chemotherapy Reports*, **50**, 163–170.

MANTEL, N. and HAENZEL, W. (1959). Statistical aspects of the analysis of data from retrospective studies of disease. *J. Nat. Cancer Inst.*, **22**, 719–748.

PETO, R. and PETO, J. (1972). Asymptotically efficient rank invariant test procedures. *J. R. Statist. Soc. A* **135**, 185–206.

PRATT, J. W. (1962). Contribution to discussion of paper by A. Birnbaum. *J. Am. Statist. Assoc.*, **57**, 314–316.

SAMPFORD, M. R. (1952). The estimation of response-time distributions, II: Multi-stimulus distributions. *Biometrics*, **8**, 307–369.

—— (1954). The estimation of response-time distribution, III: Truncation and survival. *Biometrics*, **10**, 531–561.

SAMPFORD, M. R. and TAYLOR, J. (1959). Censored observations in randomized block experiments. *J. R. Statist. Soc. B*, **21**, 214–237.

SAVAGE, I. R. (1956). Contributions to the theory of rank order statistics—the two-sample case. *Ann. Math. Statist.*, **27**, 590–615.

SHOOMAN, M. L. (1968). Reliability physics models. *IEEE Trans. on Reliability*, **17**, 14–20.

WATSON, G. S. and LEADBETTER, M. R. (1964a). Hazard analysis, I. *Biometrika*, **51**, 175–184.

WATSON, G. S. and LEADBETTER, M. R. (1964b). Hazard analysis, II. *Sankhyā*, A, **26**, 101–116.

Summary of discussion

When the paper was read to Royal Statistical Society a vote of thanks was proposed by F. Downton and seconded by R. Peto and the discussion continued by D. J. Bartholomew, D. Oakes, D. V. Lindley and P. W. Glassborow. Written contributions were received by D. E. Barton, S. Howard, B. Benjamin, J. J. Gart, L. D. Meshalkin and A. R. Kagan, M. Zelen, R. E. Barlow, J. Kalbfleisch and R. L. Prentice, and by N. Breslow. A brief reply concluded the paper.

Downton explores the connection with the theory of rank tests and Lehmann alternatives, notably in the context of an analogue of a randomized block design. Peto emphasizes the connection with score tests and develops with more care than did the paper the procedure for dealing with tied values. Bartholomew points out potential application to labour turnover issues and asks about the relation between the log-normal-based analyses traditional in that field at the time and the methods of the paper. Oakes suggests an improved method of estimating the baseline hazard, as does Breslow. Lindley examines the two-sample problem in detail and questions the appropriateness of the likelihood used in the paper. Glassborow dislikes the term 'censoring' as used in the paper and correctly questions the assumption of uninformative censoring that is so widely used in this field.

Barton discusses the property that the Kaplan–Meier estimate is a maximum likelihood estimate in which effectively a very large number of parameters had been estimated. The importance of this property had been questioned in the paper and was defended by Barton; the point remains of theoretical interest. Howard raises the important issue of computational implementation. Benjamin discusses the attitude of and contributions of actuaries to the analysis of survival data. Gart and Zelen both discuss the connection of the methods in the paper with the analysis of contingency tables. Meshalkin and Kagan discuss methods based on parametric forms of hazard function. Barlow points out the connection with the cumulative total time on test statistic. Kalbfleisch and Prentice, and separately Breslow, question the calculation of the likelihood that is a primary tool in separating the estimation of the regression part of the model from the unknown and assumed arbitrary baseline hazard. The paper was obscure on that point and in the reply the argument was clarified at a qualitative level, subsequent further clarification coming from Paper Th8 in Volume II, page 175.

Appl. Statist. (1979),
28, *No.* 1, *pp.* 73–75

Miscellanea

A Note on Multiple Time Scales in Life Testing

By V. T. Farewell   and   D. R. Cox

Fred Hutchinson Cancer Research Center,
1124 Columbia Street,
Seattle, Washington 98104, U.S.A.

Department of Mathematics,
Imperial College,
London SW 7.

[Received August 1977. Revised August 1978]

Summary

A model is proposed to study life tests where failures are recorded on more than one time scale. A single time scale which simply explains the failure pattern is sought. A medical example is given.

Keywords: PROPORTIONAL HAZARD; SURVIVAL DATA; RELIABILITY

1. Introduction

In life testing, observations, usually representing some sort of failure, may be recorded along more than one "time" axis. In a study of the failure pattern of electric light bulbs, for example, it may be possible to record, at the time of failure, both the total length of time for which the bulb has been on and the number of times it has been switched on. The incidence time for breast cancer may be recorded as chronological age or as age since some major hormonal event, such as the birth of the woman's first baby.

If one scale is clearly of primary practical importance, then others can, if necessary, be introduced into a model as time-dependent covariates. It may, however, not be clear which time scale is the more important or whether a combination of time scales could be particularly informative. This note gives a procedure for studying such a situation, assuming for simplicity that just two time scales are involved, and, again for simplicity, ignoring censoring and possible explanatory variables, including treatment effects.

The key idea is to search, in the space defined by the two scales, for a pair of axes, one of which is in the positive quadrant, such that the failure pattern along the positive quadrant axis is explained without reference to the other axis. Such a time scale does not necessarily exist nor, if one does, is it necessarily unique. That is, we look for a time scale that accounts for as much of the variation as possible, in some sense. In some contexts, it might be required to find a time scale on which failures form a Poisson process.

2. A Model

For each individual let there be two evolving time scales, t_1 and t_2, on which failure is recorded. We assume that not all individuals follow the same trajectory in the (t_1, t_2) plane and that individual trajectories are known. Some special cases are the deterministic trajectories $t_2 = a_i t_1$, $t_2 = t_1 + b_i$, $t_2 = a t_1 + b_i$ and $t_2 = a_i t_1 + b_i$, where a_i and b_i are constant for each individual varying between individuals.

Consider a new time axis $t_\theta = t_1 \cos \theta + t_2 \sin \theta$ and associate with it, to represent the second dimension, the orthogonal coordinate axis, $x_\theta = t_1 \sin \theta - t_2 \cos \theta$. For t_θ to be a meaningful time scale, we require that t_θ should increase as an individual follows a trajectory moving "northeasterly" in the (t_1, t_2) plane. For this in general we need $0 \leqslant \theta \leqslant \pi/2$ but in particular cases a wider interval of θ values may be used. Thus in the example to be treated later, where the trajectories are $t_2 = t_1 - b_i$, the condition of monotonicity is $\cos\theta + \sin\theta > 0$, i.e. $-\pi/4 < \theta < 3\pi/4$.

To find a suitable single time scale, we examine the hypothesis that there is no dependence on x_θ. There are various ways in which this might be done, one being *via* the hazard function (Cox, 1972)

$$\lambda(t_\theta; x_\theta) = \lambda_0(t_\theta)\exp(\beta x_\theta), \tag{1}$$

where $\lambda_0(.)$ is an unspecified function and β is a scalar parameter. The particular choice of the orthogonal direction for the second direction is to avoid ill-conditioning. Clearly, any distinct direction would serve for the second variable.

If $\beta = 0$ then failure along t_θ can be simply modelled, ignoring x_θ. The hypothesis that $\beta = 0$ can be examined for all relevant θ. At a significance level α a confidence region for θ can be defined which gives all linear time scales for which the failure pattern is simply explained. It is possible that no θ is in the region or that all θ are. Linear combinations of t_1 and t_2 are used for convenience only and the procedure generalizes to more than two time scales. Note that the particular approach used here treats t_θ and x_θ quite asymmetrically; it allows arbitrary dependence on t_θ and then looks for a very "smooth" form for any remaining dependence on x_θ.

3. AN EXAMPLE

As a simple illustration we examine the data of Table 1(a) for a group of 34 parous women who have developed breast cancer. We consider two time scales for defining occurrence,

TABLE 1

Data on 34 parous women developing breast cancer

(a) Age at first birth; age at development of cancer

First birth	Cancer	First birth	Cancer	First birth	Cancer
42	53	26	68	27	59
32	65	32	57	27	64
21	52	34	59	28	41
26	53	17	59	28	66
20	58	29	56	28	43
25	56	33	45	21	54
29	61	25	51	27	49
23	56	36	55	26	63
20	52	33	55	24	58
29	53	27	59	39	62
28	58	30	61	28	46
27	66				

(b) Estimated dependence on x_θ†

θ	β (yr^{-1})	Standard error of β
$-\frac{1}{3}\pi$	$-0\cdot126$	$0\cdot023$
$-\frac{1}{4}\pi$	$-0\cdot070$	$0\cdot024$
0	$0\cdot005$	$0\cdot036$
$\frac{1}{8}\pi$	$0\cdot061$	$0\cdot049$
$\frac{1}{4}\pi$	$0\cdot112$	$0\cdot054$
$\frac{3}{8}\pi$	$0\cdot140$	$0\cdot051$
$\frac{1}{2}\pi$	$0\cdot150$	$0\cdot043$
$\frac{5}{8}\pi$	$0\cdot146$	$0\cdot030$
$\frac{7}{10}\pi$	$0\cdot154$	$0\cdot026$

† t_1 is the age at cancer, t_2 is age at cancer minus age at first birth,
$$x_\theta = t_1 \sin\theta - t_2 \cos\theta.$$

chronological age denoted t_1 and age since the birth of the woman's first child denoted t_2. It is widely believed that a woman's age at first birth is important in determining the incidence of breast cancer. This example corresponds to simple deterministic trajectories $t_2 = t_1 - b_i$.

Table 1(b) gives estimates of β for a number of values of θ. For the hazard function model $\lambda_0(t_1) \exp(\beta t_2)$ which corresponds to model (1) with $\theta = 0$, the data are entirely consistent with $\beta = 0$. On the other hand, the model $\lambda_0(t_2) \exp(\beta t_1)$ corresponding to $\theta = \frac{1}{2}\pi$ produces strong evidence of non-zero β.

The results indicate that the simplest model for time of occurrence uses the chronological age scale. If age since first birth were used, we would need chronological age as a covariate in order to describe the time pattern of breast cancer. Of course it would be rash to put much weight on conclusions drawn from one small sample and, in any case, for a full interpretation it is essential to consider also the women in the study who did not develop breast cancer.

REFERENCE

Cox, D. R. (1972). Regression models and life tables (with discussion). *J. R. Statist. Soc.* B, **34**, 187–220.

Statistical Science
1993, Vol. 8, No. 3, 204-283

Linear Dependencies Represented by Chain Graphs

D. R. Cox and Nanny Wermuth

Abstract. Various special linear structures connected with covariance matrices are reviewed and graphical methods for their representation introduced, involving in particular two different kinds of edge between the nodes representing the component variables. The distinction between decomposable and nondecomposable structures is emphasized. Empirical examples are described for the main possibilities with four component variables.

Key words and phrases: Chain model, conditional independence, covariance selection, decomposable model, linear structural equation, multivariate analysis, path analysis.

1. INTRODUCTION

This paper has three broad objectives. The first is to illustrate the rich variety of special forms of association and dependence that can arise even with as few as three or four variables. The second is to show the value of graphical representation in clarifying these dependencies; for this we introduce graphs with two different kinds of edge and some further features which are also new. The third objective is to show the importance in interpretation of the distinction between decomposable and nondecomposable models.

A series of examples will be used in illustration, partly to show that many of the special structures do indeed arise in applications and partly to show in outline the implications for interpretation, although reference to the subject matter literature is necessary for a full account. Most of the examples arise from recent investigations at University of Mainz. For purposes of exposition we have chosen examples with at most four variables; that is, we have simplified by omitting mention of variables which analysis had shown to have no bearing on the points at issue.

We confine the discussion to those problems with essentially linear structure in which the interrelationships are adequately captured by the covariance matrix of the variables. Of course in applications, checks for nonlinearities and outliers are required, and these have been done for all examples whenever we had access to the raw data.

D. R. Cox is Warden, Nuffield College, Oxford OX1 1NF, United Kingdom. Nanny Wermuth is Professor, Psychologisches Institut, Johannes Gutenberg–Universität Mainz, Postfach 3980, D-55099 Mainz, Germany.

The need to discuss special structures arises partly because the relations of marginal independence and conditional independence expressed thereby are often of substantive interest and partly because in a *saturated model* with p component variables, that is, one in which the covariance matrix is unrestricted other than to being positive definite, there are $1/2p(p-1)$ correlations, and reduction of dimensionality may be desirable to avoid a superabundance of parameters.

There are strong connections with, in particular, the long history of work in path analysis in genetics, in simultaneous equations in econometrics and linear structural models in psychometrics and with the body of recent work applying graph-theoretic ideas to the study of systems of conditional independencies arising especially in the study of expert systems.

In Section 2 we review some general properties of linear regression systems as related to the covariance matrix of the variables and stress the distinction between multivariate regression and block regression and between decomposable and nondecomposable structures. In Section 3 we introduce the main conventions useful in a graph-theoretic representation of the independency relations that may hold; in Section 4 we discuss relations with previous work, and in Section 5 we give a series of empirical examples for four variables. The paper concludes in Section 6 with some general discussion. The emphasis throughout is on the structure and interpretation of the various models rather than on the procedures for fitting.

2. SOME PROPERTIES OF COVARIANCE MATRICES

It is convenient to set out some properties of systems of linear least squares regressions derivable from a covariance matrix. These are full regression equations in a multivariate normal distribution. There is throughout the usual interplay between relatively weak second-order properties of least squares regression and the strong properties derivable from an assumption of multivariate normality, such as that zero correlation or zero partial correlation implies independence or conditional independence.

We consider first the $p \times 1$ vector $Y = (Y_1, \ldots, Y_p)^T$ with mean $E(Y) = \mu$. We denote the positive definite covariance matrix by $\text{cov}(Y) = \Sigma$, and its inverse, the concentration matrix, therefore by Σ^{-1}; the diagonal elements of Σ are the variances (σ_{ii}), those of Σ^{-1} are the precisions (σ^{ii}). The off-diagonal elements of Σ are the covariances (σ_{ij}), those of Σ^{-1} are the concentrations (σ^{ij}). A marginal correlation ρ_{ij} is expressible via elements of the covariance matrix, in a way similar to that in which a partial correlation, $\rho_{ij.k}$, given all of the remaining variables $k = \{1, \ldots, p\} \setminus \{i, j\}$, is expressible via elements of the concentration matrix:

$$\rho_{ij} = \sigma_{ij} (\sigma_{ii} \sigma_{jj})^{-1/2}, \quad \rho_{ij.k} = -\sigma^{ij} (\sigma^{ii} \sigma^{jj})^{-1/2}.$$

This implies in particular that in the usual notation (Dawid, 1979a) for independence,

$$Y_i \perp\!\!\!\perp Y_j, \quad \text{if and only if } \sigma_{ij} = 0,$$
$$Y_i \perp\!\!\!\perp Y_j \mid Y_k, \quad \text{if and only if } \sigma^{ij} = 0,$$

where as above $k = \{1, \ldots, p\} \setminus \{i, j\}$.

To study regression models, we partition Y into Y_a and Y_b, $p_a \times 1$ and $p_b \times 1$, respectively, $p_a + p_b = p$, and call the two parts response and explanatory variables. Let the covariance matrix and the concentration matrix be conformally partitioned:

$$(1) \quad \Sigma = \begin{pmatrix} \Sigma_{aa} & \Sigma_{ab} \\ \cdot & \Sigma_{bb} \end{pmatrix}, \Sigma^{-1} = \begin{pmatrix} \Sigma^{aa} & \Sigma^{ab} \\ \cdot & \Sigma^{bb} \end{pmatrix},$$

then the covariance matrix Σ_{bb} of the explanatory variables and correspondingly their concentration matrix $\Sigma_{bb}^{-1} = \Sigma^{bb.a} = \Sigma^{bb} - (\Sigma^{ab})^T (\Sigma^{aa})^{-1} \Sigma^{ab}$ do not contain parameters needed to specify a standard regression model of Y_a on Y_b. Instead, their observed counterparts are taken as fixed or indeed sometimes are fixed by sampling design.

We now distinguish between a multivariate regression and a block regression. To simplify the notation we shall without essential loss of generality take often $E(Y) = 0$. We describe the distinct parameters in the two types of regression models, that is, the two ways of parametrizing the conditional distribution of Y_a given Y_b. For a *multivariate regression* of Y_a on Y_b, that is, for $Y_a = \Pi_{a|b} Y_b + \varepsilon_a$ with $E(\varepsilon_a) = 0$, $E(\varepsilon_a Y_b^T) = 0$, the regression equation parameters $\Pi_{a|b}$ and the residual variance $\text{var}(\varepsilon_a)$ can be written in a matrix as $(\Sigma_{aa.b}, \Pi_{a|b})$, where

$$(2) \quad \begin{aligned} \Pi_{a|b} &= \Sigma_{ab} \Sigma_{bb}^{-1}, \\ \text{var}(\varepsilon_a) &= \Sigma_{aa.b} = \Sigma_{aa} - \Sigma_{ab} \Sigma_{bb}^{-1} \Sigma_{ab}^T. \end{aligned}$$

In a saturated multivariate regression (2) each component of Y_a is regressed separately on the full set of components Y_b.

On the other hand in a saturated *block regression* each component of Y_a is regressed not only on Y_b but also on all remaining components of Y_a. Then the regression equation parameters are instead proportional to the elements of the matrix $(\Sigma^{aa}, \Sigma^{ab})$ (Wermuth, 1992). The reason is that the expected value of a component Y_i of Y_a given all remaining variables of Y can be obtained by taking expectations in

$$(3) \quad \Sigma^{aa} Y_a + \Sigma^{ab} Y_b = \omega_a$$

where $E(\omega_a) = 0$, $\text{var}(\omega_a) = \Sigma^{aa}$ and dividing the ith equation by the concentration σ^{ii}. Equation (3) is derived from a block triangular decomposition of the concentration matrix, $\Sigma^{-1} = A^T T^{-1} A$, where

$$(4) \quad A = \begin{pmatrix} I_{aa} & (\Sigma^{aa})^{-1} \Sigma^{ab} \\ 0 & I_{bb} \end{pmatrix},$$

$$T^{-1} = \begin{pmatrix} \Sigma^{aa} & 0 \\ 0 & \Sigma^{bb.a} \end{pmatrix},$$

as the first p_a equations of $(T^{-1} A)(Y - E(Y)) = \omega$. The residuals ω have zero mean and covariance matrix T^{-1}.

For a block regression, the resulting coefficient of variable Y_j in the ith equation is minus a partial regression coefficient given all remaining variables of Y, that is, given all remaining response and explanatory variables. On the other hand, in a multivariate regression the coefficient of Y_j in the ith equation is a partial regression coefficient given all remaining variables of Y_b, that is, given all remaining explanatory variables. To express this distinction more formally, we write a partial regression coefficient $\beta_{ij.d}$ for $\{1, \ldots, p\} = a \cup b = \{\{i, j\}, d, g\}$ in terms of elements of the conditional covariance matrix of (Y_i, Y_j) given Y_d and of elements of the concentration matrix of (Y_i, Y_j), having marginalized over Y_g, as

$$\beta_{ij.d} = \frac{\sigma_{ij.d}}{\sigma_{jj.d}} = -\frac{\sigma^{ij.g}}{\sigma^{ii.g}}.$$

Note that in the case of a block regression g is empty and d is the set of all remaining variables of Y, that is, $d = (a \cup b) \setminus \{i, j\}$, while in the case of a multivariate regression $d = b \setminus \{j\}$, and $g = a \setminus \{i\}$. Note further that

$$(5) \quad Y_i \perp\!\!\!\perp Y_j \mid Y_d, \quad \text{if and only if } \beta_{ij.d} = 0.$$

To judge the relative strength of the dependence of a response on several explanatory variables, it is sometimes useful to compare the standardized regression coefficients, that is, $\beta_{ij.d}^* = \beta_{ij.d} \sigma_{jj}^{\frac{1}{2}} \sigma_{ii}^{-\frac{1}{2}}$.

One of the major distinctions between multivariate regression and block regression lies in the meaning of the relation between two components Y_i and Y_j, both within Y_a, and in the meaning of the relation of a component Y_i from Y_a to a component Y_j from Y_b. To describe this in detail it is useful to recall how a partial regression coefficient relates to a partial correlation coefficient

$$\beta_{ij.d} = \rho_{ij.d}\sqrt{\frac{\sigma_{ii.d}}{\sigma_{jj.d}}} = \rho_{ij.d}\sqrt{\frac{\sigma^{jj.g}}{\sigma^{ii.g}}}.$$

Thus, in a block regression, that is, where $d = (a \cup b)\backslash\{i, j\}$, the relation between Y_i from Y_a and Y_j is measured essentially by the partial correlation given all remaining variables of Y, no matter whether Y_j is from Y_a or it is from Y_b. By contrast in a multivariate regression, that is, where $d = b\backslash\{j\}$, the measure of the relation of Y_i from Y_a to Y_j from Y_b is proportional to the partial correlation given the variables in Y_b other than Y_j; the correlation between Y_i and Y_j both within Y_a is given all variables in Y_b. Thus, a larger set of variables is considered simultaneously in block regression if compared with the corresponding multivariate regression. Written in matrix notation their parameters are related by

$$(6) \quad \Pi_{a|b} = -(\Sigma^{aa})^{-1}\Sigma^{ab}, \Sigma_{aa.b} = (\Sigma^{aa})^{-1}$$

$$(7) \quad \Sigma^{ab} = -(\Sigma_{aa.b})^{-1}\Pi_{a|b}, \Sigma^{aa} = (\Sigma_{aa.b})^{-1}.$$

Some of the special models we shall consider correspond to specifying some elements of regression equations to be zero, that is, to structures that appear simplified if compared with the saturated model. The choice between block regression and multivariate regression is then largely determined by the research questions and by a decision as to which of the two parametrizations permits a simpler description of the relations. For instance, in each of Examples 1, 2 and 7 of the empirical examples of Section 5 we can think of two variables as joint responses, $Y_a = (Y, X)^T$, and of two variables as explanatory, $Y_b = (V, W)^T$. A simplifying description is possible with block regression but not with multivariate regression in Example 1, while a simpler structure results with multivariate regression than with block regression in Examples 2 and 7.

If not only the conditional distribution of Y_a given Y_b is of interest, but the marginal relations among component variables within Y_b as well, we are led to a simple type of regression chain model: we specify the joint density via

$$f_{ab} = f_{a|b}f_b,$$

and make a choice for $f_{a|b}$ among a multivariate and a block regression.

A specification of the joint distribution of Y_a, Y_b by a saturated *multivariate regression chain model* has $(\Sigma_{aa.b}, \Pi_{a|b})$ as parameters for the conditional distribution of Y_a given Y_b and Σ_{bb} for the marginal distribution of Y_b. With a saturated *block regression chain model* the parameters are the regression coefficients obtained

as described above from $(\Sigma^{aa}, \Sigma^{ab})$ and the concentration matrix $\Sigma^{bb.a} = \Sigma_{bb}^{-1}$.

Considering, for instance, a multivariate regression chain model instead of a multivariate regression model can lead to a simpler structure. This is the case in Example 7 but not in Example 2 of Section 5 since the explanatory variables can be taken to be marginally uncorrelated in the former but not in the latter.

In the next more complex regression chain model the joint density of three (vector) variables Y_a, Y_b and Y_c is specified via

$$f_{abc} = f_{a|bc}f_{b|c}f_c,$$

that is, via a regression of Y_a on Y_b and Y_c, a regression of Y_b on Y_c and the marginal distribution of Y_c. This would be an adequate approach if the components of Y_a are the response variables of primary interest having Y_b and Y_c as potential explanatory variables, if Y_b plays the role of an intermediate variable containing potentially explanatory components for Y_a and possible responses to Y_c and, finally, if Y_c consists of explanatory variables whose joint distribution is to be analyzed.

A particularly important family of regression chains are the *univariate recursive regressions* in which, for a given ordering of the components of $Y = (Y_1, \ldots, Y_q)^T$, we define the model via the regression of Y_r on $Y_{r+1}, \ldots, Y_p$ for $r = 1, \ldots, q$; $q \leq p-1$. An independence hypothesis is said to be *decomposable* if it specifies one or more of the regression coefficients in such a system to be zero. Early descriptions of univariate recursive regressions have been given by Wright (1921, 1923) with an emphasis on applications in genetics and by Tinbergen (1937) for the study of business cycles.

By contrast a *nondecomposable independence hypothesis* consists of a set of k independence relations for k distinct variable pairs that cannot, in its entirety, be reexpressed in terms of vanishing coefficients in the above form: that is, no ordering of the variables would produce a decomposable independence hypothesis with the same implications from the same distributional assumption. The following arguments apply provided that there are no so-called forbidden states, that is, states of zero probability (Dawid, 1979a).

For instance, for a trivariate normal distribution of Y, Z, X the hypothesis $Y \perp\!\!\!\perp X \mid Z$ and $X \perp\!\!\!\perp Z \mid Y$ corresponds to zero concentrations for pairs (Y, X) and (X, Z) and it implies $X \perp\!\!\!\perp (Y, Z)$. This hypothesis can be reexpressed by $Y \perp\!\!\!\perp X \mid Z$ and $X \perp\!\!\!\perp Z$ corresponding to $\beta_{yx.z} = \beta_{xz} = 0$ in a univariate recursive system for $(Y, X, Z)^T$. Thus the hypothesis is decomposable even though initially not expressed in that form. On the other hand, no ordering of the variables would permit us to specify the hypothesis $Y \perp\!\!\!\perp X$ and $Z \perp\!\!\!\perp U$ as zero restrictions in a univariate recursive regression system. Thus the hypothesis is nondecomposable. Further examples for nondecomposable hypotheses are discussed in Section 5.

They arise in applications with four or more variables, as we shall see below, but suffer from a number of disadvantages both in terms of the difficulty of fitting, but more importantly, in terms of indirectness of interpretation. The need for such models was noted by Haavelmo (1943) who pointed out substantive research questions about relations which form a system of equations to be fulfilled simultaneously, but which are not a system of univariate recursive regressions. His subject matter example is as follows: consumption in an economy per year depends on total income, investment per year depends on consumption and total income is the sum of consumption and investment. A slightly simplified version of Haavelmo's argument for the simultaneous treatment of equations is given in Section 4. As a consequence of his results, the class of linear structural equations was developed to study simultaneous relations. It is mainly discussed in econometrics (Goldberger, 1964), in psychometrics (Jöreskog, 1973) and in sociology (Duncan, 1969); it includes univariate recursive regression systems and multivariate regressions as a subclass but, in general, a zero coefficient in a structural equation does not correspond to an independence relation. More generally the graphical representations to be introduced in Section 3 are equivalent to those used in path analysis and in discussions of structural equations only in rather special cases. We deal with this important point further in Section 4.

A representation in terms of univariate recursive regressions combines several advantages. First, and most importantly, it describes a stepwise process by which the observations could have been generated and in this sense may prove the basis for developing potential causal explanations. Second, each parameter in the system has a well-understood meaning since it is a regression coefficient: that is, it gives for unstandardized variables the amount by which the response is expected to change if the explanatory variable is increased by one unit and all other variables in the equation are kept constant. As a consequence, it is also known how to interpret each additional zero restriction: in the case of jointly normal variables, each added restriction introduces a further conditional independence, and it is known how parameters are modified if variables are left out of a system (Wermuth, 1989). Third, general results are available for interpreting structures, that is, for reading all implied independencies directly off a corresponding graph (Pearl, 1988; Lauritzen et al., 1990) and for deciding from the graphs of two distinct models whether they are equivalent (Frydenberg, 1990a). Fourth, an algorithm exists (Pearl and Verma, 1991; Verma and Pearl, 1992) which decides for arbitrary probability distributions and an almost arbitrary list of conditional independence statements whether the list defines a univariate recursive system; if it does, a corresponding directed acyclic

graph is drawn. Fifth, the analysis of the whole association structure can be achieved with the help of a sequence of separate univariate linear regression analyses (Wold, 1954).

The word *causal* is used in a number of different senses in the literature; for a review see Cox (1992). Glymour et al. (1987) and Pearl (1988) have developed valuable procedures for finding relatively simple structures of conditional independencies which they define to be causal. We prefer to restrict the word to situations where there is some understanding of an underlying process. From this perspective it is unrealistic to think that causality could be established from a single empirical study or even from a number of studies of similar form. We aim, however, by introducing appropriate subject matter considerations into the empirical analysis, to produce descriptions and summaries of the data which point toward possible explanations and which in some cases of univariate recursive systems could be consistent with a causal explanation.

3. SOME GRAPHICAL REPRESENTATIONS

With only three component variables, the number of possible special independency models is fairly small but with four and more components there is a quite rich and potentially confusing variety of special cases to be considered. Graphical representation helps clarify the various possibilities, and it is convenient to introduce the key ideas and conventions in terms of three variables.

A systematic account of graphical methods by Whittaker (1990) emphasizes undirected graphs, that is, systems in which all variables are treated on an equal footing. Here we use largely directed graphs to emphasize relations of response and dependence; it is fruitful also to allow two different kinds of edge between the nodes of a graph and to introduce some additional special features.

First we introduce, where appropriate, a distinction between the response variables of primary interest, one or more levels of intermediate response variables, and explanatory variables, all in general with several component variables. The distinction between variable types is usually introduced on a priori subject matter considerations, for example via the temporal ordering of the variables. Sometimes, however, there are several such provisional interpretations and some may be suggested by the data under analysis. The distinction between variable types is expressed in the graphs via (c) below.

The following conventions have been used in constructing the graphs in this paper and are illustrated in their simplest form in Figures 1–3 for three variables:

(a) each continuous variable is denoted by a node, a circle;

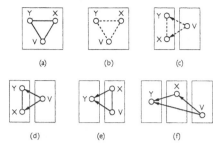

(a)   (b)   (c)

(d)   (e)   (f)

FIG 1. *Six distributionally equivalent ways of specifying a saturated model for three variables. (a) Joint distribution of Y, X, V with three substantial concentrations; (b) joint distribution of Y, X, V with three substantial covariances; (c) multivariate regression chain model with regressions of Y on V and of X on V and with correlated errors; (d) block regression chain model with regressions of Y on X, V and of X on Y, V; (e) univariate regression of Y on X, V and joint distribution of X, V; (f) univariate recursive regression system with Y as response to X, V; X as intermediate response to V. For instance, graph (e) with double lines round the right-hand box would represent the standard linear model for regression of Y on fixed explanatory variables X, V.*

(b) there is at most one connecting line between each pair of nodes, an edge;

(c) variables are graphed in boxes so that variables in one box are considered conditionally on all boxes to the right (in line with the notation $P(A \mid B)$ for the probability of A given B) so that the response variables of primary interest are in the left-hand box and its explanatory variables are in boxes to the right;

(d) if full lines are used as edges, each variable is considered conditionally on other variables in the same box (as well as those to the right), whereas if dashed lines are used variables are considered ignoring other response variables in the same box, that is, marginally with respect to response variables in the same box;

(e) the absence of an edge means that the corresponding variable pair is conditionally independent, the conditioning set being as specified in (d);

(f) variables in the same box are to be regarded

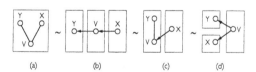

(a)   (b)   (c)   (d)

FIG. 2. *Four distributionally equivalent ways of specifying $Y \perp\!\!\!\perp X \mid V$; (a) covariance selection model for Y, X, V having parameters $\rho_{yv.x} \neq 0$, $\rho_{xv.y} \neq 0$, and $\rho_{yx.v} = 0$; (b) univariate recursive regression model with $\beta_{yv.x} \neq 0$, $\beta_{yx.v} = 0$, $\beta_{xv} \neq 0$; (c) block regression chain model with Y, V as joint responses to X and with independent parameters $\rho_{yv.x} \neq 0$, $\beta_{yx.v} = 0$, $\beta_{vx.y} \neq 0$; (d) two independent regressions of Y on V and of X on V with $\beta_{yv} \neq 0$, $\beta_{xv} \neq 0$, $\rho_{yx.v} = 0$.*

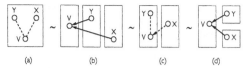

(a)   (b)   (c)   (d)

FIG. 3. *Four distributionally equivalent ways of specifying $Y \perp\!\!\!\perp X$; (a) linear structure in covariances with $\rho_{yv} \neq 0$, $\rho_{xv} \neq 0$, $\rho_{yx} = 0$; (b) univariate recursive regression model with $\beta_{vx.y} \neq 0$, $\beta_{vy.x} \neq 0$, $\beta_{yx} = 0$; (c) multivariate regression chain model with $\rho_{yv.x} \neq 0$, $\beta_{ux} \neq 0$, $\beta_{yx} = 0$; (d) multiple regression of V on two independent regressors Y, X, with $\beta_{vy.x} \neq 0$, $\beta_{vx.y} \neq 0$, $\rho_{yx} = 0$.*

in a symmetrical way, for instance as both response variables, and connected by undirected edges (lines without arrowheads, for correlations), whereas relations between variables in different boxes are shown by directed edges (arrows, for regression coefficients) such that an arrow points from the explanatory variable to the response;

(g) graphs drawn with boxes represent substantive research hypotheses (Wermuth and Lauritzen, 1990) in which the presence of an edge means that the corresponding partial correlation is large enough to be of substantive importance. This corresponds to the notion that the model being represented is the simplest appropriate one in the sense that relations considered to be unimportant are not part of the model; graphs obtained by removing the boxes represent statistical models in which a connecting edge places no such constraint on the correlation, that is, it could also be a zero correlation;

(h) a row of unstacked boxes implies an ordered sequence of (joint) responses and (joint) intermediate responses, each together with their explanatory variables. Boxes are stacked if no order is to be implied, in order to indicate independence of several (joint) variables conditionally on all boxes to the right;

(i) if the right-hand box has two lines around it, then the relations among variables in this box are regarded as fixed at their observed levels; this is to indicate a regression model instead of a regression chain model, the latter containing parameters also for those components which are exclusively explanatory.

In the present paper we use only graphs with edges of one type, that is, either all full lines or all dashed lines. It would be possible to have mixture of the two types of edge in the same graph, for example provided that all the edges within one block are of the same type and all the edges directed at a particular block are of the same type.

In a sense the distinction between full and dashed edges serves a double purpose. The distinction between full and dashed arrows from one box to another determines the different conditioning sets used in the various regression equations under consideration. The distinction between full and dashed lines within a box specifies whether it is the concentration or the covariance matrix of the residuals that is the focus of interest. In this sense the nature of the edges corresponds to the parameters of interest.

The joint distribution of all variables is in the present context specified by the vector of means and the covariance or the concentration matrix. However any such given matrix may correspond to a number of models with quite different interpretations in the light of the distinction between types of variable as response, intermediate response or explanatory variable. A complete graph, that is, one in which all edges are present, represents a saturated model, that is, in the present context a model without any specified independence relations.

To stress the distinction between the multivariate regression and block regression contained in Figure 1, we write the corresponding equations explicitly. The multivariate regression equations implied by Figure 1c are

$$E(Y \mid V = v) - \mu_y = \beta_{yv}(v - \mu_v),$$
$$E(X \mid V = v) - \mu_x = \beta_{xv}(v - \mu_v),$$

with

$$\mathrm{cov}(\varepsilon_{y.v}, \varepsilon_{x.v}) = \rho_{yx.v}(\sigma_{yy.v}\sigma_{xx.v})^{1/2}.$$

By contrast the block regression equations implied by Figure 1d are

$$E(Y \mid X = x, V = v) - \mu_y$$
$$= \beta_{yx.v}(x - \mu_x) + \beta_{yv.x}(v - \mu_v),$$
$$E(X \mid Y = y, V = v) - \mu_x$$
$$= \beta_{xy.v}(y - \mu_y) + \beta_{xv.y}(v - \mu_v),$$

with

$$\beta_{yx.v} = \rho_{yx.v}(\sigma_{yy.v}/\sigma_{xx.v})^{1/2}, \beta_{xy.v} = \rho_{yx.v}(\sigma_{xx.v}/\sigma_{yy.v})^{1/2},$$
$$\mathrm{cov}(\varepsilon_{y.xv}, \varepsilon_{x.yv}) = -\rho_{yx.v}(\sigma_{yy.xv}\sigma_{xx.yv})^{1/2},$$

where the conditional variance of the variable given all remaining variables is the reciprocal value of a precision, for example, $\sigma_{yy.xv} = 1/\sigma^{yy}$. Relations between the sets of parameters in the two types of regressions are given by Equations (6) and (7).

4. RELATIONS WITH PREVIOUS WORK

We illustrate the distinction between the graphical chain models of the present paper and structural equation models via two examples. Suppose first that X and Y are standardized to mean zero and variance one and denote their correlation coefficient by ρ. Then

$$Y = \rho X + \varepsilon_y, \quad X = \rho Y + \varepsilon_x,$$

where $(\varepsilon_y, \varepsilon_x)$ are residuals from linear regression equa-

tions. That is, the coefficients ρ in these equations have an interpretation as regression coefficients. Direct calculation shows that

$$\mathrm{cov}(\varepsilon_y, \varepsilon_x) = \mathrm{cov}(Y - \rho X, X - \rho Y) = -\rho(1 - \rho^2),$$

which is nonzero unless $\rho = 0$. That is, the two regression equations imply correlated residuals except for degenerate cases.

On the other hand, if we were to adopt

$$Y - \rho X = \varepsilon_y, \quad X - \rho Y = \varepsilon_x$$

as structural equations with uncorrelated residuals, then another direct calculation shows that the regression of Y on X is

$$E(Y \mid X = x) = \frac{E(YX)}{E(X^2)}x = \frac{\mathrm{var}(\varepsilon_y) + \mathrm{var}(\varepsilon_x)}{\mathrm{var}(\varepsilon_y) + \rho^2\mathrm{var}(\varepsilon_x)}\rho x$$

which is not ρx, again unless $\rho = 0$. That is, the coefficients in these structural equations do not have an interpretation as regression coefficients, as was noted by Haavelmo (1943).

To make the related point that missing edges in the graphical representation of linear structural equations (Van de Geer, 1971) do not in general have the independency interpretation of chain graphs, consider the following two structural equations

$$Y + \gamma_{yx}X + \gamma_{yv}V = \varepsilon_y,$$
$$\gamma_{xy}Y + X + \gamma_{xw}W = \varepsilon_x,$$

illustrated in Figure 4. For correlated errors $(\varepsilon_y, \varepsilon_x)$, a count of parameters shows that this represents a saturated model; that is, it allows an arbitrary covariance matrix for $(Y, X, V, W)^T$. That is, in particular, the missing edges between V and X, and between W and Y do not imply independencies, conditional or unconditional. For some further discussion of possibilities for interpreting the parameters in this model see Wermuth (1992) and Goldberger (1992). For linear structural equations in general, the interpretation of equation pa-

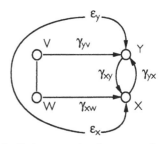

FIG. 4. *Graphical representation of two structural equations in which the missing edges for (V, X) and (W, Y) do not correspond to independencies and do not restrict the covariance matrix for (Y, X, W, V)ᵀ.*

rameters, be they present or missing, has to be derived from scratch for each model considered.

However, an interpretation in terms of independencies is available also for structural equations, whenever such a model is *distributionally equivalent* to one of the chain graph models, that is, if the same joint distribution holds for the two types of models, possibly specified in two distinct ways, *and* the parameter vectors of the two models are in one-to-one correspondence.

Three classes or families of models can be identified to have this property. These are models that have a representation by a chain graph which is:

[1] a *covariance graph*, that is, a single box graph in which all present edges are undirected dashed lines, as in Figures 1b and 3a;

[2] a *multivariate regression graph*, that is, a two-box graph in which all present edges are dashed, being lines within and arrows between boxes, as in Figures 1c and 3c and in which the right-hand box has two lines around it, the distribution of its components being fixed.

[3] a *univariate recursive regression graph*, that is, a graph of $q + 1$ boxes, q of them with a single response variable and the right-hand box with $p - q$ additional explanatory variables, as in Figures 1f, 2b and 3b. In addition the right-hand box has two lines around it to indicate that only the conditional distribution of $Y_1, \ldots, Y_q$ given the remaining variables is the model of interest.

The conventions (a) to (i) for constructing chain graphs imply for univariate recursive regression graphs that arrows have the same interpretation no matter whether they are all dashed or whether they are all full arrows. That is whenever there are no proper joint responses in a model then dashed and full edge arrows are interpreted in the same way.

To distinguish better between dashed and full-edge graphs when their interpretation differs we suggest speaking further of:

[4] a *concentration graph*, that is, a single box graph in which all edges are undirected full lines, as in Figures 1a and 2a;

[5] a *block regression graph*, that is, a two-box graph in which all present edges are full, being lines within and arrows between boxes, as in Figures 1d and 2c and in which the right-hand box has two lines around it.

Then, a *multivariate regression chain graph* can be viewed as a combination of a (sequence of) graph(s) [2] with [1] and a *block regression chain graph* as a combination of a (sequence of) graph(s) [5] with [4]. More general chain graphs with both types of edges

result as further combinations of these four building blocks.

Univariate recursive regression graphs are essentially identical to the *directed acyclic graphs* used in work on expert systems (Pearl, 1988). One of the latter results from one of the former by replacing the complete undirected graph of the explanatory variables by an acyclic orientation, that is, by a univariate recursive regression graph in arbitrary order of the nodes and by discarding all boxes.

To investigate distributional equivalence it is helpful to use the notion of a skeleton graph introduced by Verma and Pearl (1992). A *skeleton graph* is obtained from our Figures by removing boxes and arrows and ignoring the type of edge. For instance, the skeleton graphs in Figures 2a to 2d are all the same. If the skeletons differ then the corresponding models cannot be equivalent. But if the skeletons are the same, then the graphs may still imply different independencies, as in Figures 2 and 3.

Distributional equivalence to a model of univariate recursive regressions is closely tied to our notion of a nondecomposable independence hypothesis. We speak of a *decomposable model* if it is distributionally equivalent to a model of univariate recursive regressions and of a nondecomposable model otherwise. Thus, all saturated chain models for linear relations considered in this paper are decomposable, since they all specify the same joint distribution (Figure 1). A nonsaturated model is decomposable if and only if it contains *not* even one nondecomposable independence hypothesis. In complex cases, such a model may contain large sections that are decomposable and in analysis and interpretation account can be taken of that.

This notion of a decomposable model coincides with the notion of a decomposable graph when this graph has undirected full edges, that is, when it is a concentration graph. For variables with a joint normal distribution a concentration graph specifies a covariance selection model (Dempster, 1972). Such a model is decomposable if and only if the concentration graph is triangulated, that is, if it does not contain a chordless n-cycle for $n \geq 4$ (Wermuth, 1980; Speed and Kiiveri, 1986). A sequence of nodes $(a_1, \ldots, a_n)$ is said to form a *chordless n-cycle* in a chain graph if only consecutive nodes and the endpoints of the sequence are connected by edges and a chordless cycle in a sequence of four or more variables characterizes a nondecomposable independence hypothesis in concentrations. An example is Form (i) for (Y, X, V, W) discussed in Section 5. A special well-studied example of a decomposable covariance selection model is represented by a *chordless n-chain* in concentrations, that is, sequence of nodes $(a_1, \ldots, a_n)$ for which only consecutive nodes of the sequence are connected by edges. This is a Markov

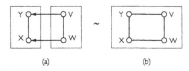

(a) (b)

Fig. 5. *Block regression chain model (a) and covariance selection model (b) both specifying the nondecomposable hypothesis (i):* $Y \perp\!\!\!\perp W \mid (X, V)$ *and* $X \perp\!\!\!\perp V \mid (Y, W)$.

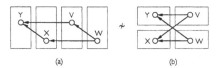

(a) (b)

Fig. 7. *Univariate recursive regressions (a) specifying (iv):* $Y \perp\!\!\!\perp W \mid (X, V)$ *and* $X \perp\!\!\!\perp V \mid W$ *and independent multiple regressions with independent explanatory variables (b) specifying (v):* $Y \perp\!\!\!\perp X \mid (V, W)$ *and* $V \perp\!\!\!\perp W$.

chain model. An example is Form (vi) for (Y, X, V, W) discussed in Section 5.

Figures 1–3 show that not only full-edge but also dashed-edge chain graph models can be decomposable, that is, distributionally equivalent to a model of univariate recursive regressions. We characterize situations in which this is not possible for four variables in the next section.

5. SOME EMPIRICAL EXAMPLES

We now introduce eight special kinds of independence hypothesis for four variables, together with their associated graphs, and illustrate most of them via empirical examples. All involve two or more independency conditions. The special structures we shall consider are as follows, the first three and the last two being nondecomposable:

(i) $Y \perp\!\!\!\perp W \mid (X, V)$ and $X \perp\!\!\!\perp V \mid (Y, W)$,

(see Figures 5a and 5b) called the chordless four-cycle in concentrations and which correspond to the vanishing of two elements in the concentration matrix, and hence to a special case of the covariance selection models (Dempster, 1972). It can also be viewed as a chordless four-cycle in a block regression chain model with joint responses Y, X and joint explanatory variables V, W. Next we consider

(ii) $Y \perp\!\!\!\perp W \mid V$ and $X \perp\!\!\!\perp V \mid W$,

called a chordless four-cycle in a multivariate regression chain model (see Figure 6a) and which contains regressions of Y and X on V and W, being a special case of the seemingly unrelated regressions of Zellner (1962);

(iii) $Y \perp\!\!\!\perp W$ and $X \perp\!\!\!\perp V$,

called the chordless four-cycle in correlations (see Figure 6b), a special case of covariance matrices with linear structure (Anderson, 1973).

These may be contrasted with a decomposable model based on a recursive sequence of univariate regressions with Y as response to X, V, W, with X as response to V, W and with V as response to W and having restrictions on the same two variable pairs (see Figure 7a)

(iv) $Y \perp\!\!\!\perp W \mid (X, V)$ and $X \perp\!\!\!\perp V \mid W$.

Four further cases, the first two decomposable, the last two not, are

(v) $Y \perp\!\!\!\perp X \mid (V, W)$ and $V \perp\!\!\!\perp W$,

two independent regressions of Y and X on two independent regressors V and W (see Figure 7b);

(vi) $Y \perp\!\!\!\perp (V, W) \mid X$ and $X \perp\!\!\!\perp W \mid V$,

called a chordless four-chain in concentrations or a Markov chain (see Figures 8a and 8b), that is, a chordless four-chain in a system of univariate recursive regressions again with Y as response to X, V, W, with X as response to V, W and with V as response to W and having response Y and explanatory variable W as chain endpoints;

(vii) $Y \perp\!\!\!\perp W$ and $X \perp\!\!\!\perp V$ and $V \perp\!\!\!\perp W$,

called a chordless four-chain in covariances (see Figures 9a and 9b) or a chordless four-chain in a multivariate regression chain model with Y, X as joint responses and having explanatory variables V, W as chain endpoints;

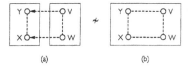

(a) (b)

Fig. 6. *Multivariate regression chain model (a) specifying the nondecomposable hypothesis (ii):* $Y \perp\!\!\!\perp W \mid V$ *and* $X \perp\!\!\!\perp V \mid W$ *and a linear in covariances structure (b) specifying the nondecomposable hypothesis (iii):* $Y \perp\!\!\!\perp W$ *and* $X \perp\!\!\!\perp V$.

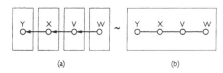

(a) (b)

Fig. 8. *Univariate recursive regressions (a) and covariance selection model both specifying the decomposable hypothesis (vi):* $Y \perp\!\!\!\perp (V, W) \mid X$ *and* $X \perp\!\!\!\perp W \mid V$.

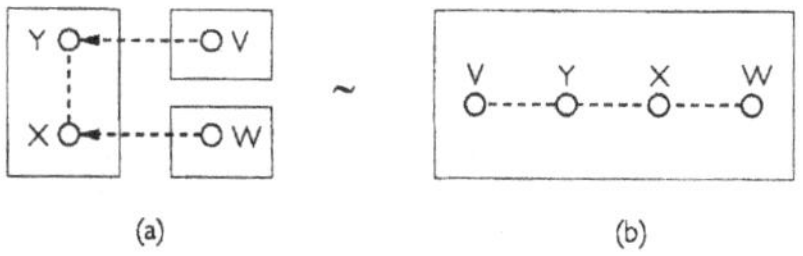

FIG. 9. *Multivariate regression chain model (a) and a linear in covariances structure (b) both specifying the nondecomposable hypothesis (vii): $Y \perp\!\!\!\perp W$ and $X \perp\!\!\!\perp V$ and $V \perp\!\!\!\perp W$.*

(viii) $Y \perp\!\!\!\perp W \mid (X, V)$ and $X \perp\!\!\!\perp V \mid (Y, W)$
        and $V \perp\!\!\!\perp W$;

called a chordless four-chain in a block regression chain model with Y, X as joint responses and having explanatory variables V, W as chain endpoints. The corresponding chain graph has the same shape as the graph in Figure 9a, but dashed lines and arrows are replaced by full lines and arrows.

For our present purpose we give for each empirical example correlations and standardized concentrations showing these as the lower and upper triangle, respectively, such as in Table 1. This allows direct detection of linear marginal independencies between pairs of variables, as shown by very small marginal correlations, that is, standardized covariances, and linear conditional independencies between pairs of variables given all remaining variables, as shown by very small partial correlations, that is, standardized concentrations.

For a formal analysis, consistency of data with a particular structure would be examined via a likelihood ratio test or its equivalent, typically comparing a maximum likelihood fit of the constrained model with that of a saturated model. For the present purposes, however, it is enough to rely on informal comparisons of marginal correlations, partial correlations or standardized regression coefficients, although such dimensionless measures are not in general appropriate for comparing different studies.

Example 1 [Table 1, Figure 5, Form (i)]. Emotions as dispositions or traits of a person and emotions as states, that is, as evoked by particular situations, are notions central to research on stress and on strategies

to cope with stressful events. Questionnaires with which the state-trait versions of the emotions anxiety and anger are measured have been developed by Spielberger et al. (1970, 1983). We obtained data for 684 female college students from C. Spielberger on the variables Y, state anxiety; X, state anger; V, trait anxiety and W, trait anger; summaries are displayed in Table 1.

The upper corner of Table 1 shows close agreement with the Form (i): $Y \perp\!\!\!\perp W \mid (X, V)$ and $X \perp\!\!\!\perp V \mid (Y, W)$, see also Figures 5a and 5b. This nondecomposable model has the simple interpretation that prediction of either state variable is not further improved by adding the other trait variable to the remaining two explanatory variables but it does not directly suggest a stepwise process by which the data might have been generated.

Example 2 [Table 2, Figure 6a, Form (ii)]. From a study of the status and reactions of patients awaiting a particular kind of operation (Slangen, Kleeman and Krohne, 1992) we obtained as basic information for 44 female patients: Y, the ratio of systolic to diastolic blood pressure; X, the diastolic blood pressure; both measured in logarithmic scale; V, body mass, that is, weight relative to height, and W, age. Table 2 shows substantial correlations except for a small marginal correlation of pair (Y, W) and a small partial correlation of pair (X, V). These are not to be directly interpreted if – as appears reasonable – each of the blood pressure variables is regarded as a potential response to body mass and age. Instead, the standardized regression coefficients in a saturated multivariate regression of Y, X on V, W display possible independencies of interest. They show close agreement with Form (ii): $Y \perp\!\!\!\perp W \mid V$ and $X \perp\!\!\!\perp V \mid W$, see also Figure 6a, with standardized regression coefficients

$$\begin{pmatrix} \hat{\beta}^*_{yv.w} & \hat{\beta}^*_{yw.v} \\ \hat{\beta}^*_{xv.w} & \hat{\beta}^*_{xw.v} \end{pmatrix} = \begin{pmatrix} 0.486 & 0.040 \\ 0.037 & -0.275 \end{pmatrix}$$

and from Table 2 correlated errors since $\hat{\beta}_{yx.vw} = -0.566$. This nondecomposable model gives as interpretation that diastolic blood pressure increases just with age

TABLE 1
Observed marginal correlations (lower half) and observed partial correlations given two remaining variables (upper half) means and standard deviations for $n = 684$ students

Variable	Y State anx	X State ang	V Trait anx	W Trait ang
Y: = State anxiety	1	0.45	0.47	−0.04
X: = State anger	0.61	1	0.03	0.32
V: = Trait anxiety	0.62	0.47	1	0.32
W: = Trait anger	0.39	0.50	0.49	1
Mean	18.87	15.23	21.20	23.42
Standard deviation	6.10	6.70	5.68	6.57

Data for Example 1 to Form (i): $Y \perp\!\!\!\perp W \mid (X, V)$ and $X \perp\!\!\!\perp V \mid (Y, W)$ and to Figures 5a and 5b.

TABLE 2

Observed marginal correlations (lower half), observed partial correlations given all remaining variables (upper half), means and standard deviations for n = 44 patients

Variable	Y Lratio bp	X Lsyst. bp	V Body mass	W Age
Y: = Log (syst/diast) bp	1	−0.566	−0.241	0.300
X: = Log diastolic bp	−0.544	1	−0.107	0.491
V: = Body mass	−0.253	0.336	1	0.572
W: = Age	−0.131	0.510	0.608	1
Mean	0.453	4.29	0.379	29.52
Standard deviation	0.091	0.13	0.060	10.59

Data for Example 2 to Form (ii): $Y \perp\!\!\!\perp W \mid V$ and $X \perp\!\!\!\perp V \mid W$ and to Figure 6a.

after controlling for an increase in body mass and that the ratio of systolic to diastolic blood pressure is higher the lower the body mass for persons of the same age. But again, the model does not directly suggest a stepwise process by which the data could have been generated.

Example 3 [Table 3, Figure 6b, Form (iii)]. In a study of strategies to cope with stressful events Kohlmann (1990) collected data for 72 students replying to a German and an American questionnaire. They are both intended to capture two similar strategies: Y, cognitive avoidance and V, blunting are thought of as strategies to reduce emotional arousal and X, vigilance and W, monitoring as strategies to reduce insecurity. The data in Table 3 agree well with Form (iii): $Y \perp\!\!\!\perp W$ and $X \perp\!\!\!\perp V$, see also Figure 6b, but not with (i) because in this case the marginal correlations but not the partial correlations are small.

It is plausible to see strong positive correlations between both pairs of similar strategies, a moderate negative correlation between each set of competing strategies measured one way and no correlation between a strategy measured with one questionnaire and the competing strategy measured with the other questionnaire. However, this structure again cannot be reexpressed with zero regression coefficients in any system of recursive univariate regressions; that is, it does not have a direct explanation as a process by which the data could have been generated.

Pairs of forms from the above special cases (i) to (iv) are mutually exclusive whenever the correlations of all variable pairs other than the two constrained pairs (Y, W) and (X, V) are substantial although with limited data it is of course possible that several different simplified structures are consistent with the data. An exception where two different sets of the above conditions may hold simultaneously is provided by (i) and (iii), that is, a chordless four-cycle in concentrations and in correlations can occur together if a very special structure is present, that is if the marginal correlations in the population satisfy orthogonalities such as

$$
\begin{aligned}
&\rho_{yw} = 0, \rho_{xv} = 0, \\
&\rho_{yv}\rho_{vw} + \rho_{yx}\rho_{xw} = 0, \\
&\rho_{yv}\rho_{yx} + \rho_{vw}\rho_{xw} = 0.
\end{aligned}
$$
(8)

The next set of data is an example of this special case.

Example 4 [Table 4, Figures 5b and 6b, Forms (i) and (iii)]. In a study of effects of working conditions on the manifestation of hypertension, Weyer and Hodapp (1979) report the correlations among the four potential influencing variables displayed in Table 4 for 106 healthy employees. The variables, which are measured with questionnaires, are Y, nervousness; X, stress at work; V, satisfaction with work and W, hierarchical status at work. The observations agree well with both (i): $Y \perp\!\!\!\perp W \mid (X,V)$ and $X \perp\!\!\!\perp V \mid (Y,W)$ (see also Figure 5b) and with (iii): $Y \perp\!\!\!\perp W$ and $X \perp\!\!\!\perp V$ (see also Figure 6b). There is no immediate interpretation; however, one

TABLE 3

Observed marginal correlations (lower half) and observed partial correlations given two remaining variables (upper half), means and standard deviations for n = 72 students

Variable	Y Cogn. avoid.	X Vigilance	V Blunting	W Monitoring
Y: = Cognitive Avoidance	1	−0.30	0.49	0.21
X: = Vigilance	−0.20	1	0.21	0.51
V: = Blunting	0.46	0.00	1	−0.25
W: = Monitoring	0.01	0.47	−0.15	1
Mean	17.49	12.57	3.71	10.40
Standard deviation	6.77	6.39	2.12	3.07

Data for Example 3 to Form (iii): $Y \perp\!\!\!\perp W$ and $X \perp\!\!\!\perp V$ and to Figure 6b.

TABLE 4
*Observed marginal correlations (lower half) and observed partial correlations given two remaining variables (upper half)
for n = 106 healthy employees*

Variable	Y Nervous	X Stress	V Satisf.	W Hier. Stat.
Y: = Nervousnous	1	0.33	0.26	**0.00**
X: = Stress at work	0.34	1	**0.06**	0.30
V: = Satisfaction with work	0.27	**0.04**	1	−0.35
W: = Hierarchical status	**0.01**	0.29	−0.34	1

Data for Example 4 to Forms (i) and (iii) and to Figures 5b and 6b simultaneously.

explanation for this special structure is that a different combination of the questionnaire items of X, V would lead to variables X^*, V^* such that the much simpler structure $(X^*, Y) \perp\!\!\!\perp (V^*, W)$ holds (Cox and Wermuth, 1992a). For the special structure (8) both the canonical correlations and the transformation matrix to obtain X^*, V^* can be expressed in closed form.

Example 5 [*Table 5, Figure 7b, Form* (v)]. For an analysis of aggregate economic data von der Lippe (1977) computed growth rates for 24 postwar years in Germany for Y, employment; X, capital gains; V, private consumption and W, exports. The correlation structure suggests that knowing the change in capital gain does not help in predicting the change in employment for given change levels of the demand side, that is, consumption and export (Wermuth, 1979); in addition, changes in consumption were not correlated with changes in exports. This implies two independent responses to two independent explanatory variables or close agreement to Form (v): $Y \perp\!\!\!\perp X \mid (V, W)$ and $V \perp\!\!\!\perp W$; see also Figure 7b.

Example 6 [*Table 6, Figure 8, Form* (vi)]. In a conditioning experiment with 48 subjects (Zeiner and Schell, 1971), one purpose was to examine discrimination between a noxious and an innocuous stimulus in two periods of a conditioning experiment with Y, a long-interval discriminatory response (6–10 seconds); X, a short-interval discriminatory response (1–5 seconds) in the light of earlier responses; V, the strongest response in the first interval and W, the response to an innocuous stimulus before the experiment itself; all responses are measured as skin resistance. The correlations displayed in Table 6 suggest (Hodapp and Wermuth, 1983, p. 384) a Markov structure (vi) in which $Y \perp\!\!\!\perp (V,W) \mid X$ and $X \perp\!\!\!\perp W \mid V$, see also Figures 8a and 8b, and thus in which the long-interval discriminatory response depends directly only on the short-interval discriminatory response; this short-interval response is directly dependent on the strongest response in the short interval and the latter is well predicted by just the response to an innocuous stimulus before the experiment.

Example 7 [*Table 7, Figure 9, Form* (vii)]. From an

TABLE 5
*Observed marginal correlations (lower half) and observed partial correlations given two remaining variables (upper half)
of growth rates for n = 24 postwar years in Germany*

Variable	Y Employment	X Capital gain	V Consumption	W Export
Y: = Employment	1	−0.11	0.68	0.55
X: = Capital gain	0.47	1	0.50	0.43
V: = Consumption	0.67	0.55	1	−0.51
W: = Export	0.44	0.39	**0.04**	1

Data for Example 5 to Form (v): $Y \perp\!\!\!\perp X \mid (V, W)$ and $V \perp\!\!\!\perp W$ and to Figure 7b.

TABLE 6
*Observed marginal correlations (lower half) and observed partial correlations given two remaining variables (upper half)
for n = 48 subjects*

Variable	Y Long	X Short	V Strong	W Innoc
Y: = Long int. discriminatory response	1	0.70	−0.04	−0.12
X: = Short int. discriminatory response	0.72	1	0.29	**0.14**
V: = Strongest short interval response	0.30	0.54	1	0.62
W: = Response to innocuous stimulus	0.19	0.43	0.71	1

Data for Example 6 to Form (vi): $Y \perp\!\!\!\perp (V,W) \mid X$ and $X \perp\!\!\!\perp V \mid W$ and to Figures 8a and 8b.

TABLE 7
Observed marginal correlations (lower half) and observed partial correlations given two remaining variables (upper half), means and standard deviations for n = 39 diabetic patients

Variable	Y GHb	X Knowledge	V Duration	W Fatalism
Y: = Glucose control, GHb	1	−0.431	−0.407	−0.262
X: = Knowledge, illness	−0.344	1	0.042	−0.517
V: = Duration, illness	−0.404	0.042	1	−0.028
W: = Fatalism, illness	−0.071	−0.460	0.060	1
Mean	10.02	33.18	147.05	20.13
Standard deviation	2.07	7.86	92.00	5.75

Data for Example 7 to Form (vii): $Y \perp\!\!\!\perp V$, $Y \perp\!\!\!\perp W$, and $X \perp\!\!\!\perp V$ and to Figures 9a and 9b.

investigation of determinants of blood glucose control (Kohlmann et al., 1991), we have data for 39 diabetic patients, who had at most 10 years of formal schooling. The variables considered are Y, a particular metabolic parameter, the glycosylated hemoglobin GHb; X, a score for particular knowledge about the illness, V, the duration of illness in months, and W, a questionnaire score measuring the patients external attribution to "chance" of the occurence of events related to the illness; an attitude called external fatalism. The correlations in Table 7 suggest a structure of the Form (vii), that is, with $Y \perp\!\!\!\perp W, X \perp\!\!\!\perp V$, and $V \perp\!\!\!\perp W$, see also Figures 9a and 9b. One interpretation is that duration of illness and external fatalism are independent explanatory variables in two seemingly independent regressions, where metabolic adjustment is better (low values of GHb) the longer the duration of the illness, knowledge about the illness is lower the higher the external fatalism of a person, and after conditioning on duration and fatalism the metabolic adjustment is still better the higher the knowledge ($\beta_{yx.vw} = -0.431$).

6. DISCUSSION

There are a number of general issues arising from the special cases discussed in the previous section, especially the extension to more than four component variables and to models with other than only linear dependencies; for the latter see Cox and Wermuth (1993).

Graphs with, in our notation, full edges have an elegant connection with the theory of Markov random fields which allows general properties to be deduced. See Lauritzen (1989) for a survey of these topics and Isham (1981) for a review of Markov random fields in a broader context. Graphs with dashed edges, or possibly graphs with mixtures of dashed and full edges, do not have the same general features, and it is an open question as to what exactly can be said about them in generality.

There are four types of nondecomposable independence hypotheses illustrated in Section 4 for four variables, namely:

(a) *Nondecomposable hypotheses in block regression chain models [Form* (i), *Example 1, Table 1, Figure 5a and Form* (viii)]. In a block regression chain model the components, even in the simplest case, are divided into responses $Y_a = (Y, X)$ and explanatory variables $Y_b = (V, W)$ with a full directed arrow unless the corresponding regression coefficient in (3) is zero and a full undirected line for the explanatory variables unless they are marginally uncorrelated. For four variables a nondecomposable independence hypothesis in a block regression chain model is characterized by a chordless four-chain in the full edge chain graph, with the two ends of the sequence being explanatory variables, that is, for (V, Y, X, W) in our examples. Figure 5a with Form (i) gives an example of the four-cycle which contains the described four-chain, while Form (viii) leads to an example of the chordless four-chain;

(b) *Nondecomposable hypotheses in concentrations [Form* (i), *Example 1, Table 1, Figure 5b].* Models of zero concentrations, that is, the covariance selection models of Dempster (1972), differ from block regression models – from (a) – in treating all variables on an equal footing, that is, having them in the same box where all edges are full undirected lines unless the corresponding variables are partially uncorrelated given the remaining component variables. For four variables a nondecomposable hypotheses in concentrations is characterized by a chordless four-cycle in the associated undirected graph of full edges, that is, in the concentration graph. Figure 5b with Form (i) gives an example of a chordless four-cycle in concentrations for (V, Y, X, W).

(c) *Nondecomposable hypotheses in multivariate regression chain models [Form* (ii), *Example 2, Table 2, Figure 6a and Form* (vii), *Example 7, Table 7, Figure 9a].* In multivariate regression chain models the components are – as for (a) – even in the simplest case divided into responses $Y_a = (Y, X)$ and explanatory variables $Y_b = (V, W)$ with a dashed directed arrow unless the corresponding regression coefficient in (2) is zero, a dashed undirected line for the responses unless they are partially uncorrelated given the explanatory variables, and a dashed undirected line for the explanatory variables unless they are marginally uncorrelated. For four variables a nondecomposable independence hypothesis in a multivariate regression chain model is characterized by a chordless four-chain in the dashed edge chain graph with the two ends of the sequence being explanatory variables, that is, for (V, Y, X, W) in our examples. Figure 6a with Form (ii) gives an example of the four-cycle which contains the described four-chain, while Figure 9a with Form (vii) gives an example of the four-chain. Both are seemingly unrelated regressions (Zellner, 1962) together with a specification for the distribution of the explanatory variables.

(d) *Nondecomposable hypotheses in covariances* [*Form* (iii), *Example 3, Table 3, Figure* 6b *and Form* (vii), *Example 7, Table 7, Figure* 9b]. Models of zero covariances, that is, models for hypotheses linear in covariances (Anderson, 1973), have—as in (b)—a single block of variables. All edges are dashed undirected lines unless the corresponding variables are marginally uncorrelated. For four variables a nondecomposable independence hypothesis in covariances is characterized by a chordless four-chain in the associated undirected graph of dashed edges, that is, in the covariance graph. Figure 6b with Form (iii) gives an example of the four-cycle which contains a chordless four-chain, while Figure 9b with Form (vii) gives an example of the four-chain.

Models which contain even a single nondecomposable independence hypothesis cannot be distributionally equivalent to a model of univariate recursive regressions. Our examples illustrate that such nondecomposable structures arise in a number of different contexts. There is need to identify them and to find explanations of how they could have been generated. Criteria for establishing nondecomposability for more than four variables are not yet published for general dashed-edge chain graphs, while for full-edge chain graphs such criteria were given by Lauritzen and Wermuth (1989) and for undirected dashed line graphs by Pearl and Wermuth (1993).

We have in this paper concentrated on the kinds of special structure that can arise, especially on their specification and interpretation, rather than on the details of fitting and assessing model adequacy. Under normal-theory assumptions maximum-likelihood fitting and testing for nondecomposable models will call for iterative procedures. A rather general asymptotically efficient noniterative procedure based on embedding the model to be fitted in a saturated model is available (Cox and Wermuth, 1990) either for direct use or as a starting point for iteration (Jensen, Johansen and Lauritzen, 1991). Several issues are important for iterative algorithms. Is there a global maximum or are there several local maxima? Which conditions guarantee the existence of maximum-likelihood estimates? What are

the convergence properties of an algorithm? Again, more is known for models represented by full-edged graphs (Speed and Kiiveri, 1986; Frydenberg and Edwards, 1989; Frydenberg and Lauritzen, 1989; Edwards, 1992) than for models with dashed edge graphs. Some of the latter may be fitted with algorithms suitable for linear structural equations; for a discussion of different alternatives see Lee, Poon and Bentler (1992).

For mixtures of discrete and continuous variables, models corresponding to chain graphs with full edges have been intensively studied (Lauritzen and Wermuth, 1989; Lauritzen, 1989; Frydenberg, 1990b; Wermuth and Lauritzen, 1990; Cox and Wermuth, 1992b; Wermuth, 1993), but for models corresponding to chain graphs with dashed edges or possibly mixtures of dashed and full edges the extensions to discrete and mixtures of discrete and continuous variables remain to be developed.

The issue of model choice in the analysis of data has too many ramifications to be discussed satisfactorily in the present paper; some different suitable strategies for analyses with a moderate number of variables are discussed in Wermuth and Cox (1992). In general, if there is sufficient substantive knowledge to give a firm indication both of the nature of the variables and of the independencies expected, then model choice consists largely of testing the adequacy of the proposed model, in particular in examining the supposedly zero correlations, concentrations and regression coefficients. The less the guidance from subject matter considerations, the more tentative will be the conclusions about model structure, but the broad principles of variable selection in empirical regression discussed, for example, by Cox (1968) and Cox and Snell (1974), will apply. In particular, where a number of different models of roughly equal complexity give satisfactory fits to the data, all should be incorporated in the conclusions, unless a choice can be made on subject matter grounds.

There are many aspects of the study of multiple dependencies and associations not addressed in the present paper. In particular the role of latent or hidden variables in clarifying the interpretation of relatively complex structures has not been dealt with, nor has the related matter of the effect of errors of observations in possibly distorting dependencies. Finally, we reemphasize the point made in Section 3 that a key argument for aiming for univariate recursive regressions consistent with subject matter knowledge is that they suggest a stepwise process by which the data might have been generated.

ACKNOWLEDGMENTS

We are grateful to Morten Frydenberg for his helpful comments, to the referees for very constructive suggestions and to the British–German Academic Research Collaboration Programme for supporting our joint work.

REFERENCES

ANDERSON, T. W. (1973). Asymptotically efficient estimation of covariance matrices with linear structure. *Ann. Statist.* 1 135–141.

COX, D. R. (1968). Regression methods; notes on some aspects of regression analysis (with discussion). *J. Roy. Statist. Soc. Ser. A* 131 265–279.

COX, D. R. (1992). Causality; some statistical aspects. *J. Roy. Statist. Soc. Ser. A* 155 291–301.

COX, D. R. and SNELL, E. J. (1974). The choice of variables in observational studies. *J. Roy. Statist. Soc. Ser. C* 23 51–59.

COX, D. R. and WERMUTH, N. (1990). An approximation to maximum-likelihood estimates in reduced models. *Biometrika* 77 747–761.

COX, D. R. and WERMUTH, N. (1992a). On the calculation of derived variables in the analyses of multivariate responses. *J. Multivariate Anal.* 42 167–173.

COX, D. R. and WERMUTH, N. (1992b). Response models for mixed binary and quantitative variables. *Biometrika* 79 441–461.

COX, D. R. and WERMUTH, N. (1993). Some recent work on methods for the analysis of multivariate observational data in the social sciences. In *Conference Proceedings of the 7th International Conference on Multivariate Analysis, Pennsylvania State Univ., May 1992.* North-Holland, Amsterdam. To appear.

DAWID, A. P. (1979a). Conditional independence in statistical theory (with discussion). *J. Roy. Statist. Soc. Ser. B* 41 1–31.

DEMPSTER, A. P. (1972). Covariance selection. *Biometrics* 28 157–175.

DUNCAN, O. D. (1969). Some linear models for two-wave, two-variable panel analysis. *Psychological Bulletin* 72 177–182.

EDWARDS, D. (1992). *Graphical Modelling with MIM.* Manual, Univ. Copenhagen.

FRYDENBERG, M. (1990a). The chain graph Markov property. *Scand. J. Statist.* 17 333–353.

FRYDENBERG, M. (1990b). Marginalization and collapsibility in graphical interaction models. *Ann. Statist.* 18 790–805.

FRYDENBERG, M. and EDWARDS, D. (1989). A modified iterative proportional scaling algorithm for estimation in regular exponential families. *Comput. Statist. Data Anal.* 8 143–153.

FRYDENBERG, M. and LAURITZEN, S. L. (1989). Decomposition of maximum-likelihood in mixed interaction models. *Biometrika* 76 539–555.

GLYMOUR, C., SCHEINES, R., SPIRTES, P. and KELLY, K. (1987). *Discovering Causal Structure.* Academic, New York.

GOLDBERGER, A. S. (1964). *Econometric Theory.* Wiley, New York.

GOLDBERGER, A. S. (1992). Models of substance; comment on "On block recursive linear regression equations," by N. Wermuth. *Revista Brasileira de Probabilidade e Estatística* 6 46–48.

HAAVELMO, T. (1943). The statistical implications of a system of simultaneous equations. *Econometrica* 11 1–12.

HODAPP, V. and WERMUTH, N. (1983). Decomposable models: A new look at interdependence and dependence structures in psychological research. *Multivariate Behavioral Research* 18 361–390.

ISHAM, V. (1981). An introduction to spatial point processes and Markov random fields. *Internat. Statist. Rev.* 49 21–43.

JENSEN, S. T., JOHANSEN, S. and LAURITZEN, S. L. (1991). Globally convergent algorithms for maximizing a likelihood function. *Biometrika* 78 867–878.

JÖRESKOG, K. G. (1973). A general method for estimating a linear structural equation system. In *Structural Equation Models in the Social Sciences* (A. S. Goldberger and O. D. Duncan, eds.) 85–112. Seminar Press, New York.

KOHLMANN, C.-W. (1990). *Streßbewältigung und Persönlichkeit.* Huber, Bern.

KOHLMANN, C. W., KROHNE, H. W., KÜSTNER E., SCHREZENMEIR, J., WALTHER, U. and BEYER, J. (1991). Der IPC-Diabetes-Fragebogen: ein Instrument zur Erfassung krankheitsspezifischer Kontrollüberzeugungen bei Typ-I-Diabetikern. *Diagnostica* 37 252–270.

LAURITZEN, S. L. (1989). Mixed graphical association models (with discussion). *Scand. J. Statist.* 16 273–306.

LAURITZEN, S. L., DAWID, A.P., LARSEN, B. and LEIMER, H. G. (1990). Independence properties of directed Markov fields. *Networks* 20 491–505.

LAURITZEN, S. L. and WERMUTH, N. (1989). Graphical models for association between variables, some of which are qualitative and some quantitative. *Ann. Statist.* 17 31–57.

LEE, S.-Y., POON, W.-Y. and BENTLER, P. M. (1992). Structural equation models with continuous and polytomous variables. *Psychometrika* 57 89–105.

PEARL, J. (1988). *Probabilistic Reasoning in Intelligent Systems.* Morgan Kaufman, San Mateo, CA.

PEARL, J. and VERMA, T. S. (1991). A theory of inferred causation. In *Principles of Knowledge Representation and Reasoning* (J. A. Allen, R. Fikes and E. Sandewall, eds.). Morgan Kaufman, San Mateo, CA.

PEARL, J. and WERMUTH, N. (1993). When can an association graph admit a causal interpretation? In *Conference Proceedings of the 4th International Workshop on Artificial Intelligence and Statistics, Fort Lauderdale, Florida.* To appear.

SLANGEN, K., KLEEMANN, P. P. and KROHNE, H. W. (1992). Coping with surgical stress. In *Attention and Avoidance; Strategies in Coping with Aversiveness* (H. W. Krohne, ed.) 321–348. Springer, New York.

SPEED, T. P. and KIIVERI, H. T. (1986). Gaussian Markov distributions over finite graphs. *Ann. Statist.* 14 138–150.

SPIELBERGER, C. D., GORSUCH, R. L. and LUSCHENE, R. E. (1970). *Manual for the State-Trait Anxiety Inventory.* Consulting Psychologists Press, Palo Alto, CA.

SPIELBERGER, C. D., RUSSELL, S. and CRANE, R. (1983). Assessment of anger. In *Advances in Personality Assessment* (J. N. Butcher and C. D. Spielberger, eds.) 2 159–187. Erlbaum, Hillsdale, NJ.

TINBERGEN, J. (1937). *An Econometric Approach to Business Cycle Problems.* Hermann, Paris.

VAN DE GEER, J. P. (1971). *Introduction to Multivariate Analysis for the Social Sciences.* Freeman, San Francisco.

VERMA, T. S. and PEARL, J. (1992). An algorithm for deciding if a set of observed independencies has a causal explanation. In *Uncertainty in Artificial Intelligence* (D. Dubois, M. P. Wellman, B. D'Ambrosio and P. Smets, eds.) 8 323–330. Morgan Kaufman, San Mateo, CA.

VON DER LIPPE, P. (1977). Beschäftigungswirkung durch Umverteilung? *WSI-Mitteilungen* 8 505–512.

WERMUTH, N. (1979). Datenanalyse und multiplikative Modelle. *Allgemeines Statistisches Archiv* 63 323–339.

WERMUTH, N. (1980). Linear recursive equations, covariance selection, and path analysis. *J. Amer. Statist. Assoc.* 75 963–972.

WERMUTH, N. (1989). Moderating effects in multivariate normal distributions. *Methodika* 3 74–93.

WERMUTH, N. (1992). On block-recursive regression equations

(with discussion). *Revista Brasileira de Probabilidade e Es-tatística* 6 1–56.

WERMUTH, N. (1993). Association structures with few variables: characteristics and examples. In *Theory and Methods for Population Health Research* (K. Dean, ed.) 181–202. Sage, London.

WERMUTH, N. and COX, D. R. (1992). Graphical models for dependencies and associations. In *Computational Statistics, Proceedings of the 10th Symposium on Computational Statistics, Neuchâtel.* (Y. Dodge and J. Whittaker, eds.) 1 235–249. Physica, Heidelberg.

WERMUTH, N. and LAURITZEN, S. L. (1990). On substantive research hypotheses, conditional independence graphs and graphical chain models (with discussion). *J. Roy. Statist. Soc. Ser. B* 52 21–72.

WEYER, G. and HODAPP, V. (1979). Job-stress and essential hyper-

tension. In *Stress and Anxiety* (I. G. Sarason and C. D. Spielberger, eds.) 6 337–349. Hemisphere, Washington, D.C.

WHITTAKER, J. (1990). *Graphical Models in Applied Multivariate Statistics.* Wiley, Chichester.

WOLD, H. O. (1954). Causality and econometrics. *Econometrica* 22 162–177.

WRIGHT, S. (1921). Correlation and causation. *Journal of Agricultural Research* 20 557–585.

WRIGHT, S. (1923). The theory of path coefficients: A reply to Niles' criticism. *Genetics* 8 239–255.

ZEINER, A. R. and SCHELL, A. M. (1971). Individual differences in orienting, conditionality, and skin resistance responsitivity. *Psychophysiology* 8 612–622.

ZELLNER, A. (1962). An efficient method of estimating seemingly unrelated regressions and tests for aggregation bias. *J. Amer. Statist. Assoc.* 57 348–368.

Summary of discussion

This paper was discussed jointly with a related paper by Spiegelhalter et al. (1993). Comments specific to the present paper were made by A. P. Dempster, C. Glymour and P. Spirtes, S.-L. Normand, J. Pearl, M. E. Sobel and J. Whittaker.

Dempster raises central issues about how informal knowledge is to be incorporated into such analyses and about how the use of simplifying models to aid interpretation fits in with the complexity of the real world. Glymour and Spirtes, among many other comments about the importance of directed acyclic graphs, suggest the class of partially ordered 'inducing path graphs' as the basis of a synthesis of the various possibilities in the paper. Normand raises various issues especially from the perspective of quite large problems in a clinical setting: in particular, how should these models be assessed for adequacy? Pearl develops the connection with his notion of intervention, and Sobel discusses the connection with simultaneous models in econometrics and the role of graphical models in social research. Finally, Whittaker argues that some at least of the special graphs in the paper are not needed and makes other comments connected with the algebra of the multivariate Gaussian case.

Reference

Spiegelhalter, D. J., Dawid, A. P., Lauritzen, S. L. and Cowell, R. G. (1993). Bayesian analysis in expert systems (with discussion). *Statistical Science* 8, 219–283.

Reprinted from JOURNAL OF MULTIVARIATE ANALYSIS
All Rights Reserved by Academic Press, New York and London

On the Calculation of Derived Variables in the Analysis of Multivariate Responses

D. R. Cox

Nuffield College, Oxford, United Kingdom

AND

Nanny Wermuth

Psychologisches Institut, Universität Mainz, Germany

Communicated by the Editors

The multivariate regression of a $p \times 1$ vector Y of random variables on a $q \times 1$ vector X of explanatory variables is considered. It is assumed that linear transformations of the components of Y can be the basis for useful interpretation whereas the components of X have strong individual identity. When $p \geqslant q$ a transformation is found to a new $q \times 1$ vector of responses Y^* such that in the multiple regression of, say, Y_1^* on X, only the coefficient of X_1 is nonzero, i.e. such that Y_1^* is conditionally independent of $X_2, ..., X_q$, given X_1. Some associated inferential procedures are sketched. An illustrative example is described in which the resulting transformation has aided interpretation. © 1992 Academic Press, Inc.

1. INTRODUCTION

Many of the standard methods of multivariate analysis derived from the multivariate normal distribution are essentially invariant under nonsingular linear transformations. A typical example is canonical correlation (or canonical regression) analysis. Here the relation between a $p \times 1$ vector response variable Y and another $q \times 1$ vector X, either of responses or of explanatory variables, is studied by finding linear combinations of the components of Y and of X that are maximally related, the resulting analysis being essentially invariant under separate linear transformations of Y and of X.

Received September 9, 1991; revised January 4, 1992.

AMS 1980 subject classification: 62J10.

Key words and phrases: canonical analysis, conditional independence, derived variable, graphical chain model, multivariate linear model.

162

This invariance is sometimes, but by no means always, appealing on subject-matter grounds. For example, linear combinations of log height and log weight may form derived variables that are entirely satisfactory for the interpretation of the effect of the "size" of individuals on various medical outcomes and the summarization of blood pressure may be best carried out via a combination of diastolic and systolic blood pressures (or their logarithms), which are themselves somewhat arbitrarily chosen summaries of the blood pressure cyclical variation. On the other hand, variables such as anger and anxiety express distinct concepts and while a linear combination of them may well arise in a multiple regression equation as an expression of their relative importance in contributing to a third variable, the formation of a new derived response variable as an arbitrary linear combination of anger and anxiety is much less appealing. These remarks could be paralleled in many fields.

In the present paper we consider problems in which arbitrary linear combinations of a $p \times 1$ variable Y are allowable but in which it is desired to preserve the distinctive individual component structure of a $q \times 1$ variable X. It is convenient to begin by treating X as random, although in some applications conditioning on the observed values will be called for, this having virtually no effect on the following arguments.

There are a number of ways in which the problem can be formalized. The simplest and the one on which we shall concentrate is as follows. Suppose that $p \geqslant q$. We seek a linear transformation from $Y = (Y_1, ..., Y_p)^T$ to new variables $Y^* = (Y_1^*, ..., Y_q^*)^T = AY$ such that in the multiple regression of Y_s^* on X only the coefficient of X_s is nonzero ($s = 1, ..., q$); that is, Y_s^* is conditionally independent of all the X_t ($t \neq s$) given X_s. That is, in a sense Y_s^* is that derived response variable especially tied to X_s. In Section 2 we discuss the relatively simple situation in which $p = q$, extending the discussion to the case $p > q$ in Section 3. The construction is possible if and only if Σ_{yx} is of full rank. Section 4 outlines the inferential problems associated with the procedure and Section 5 described a specific application.

2. EQUAL DIMENSIONALITY

Suppose first for simplicity that $p = q$, i.e., that the dimensionalities of the two vectors are the same. For simplicity assume both vectors have zero mean and partition the joint covariance matrix Σ in terms of

$$\Sigma_{xx} = \text{cov}(X) = E(XX^T), \ \Sigma_{yx} = \text{cov}(Y, X) = E(YX^T) = \Sigma_{xy}^T,$$

$$\Sigma_{yy} = E(YY^T).$$

Now $\text{cov}(Y^*, X) = A\,\Sigma_{yx}$ and the matrix of regression coefficients of Y^* on X is thus

$$B_{y^*x} = A\Sigma_{yx}\Sigma_{xx}^{-1} = AB_{yx},$$

where B_{yx} is the matrix of regression coefficients of Y on X. We require this to be diagonal and if new variables are scaled to have unit regression coefficients on the explanatory variables B_{y^*x} must be the $q \times q$ identity matrix, so that

$$A = B_{yx}^{-1} = (\Sigma_{yx}\Sigma_{xx}^{-1})^{-1} = \Sigma_{xx}\Sigma_{yx}^{-1}.$$

Thus the new variable is given by

$$Y^* = \Sigma_{xx}\Sigma_{yx}^{-1}Y. \tag{1}$$

Note that $\text{cov}(Y^*, X) = E(Y^*X^\mathrm{T}) = \Sigma_{xx}$; i.e., the new variables are such that they have the same covariance matrix with X as does X with itself. Another interpretation is via the equation $Y = B_{yx}Y^*$.

The new variable Y^* exists and is uniquely defined provided that Σ_{yx} is nonsingular. A necessary and sufficient condition for this is that no linear combination of the components of Y be uncorrelated with all the components of X. Singularity will occur for some simple patterns, as for example when all cross-correlations are equal. In the singular case certain components of Y^* may nevertheless be determined.

There are other criteria that might be used to express the notion that each component of the transformed vector is attached to a unique component of X. For example, one might require that the sth component of the new vector have zero marginal correlation with all components of X except the sth. We shall not explore this further.

The special case $q = 2$ throws some light on the above formulae. If we normalize all variables to unit standard deviation, i.e., replace covariance matrices by correlation matrices, we need the condition $\rho_{x_1y_1}\rho_{x_2y_2} - \rho_{x_1y_2}\rho_{x_2y_1} \neq 0$ for nonsingularity. Subject to this and ignoring a constant of proportionality we can take

$$Y_1^* = (\rho_{x_2y_2} - \rho_{x_1x_2}\rho_{x_1y_2})\,Y_1 - (\rho_{x_2y_1} - \rho_{x_1x_2}\rho_{x_1y_1})\,Y_2,$$
$$Y_2^* = -(\rho_{x_1y_2} - \rho_{x_1x_2}\rho_{x_2y_2})\,Y_1 + (\rho_{x_1y_1} - \rho_{x_1x_2}\rho_{x_2y_1})\,Y_2.$$

Note that, for example, Y_1^* depends only on Y_1 if and only if the partial correlation of Y_1 with X_2 given X_1 vanishes, as is clear on general grounds. Note also that if X_1 and X_2 are uncorrelated the derived variables take a simple form, namely

$$Y_1^* = \rho_{x_2y_2}Y_1 - \rho_{x_2y_1}Y_2, \qquad Y_2^* = -\rho_{x_1y_2}Y_1 + \rho_{x_1y_1}Y_2.$$

3. Unequal Dimensionality

Suppose now that $p > q$. It can be shown that there is no unique linear combination Y_1^*, say, with the required property of dependence only on X_1. A sensible approach is to first reduce Y to the $q \times 1$ vector of canonical variables that contain the regression on X; for this we use the theory of canonical correlation or regression. Then the results of Section 2 can be applied to the new variables. The resulting procedure will choose from the multiplicity of solutions that Y^* with maximal regression on X. Indeed a nonsingular transformation can be made to the canonical variables supplemented by a set of $p - q$ random variables independent of the canonical variables and moreover independent of X. Under normal theory it follows from the factorization of the likelihood and the resulting sufficiency that inference about the dependence of Y on X should involve Y only via the canonical variables.

Now the q canonical variables formed from Y for capturing the regression on X are $c_1^T Y, ..., c_q^T Y$, where the c_j are eigenvectors of $\Sigma_{yy}^{-1} \Sigma_{yx} \Sigma_{xx}^{-1} \Sigma_{xy}$ (Rao [2, Sect. 8f]), the eigenvalues being the corresponding squared canonical correlations. It is not necessary to impose any particular normalization on the c_j, although it is convenient for exposition to require that $c_j^T \Sigma_{yy} c_j = 1$; it is known that $c_j^T \Sigma_{yy} c_k = 0$ $(j \neq k)$. We now transform from Y to the $q \times 1$ vector $Z = C^T Y$, where C is the $p \times q$ matrix $(c_1 \cdots c_q)$.

Then the covariance matrix of Z is the identity and

$$\text{cov}(Z, X) = E(ZX^T) = C^T \Sigma_{yx}.$$

If now we apply the results of the previous section to the regression of Z on X we obtain a new variable given by

$$Y^* = \Sigma_{xx}(C^T \Sigma_{yx})^{-1} Z$$
$$= \Sigma_{xx}(C^T \Sigma_{yx})^{-1} C^T Y; \tag{2}$$

it is easily verified that when $p = q$ (2) reduces to (1), then Y^* does not depend on the particular normalization of C, and that, as before,

$$\text{cov}(Y^*, X) = \Sigma_{xx}.$$

We are very grateful to the referee for deriving by a different route involving the Loewner of matrices the solution

$$Y^* = \Sigma_{xx}(\Sigma_{xy} \Sigma_{yy}^{-1} \Sigma_{yx})^{-1} \Sigma_{xy} \Sigma_{yy}^{-1} Y. \tag{3}$$

To see that (2) and (3) are in fact identical, note that by the properties of canonical variables (Rao [2, Sect. 8f 1.2 and 1.5]) there exists a nonsingular matrix F such that

$$C^T \Sigma_{yx} F = \Gamma_q, \qquad \Sigma_{yy}^{-1} \Sigma_{yx} F = C \Gamma_q, \qquad F^T \Sigma_{xy} \Sigma_{yy}^{-1} \Sigma_{yx} F = \Gamma_q^2,$$

where Γ_q is the diagonal matrix of canonical correlations. Thus, on using these three results in turn, we have that the right-hand side of (2) is equal to

$$\Sigma_{xx} (\Gamma_q F^{-1})^{-1} C^T = \Sigma_{xx} (F \Gamma_q^{-1}) \Gamma_q^{-1} F^T \Sigma_{xy} \Sigma_{yy}^{-1}$$

$$= \Sigma_{xx} F F^{-1} (\Sigma_{xy} \Sigma_{yy}^{-1} \Sigma_{yx})^{-1} (F^T)^{-1} F^T \Sigma_{xy} \Sigma_{yy}^{-1}$$

which reduces to the right-hand side of (3). The form (3) has the major advantage of avoiding the eigen analysis involved in (2); on the other hand, in applications, we have found it wise always to compute the canonical correlations as some check that the smallest of them is large enough to make Y^* reasonably well defined.

It can be shown by calculating the covariance matrix of Y^* that if the explanatory variables are uncorrelated, then the derived variables also are uncorrelated.

When the number of explanatory variables exceeds the number of responses, $p < q$, it will be necessary to select p or fewer variables or combinations from the q before applying the method.

4. INFERENCE

The above results are for probability distributions. For application to data we shall assume multivariate normality, at least of Y given X, and therefore replace all population covariance matrices by the corresponding matrices of mean sums of squares and products. We regard the method as primarily a way of suggesting relatively simple derived variables and therefore to be used rather flexibly; thus elaborate discussion of formal inference procedures would be out of place. Nevertheless some simple results are available.

When $p = q$, we can obtain a confidence cone for the coefficients of, say, the first component of Y^*. Let it be hypothesized that $a^T Y$ is conditionally independent of $X_2, ..., X_p$ given X_1. This can be tested in the multiple regression of $a^T Y$ on X by a standard F test with degrees of freedom ($p - 1$, $n - p - 1$), where n is the number of observations. A confidence cone can be formed from the subset of a not rejected in such a significance test.

The hypothesis that Y_1^*, say, does not depend on a particular set of components of Y, say the last t components $Y_{p-t+1}, ..., Y_p$, can be tested by checking that the multivariate regression of $Y_1, ..., Y_{p-t}$ on X contains no contribution from $X_2, ..., X_p$ using any of the standard multivariate analysis of variance test statistics, e.g. the determinantal ratio (Rao [2, Sect. 8c.3]). If $t = 1$, i.e., only one component is hypothesized to be missing from the derived response, a standard F test is available.

The hope in using the present method will often be that one can find quite simple linear combinations of the components of the original Y that can replace the Y^* and that have a specific subject-matter interpretation. Simplicity here may mean that the coefficients defining Y have a simple interpretation or that each component Y_s^* involves only a limited number of the components of the original Y.

In some, but by no means all, cases the latter argument can be used as an alternative to the introduction of the canonical variables of Section 3. For example, suppose that $p = 3$, $q = 2$, and that on substantive grounds it is suspected that Y_1 is conditionally independent of Y_2 given (Y_3, Y_1, Y_2) that Y_1 is conditionally independent of X_2 given (Y_3, X_1), and that Y_2 is conditionally independent of X_1 given (Y_3, X_2). Suppose further that these relations are consistent with the data. Then an alternative to the use of canonical correlation as a method of reducing the dimension of Y from three to two is to restrict Y_1^* to be a combination of Y_1 and Y_3 and at the same time Y_2^* to be a combination of Y_2 and Y_3. It can then be verified that the appropriate combinations are, in the standard notation for conditional covariances,

$$Y_1^* = Y_1 - Y_3 \sigma_{Y_1 X_2 \cdot X_1} / \sigma_{Y_3 X_2 \cdot X_1},$$
$$Y_2^* = Y_2 - Y_3 \sigma_{Y_2 X_1 \cdot X_2} / \sigma_{Y_3 X_1 \cdot X_2}. \tag{3}$$

This relates the present analysis to the study of graphical models of conditional independency (Lauritzen and Wermuth [1]). We shall not explore this further here; in the example to be discussed in Section 5 this approach leads to essentially the same answer as reported there obtained by the method of Section 3.

5. An Example

Table I summarizes key aspects of observations obtained on 40 patients who have not received a preoperative treatment. There are three variables measured directly before an operation, the log concentrations of the three fatty acids palmitic acid, Y_1, linoleic acid, Y_2, and oleic acid, Y_3, and for the present purpose these form the response variable, Y. There are two explanatory variables forming the vector X, blood sugar measured the

COX AND WERMUTH

TABLE I

Observed Marginal Correlations, Means, and Standard Deviations for 40 Patients

Variable	Y_1	Y_2	Y_3	X_1	X_2
Log palmitic acid (Y_1)	1				
Log linoleic acid (Y_2)	0.90	1			
Log oleic acid (Y_3)	0.95	0.92	1		
Blood sugar (X_1)	−0.25	−0.27	−0.32	1	
Sex (X_2)	0.28	0.43	0.23	−0.03	1
Mean	4.91	4.26	4.88	80.93	0.05
Standard deviation	0.3726	0.4745	4.4073	9.1661	1.02

morning before the operation, X_1, and sex, X_2, the latter coded as 1 for females and −1 for males. Log concentrations are used partly because the concentrations themselves are positive variables with large coefficients of variation around 50% and hence with very skew distributions and partly because linear concentrations of logs with simple numerical coefficients may be hoped to have a simple interpretation.

We apply the results of Section 3 with $p = 3$, $q = 2$. The two nonzero canonical correlations between Y and X are 0.60 and 0.38, neither being near zero. This points to an appreciable relation between the derived responses Y_1^* and Y_2^* and the corresponding explanatory variables X_1 and X_2.

The transformation matrix obtained from (2) is

$$\begin{pmatrix} 110.3 & 17.5 & -163.5 \\ -3.0 & 8.1 & -9.7 \end{pmatrix}$$

and suggests taking as simple forms of derived variable $Y_1^* = Y_1 - Y_3$ and $Y_2^* = Y_2 - Y_3$. This implies that the ratio of palmitic to linoleic acid is primarily connected to blood sugar as is the ratio of linoleic to oileic acid to sex.

Table II gives the correlation matrix of Y_1^*, Y_2^*, X_1, X_2. In this particular example the derived variables turn out to be nearly uncorrelated and this, together with the negligible correlation between X_1 and X_2, yields a very simple structure in which (Y_1^*, X_1) are completely independent of (Y_2^*, X_2). A likelihood ratio test of consistency with this structure yields chi-squared of 0.59 with 4 degrees of freedom.

Thus the 9 nonnegligible correlations of the original variables have been reduced to a simple structure with just two appreciable correlations. This

TABLE II

Observed Marginal Correlations for the Derived Responses
and the Explanatory Variables of Table I

Variable	Y_1^*	Y_2^*	X_1	X_2
$Y_1^* = Y_1 - Y_3$	1			
$Y_2^* = Y_2 - Y_3$	0.09	1		
X_1	0.32	0.01	1	
X_2	0.09	0.57	−0.03	1

is a simplification special to this problem consequent on the essentially independent explanatory variables.

In general our procedure with $p = 3$, $q = 2$ imposes two conditional independencies, namely that Y_1^* is conditionally independent of X_2, given X_1, and that Y_2^* is conditionally independent of X_1 given X_2, leading usually to a nondecomposable independency structure (Wermuth and Cox, [3]) in the multivariate regression of Y^* on X requiring iterative fitting for maximum likelihood estimation.

The analysis was repeated on a different set of 40 patients for whom the same variables but quite different correlations were observed. The method based on (2) yielded essentially the same derived variables.

The computations were done using MATLAB.

6. DISCUSSION

While the method has been reasonably successful on the above example, it may often prove ineffective, even when the broad formulation in terms of linear combinations of Y that preserve the individual structure of X is quite appealing. The method will work best when the q canonical correlations are all reasonably large and there is no strong collinearity between the columns of X. In other cases only some of the components of the transformed vector may be reasonably well defined. For such reasons, it essential, as indeed with other relatively advanced methods, to have checks that the method is in some sense reasonably effective.

It should be verified that the derived variables do have appreciable regression on their target x-components, and if one or more of the canonical correlations is small, say less than 0.1–0.2, it is unlikely that all the components of the derived response will be effective.

ACKNOWLEDGMENTS

We are grateful to Dr. A. Theisen, University of Mainz, for the data of Section 5, to the referee for the result referred to in Section 3, and to the Anglo German Academic Research Collaboration Programme for supporting our joint work.

REFERENCES

[1] LAURITZEN, S. L., AND WERMUTH, N. (1989). Graphical models for associations between variables, some of which are qualitative and some quantitative. *Ann. Statist.* **17** 31–57.

[2] RAO, C. R. (1973). *Linear Statistical Inference and Its Applications.* 2nd ed. New York: Wiley.

[3] WERMUTH, N., AND COX, D. R. (1991). Explanations for nondecomposable hypotheses in terms of univariate recursive regressions including latent variables. In *Berichte zur Stochastic und verwandten Gebieten 91–2, Universität Mainz.*

Biometrika (1992), **79**, 3, *pp.* 441-61
Printed in Great Britain

Response models for mixed binary and quantitative variables

By D. R. COX

Nuffield College, Oxford OX1 1NF, U.K.

AND NANNY WERMUTH

Psychological Institute, University of Mainz, 6500 Mainz, Germany

SUMMARY

A number of special representations are considered for the joint distribution of qualitative, mostly binary, and quantitative variables. In addition to the conditional Gaussian models and to conditional Gaussian regression chain models some emphasis is placed on models derived from an underlying multivariate normal distribution and on models in which discrete probabilities are specified linearly in terms of unknown parameters. The possibilities for choosing between the models empirically are examined, as well as the testing of independence and conditional independence and the estimation of parameters. Often the testing of independence is exactly or nearly the same for a number of different models.

Some key words: Conditional Gaussian model; Graphical chain model; Linear model; Logistic function; Multivariate normal distribution; Probit model.

1. INTRODUCTION

The object of this paper is to compare a number of models for the joint distribution of quantitative and binary response variables. One role of such models is as a route for testing hypotheses of independence or conditional independence. We examine the extent to which essentially the same test arises from different models. A further important point is that for some models particular null hypotheses may be satisfied only under much stronger versions of independence than those it is desired to test, so that the models are unsuitable for the required purpose.

Two of the families of models under consideration are models based on conditional Gaussian distributions, i.e. for conditional normality of the continuous components and an arbitrary distribution for the discrete components, and on conditional Gaussian regressions. These take some conditional regression relations from a conditional Gaussian distribution and then separately assign distributions, possibly arbitrary, to the conditioning variables. The latter have been introduced as graphical chain models by Lauritzen & Wermuth (1989). In addition we examine some aspects of models in which the probabilities for the binary components are specified linearly and also of models in which there is an initial multivariate normal distribution from which the binary components, or more generally the ordinal components, are derived by forming discrete classes. Indeed the connection between multivariate continuous distributions, in particular the multivariate normal distribution, and binary and ordinal data has a long history and many facets. Probit-style models (Finney, 1952) for binary variables generated from a normal distribution of underlying observed 'tolerances' form probably the most familiar example. Another instance is Pearson's (1901) tetrachoric correlation in which the relation between two binary variables is summarized via a bivariate normal distribution fitted to the contingency table formed from the discrete responses.

We shall study first in two, then in three, dimensions aspects of the following: the general nature of the relationships between various models, the implications for estimation, and the implications for testing null hypotheses of independence or conditional independence.

2. BIVARIATE DISTRIBUTIONS

2·1. *Some distributional results*

We consider first just two response variables, i.e. we focus on the joint distribution of two random variables. It is convenient to separate the discussion into the study of a continuous response conditional on a discrete explanatory variable and an analysis the other way round. This is not a conventional study of dependence, where, even if the explanatory variable is random, conditioning on the observed values of the explanatory variable is used in inference.

Suppose first that (U, X) are bivariate normal with zero means, unit variances and correlation ρ_{xu}. Let a dichotomous variable A be formed from U via a cut-off point α; we write

$$A = \begin{cases} 1 & (U \geq \alpha), \\ 0 & (U < \alpha). \end{cases} \tag{2·1}$$

Then derivatives of the moment generating function of a truncated normal distribution (Tallis, 1961) or direct calculations show that

$$\mu_u^+(\alpha) = E(U \mid A = 1) = \phi(\alpha)/\Phi(-\alpha),$$
$$\mu_u^-(\alpha) = E(U \mid A = 0) = -\phi(\alpha)/\Phi(\alpha), \tag{2·2}$$

where $\phi(.)$ and $\Phi(.)$ are respectively the standard normal density and integral. Further

$$\sigma_{uu}^+(\alpha) = \text{var}(U \mid A = 1) = 1 + \alpha\mu_u^+(\alpha) - \{\mu_u^+(\alpha)\}^2,$$
$$\sigma_{uu}^-(\alpha) = \text{var}(U \mid A = 0) = 1 + \alpha\mu_u^-(\alpha) - \{\mu_u^-(\alpha)\}^2, \tag{2·3}$$

and the third cumulants, or third moments about the mean, are

$$\kappa_{3,u}^+(\alpha) = (\alpha^2 - 1)\mu_u^+(\alpha) - 3\alpha\mu_u^+(\alpha)^2 + 2\mu_u^+(\alpha)^3,$$

with a corresponding formula for $\kappa_{3,u}^-(\alpha)$, and from these the standardized third cumulants

$$\gamma_{1,u}^+(\alpha) = \kappa_{3,u}^+(\alpha)/\{\sigma_{uu}^+(\alpha)\}^{3/2} \tag{2·4}$$

and $\gamma_{1,u}^-(\alpha)$ are calculated directly.

Because $X = \rho_{xu}U + \varepsilon_{x.u}$, where $E(\varepsilon_{x.u}) = 0$, $\text{cov}(U, \varepsilon_{x.u}) = 0$, $\text{var}(\varepsilon_{x.u}) = \sigma_{xx.u} = 1 - \rho_{xu}^2$, we have that

$$\mu_{x.a}^+ = E(X \mid A = 1) = \rho_{xu}\mu_u^+(\alpha), \quad \sigma_{xx.a}^+ = \text{var}(X \mid A = 1) = \rho_{xu}^2\sigma_{uu}^+(\alpha) + (1 - \rho_{xu}^2),$$

with corresponding formulae for $\mu_{x.a}^-$, $\sigma_{xx.a}^-$ obtained by replacing $+$ by $-$ everywhere. Further

$$\kappa_{3,x.a}^+ = E\{(X - \mu_{x.a}^+)^3 \mid A = 1\} = \rho_{xu}^3\kappa_{3,\mu}^+(\alpha).$$

Note that if U is not observed there is no loss of generality in taking it in standardized form. If Y has mean μ_y and variance σ_{yy}, the above formulae for X correspond to the moments of $(Y - \mu_y)/\sqrt{\sigma_{yy}}$ so that in this general case

$$\mu_{y.a}^+ = \mu_y + \rho_{yu}\sigma_{yy}^{\frac{1}{2}}\mu_u^+(\alpha), \quad \sigma_{yy.a}^+ = \sigma_{yy}\{\rho_{yu}^2\sigma_{uu}^+(\alpha) + (1 - \rho_{yu}^2)\},$$

$$\kappa_{3.y.a}^+ = \sigma_{yy}^{3/2}\rho_{yu}^3\kappa_{3.u}^+(\alpha). \tag{2.5}$$

For studying dependencies of A on Y we start from the conditional distribution of U given $Y = y$ which is normal with mean $\rho_{yu}(y - \mu_y)/\sqrt{\sigma_{yy}}$ and variance $(1 - \rho_{yu}^2)$ so that

$$\text{pr}\,(A = 1 \mid Y = y) = \Phi\left\{\frac{\rho_{yu}(y - \mu_y)/\sqrt{\sigma_{yy}} - \alpha}{\sqrt{(1 - \rho_{yu}^2)}}\right\}. \tag{2.6}$$

In many practical cases this will be virtually indistinguishable from a linear logistic regression (Cox, 1966; Cox & Snell, 1989, p. 22), i.e. the simplest form of a conditional Gaussian regression with discrete response, and provided the probabilities are not too extreme, say $0.2 \leqslant \text{pr}\,(A = i \mid Y = y) \leqslant 0.8$, this implies near linearity of the log odds, i.e.

$$\log\frac{\text{pr}\,(A = 1 \mid Y = y)}{\text{pr}\,(A = 0 \mid Y = y)} \simeq d\frac{\rho_{yu}(y - \mu_y)/\sqrt{\sigma_{yy}} - \alpha}{\sqrt{(1 - \rho_{yu}^2)}}, \tag{2.7}$$

where the most suitable value of the constant d depends on the range over which the approximation is required. To match the functions at the 20%, 50% and 80% points, we take $d = 1.65$.

By contrast, in the conditional Gaussian distribution the conditional distributions of Y given $A = 0, 1$ are normal. In the homogeneous case, the conditional variances are the same at both levels of A, while the marginal probabilities $\text{pr}\,(A = i)$ are positive but otherwise arbitrary. The relationship $\text{pr}\,(A = 1 \mid Y = y)$ derived from this joint distribution is linear logistic in the homogeneous and quadratic logistic in the nonhomogeneous case, while the marginal distribution of Y is a mixture of normals. Any conditional Gaussian regression looks like a conditional distribution derived from a joint conditional Gaussian distribution: with a dichotomous response, A, it is a logistic regression and with a continuous response, Y, it is a linear regression. A joint distribution defined by a sequence of conditional Gaussian regressions and a marginal conditional Gaussian distribution of the variables which are not responses is called a conditional Gaussian regression chain model. With just two variables, A and Y, the conditional Gaussian regression chain model with Y as response to A defines a joint conditional Gaussian distribution while, in general, this is not the case for a conditional Gaussian regression chain model with A as response to Y.

Both the conditional Gaussian regression chain model for A as response to Y and the dichotomized normal model are special cases of one in which Y is marginally normal and in which

$$\pi_{1|y}^{A|Y} = \text{pr}\,(A = 1 \mid Y = y) = G(\theta_0^{(g)} + \theta_1^{(g)}y), \tag{2.8}$$

where $G(x)$ is the logistic function $L(x) = e^x/(1 + e^x)$ for the conditional Gaussian regression chain model, and $G(x)$ is the standardized normal distribution function $\Phi(x)$ for the dichotomized normal. Clearly other choices of G are possible, conceivably containing additional nuisance parameters to allow a data-based choice of G (Aranda-Ordaz, 1981). Despite some obvious limitations, the choice $G(x) = x$, leading to a linear model for probabilities, is useful in particular as an approximation to the logistic and probit functions wherever, as noted above, the conditional probabilities are largely confined to the range $(0 \cdot 2, 0 \cdot 8)$. We therefore write

$$\pi_{1|y}^{A|Y} = \pi_1^A + \beta_{ay}(y - \mu_y), \qquad \pi_{0|y}^{A|Y} = \pi_0^A - \beta_{ay}(y - \mu_y), \qquad (2\cdot9)$$

provided that the values so defined are in $[0, 1]$. In some applications the assumption that the probabilities are confined to a central range is entirely reasonable. An example are probabilities of changing a field of study investigated by Weck (1991) under a wide range of different conditions in Germany; see § 3·2.

One advantage of the linear representation is the very direct interpretation of the parameters and another is that marginalization over Y is immediate. Provided only that $E(Y) = \mu_y$, we recover the stated marginal probability. Under the dichotomized normal model, the corresponding probability is $\Phi(-\alpha)$, whereas under the conditional Gaussian regression chain model the marginal probability is $E_Y\{L(\theta_0^{(\lambda)} + \theta_1^{(\lambda)} Y)\}$, where the expectation is over the normal distribution of Y, $N(\mu_y, \sigma_y^2)$. While this cannot be evaluated exactly in closed form, a good approximation is obtained by writing $L(x) \simeq \Phi(xc)$, where $c = 0 \cdot 607$. Then, on omitting the superscript λ for convenience, we have

$$E\{L(\theta_0 + \theta_1 y)\} \simeq \int_{-\infty}^{\infty} \Phi(c\theta_0 + c\theta_1 y)\phi(y; \mu_y, \sigma_y^2) \, dy,$$

where $\phi(y; \mu_y, \sigma_y^2)$ is the density of $N(\mu_y, \sigma_y^2)$. This integral can be evaluated in closed form to give

$$\Phi\left\{\frac{c(\theta_0 + \theta_1\mu_y)}{\sqrt{(1 + c^2\theta_1^2\sigma_y^2)}}\right\} \simeq L\left\{\frac{\theta_0 + \theta_1\mu_y}{\sqrt{(1 + c^2\theta_1^2\sigma_y^2)}}\right\}. \qquad (2\cdot10)$$

The advantages of the simpler marginalization of $(2\cdot9)$ are particularly strong in a larger number of dimensions.

A model in which, for example, the conditional distribution of Y given $A = i$ is normal and the marginal distribution of Y therefore nonnormal is distinct from a model in which the marginal distribution is normal. Nevertheless the separation of the two normal components in the first model must be appreciable if the distinction is to be detectable with realistic amounts of data. This can be verified numerically or seen analytically by noting that a mixture with probabilities $(\frac{1}{2} + \frac{1}{2}\xi), (\frac{1}{2} - \frac{1}{2}\xi)$ of normal distributions of means $\delta, -\delta$ and unit variances has density

$$f_y^Y = \frac{1}{\sqrt{(2\pi)}} \{\tfrac{1}{2}(1 + \xi) e^{-\frac{1}{2}(y-\delta)^2} + \tfrac{1}{2}(1 - \xi) e^{-\frac{1}{2}(y+\delta)^2}\}$$

$$= \frac{1}{\sqrt{\{2\pi(1 + \delta^2 - \xi^2\delta^2)\}}} \exp\{-\tfrac{1}{2}(1 + \delta^2 - \xi^2\delta^2)^{-1}(y - \xi\delta)^2\}\{1 + O(\delta^3)\}, \qquad (2\cdot11)$$

after some manipulation, thus showing that nonnormality enters only in the term of order δ^3. A similar argument shows that if the marginal distribution of Y is normal and the relation between A and Y probit or logistic confined to the range $(0 \cdot 2, 0 \cdot 8)$ and hence effectively linear, then the conditional distribution of Y given $A = i$ is very close to normality. To see this analytically note that, with $i^* = 2i - 1$,

$$f_{y|i}^{Y|A} = f_y^Y f_{i|y}^{A|Y} / f_i^A \simeq \frac{1}{\sqrt{(2\pi)\sigma_y}} \exp\left\{-\frac{(y - \mu_y)^2}{2\sigma_y^2}\right\}\{f_i^A + \gamma_{ay}(y - \mu_y)i^*\}/f_i^A. \qquad (2\cdot12)$$

The limitation on the probabilities implies that $\gamma_{ay}\sigma_y$ is small. If we write $\gamma_{ay} = \varepsilon/\sigma_y$ and incorporate the linear term into the exponential we have that with an error of order ε^3 the conditional distributions are normal with the same variance.

2·2. General statistical interpretation

A number of broad conclusions can be drawn from the above results, in particular from the numerical results in Tables 1 and 2 which are based directly on the formulae of § 2·1. By the symmetry of the problem it is enough to suppose that $\alpha \geqslant 0$.

Table 1. *Variance ratios* $\sigma^+_{yy.a}/\sigma^-_{yy.a}$ *of* Y *given* $(A = i)$
in a dichotomized bivariate normal distribution; α, *cut-off point for dichotomized* U; ρ_{yu}, *correlation*

α	$\rho_{yu} = 0\cdot2$	$\rho_{yu} = 0\cdot5$	$\rho_{yu} = 0\cdot8$	$\rho_{yu} \simeq 0\cdot99$
0	1·000	1·000	1·000	1·000
0·5	0·991	0·938	0·639	0·552
1·5	0·975	0·835	0·533	0·193
2·5	0·965	0·781	0·429	0·093

Table 2. *Standardized skewnesses* $\gamma^+_{1y.a}$ *and* $\gamma^-_{1y.a}$ *of* Y
given $(A = i)$ *in a dichotomized bivariate normal distri-*
bution; α, *cut-off point;* ρ_{yu}, *correlation*

α	Level of A	$\rho_{yu} = 0\cdot2$	$\rho_{yu} = 0\cdot5$	$\rho_{yu} = 0\cdot8$	$\rho_{yu} \simeq 0\cdot99$
0·0	+	0·002	0·035	0·245	0·995
	−	−0·002	−0·035	−0·245	−0·995
0·5	+	0·001	0·028	0·215	1·169
	−	−0·002	−0·042	−0·252	−0·800
1·5	+	0·001	0·015	0·139	1·444
	−	−0·002	−0·036	−0·172	−0·391
2·5	+	0·000	0·008	0·084	1·676
	−	−0·001	−0·012	−0·051	−0·102

If emphasis is on the conditional distribution of the continuous component Y given the discrete component A, then, under the model based on a bivariate normal distribution, unequal variances combined with skewness will be encountered in the two groups if α and ρ_{yu} are not equal to zero. These effects are likely to be important and empirically detectable only if the correlation between Y and U is fairly high, and, for the inequality of variances, if the dichotomy of U is into quite unequal groups. By contrast in the conditional Gaussian distribution model, the conditional distributions of Y given A are normal and in its homogeneous form have equal variances.

In the dichotomized normal model the conditional distribution of A given Y has probit form: clear departure from that would be evidence against an underlying bivariate normal distribution. The joint distribution of (A, Y) has four independent parameters which can be taken in various forms; $(\mu_y, \sigma_{yy}, \rho_{yu}, \alpha)$ is one natural choice. The parameters can be estimated in several ways; see § 2·3.

Broadly similar results apply when U is divided into three groups. If the trichotomy is symmetrical, the variance of Y within groups will be the same in the outer groups and different, in general, for the central group. Furthermore, the conditional distributions of

A given Y will have probit form in the outer groups but not for the central group. As the number of groups increases we quite rapidly approach recovery of the information about correlation that would be available were the underlying continuous variables to the observed (Cox, 1958).

2·3. Estimation

The estimation of parameters from a conditional Gaussian distribution model and from a conditional Gaussian regression chain model with discrete response follows standard maximum likelihood methods. For n independent observations $(i_1, y_1), \ldots, (i_n, y_n)$ from the dichotomized bivariate distribution, where $i_r = 1$ if $U_r > \alpha$ and $i_r = 0$ if $U_r \leqslant \alpha$, the log likelihood is best written as the sum of the marginal log likelihood from $(y_1, \ldots, y_n)$ and the log conditional likelihood for A given $Y = y$ writing $\text{pr}(A = 1 | Y = y)$ in the form $\Phi(\theta_0 + \theta_1 y)$, where

$$\theta_0 = -(\rho_{yu}\mu_y/\sigma_{yy}^{\frac{1}{2}} + \alpha)(1 - \rho_{yu}^2)^{-\frac{1}{2}}, \quad \theta_1 = \rho_{yu}\{\sigma_{yy}(1 - \rho_{yu}^2)\}^{-\frac{1}{2}}.$$

Thus, $\hat{\mu} = \bar{y}$, $\hat{\sigma}_{yy} = \sum (y_r - \bar{y})^2/n$ from marginal normality and they are asymptotically uncorrelated with $\hat{\theta}_0, \hat{\theta}_1$ derived via a probit analysis (Mailtz, 1955) of $(i_1, \ldots, i_n)$ on $(y_1, \ldots, y_n)$.

An older and computationally simpler method proceeds by first noting that because $\text{pr}(U > \alpha) = \text{pr}(A = 1) = \Phi(-\alpha)$ we can estimate α by $\tilde{\alpha} = \Phi^{-1}(\bar{i})$, where $\bar{i}$ is the overall proportion of 1's. Further

$$E(Y|A=1) - E(Y|A=0) = \rho_{yu}\sigma_{yy}^{\frac{1}{2}}\phi(\alpha)\{\Phi(\alpha)\Phi(-\alpha)\}^{-1}, \tag{2·13}$$

so that in a self-explanatory notation we can write

$$\tilde{\rho}_{yu} = \frac{(\bar{Y}^+ - \bar{Y}^-)\bar{i}(1 - \bar{i})}{\hat{\sigma}_{yy}^{\frac{1}{2}}\phi\{\Phi^{-1}(\bar{i})\}}, \tag{2·14}$$

where $\hat{\sigma}_{yy} = \sum (y_r - \bar{y})^2/n$. This is Pearson's (1903) biserial correlation coefficient. The dependence of $\tilde{\rho}_{yu}$ on $\bar{i}$ near $\bar{i} = \frac{1}{2}$, i.e. median dichotomy, is very slow; as $\bar{i}$ varies from 0·5 to 0·3 the $\bar{i}$-dependent factor in (2·14) changes only from $0·624 = \sqrt{(\pi/8)}$ to 0·604. Then, for some purposes it is sensible to replace (2·14) by

$$\tilde{\rho}_{yu}^* = \frac{(\bar{Y}^+ - \bar{Y}^-)}{\hat{\sigma}_{yy}^{\frac{1}{2}}}\sqrt{\left(\frac{\pi}{8}\right)}. \tag{2·15}$$

Comparison of the asymptotic variances of $\tilde{\rho}_{yu}$ and $\hat{\rho}_{yu}$ shows that the asymptotic efficiency of $\tilde{\rho}_{yu}$ relative to $\hat{\rho}_{yu}$ is 1·00 if $\rho = 0$, a result related to a general result about testing independence to be discussed in § 2·4. The efficiency is, however, appreciably less than one when $\rho > \frac{1}{2}$, say (Tate, 1955). For this result one needs the asymptotic variance of $\tilde{\rho}_{yu}$ (Soper, 1915), calculated most directly by finding the asymptotic covariance matrix of $\sum I_r$, $\sum Y_r$, $\sum I_r Y_r$, $\sum (Y_r - \bar{Y})^2$ in terms of which $\tilde{\rho}_{yu}$ can be expressed. Note that to the required accuracy $\sum (Y_r - \bar{Y})^2$ can be replaced by $\sum (Y_r - \mu_y)^2$. The asymptotic variance of $\hat{\rho}_{yu}$ is obtained from the Fisher information matrix (Tate, 1955; Prince & Tate, 1966).

Fitting of the model linear in probabilities (2·9) is most conveniently done by unweighted least squares applied to the (0, 1) responses, comparing the residual mean square to $\bar{i}(1 - \bar{i})$ for an approximate test of adequacy. It would also be possible to fit this model by maximum likelihood. Approximate calculations in Appendix 1 imply that, provided the fitted probabilities lie in the range (0·2, 0·8), the point estimates will be nearly the same and the reduction in variance is small, at most 5% and usually much less than this.

An example in which least squares and maximum likelihood estimation show these properties is given with the following data, collected by Dr N. Schmitt in connection with a medical dissertation at the pain clinic in Mainz. Success of treatment is predicted from a score for stage of chronic pain for $n = 58$ male patients treated for three weeks in the pain clinic. Table 3 shows predicted probabilities of successful treatment under a logit, a probit and a linear-in-probabilities regression. The only notable difference is that, in fitting a linear model by maximum likelihood, relatively greater weight is attached to the two individuals at the extreme level of chronic pain $y = 11$.

Table 3. *Different estimates for regression of treatment success on stage of chronic pain, y*

			Probabilities of treatment success, estimated by				
			observed				
		Number	relative	linear	linear	linear regression in	
	Total	of	frequen-	logistic	probit	binary responses via	
Stage	count	successes	cies	regression	regression	LS	ML
y	n_y	n_{1y}	n_{1y}/n_y	(a)	(b)	(c)	(d)
6	8	7	0·88	0·75	0·76	0·75	0·78
7	9	5	0·56	0·64	0·64	0·63	0·64
8	15	6	0·40	0·50	0·50	0·50	0·51
9	14	6	0·43	0·36	0·36	0·37	0·37
10	10	3	0·30	0·24	0·24	0·24	0·23
11	2	0	0·00	0·15	0·15	0·11	0·01

(a) $\log(\hat{\pi}_{1|y}/\hat{\pi}_{0|y}) = 4\cdot52 - 0\cdot57y,$ (b) $\Phi^{-1}(\hat{\pi}_{1|y}/\hat{\pi}_{0|y}) = 2\cdot82 - 0\cdot35y,$

(c) $\hat{\beta}_{LS} = (1\cdot52, -0\cdot13),$ $\mathrm{cov}(\hat{\beta}_{LS}) = \begin{pmatrix} 0\cdot1205 & -0\cdot0140 \\ \cdot & 0\cdot0017 \end{pmatrix}$

(d) $\hat{\beta}_{ML} = (1\cdot61, -0\cdot14),$ $\mathrm{cov}(\hat{\beta}_{ML}) = \begin{pmatrix} 0\cdot1170 & -0\cdot0136 \\ \cdot & 0\cdot0016 \end{pmatrix}$

2·4. *Tests of independence*

In § 2·3 the emphasis is on estimation, for example of the correlation coefficient in an underlying bivariate normal distribution. Sometimes, however, there is special interest in testing the null hypothesis of independence between discrete and continuous components. Here there is a certain robustness to formulation which arises also in purely continuous and purely discrete cases.

For example, the optimal test for independence in a bivariate normal distribution of (X, Y) can be regarded as a test of the correlation coefficient, treating both variables symmetrically, or as a test of linear regression either of Y on X or of X on Y, which in turn can be regarded as arising in a number of ways. Very similar remarks apply to the purely binary case where Fisher's exact test for 2×2 table can be derived from several viewpoints, and analogous results are available for $r \times s$ tables (Birch, 1963).

Faced with a random sample from a mixed binary and continuous distribution, one directly appealing test of independence is based on the difference between the means of the continuous variable Y at the two levels of the binary variable A, standardized similarly to the Student t statistic to produce under the null hypothesis a statistic with approximately a standard normal distribution. Note that the unequal variances examined in § 2·1 arise only when $\rho_{yu}^2 \neq 0$. The resulting test is exactly or asymptotically optimal under the following sets of assumptions.

(i) For each $A = i$, Y is normal with mean μ_i $(i = 0, 1)$ and constant variance, the null hypothesis of independence being equivalent to $\mu_1 = \mu_0$. Here the t test of a difference is directly appropriate.

(ii) To cover several possible types of departure from independence, suppose that the joint distribution of (A, Y) is specified by a marginal density $f_y^Y = g(y; \theta)$ for Y and a conditional distribution for A given $Y = y$ of the form

$$\pi_{i|y}^{A|Y} = \mathrm{pr}\,(A = i \mid Y = y) = \{h(\alpha, \beta y)\}^i \{1 - h(\alpha, \beta y)\}^{1-i}, \qquad (2\cdot16)$$

where g, h are known functions, and α, β, θ are unknown parameters with independent parameter spaces. The null hypothesis of independence is $\beta = 0$. A crucial point is that y appears linearly in $(2\cdot16)$ multiplying the parameter β, so that probit and logistic regressions are among the many special cases. On evaluating the β-component of the derivative of the log likelihood function at the null hypothesis, we obtain

$$\left[\frac{\partial h(\alpha, \phi)}{\partial \phi}\right]_{\phi=0} \left[\sum_{i:i_j} y_{ij}/h(\alpha, 0) - \sum_{r:i_j=0} v_{ir}/\{1 - h(\alpha, 0)\}\right]. \qquad (2\cdot17)$$

Now under the null hypothesis, the maximum likelihood estimate of $h(\alpha, 0)$ is the proportion of observations with $i_r = 1$ so that $(2\cdot17)$ can be replaced by the difference of the two sample means, leading after standardization to the Student t statistic as having the usual asymptotic properties of the score test; see for example Cox & Hinkley (1974, pp. 315, 324). Note that, if y in $(2\cdot16)$ is replaced by suitable nonlinear functions of y, robust tests of location can be generated.

(iii) Finally, note that, under the null hypothesis $\beta = 0$, very generally $(y_1, \ldots, y_n, \Sigma\, i_r)$ or some reduction thereof is sufficient for (α, θ). If then a test is based on $\Sigma\, i_r y_r$, the sample total in the $A = 1$ group, an 'exact' test can be obtained from the permutation distribution of the sample total and will typically be close to the Student t test.

Thus as in the purely continuous or purely binary cases, essentially the same test of independence can be derived from various viewpoints.

2·5. Bivariate response plus explanatory variable

Now suppose that, in addition to the bivariate response variable (A, Y), there is on each individual a vector z of explanatory variables which are either not random or, if random, are treated as fixed at their observed values for the purpose of analysis. In the spirit of the previous discussion, some relatively simple models for interpretation of such data are as follows, with $\bar{z}$ denoting the mean of z over the data.

(i) We have that Y given Z is normal with mean $\mu_y + \beta_{yz}^T(z - \bar{z})$, variance $\sigma_{yy.z}$ and, conditionally on $Y = y$ and z, A is governed by a probit law:

$$\mathrm{pr}\,(A = 1 \mid Y = y, z) = \Phi\{\gamma_{a.yz}^{(p)} + \gamma_{ay.z}^{(p)}(y - \mu_y) + \gamma_{az.y}^{T(p)}(z - \bar{z})\}. \qquad (2\cdot18)$$

This is a direct generalization of the model discussed above and can be derived via an underlying bivariate normal distribution of (U, Y) with U dichotomized to form A.

(ii) The probit relation $(2\cdot18)$ can be replaced by a logistic relation. This specifies a homogeneous conditional Gaussian regression chain model if Z has marginally a normal distribution.

(iii) We have that A is governed by a probit law conditionally on z,

$$\mathrm{pr}\,(A = 1 \mid z) = \Phi\{\gamma_{a.z}^{(p)} + \gamma_{az}^{\mathrm{T}(p)}(z - \bar{z})\}, \qquad (2\cdot19)$$

and, given $A = i$ and z, Y is conditionally normal with constant variance $\sigma_{yy.za}$ and mean

$$\beta_{y.az} + \beta_{ya.z}i + \beta_{yz.a}^{\mathrm{T}}(z - \bar{z}). \qquad (2\cdot20)$$

(iv) The probit relation (2·19) can be replaced by a logistic relation. Normality of Y can then take on at least two forms: (a) if in this case Z is conditionally normal given $A = i$ and has constant variance at both levels of A this specifies a homogeneous conditional Gaussian distribution for A, Y, Z; (b) if, however, Y is marginally normal then a homogeneous conditional Gaussian regression chain model with discrete response results which is different from the one under (ii).

The above models have strong assumptions not only of linearity but, at least as importantly, of parallelism of regression lines. Nonparallelism, at least in extreme cases, can have major substantive implications. Therefore checks of parallelism are necessary in applications.

For instance in (2·20) we may allow $\beta_{yz.a}^{\mathrm{T}}$ to depend on levels of A, for example by inserting $i(z - \bar{z})$ as an additional explanatory variable, possibly constraining the non-parallelism to certain components of z. This gives generalized linear models (McCullagh & Nelder, 1989). If in addition the variance of Y $\sigma_{yy.za}$ is allowed to depend on the level of A, then this specifies not a generalized linear model, but, in the case of marginally normal Y, that is, case (b) under (iv), a nonhomogeneous graphical chain model.

2·6. *Comparison of models*

The results of §§ 2·2 and 2·4 suggest that empirical choice between the various models studied here is likely to be feasible only when substantial correlation is present between the binary and continuous components. The difficulties are likely to be compounded when 'fixed' explanatory variables are present.

A key distinction is between the conditional Gaussian distribution in which for each $A = i$ the continuous variable Y has a 'simple' form and those models in which the marginal distribution of Y is of 'simple' form. Typically, simplicity in the marginal distribution of Y corresponds to fairly complicated conditional distributions and vice versa; see Table 4. However, whenever the linear-in-probability model is a suitable approximation to the conditional dependence of A on Y, then marginal normality of Y corresponds to approximately normal conditional distributions of Y given $A = i$ by (2·12), while conditional normal distributions of Y given $A = i$ correspond to an approximately normal marginal distribution of Y by (2·11). If in this latter case the conditional distribution of A given $Y = y$ is of probit form, it corresponds to an underlying bivariate normal distribution; if instead the conditional distribution of A given $Y = y$ is of logistic form, the specifications correspond to those of a homogeneous conditional Gaussian regression chain model, and, as noted before, these two models are likely to be close.

Dr G. K. Reeves, in work as yet unpublished, has confirmed these qualitative conclusions by imbedding the conditional Gaussian distribution and the conditional Gaussian regression chain models in a single family containing an additional parameter taking values 0 and 1 for the two families in question and showing that the profile likelihood for that parameter is typically very flat.

Table 4. *Distributional properties of models for A, Y*

Specification of model	A given $Y = y$	Y	Y given $A = i$	A
(i) (U, Y) bivariate normal, A results from partitioning U	Probit regression	Normal	Special skewed distributions of unequal variances (Tables 1, 2)	Arbitrary
(ii) Homogeneous CG-regression chain model with Y as response*	Logistic regression with $\gamma_{ay}^{(l)} = (\mu_1 - \mu_0)/\sigma^2$	Mixture of normals	Normal	Arbitrary
(iii) Homogeneous CG-regression chain model with A as response	Logistic regression with arbitrary γ_{ay}	Normal	Approximately like (i) (2·10)	Arbitrary
(iv) Y normal, $0 \cdots$ depends linearly on y	Linear-in-probabilities regression	Normal	Approximately normal (2·12)	Requires $0\cdot2 \leqslant \pi_i^A \leqslant 0\cdot8$
(v) $Y \mid (A = i)$ normal $0\cdot2 \leqslant \pi_{i\mid y}^{A\mid Y} \leqslant 0\cdot8$ depends linearly on y	Linear-in-probabilities regression	Approximately normal (2·11)	Normal	Requires $0\cdot2 \leqslant \pi_i^A \leqslant 0\cdot8$

* Equivalant to a CG, conditional Gaussian, distribution for A, Y.

3. Trivariate distributions

3·1. *Preliminaries*

We now consider three response variables, at least one discrete, usually binary. There are many ways of specifying the joint distribution and to some extent the most suitable formulation for a particular application depends on the questions to be asked, for example the kinds of conditional independencies under investigation. There are further questions as to the extent that different models can in practice be distinguished empirically and, corresponding to the discussion of § 2·4, the extent to which tests of independence derived from different models are essentially the same.

3·2. *Three binary response variables*

It is necessary to begin by reviewing methods for economical representation of three binary variables

$$\pi_{ijk}^{ABC} = \mathrm{pr}\,(A = i, B = j, C = k) \quad (i, j, k = 0, 1).$$

Some can be derived by specializing the saturated model

$$\pi_{ijk}^{ABC} = H(\mu^{(h)} + \xi_A^{(h)} i^* + \xi_B^{(h)} j^* + \xi_C^{(h)} k^* + \xi_{AB}^{(h)} i^* j^* + \xi_{AC}^{(h)} j^* k^* + \xi_{BC}^{(h)} j^* k^* + \xi_{ABC}^{(h)} i^* j^* k^*),$$

$$(3\cdot1)$$

where $i^* = 2i - 1$, etc. and so takes values $(-1, 1)$ as i takes values $(0, 1)$; $\mu^{(h)}$ is a normalizing constant and $H(x)$ is a suitable function. The choices $H(x) = e^x$, $H(x) = x$, $H(x) = \Phi(x)$ are the most common, the first having advantages for the expressions of conditional independencies, the second allowing the simple calculation of marginal distributions, and the last being closely related to tetrachoric correlations, having the longest history. The corresponding parameters are denoted by (l) for log linear, are written without superscript in the linear case, and get the superscript (p) for joint probit in the last case.

The representations above arise when the joint distribution of the variables A, B, C is a natural starting point, treating the three variables on an equal footing. If, however, there is an univariate recursive system with A being a response to B, C, and B being a response to C, a different representation suggests itself. In this case we write

$$\pi_{1|jk}^{A|BC} = G(\gamma_{a.bc(bc)}^{(g)} + \gamma_{ab.c(bc)}^{(g)} j^* + \gamma_{ac.b(bc)}^{(g)} k^* + \gamma_{a(bc).bc}^{(g)} j^* k^*),$$

$$\pi_{1|k}^{B|C} = G(\gamma_{b.c}^{(g)} + \gamma_{bc}^{(g)} k^*), \quad \pi_1^C = G(\gamma_c^{(g)}),$$

(3·2)

where, as before, $j^* = 2j - 1$, $k^* = 2k - 1$ and $G(x)$ is a suitable function. The choices $G(x) = L(x)$, $G(x) = x$ and $G(x) = \Phi(x)$ correspond most directly to those of $H(x)$ discussed for (3·1). The corresponding parameters are denoted by (l) for logit regression, are written without a superscript in the linear regression case and have superscript (p) for probit regression, in the last case.

If $G(x) = L(x)$, the equations in (3·2) are logit regressions with discrete explanatory variables. The parameters in such a regression relate in a simple way to the parameters in a log linear model for the corresponding joint probabilities, since from $G(x) = L(x)$ we get e.g.

$$\log(\pi_{1|jk}^{A|BC} / \pi_{0|jk}^{A|BC}) = \log \pi_{1jk}^{ABC} - \log \pi_{0jk}^{ABC}.$$

Conditional independencies correspond to vanishing logit regression coefficients, for instance $A \perp B \mid C$ is expressed by $0 = \gamma_{a(bc).bc}^{(l)} = \gamma_{ab.c(bc)}^{(l)}$ or $B \perp C$ is expressed by $0 = \gamma_{bc}^{(l)}$. The question of when the same independence structure results from restrictions on systems of recursive logit regressions like (3·2) or more complex ones and from restrictions on a corresponding log linear model has been answered by Wermuth & Lauritzen (1983).

If $G(x) = \Phi(x)$ then the equations in (3·2) are probit regressions with discrete explanatory variables. The parameters in these probit regressions do not relate in a simple way to the parameters in probits for the corresponding joint probabilities except in the case of median dichotomized bivariate normal variates. Note that for a trivariate normal distribution the joint probit model of (3·1) and the corresponding system of recursive probit regressions (3·2) are not equivalent even in the case of median-dichotomizing all three variables. Similarly, the parameters of the linear-in-probabilities regressions, that is $G(x) = x$, do not in general connect simply to the parameters of a linear model for the corresponding joint probabilities; see (A2·14). However, they mimic relations connecting total with partial regression coefficients. For instance, we can write in the case of $\gamma_{a(bc).bc} = 0$:

$$E_{A|C}(\pi_{1|k}^{A|C}) = E_{B|C} E_{A|BC}(\pi_{1|jk}^{A|BC}) = \gamma_{a.c} + \gamma_{ac} k^*,$$

where

$$\gamma_{a.c} = \gamma_{a.bc} + \gamma_{ab.c} \gamma_{b.c}, \quad \gamma_{ac} = \gamma_{ac.b} + \gamma_{ab.c} \gamma_{bc}.$$

Further conditional probabilities such as $\pi_{j|ik}^{B|AC}$ can of course be calculated but involve ratios of combinations of the original parameters. See Appendix 2 for more detailed discussion.

A data set in which recursive linear-in-probabilities regressions give an appropriate description is taken from Weck (1991, p. 182) for $n = 2026$ German students. The three binary variables ($1 = $ yes, $0 = $ no) are: A, change of field study; B, poor integration in high school classes; C, change of primary school. The counts n_{ijk} are

$$(n_{111}, n_{011}, n_{101}, n_{001}, n_{110}, n_{010}, n_{100}, n_{000}) = (15, 33, 84, 278, 40, 113, 246, 1217).$$

The saturated model of the type (3·2) with $G(x) = x$, as in (A2·12), is

$$\hat{\pi}_{1|jk}^{A|BC} = 0{\cdot}244 + 0{\cdot}043j^* + 0{\cdot}029k^* + 0{\cdot}0003j^*k^*,$$

$$\hat{\pi}_{1|k}^{B|C} = 0{\cdot}106 + 0{\cdot}011k^*, \quad \hat{\pi}_1^C = 0{\cdot}202.$$

It is well reproduced assuming $B \perp C$ that is B independent of C, and no interaction effect of B and C on A, that is by taking $\pi_{ijk}^{ABC} = \pi_{i|jk}^{A|BC} \pi_j^B \pi_k^C$ and least squares estimates

$$\tilde{\pi}_{1|jk}^{A|BC} = 0{\cdot}245 + 0{\cdot}045j^* + 0{\cdot}031k^*, \quad \tilde{\pi}_1^B = 0{\cdot}109, \quad \tilde{\pi}_1^C = 0{\cdot}202.$$

This gives as highest risk to change the field of study $\tilde{\pi}_{1|11}^{A|BC} = 0{\cdot}32$ and as lowest $\tilde{\pi}_{1|00}^{A|BC} = 0{\cdot}17$.

3·3. Two continuous and one binary variable

In studying two continuous variables X, Y and one binary variable A, there are four types of independence which may be of interest, exemplified by $X \perp Y | A$, $A \perp X | Y$, $X \perp Y$, $A \perp X$. We shall consider five families of models.

(i) We have a homogeneous conditional Gaussian distribution model in which the distribution of A is arbitrary and in which the conditional distribution of (X, Y) given $A = i$ is bivariate normal with vector mean $\mu_i = (\mu_x(i), \mu_y(i))$ and covariance matrix Σ.

(ii) Secondly we have a homogeneous conditional Gaussian regression chain model with A as response in which the marginal distribution of (X, Y) is bivariate normal with vector mean $\mu = (\mu_x, \mu_y)$ and covariance matrix Σ, and in which given $X = x$, $Y = y$, A has linear logistic regression

$$\pi_{1|xy}^{A|XY} = L\{\gamma_{a.xy}^{(l)} + \gamma_{ax.y}^{(l)}(x - \mu_x) + \gamma_{ay.x}^{(l)}(y - \mu_y)\}. \tag{3·3}$$

(iii) Thirdly we have a dichotomized normal model in which (U, X, Y) are trivariate normal and in which A is formed by dichotomizing U thus producing a model differing from (ii) only in replacing (3·3) by

$$\pi_{1|xy}^{A|XY} = \Phi\{\gamma_{a.xy}^{(p)} + \gamma_{ax.y}^{(p)}(x - \mu_x) + \gamma_{ay.x}^{(p)}(y - \mu_y)\}. \tag{3·4}$$

(iv) Fourthly we have a homogeneous conditional Gaussian regression chain model with (A, X) as joint responses to Y in which Y is marginally normal, and in which A given $Y = y$ has linear logistic form

$$\pi_{1|y}^{A|Y} = L\{\gamma_{a.y}^{(l)} + \gamma_{ay}^{(l)}(y - \mu_y)\}, \tag{3·5}$$

and in which X given both of $A = i$, $Y = y$ has a normal distribution with parallel regression lines on y at the two levels of A is denoted by

$$E(X | A = i, Y = y) = \mu_x(i) + \beta_{xy.a}(y - \mu_y(i)) = \beta_{x.ay}(i) + \beta_{xy.a}y. \tag{3·6}$$

Thus while Y is marginally normal, X is not unless $\mu_x(1) = \mu_x(0)$, but it is close to normality if the standardized difference in means is small, as in (2·11).

(v) Finally we have a linear representation of probabilities in which, over an inevitably restricted range,

$$\pi_{i|xy}^{A|XY} = \pi_i^A + \gamma_{ax.y}(x - \mu_x)i^* + \gamma_{ay.x}(y - \mu_y)i^*, \tag{3.7}$$

and in which (X, Y) is bivariate normal. Marginalization in this model gives

$$\pi_{i|x}^{A|X} = \pi_i^A + (\gamma_{ax.y} + \beta_{yx}\gamma_{ay.x})(x - \mu_x)i^* = \pi_i^A + \gamma_{ax}(x - \mu_x)i^*,$$

say, where β_{yx} is the linear regression coefficient of Y on X. Note that γ_{ax} is determined by a relation of the same form as used in marginalizing least squares regression coefficients.

Choice between these models is partly an empirical matter, although (ii) and (iii) are known to be distinguishable only from very large amounts of data, and model (v), which is subject to the restriction that the right-hand side lies in $(0, 1)$, is essentially the same as models (ii) and (iii) if X and Y are such that with high probability $\pi_{i|xy}^{A|XY}$ is in the range $(0.2, 0.8)$. Marginal nonnormality of both or one of Y or X points toward (i) or (iv). Also, if we wish to test a particular conditional independence, certain models will be virtually excluded, because under some models particular hypotheses may arise only in a way that demand independencies additional to the one to be tested, implying that such a model is unsuitable for the required purpose.

The hypothesis $X \perp Y \mid A$ is tested from (i) and (iv) via the vanishing of the regression coefficient, say of Y on X, within the two groups of observations with respectively $A = 0, 1$, so that a standard normal theory test is available. On the other hand, under the models (ii), (iii), (v), that is those having A as univariate response, $X \perp Y \mid A$ if and only if either $X \perp (A, Y)$ or $Y \perp (A, X)$, and testing either of these would amount to examining a hypothesis much more stringent than the hypothesis of interest initially.

The hypothesis $A \perp X \mid Y$ can be tested from (ii), (iii), (v) via the vanishing of the $\gamma_{ax.y}^{(g)}$ in the binary regression of A on X, Y. Comparison of the tests under the different models is no longer straightforward, in part because the null hypotheses being tested which allow dependence of A on the second variable Y are not the same under the different models. If we take as the model

$$\pi_{1|xy}^{A|XY} = G(\mu + \gamma_x x + \gamma_y y), \tag{3.8}$$

where we have simplified the notation slightly as compared with (3.2), a component of the log likelihood from independent observations (i_j, x_j, y_j) $(j = 1, \ldots, n)$ is

$$\mathcal{L} = \sum [i_j \log G(\mu + \gamma_x x_j + \gamma_y y_j) + (1 - i_j) \log \{1 - G(\mu + \gamma_x x_j + \gamma_y y_j)\}],$$

and the score statistic for testing $\gamma_x = 0$ is based on $U_{x0} = (\partial \mathcal{L}/\partial \gamma_x)$ evaluated at $\gamma_x = 0$ and at $\mu = \hat{\mu}_0$, $\gamma_y = \hat{\gamma}_{y0}$, the maximum likelihood estimates of (μ, γ_y) at $\gamma_x = 0$. In fact

$$U_{x0} = \sum \hat{W}_{j0} x_j (i_j - \hat{G}_{j0}),$$

where $\hat{G}_{j0} = G(\hat{\mu}_0 + \hat{\gamma}_{y0} y_j)$ and $\hat{W}_{j0} = W(\hat{\mu}_0 + \hat{\gamma}_{y0} y_j)$ with

$$W(x) = G'(x)/[G(x)\{1 - G(x)\}].$$

When $G(x) = L(x)$, the unit logistic function, $W(x) = 1$, whereas, when $G(x) = \Phi(x)$, $1 \leqslant W(x)/W(0) < 1.22$ over the range in which $0.1 \leqslant \Phi^{-1}(x) \leqslant 0.9$ with 1.22 replaced by 1.10 in the narrower range $0.2 \leqslant \Phi^{-1}(x) \leqslant 0.8$. For the linear form $G(x) = x$ the weight function varies more strongly with 1.22 and 1.10 replaced by 2.18 and 1.56 respectively.

The expected information matrix for $(\mu, \gamma_x, \gamma_y)$ evaluated at $(\hat\mu_0, 0, \hat\gamma_{y0})$ is, on differentiating $\mathcal{L}$ twice with respect to the parameters and taking expectations, equal to

$$\begin{bmatrix} \sum \hat V_{j0} & \sum \hat V_{j0}x_j & \sum \hat V_{j0}y_j \\ \cdot & \sum \hat V_{j0}x_j^2 & \sum \hat V_{j0}x_jy_j \\ \cdot & \cdot & \sum \hat V_{j0}y_j^2 \end{bmatrix}, \tag{3.9}$$

where $\hat V_{j0} = V(\hat\mu_0 + \hat\gamma_{y0}y_j)$ with $V(x) = G'(x)W(x)$ and the null hypothesis variance of V_{x0} is the reciprocal of the $(2,2)$ element in the inverse of (3.9). In these calculations the $\{x_j, y_j\}$ are regarded as fixed, as would be appropriate in analyzing a given set of observations. For some theoretical purposes, however, we may take expectations over the marginal bivariate normal distribution of (X, Y). If in the fitting we measure x_j and y_j as deviations from their sample means, μ becomes orthogonal to (γ_x, γ_y) and the 2×2 information matrix for the latter is

$$n\begin{bmatrix} E\{V(\mu + \gamma_y Y)X^2\} & E\{V(\mu + \gamma_y Y)XY\} \\ \cdot & E\{V(\mu + \gamma_y Y)Y^2\} \end{bmatrix}$$

In the not very interesting case where $\gamma_y \simeq 0$, it follows immediately that the test based on U_{x0} is asymptotically independent of the form of the function $G(.)$. More generally, even if strong regression of A on y is present, the tests based on two different $G(.)$'s that give essentially the same fit to the dependence on y are unlikely to differ appreciably. For in (3.9), the $\hat G_{j0}$ will not differ much and the weights $\hat W_{j0}$ vary in a limited way as noted above. If, however, tests use forms of $G(.)$ that differ notably in their fit to the dependence of A on y, then of course the tests are different, especially if X and Y are strongly correlated. The reason is that the use of an inappropriate function for 'adjusting for' the dependence on y could induce bias in the test of $\gamma_x = 0$, quite apart from questions of efficiency.

Under the homogeneous conditional Gaussian distribution model (i), $A \perp X \mid Y$ requires that the regression lines of X on Y for $A = 0, 1$ coincide.

The hypothesis $X \perp Y$ is directly tested in (ii), (iii) and (v) via a test of independence in the postulated bivariate normal distribution of (X, Y), whereas, under (i) and (iv), the hypothesis is satisfied only if $(A, X) \perp Y$ or $(A, Y) \perp X$.

Finally $A \perp X$ is directly tested in (i) via the equality of the means of X at $A = 0, 1$. The linear form (v) is marginalized by taking expectations conditionally on $X = x$ to give (2.9), so that the required independence $A \perp X$ is directly tested from the regression of A on X or of X on A giving tests which, as discussed in § 2.4, are asymptotically equivalent to that from (i).

The same test can also be derived from the probit model (iii). Because of the close numerical equivalence of probit and logistic forms, the same test will also be effective under the logistic model in (iv) and in (ii) although in the latter case only as a result of a mathematical approximation; compare (2.10). Thus whenever a hypothesis can be tested via a number of different models, the resulting test is usually approximately the same regardless of the model, confirming the more detailed analysis of § 2.4 for two variables.

In summary, therefore, the various models give at least roughly equivalent tests for the various independence hypotheses, provided the model considered does not make the hypothesis in question collapse into a stronger form of independence; see Table 5. Typically, such problems will not occur if only hypotheses are to be tested which

Table 5. *Null hypotheses under different distributional assumptions for A, X, Y*

Specification of model (§ 3·3)	Hypothesis			
	$A\perp X \mid Y$	$X\perp Y \mid A$	$A\perp Y$	$X\perp Y$
(i) Homogeneous CG-regression chain model with (X, Y) as response to A^*	$\beta_{x.ay}(1)=\beta_{x.ay}(0)$ in (3·6)	$\beta_{xy.a}=0$ in (3·6)	$\mu_y(1)=\mu_y(0)$	Only if either $(A, X)\perp Y$ or $(A, Y)\perp X$
(ii) Homogeneous CG-regression chain model with A as response to (X, Y)	$\gamma^{(l)}_{ax.y}=0$ in (3·3)	Only if either $X\perp(A, Y)$ or $Y\perp(A, X)$	Approximately like (iii)	$\rho_{xy}=0$
(iii) Trivariate normal for (X, Y, U) with U dichotomized to form A	$\gamma^{(p)}_{ax.y}=0$ in (3·4)	Same as (ii)	$\gamma^{(p)}_{ay}=0$	Same as (ii)
(iv) Homogeneous CG-regression chain model with (X, A) as response to $Y\dagger$	Same as (i)	Same as (i)	$\gamma^{(l)}_{ay}=0$ in (3·5)	Same as (i)

* Equivalent to a CG, conditional Gaussian, distribution for A, X, Y.
† Marginal independence $A\perp X$ in this model and in a corresponding model having logistic dependence of A on Y replaced by a probit dependence requires $(A, Y)\perp X$ or $A\perp(X, Y)$, while $A\perp Y \mid X$ requires $\gamma^{(l)}_{ay}=\beta_{xy.a}\{\mu_x(1)-\mu_x(0)\}/\sigma^2_{x.ya}$.

correspond to the ordering, the conditioning of variables, implied by the dependence chain used to specify the model, i.e. whenever the dependence chain and where the independencies to be tested result from substantive considerations (Wermuth & Lauritzen, 1990).

3·4. *Two binary and one continuous variable*

We now carry out a broadly parallel analysis to §§ 3·2, 3·3 when there are two binary variables A, B and one continuous variable, X. The independency relations of interest are exemplified by $A\perp B \mid X, A\perp X \mid B, A\perp B$ and $A\perp X$. We consider five families of models.

(i) We have a homogeneous conditional Gaussian distribution model in which the distribution of (A, B) is arbitrary, with probabilities π^{AB}_{ij}, and in which X given $A=i$, $B=j$ is normally distributed with mean μ_{ij} and variance σ^2.

(ii) We have a homogeneous conditional Gaussian regression chain model with (A, B) as responses to X, with X having a marginal normal distribution of mean μ and variance σ^2, and A, B having a joint log linear model given X written conveniently in the form

$$\pi^{AB\mid X}_{ij\mid x} \propto \exp\{\mu^{(l)}+\xi^{(l)}_A i^* + \xi^{(l)}_B j^* + \xi^{(l)}_{AB} i^* j^* + \xi^{(l)}_{AX} i^*(x-\mu_x)$$
$$+ \xi^{(l)}_{BX} j^*(x-\mu_x) + \xi^{(l)}_{ABX} i^* j^*(x-\mu_x)\}. \qquad (3\cdot10)$$

(iii) We have a homogeneous conditional Gaussian regression chain model with A as a response to (B, X), that is with X given $B=j$ normal with mean μ_{xj} and variance σ^2_{xb} and with A given $X=x$, $B=j$ logistic

$$\pi^{A\mid XB}_{1\mid xj} = L\{\gamma^{(l)}_{a.xb(xb)} + \gamma^{(l)}_{ax.b(xb)}(x-\mu_{xj}) + \gamma^{(l)}_{ab.x(xb)}j^* + \gamma^{(l)}_{a(xb).bx}(x-\mu_{xj})j^*\}. \qquad (3\cdot11)$$

(iv) A number of models can be formed from underlying multivariate normal distributions, one by replacing $L(x)$ in (3·11) by $\Phi^{-1}(x)$, this corresponding to bivariate normal distributions for (U, X) given $B = j$, and another corresponding to a trivariate normal distribution for (U, V, X), the first two variables being dichotomized to form (A, B).

(v) Finally, there are representations linear in the probabilities which can be written as

$$\pi^{AB|X}_{ij|x} = \tfrac{1}{4}\{1 + \xi_A i^* + \xi_B j^* + \xi_{AB} i^* j^* + \xi_{AX} i^* (x - \mu_x) + \xi_{BX} j^* (x - \mu_x) + \xi_{ABX} i^* j^* (x - \mu_x)\},$$

(3·12)

or as (3·11) with L and superscript (l) deleted. They have advantages of simple marginalization and direct interpretation but the disadvantages of inevitably constrained parameters and give only indirect representations of some conditional independencies.

Discussion of the tests of various kinds of independency parallels that of § 3·3. Programmed algorithms for estimation in conditional Gaussian distributions are due to Edwards (1990).

Thus, for example, $A \perp B \mid X$ is tested quite directly in the conditional Gaussian regression chain models (ii) and (iii), for example in (iii) by testing $\gamma^{(l)}_{ab.x(xb)} = \gamma^{(l)}_{a(xb).bx} = 0$ in the linear logistic regression of A on X, B and XB associated with (3·11), and in (ii) the same test is appropriate, because (3·10) implies a relation of the type (3·11) for the conditional distribution of A given X and B. On the other hand, under the conditional Gaussian distribution model (i), $A \perp B \mid X$ requires both that the normal means μ_{ij} have an additive structure $\mu_{ij} = \mu + \xi_A i^* + \xi_B j^*$ and that the marginal odds ratio takes a special value

$$(\pi^{AB}_{11} \pi^{AB}_{00}) / (\pi^{AB}_{10} \pi^{AB}_{01}) = \exp(4 \xi_A \xi_B / \sigma^2)$$

(Wermuth, 1989). A likelihood ratio test can be set up for this hypothesis, but the precise relation between it and the tests associated with models (i) and (iii) is unclear. An exception is the case in which the linear-in-probability regressions approximate both of $\pi^{A|X}_{i|x}$ and $\pi^{B|X}_{j|x}$ well. Then the results of (2·11), (2·12) imply that the models are virtually indistinguishable under the hypothesis $A \perp B \mid X$, no matter whether the distributional assumptions (i), (ii) or (iii) hold.

For $A \perp X \mid B$, the conditional Gaussian distribution model (i) is immediately applicable via a test of no interaction in the two-way analysis of the cell means. Under a conditional Gaussian regression chain model with discrete responses, $A \perp X \mid B$ requires $\xi^{(l)}_{AX} = \xi^{(l)}_{ABX} = 0$ for (3·10) and $\gamma^{(l)}_{ab.x(xb)} = \gamma^{(l)}_{a(xb).bx} = 0$ for (3·11) and the two tests are equivalent since, as mentioned before, (3·10) implies a relation of the type (3·11).

3·5. Concluding remarks

Most of the models discussed above treat the three involved variables asymmetrically. This can lead to particularly simple and appealing interpretations if single variables are responses and it is especially important when it can be given a substantive interpretation. Nevertheless all the models are to be regarded as specifying the joint distribution of three random variables involved. A particular conditional hypothesis, which can be directly specified and tested as independence of a response from one of the explanatory variables, may not be satisfied in some joint distribution unless a stronger independence holds. One example is $X \perp Y \mid A$ which is a common hypothesis in a linear regression of X or Y on the remaining variables, but which cannot hold if A arises as a dichotomized variable from U, where X, Y, U have a joint normal distribution. A similar but not completely analogous case is $A \perp B \mid Y$, which is a common hypothesis in a probit regression of A or B on the remaining variables, but an unstable hypothesis in a conditional Gaussian distribution of Y, A, B, that is, even though the hypothesis can be satisfied by some expected counts, the sample size has to be very large to distinguish it from one of the stronger hypotheses $A \perp (B, Y)$ or $B \perp (A, Y)$.

If in addition there is for each individual a vector z of explanatory variables which can be treated as if fixed, the addition to the models of a term for linear dependence on z is in most cases as discussed in § 2·5.

An important qualitative conclusion is that there is a variety of models for representing this kind of data and that to a considerable extent tests of conditional independence do not depend strongly on model choice. This allows some flexibility of choice in selecting models that are convenient for substantive interpretation and for probability calculations.

Broadly similar results apply to situations with more than three variables, but conceptually new problems may also arise if distributions of four or more variables are studied. For instance, some models will contain so-called nondecomposable independence hypotheses, i.e. independencies which cannot be conveniently specified by zero restrictions on individual parameters of recursive systems such as (3·2), but which involve associated joint responses instead. As a consequence not only the interpretation can be more difficult, but the available estimation procedures for sequences of univariate recursive regressions have to be extended to obtain estimates. Noniterative approximations may be utilized in some situations (Cox & Wermuth, 1990, 1991), but typically iterative algorithms are required to obtain maximum likelihood estimates under such more complex models.

ACKNOWLEDGEMENT
We are grateful to the British German Academic Research Collaboration Programme for supporting our work.

APPENDIX 1

Least squares analysis of linear representations of probabilities

Let $Y_1, \ldots, Y_n$ be independent binary random variables with

$$E(Y_j) = \mathrm{pr}\,(Y_j = 1) = \pi_j, \quad \mathrm{var}\,(Y_j) = \pi_j(1 - \pi_j) = \bar{\nu}(1 + \delta_j),$$

where $\bar{\nu} = \mathrm{ave}\,\{\pi_j(1 - \pi_j)\}$ and $\Sigma\,\delta_j = 0$. Consider a linear representation of the probabilities, $\pi_j = x_j/\beta$, where x_j is a $1 \times p$ and β is a $p \times 1$ vector of parameters. If Y is the $n \times 1$ vector of the y_j, then $E(Y) = x\beta$, where x is $n \times p$ and assumed to be of full rank.

We suppose that the $\{\pi_j\}$ are confined to a central range such as $(0·2, 0·8)$, so that the constraints $0 \le \pi_j \le 1$ can be ignored. Also the $\{\delta_j\}$ are then small.

The ordinary least squares estimates $\hat{\beta}_{\mathrm{LS}} = (x^{\mathrm{T}}x)^{-1}x^{\mathrm{T}}Y$ have covariance matrix

$$\mathrm{cov}\,(\hat{\beta}_{\mathrm{LS}}) = \bar{\nu}(x^{\mathrm{T}}x)^{-1}\{I + (x^{\mathrm{T}}\Delta x)(x^{\mathrm{T}}x)^{-1})\}, \tag{A1·1}$$

where $\Delta = \mathrm{diag}\,(\delta_1, \ldots, \delta_n)$. Direct calculation shows that the expected value of the residual mean square is

$$\bar{\nu}[1 - (n-p)^{-1}\,\mathrm{tr}\,\{(x^{\mathrm{T}}\Delta x)(x^{\mathrm{T}}x)^{-1}\}], \tag{A1·2}$$

so that some adjustment is in principle desirable in attaching standard errors to the components of $\hat{\beta}_{\mathrm{LS}}$.

The log likelihood function is

$$L(\beta) = \sum_j \{y_j \log(x_j\beta) + (1 - y_j) \log(1 - x_j\beta)\}$$

and, on differentiating twice with respect to β and taking expectations, it follows that the Fisher information matrix for β is

$$\bar{\nu}^{-1}x^\mathrm{T} \operatorname{diag}\{(1 + \delta_j)^{-1}\}x = \bar{\nu}^{-1}x^\mathrm{T}x\{I - (x^\mathrm{T}x)^{-1}(x^\mathrm{T}\Delta x) + (x^\mathrm{T}x)^{-1}(x^\mathrm{T}\Delta^2 x) + O(\Delta^3)\}. \quad (\text{A1·3})$$

Thus the asymptotic covariance matrix of $\hat{\beta}_\mathrm{ML}$, the maximum likelihood estimate, is on inversion

$$\bar{\nu}(x^\mathrm{T}x)^{-1}[I + (x^\mathrm{T}\Delta x)(x^\mathrm{T}x)^{-1} - x^\mathrm{T}\Delta^\mathrm{T}\{I - x(x^\mathrm{T}x)^{-1}x^\mathrm{T}\}\Delta x(x^\mathrm{T}x)^{-1} + O(\Delta^3)]. \quad (\text{A1·4})$$

Comparison with (A1·1) shows that the inflation of variance by using $\hat{\beta}_\mathrm{LS}$ rather than $\hat{\beta}_\mathrm{ML}$ is of order Δ^2. This could have been anticipated from the identity between maximum likelihood estimation and weighted least squares with appropriately iterated weights and the known insensitivity of weighted least squares to perturbations in the weights.

As a rather extreme case consider the model with

$$D(Y_j) \quad \rho_0 \quad (j = 1, \quad , n_j) \quad F(Y_j) = \beta_0 - \beta_1 \quad (j = n_1 + 1, \ldots, n_1 + \tfrac{1}{2}n_2),$$

$$E(Y_j) = \beta_0 + \beta_1 \quad (j = n_1 + \tfrac{1}{2}n_2 + 1, \ldots, n_1 + n_2),$$

with in fact $\beta_0 = \tfrac{1}{2}$, $\beta_1 = 0.3$, so that maximal changes of variance are encountered. Then

$$\operatorname{var}(\hat{\beta}_{0,\mathrm{LS}})/\operatorname{var}(\hat{\beta}_{0,\mathrm{ML}}) = (n_1 + 2.2025 n_1 n_2 + n_2^2)/(n_1 + n_2)^2 < 1.051.$$

It thus seems likely that the loss of efficiency is typically less than, and often much less than, 5%.

These arguments can be extended to any simple exponential family problem involving independent observations in which the mean parameter is specified by a linear model.

Appendix 2

Models linear in probabilities

In this appendix we develop further some aspects of models linear in probabilities as set out in §3. The advantages of such models are that parameters are directly interpreted via differences or contrasts of probabilities, that simple marginalization is available by addition of probabilities and that fitting by ordinary least squares is often highly efficient. The disadvantages are that independence is a multiplicative rather than an additive concept and that constraints on the parameters are unavoidable unless the probabilities are restricted to a central range.

We discuss separately the linear models based on the joint distribution (3·1) and the linear models formulated recursively (3·2).

First note that it is possible to augment (3·1) by terms depending on one or more quantitative variables as in (3·12) and indeed we can include terms in $(X - \mu_x)^2$ if desired. Taking expectations over X in (3·12) leads back to the model for π_{ij}^{AB}.

The relative clumsiness of the linear representation for π_{ij}^{AB} in dealing with independence and conditional distributions is shown by formulae like

$$\pi_i^A = \tfrac{1}{2}(1 + \xi_A i^*), \quad \pi_j^B = \tfrac{1}{2}(1 + \xi_B j^*),$$
$$\pi_{i|j}^{A|B} = \tfrac{1}{4}(1 + \xi_A i^* + \xi_B j^* + \xi_{AB} i^* j^*)/\{\tfrac{1}{2}(1 + \xi_B j^*)\}. \quad (\text{A2·1})$$

Thus, A and B are independent if and only if $\xi_{AB} = \xi_A \xi_B$ so that independence can be assessed via the nonlinear combination

$$\eta_{AB} = \xi_{AB} - \xi_A \xi_B. \quad (\text{A2·2})$$

This is equivalent to $\xi_{AB} = 0$ if and only if at least one of A, B is equally likely to take values zero, one. Introduction of the categories i^*, etc. taking values $(1, -1)$ instead of i taking values

(0, 1) is not essential but does symmetrize the formulae. Note that, if I^* is the random variable corresponding to A, then $E(I^*) = \xi_A$, var $(I^*) = 1 - \xi_A^2$.

Similarly for three variables A, B, C starting from

$$\pi_{ijk}^{ABC} = \tfrac{1}{8}(1 + \xi_A i^* + \xi_B j^* + \xi_C k^* + \xi_{AB} i^* j^* + \xi_{AC} i^* k^* + \xi_{BC} j^* k^* + \xi_{ABC} i^* j^* k^*), \qquad \text{(A2·3)}$$

conditional independence $A \perp B \mid C$ involves two conditions

$$\pi_{111}^{ABC} / \pi_{11}^{BC} = \pi_{110}^{ABC} / \pi_{10}^{BC}, \quad \pi_{101}^{ABC} / \pi_{01}^{BC} = \pi_{100}^{ABC} / \pi_{01}^{BC},$$

leading to

$$\eta_{AC}\eta_{BC} = \eta_{AB}(1 - \xi_C^2), \quad \xi_{ABC} - \xi_A\xi_B\xi_C = \xi_A\eta_{BC} + \xi_B\eta_{AC} + \xi_C\eta_{AB}. \qquad \text{(A2·4)}$$

Under complete independence $A \perp B \perp C$ we have that

$$\eta_{AB} = \eta_{AC} = \eta_{BC} = 0, \quad \xi_{ABC} = \xi_A\xi_B\xi_C, \qquad \text{(A2·5)}$$

suggesting that it may sometimes be convenient to define

$$\eta_{ABC} = \xi_{ABC} - \xi_A\xi_B\xi_C. \qquad \text{(A2·6)}$$

Note that when the marginal probabilities are equal or close to $\tfrac{1}{2}$ as when binary variables are produced by median dichotomizing of continuous variables, then $\xi_A = \xi_B = \xi_C = 0$ and equations (A2·4)–(A2·6) simplify appreciably.

Conditional distributions can be written down in forms exemplified by

$$\pi_{i|jk}^{A|BC} = \pi_{ijk}^{ABC} / \{\tfrac{1}{4}(1 + \xi_B j^* + \xi_C k^* + \xi_{BC} j^* k^*)\}, \quad \pi_{ij|k}^{AB|C} = \pi_{ijk}^{ABC} / \{\tfrac{1}{2}(1 + \xi_C k^*)\}. \qquad \text{(A2·7)}$$

A second set of useful linear representations can be obtained when the variables are ordered to have A as response to (B, C) and B as response to C, that is in such a way that it is sensible to build up the joint distribution from the marginal distribution of C, the conditional distribution of B given C and the conditional distribution of A given B and C. We write, in a notation chosen to stress the relation with least squares regression formulae,

$$\pi_k^C = \tfrac{1}{2}(1 + \xi_C k^*), \quad \pi_{j|k}^{B|C} = \tfrac{1}{2}(1 + \gamma_{b.c} j^* + \gamma_{bc} j^* k^*). \qquad \text{(A2·8)}$$

Thus the marginal distribution of B has

$$\pi_j^B = \tfrac{1}{2}(1 + \gamma_{b.c} j^* + \gamma_{bc}\xi_C j^*) = \tfrac{1}{2}(1 + \xi_B j^*),$$

where $\xi_B = \gamma_{b.c} + \gamma_{bc}\xi_C$. Note also that cov $(J^*, K^*) = \gamma_{bc}(1 - \xi_C^2)$ so that the regression coefficient of J^* on K^* is γ_{bc}. This can be found, for example, from the joint distribution of B and C obtained by multiplying the two equations (A2·8) using $k^{*2} = 1$, that is

$$\pi_{jk}^{BC} = \tfrac{1}{4}\{1 + (\gamma_{b.c} + \gamma_{bc})j^* + \xi_C k^* + \gamma_{bc} j^* k^*\} = \tfrac{1}{4}(1 + \xi_B j^* + \xi_C k^* + \xi_{BC} j^* k^*), \qquad \text{(A2·9)}$$

where

$$\xi_B = \gamma_{b.c} + \gamma_{bc}\xi_C, \quad \xi_{BC} = \gamma_{bc}, \qquad \text{(A2·10)}$$

establishing a connection with the direct specification via the joint distribution. Also

$$E(J^* \mid K^* = k^*) = \gamma_{b.c} + \gamma_{bc} k^*. \qquad \text{(A2·11)}$$

Next write

$$\pi_{i|jk}^{A|BC} = \tfrac{1}{2}\{1 + \gamma_{a.bc} i^* + \gamma_{ab.c(bc)} i^* j^* + \gamma_{ac.b(bc)} i^* k^* + \gamma_{a(bc).bc} i^* j^* k^*\}. \qquad \text{(A2·12)}$$

On taking expectations over the levels of B given C, using (A2·11), we have that

$$\begin{aligned}\pi_{i|k}^{A|C} &= \tfrac{1}{2}\{1 + (\gamma_{a.bc} + \gamma_{ab.c(bc)}\gamma_{b.c} + \gamma_{a(bc).bc}\gamma_{bc})i^* + (\gamma_{ac.b(bc)} + \gamma_{ab.c(bc)}\gamma_{bc} + \gamma_{a(bc).bc}\gamma_{b.c})i^* k^*\} \\ &= \tfrac{1}{2}(1 + \gamma_{a.c} i^* + \gamma_{ac} i^* k^*), \qquad \text{(A2·13)}\end{aligned}$$

say, and again the relations linking the coefficients in (A2·13) with those in (A2·8), (A2·9) and (A2·12) are ordinary regression ones.

Finally, the joint distribution of A, B, C is obtained by multiplying (A2·9) and (A2·12) to form (A2·3) with

$$\xi_A = \gamma_{a.bc} + \xi_B\gamma_{ab.c(bc)} + \xi_C\gamma_{ac.b(bc)} + \xi_{BC}\gamma_{a(bc).bc},$$
$$\xi_{AB} = \gamma_{ab.c(bc)} + \xi_B\gamma_{a.bc} + \xi_C\gamma_{a(bc).bc} + \xi_{BC}\gamma_{ac.b(bc)},$$
$$\xi_{AC} = \gamma_{ac.b(bc)} + \xi_C\gamma_{a.bc} + \xi_B\gamma_{a(bc).bc} + \xi_{BC}\gamma_{ab.c(bc)}, \quad (A2·14)$$
$$\xi_{ABC} = \gamma_{a(bc).bc} + \xi_B\gamma_{ac.b(bc)} + \xi_C\gamma_{ab.c(bc)} + \xi_{BC}\gamma_{a.bc},$$

with the ξ's on the right-hand side having already been defined. These simplify considerably if there is no three-factor interaction, leading, for instance, to

$$\gamma_{a.bc} = \xi_A - \gamma_{ab.c}\xi_B - \gamma_{ac.b}\xi_C, \quad (A2·15)$$

and if, in addition, the marginal probabilities are all equal to $\frac{1}{2}$ we have

$$\gamma_{ab.c} = \frac{\xi_{AB} - \xi_{AC}\xi_{BC}}{1 - \xi_{BC}^2}, \quad \gamma_{ac.b} = \frac{\xi_{AC} - \xi_{AB}\xi_{BC}}{1 - \xi_{BC}^2}. \quad (A2·16)$$

If, however, conditional relations are required in which the order with A as response to B, C and B as response to C is not preserved, then in the present parameterization the linear structure is lost and we return to the form (A2·1).

References

Aranda-Ordaz, F. J. (1981). On two families of transformations to additivity for binary response data. *Biometrika* **68**, 357-63.
Birch, M. W. (1963). Maximum-likelihood in three-way contingency tables. *J. R. Statist. Soc.*, Suppl. **5**, 171-6.
Cox, D. R. (1958). The regression analysis of binary sequences (with discussion). *J. R. Statist. Soc.* B **20**, 215-42.
Cox, D. R. (1966). Some procedures connected with the logistic response curve. In *Research Papers in Statistics, Essays in Honour of J. Neyman's 70th birthday*, Ed. F. N. David, pp. 55-71. London: Wiley.
Cox, D. R. & Hinkley, D. V. (1974). *Theoretical Statistics*. London: Chapman and Hall.
Cox, D. R. & Snell, E. J. (1989). *Analysis of Binary Data*, 2nd ed. London: Chapman and Hall.
Cox, D. R. & Wermuth, N. (1990). An approximation to maximum likelihood estimates in reduced models. *Biometrika* **77**, 747-61.
Cox, D. R. & Wermuth, N. (1991). A simple approximation for bivariate and trivariate normal integrals. *Int. Statist. Rev.* **59**, 263-9.
Edwards, D. (1990). Hierarchical mixed interaction models (with discussion). *J. R. Statist. Soc.* B **52**, 3-20.
Finney, D. J. (1952). *Probit Analysis*, 2nd ed. Cambridge University Press.
Lauritzen, S. L. & Wermuth, N. (1989). Graphical models for associations between variables, some of which are qualitative and some quantitative. *Ann. Statist.* **17**, 31-57.
Maritz, J. S. (1953). Estimation of the correlation coefficient in the case of a bivariate normal population when one of the variables is dichotomized. *Psychometrika* **18**, 97-110.
McCullagh, P. & Nelder, J. A. (1989). *Generalized Linear Models*, 2nd ed. London: Chapman and Hall.
Person, K. (1901). Mathematical contributions to the theory of evolution—VII. On the correlation of characters not quantitatively measurable. *Phil. Trans. R. Soc. Lond.* A **195**, 1-47.
Pearson, K. (1903). Mathematical contributions to the theory of evolution—XI. On the influence of natural selection on the variability and correlation of organs. *Phil. Trans. R. Soc. Lond.* A **200**, 1-66.
Prince, J. & Tate, R. F. (1966). Accuracy of maximum likelihood estimates of correlation for a biserial model. *Psychometrika* **31**, 85-92.
Soper, H. E. (1915). On the probable error for the bi-serial expression for the correlation coefficient. *Biometrika* **10**, 384-90.
Tallis, G. M. (1961). The moment generating function of the truncated multinormal distribution. *J. R. Statist. Soc.* B **23**, 233-9.
Tate, R. F. (1955). The theory of correlation between two continuous variables when one variable is dichotomized. *Biometrika* **42**, 205-16.
Weck, M. P. (1991). *Der Studienfachwechsel. Eine Längsschnittanalyse der Interaktionsstruktur von Bedingungen des Studienverlaufs*. Frankfurt: Lang.
Wermuth, N. (1989). Moderating effects of subgroups in linear models. *Biometrika* **76**, 81-92.

WERMUTH, N. & LAURITZEN, S. L. (1983). Graphical and recursive models for contingency tables. *Biometrika* **70**, 537-52.

WERMUTH, N. & LAURITZEN, S. L. (1990). On substantive research hypotheses, conditional independence graphs and graphical chain models (with discussion). *J. R. Statist. Soc.* B **52**, 21-72.

[*Received March* 1991. *Revised October* 1991]

An Analysis of Transformations

By G. E. P. Box   and   D. R. Cox

University of Wisconsin   *Birkbeck College, University of London*

[Read at a Research Methods Meeting of the Society, April 8th, 1964,
Professor D. V. Lindley in the Chair]

Summary

In the analysis of data it is often assumed that observations $y_1, y_2, ..., y_n$ are independently normally distributed with constant variance and with expectations specified by a model linear in a set of parameters θ. In this paper we make the less restrictive assumption that such a normal, homoscedastic, linear model is appropriate after some suitable transformation has been applied to the y's. Inferences about the transformation and about the parameters of the linear model are made by computing the likelihood function and the relevant posterior distribution. The contributions of normality, homoscedasticity and additivity to the transformation are separated. The relation of the present methods to earlier procedures for finding transformations is discussed. The methods are illustrated with examples.

1. Introduction

The usual techniques for the analysis of linear models as exemplified by the analysis of variance and by multiple regression analysis are usually justified by assuming

(i) simplicity of structure for $E(y)$;
(ii) constancy of error variance;
(iii) normality of distributions;
(iv) independence of observations.

In analysis of variance applications a very important example of (i) is the assumption of additivity, i.e. absence of interaction. For example, in a two-way table it may be possible to represent $E(y)$ by additive constants associated with rows and columns.

If the assumptions (i)–(iii) are not satisfied in terms of the original observations, y, a non-linear transformation of y may improve matters. With this in mind, numerous special transformations for use in the analysis of variance have been examined in the literature; see, in particular, Bartlett (1947). The main emphasis in these studies has tended to be on obtaining a constant error variance, especially when the variance of y is a known function of the mean, as with binomial and Poisson variates.

In multiple regression problems, and in particular in the analysis of response surfaces, assumption (i) might be that $E(y)$ is adequately represented by a rather simple empirical function of the independent variables $x_1, x_2, ..., x_l$ and we would want to transform so that this assumption, together with assumptions (ii) and (iii), is approximately satisfied. In some cases transformation of independent as well as of dependent variables might be desirable to produce the simplest possible regression model in the transformed variables. In all cases we are concerned not merely to find a transformation which will justify assumptions but rather to find, where possible, a metric in terms of which the findings may be succinctly expressed.

Each of the considerations (i)–(iii) can, and has been, used separately to select a suitable candidate from a parametric family of transformations. For example, to achieve additivity in the analysis of variance, selection might be based on

(a) minimization of the F value for the degree of freedom for non-additivity (Tukey, 1949); or

(b) minimization of the F ratio for interaction versus error; or

(c) maximization of the F ratio for treatments versus error (Tukey, 1950).

Tukey and Moore (1954) used method (a) in a numerical example, plotting contours of F against (λ_1, λ_2) for transformations in the family $(y + \lambda_2)^{\lambda_1}$. They found that in their particular example the minimizing values were very imprecisely determined.

In both (a) and (b) the general object is to look for a scale on which effects are additive, i.e. to see whether an apparent interaction is removable by a transformation. Of course, only a particular type of interaction is so removable. Whereas (a) can be applied, for example, to a two-way classification without replication, method (b) requires the availability of an error term separated from the interaction term. Thus, if applied to a two-way classification, method (b) could only be used when there was some replication within cells. Finally, method (c) can be used even in a one-way analysis to find the scale on which treatment effects are in some sense most sensitively expressed. In particular, Tukey (1950) suggested multivariate canonical analysis of (y, y^2) to find the linear combination $y + \lambda y^2$ most sensitive to treatment effects. Incidentally, care is necessary in using $y + \lambda y^2$ over the wide ranges commonly encountered with data being considered for transformation, for such a transformation is sensible only so long as the value of λ and the values of y are such that the transformation is monotonic.

For transformation to stabilize variance, the usual method (Bartlett, 1947) is to determine empirically or theoretically the relation between variance and mean. An adequate empirical relation may often be found by plotting log of the within-cell variance against log of the cell mean. Another method would be to choose a transformation, within a restricted family, to minimize some measure of the heterogeneity of variance, such as Bartlett's criterion. We are grateful to a referee for pointing out also the paper of Kleczkowski (1949) in which, in particular, approximate fiducial limits for the parameter λ in the transformation of y to $\log(y + \lambda)$ are obtained. The method is to compute fiducial limits for the parameters in the linear relation observed to hold when the within-cell standard deviation is regressed on the cell mean.

Finally, while there is much work on transforming a single distribution to normality, constructive methods of finding transformations to produce normality in analysis of variance problems do not seem to have been considered.

While Anscombe (1961) and Anscombe and Tukey (1963) have employed the analysis of residuals as a means of detecting departures from the standard assumptions, they have also indicated how transformations might be constructed from certain functions of the residuals.

In regression problems, where both dependent and independent variables can be transformed, there are more possibilities to be considered. Transformation of the independent variables (Box and Tidwell, 1962) can be applied without affecting the constancy of variance and normality of error distributions. An important application is to convert a monotonic non-linear regression relation into a linear one. Obviously it is useless to try to linearize a relation which is not monotonic, but a transformation is sometimes useful in such cases, for example, to make a regression relation more nearly quadratic around its maximum.

2. General Remarks on Transformations

The main emphasis in this paper is on transformations of the dependent variable. The general idea is to restrict attention to transformations indexed by unknown parameters λ, and then to estimate λ and the other parameters of the model by standard methods of inference. Usually λ will be a one-, or at most two-, dimensional parameter, although there is no restriction in principle. Our procedure then leads to an interesting synthesis of the procedures reviewed in Section 1. It is convenient to make first a few general points about transformations.

First, we can distinguish between analyses in which either (a) the particular transformation, λ, is of direct interest, the detailed study of the factor effects, etc., being of secondary concern; or (b) the main interest is in the factor effects, the choice of λ being only a preliminary step. Type (b) is likely to be much the more common. Nevertheless, (a) can arise, for example, in the analysis of a preliminary set of data. Or, again, we may have two factors, A and B, whose main effects are broadly understood, it being required to study the λ, if any, for which there is no interaction between the factors. Here the primary interest is in λ. In case (b), however, we shall need to fix one, or possibly a small number, of λ's and go ahead with the detailed estimation and interpretation of the factor effects on this particular transformed scale. We shall choose λ partly in the light of the information provided by the data and partly from general considerations of simplicity, ease of interpretation, etc. For instance, it would be quite possible for the formal analysis to show that say $\sqrt{y}$ is the best scale for normality and constancy of variance, but for us to decide that there are compelling arguments of ease of interpretation for working say with $\log y$. The formal analysis will warn us, however, that changes of variance and non-normality may need attention in a refined and efficient analysis of $\log y$. That is, the method developed below for finding a transformation is useful as a guide, but is, of course, not to be followed blindly. In Section 7 we discuss briefly some of the consequences of interpreting factor effects on a scale chosen in the light of the data.

In regression studies, it is sometimes necessary to take an entirely empirical approach to the choice of a relation. In other cases, physical laws, dimensional analysis, etc., may suggest a particular functional form. Thus, in a study of a chemical system one would expect reaction rate to be proportional to some power of the concentration and to the antilog of the reciprocal of absolute temperature. Again, in many fields of technology relationships of the form

$$y \propto x_1^{\beta 1} \ldots x_l^{\beta l}$$

are very common, suggesting a log transformation of all variables. In such cases the reasonable thing will often be first to apply the transformations suggested by the prior reasoning, and after that consider what further modifications, if any, are needed. Finally, we may know the behaviour of y when the independent variables x_i tend to zero or infinity, and certainly, if we are hopeful that the model might apply over a wide range, we should consider models that are consistent with such limiting properties of the system.

We can distinguish broadly two types of dependent variable, extensive and non-extensive. The former have a relevant property of physical additivity, the latter not. Thus yield of product per batch is extensive. The failure time of a component would be considered extensive if components are replaced on failure, the main thing of interest being the number of components used in a long time. Properties like temperature, viscosity, quality of product, etc., are not extensive. In the absence of the sort of prior consideration mentioned in the previous paragraph there is no reason to prefer the initial form of a non-extensive variable to any monotonic function of it. Hence, transformations can be applied freely to non-extensive variables. For extensive variables, however, the population mean of y is the parameter determining the long-run behaviour of the system. Thus in the two examples mentioned above, the total yield of product in a long period and the total number of components used in a very long time are determined respectively by the population mean of yield per batch and the mean failure time per component, irrespective of distributional form.

In a narrowly technological sense, therefore, we are interested in the population mean of y, not of some function of y. Hence we either analyse linearly the untransformed data or, if we do apply a transformation in order to make a more efficient and valid analysis, we convert the conclusions back to the original scale. Even in circumstances where, for immediate application, the original scale y is required, it may be better to think in terms of transformed values in which, say, interactions have been removed.

In general, we can regard the usual formal linear models as doing two things:

(a) specifying the questions to be asked, by defining explicitly the parameters which it is the main object of the analysis to estimate;

(b) specifying assumptions under which the above parameters can be simply and effectively estimated.

If there should be conflict between the requirements for (a) and for (b), it is best to pay most attention to (a), since approximate inference about the most meaningful parameters is clearly preferable to formally "exact" inference about parameters whose definition is in some way artificial. Therefore in selecting a transformation we might often give first attention to simplicity of the model structure, for example to additivity in the analysis of variance. This allows simplicity of description and also the main effect of a factor **A**, measured on a scale for which there appears to be no interaction with a factor **B**, often has a reasonable possibility of being valid for levels of **B** outside those of the initial experiment.

3. Transformation of the Dependent Variable

We work with a parametric family of transformations from y to $y^{(\lambda)}$, the parameter λ, possibly a vector, defining a particular transformation. Two important examples considered here are

$$y^{(\lambda)} = \begin{cases} \dfrac{y^\lambda - 1}{\lambda} & (\lambda \neq 0), \\[2mm] \log y & (\lambda = 0), \end{cases} \tag{1}$$

and

$$y^{(\lambda)} = \begin{cases} \dfrac{(y+\lambda_2)^{\lambda_1} - 1}{\lambda_1} & (\lambda_1 \neq 0), \\[2mm] \log(y+\lambda_2) & (\lambda_1 = 0). \end{cases} \tag{2}$$

The transformations (1) hold for $y > 0$ and (2) for $y > -\lambda_2$. Note that since an analysis of variance is unchanged by a linear transformation (1) is equivalent to

$$y^{(\lambda)} = \begin{cases} y^\lambda & (\lambda \neq 0), \\ \log y & (\lambda = 0); \end{cases} \tag{3}$$

the form (1) is slightly preferable for theoretical analysis because it is continuous at $\lambda = 0$. In general, it is assumed that for each λ, $y^{(\lambda)}$ is a monotonic function of y over the admissible range. Suppose that we observe an $n \times 1$ vector of observations $\mathbf{y} = \{y_1, \dots, y_n\}$, and that the appropriate linear model for the problem is specified by

$$E\{\mathbf{y}^{(\lambda)}\} = \mathbf{a}\boldsymbol{\theta}, \tag{4}$$

where $\mathbf{y}^{(\lambda)}$ is the column vector of *transformed* observations, $\mathbf{a}$ is a known matrix and $\boldsymbol{\theta}$ a vector of unknown parameters associated with the transformed observations.

We now assume that for some unknown λ, the transformed observations $y_i^{(\lambda)}$ $(i = 1, \dots, n)$ satisfy the full normal theory assumptions, i.e. are independently normally distributed with constant variance σ^2, and with expectations (4). The probability density for the untransformed observations, and hence the likelihood *in relation to these original observations*, is obtained by multiplying the normal density by the Jacobian of the transformation.

The likelihood in relation to the original observations $\mathbf{y}$ is thus

$$\frac{1}{(2\pi)^{\frac{1}{2}n}\,\sigma^n} \exp\left\{-\frac{(\mathbf{y}^{(\lambda)} - \mathbf{a}\boldsymbol{\theta})'\,(\mathbf{y}^{(\lambda)} - \mathbf{a}\boldsymbol{\theta})}{2\sigma^2}\right\} J(\lambda; \mathbf{y}), \tag{5}$$

where

$$J(\lambda; \mathbf{y}) = \prod_{i=1}^{n} \left|\frac{dy_i^{(\lambda)}}{dy_i}\right|.$$

We shall examine two ways in which inferences about the parameters in (5) can be made. In the first, we apply "orthodox" large-sample maximum-likelihood theory to (5). This approach leads directly to point estimates of the parameters and to approximate tests and confidence intervals based on the chi-squared distribution.

In the second approach, via Bayes's theorem, we assume that the prior distributions of the θ's and $\log \sigma$ can be taken as essentially uniform over the region in which the likelihood is appreciable and we integrate over the parameters to obtain a posterior distribution for λ; for general discussion of this approach, see, in particular, Jeffreys (1961).

We find the maximum-likelihood estimates in two steps. First, for given λ, (5) is, except for a constant factor, the likelihood for a standard least-squares problem. Hence the maximum-likelihood estimates of the θ's are the least-squares estimates for the dependent variable $y^{(\lambda)}$ and the estimate of σ^2, denoted for fixed λ by $\hat{\sigma}^2(\lambda)$, is

$$\hat{\sigma}^2(\lambda) = \mathbf{y}^{(\lambda)'}\mathbf{a}_r\,\mathbf{y}^{(\lambda)}/n = S(\lambda)/n \tag{6}$$

where, when $\mathbf{a}$ is of full rank,

$$\mathbf{a}_r = \mathbf{I} - \mathbf{a}(\mathbf{a}'\mathbf{a})^{-1}\mathbf{a}', \tag{7}$$

and $S(\lambda)$ is the residual sum of squares in the analysis of variance of $y^{(\lambda)}$.

Thus for fixed λ, the maximized log likelihood is, except for a constant,

$$L_{\max}(\lambda) = -\tfrac{1}{2}n \log \hat{\sigma}^2(\lambda) + \log J(\lambda; \mathbf{y}). \tag{8}$$

In the important special case (1) of the simple power transformation, the second term in (8) is

$$(\lambda - 1)\Sigma \log y_i. \tag{9}$$

In (2), when an unknown origin λ_2 is included, the term becomes

$$(\lambda_1 - 1)\Sigma \log(y_i + \lambda_2). \tag{10}$$

It will now be informative to plot the maximized log likelihood $L_{\max}(\lambda)$ against λ for a trial series of values. From this plot the maximizing value $\hat{\lambda}$ may be read off and we can obtain an approximate $100(1-\alpha)$ per cent confidence region from

$$L_{\max}(\hat{\lambda}) - L_{\max}(\lambda) < \tfrac{1}{2}\chi^2_{\nu_\lambda}(\alpha), \tag{11}$$

where ν_λ is the number of independent components in λ. The main arithmetic consists in doing the analysis of variance of $\mathbf{y}^{(\lambda)}$ for each chosen λ.

If it were ever desired to determine $\hat{\lambda}$ more precisely this could be done by determining numerically the value $\hat{\lambda}$ for which the derivatives with respect to λ are all zero. In the special case of the one parameter power transformation $y^{(\lambda)} = (y^\lambda - 1)/\lambda$,

$$\frac{d}{d\lambda} L_{\max}(\lambda) = -n \frac{\mathbf{y}^{(\lambda)\prime}\mathbf{a}_r\mathbf{u}^{(\lambda)}}{\mathbf{y}^{(\lambda)\prime}\mathbf{a}_r\mathbf{y}^{(\lambda)}} + \frac{n}{\lambda} + \Sigma \log y_i, \tag{12}$$

where $\mathbf{u}^{(\lambda)}$ is the vector of components $\{\lambda^{-1}y_i^\lambda \log y_i\}$. The numerator in (12) is the residual sum of products in the analysis of covariance of $\mathbf{y}^{(\lambda)}$ and $\mathbf{u}^{(\lambda)}$

The above results can be expressed very simply if we work with the normalized transformation

$$\mathbf{z}^{(\lambda)} = \mathbf{y}^{(\lambda)}/J^{1/n},$$

where $J = J(\lambda; \mathbf{y})$. Then

$$L_{\max}(\lambda) = -\tfrac{1}{2}n \log \hat{\sigma}^2(\lambda; \mathbf{z}),$$

where

$$\hat{\sigma}^2(\lambda; \mathbf{z}) = \frac{\mathbf{z}^{(\lambda)\prime}\mathbf{a}_r\mathbf{z}^{(\lambda)}}{n} = \frac{S(\lambda; \mathbf{z})}{n},$$

where $S(\lambda; \mathbf{z})$ is the residual sum of squares of $\mathbf{z}^{(\lambda)}$. The maximized likelihood is thus proportional to $\{S(\lambda; \mathbf{z})\}^{-n}$ and the maximum-likelihood estimate is obtained by minimizing $S(\lambda; \mathbf{z})$ with respect to λ.

For the simple power transformation

$$z^{(\lambda)} = \frac{y^\lambda - 1}{\lambda \dot{y}^{\lambda-1}},$$

where $\dot{y}$ is the geometric mean of the observations.

For the power transformation with shifted location

$$z^{(\lambda)} = \frac{(y+\lambda_2)^{\lambda_1} - 1}{\lambda_1 \{\mathrm{gm}\,(y+\lambda_2)\}^{\lambda_1 - 1}},$$

where $\mathrm{gm}\,(y+\lambda_2)$ is the sample geometric mean of the $(y+\lambda_2)$'s.

Consider now the corresponding Bayesian analysis. Let the degrees of freedom for residual be $\nu_r = n - \mathrm{rank}\,(\mathbf{a})$, and let

$$s^2(\lambda) = \frac{\mathbf{y}^{(\lambda)\prime}\mathbf{a}_r\mathbf{y}^{(\lambda)}}{\nu_r} = \frac{S(\lambda)}{\nu_r} \tag{13}$$

be the residual mean square in the analysis of variance of $\mathbf{y}^{(\lambda)}$; note the distinction between $\hat{\sigma}^2(\lambda)$, the maximum-likelihood estimate with divisor n, and $s^2(\lambda)$ the "usual" estimate, with divisor the degrees of freedom ν_r. We first rewrite the likelihood (5), i.e. the conditional probability density function of the y's given $\boldsymbol{\theta}$, σ^2, λ, in the form

$$p(\mathbf{y} \mid \boldsymbol{\theta}, \sigma^2, \lambda) = \frac{1}{(2\pi)^{\frac{1}{2}n} \sigma^n} \exp \left\{ -\frac{\nu_r s^2(\lambda) + (\boldsymbol{\theta} - \hat{\boldsymbol{\theta}}_\lambda)' \mathbf{a}' \mathbf{a} (\boldsymbol{\theta} - \hat{\boldsymbol{\theta}}_\lambda)}{2\sigma^2} \right\} J(\lambda; \mathbf{y}), \qquad (14)$$

where $\hat{\boldsymbol{\theta}}_\lambda$ is the least-squares estimate of $\boldsymbol{\theta}$ for given λ.

Now consider the choice of the joint prior distribution for the unknown parameters. We first parametrize so that the θ's are linearly independent and hence $n - \nu_r$ in number. Let $p_0(\lambda)$ denote the marginal prior density of λ. We assume that it is reasonable, when making inferences about λ, to take the conditional prior distribution of the θ's and $\log \sigma$, given λ, to be effectively uniform over the range for which the likelihood is appreciable. That is, the conditional prior element given λ is

$$g(\lambda) \, d\boldsymbol{\theta}_\lambda \, d(\log \sigma_\lambda), \qquad (15)$$

where, for definiteness, we for the moment denote the effects and variance measured in terms of $y^{(\lambda)}$ by a suffix λ. The factor $g(\lambda)$ is included because the general size and range of the transformed observations $y^{(\lambda)}$ may depend strongly on λ. If the conditional prior distribution (15) were assumed independent of λ, nonsensical results would be obtained.

To determine $g(\lambda)$ we argue as follows. Fix a standard reference value of λ, say λ_1. Suppose provisionally that, for fixed λ, the relation between $y^{(\lambda)}$ and $y^{(\lambda_1)}$ over the range of the observations is effectively linear, say

$$y^{(\lambda)} = \text{const} + l_\lambda y^{(\lambda_1)}. \qquad (16)$$

We can then choose $g(\lambda)$ so that when (16) holds, the conditional prior distributions (15) are consistent with one another for different values of λ. In fact, we shall need to apply the answer when the transformations are appreciably non-linear, so that (16) does not hold. There may be a better approach to the choice of a prior distribution than the present one.

It follows from (16) that

$$\log \sigma_\lambda^2 = \text{const} + \log \sigma_{\lambda_1}^2 \qquad (17)$$

and hence, to this order, the prior density of σ_λ^2 is independent of λ. However, the θ_λ's are linear combinations of the expected values of the $y^{(\lambda)}$'s, so that

$$\frac{d\theta_\lambda}{d\theta_{\lambda_1}} = l_\lambda.$$

Since there are $n - \nu_r$ independent components to $\boldsymbol{\theta}$, it follows that $g(\lambda)$ is proportional to $1/l_\lambda^{n-\nu_r}$.

Finally we need to choose l_λ. In passing from λ_1 to λ, a small element of volume of the n dimensional sample space is multiplied by $J(\lambda; \mathbf{y})/J(\lambda_1; \mathbf{y})$. An average scale change for a single y component is the nth root of this and, since λ_1 is only a standard reference value, we have approximately

$$l_\lambda = \{J(\lambda; \mathbf{y})\}^{1/n}. \qquad (18)$$

Thus, approximately, the conditional prior density (15) is

$$\frac{d\boldsymbol{\theta}_\lambda \, d(\log \sigma_\lambda)}{\{J(\lambda; \mathbf{y})\}^{(n-\nu_r)/n}}.$$

The combined prior element of probability is thus

$$\frac{d\theta \, d(\log \sigma)}{\{J(\lambda; \mathbf{y})\}^{(n-\nu_r)/n}} p_0(\lambda) \, d\lambda, \tag{19}$$

where we now suppress the suffix λ on θ and σ.

This is only an approximate result. In particular, the choice of (18) is somewhat arbitrary. However, when a useful amount of information is actually available from the data about the transformation, the likelihood will dominate and the exact choice of (19) is not critical. The prior distribution (19) is interesting in that the observations enter the approximate standardizing coefficient $J(\lambda; \mathbf{y})$.

We now have the likelihood (14) and the prior density (19) and can apply Bayes's theorem to obtain the marginal posterior distribution of λ in the form

$$K'_y \frac{I(\lambda \mid y) \, p_0(\lambda)}{\{J(\lambda; \mathbf{y})\}^{(n-\nu_r)/n}}, \tag{20}$$

where K'_y is a normalizing constant independent of λ, chosen so that (20) integrates to one with respect to λ, and

$$I(\lambda \mid y) = \int_{-\infty}^{\infty} d(\log \sigma) \int_{-\infty}^{\infty} d\theta \, p(\mathbf{y} \mid \theta, \sigma^2, \lambda). \tag{21}$$

The integral (21) can be evaluated to give

$$I(\lambda \mid y) = \frac{|\mathbf{a}'\mathbf{a}|^{-\frac{1}{2}} 2^{\frac{1}{2}\nu_r} \Gamma(\frac{1}{2}\nu_r)}{(2\pi)^{\frac{1}{2}\nu_r} \{s^2(\lambda)\}^{\frac{1}{2}\nu_r} \nu_r^{\frac{1}{2}\nu_r}} J(\lambda; \mathbf{y}).$$

Substituting into (20), we have that the posterior distribution of λ is

$$K_y \frac{\{J(\lambda; \mathbf{y})\}^{\nu_r/n}}{\{s^2(\lambda)\}^{\frac{1}{2}\nu_r}} p_0(\lambda),$$

where K_y is a normalizing constant independent of λ.

Thus the contribution of the observations to the posterior distribution of λ is represented by the factor

$$\{J(\lambda; y)\}^{\nu_r/n} / \{s^2(\lambda)\}^{\frac{1}{2}\nu_r}$$

or, on a log scale, by the addition of a term

$$L_b(\lambda) = -\tfrac{1}{2}\nu_r \log s^2(\lambda) + (\nu_r/n) \log J(\lambda; y) \tag{22}$$

to $\log p_0(\lambda)$.

Once again if we work with the normalized transformation $z^{(\lambda)} = y^{(\lambda)}/J^{1/n}$, the result is expressed with great simplicity, for

$$L_b(\lambda) = -\tfrac{1}{2}\nu_r \log s^2(\lambda; \mathbf{z}) \tag{23}$$

and the posterior density is

$$p(\lambda) = \text{const} \times p_0(\lambda) \times \{S(\lambda; \mathbf{z})\}^{-\frac{1}{2}\nu_r}.$$

In practice we can plot $\{S(\lambda; \mathbf{z})\}^{-\frac{1}{2}\nu_r}$ against λ, combining it with any prior information about λ. When the prior density of λ can be taken as locally uniform, the posterior distribution is obtained directly by plotting

$$p_u(\lambda) = k\{S(\lambda; \mathbf{z})\}^{-\frac{1}{2}\nu_r}, \tag{24}$$

where k is chosen to make the total area under the curve unity.

We normally end by selecting a value of λ in the light both of this plot and of other relevant considerations discussed in Section 2. We then proceed to a standard analysis using the indicated transformation.

The maximized log likelihood and the log of the contribution to the posterior distribution of λ may be written respectively as

$$L_{\max}(\lambda) = -\tfrac{1}{2}n\log\{S(\lambda; \mathbf{z})/n\}, \quad L_b(\lambda) = -\tfrac{1}{2}\nu_r\log\{S(\lambda; \mathbf{z})/\nu_r\}.$$

They differ only by substitution of ν_r for n. They are both monotonic functions of $S(\lambda; \mathbf{z})$ and their maxima both occur when the sum of squares $S(\lambda; \mathbf{z})$ is minimized. For general description, $L_{\max}(\lambda)$ and $L_b(\lambda)$ are substantially equivalent. However, it can easily happen that ν_r/n is appreciably less than one, even when n is quite large. Therefore, in applications, the difference cannot always be ignored, especially when a number of models are simultaneously considered.

There are some reasons for thinking $L_b(\lambda)$ preferable to $L_{\max}(\lambda)$ from a non-Bayesian as well as from a Bayesian point of view; see, for example, the introduction by Bartlett (1937) of degrees of freedom into his test for the homogeneity of variance. The general large-sample theorems about the sampling distributions of maximum-likelihood estimates, and the maximum-likelihood ratio chi-squared test, apply just as much to $L_b(\lambda)$ as to $L_{\max}(\lambda)$.

4. Two Examples

We have supposed that after suitable transformation from y to $y^{(\lambda)}$, (a) the expected values of the transformed observations are described by a model of simple structure; (b) the error variance is constant; (c) the observations are normally distributed. Then we have shown that the maximized likelihood for λ, and also the approximate contribution to the posterior distribution of λ, are each proportional to a negative power of the residual sum of squares for the variate $z^{(\lambda)} = y^{(\lambda)}/J^{1/n}$.

The "overall" procedure seeks a set of transformation parameters λ for which (a), (b) and (c) are simultaneously satisfied, and sample information on all three aspects goes into the choice. In this Section we now apply this overall procedure to two examples. In Section 5 we shall show how further analysis can show the separate contributions of (a), (b) and (c) in the choice of the transformation. We shall then illustrate this separation using the same two examples.

The above procedure depends on specific assumptions, but it would be quite wrong for fruitful application to regard the assumptions as final. The proper attitude of sceptical optimism is accurately expressed by saying that we tentatively entertain the basis for analysis, rather than that we assume it. The checking of the plausibility of the present procedure will be discussed in Section 5.

A Biological Experiment using a 3×4 Factorial Design with Replication

Table 1 gives the survival times of animals in a 3×4 factorial experiment, the factors being (a) three poisons and (b) four treatments. Each combination of the two factors is used for four animals, the allocation to animals being completely randomized.

We consider the application of a simple power transformation $y^{(\lambda)} = (y^\lambda - 1)/\lambda$. Equivalently we shall actually analyse the standardized variate $z^{(\lambda)} = (y^\lambda - 1)/(\lambda \dot{y}^{\lambda-1})$.

TABLE 1

Survival times (unit, 10 hr) of animals in a 3×4 factorial experiment

Poison	Treatment			
	A	B	C	D
I	0·31	0·82	0·43	0·45
	0·45	1·10	0·45	0·71
	0·46	0·88	0·63	0·66
	0·43	0·72	0·76	0·62
II	0·36	0·92	0·44	0·56
	0·29	0·61	0·35	1·02
	0·40	0·49	0·31	0·71
	0·23	1·24	0·40	0·38
III	0·22	0·30	0·23	0·30
	0·21	0·37	0·25	0·36
	0·18	0·38	0·24	0·31
	0·23	0·29	0·22	0·33

We are tentatively entertaining the model that after such transformation

(a) the expected value of the transformed variate in any cell can be represented by additive row and column constants, i.e. that no interaction terms are needed,
(b) the error variance is constant,
(c) the observations are normally distributed.

The maximized likelihood and the posterior distribution are functions of the residual sum of squares for $z^{(\lambda)}$ after eliminating row and column effects. This sum of squares is denoted $S(\lambda; z)$. It has 42 degrees of freedom and is the result of pooling the "within groups" and the "interaction" sums of squares.

Table 2 gives $S(\lambda; z)$ together with $L_{\max}(\lambda)$ and $p_u(\lambda)$ over the interesting ranges. The constant k in $k e^{L_b(\lambda)} = p_u(\lambda)$ is the reciprocal of the area under the curve $Y = e^{L_b(\lambda)}$ determined by numerical integration. Graphs of $L_{\max}(\lambda)$ and of $p_u(\lambda)$ are shown in Fig. 1. This analysis points to an optimal value of about $\hat{\lambda} = -0.75$. Using (11) the curve of maximized likelihood gives an approximate 95 per cent confidence interval for λ extending from about -1.13 to -0.37.

The posterior distribution $p_u(\lambda)$ is approximately normal with mean -0.75 and standard deviation 0.22. About 95 per cent of this posterior distribution is included within the limits -1.18 and -0.32.

The reciprocal transformation has a natural appeal for the analysis of survival times since it is open to the simple interpretation that it is the *rate of dying* which is to be considered. Our analysis shows that it would, in fact, embody most of the advantages obtainable. The complete analysis of variance for the untransformed data and for the reciprocal transformation (taken in the z form) is shown in Table 3.

Whereas no great change occurs on transformation in the mean squares associated with poisons and treatments, the within groups mean square has shrunk to a third of

TABLE 2

Biological data. Calculations based on an additive, homoscedastic, normal model in the transformed observations

λ	$S(\lambda; \mathbf{z})$	$L_{\max}(\lambda)$	λ	$S(\lambda; \mathbf{z})$	$L_{\max}(\lambda)$
1·0	1·0509	91·72	−1·0	0·3331	119·29
0·5	0·6345	103·83	−1·2	0·3586	117·52
0·0	0·4239	113·51	−1·4	0·4007	114·86
−0·2	0·3752	116·44	−1·6	0·4625	111·43
−0·4	0·3431	118·58	−2·0	0·6639	102·74
−0·6	0·3258	119·82	−2·5	1·1331	89·91
−0·8	0·3225	120·07	−3·0	2·0489	75·69

λ	$p_u(\lambda)$	λ	$p_u(\lambda)$
0·0	0·01	−0·8	1·82
−0·1	0·02	−0·9	1·42
−0·2	0·08	−1·0	0·92
−0·3	0·26	−1·1	0·47
−0·4	0·49	−1·2	0·19
−0·5	0·94	−1·3	0·07
−0·6	1·46	−1·5	0·01
−0·7	1·82		

$$L_{\max}(\lambda) = -24 \log \hat{\sigma}^2(\lambda; \mathbf{z}) = \log\{S(\lambda; \mathbf{z})\}^{-24} + 92\cdot91; \ p_u(\lambda) = k\, e^{L_b(\lambda)} = 0\cdot866 \times 10^{-10} \{S(\lambda; \mathbf{z})\}^{-21}.$$

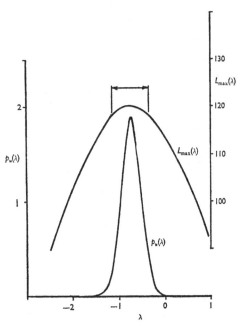

FIG. 1. Biological data. Functions $L_{\max}(\lambda)$ and $p_u(\lambda)$. Arrows show approximate 95 per cent. confidence interval for λ.

9

its value and the interaction mean square is now much closer in size to that within groups. Thus, in the transformed metric, not only is greater simplicity of interpretation possible but also the sensitivity of the experiment, as measured by the ratios

TABLE 3

Analyses of variance of biological data

	Degrees of freedom	Mean squares × 1000	
		Untransformed	Reciprocal transformation (z form)
Poisons . .	2	516·5	568·7
Treatments . .	3	307·1	221·9
$P \times T$ . . .	6	41·7	8·5
Within groups .	36	22·2	7·8

of the poisons and the treatments mean squares to the residual square, has been increased almost threefold. We shall not here consider the detailed interpretation of the factor effects.

A Textile Experiment using a Single Replicate of a 3^3 Design

In an unpublished report to the Technical Committee, International Wool Textile Organization, Drs A. Barella and A. Sust described some experiments on the behaviour of worsted yarn under cycles of repeated loading. Table 4 gives the numbers of cycles to failure, y, obtained in a single replicate of a 3^3 experiment in which the factors are

x_1: length of test specimen (250, 300, 350 mm.),
x_2: amplitude of loading cycle (8, 9, 10 mm.),
x_3: load (40, 45, 50 gm.).

In Table 4 the levels of the x's are denoted conventionally by $-1, 0, 1$.

It is useful to describe first the results of a rather informal analysis of Table 4. Barella and Sust fitted a full equation of second degree in x_1, x_2 and x_3, but the conclusions were very complicated and messy. In view of the wide relative range of variation of y, it is natural to try analysing instead log y, and there results a great simplification. All linear regression terms are very highly significant and all second-degree terms are small. Further, it is natural to take logs also for the independent variables, i.e. to think in terms of relationships like

$$y \propto x_1^{\beta_1} x_2^{\beta_2} x_3^{\beta_3}. \tag{25}$$

The estimates of the β's, from the linear regression coefficients of log y on the log x's, are, with their estimated standard errors,

$$\hat{\beta}_1 = 4{\cdot}96 \pm 0{\cdot}20, \quad \hat{\beta}_2 = -5{\cdot}27 \pm 0{\cdot}30, \quad \hat{\beta}_3 = -3{\cdot}15 \pm 0{\cdot}30.$$

Since $\hat{\beta}_1 \simeq -\hat{\beta}_2$, the combination $\log x_1 - \log x_2 = \log (x_1/x_2)$ is suggested by the data as of possible importance. In fact, x_2/x_1 is just the fractional amplitude of the loading cycle; indeed, naïve dimensional considerations suggest this as a possible factor, although there are in fact other relevant lengths, so that dependence on x_1

and x_2 separately is not inconsistent with dimensional considerations. If, however, we write $x_2/x_1 = x_4$ and round the regression coefficients, we have the simple formula

$$y \propto x_4^{-5} x_3^{-3}$$

which fits the data remarkably well.

TABLE 4

Cycles to failure of worsted yarn: 3^3 factorial experiment

Factor levels			Cycles to failure, y
x_1	x_2	x_3	
−1	−1	−1	674
−1	−1	0	370
−1	−1	+1	292
−1	0	−1	338
−1	0	0	266
−1	0	+1	210
−1	+1	−1	170
−1	+1	0	118
−1	+1	+1	90
0	−1	−1	1,414
0	−1	0	1,198
0	−1	+1	634
0	0	−1	1,022
0	0	0	620
0	0	+1	438
0	+1	−1	442
0	+1	0	332
0	+1	+1	220
+1	−1	−1	3,636
+1	−1	0	3,184
+1	−1	+1	2,000
+1	0	−1	1,568
+1	0	0	1,070
+1	0	+1	566
+1	+1	−1	1,140
+1	+1	0	884
+1	+1	+1	360

In this case, there seem strong general arguments for starting with a log trans-formation of all variables. Power laws are frequently effective in the physical sciences; also, provided that the signs of the β's are right, (25) has sensible limiting behaviour for $x_2, x_3 \to 0, \infty$; finally, the obvious normal theory model based on transforming (25) gives distributions over positive values of y only.

Nevertheless, it is interesting to see whether the method of the present paper applied directly to the data of Table 4 produces the log transformation. In this paper, transformations of the dependent variable alone are considered; in fact, since the relative range of the x's is not very great, transformation of the x's does not have a big effect on the linearity of the regression.

We first consider the application of a simple power transformation in terms, as before, of the standardized variate $z^{(\lambda)} = (y^\lambda - 1)/(\lambda \dot{y}^{\lambda-1})$. We tentatively suppose that after such transformation

  (a) the expected value of the transformed response can be represented merely by a model *linear* in the x's,
  (b) the error variance is constant,
  (c) the observations are normally distributed.

The maximized likelihood and the posterior distribution are functions of the residual sum of squares for $z^{(\lambda)}$ after fitting only a linear model to the x's. Since there are four constants in the linear regression model this residual sum of squares has $27 - 4 = 23$ degrees of freedom; we denote it by $S(\lambda; z)$.

Table 5 shows $S(\lambda; z)$ together with $L_{max}(\lambda)$ and $p_u(\lambda)$ over the interesting ranges and the results are plotted in Fig. 2. The optimal value for the transformation parameter is $\hat{\lambda} = -0.06$. The transformation is determined remarkably closely in this

TABLE 5

Textile data. Calculations based on normal linear model in the transformed observations

λ	$S(\lambda; z)$	$L_{max}(\lambda)$	λ	$S(\lambda; z)$	$L_{max}(\lambda)$
1·00	5·4810	21·52	−0·20	0·2920	61·11
0·80	2·9978	29·67	−0·40	0·5478	52·61
0·60	1·5968	38·17	−0·60	1·1035	43·16
0·40	0·8178	47·21	−0·80	2·1396	34·22
0·20	0·4115	56·48	−1·00	3·9955	25·79
0·00	0·2519	63·10			

λ	$p_u(\lambda)$	λ	$p_u(\lambda)$
0·20	0·02	−0·10	4·66
0·15	0·09	−0·15	2·36
0·10	0·42	−0·20	0·77
0·05	1·58	−0·25	0·19
0·00	4·18	−0·30	0·04
−0·05	5·64	−0·35	0·01

$L_{max}(\lambda) = -13.5 \log \hat{\sigma}^2(\lambda; z) = \{S(\lambda; z)\}^{-13.5} + 44.49.$
$p_u(\lambda) = k e^{L_b(\lambda)} = 0.540 \times 10^{-6} \{S(\lambda; z)\}^{-11.5}.$

example, the approximate 95 per cent confidence range extending only from −0·18 to +0·06. The posterior distribution $p_u(\lambda)$ has its mean at −0·06. About 95 per cent of the distribution is included between −0·20 and +0·08. As we have mentioned, the advantages of a log transformation corresponding to the choice $\lambda = 0$ are very great and such a choice is now seen to be strongly supported by the data.

The complete analysis of variance for the untransformed and the log transformation, taken in the z form, is shown in Table 6.

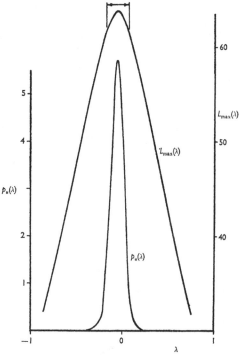

FIG. 2. Textile data. Functions $L_{\max}(\lambda)$ and $p_u(\lambda)$. Arrows show approximate
95 per cent confidence interval for λ.

TABLE 6

Analyses of variance of textile data

	Degrees of freedom	Mean squares × 1000	
		Untransformed	Logarithmic transformation (z form)
Linear . .	3	4,916·2	2,374·4
Quadratic . .	6	704·1	8·1
Residual . .	17	73·9	11·9

The transformation eliminates the need for second-order terms in the regression
equation while at the same time increasing the sensitivity of the analysis by about
three, as judged by the ratio of linear and residual mean squares.

For this example we have also tried out the procedures we have discussed using
the two parameter transformation $y^{(\lambda)} = \{(y+\lambda_2)^{\lambda_1}-1\}/\lambda_1$ or in the z form actually

used here $z^{(\lambda)} = \{(y+\lambda_2)^{\lambda_1} - 1\}/\{\lambda_1 \operatorname{gm}(y+\lambda_2)\}^{\lambda_1-1}$. Incidentally the calculation and print out of 77 analysis of variance tables, involving in each case the fitting of a general equation of second degree, and calculation of residuals and fitted values took 2 min. 6 sec. on the C.D.C. 1604 electronic computer. The full numerical results can be obtained from the authors, but are not given here. Instead approximate contours of $-11\cdot5 \log S(\lambda; \mathbf{z})$, and hence of $S(\lambda; \mathbf{z})$ itself, of the maximized likelihood and of $p_u(\lambda_1, \lambda_2)$, are shown in Fig. 3. If the joint posterior distribution $p_u(\lambda_1, \lambda_2)$ were normal then a region which excluded 100α per cent of the total posterior probability could be given by

$$L_b(\hat{\lambda}_1, \hat{\lambda}_2) - L_b(\lambda_1, \lambda_2) = \chi_2^2(\alpha). \tag{26}$$

The shape of the contours indicates that the normal assumption is not very exact. Nevertheless, the quantity 100α obtained from (26) has been used to label the contours in Fig. 3 which thus roughly indicates the posterior probability distribution. For this example no appreciable improvement results from the addition of the further transformation parameter λ_0.

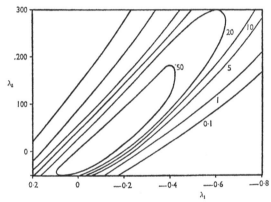

FIG. 3. Textile data. Transformation to $(y+\lambda_2)^{\lambda_1}$. Contours of $p_u(\lambda_1, \lambda_2)$ labelled with approximate percentage of posterior distribution excluded.

5. FURTHER ANALYSIS OF THE TRANSFORMATION

5.1. *General Procedure for Further Analysis*

The general procedure discussed above seeks to achieve simultaneously a model with (a) simple structure for the expectations, (b) constant variance and (c) normal distributions. Further analysis is sometimes profitable to see the separate contributions of these three elements to the transformation. Such analysis may indicate

(i) how simple a model we are justified in using;
(ii) what weight is given to the considerations (a) – (c) in choosing λ;
(iii) whether different transformations are really needed to achieve the different objectives and hence whether or not the value of λ chosen using the overall procedure is a compatible compromise.

Of course, quite often careful inspection of the data will answer (i)–(iii) adequately for practical purposes. Nevertheless, a further analysis is of interest.

We aim at simplicity both to achieve ease of understanding and to allow an efficient analysis. Validity of the formal tests associated with analysis of variance may, in virtue of the robustness of these tests, often hold to a good enough approximation even with the untransformed data. We stress, however, that such approximate validity is not by itself enough to justify an analysis; sensitivity must be considered as well as robustness. Thus in the biological example we have about one-third the sensitivity on the original scale as on the transformed scale. The approximate validity of significance tests on the original scale would be very poor consolation for the substantial loss of information involved in using the untransformed analysis. In any case even such validity is usually only preserved under the null hypothesis that all treatment effects are zero.

For the further analysis we again explore two approaches, one via maximum likelihood and the other via Bayes's theorem. Consider a general model to which a constraint C can be applied or relaxed, so that the relative merits of the simple and of the more complex model can be assessed. For example, the general model may include interaction terms, the constraint C being that the interaction terms are zero.

If $L_{max}(\lambda)$ and $L_{max}(\lambda|C)$ denote maximized log likelihoods for the general model and for the constrained model, then

$$L_{max}(\lambda|C) = L_{max}(\lambda) + \{L_{max}(\lambda|C) - L_{max}(\lambda)\}. \tag{27}$$

Here the second term on the right-hand side is a statistic for testing for the presence of the constraint.

More generally, with a succession of constraints, we have

$$L_{max}(\lambda|C_1, C_2) = L_{max}(\lambda) + \{L_{max}(\lambda|C_1) - L_{max}(\lambda)\}$$
$$+ \{L_{max}(\lambda|C_1, C_2) - L_{max}(\lambda|C_1)\}, \tag{28}$$

and the three terms on the right of (28) can be examined separately. The detailed procedure should be clear from the examples to follow.

To apply the Bayesian approach, we write the posterior density of λ

$$p(\lambda|C) = p(\lambda) \times \frac{p(C|\lambda)}{p(C)}, \tag{29}$$

where $p(C) = E_\lambda\{p(C|\lambda)\}$ is a constant independent of λ. That is, the posterior density of λ under the constrained model is the posterior density under the general model multiplied by a factor proportional to the conditional probability of the constraint given λ. Successive factorization can be applied when there is a series of successively applied constraints, giving, for example,

$$p(\lambda|C_1, C_2) = p(\lambda) \times \frac{p(C_1|\lambda)}{p(C_1)} \times \frac{p(C_2|\lambda, C_1)}{p(C_2|C_1)}, \tag{30}$$

where $p(C_2|C_1) = E_\lambda\{p(C_2|\lambda, C_1)\}$ is a further constant independent of λ. Note that we are concerned here not with the probabilities that the constraints are true, but with the contributions of the constraints to the final function $p(\lambda|C_1, C_2)$.

5.2. Structure of the Expectation

Now very often the most important question is: how simple a form can we use for $E\{y^{(\lambda)}\}$? Thus in the analysis of the biological example in Section 4, we assumed, among other things, that additivity can be achieved by transformation. In fact, interaction terms may or may not be needed. Similarly, in our analysis of the textile example we took a linear model with four parameters; the full second-degree model with ten parameters may or may not be necessary.

Now let A, H and N denote respectively the constraints to the simpler linear model (without interaction or second-degree terms), to a heteroscedastic model and to a model with normal distributions. Then,

$$L_{\max}(\lambda|A,H,N) = L_{\max}(\lambda|H,N) + \{L_{\max}(\lambda|A,H,N) - L_{\max}(\lambda|H,N)\}. \quad (31)$$

Let the parameter θ in the expectation under the general linear model be partitioned (θ_1, θ_2) where $\theta_2 = 0$ is the constraint A. Denote the degrees of freedom associated with θ_1 and θ_2 by ν_1 and ν_2. If ν_r is the number of degrees of freedom for residual in the complex model, the number in the simpler model is thus $\nu_r + \nu_2$.

As before, we work with the standardized variable $z^{(\lambda)} = y^{(\lambda)}/J^{1/n}$. If we identify residual sums of squares by their degrees of freedom, we have

$$L_{\max}(\lambda|\theta_2 = 0, H, N) = -\tfrac{1}{2}n\log\{S_{\nu_r+\nu_2}(\lambda; z)/n\}, \quad (32)$$

whereas

$$L_{\max}(\lambda|H,N) = -\tfrac{1}{2}n\log\{S_{\nu_r}(\lambda; z)/n\}. \quad (33)$$

Thus, in the textile example, S_{ν_r} refers to the residual sum of squares from a second-degree model and $S_{\nu_r+\nu_2}$ refers to the residual sum of squares from a first-degree model. Quite generally

$$S_{\nu_r+\nu_2}(\lambda; z) = S_{\nu_r}(\lambda; z) + S_{\nu_2.\nu_1}(\lambda; z),$$

where $S_{\nu_2.\nu_1}(\lambda; z)$ denotes the extra sum of squares of $z^{(\lambda)}$ for fitting θ_2, adjusting for θ_1, and has ν_2 degrees of freedom.

Thus with (32) and (33) the decomposition (31) becomes

$$L_{\max}(\lambda|\theta_2 = 0, H, N) = L_{\max}(\lambda|H,N) - \tfrac{1}{2}n\log\left\{1 + \frac{\nu_2}{\nu_r}F(\lambda; z)\right\}, \quad (34)$$

where

$$F(\lambda; z) = \frac{S_{\nu_2.\nu_1}(\lambda; z)/\nu_2}{S_{\nu_r}(\lambda; z)/\nu_r} \quad (35)$$

is the standard F ratio, in the analysis of variance of $z^{(\lambda)}$, for testing the restriction to the simpler model.

Equation (34) thus provides an analysis of the overall criterion into a part taking account only of homoscedasticity (H) and normality (N) plus a part representing the additional requirement of a simple linear model, given that H and N have been achieved.

In the corresponding Bayesian analysis (30) gives

$$p(\lambda|\theta_2 = 0, H, N) = p(\lambda|H,N) \times k_A\, p(\theta_2 = 0|\lambda, H, N), \quad (36)$$

where

$$1/k_A = E_{\lambda|H,N}\{p(\theta_2 = 0|\lambda, H, N)\},$$

the expectation being taken over the distribution $p(\lambda|H,N)$.

Note that since the condition $\theta_2 = 0$ is given, there is no component for these parameters in the prior distribution, so that the left-hand side of (36) is the posterior density obtained previously assuming A. Thus, in terms of the standardized variable $z^{(\lambda)}$, the left-hand side is

$$p_0(\lambda)\, C_{\nu_r+\nu_2}\{S_{\nu_r+\nu_2}(\lambda; z)\}^{-\frac{1}{2}(\nu_r+\nu_2)}, \quad (37)$$

where the normalizing constant is given by

$$C_{\nu_r+\nu_2}^{-1} = \int p_0(\lambda)\{S_{\nu_r+\nu_2}(\lambda;\,\mathbf{z})\}^{-\frac{1}{2}(\nu_r+\nu_2)}\,d\lambda.$$

Similarly, in the general model with $\boldsymbol{\theta}_1$ and $\boldsymbol{\theta}_2$ both free to vary, we obtain the first factor on the right-hand side of (36) as

$$p(\lambda\,|\,H, N) = p_0(\lambda)\,C_{\nu_r}\{S_{\nu_r}(\lambda;\,\mathbf{z})\}^{-\frac{1}{2}\nu_r}, \tag{38}$$

with

$$C_{\nu_r}^{-1} = \int p_0(\lambda)\{S_{\nu_r}(\lambda;\,\mathbf{z})\}^{-\frac{1}{2}\nu_r}\,d\lambda.$$

Thus, from (37) and (38), the second factor on the right-hand side of (36) must be

$$\frac{C_{\nu_r+\nu_2}}{C_{\nu_r}}\frac{\{S_{\nu_r}(\lambda;\,\mathbf{z})\}^{\frac{1}{2}\nu_r}}{\{S_{\nu_r+\nu_2}(\lambda;\,\mathbf{z})\}^{\frac{1}{2}(\nu_r+\nu_2)}}. \tag{39}$$

Now the general equation (36) shows that this last expression must be proportional to $p(\boldsymbol{\theta}_2 = 0\,|\,\lambda, H, N)$. It is worth proving this directly. To do this, consider a transformed scale on which constant variance and normality have been attained and the standard estimates $\hat{\boldsymbol{\theta}}_2$ and s^2 calculated. For the moment, we need not indicate explicitly the dependence on λ and $\mathbf{z}$. We denote the matrix of the reduced least-squares equations for $\boldsymbol{\theta}_2$, eliminating $\boldsymbol{\theta}_1$, by $\mathbf{b}$, so that the covariance matrix of $\boldsymbol{\theta}_2$ is $\sigma^2\mathbf{b}^{-1}$. The elements of $\mathbf{b}$ and $\mathbf{b}^{-1}$ are denoted b_{ij} and b^{ij}. Also we write $\rho_{ij} = b^{ij}/\sqrt{(b^{ii}\,b^{jj})}$ and $\{\rho^{ij}\}$ for the matrix inverse to $\{\rho_{ij}\}$. Then the joint distribution of

$$t_i = \frac{\theta_{2i} - \hat{\theta}_{2i}}{s\sqrt{b^{ii}}}$$

is (Cornish, 1954; Dunnett and Sobel, 1954)

$$\text{const} \times \left(1 + \frac{\sum\rho^{ij}\,t_i\,t_j}{\nu_r}\right)^{-\frac{1}{2}(\nu_r+\nu_2)}$$

where here and later the constant involves neither the parameters nor the observations. With uniform prior distributions for the θ's and for $\log\sigma$, this is also the posterior distribution of the quantities $(\theta_{2i} - \hat{\theta}_{2i})/(s\sqrt{b^{ii}})$, where now the θ_{2i} are the random variables. Transforming from the t_i's to the θ_{2i}'s we have that

$$p(\boldsymbol{\theta}_2\,|\,\lambda, H, N) = \text{const} \times (s_{\nu_r}^2)^{-\frac{1}{2}\nu_2}\left\{1 + \frac{(\boldsymbol{\theta}_2 - \hat{\boldsymbol{\theta}}_2)'\mathbf{b}(\boldsymbol{\theta}_2 - \hat{\boldsymbol{\theta}}_2)}{\nu_r\,s_{\nu_r}^2}\right\}^{-\frac{1}{2}(\nu_r+\nu_2)}$$

whence

$$p(\boldsymbol{\theta}_2 = 0\,|\,\lambda, H, N) = \text{const} \times (S_{\nu_r})^{-\frac{1}{2}\nu_2}\left\{1 + \frac{\hat{\boldsymbol{\theta}}_2'\mathbf{b}\hat{\boldsymbol{\theta}}_2}{S_{\nu_r}}\right\}^{-\frac{1}{2}(\nu_r+\nu_2)}$$

$$= \text{const} \times \frac{S_{\nu_r}^{\frac{1}{2}\nu_r}}{(S_{\nu_r+\nu_2})^{\frac{1}{2}(\nu_r+\nu_2)}}. \tag{40}$$

If now we restore in our notation the dependence on λ, comparison of (40) with (39) proves the required result; the appropriateness of the constant is easily checked.

Thus (36) provides an analysis of the overall density into a part $p(\lambda\,|\,H, N)$ taking account only of homoscedasticity and normality, and a second part, (39), in which the influence of the simplifying constraint is measured.

Equation (39) can be rewritten

$$\text{const} \times \{S_{\nu_r}(\lambda; \mathbf{z})\}^{-\frac{1}{2}\nu_2} \left\{1 + \frac{\nu_2}{\nu_r} F(\lambda; \mathbf{z})\right\}^{-\frac{1}{2}(\nu_r + \nu_2)}. \tag{41}$$

Now, by (34), the corresponding expression in the maximum-likelihood approach is given, in a logarithmic version, by

$$-\tfrac{1}{2} n \log\left\{1 + \frac{\nu_2}{\nu_r} F(\lambda; \mathbf{z})\right\}. \tag{42}$$

The essential difference between (41) and (42) is the occurrence of the term in $S_{\nu_r}(\lambda; \mathbf{z})$ in (41). In conventional large sample theory, ν_r is supposed large compared with ν_2 and then in the limit the variation with λ of the additional term is negligible, and the effect of both terms can be represented by plotting the standard F ratio as a function of λ. In applications, however, ν_2/ν_r may well be appreciable; thus in the textile example $\nu_2/\nu_r = 6/17$.

Hence (41) and (42) could lead to appreciably different conclusions, for example, if we found a particular value of λ giving a low value of $F(\lambda; \mathbf{z})$ but a relatively high value of $S_{\nu_r}(\lambda; \mathbf{z})$.

The distinction between (41) and (42) from a Bayesian point of view can be expressed as follows. In (41) there occurs the *ordinate* of the posterior distribution of $\boldsymbol{\theta}_2$ at $\boldsymbol{\theta}_2 = 0$. On the other hand, the F ratio, which determines (42), is a monotonic function of the *probability mass* outside the contour of the posterior distribution passing through $\boldsymbol{\theta}_2 = 0$. Alternatively, a calculation of the posterior probability of a small region near $\boldsymbol{\theta}_2 = 0$ having a length proportional to σ_z in each of the ν_2 component directions gives an expression equivalent to (42). The difference between (41) and (42) will be most pronounced if there exists an extreme transformation producing a low value of $F(\lambda; \mathbf{z})$ but a large value of $S_{\nu_r}(\lambda; \mathbf{z})$, corresponding to a large spread of the posterior distribution of $\boldsymbol{\theta}_2$. Expression (42) would give an answer tending to favour this transformation, whereas (41) would not.

5.3. *Application to Textile Example*

We now illustrate the above analysis using the textile data. The calculations are set out in Table 7 and displayed in Figs. 4 and 5. We discuss the conclusions in some detail here. In practice, however, the most useful aspect of this approach is the opportunity for graphical assessment.

Fig. 4 shows that the curvature of $L_{\max}(\lambda|H,N)$ is much less than that of $L_{\max}(\lambda|A,H,N)$ previously given in Fig. 2, the constraint A here being that the second-degree terms are supposed zero. The inequality

$$L_{\max}(\hat{\lambda}|H,N) - L_{\max}(\lambda|H,N) < \tfrac{1}{2}\chi_1^2(\alpha) \tag{43}$$

thus gives the much wider approximate 95 per cent confidence interval $(-0\cdot48, 0\cdot13)$ for λ indicated by HN in Fig. 4 and compared with the previous interval, marked AHN. Since the constraint has 6 degrees of freedom the sampling distribution of

$$-2\{L_{\max}(\lambda|A,H,N) - L_{\max}(\lambda|H,N)\} \tag{44}$$

for fixed normalizing λ is asymptotically χ_6^2. Alternatively, (44), being a monotonic function of F, can be tested exactly. Thus we can decide for which λ's, if any, the inclusion of the constraint is compatible with the data. In Fig. 5, $F(\lambda; \mathbf{z})$ is close to unity over the interesting range of λ close to zero, so that we can use the simpler model in this neighbourhood. The range indicated by C in Fig. 4 is that for which F is less than 2·70, the 5 per cent significance point.

TABLE 7

Textile data. Calculations for the analysis of the transformation

λ	$L_{\max}(\lambda \mid A, H, N)$	$L_{\max}(\lambda \mid H, N)$	*Difference* $= -13\cdot5 \times$	
			$\log(1 + \tfrac{6}{17}F(\lambda; z))$	$F(\lambda; z)$
1·00	21·52	41·41	$-19\cdot89$	9·52
0·80	29·67	49·14	$-19\cdot47$	9·15
0·60	38·17	55·65	$-17\cdot48$	7·50
0·40	47·21	60·59	$-13\cdot38$	4·80
0·20	56·48	63·99	$-7\cdot51$	2·09
0·00	63·10	66·02	$-2\cdot92$	0·68
$-0\cdot20$	61·11	66·89	$-5\cdot78$	1·51
$-0\cdot40$	52·61	66·07	$-13\cdot46$	4·84
$-0\cdot60$	43·16	62·68	$-19\cdot52$	9·19
$-0\cdot80$	34·22	56·44	$-22\cdot22$	11·85
$-1\cdot00$	25·79	48·18	$-22\cdot39$	12·03

λ	$p_u(\lambda \mid A, H, N)$	$p_u(\lambda \mid H, N)$	$k_A p_u(A \mid \lambda, H, N)$
0·20	0·02	0·32	0·05
0·15	0·09	0·49	0·18
0·10	0·42	0·69	0·62
0·05	1·58	0·93	1·71
0·00	4·18	1·19	3·51
$-0\cdot05$	5·64	1·47	3·84
$-0\cdot10$	4·66	1·76	2·65
$-0\cdot15$	2·36	1·96	1·20
$-0\cdot20$	0·77	2·06	0·37
$-0\cdot25$	0·19	2·03	0·09
$-0\cdot30$	0·04	1·88	0·02
$-0\cdot35$	0·01	1·59	0·01

The Bayesian analysis follows parallel lines. In Fig. 4, $p_u(\lambda \mid H, N)$ has a much greater spread than $p_u(\lambda \mid A, H, N)$. Fig. 5 shows $p_u(\lambda \mid H, N)$ with the component $k_A p(A \mid \lambda, H, N)$ from the constraint. When multiplied together they give the overall density $p_u(\lambda \mid A, H, N)$. A value of λ near zero maximizes the posterior density assuming the constraint and is consistent with the information in $p_u(\lambda \mid H, N)$.

There is, however, nothing in our Bayesian analysis itself to tell us whether the simplified model with the constraint is compatible with the data, even for the best possible λ. There is an important general point here. All probability calculations in statistical inference are conditional in one way or another. In particular, Bayesian posterior distributions such as $p_u(\lambda \mid A, H, N)$ are conditional on the model, in particular here on assumption A. It could easily happen that there is no value of λ for which A is at all reasonable, but to check on this we need to supplement the

Bayesian argument (Anscombe, 1961). Here we can do this by a significance test based on the sampling distribution of a suitable function of the observations, namely $F(\lambda; \mathbf{z})$. For λ around zero the value of $F(\lambda; \mathbf{z})$ is, in fact, well within the significance limits, so that we can reasonably use the posterior distribution of λ in question.

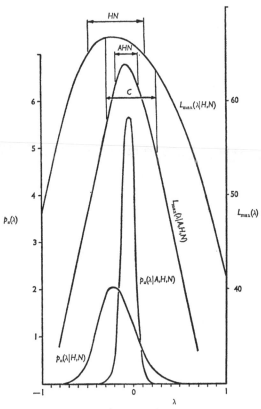

FIG. 4. Textile data. Functions $L_{max}(\lambda)$ and $p_u(\lambda)$ under different models. A: additivity. H: homogeneity of variance. N: normality. Arrows HN, AHN show approximate 95 per cent confidence intervals for λ. Arrows C show range for which F for second-degree terms is not significant at 5 per cent level.

5.4. *Homogeneity of Variance*

Suppose that we have k groups of data, the expectation and variance being constant within each group. In the lth group, let the variance be σ_l^2 and let $S^{(l)}$ denote the sum of squares of deviations, having $\nu_l = n_l - 1$ degrees of freedom. Write $\Sigma n_l = n$, $\Sigma \nu_l = n - k$. Thus in our biological example, $k = 12$, $\nu_1 = \ldots = \nu_{12} = 3$, $n_1 = \ldots = n_{12} = 4$ and $\nu = 36$, $n = 48$.

Now suppose that a transformation to $y^{(\lambda)}$ exists which induces normality simultaneously in all groups. Then in terms of the standardized variable $z^{(\lambda)}$, the maximized log likelihood is

$$L_{max}(\lambda \mid N) = -\tfrac{1}{2}\Sigma n_l \log\{S^{(l)}(\lambda; \mathbf{z})/n_l\}, \qquad (45)$$

where $S^{(l)}(\lambda; \mathbf{z})$ is the sum of squares $S^{(l)}$, considered as a function of λ and calculated from the standardized variable $z^{(\lambda)}$.

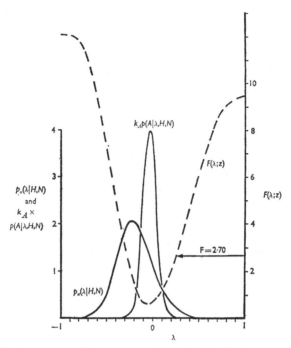

Fig. 5. Textile data. ——— Components of posterior distribution. – – – – – Variance ratio, $F(\lambda; \mathbf{z})$. Arrow gives 5 per cent significance level.

We now consider the constraint $H, \sigma_1^2 = \ldots = \sigma_k^2$, i.e. look at the possibility that a transformation exists simultaneously achieving normality and constant variance. Then if $S_\nu = \Sigma S^{(l)}$ is the pooled sum of squares within groups

$$L_{\max}(\lambda | H, N) = -\tfrac{1}{2}n \log\{S_\nu(\lambda; \mathbf{z})/n\}. \qquad (46)$$

Therefore

$$L_{\max}(\lambda | H, N) = L_{\max}(\lambda | N) + \log\left[\frac{\Pi\{S^{(l)}(\lambda; \mathbf{z})/n_l\}^{\frac{1}{2}n_l}}{\{S_\nu(\lambda; \mathbf{z})/n\}^{\frac{1}{2}n}}\right]$$

$$= L_{\max}(\lambda | N) + \log L_1(\lambda; \mathbf{z}), \qquad (47)$$

say. Here the second factor is the log of the Neyman–Pearson L_1 criterion for testing the hypothesis $\sigma_1^2 = \ldots = \sigma_k^2$.

In the corresponding Bayesian analysis, (29) gives

$$p(\lambda | H, N) = p(\lambda | N) \times k_H \, p(\sigma_1^2 = \ldots = \sigma_k^2 | \lambda, N), \qquad (48)$$

where

$$k_H^{-1} = E_{\lambda|N}\{p(\sigma_1^2 = \ldots = \sigma_k^2 | \lambda, N)\}.$$

For the general model in which $\sigma_1^2, \ldots, \sigma_k^2$ may be different, the prior distribution is

$$p_0(\lambda)\,(\Pi d\theta_i)\,(\Pi d\log \sigma_l)\,J^{-\nu/n}$$

and

$$p(\lambda \mid N) = p_0(\lambda)\, c\Pi\{S^{(l)}(\lambda;\, \mathbf{z})\}^{-\frac{1}{2}\nu_l}, \tag{49}$$

with

$$c^{-1} = \int p_0(\lambda)\, \pi\{S^{(l)}(\lambda;\, \mathbf{z})\}^{-\frac{1}{2}\nu_l} d\lambda.$$

For the restricted model in which the variances are all equal to σ^2, the appropriate prior distribution is

$$p_0(\lambda)\,(\Pi d\theta_l)\,(d\log\sigma) J^{-\nu/n}$$

and

$$p(\lambda \mid H, N) = \{p_0(\lambda)\, c_\nu(\lambda;\, \mathbf{z})\}^{-\frac{1}{2}\nu}. \tag{50}$$

Hence, on dividing (50) by (49), we have that the second factor in (48) is

$$\frac{c_\nu\,\Pi\{S^{(l)}(\lambda;\, \mathbf{z})\}^{-\frac{1}{2}\nu_l}}{c\;\{S_\nu(\lambda;\, \mathbf{z})\}^{-\frac{1}{2}\nu}} = \frac{c_\nu}{c}\,\frac{\Pi\nu_l^{\frac{1}{2}\nu_l}}{\nu^{\frac{1}{2}\nu}}\, e^{-\frac{1}{2}M(\lambda;\,\mathbf{z})}, \tag{51}$$

where (Bartlett, 1937)

$$M(\lambda;\, \mathbf{z}) = \nu\log\left\{\frac{S_\nu(\lambda;\, \mathbf{z})}{\nu}\right\} - \Sigma\nu_l\log\left\{\frac{S^{(l)}(\lambda;\, \mathbf{z})}{\nu_l}\right\}$$

is the modification of the L_1 statistic for testing homogeneity of variance, replacing sample sizes by degrees of freedom.

From our general argument, (51) must be proportional to $p(\sigma_1^2 = \ldots = \sigma_k^2 \mid \lambda, N)$. This can be verified directly by finding the joint posterior distribution of $\sigma_1^2, \ldots, \sigma_k^2$, transforming to new variables $\sigma_1^2, \sigma_2^2/\sigma_1^2, \ldots, \sigma_k^2/\sigma_1^2$, integrating out σ_1^2, and then taking unit values of the remaining arguments.

5.5. *Application to Biological Example*

In the biological example, we can now factorize the overall criterion into three parts. These correspond to the possibilities that in addition to normality within each group, we may be able to get constant variance and that it may be unnecessary to include interaction terms in the model, i.e. that additivity is achievable.

In terms of maximized likelihoods,

$$L_{\max}(\lambda \mid A, H, N) = L_{\max}(\lambda \mid N) + \log L_1(\lambda;\, \mathbf{z})$$
$$- \tfrac{1}{2}n\log\left\{1 + \frac{\nu_2}{\nu_r} F(\lambda;\, \mathbf{z})\right\}, \tag{52}$$

where $L_1(\lambda;\, \mathbf{z})$ is the criterion for testing constancy of variance given normality and $F(\lambda;\, \mathbf{z})$ is the criterion for absence of interaction given normality and constancy of variance.

The corresponding Bayesian analysis is

$$p(\lambda \mid A, H, N) = p(\lambda \mid N) \times k_H\, p(\sigma_1^2 = \ldots = \sigma_k^2 \mid \lambda, N) \times k_A\, p(\boldsymbol{\theta}_2 = 0 \mid \lambda, N, H). \tag{53}$$

The results are set out in Table 8 and in Figs. 6–8. The graphs of $L_{\max}(\lambda \mid N)$ and $p_u(\lambda \mid N)$ in Fig. 6 show that the information about λ coming from within group normality is very slight, values of λ as far apart as -1 and 2 being acceptable on this basis. The requirement of constant variance, however, has a major effect on the choice of λ; further, some information is contributed by the requirement of additivity.

TABLE 8

Biological data. Calculations for analysis of the transformation

λ	$L_{\max}(\lambda \mid A, H, N)$	$L_{\max}(\lambda \mid H, N)$	$L_{\max}(\lambda \mid N)$	$M(\lambda; \mathbf{z})$	$F(\lambda; \mathbf{z})$
4·0			125·33		1·17
3·0			128·50		1·48
2·0	62·97	69·36	130·78	92·13	1·83
1·0	91·72	98·24	131·93	50·54	1·88
0·5	103·83	109·55	132·15	33·90	1·62
0·0	113·51	117·96	131·95	20·99	1·22
−0·2	116·44	120·37	131·79	17·13	1·07
−0·4	118·58	122·13	131·59	14·19	0·95
−0·6	119·82	123·21	131·35	12·21	0·90
−0·8	120·07	123·60	131·04	11·16	0·94
−1·0	119·29	123·30	130·69	11·09	1·09
−1·2	117·52	122·35	130·29	11·91	1·33
−1·4	114·86	120·76	129·85	13·64	1·67
−1·6	111·43	118·55	129·37	16·23	2·08
−2·0	102·74	112·50	128·27	23·66	3·01
−2·5	89·91	102·46	126·68	36·33	4·12
−3·0	75·69	90·10	124·84	52·11	4·93

λ	$p_u(\lambda \mid A, H, N)$	$p_u(\lambda \mid H, N)$	$p_u(\lambda \mid N)$	$k_H p(H \mid \lambda, N)$	$k_A p(A \mid \lambda, H, N)$
1·0			0·335		
0·5		0·006	0·398		0·03
0·0	0·006	0·021	0·342	0·06	0·28
−0·1	0·023	0·055	0·324	0·17	0·39
−0·2	0·076	0·127	0·304	0·42	0·60
−0·3	0·257	0·261	0·283	0·92	0·98
−0·4	0·492	0·471	0·261	1·80	1·04
−0·5	0·942	0·754	0·240	3·14	1·25
−0·6	1·462	1·059	0·218	4·85	1·38
−0·7	1·823	1·320	0·196	6·73	1·38
−0·8	1·823	1·430	0·173	8·27	1·27
−0·9	1·419	1·360	0·153	8·88	1·04
−1·0	0·923	1·136	0·134	8·47	0·81
−1·1	0·468	0·850	0·116	7·33	0·55
−1·2	0·194	0·558	0·099	5·64	0·35
−1·3	0·067	0·329	0·083	3·96	0·20
−1·4	0·019	0·170	0·069	2·46	0·11
−1·5	0·005	0·078	0·058	1·34	0·06
−1·6	0·001	0·032	0·050	0·64	0·03
−1·7		0·009			

From Fig. 7, which shows the detailed separation of the maximum-likelihood and Bayesian components, any transformation in the region y^{-1} to $y^{-\frac{1}{2}}$ gives a compatible compromise.

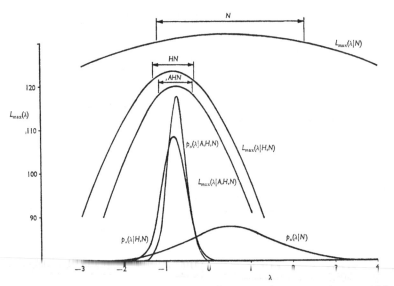

FIG. 6. Biological data. Functions $L_{max}(\lambda)$ and $p_u(\lambda)$ under different models. A: additivity. H: homogeneity of variance. N: normality. Arrows N, HN, AHN show approximate 95 per cent confidence intervals for λ.

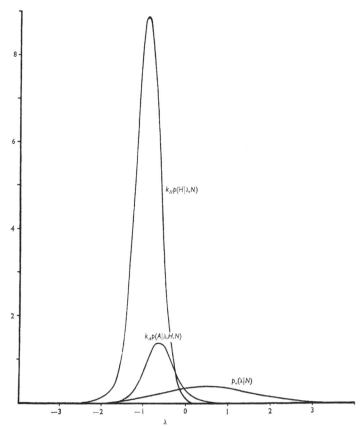

FIG. 7. Biological data. Components of posterior distribution.

Since the groups all contain four observations

$$-2\log L_1(\lambda; \mathbf{z}) = \tfrac{4}{3} M(\lambda; \mathbf{z})$$

and the graph of $M(\lambda; \mathbf{z})$ in Fig. 8 is equivalent to one of $L_1(\lambda; \mathbf{z})$. Since on the null hypothesis the distribution of $M(\lambda; \mathbf{z})$ is approximately χ^2_{11}, we can use Fig. 8 to

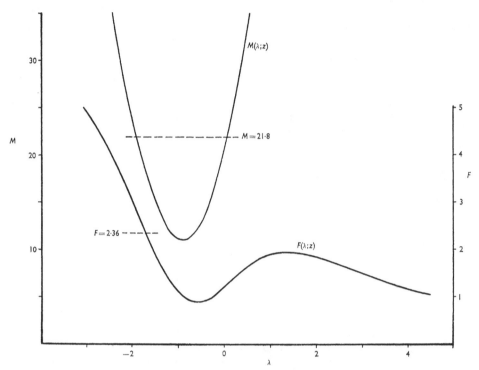

FIG. 8. Biological data. Variance ratio, $F(\lambda; \mathbf{z})$, for interaction against error as a function of λ. Bartlett's criterion, $M(\lambda; \mathbf{z})$, for equality of cell variances as a function of λ. Dotted lines give 5 per cent significance limits.

find the range in which the data are consistent with homoscedasticity. Similarly the graph of $F(\lambda; \mathbf{z})$ indicates the range within which the data are consistent with additivity. The dotted lines indicate the 5 per cent significance levels of M and of F.

The minimum of $M(\lambda; \mathbf{z})$ is very near $\lambda = -1$. It is of interest that the regression coefficient of $\log$(sample variance) on $\log$(sample mean) is nearly 4, so that the reciprocal transformation is suggested also by the usual approximate argument for stabilizing variance.

6. ANALYSIS OF RESIDUALS†

We now examine briefly a connection between the methods of the present paper and those based on the analysis of residuals. The analysis of residuals is intended

† We are greatly indebted to Professor F. J. Anscombe for pointing out an error in the approximation for α as we originally gave it. In the present modified version terms originally neglected in this Section have been included to correct the discrepancy.

primarily to examine what happens on one particular scale, although its use to indicate a transformation has been suggested (Anscombe and Tukey, 1963). Corresponding to an observation y, let Y be the deviation $\hat{y} - \bar{y}$ of the fitted value $\hat{y}$ from the sample mean and let $r = y - \hat{y}$ be the residual. If the ideal assumptions are satisfied r and Y will be distributed independently. Different sorts of departures from ideal assumptions can be measured, therefore, by studying the deviations of the statistics $T_{ij} = \Sigma r^i Y^j$ from $nE(r^i) E(Y^j)$. In addition to graphical analysis, a number of such functions have indeed been proposed for particular study (Anscombe, 1961; Anscombe and Tukey, 1963).

Specifically, the statistics

$$T_{30} = \Sigma r^3, \quad T_{40} = \Sigma r^4, \quad T_{21} = \Sigma r^2 Y, \quad T_{12} = \Sigma r Y^2 \tag{54}$$

were put forward as measures respectively of skewness, kurtosis, heterogeneity of variance and non-additivity. Tukey's degree of freedom for non-additivity (Tukey, 1949) involves the sum of squares corresponding to T_{12} considered as a contrast of residuals with "fixed" coefficients $r^{(i)}$.

Suppose now that we consider the family of power transformations and, writing $z = y/\dot{y}$, and $w = z - 1$, make the expansion

$$z^{(\lambda)} = \frac{z^\lambda - 1}{\lambda} = w + \tfrac{1}{2}(\lambda - 1) w^2 + \tfrac{1}{6}(\lambda - 1)(\lambda - 2) w^3 + O(w^4)$$

$$= w - \alpha w_2 + \tfrac{2}{3}\alpha(\alpha + \tfrac{1}{2}) w_3 + O(w^4), \tag{55}$$

where $w_2 = w^2$, $w_3 = w^3$ and $\alpha = 1 - \lambda$.

Now, $L_{\max}(\lambda)$ and $L_b(\lambda)$ are determined by the residual sum of squares of $z^{(\lambda)}$, which is approximately

$$\{w - \alpha w_2 + \tfrac{2}{3}\alpha(\alpha + \tfrac{1}{2}) w_3\}' a_r \{w - \alpha w_2 + \tfrac{2}{3}\alpha(\alpha + \tfrac{1}{2}) w_3\}. \tag{56}$$

If we take terms up to the fourth degree in w and then differentiate with respect to α, we have that the maximum-likelihood estimate of α is approximately

$$\hat{\alpha} = \frac{3w' a_r w_2 - w' a_r w_3}{3w_2' a_r w_2 + 4w' a_r w_3}. \tag{57}$$

If we write $y_1 = y - \dot{y}$, $y_2 = (y - \dot{y})^2$, $y_3 = (y - \dot{y})^3$ and denote by $\hat{y}_1, \hat{y}_2, \hat{y}_3$ the values obtained by fitting y_1, y_2 and y_3 to the model, the above approximation may be expressed in terms of the original observations as

$$\hat{\alpha} = \frac{3\dot{y}(y_1' a_r y_2) - y_1' a_r y_3}{3y_2' a_r y_2 + 4y_1' a_r y_3} = \frac{3\dot{y}\Sigma(y_1 - \hat{y}_1)(y_2 - \hat{y}_2) - \Sigma(y_1 - \hat{y}_1)(y_3 - \hat{y}_3)}{3\Sigma(y_2 - \hat{y}_2)^2 + 4\Sigma(y_1 - \hat{y}_1)(y_3 - \hat{y}_3)}. \tag{58}$$

To see the relation between this expression and the T statistics, write $d = \bar{y} - \dot{y}$. Then $y_1 = y - \dot{y} = r + Y + d$. Bearing in mind that $a_r Y = 0$, $a_r r = r$, $Y'r = 0$, $a_r 1 = 0$, $1'r = 0$, where 1 denotes a vector of ones, terms such as $y_1' a_r y_2$ can easily be expressed in terms of sums of powers and products of r, Y and d. In particular, on writing S for Σr^2, we find the numerator of (58) to be

$$3(\bar{y} - 3d)(T_{30} + 2T_{21} + T_{12}) - (T_{40} + 3T_{31} + 3T_{22} + T_{13}) + 3d(2\bar{y} - 3d)S. \tag{59}$$

To this order of approximation the maximum-likelihood estimate of α thus involves all the T statistics of orders 3 and 4.

As a very special case, for data assumed to form a single random sample

$$\hat{\alpha} = \frac{3\bar{y}\Sigma(y-\bar{y})(y_2-\bar{y}_2)+\Sigma(y-\bar{y})(y_3-\bar{y}_3)}{3\Sigma(y_2-\bar{y}_2)^2+4\Sigma(y-\bar{y})(y_3-\bar{y}_3)}.$$

Here questions such as non-additivity and non-constancy of variance do not arise and the transformation is attempting only to produce normality. Correspondingly in (59), $T_{21} = T_{12} = T_{31} = T_{22} = T_{13} = 0$, since $Y = \hat{y}-\bar{y} = 0$. In fact if we write $m_1 = \bar{y}$, $m_p = n^{-1}\Sigma(y-\bar{y})^p$ ($p = 2, 3, ...$) and make the approximation $d = \frac{1}{2}m_2/m_1$, we have that

$$\hat{\alpha} = \frac{m_1 m_3 - \frac{1}{3}\left\{(m_4 - 3m_2^2)+\frac{3m_2 m_3}{m_1}+\frac{9}{4}\frac{m_2^3}{m_1^2}\right\}}{6m_2^2+\frac{1}{3}\left\{7(m_4-3m_2^2)+12\frac{m_2 m_3}{m_1}+6\frac{m_2^3}{m_1^2}\right\}}. \tag{60}$$

For distributions in which m_1, m_2, m_3 and $m_4 - 3m_2^2$ are of the same order of magnitude, the terms in curly brackets are of one order higher in $1/m_1$ than are the other terms of the numerator and denominator. If we ignore the higher-order terms, we have

$$\hat{\alpha} \simeq \frac{m_1 m_3}{6m_2^2}.$$

A useful check suggested by Anscombe is to consider the χ^2 distribution for moderate degrees of freedom and the Poisson distribution for not too small a mean. For χ^2 we find $\alpha \simeq \frac{1}{6}$, whence $\lambda \simeq \frac{1}{3}$, corresponding to the well-known Wilson–Hilferty transformation. For the Poisson distribution, $\alpha \simeq \frac{1}{3}$, whence $\lambda \simeq \frac{2}{3}$.

7. ANALYSIS OF EFFECTS AFTER TRANSFORMATION

In Section 2 we suggested that, having chosen a suitable λ, we should make the usual detailed estimation and interpretation of effects on this transformed scale. Thus in our two examples we recommended that the detailed interpretation should be in terms of a standard analysis of respectively $1/y$ and $\log y$. Since the value of λ used is selected at least partly in the light of the data, the question arises of a possible need to allow for this selection when interpreting the factor effects.

To investigate an appropriate allowance, we regard λ as an unknown parameter with "true" value λ_0, say, and suppose the true factor effects to be measured in terms of the scale λ_0. If we were, for instance, to analyse the factor effects on the scale corresponding to the maximum-likelihood estimate $\hat{\lambda}$, we might expect some additional error arising from the difference between $\hat{\lambda}$ and λ_0. We now investigate this matter, although the present formulation of the problem is not always completely realistic. For example, in our biological example, having decided to work with $1/y$, we shall probably be interested in factor effects measured on this scale and not those measured in some unknown scale corresponding to an unknown "true" λ_0. On the other hand, if we are interested in whether there is interaction between two factors, it is possibly dangerous to answer this by testing for interaction on the scale $\hat{\lambda}$, since $\hat{\lambda}$ may be selected at least in part to minimize the sample interaction. A more reasonable formulation here may often be: on some unknown "true" scale λ_0, are interaction terms necessary in the model?

From the maximum-likelihood approach, the most useful result is that significance tests for null hypotheses, such as that just mentioned about the absence of interaction, can be obtained in a straightforward way in terms of the usual large-sample chi-squared test. Thus, in the textile example, we could test the null hypothesis that second-degree terms are absent for some unknown "true" λ_0, by testing twice the difference of the maxima of the two curves of $L_{\max}(\lambda)$ in Fig. 4 as χ_6^2. Note that the maxima occur at different values of λ. In this particular example, such a test is hardly necessary.

It would be possible to obtain more detailed results by evaluating the usual large-sample information matrix for the joint estimation of λ, σ^2 and $\boldsymbol{\theta}$. Since, however, more specific results can be obtained from the Bayesian analysis, we shall present only those. The general conclusion will be that to allow for the effect of analysing in terms of $\hat{\lambda}$ rather than λ_0, the residual degrees of freedom need only be reduced by ν_λ, the number of component parameters in λ. This result applies provided that the population and sample effects are measured in terms of the normalized variables $\mathbf{z}^{(\lambda)}$.

Consider locally uniform prior densities for $\boldsymbol{\theta}$, $\log \sigma$ and λ. Then the posterior density for $\boldsymbol{\theta}$ is

$$\frac{\int \{(\mathbf{z}^{(\lambda)} - \mathbf{a}\boldsymbol{\theta})' (\mathbf{z}^{(\lambda)} - \mathbf{a}\boldsymbol{\theta})\}^{-\frac{1}{2}n} \, d\lambda}{\int \{\nu_r s^2(\lambda; \mathbf{z})\}^{-\frac{1}{2}\nu_r} \, d\lambda}. \tag{61}$$

Approximate evaluation of the integral in (61) is done by expansion around the maxima of the integrands. The maximum of the integrand in the denominator is at the maximum-likelihood estimate $\hat{\lambda}$, and that of the numerator is near $\hat{\lambda}$, so long as $\boldsymbol{\theta}$ is near its maximum-likelihood value. The answer is that (61) is approximately

$$\frac{\{(\mathbf{z}^{(\hat{\lambda})} - \mathbf{a}\boldsymbol{\theta})' (\mathbf{z}^{(\hat{\lambda})} - \mathbf{a}\boldsymbol{\theta})\}^{-\frac{1}{2}(n-\nu_\lambda)}}{\{\nu_r s^2(\hat{\lambda}; \mathbf{z})\}^{-\frac{1}{2}(\nu_r - \nu_\lambda)}}. \tag{62}$$

This is exactly the posterior density of $\boldsymbol{\theta}$ for some known fixed λ with the degrees of freedom reduced by ν_λ.

To derive (62) from (61), we need to evaluate integrals of the form

$$I = \int \{q(\lambda)\}^{-\frac{1}{2}\nu} \, d\lambda, \tag{63}$$

where ν is large, and $q(\lambda)$ is assumed positive and to have a unique minimum at $\lambda = \hat{\lambda}$, with a finite Hessian determinant Δ_q at the minimum. We can then make a Laplace expansion, writing

$$I = \int \exp\left[-\frac{\nu}{2} \log q(\hat{\lambda}) - \frac{\nu}{2} \log\left\{ 1 + \frac{q(\lambda) - q(\hat{\lambda})}{q(\hat{\lambda})} \right\} \right] d\lambda$$

$$\simeq \frac{\{q(\hat{\lambda})\}^{-\frac{1}{2}\nu - \frac{1}{2}\nu_\lambda}}{\Delta_q^{\frac{1}{2}}} \times \text{const}; \tag{64}$$

for this we expand the second logarithmic term as far as the quadratic terms and then integrate over the whole ν_λ-dimensional space of λ. In our application the terms $\Delta_q^{\frac{1}{2}}$ in numerator and denominator are equal to the first order.

Finally, we can obtain an approximation to the posterior distribution $p_u(\lambda)$ of λ that is better than the usual type of asymptotic normal approximation. For an expansion about $\hat{\lambda}$ gives that

$$p_u(\lambda) = \frac{\{s^2(\lambda; \mathbf{z})\}^{-\frac{1}{2}\nu_r}}{\int \{s^2(\lambda; \mathbf{z})\}^{-\frac{1}{2}\nu_r} d\lambda}$$

$$\simeq \frac{\text{const}}{\left\{1 + \dfrac{(\lambda - \hat{\lambda})' \mathbf{b} (\lambda - \hat{\lambda})}{\nu_r s^2(\hat{\lambda}; \mathbf{z})}\right\}^{\frac{1}{2}\nu_\lambda}}. \qquad (65)$$

Here

$$\mathbf{b} = \mathbf{d}'(\hat{\lambda}) \, \mathbf{a}_r \, d(\hat{\lambda}), \qquad (66)$$

with $\mathbf{d}(\lambda)$ being the $n \times \nu_\lambda$ matrix with elements

$$\frac{\partial z_i^{(\lambda)}}{\partial \lambda_j} \quad (i = 1, ..., n; \; j = 1, ..., \nu_\lambda).$$

The matrix $\mathbf{b}$ determines the quadratic terms in the expansion of $s^2(\lambda; \mathbf{z})$ around $\hat{\lambda}$.

Thus the quantities $(\lambda_j - \hat{\lambda}_j)/\{s(\hat{\lambda}; \mathbf{z})\sqrt{b^{ii}}\}$ have approximately a posterior multivariate t distribution and

$$\frac{(\lambda - \hat{\lambda})' \mathbf{b} (\lambda - \hat{\lambda})}{\nu_r s^2(\hat{\lambda}; \mathbf{z})}$$

a posterior F distribution. In fact, however, it will usually be better to examine the posterior distribution of λ directly, as we have done in the numerical examples.

8. Further Developments

We now consider in much less detail a number of possible developments of the methods proposed in this paper. Of these, the most important is probably the simultaneous transformation of independent and dependent variables in a regression problem. Some general remarks on this have been made in Section 1.

Denote the dependent variable by y and the independent variables by $x_1, ..., x_l$. Consider a family of transformations from y into $y^{(\lambda)}$ and $x_1, ..., x_l$ into $x_1^{(\kappa_1)}, ..., x_l^{(\kappa_l)}$, the whole transformation being thus indexed by the parameters $(\lambda; \kappa_1, ..., \kappa_l)$. It is not necessary that the family of transformations of say x_1 into $x_1^{(\kappa_1)}$ and x_2 into $x_2^{(\kappa_2)}$ should be the same, although this would often be the case.

We now assume that for some unknown $(\lambda; \kappa_1, ..., \kappa_l)$ the usual normal theory assumptions of linear regression theory hold. We can then compute say the maximized log likelihood for given $(\lambda; \kappa_1, ..., \kappa_l)$, obtaining exactly as in (8)

$$L_{\max}(\lambda; \kappa_1, ..., \kappa_l) = -\tfrac{1}{2} \log \hat{\sigma}^2(\lambda; \kappa_1, ..., \kappa_l) + \log J(\lambda; \mathbf{y}), \qquad (67)$$

where $\hat{\sigma}^2(\lambda; \kappa_1, ..., \kappa_l)$ is the maximum-likelihood estimate of residual variance in the standard multiple regression analysis of the transformed variable. The corresponding expression from the Bayesian approach is

$$L_b(\lambda; \kappa_1, ..., \kappa_l) = -\tfrac{1}{2}\nu_r \log s^2(\lambda; \kappa_1, ..., \kappa_l) + \frac{\nu_r}{n} \log J(\lambda; \mathbf{y}). \qquad (68)$$

The straightforward extension of the procedure of Section 3 is to compute (67) or (68) for a suitable set of $(\lambda; \kappa_1, ..., \kappa_l)$ and to examine the resulting surface especially near its maximum. This is, however, a tedious procedure, except perhaps for $l = 1$. Further, graphical presentation of the conclusions will not be easy if $l > 1$; for $l = 1$ we can plot contours of the functions (67) and (68).

When λ is fixed, i.e. transformations of the independent variables only are involved, Box and Tidwell (1962) developed an iterative procedure for the corresponding non-linear least-squares problem. In this the independent variables are, if necessary, first transformed to near the optimum form. Then two terms of the Taylor expansion of $x_1^{(\kappa_1)}, ..., x_l^{(\kappa_l)}$ are taken. For example if $x_1^{(\kappa_1)} = x^{\kappa_1}$ and the best value for κ_1 is thought to be near 1, we write

$$x_1^{\kappa_1} = x_1 + (\kappa_1 - 1) x_1 \log x_1. \tag{69}$$

A linear regression term $\beta_1 x_1^{\kappa_1}$ can then be written approximately

$$\beta_1 x_1 + \beta_1 (\kappa_1 - 1) x_1 \log x_1 = \beta_1 x_1 + \gamma_1 x_1 \log x_1,$$

say. If the linear model involves linear regression on $x_1, ..., x_l$ and if all the transformations of the independent variable are to powers, we can therefore take the linear regression on $x_1, ..., x_l$, $x_1 \log x_1, ..., x_l \log x_l$ in order to estimate the β's and γ's and hence also the κ's. The procedure can then be iterated. Transformation of the dependent variable will usually be the more critical. Therefore, a reasonable practical procedure will often be to combine straightforward investigation of transformation of the dependent variable with Box and Tidwell's method applied to the independent variables.

It is possible also to consider simplifications of the procedure for determining a transformation of the dependent variable. The main labour in straightforward application of the method of Section 3 is in applying the transformation for various values of λ and then computing the standard analysis of variance for each set of transformed data. Such a sequence of similar calculations is straightforward on an electronic computer. It is perfectly practicable also for occasional desk calculation, although probably not for routine use. There are a number of possible simplifications based, for example, on expansions like (69) or even (55), but they have to be used very cautiously.

In the present paper we have concentrated largely on transformations for those standard "fixed-effects" analysis of variance situations where the response can be treated as a continuous variable. The same general approach could be adopted in dealing with "random-effects" models, and with various problems in multivariate analysis and in the analysis of time series. We shall not go into these applications here.

An important omission from our discussion concerns transformations specifically for data suspected of following the Poisson or binomial distributions. There are two difficulties here. One is purely computational. Suppose we assume that our observations, y, follow, for example, Poisson distributions with means that obey an additive law on an unknown transformed scale. Thus, in a row–column arrangement, it might be assumed that the Poisson mean in row i and column j has the form

$$(\mu + \alpha_i + \beta_j)^{1/\lambda} \quad (\lambda \neq 0),$$

$$\mu \alpha_i \beta_j \quad (\lambda = 0),$$

where λ is unknown. Then λ and the other parameters of the model can be estimated by maximum likelihood (Cochran, 1940). It would probably be possible to develop reasonable approximations to this procedure although we have not investigated this matter.

An essential distinction between this situation and the one considered in Section 3 is that here the untransformed observations y have known distributional properties. The analogous normal theory situation would involve observations y normally distributed with constant variance on the untransformed scale, but for which the population means are additive on a transformed scale. The maximum-likelihood solution in this case would involve, at least in principle, a straightforward non-linear least-squares problem. However, this situation does not seem likely to arise often; certainly, it is inappropriate in our examples.

An important possible complication of the analysis of data connected with Poisson and binomial distributions has been particularly stressed by Bartlett (1947). This is the presence of an additional component of variance of unknown form on top of the Poisson or binomial variation. If inspection of the data shows that such additional variation is substantial, it may be adequate to apply the methods of Section 3. For integer data with range $(0, 1, ...)$ it will often be reasonable to consider power transformations. For data in the form of proportions of "successes" in which "successes" and "failures" are to be treated symmetrically, Professor J. W. Tukey has, in an unpublished paper, suggested the family of transformations from y to

$$y^\lambda - (1-y)^\lambda.$$

For suitable λ's this approximates closely to the standard transforms of proportions, the probit, logistic and angular transformations. The methods of the present paper could be applied with this family of transformations.

ACKNOWLEDGEMENT
We thank many friends for remarks leading to the writing of this paper.

REFERENCES
ANSCOMBE, F. J. (1961), "Examination of residuals", *Proc. Fourth Berkeley Symp. Math. Statist. and Prob.*, **1**, 1–36.
—— and TUKEY, J. W. (1963), "The examination and analysis of residuals", *Technometrics*, **5**, 141–160.
BARTLETT, M. S. (1937), "Properties of sufficiency and statistical tests", *Proc. Roy. Soc.* A, **160**, 268–282.
—— (1947), "The use of transformations", *Biometrics*, **3**, 39–52.
BOX, G. E. P. and TIDWELL, P. W. (1962), "Transformation of the independent variables", *Technometrics*, **4**, 531–550.
COCHRAN, W. G. (1940), "The analysis of variance when experimental errors follow the Poisson or binomial laws", *Ann. math. Statist.*, **11**, 335–347.
CORNISH, E. A. (1954), "The multivariate t distribution associated with a set of normal sample deviates", *Austral. J. Physics*, 7, 531–542.
DUNNETT, C. W. and SOBEL, M. (1954), "A bivariate generalization of Student's t distribution", *Biometrika*, **41**, 153–169.
JEFFREYS, H. (1961), *Theory of Probability*, 3rd ed. Oxford University Press.
KLECZKOWSKI, A. (1949), "The transformation of local lesion counts for statistical analysis", *Ann. appl. Biol.*, **36**, 139–152.
TUKEY, J. W. (1949), "One degree of freedom for non-additivity", *Biometrics*, **5**, 232–242.
—— (1950), "Dyadic anova, an analysis of variance for vectors", *Human Biology*, **21**, 65–110.
—— and MOORE, P. G. (1954), "Answer to query 112", *Biometrics*, **10**, 562–568.

Summary of discussion

When the paper was read to the Royal Statistical Society a vote of thanks was proposed by J. A. Nelder and seconded by J. Hartigan and the discussion continued by J. W. Tukey, R. L. Plackett, M. S. Bartlett, M. R. Sampford, C. A. B. Smith, D. Kerridge and E. M. L. Beale. There was a written contribution by F. J. Anscombe.

Nelder, Tukey, Plackett and Bartlett all raise concerns about the possible overcomplication of the method in applications. Nelder points out that the difference between formulations involving sample sizes and those with degrees of freedom could be obviated essentially by a REML-like device. Hartigan proposes a nonparametric solution of the issue. Sampford is concerned about the apparent emphasis on a normality assumption. Beale points out possible difficulties when a form of transformation with two unknown parameters is involved. Kerridge remarks that the Bayesian treatment did not rely on personalistic assessment of prior beliefs. Anscombe elaborates the connection with the analysis of residuals. Several speakers, especially Tukey, ask about the balance between additivity, variance stabilization and achievement of normal error distributions,

THE ROLE OF STATISTICAL METHODS IN SCIENCE AND TECHNOLOGY

1. INTRODUCTION

This lecture is concerned more with vague generalities than with a technical account of recent research. It is fairly well established that statistical ideas and methods are important in a wide variety of technological investigations. Illustrations are the design and analysis of clinical trials for new drugs, or of agricultural field trials, or of industrial investigations of many types. Broadly speaking, the nearer one gets to the ultimate practical application, the more immediately applicable statistical ideas are likely to be.

The role of statistics in more fundamental research is less clear. Of course, statistical concepts have been important in some fields, for example in theoretical physics and in certain aspects of genetics. Again, to take a rather different example, Gauss's work on the method of least squares, done in connexion with errors of observation in astronomy, is still of fundamental importance for an essential group of modern statistical methods.

However, there is current a feeling that in some fields of fundamental research, statistical ideas are sometimes not just irrelevant, but may actually be harmful as a symptom of an over-empirical approach. This view, while understandable, seems to me to come from too narrow a concept of what statistical methods are about. In this lecture, I want to illustrate with examples a number of ways in which statistical ideas can be useful in science and technology.

2. THE MEASUREMENT OF UNCERTAINTY

The aspect that receives most attention in many text-books and courses on statistics is the measurement of the uncertainty involved in drawing conclusions from data. More precisely, what is measured is that part of the uncertainty stemming from the presence of random variation.

As an example, the first column of Table 1 gives measurements of the velocity of light up to 1932, taken from a review paper by von Friesen (1937). On the basis of these figures, Gheury de Bray (1934) had earlier suggested a periodic variation in the velocity of light ; the observations are remarkably well fitted by the formula

$$\text{vel. of light} = 299,885 + 115 \sin \left\{ \frac{2\pi}{40} (T - 1901) \right\} \text{ km/sec, (1)}$$

T being measured in years.

However, even if we ignore any possible systematic differences between experiments and between methods, each value is subject to uncertainty, as indicated approximately by the *confidence limits* in the third column of Table 1 ; these include some allowance for systematic errors. These are so wide that it is clear from a graph that the evidence for a periodic variation is entirely inadequate (von Friesen, 1937). This type of conclusion can be put quantitatively and objectively in terms of a *significance test*.

Table 1

Determinations of velocity of light by "direct" methods

Year	Velocity (km/sec)	Fit by (1)	Stated error
1874	299,990*	299,987	±300
1879	299,910	299,921	±50
1882	299,860	299,867	±30
1882	299,853	299,867	±60
1902	299,901	299,903	±84
1924	299,802	299,833	±30
1926	299,796	299,804	±4
1928	299,778	299,783	±20
1932	299,774	299,771	±11

* There is some doubt about the correct value here. Von Friesen recommended 300,400 km/sec.

There are at least three general purposes underlying the use of significance tests and confidence intervals:

(i) as a control on one's natural optimism (or pessimism) in interpreting data;

(ii) to give a fairly objective method for deciding what size of investigation is required;

(iii) to help convince other people that one's conclusions are justified.

While this last aspect is important, I have the impression that some published applications of significance tests are more directed at avoiding criticism from editors of journals and from Ph.D. examiners than in adding anything fundamental to the interpretation of the data.

There remains fairly substantial scope for developing new statistical tests, especially in connexion with more subtle questions, for example in connexion with apparently discontinuous relations.

3. COMPLEX PATTERNS OF VARIATION AND THE STRUCTURE OF ERROR

An important section of statistical work is concerned with complex patterns of haphazard variation. As a first example,

4

Table 2 refers to part of a recent investigation by Mr D. Legge, Psychology Department, University College. In this, eight subjects were tested in four sessions, with four doses of a drug, and under four perceptual conditions.

Table 2

Part of the design of Mr Legge's experiment

(a) *Assignment of drug doses* (D_1, D_2, D_3, D_4) *to sessions and subjects*

| | Sessions | | | | | | Sessions | | | |
Subjects	I	II	III	IV	Subjects	I	II	III	IV
S_1	D_1	D_2	D_4	D_3	S_5	D_1	D_2	D_4	D_3
S_2	D_4	D_1	D_3	D_2	S_6	D_3	D_4	D_2	D_1
S_3	D_3	D_4	D_2	D_1	S_7	D_2	D_3	D_1	D_4
S_4	D_2	D_3	D_1	D_4	S_8	D_4	D_1	D_3	D_2

(b) *Assignment of perceptual conditions* (1, 2, 3, 4) *for drug-dose* D_1

| | Period | | | | | | Period | | | | |
	A	B	C	D	E		A	B	C	D	E
S_1 I	4	1	3	2	2	S_5 I	3	4	2	1	1
S_2 II	3	4	2	1	1	S_6 IV	4	1	3	2	2
S_3 IV	1	2	4	3	3	S_7 III	1	2	4	3	3
S_4 III	2	3	1	4	4	S_8 II	2	3	1	4	4

The following effects may be present :
systematic differences between subjects,
systematic differences between sessions,
systematic differences between periods within sessions,
effect of drug doses,
effect of perceptual conditions,
residual effects of the last two from one observation to the next, interactions of the above.

The objects of statistical investigation here are to consider the following:

(i) For a given pattern of observations, which of the above sources of variation can be estimated (this has, of course, bearing on the best experimental design to use)?

(ii) How can each set of effects be most efficiently measured and estimated?

(iii) Suppose that we are primarily interested in one set of effects, say in how the effect of perceptual conditions changes with

5

dose. How is the "error" of these comparisons built up and how can it be estimated?

The last is important in any context in which a careful study of the structure of "error" is desirable, for example in connexion with methods of chemical or biological analysis in which several operations, each introducing error, are involved.

There is available a quite highly developed set of techniques, centring on analysis of variance, for tackling these problems.

A rather different type of application is illustrated by some recent work of Dr M. G. Kendall and mine for the Atomic Energy Authority on the analysis of temperature variations in nuclear reactors. Here some of the main sources of variation are spatial across the reactor and between different can positions down a channel. A special aspect here is that it is required to use the estimated components of variance to infer the highest temperatures likely to be attained in the reactor.

4. STATISTICAL ASPECTS OF THE PLANNING OF INVESTIGATIONS

This is conventionally divided into two sections, the design of experiments and the design of sample surveys, a rather rigid distinction being drawn between them. Roughly, this is that a sample survey is concerned with estimating by sampling the properties of a system already in existence, whereas, in an experiment, the experimenter is assumed to have complete control over the setting up of the system for investigation. There is no doubt, however, that there are important intermediate situations, and that much remains to be done in studying these.

Consider first experiments. Here we can distinguish at least three themes:
  (i) the avoidance of ambiguities of interpretation;
  (ii) methods for the simultaneous investigation of many factors and their interactions;
  (iii) the economy of investigations.

An example of (ii) is provided by some work of Dr J. D. Biggers, Wistar Institute, Philadelphia. In this, chicken bones were grown over a nutrient medium whose chemical composition was precisely defined. By having a number of solutions with different constitutions, the effect on growth rate could be studied. In this particular investigation, main effects and simple interactions between 11 amino acids were to be investigated. If in each trial each amino acid is either absent or present in a standard amount, there are $2^{11} = 2,048$ possible nutrient media. In fact, however, there is a subset of 128, divided into 4 sets of 32, which enables the effect of each amino acid

6

and of the interactions between pairs of amino acids to be estimated easily and efficiently.

The technique applied here, fractional replication, was first used in an experiment connected with agricultural botany, and has been extensively developed for applications to do with chemical processing. Fractional replication is a powerful tool for investigating complex systems and especially for preliminary investigations to select effects for more detailed study. However, I want to emphasize strongly that statistical ideas on experimental design are not confined to combinatorial trickery connected with complex experiments. The idea, held in some quarters, that statisticians inevitably favour complicated experiments rather than simple experiments is quite wrong.

Questions raised by Dr D. K. Butt, Physics Department, Birkbeck College, about angular correlation studies, illustrate methods to economise on observations. In this work observations are made on the rate of emission of pairs of particles whose paths subtend an angle θ at a radioactive source. There are strong reasons for expecting that the rate of emission at angle θ will be

$$\alpha_0 + \alpha_2 P_2 (\cos \theta) + \alpha_4 P_4 (\cos \theta)$$

$$= \alpha_0 + \alpha_2 \left\{ \frac{3 \cos^2 \theta - 1}{2} \right\} + \alpha_4 \left\{ \frac{35 \cos^4 \theta - 30 \cos^2 \theta + 3}{8} \right\}, \qquad (2)$$

where $P_2 (\cos \theta)$, $P_4 (\cos \theta)$ are Legendre polynomials and α_0, α_2, α_4 are unknown parameters to be estimated. Observations can thus be confined to be range $90° \leqslant \theta \leqslant 180°$. At how many angles should observations be made? There is an elegant general theorem that only three values of θ should be used, equally spaced in terms of $\cos^2 \theta$; this means $90°$, $135°$, $180°$. The distribution of observation times over these three angles depends on the relative importance of the different α's. One plausible choice is to distribute the observational times equally between the three angles.

If the restriction to three angles seems intaitively wrong, this is because it leaves no way of checking the validity of (2). It might be more reasonable to insert a term $\alpha_6 P_6 (\cos \theta)$ and to take observations at $\cos^2 \theta = 0, \frac{1}{3}, \frac{2}{3}, 1$, i.e., at $\theta = 90°$, $125°$, $145°$, $180°$. Goodness of fit of (2) is then tested by checking that the estimate of α_6 is small, but there is some loss of efficiency in estimating α_0, α_2 and α_4.

Sample surveys can be important because
(i) the field of study is not easily accessible by experiment, an example being criminology;
(ii) it is required to point to problems for detailed experimental study, there being many medical examples. There is much scope for the development of analytical survey methods.

7

Sometimes, what should be regarded as suggestive associations obtained from survey data, get inflated importance. An example is the complex "theory" of kinship relations constructed by Murdock (1949) on the basis largely of 2 x 2 contingency tables assembled from the Yale Cross-Cultural survey.

5. REDUCTION AND ANALYSIS OF COMPLEX DATA

One further important group of problems concerns complex responses. For each individual subject, animal, batch of material, etc., a whole set of observations, possibly of different types, is obtained. The following examples illustrate some of the problems of analysis.

First, we may have a series of observations in time or space which have to be reduced into simpler form before they can be interpreted. For instance, we may have for a number of looms the instants of occurrence of stops of various types. A closely related example occurs in plant ecology (Greig-Smith, 1957) where one studies the distribution over an area of plants of different species.

Another example is provided by some work of Dr J. D. Biggers, Dr C. Finn and Dr Anne McLaren on the reproductive performance of female mice (Biggers *et al.*, 1962); see also McLaren and Michie (1959). In this, two groups of mice were compared, the mice in one group having had one ovary removed. Observations for each mouse consisted of the size of litter for each litter up to the apparent end of reproductive life. The full interpretation of the data depends on correctly combining and comparing the curves for individual mice. Biggers *et al.* did this by means of a "standardised curve of reproductive performance"; just to have pooled all the results for mice in one group would not have been adequate.

A second group of problems arises when we have measurements of many types on each individual. The general set of techniques called multivariate analysis have been admirably reviewed recently by Professor C. R. Rao (1961). It is likely that with the general availability of electronic computers, this field will develop rapidly. For even techniques, like the method of principal components, which are conceptually naïve, lead to formidable calculations. What seems needed are methods for introducing specific technical information about the types of relation likely to hold between the variables; the method of path coefficients used in genetics may provide a clue. An interesting example of an essentially multivariate problem is that of bacterial classification (Sneath, 1957).

6. DECISION THEORY AND THE STATISTICS OF ROUTINE PROCESSES

There has been much emphasis recently on statistics as the

8

study of decision making in the face of uncertainty. In one approach to these problems, the stress is on subjective ideas of probability and utility and the object is to set up rules for an internally consistent scheme of decision making for any one individual. This sort of work is often discussed in terms of homely anecdotes, such as the following (Savage, 1954, p. 13).

Your wife has just broken five good eggs into a bowl, when you volunteer to finish making the omelette. A sixth egg lies unbroken on the table. There are three decisions: to break the egg into the bowl, to break into a saucer and if good transfer to the bowl, and to throw the last egg away. The "states of nature" are that the egg is (a) good, and (b) bad. The point of the discussion is to show that a decision should be reached as if a numerical value was assigned to the prior probability that the egg is bad and numerical values to the utilities of the various outcomes, and to study how the personal probabilities change as observations are obtained. In this approach the emphasis is on probability as a personal measure of (hypothetical) behaviour. Of course, personal judgment enters into all aspects of research, including the interpretation of data, and "conventional" statistical methods leave ample scope for this. However, the idea that the whole of statistical inference can be formulated as the study of consistent personal whim strikes me as a bold and imaginative step, backwards.

There is a very important collection of "objective" decision problems connected with the running of routine, or nearly routine, processes, for example:

(1) industrial inspection and process control;
(2) selection, including educational selection, plant-breeding, drug-screening (these are, of course, different in essential respects);
(3) stock control, production scheduling and replacement problems;
(4) medical diagnosis.

While it is probably rare for any of these to be run rigidly as a system with preset rules, one can still construct fairly realistic models.

7. Special Probabilistic Models

In many statistical analyses in which the observations are represented by random variables, the probability model is a quite empirical one, in which no attempt is made to build up a special probabilistic theory for the problem. A rather different type of statistical work is the development of mathematical theories of phenomena in which probabilistic aspects are important.

9

As one example, I should like to mention some recent work by Professor C. T. Ingold and Miss S. A. Hadland, Botany Department, Birkbeck College (Ingold and Hadland, 1959). Spores of *Sordaria* are grouped in sets of eight. When they escape, the set of eight may remain together, or may divide into two or more projectiles, each of fewer spores. Table 2 gives one set of observations of the number of spores per projectile. The simplest theoretical model for this is that for each of the seven links in a chain of eight spores, there is a constant probability θ of breakage, independently for all links. The theoretical frequencies of projectiles of $1, 2, \ldots, 8$ spores can be then calculated by a combinatorial probability argument and θ estimated ($\theta = 0.168$ for Table 3). The general agreement between observed and theoretical frequencies was good; rigorous testing of the adequacy of fit would have been difficult because of the complicated correlations between the different frequencies.

Table 3

Number of Spores per projectile in Sordaria

Number	Observed frequency	Theoretical frequency
1	490	459
2	343	361
3	265	282
4	199	221
5	200	171
6	134	132
7	72	101
8	272	249

This is an example of a "static" probability problem. Often, however, one has a "dynamic" problem in which a system is changing with time in a probabilistic way. This leads to a field of great mathematical interest, the theory of stochastic processes. Applications include statistical problems in theoretical physics, and in cosmology, the theory of epidemics and of competition between species, and a range of industrial problems including studies of congestion and of inventory control and production planning.

In some fields at least, the most valuable mathematical theory is the very simple formula that brings together approximately data that are apparently unrelated. For many of the topics just mentioned, the construction of anything approaching a realistic theory leads to great complexity. Simulation on an electronic computer is then likely to be the best method of investigation.

10

8. DISCUSSION

I have dealt with a number of types of application of statistical ideas. However, the last thing I want to do is to suggest that the subject could, or should, be divided into separate compartments; few things could be more arid. For, in many investigations, several of the above general themes will be involved simultaneously. Further, several of the types of problem are closely related mathematically. Again, I hope that I have not given the impression that there are simple routine answers available for all problems like those discussed. Active research is going on in all these general topics. What can reasonably be claimed is that the general ideas are available for thinking about these problems qualitatively, and that quantitative answers are available for the simpler problems. But much remains to be done.

There are many further fields in which statistical ideas are applicable, in addition to those I have mentioned. Important examples are in economics, administration, and in commercial contexts.

One difficult choice that has to be made in teaching statistics is of emphasis between general principles that can be adapted to meet new problems and the details of special methods that are found to be commonly useful. The view that no two statistical problems are really the same, is expressed in a well-known aphorism: there is no such thing as a routine statistical question, there are only questionable statistical routines. While this is too extreme, it is clear that one main direction in which the subject will develop is by our learning how to incorporate efficiently into statistical analyses special theoretical and other knowledge peculiar to particular applications.

ACKNOWLEDGEMENTS

I am grateful to Dr J. D. Biggers, Dr D. K. Butt, Professor C. T. Ingold, Dr M. G. Kendall, Mr D. Legge, Dr A. McLaren and the Atomic Energy Authority for permission to quote illustrative examples.

REFERENCES

BIGGERS, J. D., FINN, C. and McLAREN, A. (1962). "The long-term reproductive performance of female mice, I, II." *J. Reproduction and Fertility*, to appear.

FRIESEN, S. VON (1937), "On the values of fundamental atomic constants," *Proc. Roy. Soc.* A, 160, 424-439.

11

GHEURY DE BRAY, M. E. J. (1934), "Velocity of light," *Nature, Lond.*, 137, 948-949.

GREIG-SMITH, P. (1957), *Quantitative plant ecology*. London: Butterworth.

INGOLD, C. T. and HADLAND, S. A. (1959), "The ballistics of *Sordaria*," *New Phytologist*, 58, 46-57.

McLAREN, A., and MICHIE, D. (1959), "The spacing of implantations in the mouse uterus," *Mem. Soc. Endocrinol.*, 6, 65-74.

MURDOCK, G. P. (1949), *Social structure*. New York: Macmillan.

RAO, C. R. (1960), "Multivariate analysis: an indispensible statistical aid in applied research," *Sankhya*, 22, 317-338.

SAVAGE, L. J. (1954), *The foundations of statistics*. New York: Wiley.

SNEATH, P. H. A. (1957), "Some thoughts on bacterial classification," *J. Gen. Microbiol.*, 17, 184-200.

12

one should include the loss arising from the fact that n individuals will be given the worse treatment, so that the total loss will be equal to

$$n + (N - 2n)p(n, \boldsymbol{\theta})$$

multiplied by some function $g(\boldsymbol{\theta})$ of θ_1 and θ_2. Sampling costs (e.g., $2nc$) would be added to this. Maurice formulated this model in terms of industrial and agricultural settings where "treatment" could mean production method, fertilizer, etc.

Colton [2] adapted the model to clinical trials for the selection of one out of two medical treatments.

This model is used to assist in assessing a desirable value for n—the size of the trial with each treatment. There is generally no minimax* solution if the function $g(\boldsymbol{\theta})$ is unbounded. Maurice [4] shows how to determine a minimax solution when $g(\cdot)$ is bounded, in particular when (1) the observed variables are normally distributed with standard deviation σ and expected values θ_1 and θ_2 in π_1 and π_2, respectively, and (2) $g(\boldsymbol{\theta})$ is proportional to $|\theta_1 - \theta_2|$. She also obtains an (approximately) optimal sequential procedure. Colton [2] also uses a different criterion to obtain an optimal value for n —"maximin expected net gain". "Gain" is defined as $+(-)g(\boldsymbol{\theta})$ if the better (worse) treatment is received, so that expected net gain is proportional to

$$(N - 2n)\big[(1 - p_n(\boldsymbol{\theta})) - p_n(\boldsymbol{\theta})\big]g(\boldsymbol{\theta}).$$

Colton [2] also obtains an (approximately) optimal sequential procedure (see also Anscombe [1]), and later [3] describes optimal two-stage sampling procedures based on the same model.

In more recent investigations, methods of utilizing early results in a sequential trial to assist in determining treatment assignments so as to reduce the expected number of individuals given the worse treatment during the trial has been given considerable attention (*see* PLAY-THE-WINNER).

Oudin and Lellouch [5] consider another aspect, in which it is desired to combine the results of parallel trials on groups of individuals in different categories according to values of some concomitant variables* (age, sex, etc.).

References

[1] Anscombe, F. J. (1963). *J. Amer. Statist. Ass.*, **58**, 365–383.
[2] Colton, T. (1963). *J. Amer. Statist. Ass.*, **58**, 388–400.
[3] Colton, T. (1965). *Biometrics*, **21**, 169–180.
[4] Maurice, R. J. (1959). *J. R. Statist. Soc. B*, **21**, 203–213.
[5] Oudin, C. and Lellouch, J. (1975). *Rev. Statist. Appl.*, **23**, 35–41.

(CLINICAL TRIALS
DECISION THEORY
PLAY-THE-WINNER
SEQUENTIAL ANALYSIS)

COMBINATION OF DATA

"Combination of Observations" is an old term for the numerical analysis of data. In a sense, the words "combination of data" encompass the whole of the statistical analysis of data, with an implied emphasis on condensation and summarization, two important general themes. The present article, however, concentrates on the combination of information from separate sets of data, each of which has a viable analysis in its own right. What issues arise in putting together conclusions from separate analyses?

Such combination can arise in two rather different ways. First, the different sets of data may be obtained from quite distinct investigations, possibly even using dissimilar designs and different experimental techniques. Secondly, a sensible strategy for handling data of relatively complex structure is often to divide the data into simpler subsections, to analyze these separately, and then, in a second stage of analysis, to merge the conclusions from the component analyses. While the technical statistical problems in these two kinds of application may be identical, assumptions of homogeneity, for example of error variance, may be made with greater confidence in the second situation and the need to produce a single final conclusion is greater.

In general, the drawing together and comparison of interrelated information from different sources is very important. Often, however, it is enough to proceed qualitatively.

WEIGHTED MEANS

Suppose first that a parameter θ of clear interest can be estimated from m independent sets of data. Thus θ might be a mean, a difference of means, a factorial contrast*, a log odds ratio* in a comparison of binomial probabilities, etc., and the m sets might correspond to m independent studies by different investigators. Let $t_1, \ldots, t_m$ be the estimates, assumed for the moment normally distributed around θ with known variances $v_1, \ldots, v_m$, calculated from the internal variability within the separate sets of data.

If these assumptions are a reasonable basis for the analysis, the "best" combined estimate of θ is

$$\tilde{t} = \left(\sum t_j/v_j\right)/\left(\sum 1/v_j\right); \qquad (1)$$

this is called a weighted mean of the t_j's and $1/v_j$ is called the weight attached to t_j. Then $\tilde{t}$ is normally distributed around θ with standard error

$$\left(\sum 1/v_j\right)^{-1/2}, \qquad (2)$$

from which confidence limits for θ can be calculated. Equation (1) is summarized by the slogan "Weight a value inversely as its variance."

Contrast (1) with the unweighted mean,

$$\bar{t} = \left(\sum t_j\right)/m, \qquad (3)$$

the two formulae being the same if and only if all the variances v_j are equal. It can be shown that the general weighted mean, with weights w_j,

$$\tilde{t}_w = \sum w_j t_j / \sum w_j, \qquad (4)$$

has variance $\sum w_j^2 v_j/(\sum w_j)^2$, close to that of its minimum $(\sum 1/v_j)^{-1}$, achieved for $\tilde{t}$, for

a fairly wide range of weights. Thus the choice of weights is usually not critical provided that extreme weights are avoided. This is some comfort when there is doubt about the v_j.

A number of assumptions are made in (1) and (2). The principal one is that the parameter θ is indeed the same for all sets of data. It is an important general principle that, before merging information from different sources, mutual consistency should be checked. Sometimes informal inspection is enough, but if a formal test is required,

$$\tilde{d} = \sum v_j^{-1}(t_j - \tilde{t})^2$$
$$= \sum t_j^2/v_j - \left(\sum t_j/v_j\right)^2\left(\sum 1/v_j\right)^{-1} \qquad (5)$$

can be treated as chi-squared* with $m - 1$ degrees of freedom.

Incidentally, the quickest derivation of (1), (2), and (5), and their numerous generalizations, is via a standard least-squares analysis of the linear model

$$E\left(t_j/\sqrt{v_j}\right) = \left(1/\sqrt{v_j}\right)\theta, \qquad \operatorname{var}\left(t_j/\sqrt{v_j}\right) = 1. \qquad (6)$$

Then (1) is the "ordinary" least-squares* estimate of θ and (5) is the "ordinary" residual sum of squares corresponding under the model to a true variance of 1.

Too extreme a value of (5) suggests that something is wrong; usually, it will be that large values of (5) are obtained, indicating either that the v_j underestimate the real error in the t_j or that $E(t_j)$ is not constant.

POOLING* IN THE PRESENCE OF OVERDISPERSION

Suppose that (5) shows that the separate estimates t_j differ by more than they should under the assumptions of homogeneity outlined above (1). Here are some possible next steps:

1. It may happen that the v_j are based on unrealistic assumptions about the individual sets of data. Revision of the val-

ues of the v_j will in general give a new estimate (1), although as noted above, minor adjustments to weights are usually unimportant. If, however, all the v_j are assumed in error by an unknown common factor, the estimate (1) is still optimal, although its standard error (2) has to be multiplied by $\sqrt{\{\tilde{d}/(m-1)\}}$, a reasonable thing to do only if m is not very small. This assumption about variances may sometimes be reasonable in connection both with superficially Poisson or binomial variation and also with cluster sampling*.

2. Interest may switch from providing a single estimate to a study of the differences between the individual estimates. Such a move led to the discovery of the inert gases.

3. It may be reasonable to suppose that the property measured by θ is the same for all sets of data, but that some or all of the determinations are subject to systematic errors. For example, θ may be a fundamental constant, such as the velocity of light, or the rate constant of a well-defined chemical reaction. Then

$$E(t_j) = \theta + \Delta_j, \qquad (7)$$

where Δ_j is an unknown constant. Sometimes bounds for the Δ_j, can be calculated from detailed examination of the measurement techniques. We can, by adding suitable constants to the t_j's, take such bounds in the form $|\Delta_j| \leqslant a_j$. Then for the weighted mean $\tilde{t}_w = \sum w_j t_j / \sum w_j$, cautious $1 - 2\epsilon$ confidence limits are

$$- \frac{\sum w_j a_j}{\sum w_j} - k_\epsilon^* \frac{\left(\sum w_j^2 v_j\right)^{1/2}}{\sum w_j} + \tilde{t}_w,$$

$$\frac{\sum w_j a_j}{\sum w_j} + k_\epsilon^* \frac{\left(\sum w_j^2 v_j\right)^{1/2}}{\sum w_j} + \tilde{t}_w, \qquad (8)$$

where k_ϵ^* is the upper ϵ-point of the standardized normal distribution. The adjustment for systematic error covers the possibility that all systematic errors are simultaneously extreme. For given

a_j, v_j, and ϵ, the w_j's can be chosen to minimize the width of the interval (8); in the extreme case where systematic errors predominate, concentrate on the estimates with smallest a_j.

Some of the sets of data, for example "high-quality" determinations, have, or can be assumed to have, or can even be defined to have, zero Δ_j. For example, some sets of data might compare a new treatment with a concurrent control, with due randomization*, whereas other sets might use historical controls*. One reasonable procedure is then as follows. Examine the "bias-free" estimates for mutual consistency. Subject to this, pool them by (1) to give $\tilde{t}_{BF}$. Then examine the remaining t_j for consistency with $\tilde{t}_{BF}$, pooling into $\tilde{t}_{BF}$ those clearly consistent with it. The remaining values are studied and if possible the biases in them explained, for example by adjustment for discrepant concomitant variables. Care is needed in case, for example, all the possibly biased values in fact err in the same direction. A final statement of conclusions should include both $\tilde{t}_{BF}$ and the final pooled estimate.

If the Δ_j can be modeled as random variables, analyses similar to those discussed in step 6 below become available, but such assumptions about systematic errors should be treated with due scepticism. An assumption that systematic errors in different sections of the data are independent random variables of zero mean deserves extreme scepticism.

4. Inspection, or more detailed analysis, may show that the parameter θ is inappropriately defined. Thus θ might be taken initially as the difference in mean response between two treatments. Yet the data may show that while the difference between means varies between the sets of data, the ratio of means is stable. This suggests redefining θ before combination.

5. It may be necessary to abandon the notion of a single true θ, supposing then

that

$$E(t_j) = \theta_j, \qquad (9)$$

where θ_j is the parameter value of interest in the jth set of data. Note that although (9) is superficially similar to (7), the interpretation is different, in that in (7) only θ is really of concern. If m is very small (e.g., 2) and the discrepancy cannot be explained, it may be best to abandon the notion of combining the sets of data. With larger values of m one would normally first attempt an empirical explanation of the variation of θ_j. For example, if z_j is a scalar or vector explanatory variable attached to the whole of the jth set of data (e.g., a characteristic of the center or laboratory in which the jth set of data is obtained), one might hope that

$$\theta_j = \theta + \boldsymbol{\beta}^T(\mathbf{z}_j - \mathbf{z}_0), \qquad (10)$$

where θ is the value at some reference level $\mathbf{z}_0$, possibly $\bar{\mathbf{z}} = \sum \mathbf{z}_j / m$, and $\boldsymbol{\beta}$ is a vector of unknown parameters to be estimated, for example via the generalization of (6).

A less ambitious representation is

$$\theta_j = \theta + \boldsymbol{\beta}^T(\mathbf{z}_j - \mathbf{z}_0) + \eta_j, \qquad (11)$$

where $\eta_1, \ldots, \eta_m$ are independent random variables of zero mean and variance σ_η^2, in which only a portion of the variability of θ_j can be explained. Equations (10) and (11) lead in effect to empirically estimated "corrections" to apply to the t_j to adjust to the reference level, $\mathbf{z} = \mathbf{z}_0$.

Another possibility is that variation among the θ_j can be explained by one, or a small number, of anomalous data sets. Then the inconsistent set or sets can be studied separately and the remainder combined by (1). Occasionally, it may be wise to replace the mean (1) by some robust* estimate of location in which isolated extreme values are automatically discounted.

6. Finally, in the absence of a specific explanation of the variation in θ, it may sometimes be taken as random; i.e., we may take a representation

$$\theta_j = \theta + \eta_j, \qquad (12)$$

where $\eta_1, \ldots, \eta_m$ are, as in (11), independent random variables normally distributed with zero mean and variance σ_η^2. This is equivalent to supposing $t_1, \ldots, t_m$ to be independently normally distributed around θ with variances $v_1 + \sigma_\eta^2, \ldots, v_m + \sigma_\eta^2$. For the moment assume, usually quite unrealistically, that σ_η^2 is known. Then the discussion of (1) and (2) holds, with changed variances, so that the weighted mean (4) is indicated with $w_j = (v_j + \sigma_\eta^2)^{-1}$. Note that the new mean is intermediate between the original weighted mean t of (1) and the unweighted mean $\bar{t}$ of (3). Of course, if the v_j are all equal, the estimate is unchanged, although its standard error now includes a contribution from the variation between sets.

Representation (12) is a version of a components of variance* or random-effects model, unbalanced if the v_j are unequal, balanced if they are equal. The difference from the standard form of such models is that the "within-group" variances v_j are possibly different but assumed known, i.e., estimated with relatively high precision. Maximum likelihood estimation* of θ and σ^2 from $t_1, \ldots, t_m$ is straightforward. The maximum likelihood estimate $\hat{\theta}$ is a weighted mean with weights obtained by replacing σ_η^2 by its maximum likelihood estimate $\hat{\sigma}_\eta^2$. Asymptotically, $\hat{\theta}$ and $\hat{\sigma}_\eta^2$ are independent.

Alternatively, σ_η^2 can be estimated by equating d of (5) to its expectation, or indeed by equating other quadratic forms*, such as $\bar{d} = \sum (t_j - \bar{t})^2$, to their expectations. Of course, m must not be too small or this will be ineffective. Having estimated σ_η^2 in some such way, we may use a weighted mean with estimated weights.

The use of a random-effects representation (12) needs careful consideration. If the sets of data arise say from different centers or laboratories or batches of material randomly sampled from a well-defined universe of centers or laboratories or batches, a model of this broad type is appropriate; θ can be regarded as an average applying to the universe in question. Unfortunately, such clear-cut justification and interpretation is relatively rare. More commonly, we have a number of estimates that are inexplicably different and if, as is frequently the case, a single estimate is required, we have little choice but to assume that the additional variation is in some loose sense random and to adopt (12): note that (12) is in effect a definition of the target parameter θ as the value achieved when this additional source of variability averages out to zero, a somewhat hypothetical construct.

SUMMARY DISCUSSION

If the separate estimates are mutually consistent, to merge results, calculate a weighted mean. If the separate estimates are not consistent in the light of the internal estimates of error, explain the discrepancies, if possible. If this is not possible and merging is desirable, with due caution treat the superfluous variation as random. If m is reasonably small, it will usually be desirable to report the individual estimates as well as the pooled one.

RELATION WITH SIGNIFICANCE TESTS

The discussion above has concentrated on estimation rather than significance testing, in the belief that this is normally more fruitful. If a particular value of θ, say θ_0, is of special interest this can be examined via the final estimate and its standard error.

The practice of testing the hypothesis $\theta = \theta_0$ separately for each set of data, dividing the sets into two sections, those consistent with θ_0 and those inconsistent at some selected level, is in general bad. If, in fact, all sets of data correspond to the same value of θ deviating modestly from θ_0, quite mis-

leading conclusions result from such a split of the data.

There is sometimes the following related difficulty of interpretation. Suppose that essentially the same topic is investigated independently in a number of centers and the hypothesis $\theta = \theta_0$ tested in each: the hypothesis might, for example, correspond to a new treatment being equivalent to a control. Suppose that one of the centers finds significant evidence against $\theta = \theta_0$ and the others do not. It is clear that all centers should report their results. Next suppose that the results from the different centers are mutually consistent and that the pooled estimate leads to confidence limits for θ containing and sharply concentrated around θ_0. What should be concluded?

The significance level attaching to the results from the center reporting a significant effect is concerned with what can reasonably be concluded from that center's results in isolation. Because interpretation should be based, as far as reasonably feasible, on all relevant available information, this one significance level is unimportant when all the data become available, unless strong rational arguments can be produced for singling out the one center as different in some crucial respect from the others. Of course, it may be hard to get agreement on this last point! The morals are partly the need to report all results significant or not, partly the desirability of avoiding undue emphasis on significance tests and partly the importance of assessing all available relevant information.

MORE TECHNICAL COMMENTS

The discussion above has focused on general ideas, and by concentrating on the simplest situations in which the t_j are normally distributed with known variances, detailed formulae have been largely avoided. Broadly similar problems arise when the t_j have specific nonnormal distributions and when there are nuisance parameters present, in particular in normal theory when the variances of the t_j are unknown, but can be estimated.

Theoretical discussion of such problems can be hard. For example, suppose that $t_1, \ldots, t_m$ are independently normally distributed with mean θ and unknown variances; let these variances be estimated by $v_1, \ldots, v_m$, which are now normal theory estimates of variance with $f_1, \ldots, f_m$ degrees of freedom, independently of one another and of the t_j. Thus if t_j is the mean of a random sample of n_j observations, $f_j = n_j - 1$ and v_j is the usual sample estimate of population variance divided by n_j.

The simplest solution is to use (1), which we now call a weighted mean with empirically estimated weights. This is satisfactory if all the degrees of freedom are reasonably large (e.g., provided that none is smaller than 10). If m is large and some or all of the f_j are small, an asymptotically efficient estimate is constructed by solving for $\tilde{\theta}$ the estimating equation

$$\sum \frac{(f_j - 1)(t_j - \tilde{\theta})}{f_j v_j + (t_j - \tilde{\theta})^2} = 0 \qquad (13)$$

In principle, exact confidence limits can be derived. Note that (13) has the odd feature that samples with $f_j = 1$, and indeed those with $f_j = 0$, do not contribute. If $f_j \gg 1$ and the second term in the denominator of (13) is ignored, (13) defines the weighted mean (1).

The formal optimal properties of $\tilde{\theta}$ arise from a limiting operation as $m \to \infty$, involving a model with $m + 1$ parameters. Standard maximum likelihood theory is inapplicable and indeed (13) is not quite the maximum likelihood estimating equation, which, when the t_j are means of random samples and the v_j corresponding estimated variances, has $(f_j + 1)$ instead of $(f_j - 1)$ in the numerator. Models with large numbers of incidental parameters are in any case best avoided, and if the variation in the true variances cannot be rationally explained, a reasonable procedure for large m would often be to take the true variances as random

variables (i.e., to have a random-effects model for the variances). Under suitable special assumptions this leads to a modification of (13) in which the weights are shrunk toward their mean.

POOLING OF DATA AND EMPIRICAL BAYES* ESTIMATION

A different application may be made of the random-effects model (12). Instead of combining data to reach conclusions about θ, one may wish to study θ_m, the parameter value in one of the sets, taken without loss of generality to be the last. In particular, if the sets are ordered in time, we may wish to study the current set, making some use of historical data.

Because θ_m is a random variable, Bayesian arguments are applicable, provided that the prior distribution is known or can be estimated. With the new focus of interest, (12) is called an empirical Bayes model. If θ, v_m, and σ_η^2 are all known, and normality assumptions hold, the posterior distribution is normal with mean

$$\frac{t_m / v_m + \theta / \sigma_\eta^2}{1/v_m + 1/\sigma_\eta^2} \qquad (14)$$

and variance

$$\left(1/v_m + 1/\sigma_\eta^2\right)^{-1}. \qquad (15)$$

This involves the data only through t_m, and (14) is formally a weighted mean of two "observations," t_m and θ with variances v_m and σ_η^2, respectively. In parametric empirical Bayes estimation θ, σ_η^2, and, if necessary, v_m are replaced by estimates, so that all the data become involved. The proportional "shrinkage"* of even extreme values in (14) has to be viewed critically because it depends strongly on the normality of η_m in (12).

More elaborate representations of the η_j's, for example involving serial dependence, may be useful.

COMBINATION OF DATA ABOUT NUISANCE PARAMETERS*

Sometimes, for example in repetitive routine tests, sets of data are obtained each with its own value of the parameter, θ, of interest (e.g., a mean). The possibility of an empirical Bayes analysis of the θ's then arises. Suppose that this is undesirable, for example because of doubts about an appropriate form of prior distribution. It may nevertheless be useful to use previous information about a nuisance parameter, for example a variance, especially if this is estimated poorly from each individual set of data. Such an approach is sometimes called partially empirical Bayes. Detailed formulae will not be given here. An extreme case is when a fixed "historical" value is used for, say, the variance, the "fixed" value preferably being updated from time to time.

Even in contexts, such as agricultural field trials, with satisfactory internal estimates of error, it will be wise to compare qualitatively with historical values. Often the coefficient of variation of yield for a particular crop and plot size will be known roughly, and comparison with this gives some check on reliability.

MORE COMPLEX CONTEXTS

The discussion so far has assumed one or more clearly defined parameters of interest. A conceptually more difficult problem arises if we have a number of sets of similar data and wish to describe as concisely as possible the common features of and differences between the sets.

First, if separate fits are made to the individual sets of data, it is normally desirable, unless there are specific arguments to the contrary, to fit all sets in compatible fashion. To take a simple example, suppose that a regression of y on x is fitted to each set and that judged in isolation some sets are satisfactorily fitted by a straight line and that others need a quadratic. Then for comparative purposes a quadratic should be fitted to

all sets. As pointed out above in a slightly different context, it is possible that all sets are consistent with the same modest curvature.

Having fitted the separate sets in some compatible fashion, we look in principle for a parameterization in which as many component parameters as possible are the same for all sets of data. Thus, if the data are not totally homogeneous, one might hope that all except one of the component parameters are constant and that the variation in the anomalous component can be explained or modeled as in the section "Pooling in the Presence of Overdispersion". For example, if separate straight lines are fitted to the separate sets of data, are the lines either parallel or concurrent? Canonical regression analysis might be applied to the estimates in the original parameterization to discover combinations that are nearly constant, although often a more informal approach will do.

SERIES OF EXPERIMENTS OF SIMILAR DESIGN

A special problem that has been studied in considerable detail is the combination of data from a series of experiments of similar or even identical design, methods based on analysis of variance being appropriate. For simplicity, suppose that the same design, say in b randomized blocks, is used to compare t treatments in m "places." Very often the treatments will have factorial structure.

For one set of data, the analysis of variance has the form shown in Table 1a, where a decomposition of treatment comparisons into component contrasts is hinted at. Subject to homogeneity of error, which can be examined, the composite data have the conventional analysis of variance shown in Table 1b. Note that the treatments × places interaction can be partitioned as in Table 1c.

For a particular single-degree-of-freedom contrast, the analysis suggested in the first two sections corresponds in effect to Table 1c. That is, the constancy of a particular

Table 1   Some Analysis-of-Variance Tables for a Series of Similar Experiments: (a) Single Place; (b) Several Places; (c) Several Places, with Decomposition of Interaction[a]

(a)		(b)		(c)
Mean	1	Mean	1	
		Places	$m-1$	
Blocks	$b-1$	Blocks within		
		places	$m(b-1)$	As for
Treatments	$t-1:C_1$	Treatments	$t-1:C_1$	(b)
	$\vdots$		$\vdots$	with
	$\dot{C}_k$		$\dot{C}_k$	Treatments $\times$ places:
		Treatments $\times$ places		$C_1 \times$ places
			$(t-1)(m-1)$	$\vdots$
Residual	$(b-1)(t-1)$	Residual	$m(b-1)(t-1)$	$\dot{C}_k \times$ places

[a]$C_1, \ldots, C_k$ treatment contrasts.

contrast C is examined by comparing the component of interaction $C \times$ places with error. If interaction is present, cannot be explained, and a random-effects representation is adopted, the standard error of the overall mean contrast is estimated via the interaction mean square, $C \times$ places.

An alternative is to use the analysis of Table 1b, in which homogeneity of the interaction mean squares is assumed. If m is small, one must use either this or something intermediate between Tables 1b and 1c. In a general way, it is likely that larger effects will show larger interactions and hence, for example, treatment main effects $\times$ places are likely to be bigger than high-order-treatment interactions $\times$ places, so that the analysis of Table 1c is preferable when enough degrees of freedom are available.

There are numerous generalizations of the above, for example to experiments with different error variances and different designs. If a second factor (e.g., times) is included with levels regarded as random, the error variance for treatment contrasts has to be estimated by synthesis of variance.

DIVISION OF DATA FOR ANALYSIS

Most of the previous discussion assumes a number of independent sets of data for pos-

sible comparison. As indicated in the introduction, the sets may arise, not from separate investigations, but as part of a plan of statistical analysis in which the data are divided into rational sections for separate analysis and ultimately combination.

It is impossible to say at all precisely when such split analysis is advisable. Computational considerations may demand it. The investigators' lack of experience with complex techniques of analysis may make splitting an aid to comprehension. Lack of homogeneity, especially with time series data, may be best detected by splitting. Finally, a direct empirical demonstration of reproducibility of conclusions can often be nicely displayed by splitting the data.

Bibliography

Cochran, W. G. (1954). *Biometrics*, **10**, 101–129.

Cochran, W. G. and Cox, G. M. (1957). *Experimental Design*, 2nd ed. Wiley, New York. (Chapter 14 concerns series of experiments.)

Cox, D. R. (1975). *Biometrika*, **62**, 651–654.

James, G. S. (1956). *Biometrika*, **43**, 304–321. (Theoretical discussion.)

Patterson, H. D. and Silvey, V. (1980). *J. R. Statist. Soc. A*, **143**. (Discussion of specific application and numerous practical complications.)

Yates, F. and Cochran, W. G. (1938). *J. Agric. Sci.*, **28**, 556–580. (General discussion, methods, and practical examples.)

D. R. Cox

International Statistical Review, (1984), **52**, 1, pp. 1–31. Printed in Great Britain
© International Statistical Institute

Interaction

D.R. Cox

Department of Mathematics, Imperial College, London SW7 2BZ, U.K.

Summary

A broad review is given of various aspects of interaction. After a brief discussion of the relation between absence of interaction and constancy of variance, the main part of the paper considers interaction in the usual sense, connected with the structure of an expected value in multiply classified data. Interactions are first classified depending on the character of the factors involved. Techniques for detecting interactions are reviewed, including aspects of model formulation, especially for unbalanced data. Special methods for interpreting interactions are described. Finally, the relation between interactions and error estimation is considered and some short remarks made on extensions to binary and other essentially nonnormal models.

Key words: Balanced design; Component of variance; Contrast; *F* test; Factorial experiment; Interaction; Latin square; Log linear model; Main effect; Transformation; Unbalanced design.

1 Introduction

The notion of interaction and indeed the very word itself are widely used in scientific discussion. This is largely because of the relation between interaction and causal connexion. Interaction in the statistical sense has, however, a more specialized meaning related, although often in only a rather vague way, to the more general notion.

The object of the present paper is to review the statistical aspects of interaction. The main discussion falls under three broad headings:

(i) the definition of interaction and the reasons for its importance;
(ii) the detection of interaction;
(iii) the interpretation and application of interaction.

For the most part the discussion is in terms of the linear and quadratic statistics familiar in the analysis of variance and linear regression analysis, although most of the ideas apply much more generally, for instance to various analyses connected with the exponential family, including logistic and other analyses for binary data. First, however, in § 2 some brief comments are made on the connexion between constancy of distributional shape and constancy of variance and the absence of interaction.

The paper presupposes some familiarity with the broad issues involved. Emphasis is placed on matters, for example of interpretation, that, however widely known, are not extensively discussed in the literature. References are confined to key ones.

Although some emphasis will be put on data from balanced experimental designs, in principle the ideas apply equally to observational data and to unbalanced data.

2 Distributional form and interaction

In the majority of the discussion of interaction we concentrate on the structure of the expected response, Y, or at least on contrasts of location. That is, any nonconstancy of variance may suggest that a more elaborate method of analysis involving weighting is desirable, but does not by itself change the central objectives of the analysis. In some contexts, however, the form of the frequency distribution is of some primary interest and there is then a relation with notions of interaction in the following sense.

Consider first an experiment with just two treatments. Under the simplest assumption of unit-treatment additivity, the response obtained on a particular experimental unit under treatment 2 differs by a constant τ, say, from the observation that would have been obtained on that unit under treatment 1. This implies that if $F_2(.)$ and $F_1(.)$ are the corresponding cumulative distribution functions of response then

$$F_2(y) = F_1(y - \tau). \tag{2.1}$$

That is, the two distributions differ only by translation and in particular have the same variance. Any departure from (2.1) implies that simple unit-treatment additivity cannot hold; in other words, to match the two distributions in (2.1), τ must be a function of y, $\tau(y)$ say. That is, the treatment effect depends on the initial level of response and this is one form of interaction.

Consistency with (2.1) can be examined by inspecting histograms or cumulative distribution functions, but preferably via a Q–Q plot (Wilk & Gnanadesikan, 1968) in which corresponding quantiles of the two distributions are plotted against one another to produce under (2.1) a line of unit slope.

Doksum & Sievers (1976) have given formal procedures based on this and in particular a confidence band for $\tau(y)$ defined by

$$F_2(y) = F_1\{y - \tau(y)\}, \quad \tau(y) = y - F_1^{-1}F_2(y).$$

If (2.1) does not hold, it is natural to consider whether a monotonic transformation from y to $h(y)$ is available such that $h(Y)$ satisfies (2.1). G.E.H. Reuter (unpublished) has shown that such a transformation exists if F_1 and F_2 do not intersect. If there are more than two 'treatments' transformation to a location family representation exists only exceptionally.

A more traditional approach, very reasonable with modest amounts of data insufficient for detailed examination of distributional shape and with distributions that are not very long-tailed, is to consider only the variance. That is, we replace (2.1) by the condition of constant variance. The condition for the existence of an approximate variance stabilizing transformation is, for any number of treatment groups, that the variance is determined by the mean. Then if

$$\mathrm{var}\,(Y) = v(\mu), \quad E(Y) = \mu$$

a familiar argument yields

$$h(y) = \int_0^y dx/\sqrt{v(x)} \tag{2.2}$$

as the appropriate transformation. That is, evidence of the kind of interaction under discussion here could come only by showing that

(i) two or more distributions with the same mean have different variances; or
(ii) aspects of distributional shape, e.g. third moments, are not adequately stabilized by (2.2); or
(iii) the linearizing approximations involved in deriving (2.2) are inadequate and that a more careful examination of variance shows that it cannot be reasonably stabilized.

Points (ii) and (iii) are probably best studied numerically in a particular application, although some results can be given by higher order expansion. Thus, if $\rho_{Y3}(\mu)$ denotes the standardized third cumulant of Y, a higher-order approximation is

$$\text{var}\{h(Y)\} = v(\mu)\{h'(\mu)\}^2 + \rho_{Y3}(\mu)\{v(\mu)\}^{\frac{3}{2}}h'(\mu)h''(\mu). \tag{2.3}$$

Thus, on writing $g(\mu) = \{h'(\mu)\}^2$, we have a first-order linear differential equation for the variance-stabilizing transformation. If the solution does not lead to a monotonic h, variance cannot be stabilized. An alternative use of (2.3) is to retain the transformation (2.2) when we have that

$$\text{var}\{h(Y)\} = 1 - \tfrac{1}{2}\rho_{Y3}(\mu)v'(\mu)/\sqrt{v(\mu)}, \tag{2.4}$$

and the magnitude of the correction term can be examined.

For the special choice (2.2) we have that

$$\rho_{h(Y),3}(\mu) = \rho_{Y3}(\mu) - \frac{3v'(\mu)}{4\sqrt{v(\mu)}}\{\rho_{Y4}(\mu) - \rho_{Y3}^2(\mu) + 2\}$$

to the next order and if this is strongly dependent on μ there is evidence that distributional shape is not adequately stabilized, point (ii) above.

For multivariate responses the formulation analogous to that leading to (2.2) is that for each vector μ there is a covariance matrix $\Omega(\mu)$. It is known (Holland, 1973; Song, 1982) that only exceptionally does there exist a vector transformation from y to $h(y)$ such that the covariance matrix of $h(Y)$ is, in the linearized theory, a constant matrix, say the identity matrix; certain compatability conditions have to be satisfied. Practical techniques for finding suitable approximate transformations do not seem to be available.

In the remainder of the paper we concentrate on the structure of $E(Y)$, making the simplest assumptions about variances and distributions. The essence of the present section is that constancy of variance is a special form of absence of interaction. If variance or distributional shape vary to an extent that is judged scientifically important, the description of the effect of 'treatments' on response should not be confined to the examination of means.

3   Some key ideas

We consider a response variable Y whose expectation depends on explanatory variables, or factors, which it is convenient to classify into three types as follows.

First there are *treatment* variables, which we denote by $x_1, \ldots, x_t$. In experiments these represent the levels of various treatment factors under the control of the experimenter. In observational studies, they represent explanatory variables which for each individual could conceivably have taken different levels and which in an experimental study of the same phenomenon would have been allocated by the investigator, preferably with randomization. The levels of such variables may be binary, nominal, ordinal or quantitative. The central objective of the investigation is to study the effect of some or all of the treatment variables on the response.

Secondly, there are *intrinsic* variables, to be denoted by $z_1, \ldots, z_q$, and which are of two different kinds. First there are those that describe properties of the individuals under study that one regards, in the current context, as in principle not capable of variation by the investigator, but as defining the 'material' under study. Examples are age and sex in an observational or experimental investigation and, more broadly in experiments, concomitant observations measured before the allocation of treatments. A second type of intrinsic variable is concerned with that part of the environment of the investigation outside the investigator's control and which it is inappropriate to consider as a treatment. In an industrial experiment, the uncontrolled meteorological conditions holding during the processing of a particular batch exemplify this kind of observation. The z's may be binary, nominal, ordinal or quantitative.

Finally, there are variables whose levels are not uniquely characterized. Such variables, defining blocks, litters, replicates and so on, are sometimes called random factors, but we wish to avoid any implication that the levels are randomly sampled from a well-defined population, or that the effects associated with such variables are automatically best represented by random variables. We therefore call such variables *nonspecific*. They will be denoted by $u_1, \ldots, u_r$; the levels of each u are usually nominal. If, however, replication in time or space is involved, the levels of u are quantitative, although special care is needed if, for example, the same individual is observed at a series of time points.

We write

$$E(Y) = \eta(x_1, \ldots, x_t; z_1, \ldots, z_q; u_1, \ldots, u_r) = \eta(x, z, u), \qquad (3.1)$$

say.

A key assumption in the discussion is that interest focuses on the effect of the x's on the response; the z's and u's must be considered, but only in so far as they bear on the effect of the treatment variables, the x's.

Even if we restrict our attention for simplicity to two-factor interactions, three different kinds have to be considered:

    (a) treatment × treatment;
    (b) treatment × intrinsic;
    (c) treatment × nonspecific.

While the formal definitions are common, interpretation is rather different. We discuss briefly the simplest cases.

With just two treatment factors x_1 and x_2, and no intrinsic or nonspecific factors, absence of interaction of type (a) means that

$$\eta(x_1, x_2) = \eta^{(1)}(x_1) + \eta^{(2)}(x_2). \qquad (3.2)$$

The importance of this is partly the empirical one that, to the extent that (3.2) is adequate, the function of two variables $\eta(x_1, x_2)$ can be replaced by the much simpler form of two functions of one variable: for nominal x_1 and x_2 a two-way table is reduced to two one-way tables. In some ways, a more basic reason for the importance of (3.2) is that, because the effect of changing, say, x_1 does not depend on the level of x_2, the two treatments act in a way that appears causally independent, at least over the range of levels examined. The converse conclusion, that interaction, in the sense of failure of (3.2), implies causal dependence is in general unsound. First, transformation of the expected response scale may remove the interaction: see Scheffé (1959, p. 95) for the conditions for this when x_1 and x_2 are quantitative. Secondly, it is possible that some simple mechanism in which in a reasonable sense the effects of x_1 and x_2 are physically separate nevertheless leads to a function not satisfying (3.2). A simple example over the region $x_2 \geq 0$ is

$$\eta(x_1, x_2) = \eta_{11}(x_1) + e^{-\beta x_2} \eta_{12}(x_1),$$

which as a function of x_2 moves between levels controlled solely by x_1.

The clearest evidence of the failure of (3.2), in a way not removable by transformation, arises when the rank order of the responses at a series of levels of, say, x_1, changes with x_2. It might be worth using a special term, such as order-based interaction of x_1 with x_2, for this, an asymmetric relation.

When we turn to treatment × intrinsic interactions, the mathematical condition

$$\eta(x, z) = \eta_1(x) + \eta_2(z) \qquad (3.3)$$

appears identical to (3.2) but the interpretation is different, essentially because the roles of x and z are now asymmetrical. Here the concern is with stability of the x-effect as z varies, such stability both enabling the treatment effect to be simply specified in a single function and also providing a basis for rational extrapolation of conclusions to new individuals and environments. The function $\eta_2(z)$ itself is in the present context of no direct interest.

Finally, there are treatment $\times$ nonspecific interactions. We discuss these in more detail in § 7. Briefly, however, such interactions have either to be given a rational explanation or regarded as extra sources of unexplained variation, i.e. error.

Similar remarks apply to higher order interactions.

4   Remarks on model formulation

4.1   Some general principles

To an important extent, the interpretation of balanced data can be made via one-way, two-way, . . . tables of mean values, using simple contrasts with a tangible practical meaning, and hence not depending critically on formal probabilistic models. Such models are, however, needed either for assessments of precision, including significance tests, or as a guide for the analysis of unbalanced sets of data. While the definition and interpretation of interaction is in principle the same for unbalanced as for balanced data, the strategies of analysis may well be different.

There are two extreme approaches to the initial choice of such a model. The first is to start from the most general model estimable from the data, i.e. writing into the analysis of variance table all interactions that can be found, given the structure of the data. The second is to start from a simple representation, possibly only involving main effects, and to introduce further effects only when the data dictate a need to do so. In both cases the objective is to end with an incisive interpretation. For balanced data, the use of the full analysis of variance table has the major advantages that critical inspection of the table will reveal any unanticipated effects or anomalies and that the magnitude of the lower order interactions will indicate which interactions have to be directly considered in the final interpretation. Further, the form of the analysis of variance table is a valuable concise summary of the structure of the data, essentially equivalent to, but in many ways easier to assimilate than, the associated linear model.

For the analysis of balanced or nearly balanced data, it will normally be a good idea to start from the 'full' analysis of variance table. For unbalanced 'factorial' data, it will often be sensible to look at the structure of the analysis of variance table for the corresponding balanced data, this indicating the contrasts that may in principle be examined: of course, the presence of empty cells affects estimability. In principle, it is possible to find by least squares, corresponding to each mean square in the balanced analysis of variance table, a mean square adjusting for all other 'relevant' contrasts, although in complex cases this can be cumbersome to implement and interpret. We discuss below the meaning of the restriction to 'relevant' contrasts.

The comments above concern the initial analysis. Normally, final presentation of conclusions will be via some simpler specification, for example one involving rather few interaction terms. For balanced data, a simplification of the model will not usually change the estimates of the contrasts of primary interest. Indeed, as noted above, it is an attraction of balanced data that simple contrasts of one-way and two-way tables of means have a very direct interpretation. The error variance under the simplified model can be estimated via the error estimate in the original model, or via that in some simplified model, often but not necessarily that used to present the conclusions, i.e. in effect by some pooling of sums of squares. Unless the degrees of freedom for error in the 'full' model are rather small, decisions on pooling error mean squares usually have relatively little effect on the conclusions, and use of the error estimate from the 'full' model will often be simplest. Of course, where there are several layers of error, choice of the appropriate one can be of critical importance.

In unbalanced data, explicit formulation and fitting of a model will normally be required both for parameter estimation and for estimating error variance. In this it is usually reasonable to err on the side of overfitting; for instance, treatment main effects should not be omitted just because they are insignificant.

In formulating a 'full' model, the broad procedure is, roughly speaking, to identify all the factors specifying the explanatory variables for each observation and then to set out the 'types' of these factors (treatment, intrinsic and nonspecific) and the relationships between the factors in the design (completely cross-classified, hierarchical, incomplete): all interactions among completely cross-classified variables are included in the 'full' model, with the exception of those with nonspecific factors which are often amalgamated for error; see, however, § 7.

4.2 Two examples

Example. Consider an experiment in which in each of a number of blocks (nonspecific factor) all combinations of two treatment factors (x_1, x_2) occur with all levels of an intrinsic factor z. Table 1 outlines the 'full' analysis of variance table. Note especially the grouping according to the type of explanatory variable involved and that the interactions $x_1 \times$ blocks, ... have not been isolated.

In contemplating simplified models an almost invariable rule is that if one term is omitted, so are all 'below' it in the analysis of variance table; see, for example, Nelder (1977). For instance, if $x_1 \times x_2$ is omitted, it would rarely be sensible to include $z \times x_1 \times x_2$. The reason for this is that if some contrast interacts with, say, z, and is therefore nonzero at some levels of z, it would normally be very artificial to suppose that the value averaged out exactly to zero over the levels of z involved in defining the 'main effect' for the contrast. The implications for the analysis of unbalanced data will be developed in more detail in § 4.3.

Table 1

Outline analysis of variance table

Source of variation		Description
I	$x_1\ddagger$	Treatment*, main effec.
	$x_2\ddagger$	
	$x_1 \times x_2\ddagger$	Interaction
II	z	Intrinsic, main effect
	$z \times x_1\ddagger$	Intrinsic × treatment interaction
	$z \times x_2\ddagger$	
	$z \times x_1 \times x_2\ddagger$	
III	u	Nonspecific main effect, blocks
	$u \times$ all other contrasts	Error†

  * Treatment effects will often be subdivided, especially if one or both x's are quantitative.
  † Under some circumstances this term should be subdivided to provide separate errors for different contrasts of interest.
  ‡ Contrast of interest.

Example. As a rather more complicated example of model formulation, we consider next the two-treatment two-period cross-over design. The simplest such design is that shown in Table 2a, having just two periods; between the two periods a 'washout' period will normally be inserted. Subjects are randomized into two groups, usually but not necessarily in equal numbers. The explanatory variables to be considered are:

(i) treatment factors: direct treatment effect; residual effect of treatment (in period 2 only);
(ii) intrinsic factor: periods;
(iii) nonspecific factor: subjects.

First consider the role of the nonspecific factor, subjects. If a separate parameter is inserted for each level, i.e. for each individual subject, it is clear that analysis free of these parameters is possible only from the difference in response, period 2 minus period 1, calculated for each subject. On the other hand, subjects have been randomized to subject groups and therefore a second and independent analysis can be made of the subject totals, period 2 plus period 1, having a larger, and possibly very much larger, error variance than the analysis of differences. This device, and its obvious generalizations, avoid the need to write elaborate models containing subject terms. Interaction with the nonspecific factor is regarded as error and not entered explicitly in the model.

There remain three factors. Interaction between direct and residual treatment effect is undefined because, regarded as factors, they are an incomplete cross-classification. Similarly, residual treatment effects are defined only in the second period and so cannot lead to an interaction with periods. Thus the only interaction for consideration is direct treatment effect × periods.

Because the factors all occur at just two levels, it is convenient, at least initially, to define parameters as follows: μ, general mean; direct treatment $-\tau$ for T_1, τ for T_2; residual treatment $-\rho$ following T_1, ρ following T_2; period $-\pi$ for period 1, π for period 2; direct treatment × period interaction $-\gamma$ for T_2 in period 1 and T_1 in period 2, γ otherwise. It follows that, omitting subject parameters, the expected values are those shown in Table 2a. The parameterization is that convenient for model formulation: for interpretation the treatment effect in period 1, namely $2\tau - 2\gamma$, might be taken as one of the parameters instead of γ.

For the reasons indicated above, we replace observations by differences obtaining

Table 2

Two-period cross-over design

(a) Simple design Subject group	Period 1	Period 2	(b) Extended design Subject group	Period 1	Period 2
I	T_1 $\mu - \pi - \tau + \gamma$	T_2 $\mu + \pi + \tau - \rho + \gamma$	I	T_1 $\mu - \pi - \tau + \gamma$	T_2 $\mu + \pi + \tau - \rho - \delta + \gamma$
II	T_2 $\mu - \pi + \tau - \gamma$	T_1 $\mu + \pi - \tau + \rho - \gamma$	II	T_2 $\mu - \pi + \tau - \gamma$	T_1 $\mu + \pi - \tau + \rho - \delta - \gamma$
			A	T_1 $\mu - \pi - \tau + \gamma$	T_1 $\mu + \pi - \tau - \rho + \delta - \gamma$
			B	T_2 $\mu - \pi + \tau - \gamma$	T_2 $\mu + \pi + \tau + \rho + \delta + \gamma$

General mean, μ; period effect, π; direct treatment effect, τ; residual treatment effect, ρ; direct treatment × period interaction, γ; direct × residual treatment interaction, δ.

respectively for group I and II expected values:

> differences: $2\pi+2\tau-\rho,\ 2\pi-2\tau+\rho$;
> sums: $2\mu-\rho+2\gamma,\ 2\mu+\rho-2\gamma$.

It follows immediately that from the within-subject analysis, the only treatment parameter that can be estimated is $\tau-\frac{1}{2}\rho$, which is of interest only if ρ can be assumed negligible. If information from between-subject comparisons is added, $\gamma-\frac{1}{2}\rho$ can be estimated, so that residual effects and direct treatment $\times$ period interaction cannot be separated except by *a priori* assumption. Thus from the combined analysis only linear combinations of $\tau-\frac{1}{2}\rho$ and $\gamma-\frac{1}{2}\rho$ can be estimated and the only such combination of general interest is $\tau-\gamma$, the treatment effect in period 1. The conclusion that the design is advisable for 'routine' comparison of treatment only when there is strong *a priori* reason for thinking ρ negligible seems fairly widely accepted (Brown, 1980; Hills & Armitage, 1979; Armitage & Hills, 1982).

If we use the extended design of Table 2b, subjects are randomized between the four groups, typically with equal numbers in groups I and II and equal numbers in groups A and B. The new feature is that residual effect is cross-classified with direct effect and hence we should introduce for the second period a further parameter for direct $\times$ residual treatment interaction, $-\delta$ if the two treatments are different, δ if the two treatments are the same. Table 2b gives the expected responses. It follows that in the four groups the expected values are:

> differences: $2\pi+2\tau-\rho-\delta,\quad 2\pi-2\tau+\rho-\delta,$
> $\qquad\qquad 2\pi-\rho+\delta-2\gamma,\quad 2\pi+\rho+\delta+2\gamma$;
> sums: $2\pi-\rho-\delta+2\gamma,\quad 2\mu+\rho-\delta-2\gamma,$
> $\qquad\qquad 2\mu-\rho+\delta-2\tau,\quad 2\mu+\rho+\delta+2\tau.$

Thus if only within-subject comparisons are used, i.e. differences analysed, the parameters that can be estimated are τ, ρ, $\gamma+\frac{1}{2}\delta$, π. That is, a direct treatment effect and a residual treatment effect can be separately estimated but direct treatment $\times$ period interaction and direct $\times$ residual treatment interaction cannot be disentangled. If, however, between-subject comparisons are introduced all contrasts become estimable.

Three and four period designs are dealt with similarly, although of course the details get more complicated. The precision of between-subject comparisons can often be increased by an initial period of standardization supplemented by the measurement of a concomitant variable. We return later to the questions of interpretation; here the initial model formulation is the aspect of concern. In many applications, use of such designs is advisable only when residual effects and treatment $\times$ period interactions are very likely to be negligible. There are contexts, however, where the occurrence and nature of residual effects are of direct interest.

4.3   Unbalanced data

There are three difficulties in applying the above ideas to unbalanced data. The first is that in isolating a sum of squares for any particular contrast, we must specify which other parameters are to be included and which are to be set at null values. In a complex analysis many such specifications are involved. Secondly, the amount of computation is greatly increased and, while each individual fitting operation is simple, easy methods of specification and easily understood arrangements of answers are needed. Thirdly, and most seriously, interpretation is no longer via easily assimilated tables of means, but via corresponding sets of derived values less directly related to the 'raw' data and having a more complicated error structure.

There is probably no uniquely optimal way of addressing the problem of formulation, but the following seems a reasonable general procedure.

(i) Set out the skeleton analysis of variance table for the balanced form analogous to the data under study. Consider whether 'empty cells' will make any of the contrasts nonestimable.

(ii) Formulate a baseline model containing, usually, all main effects and any other contrasts likely to be important, or found in preliminary analysis to be important.

(iii) In isolating the sums of squares for each set of contrasts, not in the baseline model, include the parameters of the baseline model, the parameters under test, and any parameters 'above' the parameters under test in the natural hierarchy. Thus, in the notation used above, if we are examining the three-factor interaction, all the associated two-factor interactions and, of course, main effects should be included, as well as all parameters in the baseline model.

Because of the possibility mentioned in (ii) that the initial baseline model will need modification, the procedure is in general iterative. The final baseline model will usually be the one underpinning the presentation of conclusions. The procedure is close to a severely constrained form of stepwise regression.

5 Detection of interaction

5.1 *General remarks*

We now discuss some aspects of the detection of interaction. The literature on the formal use of F tests is large and we concentrate here on the appropriate choice of the upper mean square. Graphical techniques are discussed in § 5.3 and § 5.4 deals with the important issues involved when the interaction of interest is not completely estimable.

By far the most powerful device for detecting interaction is, however, the critical inspection of one-way, two-way, ... tables of means.

5.2 *Some comments on F tests*

The setting up of an F test for an interaction requires the choice of an upper mean square associated with the effect under study and a lower mean square estimating error. The former choice is normally the more critical because it amounts to a specification of the kind of interaction under examination. Where there are initially several degrees of freedom in the upper mean square, it is usually important to partition the interaction sum of squares either into single or small numbers of degrees or at least to separate off the most important degrees of freedom. The decomposition will nearly always be based on that of the corresponding treatment main effect. Choice of error mean square is usually critical only when hierarchical error structure is present.

A simple example is of a treatment × intrinsic interaction in which the treatment variable, x, is quantitative and the intrinsic variable, z, nominal with a small number of levels. If it is fruitful to decompose the main effect into linear, quadratic and remaining components, there is a corresponding decomposition of interaction

$$x_L \times z, \quad x_Q \times z, \quad x_{\text{Rem}} \times z, \tag{5.1}$$

is a self-explanatory notation.

If the treatment factor is nominal it will be sensible to decompose the interaction on the basis either of *a priori* considerations or after inspection of the treatment main effect. To take a simple example with 4 levels, suppose that inspection of main effects shows that level 3 differs from levels 1, 2, 4 which differ rather little among themselves. It is then reasonable to divide the interaction into

$$(3 \text{ versus } 1, 2, 4) \times z, \quad (\text{Within } 1, 2, 4) \times z. \qquad (5.2)$$

Tukey (1949) pointed out that the selection of the component of interaction after examining the main effect does not change the distribution of the resulting F statistic, although his use of the idea is slightly different from that here.

Examination of the various components of (5.1) and (5.2) serves rather different purposes. Most attention will usually apply to the leading components which, if they are appreciable, hopefully point to some fairly direct interpretation: in terms of power these statistics correspond to testing against quite specific and meaningful alternatives. We may call such tests directly interpretable. The final tests involved in (5.1) and (5.2) serve a rather different purpose. A large final component in (5.1) is incompatible with the viewpoint implied in the whole formulation and so would be evidence of an unexpected occurrence in need of detection and explanation, such as by the isolation of an outlying observation or group of observations or by some radical reformulation. Such tests may be called ones of specification; the distinction from directly interpretable tests is not always clearcut. Tests of higher order interactions are usually ones of specification.

The calculation of P-values, while by no means always necessary, is often a useful guide to interpretation; usually there will be several P-values from each analysis. Elaborate formalization of the process of analysis is probably counterproductive, but often the following possibilities can usefully be distinguished.

Suppose for simplicity that just two F statistics are being examined. The following possibilities may arise.

First, underlying the two statistics may be a two-dimensional space of possibilities all on an equal footing. Then a combined mean square can be calculated in the usual way.

Secondly, the two possibilities may correspond to quite separate effects, none, either one or both of which may be present. An example would be the interaction of a treatment effect with two quite different intrinsic variables. Then the separate F statistics should be examined and, as some protection against selection, the smaller P-value doubled. For balanced data the two statistics are, under the global null hypothesis, slightly positively correlated because of the common denominator, so that the standard allowance for selection slightly overcorrects.

Thirdly, the two possibilities underlying the F statistics may be nested in that, while alternative 1 without alternative 2 is the basis of a sensible interpretation, alternative 2 is sensible only in conjunction with alternative 1. The decomposition $x_L \times z$ and $x_Q \times z$ of (5.1) is a typical example: in most circumstances it would not be sensible to have an interpretation in which the slope is the same for all z whereas the curvature varies with z, although interpretation with varying slopes but constant, especially zero, curvature would be acceptable. Another example connected with the same situation is where the two F statistics are associated with the main effect x_Q and with $x_Q \times z$.

In this third case, it will be thus reasonable to examine the statistics F_1 and F_{12} corresponding to the first alternative and the merged alternatives. If P_1 and P_{12} are the significance levels associated with the marginal distributions of these statistics, i.e. calculated from the standard F tables, we may consider $P^* = \min(P_1, P_{12})$ as a test statistic. The 'allowance for selection' is a factor between 1 and 2 depending on the associated degree of freedom (and level of significance) concerned; if the degrees of freedom of F_1

and F_2 are not too disparate a factor $\frac{3}{2}$ is a rough approximation adequate for most purposes (Azzalini & Cox, 1984).

It is important that checks of specification should be carried out, not necessarily by formal calculation of significance levels, although these are needed in the background for calibration. Analysis of variance is, among other things, a powerful exploratory and diagnostic technique and total commitment to an analysis in the light of *a priori* assumptions about the form of the response can lead to the overlooking of important unanticipated effects, defects in the data or misconceptions in formulation.

There are many generalizations of the above discussion, for instance to models for nonlinear treatment effects: here the interaction with intrinsic variables should, in principle, be decomposed corresponding to the meaningful parameters in the nonlinear model.

5.3 Graphical methods

When, as in a highly factorial treatment system, there are a considerable number of component interactions, graphical representation may be helpful:

(a) in presentation of conclusions,
(b) in isolating large contrasts for detailed examination,
(c) in determining which contrasts can reasonably be used to estimate error,
(d) in detecting initially unsuspected error structure, such as the hierarchical error involved in a split plot design (Daniel, 1976, Chapter 16), or important correlations between errors.

Graphical methods work most easily when all mean squares under study have the same number of degrees of freedom. If most of the contrasts have d degrees of freedom, it may be worth transforming a mean square with d' degrees of freedom, via the probability integral transformation and an assumed value $\tilde{\sigma}^2$ for variance, into the corresponding point of the distribution for d degrees of freedom. Pearson & Hartley (1972, Tables 19 and 20) give expected order statistics corresponding to degrees of freedom $1, 2, \ldots, 7$ and this covers most of the commonly occurring cases. As a simple example, suppose that all but one of the mean squares has one degree of freedom and that a preliminary estimate of variance is $\tilde{\sigma}^2$. Suppose that there is in addition a mean square of $1 \cdot 2\tilde{\sigma}^2$ with 3 degrees of freedom. The corresponding value of $\chi_3^2 = 3 \cdot 6$ is at the $0 \cdot 69$ quantile and the corresponding point of χ_1^2 is $1 \cdot 05$ so that the single mean square is assigned value $1 \cdot 05\tilde{\sigma}^2$.

It is good to show main effects distinctively: if a primary object is (c), the estimation of error, it is best to omit the main effects, and any other important effects, in the calculation of effective sample size for finding plotting positions. That is, the expected order statistics used for plotting refer to a sample size equal to the number of contrasts used for estimating error, not to the total number of contrasts. When the contrasts have single degrees of freedom, there is a choice between plotting mean squares, or absolute value of contrasts, the half normal plot (Daniel, 1959) or signed values of contrasts. For many purposes the last seems preferable, partly because the signs of important contrasts can be given a physical interpretation and partly because if levels are defined so that all sample main effects are positive, interactions have an interpretable sign, a positive interaction indicating a reinforcement of the separate effects. Examples of the use of these plots are to be found in the books of Daniel (1976) and Daniel & Wood (1971).

It is well known that in some approximate large sample sense probability plotting is not systematically affected by dependence between the points plotted, although the precision of the points will be altered. Because the shape of 'reasonable' distributions is settled by the first four moments, a condition for the validity of probability plotting methods is that

12   D.R. Cox

the dependence shall be sufficiently weak that the first four sample moments converge in probability to their population values. Thus, in unbalanced analyses independence of the various mean squares to be plotted is not necessary.

A rather extreme example of the case of nonindependent contrasts concerns factors with nominal levels. For simplicity, consider unreplicated observations y_{ij} $(i = 1, \ldots, m_1; j = 1, \ldots, m_2)$.

Every elementary interaction contrast

$$y_{i_1 j_1} - y_{i_1 j_2} - y_{i_2 j_1} + y_{i_2 j_2} \quad (i_1 \neq i_2; j_1 \neq j_2) \tag{5.3}$$

has, under the hypothesis of no two-factor interaction and homogeneous variance, zero mean and constant variance. This suggests that all $\frac{1}{4}m_1(m_1-1)m_2(m_2-1)$ contrasts (5.3) should be plotted against normal order statistics, this being appropriate for detection of interaction confined to a relatively small number of cells. Interaction produced by outlying observations, or outlying rows or columns may be detected in this way, although the plot is highly redundant, especially if m_1 and m_2 are large; the redundancy measured by the ratio of the number of points plotted to the number of independent contrasts is $\frac{1}{4}m_1 m_2$ and so is large if either or both m_1 and m_2 are large. This technique has been studied in detail by Bradu & Hawkins (1982) and in an as yet unpublished report by R. Denis and E. Spjøtvoll.

5.4 Some special procedures

The graphical methods of § 5.3 as well as the tests of § 5.2 depend directly on or are closely related to standard linear model theory. There are, however, other tests of interaction which may sometimes be useful.

Particularly with a treatment factor and an intrinsic factor both with nominal levels one might wish to test for an order-based interaction of treatment with intrinsic factor, in the sense defined above (3.3). Suppose that the two-way array y_{ij} $(i = 1, \ldots, m_1; j = 1, \ldots, m_2)$ of observations have variance reasonably well estimated by $\tilde{\sigma}^2$. For example, there may be replication within cells and the y_{ij} may be cell means. An order-based interaction of the kind under study is revealed if for some suitably large $c > 0$ and for four cells $(i_1, j_1), (i_2, j_1), (i_1, j_2), (i_2, j_2) = \frac{1}{3}$ $(i_1 \neq i_2, j_1 \neq j_2)$, we have that

$$y_{i_1 j_2} - y_{i_1 j_1} \geq c\tilde{\sigma}\sqrt{2}, \quad y_{i_2 j_1} - y_{i_2 j_2} \geq c\tilde{\sigma}\sqrt{2}, \tag{5.4}$$

so that column effects exceeding c standard errors and in opposite directions occur in rows i_1 and i_2; if we have one treatment and one intrinsic factor, the treatment factor is assigned to columns. We take as test statistic the maximum c for which (5.4) holds for some four cells. The associated significance level is (Azzalini & Cox, 1984) approximately

$$1 - \exp\left[-\tfrac{1}{2}m_1(m_1-1)m_2(m_2-1)\{\Phi(-c)\}^2\right].$$

A second set of special techniques is connected with outliers. Especially where a single outlier is involved, the techniques for outliers in a general linear model can be used; in effect the depression in the residual sum of squares following from the omission of an observation, in turn related to a square residual, gives a diagnostic statistic. For special studies of two-factor systems, see Brown (1975), Gentleman & Wilk (1975), Gentleman (1980) and Bradu & Hawkins (1982). Similar ideas apply if a whole row or column is deleted. For unbalanced analyses, recent work on the detection of influential observations (Cook & Weisberg, 1982; Atkinson, 1982) is relevant.

Similar remarks apply if one or more whole 'rows' are regarded as outliers corresponding usually to particular levels of an intrinsic factor.

5.5   *Incompletely identified interactions*

A quite common occurrence, especially with treatments with nominal levels, is that an interaction or set of interactions are of interest having a substantial number of degrees of freedom and yet only a very small number of component degrees of freedom can be isolated at a time, because of the incomplete nature of the data. Sometimes the components for study are determined *a priori* but in other cases, especially where the interaction is being studied primarily as a check on specification, it will be required to isolate special components in a way that depends on the data. One general principle that can be used in such cases is that large component main effects are more likely to lead to appreciable interactions than small components. Also, the interactions corresponding to larger main effects may be in some sense of more practical importance. The application of this is best seen via a number of examples.

Example. Cox & Snell (1981, Example P) reanalysed some data of Fedorov, Maximov & Bogorov (1968) consisting of 16 observations on a 2^{10} treatment system, the factor levels corresponding to low and high concentrations of components in a nutrient medium. The design is a partly randomized one. The full data are given in Table 3a. A crucial aspect is that an independent estimate of error variance, $\tilde{\sigma}^2 = 14\cdot44$, is available. The following is a minor development of the earlier analysis.

Tables 3b, c summarize some aspects of the analysis:

(a) an initial fit of only main effects gives a residual mean square far in excess of $\tilde{\sigma}^2$;

(b) therefore a model including interaction terms is needed, but it is clearly impossible to include all 45 two-factor interactions;

(c) there is no uniquely appropriate procedure, but one sensible approach is to introduce main effects one at a time starting with the largest and with each pair of large main effects to introduce the corresponding two-factor interaction;

(d) after some trial and suppression of small contributions a representation is found producing a residual mean square close to $\tilde{\sigma}^2$;

(e) as a further check, addition to the model of the remaining main effects, one at a time, gives estimates that are both individually small and also of both signs: 'real' main effects are very likely to be positive, corresponding to the higher level of a component increasing response.

Example. The examination of interaction in a Latin square provides a simple illustration of the same idea. We consider a Latin square in which the letters represent a treatment factor with nominal levels; the columns represent an intrinsic factor with equally spaced quantitative levels, such as serial order; the rows represent a nonspecific factor such as subjects. Of course, there are many other versions of the Latin square design depending on the character of the three defining factors.

In the present instance, the interaction treatments $\times$ columns is the one of interest. In an $m \times m$ square the isolation of all $(m-1)^2$ degrees of freedom for that interaction is not possible and one approach is to take the linear component of the column effect and some single contrast of treatments based either on *a priori* considerations or on inspection of main effects. Table 4 illustrates the method for a 4×4 Latin square; we take C versus A, B, D as the component of treatment main effect. A derived concomitant variable is formed as a product of the column variable $-3, -1, 1, 3$ representing linear trend with the variable taking values $-1, -1, 3, -1$ for respectively A, B, C, D.

The standard Latin square analysis is augmented by including in the model a term proportional to the derived variable above. The analysis is conveniently done by the standard calculations of analysis of covariance. The derived variable, shown explicitly in

Table 3

Data and outline analysis of 2^{10} system (Fedorov et al., 1968)

(a) Yields of bacteria

						Factors					
	x_1 NH$_4$Cl	x_2 KH$_2$PO$_4$	x_3 MgCl$_2$	x_4 NaCl	x_5 CaCl$_2$	x_6 Na$_2$S 9H$_2$O	x_7 Na$_2$S$_2$O$_3$	x_8 NaHCO$_3$	x_9 FeCl$_3$	x_{10} micro-elements	Y
Levels +	1500	450	900	1500	350	1500	5000	5000	125	15	Yield
−	500	50	100	500	50	500	1000	1000	25	5	
1	−	+	+	+	−	+	−	+	−	+	14·0
2	−	−	+	+	−	+	+	−	−	+	4·0
3	+	−	−	+	+	+	−	−	−	−	7·0
4	−	−	+	−	+	+	−	+	+	+	24·5
5	+	−	+	+	+	+	+	−	−	−	14·5
6	+	−	+	−	+	+	+	+	+	+	71·0
7	−	−	−	−	−	−	−	−	−	−	15·5
8	+	+	−	+	+	−	−	+	+	−	18·0
9	−	+	−	+	−	−	+	−	−	+	17·0
10	+	+	+	+	−	−	−	−	+	−	13·5
11	−	+	+	−	+	−	+	+	+	+	52·0
12	+	+	+	−	−	−	+	+	−	−	48·0
13	+	+	−	−	+	−	+	−	+	−	24·0
14	−	+	−	−	−	+	−	−	+	−	12·0
15	+	−	−	−	−	−	−	+	+	+	13·5
16	−	−	−	+	−	+	+	+	−	+	63·0

All the concentrations are given in mg/l, with the exception of factor 10, whose central level (10 ml of solution of micro-elements per 11 of medium) corresponds to 10 times the amount of micro-element in Larsen's medium. The yield has a standard error of 3.8.

(b) Some residual mean squares

Model fitted	Residual mean squares	Degrees of freedom
All main effects	305·7	5
x_8	287·9	14
x_7, x_8	162·0	13
x_7, x_8, x_7x_8, x_9	48·8	11
$x_7, x_8, x_6, x_9, x_7x_8, x_6x_8$	19.1	9
External	14·4	

(c) Some estimates and standard errors

Final model		Additional main effects, added singly	
x_6	1·35±0·99	x_1	1·37±1·04
x_7	11·78±0·99	x_2	−1·43±1·11
x_8	11·47±0·99	x_3	−2·08±1·06
x_9	3·26±1·05	x_4	−0·47±1·19
x_6x_8	4·59±0·95	x_5	1·03±1·05
x_7x_8	9·53±0·95	x_{10}	−2·25±1·34

Table 4, is not orthogonal to the rows of the square. Because the analysis of variance of the derived variable has sums of squares:

$$\text{columns and treatments, } 0; \text{ rows, } 80; \text{ residual, } 160; \text{ total, } 240,$$

the variance of the estimated interaction is $\sigma^2/160$, compared with $\sigma^2/240$ for a corresponding orthogonal analysis. That is, in the present instance interaction is estimated with $\frac{2}{3}$

Table 4

Interaction detection in a 4×4 Latin square

Subjects	'Time' 1	2	3	4
1	B, 3	C, −3	A, −1	D, −3
2	D, 3	A, 1	B, −1	C, 9
3	A, 3	D, 1	C, 3	B, −3
4	C, −9	B, 1	D, −1	A, −3

The derived variable is formed from the product of the treatment effect C versus A, B, D $(-1, -1, 3, -1)$ and the linear column effect $(-3, -1, 1, 3)$.

efficiency. If the interaction component of interest were known *a priori* a particular Latin square for optimal estimation could be used. This form of test is best regarded as one of specification. If an interaction were detected the whole basis of the analysis would need reconsideration.

The principle involved here is that of Tukey's (1949) degree of freedom for nonadditivity, except that here the isolation of defining contrasts is based partly on *a priori* considerations rather than being entirely data-dependent.

6 Interpretation of interaction

6.1 General remarks

We now suppose that the existence of interaction has been established and that its magnitude is sufficient that explanation or interpretation is needed.

If the interactions in question are two-factor ones, treatment × treatment or treatment × intrinsic, it will often be enough to make a qualitative or graphical summary of the appropriate two-way tables of means, with associated standard errors; occasionally this is adequate also for three-factor interactions using three-way tables.

It will, however, rarely be adequate to examine expressions for an expected value of the form

$$\mu + \xi_i + \eta_j + \zeta_{ij} \quad \left(\sum \xi_i = \sum \eta_j = \sum_i \zeta_{ij} = \sum_j \zeta_{ij} = 0 \right)$$

and to use directly estimates $\hat{\zeta}_{ij}$ to characterize the interaction.

Some other simple but important possibilities will now be outlined. If all or specified interactions can be removed by transformation, part at least of the interpretation will usually exploit this.

A suitable transformation, if one exists, is often best found by informal arguments. The technique suggested by Box & Cox (1964) may yield a transformation that meets all the requirements of a normal theory linear model without the interactions in question. Otherwise, if removal of the interactions is the primary requirement, it may be better to calculate the relevant F ratio for various transformations, in particular for some parametric family of transformations; a convenient transformation with acceptably small value of F may then be used for interpretation, provided such a transformation exists. See Draper & Hunter (1969).

If there is an interaction between a treatment factor with quantitative levels and a treatment or intrinsic factor with nominal levels, one aims to characterize the effect of the

16 D.R. Cox

quantitative factor in a parametric fashion via a linear or nonlinear model and then to describe how these parameters change, preferably concentrating the change into a single parameter.

We concentrate here, as indeed throughout the paper, on the concise description and interpretation of contrasts of means. Another possibility, however, is that one wishes to predict the value of a future observation at a specified combination of levels. If the model for such prediction is clearcut, no essentially new problem arises. Particularly, however, if point estimates are required and interaction effects are on the borderline of detectability, it may be sensible in large systems to adopt an empirical Bayes approach in which, for instance, estimates obtained assuming interaction are 'shrunk' towards those appropriate in the absence of interaction (Lindley, 1972). It is a bit curious that in the literature on multiple regression explicit response prediction, as contrasted with parameter estimation, has received relatively much more attention than in the literature on analysis of variance.

Section 6.2 deals with some general aspects of the relation between interactions and main effects. Sections 6.3 and 6.4 then discuss various rather specialized ideas that are sometimes helpful in interpretation when interactions are present.

6.2 *Main effects in the presence of interaction*

In the following discussion we consider a single treatment factor and a single intrinsic factor, and suppose that interaction is clearly established; the general ideas below apply, however, much more broadly.

In § 4.1 it was stressed that models with interaction present but an associated main effect set zero were very rarely of interest; for an exception, see Cox (1977). This applied at the model fitting and interaction detection stage. In interpretation, however, main effects are fairly rarely of direct concern in the presence of appreciable interaction.

A very simple example will illustrate this. Suppose that in comparing two diets 0, 1 the mean live weight increases are, in kilograms,

    Males: treatment 0,   100;   treatment 1,   120
    Females: treatment 0,   90;   treatment 1,   80

with small standard errors of order 1 kg. The treatment effect (1 versus 0) is thus 20 kg for males and -10 kg for females and interaction, in fact qualitative interaction in the sense of § 5.2, is present. The interpretation in such cases is normally via these separate effects. The main effect, i.e. the treatment effect averaged over the levels of the intrinsic factor, is of interest only in the following circumstances.

(a) The treatment effects at the various levels of the intrinsic factor are broadly similar, so that an average treatment effect gives a convenient overall qualitative summary. Clearly, this is not the case in the illustrated example, nor in any application with qualitative interaction.

(b) An application is envisaged in which the same treatment is to be applied to all experimental units, e.g. to all animals regardless of sex. Then an average treatment effect is of some interest in which the weights attached to the different levels of the intrinsic factor are proportions in the target population. In the particular example these might be $(\frac{1}{2}, \frac{1}{2})$ regardless of the relative frequencies of males and females in the data. Note also that if the data frequencies are regarded as estimating unknown population proportions the standard error of the estimated main effect should include a contribution from the sampling error in the weights.

Note especially that in case (b) the precise definition of main effect depends in an essential way on having a physically meaningful system of weights. This formulation is in any case probably fairly rarely appropriate and there is some danger of calculating standard errors for estimates of artificial parameters.

6.3 *Product models*

Product models for interactions were in effect introduced by Fisher & Mackenzie (1923) and as a basis for a so-called joint regression approach by Yates & Cochran (1938). J. Mandel (in a series of papers) studied such models, in particular in the context of calibration studies; see, especially, Mandel (1971). For reviews, see Freeman (1973) for genetic applications and Krishnaiah & Yochmowitz (1980) for the more mathematical aspects.

In a two-factor system with nominal levels for both factors, suppose that the expected value η_{ij} in 'cell' (i, j) can be written in the special form

$$\eta_{ij} = \mu + \theta_i + \phi_j + \lambda \gamma_i \kappa_j, \tag{6.1}$$

where the normalizing conditions

$$\sum \theta_i = \sum \phi_j = 0, \quad \sum \gamma_i = \sum \kappa_j = 0, \quad \sum \gamma_i^2 = \sum \kappa_j^2 = 1$$

can be applied. One way of viewing (6.1) is that, with fitting by least squares, the model leads to a rank one approximation to the two-way matrix of interaction terms, and so on if further terms are added; this establishes a connexion with the singular-value decomposition of a matrix (Rao, 1973, p. 42).

Important restricted cases of this form of model are obtained by replacing (6.1) by one of

$$\eta_{ij} = \mu + \theta_i + \phi_j + \gamma_i \phi_j, \quad \eta_{ij} = \mu + \theta_i + \phi_j + \theta_i \kappa_j, \tag{6.2}$$

where $\sum \theta_i = \sum \phi_j = \sum \kappa_j = \sum \gamma_i = 0$, or by the symmetrical form

$$\eta_{ij} = \mu + \theta_i + \phi_j + \lambda \theta_i \phi_j. \tag{6.3}$$

The score test for $\lambda = 0$ is essentially Tukey's (1949) degree of freedom for nonadditivity.

Note that under the first of (6.2) $\bar{\eta}_{.j} = \mu + \phi_j$, so that for fixed i a plot of η_{ij} versus $\bar{\eta}_{.j}$ as j varies gives a straight line of slope $(1 + \gamma_i)$. The straight lines pass through a common point only under (6.3). Yates & Cochran (1938) gave an empirical example where the suffix i represented varieties and the suffix j places.

Snee (1982) points out the difficulty in unreplicated data of distinguishing a product structure of the above form from row or column based heterogeneity of variance; see also Yates (1972).

One particular motivation of these models can best be stated in terms of interlaboratory calibration experiments, although the idea applies much more widely. Suppose that a number of batches of material labelled by j have unobserved true values ξ_j' of some property. Samples of each batch are sent to various laboratories, labelled by i, and the ith laboratory has a calibration equation $\alpha_i' + \beta_i' \xi_j'$ converting the true value into an expected response. Thus $\eta_{ij} = \alpha_i' + \beta_i' \xi_j'$ and writing $\alpha_i' = \bar{\alpha} + \alpha_i$, $\beta_i' = \bar{\beta} + \beta_i$, $\xi_j' = \bar{\xi} + \xi_j$, with $\sum \alpha_i = \sum \beta_i = \sum \xi_j = 0$, we have that

$$\eta_{ij} = (\bar{\alpha} + \bar{\beta}\bar{\xi}) + (\alpha_i + \beta_i\bar{\xi}) + \bar{\beta}\xi_j + \beta_i\xi_j = \mu + \theta_i + \phi_j + \gamma_i\phi_j,$$

say, where $\gamma_i = \beta_i/\bar{\beta}$ is the fractional deviation of slope in the ith laboratory from the mean slope over laboratories. If the laboratories were all calibrated to give the same expected response at the mean $\bar{\xi}$, then $\theta_i \equiv 0$. Note, however, that it would rarely make sense to test the null hypothesis $\theta_i \equiv 0$ when the β_i differ, for the reasons given in § 4.1.

The above discussion is appropriate when the levels of both factors are nominal. If, for example, the rows in (6.1) have quantitative levels and models with a linear effect are under examination, it will be natural to take both the θ_i and the γ_i to be linear in the carrier variable, and this leads to the decomposition corresponding to (5.1).

6.4   Interpretation with appreciable higher-order interaction

In an investigation with many factors the presence of appreciable three factor or higher order interactions can cause serious problems of interpretation. A direct interpretation in terms of the interactions as such is rarely enlightening.

In relatively simple cases the interactions may be used just as a guide as to which tables of means should be considered. Thus, if the only important three factor or higher interaction is of the form treatment$_1 \times$ treatment$_2 \times$ intrinsic, it may be enough to set out separate treatment$_1 \times$ treatment$_2$ tables at each level of the intrinsic factor.

In more complex cases, especially where interactions among several treatment factors occur, other techniques are needed. These include the following:

(i)   transformation of the response variable may be indicated;

(ii) if the treatment factors are quantitative, and a high degree polynomial appears necessary to fit the data, the possible approaches to a simpler interpretation include transformation of response and/or factor variables and the fitting of simple nonlinear models;

(iii) the abandonment of the factorial representation of the treatments in favour of the isolation of a few possibly distinctive factor combinations;

(iv) the splitting of the factor combinations on the basis of one or perhaps more factors;

(v)   the adoption of a new system of factors for the description of the treatment combinations.

Approaches (i) and (ii) call for no special comment here. There follow some remarks on (iii)–(v).

For 2^m systems it is easy to see that if all responses except one are equal then all the standard factorial contrasts are equal except for sign. Thus the occurrence of a number of appreciable contrasts, the higher-order interactions being broadly comparable in magnitude with the main effects, suggests that there are one, or perhaps a few, factor combinations giving radically different responses. The analysis of the data then consists of specifying the anomalous combinations and their responses, and checking the homogeneity of the remaining combinations. That is, the factorial character is abandoned in the final analysis, the decomposition into main effects and interactions being a step in the preliminary analysis. For an example, see Cox & Snell (1981, Example N).

The next possibility, (iv), is that all or most of the appreciable contrasts, including higher-order interactions, involve a particular factor. It may then be possible to find a simple interpretation by decomposing the data into separate sets, one for each level of the 'special' factor. Each set may then have a relatively simple interpretation. This device is particularly appealing if the 'special' factor is an intrinsic factor, not a treatment factor, although the method is not restricted to intrinsic factors. For an example, see the comments by Cox (1974) on the data in the book of Johnson & Leone (1964).

Finally, point (v), there is the possibility (Cox, 1972; Bloomfield, 1974) that redefinition of the factors will lead to a simpler explanation or representation. For two factors each at two levels the four combinations conventionally labelled

$$1 \quad a \quad b \quad ab$$

can be relabelled

$$1 \quad a'b' \quad b' \quad a'$$
$$1 \quad b'' \quad a''b'' \quad a''$$

as well as by systems obtained by interchanging upper and lower levels of one or both factors. If, for example, in the original system the main effect A and the interaction $A \times B$ are appreciable but the main effect of B negligible, the first transformation mentioned will switch the interaction $A \times B$ into the main effect B', leaving the main effect A as the main effect A'. In general in a 2^m system the m largest contrasts that are linearly independent in the prime power commutative group of contrasts can be redefined as main effects. A more complex transformation scheme would be appropriate if it is required to control the position of more than m independent contrasts.

In all these rather special methods there is the possibility of alternative representations, possibly of different types or possibly variants of the same type, giving approximately equally good fits to the data. Formal theory is not much guide; the maximized log likelihood achieved is determined by the residual sum of squares, but the number of parameters effectively being fitted is not clearly defined and the various descriptions being compared are not nested. The safest approach in principle is to list all those relatively simple representations consistent with the data and to make any choice between them on grounds external to the data.

7 Role of nonspecific factors and the estimation of error

7.1 General remarks

So far we have not discussed in detail the estimation of error. In some situations, error estimation is based directly on variation between replicate observations within 'cells'; the possibility of hierarchical error structure has often to be considered, both in the context of split plot experiments and also, in observational data, in connexion with multistage sampling.

If there is no direct replication and all factors are either treatment or intrinsic, error will be estimated from interactions, often by pooling all those above a certain order, although there may be some case for using interactions among the intrinsic factors only, when that is feasible. Comparison with the variance found in previous experience with similar material gives one important check. Where the mean squares for several different contrasts are pooled, rough check of homogeneity is desirable, more to avoid the overlooking of an important unexpected contrast than to avoid overestimation of error.

In the present section, however, we concentrate on the role of nonspecific factors, such as blocks, replicates, etc., factors whose levels are not defined by uniquely identified features. For instance, if an investigation is repeated at a number of sites there will usually be many ways, often rather ill defined, in which the sites differ from one another.

7.2 Interpretation of interactions with nonspecific factors

Suppose for simplicity that there is a single treatment factor, x, and a single nonspecific factor, u. If there is no replication within (x, u) combinations, or other ways of estimating error, there may be little option but to treat $x \times u$ as error; in effect, we say that if the treatment effect is different at different levels of u, there is no basis for the detailed explanation or representation of this interaction and that it is to be regarded as in some sense random. This is particularly appropriate if the distinction between the different levels of u is of little direct interest, such as when u represents blocks, litters, etc.

Suppose, however, that a separate estimate of error is available and that the different levels of u are in some sense important. For example, a major study may be repeated in more or less identical form in a number of centres. If an interaction $x \times u$ is found, it should if possible be 'explained', for example,

 (i) by removal by transformation,
 (ii) by relating treatment effects to concomitant variables attached to the different levels of u,
 (iii) by regarding certain levels of u as anomalous,
 (iv) by establishing that the nominal estimate of error is an underestimate of the 'true' variability present.

Failing all these, it seems necessary to regard the interaction as a source of uncontrolled variability, to be represented in a mathematical model by an additional random variable. The corresponding treatment main effect is then an average over an infinite population of levels, usually hypothetical. Of course, where there is a natural infinite or finite population of u levels, it is more defensible to take this as the basis.

In simple balanced data, the conclusion is that with a single nonspecific factor, u, showing interaction, the error of treatment contrasts is estimated from the interaction treatments $\times u$. With unbalanced data, either maximum likelihood analysis of a complex error structure is required or some simpler approximate method based on reduction to a balanced system.

With more than one nonspecific factor, synthesis of variance is needed, even in balanced systems, to estimate error (Cochran & Cox, 1957, § 14.5):

$$\text{MS}_{\text{treat} \times u_1} + \text{MS}_{\text{treat} \times u_2} - \text{MS}_{\text{treat} \times u_1 \times u_2}$$

is that normally used, in a self-explanatory notation, and this is assigned an effective degrees of freedom. It would be worthwhile to have a more formal development of the theory of the approximate inference so involved, one approach being via maximized log likelihoods, irrelevant effects having been removed by orthogonal transformation.

If the treatment effect is decomposed into components with very different magnitudes, it will, provided that enough degrees of freedom are available, be necessary to make an analogous decomposition of the interaction with a nonspecific factor. For example, if a treatment effect were linear at each level of u, but with different slopes, the standard error of the 'overall' slope could be determined from linear treatments $\times u$, which might well be appreciably greater than quadratic treatments $\times u$, which should equal the internal error if indeed each separate relation is linear. Similar remarks apply if there are several treatment factors. In principle, the error of each component of the treatment effect is found from the interaction of that component with u.

Special care is needed if 'time' is a factor. If an investigation is repeated over time on physically independent material, and if external conditions such as meteorological ones can be treated as totally random, then it may be sensible to treat 'time' as a nonspecific factor, as above. If, however, the same individual is involved at each time point, assumptions of independence are best avoided and any made need critical attention. A commonly useful device (Wishart, 1938) is to produce for each individual one or more summary statistics measuring, for example, trend, cyclical variation, treatment effects, serial correlation, etc. and then to analyse these as derived response variables. This approach usually requires that the number of time points is not too small, although the development in § 4 of an analysis for the two-period, two-treatment cross-over design could be regarded as an illustration. Thus the derived variables are the difference and sum of the two observations on each subject.

If there are several nonspecific factors and their main effects and interactions are of intrinsic interest, the detailed study of components of variance is involved. Particular applications include biometrical genetics, the study of sources of variability in industrial processes and the evaluation of interlaboratory trials. This is a special (and extremely interesting) topic which, with regret, is omitted from the present paper.

8 Generalizations

Most of the above discussion generalizes immediately to any analysis in which a parameter describing the distribution of a response variable is related to explanatory variables via a linear expression. The generalized linear models of Nelder & Wedderburn (1972) embrace many of the commonly occurring cases. The more indirect the relation between the original data and the linear representation, the less the interpretative value of the absence of interaction.

For such generalizations, the method of analysis will usually be either maximum likelihood or a preliminary transformation followed by least squares with empirical weights. The main technical difficulty concerns systems with several layers of error. For these the direct use of weighted least squares will often be simpler, although no generally useful methods are available at the time of writing.

The definition of interaction within exponential family models involves a linear representation of some function of the defining parameter. Thus, for binary data the defining parameter is the probability of 'success', in this case the moment parameter of the exponential family. The canonical parameter is the logistic transform. In this case the canonical parameter has unrestricted range. For exponential distributions, the canonical parameter is the reciprocal of the mean; the logarithm is the simplest function of the moment parameter (or of the canonical parameter) having unrestricted range. Interaction can thus be defined in various ways, in particular relative to linear representations for the moment parameter, for the canonical parameter, or for the simplest unconstrained parameter. Especially when the interaction between a treatment and an intrinsic factor is involved, use of the unconstrained parameter will often be the most appropriate, since such representations have the greatest scope for extrapolation. Representations in terms of the canonical parameter lead to the simplest statistical inference, and those in terms of the moment parameter have the most direct physical interpretation. Of course, often there will not be enough data to show empirically that one version is to be preferred to another; with extensive data there is the possibility of finding that function, if any, of the defining parameter leading to absence of interaction but often familiarity and ease of interpretation will restrict the acceptable representations.

Darroch (1974) has discussed the relative advantages of so-called additive and multiplicative definitions of interaction for nominal variables. His account in terms of association between multidimensional responses is easily adapted to the present context of a single response and several explanatory variables.

Throughout the paper only univariate response variables have been considered. The formal procedures of multivariate analysis of variance are available for examining interactions affecting several response variables simultaneously. While such procedures may be used explicitly only relatively rarely, any demonstration that a particular type of simple representation applied to several response variables simultaneously makes such a representation more convincing.

Acknowledgements

This paper contains the union of invited papers given at the European Meeting of Statisticians, Wroclaw, the York Conference of the Royal Statistical Society, the Gordon Research Conference, and the Highland Group, Royal Statistical Society. I am very grateful for various interesting points made in discussion and thank also E.J. Snell and O. Barndorff-Nielsen for helpful and encouraging comments on a preliminary written version.

Appendix

Some open technical issues

The discussion in the main part of the paper has concentrated on general issues and deliberately put rather little emphasis on details of technique. Nevertheless there are unsolved technical matters involved and there follows a list of a few problems. They are open so far as I am aware, although any reader thinking of working on any of them is strongly advised to look through the recent (and not so recent) literature not too long after starting.

Problem 1 [§ 2]. Given random samples from m populations, test the hypothesis that there exists a translation-inducing transformation. Estimate the transformation and associated differences.

Problem 2 [§ 2]. Calculate the efficiency of your procedure for 1 as compared with some simple fully parametric formulation.

Problem 3 [§§ 2, 5]. Extend the discussion of 1 and 2 to the examination of interactions.

Problem 4 [§ 3]. Find some interesting special cases satisfying Scheffé's condition for transformation to additivity.

Problem 4 [§ 4]. Extend the notion of order-based interactions to higher order interactions.

Problem 6 [§ 4]. Develop suitable designs and methods of analysis, and make practical recommendations concerning the use of, p period, t treatment crossover designs (i) for $p = t$, (ii) for $p > t$, (iii) for $p < t$, when the treatments are (a) unstructured, (b) factorial with quantitative levels, (c) factorial with qualitative levels and when interest focuses primarily on (I) direct treatment effects with some possibility of examining residual effects, (II) estimating and testing residual effects, (III) estimating 'total' treatment effects.

Problem 7 [§§ 4, 5]. What is the role of AIC, C_p and so on in the study of interactions, especially from unbalanced data?

Problem 8 [§§ 4, 5]. What is the role of rank invariant procedures in the study of interactions?

Problem 9 [§ 5]. What is the effect of nonnormality on the plot of ordered mean squares?

Problem 10 [§ 5]. Are there useful significance tests associated with plots of ordered mean squares?

Problem 11 [§ 5]. Examine in some generality the effect of dependence on probability plotting methods.

Problem 12 [§ 5]. In the two-way arrangement with nominal factors, is there a useful set of primitive contrasts less redundant than the full set of tetrad differences?

Problem 13 [§ 5]. Develop a systematic theory of testing for treatment × column interaction in an $m \times m$ Latin square when both, one or none of the defining contrasts are known *a priori*. What is the efficiency of a randomized design for detecting interaction: what is the lowest efficiency that could be encountered? What are the implications, if any, for design?

Problem 14 [§ 5]. Develop and implement an IKBS (intelligent knowledge-based system) that would deal with the 2^{10} example of § 5 and similar but much larger problems.

Problem 15 [§ 6]. Suppose that $Y_j (j = 1, \ldots, n)$ are independently normally distributed with $E(Y_j) = \mu_j$ and unit variance. Initially interest lies in the null hypothesis $\mu_1 = \ldots = \mu_n = 0$ but overwhelming evidence against this is found. Two alternative explanations are contemplated:

(a) $\mu_j = 0 \ (j \neq \tau)$, $\mu_\tau \neq 0$, τ being unknown;
(b) $\mu_1, \ldots, \mu_n$ are independently normally distributed with zero mean and unknown variance σ^2.

Develop procedures for examining which if either of (a) and (b) is satisfactory and which is preferable.

Problem 16 [§ 6]. Extend the discussion of Problem 15 to the examination of interactions.

Problem 17 [§ 7]. Develop a simple and elegant theory for finding confidence limits for contrasts where error has to be estimated via synthesis of variance.

Problem 18 [§ 7]. Develop from first principles confidence limits for an upper variance component from balanced and from unbalanced data.

Problem 19 [§ 7]. Examine critically the role of time series models for experiments in which time is a nonspecific factor.

Problem 20 [§ 8]. Give a general discussion for analysing interactions in log linear models for Poisson and binomial data in the presence: (a) of overdispersion, (b) of underdispersion.

References

Armitage, P. & Hills, M. (1982). The two period crossover trial. *Statistician* **31**, 119–131.
Atkinson, A.C. (1982). Regression diagnostics, transformations and constructed variables (with discussion). *J. R. Statist. Soc.* B **44**, 1–36.
Azzalini A. & Cox, D.R. (1984). Two new tests associated with analysis of variance. *J. R. Statist. Soc.* B **46**. To appear.
Bloomfield, P. (1974). Linear transformations for multivariate binary data. *Biometrics* **30**, 609–617.
Box, G.E.P. & Cox, D.R. (1964). An analysis of transformations (with discussion). *J. R. Statist. Soc.* B **26**, 211–252.
Bradu, D. & Hawkins, D.M. (1982). Location of multiple outliers in two-way tables, using tetrads. *Technometrics* **24**, 103–108.
Brown, M.B. (1975). Exploring interaction effects in the analysis of variance. *Appl. Statist.* **24**, 288–298.
Brown, B.W. (1980). The crossover experiment for clinical trials. *Biometrics* **36**, 69–79.
Cochran, W.G. & Cox, G.M. (1957). *Experimental Designs*, 2nd edition. New York: Wiley.
Cook, P.D. & Weisberg, S. (1982). *Residuals and Influence in Regression*. London: Chapman and Hall.
Cox, D.R. (1972). The analysis of multivariate binary data. *Appl. Statist.* **21**, 113–120.
Cox, D.R. (1974). Discussion of paper by M. Stone. *J. R. Statist. Soc.* B **36**, 140–141.
Cox, D.R. (1977). Discussion of paper by J.A. Nelder. *J. R. Statist. Soc.* A **140**, 71–72.
Cox, D.R. & Snell, E.J. (1981). *Applied Statistics*. London: Chapman and Hall.
Daniel, C. (1959). Use of half-normal plots in interpreting factorial two-level experiments. *Technometrics* **1**, 311–341.
Daniel, C. (1976). *Applications of Statistics to Industrial Experimentation*. New York: Wiley.

Daniel, C. & Wood, F.S. (1971). *Fitting Equations to Data*, New York: Wiley.

Darroch, J.N. (1974). Multiplicative and additive interaction in contingency tables. *Biometrika* **61**, 207–214.

Draper, N.R. & Hunter, W.G. (1969). Transformations: Some examples revisited. *Technometrics* **11**, 23–40.

Doksum, K.A. & Sievers, G.L. (1976). Plotting with confidence: graphical comparisons of two populations. *Biometrika* **63**, 421–434.

Fedorov, V.D., Maximov, V.N. & Bogorov, V.G. (1968). Experimental development of nutritive media for micro-organisms. *Biometrika* **55**, 43–51.

Fisher, R.A. & Mackenzie, W.A. (1923). Studies in crop variation. II. The manurial response of different potato varieties. *J. Agric. Sci.* **13**, 311–320.

Freeman, G.H. (1973). Statistical methods for the analysis of genotype-environment interactions. *Heredity* **31**, 339–354.

Gentleman, J.F. (1980). Finding the k most likely outliers in two-way tables. *Technometrics* **22**, 591–600.

Gentleman, J.F. & Wilk, M.B. (1975). Detecting outliers in a two-way table. I. Statistical behaviour of residuals. *Technometrics* **17**, 1–14.

Hills, M. & Armitage, P. (1979). The two-period crossover clinical trial. *Brit. J. Clinical Pharmacol.* **8**, 7–20.

Holland, P.W. (1973). Covariance stabilizing transformation. *Ann. Statist.* **1**, 84–92.

Johnson, N.L. & Leone, F. (1964). *Statistics and Experimental Design*, **2**. New York: Wiley.

Krishnaiah, P.R. & Yochmowitz, M.G. (1980). Inference on the structure of interaction in two-way classification models. *Handbook of Statist.* **1**, 973–994.

Lindley, D.V. (1970). A Bayesian solution for two-way analysis of variance. *Col. Math. Soc. János Bolyai* **9**, 475–496.

Mandel, J. (1971). A new analysis of variance for non-additive data. *Technometrics* **13**, 1–18.

Nelder, J.A. (1977). A representation of linear models (with discussion). *J. R. Statist. Soc.* A **140**, 48–76.

Nelder, J.A. & Wedderburn, R.W.M. (1972). Generalized linear models, *J. R. Statist. Soc.* A **135**, 370–384.

Pearson, E.S. & Hartley, H.O. (1972). *Biometrika Tables for Statisticians*, **2**. Cambridge University Press.

Rao, C.R. (1973). *Linear Statistical Inference and its Applications*, 2nd edition. New York: Wiley.

Scheffé, H. (1959). *The Analysis of Variance*. New York: Wiley.

Snee, R.D. (1982). Nonadditivity in a two-way classification: is it interaction or nonhomogeneous variance? *J. Am. Statist. Assoc.* **77**, 515–519.

Song, C.C. (1982). Covariance stabilizing transformations and a conjecture of Holland. *Ann. Statist.* **10**, 313–315.

Tukey, J.W. (1949). One degree of freedom for non-additivity. *Biometrics* **5**, 232–242.

Wilk, M.B. & Gnanadesikan, R. (1968). Probability plotting methods for the analysis of data. *Biometrika* **55**, 1–17.

Wishart, J. (1938). Growth-rate determinations in nutrition studies with the bacon pig, and their analysis. *Biometrika* **30**, 16–28.

Yates, F. (1972). A Monte-Carlo trial on the behaviour of the nonadditivity test with nonnormal data. *Biometrika* **59**, 253–261.

Yates, F. & Cochran, W.G. (1938). The analysis of groups of experiments. *J. Agric. Sci.* **28**, 556–580.

Résumé

Le rôle en la statistique de l'action réciproque (interaction) est discuté. On classifie les interactions en fonction de la nature des facteurs qui sont engagés. Les tests pour l'interaction sont passés en revue. Les méthodes d'interprétation d'une interaction sont discutés et la connexion avec l'estimation d'erreur est décrite. Enfin problèmes non resolus sont catalogués.

[Paper received January 1983, revised May 1983]

Summary of discussion
The paper was supplemented with discussion by A. C. Atkinson, G. E. P. Box, J. N. Darroch, E. Spjøtvoll, and J. Wahrendorf. Atkinson makes a number of detailed points about the meaning of interaction in not-so-simple situations and about some associated graphical methods of analysis. Box describes a situation in which large interactions have an important effect and emphasizes the arbitrariness in the choice of variables defining factors. Darroch emphasizes the care needed in extending ideas about interactions to models for contingency tables. Spjøtvoll asks about the level of formality desirable in searching for models to represent data and outlines two pieces of work then current on nonstandard models for situations with interaction. Finally, Wahrendorf considers the interpretational aspects of interaction, especially in a medical context and gives useful references.

Regression Methods

Notes on Some Aspects of Regression Analysis

By D. R. Cox

Imperial College

[Read before the ROYAL STATISTICAL SOCIETY on Wednesday, March 20th, 1968, the President, Dr F. YATES, C.B.E., F.R.S., in the Chair]

SUMMARY

Miscellaneous comments are made on regression analysis under four broad headings: regression of a dependent variable on a single regressor variable; regression on many regressor variables; analysis of bivariate and multivariate populations; models with components of variation.

1. INTRODUCTION

THIS is an expository paper consisting not of new results but of miscellaneous and isolated comments on the theory of regression. The subject is a very broad one and the paper is in no sense comprehensive. In particular, the ideas of regression are the basis of much work in time series analysis and in multivariate analysis and these specialized subjects are barely mentioned; nor are experimental design and sampling theory problems associated with regression considered. Another serious limitation to the paper is the omission of relevant parts of the econometric literature.

Two general situations are distinguished and in their simplest forms are:

(i) a dependent variable Y has a distribution depending on a regressor variable x and it is required to assess this dependence;

(ii) there is a bivariate population of pairs (X, Y) and the joint distribution is to be analysed.

Capital letters are used for observations represented by random variables and lower-case letters for other observations.

Most theoretical discussion of regression starts from a quite tightly specified model in which some observations are regarded as corresponding to random variables with probability distributions depending in a given way on unknown parameters. Many of the difficulties of regression analysis, however, concern such questions as which observations should be treated as random variables, what are suitable families of models, and what is the practical interpretation of the conclusions.

The special computational and other problems associated with fitting non-linear models will not be considered explicitly, although much of the discussion applies as much to such problems as to the simpler linear models with which the paper is overtly concerned. Particular applications will not be discussed in detail but one or two typical but hypothetical situations will be outlined later for illustration. Through-out the discussion the measurement of uncertainty by significance tests and confidence limits is important but not paramount.

Inevitably the paper tends to emphasize difficulties likely to be encountered; of course awareness of potential difficulties is a good thing, but only as one facet of constructive scepticism.

The methodology of regression is described by Williams (1959) and by Draper and Smith (1966) and the more theoretical aspects by Plackett (1960), Kendall and Stuart (1967, Chapters 26–29) and Rao (1965).

2. Regression on a Single Regressor Variable

Suppose that there are n pairs of observations $(x_1, Y_1), ..., (x_n, Y_n)$ and that for each value of the regressor variable x there is a population of values of the dependent variable from which the observed Y_i's are randomly chosen. This is often taken for granted as the starting point for a theoretical discussion of regression; in (i) and (ii) which follow, some of its implications are discussed and then in (iii)–(vi) comments on some more advanced matters are made.

(i) *Choice of dependent and regressor variables.* In some situations the x values may be chosen deliberately by the experimenter and Y is a response dependent on x. The more difficult situation is when both types of observation can be regarded as random. Then we take as dependent variable:

(a) the "effect", the regressor variable being the explanatory variable; for a given value of the explanatory variable, we ask what is the distribution of possible responses;

(b) the variable to be predicted, the regressor variable being the variable on which the prediction is to be based. A full solution to the prediction problem is to give the conditional distribution of the variable to be predicted given all available information on the individual concerned.

Suppose that it is reasonable to consider an existing or hypothetical population of Y values for each x. Now whether or not the x's can be regarded as random may well affect the interpretation and application of the conclusions. For analysis of the regression coefficient in the model, however, we argue conditionally on the x values actually observed (Fisher, 1956, p. 156), provided that the x values by themselves would have given no information about the parameter of interest.

Example 1. A random sample of fibre segments of fixed lengths is taken from a homogeneous source and for each segment the mean diameter is measured accurately and the breaking load determined. Both observations can be regarded as random variables but in view of (a) it is reasonable to take breaking load, or rather its log, as dependent variable and log diameter as regressor variable. The same model would be appropriate if, for example, a fixed number of fibres is selected randomly from each of a number of diameter groups. Whether the regression is a fruitful thing to consider depends on its stability, i.e. on whether there is a reproducible relationship involved; see (iii).

Example 2. A more difficult case is illustrated by a calibration experiment in which for n individuals (a) a "slow" measurement and (b) a "quick" measurement are made of some property. Often (a) is a definitive determination, for example an optical measurement of fibre diameter, and (b) is the result of some indirect and much easier method. In future, only the "quick" measurement will be obtained and it is required to predict from this the corresponding "slow" measurement. If the n individuals initially observed are a random sample from the same population as the individuals for which future predictions are to be made, we take the "slow" measurement to be the dependent variable, since this is the one to be predicted. Suppose, however, that the n individuals were chosen systematically, for example to have "slow" values approximately evenly distributed over the range of interest.

Usually it would be reasonable to regard the "quick" value as having a random component and, provided that a physically stable random system is involved and the relationship linear to write

$$\text{"quick"} = \alpha + \beta \text{ "slow"} + \text{"error"},$$

where the "error" does not depend on the "slow" value. Given a new "quick" measurement, we have to estimate a non-random variable by inverse estimation (Williams, 1959, pp. 91, 95). If both variables can in fact be regarded as random, the second approach is inefficient because it ignores the information about the marginal distribution of the "slow" measurement.

(ii) *The omitted variables.* Suppose that z is a regressor variable that might have been included in the regression analysis but in fact is not, for example because no observations of it are available. What assumptions about z are made when we consider the regression of Y on x alone? Box (1966) has given an illuminating discussion of the dangers of omitting a relevant variable. The relationship ignoring z will be meaningful if:

(a) changes in z have no effect on Y;

(b) in a randomized experiment in which x corresponds to a treatment, there may be a unit-treatment additivity. Then the usual analysis will give an estimate of the effect of changing x, and an estimate of the standard error. The estimate refers to the difference between the response on one unit with certain values for x and for the omitted variable z and what would have been observed on that same unit with a different value of x and the same z;

or (c) z is a random variable, say Z, and the distributions of Z, given x, and of Y, given $Z = z$ and x, are well defined. If x is a random variable X, this amounts to the requirement that (X, Y, Z) have a well-defined three-dimensional distribution. Then the regression of Y on x is well defined and includes a contribution associated with changes in z.

Example 3. Consider observational data for individuals with an ultimately fatal disease, Y being the log time to death and x some aspect of the treatment applied, called the dose, and that the regression of Y on x is analysed. The missing variable z is the initial severity of the disease. If, as might well be the case, z largely determines x, the regression of Y on x, although well defined under the circumstances of (c), would be of very limited usefulness. In particular it would not give for a particular individual an estimate of the effect on his Y of changing dose. This is an extreme example of a difficulty that applies to many regression studies based on observational data.

(iii) *Stability of regression.* While a fitted regression equation may often be useful simply as a concise summary of data, it is obviously desirable that the relation should be stable and reproducible. This is stressed by Ehrenberg (1968) and Nelder (1968); see also Tukey (1954). Stability might mean that when the experiment is repeated under different conditions:

(a) the same regression equation holds, even though other aspects of the data change;

or (b) parallel regression equations are obtained;

or (c) satisfactory regression lines are always obtained but with different positions and slopes.

In cases (b) and (c) the fitting of regression lines will be an important first step in the analysis. The second step will be to try to account for the variation in the parameters whose estimates do vary appreciably, possibly by a further regression analysis on regressor variables characterizing the different groups of observations and taking the initial regression coefficients as dependent variables. In testing the significance of the differences between the regression coefficients in different groups it will often be important to allow for correlations between the groups of data (Yates, 1939).

(iv) *Choice of relation to be fitted.* This choice will depend on preliminary plotting and inspection of the data and possibly on the outcome of earlier unsuccessful analyses. In addition, the model may take account of:

(a) conclusions from previous sets of data;

(b) theoretical analysis, including dimensional analysis, of the system;

(c) limiting behaviour.

Further, any given model can be parametrized in various ways and, in choosing a parametrization, the following considerations may be relevant:

(a)′ individual parameters should have a physical interpretation, say in terms of components in a theoretical model or in terms of combinations of regressor variables of physical meaning;

(b)′ individual parameters and estimates should have a descriptive interpretation, for example in terms of the average slope and curvature of response over the range considered;

(c)′ interpretations, such as those of (a)′ and (b)′, should be insensitive to secondary departures from the model;

(d)′ any instability between groups should be confined to as few parameters as possible;

(e)′ the sampling errors of estimates of different parameters should not be highly correlated.

These requirements may be to some extent mutually conflicting.

There is not space here to discuss and exemplify all these points. As just one example, preliminary analysis of data of Example 1 might, if the x's have relatively little variation, suggest that linear regressions of breaking load on diameter and of log breaking load on log diameter would fit about equally well. The second would in general be preferable because, with respect to the above conditions:

(b) it permits easier comparison with the theoretical model breaking load $\propto (\text{diameter})^2$;

(c) it ensures that breaking load vanishes with diameter;

(a, b)′ the regression coefficient, being a dimensionless power, is easier to think about than a coefficient having the dimensions of load/diameter.

(v) *Goodness of fit.* This can be examined in a number of ways:

(a) by a non-probabilistic graphical or tabular analysis, for example of residuals;

(b) by a significance test, using as a test statistic some aspect of the data thought to be a reasonable measure of departure from the model. Thus, the standardized third moment of the residuals could be used, if possible skewness is of interest;

(c) by the fitting of an extended model reducing to the given model for particular parameter values. The most familiar example is the inclusion of the new regressor variable, possibly a power of the first variable, or a product of variables, when multiple regression is being considered;

(d) by the fitting of a quite different model, seeing whether it fits better than the initial one.

Such examination of the adequacy of the model is important if models are to be refined and improved. Often, but not always, the primary aspect of the model will be the form of the regression equation and the adequacy of what may be secondary assumptions about constancy of variance, normality of distribution, etc. will be of rather less importance. Formal significance tests are valuable but, of course, need correct interpretation. A very significant lack of fit means that there is decisive evidence of systematic departures from the model; nevertheless the model may account for enough of the variation to be very valuable (Nelder, 1968). A non-significant test result means that in the respect tested the model is reasonably consistent with the data; nevertheless there may be other reasons for regarding the model as inadequate.

Of (a)–(d) the least standard is (d). It is closely related to the problem of choosing between alternative regression equations (Williams, 1959, Chapter 5), for example between the regression of Y on x_1 alone and that of Y on x_2 alone. The more usual procedure in such a case will, however, be to fit both variables to cover the possibility that the joint regression is appreciably better than either separate one.

As a rather different example, suppose that normal theory linear regressions of (α) Y on x, (β) Y on $\log x$, (γ) $\log Y$ on $\log x$ are considered. The goodness of fit of (α) and (β) can be compared descriptively by the residual sums of squares, but to compare say (α) with (γ) the residual sum of squares cannot be used directly. The most usual procedure is probably then to compare squared correlation coefficients, but for comparison of the full models it is probably better to compare the maximized log likelihoods of $Y_1, ..., Y_n$ under the two models. Cox (1961, 1962) has discussed the construction of significance tests in such situations.

An alternative and in many ways preferable approach is to consider a comprehensive model containing (α), (β), (γ) as special cases. For example the normal theory linear regression of

$$\frac{Y^{\lambda_2}-1}{\lambda_2} \quad \text{on} \quad \frac{x^{\lambda_1}-1}{\lambda_1}$$

could be taken and all parameters, including (λ_1, λ_2), estimated and tested by maximum likelihood (Box and Tidwell, 1962; Box and Cox, 1964). This is computationally formidable if there are several regressor variables.

(vi) *More complex dependence on the regressor variable.* In most regression analyses it is assumed that the dependence on the regressor variable x is confined to changes in the conditional mean of Y. Transformation of Y may be necessary to achieve this; if different transformations are required to linearize the regression of the mean and to stabilize variance the first will usually have preference, simply because primary interest will usually lie in the mean. To study, for example, changes in the conditional variance of Y we can:

(a) plot residuals;

(b) group on x, calculate variances within groups, if necessary applying an adjustment for changes of mean within groups, and then consider the regression on x of log variance;

(c) fit, for example by maximum likelihood, a model in which parameters are added to account for changes in variance. For example, the variance might be taken to be $\sigma^2 \exp\{\gamma(x-\bar{x})\}$. It would nearly always be right to precede any such fitting by (a) or (b) in order to get some idea of an appropriate model and of whether the more complex fitting is likely to be fruitful.

Similar remarks apply to the study of changes of distributional shape.

If the regression on the mean is linear but there are substantial changes in variance a weighted analysis will often be required, although the changes in variance have to be quite substantial before there is appreciable gain in precision in estimating the regression coefficient. Of course the changes in variance may be of intrinsic interest, or need separate study in order to specify how the precision of prediction depends on x.

3. REGRESSION ON SEVERAL REGRESSOR VARIABLES

Suppose now that for each individual several regressor variables are available, i.e. that for the ith individual we observe $(Y_i, x_{i1}, ..., x_{ip})$. We consider mainly the case where $x_1, ..., x_p$ are physically distinct measurements rather than, for example, powers of a single x. Virtually all the discussion of Section 2 is relevant, but there are new points mostly connected with the choice of the regressor variables and the interpretation of situations in which there is appreciable non-orthogonality among the regressor variables.

There are two extreme situations to consider. In the first the number of regressor variables is quite small, say not more than three or four. It is then perfectly feasible both to fit the 2^p possible regression equations and to examine them individually. Those regressor variables the nature of whose effects is clearly established can be isolated and ambiguities arising from non-orthogonality of other variables listed and, as far as possible, interpreted. Also further regression equations involving, say, squares and cross-products of some of the original regressor variables can be fitted, if required. In the second case the number p of regressor variables is larger. It may still be computationally feasible to fit all 2^p equations, but unless all pairs of regressor variables are nearly orthogonal, the interpretation is likely to be difficult and, at the least, some further techniques are required for handling the information from the fits. In many applications of this type there is a reasonable hope that only a fairly small number of regressor variables have important effects over the region studied. The broad distinction between these two cases should be borne in mind in the following discussion.

(i) *Interpretation and objectives.* Lindley (1968) has emphasized that the choice between alternative equations depends on the purpose of the analysis and has discussed two cases in detail from a decision-theoretic viewpoint, one a prediction problem and one a control problem. His results show very explicitly the consequences of strong assumptions about the problem and are likely to be useful guidance in other cases too. The following remarks refer to cases where less explicit assumptions are possible about the nature of the problem and the objectives of the analysis.

Suppose first that the objective is to predict Y for future individuals in the region of x-space covered by the data. In particular, the x's may be random variables and the new individuals be drawn from the same population. Then any regression equation that fits the data adequately will be about equally effective on the average over a series of x-values. If, however, it is thought that not all regressor variables contribute, there is likely to be a gain from excluding regressor variables with an insignificant effect. Note that a Bayesian analysis of this situation suggests reducing the contribution of, rather than eliminating, such variables and this is sensible also from a sampling theory viewpoint. So long as p is small compared with n, it is not likely to make a major difference which of these various possibilities is taken. The algorithm of Beale *et al.* (1967) for selecting the "best" equation with a specified number of regressor variables and the various automatic stepwise procedures described by Draper and Smith (1966, Chapter 6) will be relevant.

Suppose next that the prediction is to be made for an individual in a new region of x-space. Things are now different. For example, suppose that x_1 and x_2 are almost linearly related in the initial sample of observations and that the partial regression coefficients are insignificant, the combined regression being very highly significant. It is thus known that at least one of x_1 and x_2 has an important contribution, but there will be many regression equations fitting the data about equally well. Under the circumstances of the previous paragraph this is immaterial, but if prediction of Y is attempted for (x_1, x_2) far from the original linear relation, extremely different results will be obtained from the different fits. In such cases the possibilities are:

(a) to postpone setting up a prediction equation until better data are available for estimation;

(b) to use external information to decide which is "really" the appropriate equation;

(c) to use the formal variance of prediction from the full equation as a means of detecting individuals for which prediction from *any* regression equation is hazardous.

The next, and in many ways the most important, case is where we hope that there is a unique dependence of Y on some, or all, of the regressor variables that will remain stable over a range of conditions and we wish to estimate this relation and in particular to identify the regressor variables that occur in it. In a randomized experiment it may ideally be possible to estimate the contrasts of primary interest separately and efficiently, to show that they do not interact with external factors and they account for most of the variability. Even here there are difficulties, particularly if the response surface is relatively complicated. For observational data, there are two major difficulties:

(a) the possibility of important omitted variables (see Section 2, point ii);

(b) ambiguities arising from appreciable non-orthogonality of regressor variables.

There is discussion below of some of the devices that can be used to try to overcome (b).

In the situation contemplated in the previous paragraph, the objective is essentially the same as that in a randomized experiment. A more limited objective is to analyse preliminary data in order to suggest which factors would be worth including in a subsequent experiment and to suggest appropriate spacing for the levels. It would be interesting to examine the performance of some simple strategies, even though there will always be further information to be taken into account.

(ii) *Aids to interpretation.* In some cases the main interest may lie in the regression on x_1, the variable x_2 being included as characterizing say different groups of observations, or some potentially important aspect of secondary interest in the particular investigation. If x_2 can conveniently be grouped, it will often be good to fit separate regressions on x_1 within each x_2 group and then to relate the estimated parameters to x_2. This leads to an analysis of the stability of the regression equation and possibly to the construction of models containing interaction. More generally x_1 and x_2 may be sets of regressor variables.

If there is a property x that is thought not to have an effect on Y, it will often be good to include x as a regressor variable. Significant regression on x would then be a warning, for example of an important omitted variable.

The next set of remarks refer to ambiguities arising from non-orthogonality and all depend upon introducing further information in some form.

(a) It may be thought that the regression coefficient on say x_1 should be non-negative. In some special cases this may resolve an apparent ambiguity. For instance, suppose that x_1 and x_2 are closely positively related, that the combined regression is large, but the partial regressions are insignificant, that on x_1 being negative. Incidentally the attitude to assumptions such as that about the sign of a regression coefficient needs comment. That taken here is that any such assumption should, so far as possible, be tested on the data and, if consistent with the data, its consequences should be analysed and compared with the conclusions without the assumption. It might be argued from a Bayesian viewpoint that a prior probability should be attached to the assumption and a single conclusion obtained, but, even apart from the difficulty of doing this quantitatively in a meaningful way, it seems likely that the more cautious approach will be more informative.

(b) There may be sets of regressor variables which are to a large extent physically equivalent. For example, in a textile experiment yarn strength can be measured by several different methods. Quite often the measurements may be expected to be highly correlated and equivalent as regressor variables, although the data may show this expectation to be false. In applications like this it will be natural to try to use throughout one regressor variable, possibly a simple combination of the separate variables, provided that this does not give an appreciably worse fit than full fitting.

(c) Kendall (1957, p. 75) suggested applying principal component analysis to the regressor variables and then taking new regressor variables specified by the first few principal components. Jeffers (1967) and Spurrell (1963) have given interesting applications. A difficulty seems to be that there is no logical reason why the dependent variable should not be closely tied to the least important principal component. The following modification is worth considering. The principal components may suggest simple combinations of regressor variables with physical meaning. These simple combinations, not the principal components, can be used as regressor variables and if a good fit is obtained a constructive, although not necessarily unique, simplification has emerged. If the regressor variables can be divided into meaningful sets, e.g. into physical measurements and chemical measurements, separate principal component analyses could be considered for the two sets.

(d) In some situations, especially in the physical sciences, the method of dimensional analysis may lead to a reduction in the effective number of regressor variables.

(e) Another general way of clarifying the regression relation when some of the regressor variables are random variables is to examine plausible special models for the interrelationships between *all* the variables. There are two rather different cases. If the additional assumptions cannot be tested from the data, then parameters not previously estimable may become so, and those previously estimable may have the precision of estimation increased. On the other hand, if the additional assumptions can be tested, then the gain is confined to improved precision. Sewall Wright's method of path coefficients is essentially a device for handling complex systems of interrelations. For general discussion of path coefficients not specifically in genetic terms, see Tukey (1954), Turner and Stevens (1959), Turner *et al.* (1961) and, particularly for the connection with multiple regression, Kempthorne (1957, Chapter 14). The most familiar example of the second type of situation is the use of a concomitant variable to increase the precision of treatment contrasts in controlled experiments. When the concomitant variable is measured before the treatments are applied, the special model is justified by the randomization of treatments. Another simple example is the use of an intermediate variable (Cox, 1960). Here the regression of Y on X_1 is of interest and the supposition is that X_2 is a further variable such that, given $X_2 = x_2$, Y is independent of X_1. Then, under some circumstances, observation of X_2 can lead to appreciable increase in the precision of the estimated regression of Y on X_1. In other applications, analysis of covariance is used to see whether the data are in accord with the hypothesis that Y is affected by X_1 only via X_2.

(f) A very special case is when the regressor variables can be arranged in order of priority. The main cases are the fitting of polynomials and Fourier series.

(iii) *Analysis of a set of fitted regressions.* For problems in which many alternative equations, for example all 2^p linear regressions, are fitted to the same data, the handling of the resulting information needs comment. In a prediction problem in which the predictions are to be made over a set of x values distributed in much the same way as the data, an average variance of prediction, or better the corresponding standard deviation, will often be a reasonable measure of adequacy; of course, in some applications there may be particular points in x-space at which prediction is required. Those equations significantly worse than the overall fit can be identified in some way. Note that an equation significantly in conflict with the data may be used, for example because it involves substantial economy in the number of variables to be measured. This would be reasonable if the standard deviation of prediction is thought satisfactory, but the use of this equation in a new region of x-space is likely to be especially hazardous.

Where we are looking for a (hopefully) unique relation, the first step will often be to list all equations of a particular type that are not significantly contradicted by the data, as a preliminary to trying to narrow down the choice by some of the arguments sketched in (ii). Automatic devices for selecting equations would be used with great caution if at all. Gorman and Toman (1966) and Hocking and Leslie (1967) have discussed some further methods and in particular have outlined some of the recent unpublished work of Dr C. L. Mallows. A different approach is taken by Newton and Spurrell (1967a, b) who introduce quantities called elements to summarize the set of all 2^p regression sums of squares.

Particular caution is necessary in examining the effect of regressor variables which vary much less in the data than would be expected in future applications. The standard errors of the regression coefficients will be high and there is an obvious danger in judging the potential importance of such variables solely from the statistical significance of their regression coefficients.

4. BIVARIATE POPULATIONS

Consider now situations in which the observations are pairs $(X_1, Y_1), ..., (X_n, Y_n)$ drawn from a bivariate population and in which there is no particular reason for studying the dependence of Y on X rather than that of X on Y. The example concerning heights and weights of schoolchildren discussed by Ehrenberg (1968) is an instance. Either or both regressions could legitimately be considered, but the question is whether it is fruitful to do so.

With one homogeneous set of data the concise description of the joint distribution is all that can be attempted, in the absence of a more specific objective. This may be done by a frequency table or by an estimate of the joint cumulative distribution function, or some parametric bivariate distribution can be fitted. While there has been discussion of special families of bivariate distributions other than the bivariate normal (Plackett, 1965; Moran, 1967) the bivariate normal distribution is nevertheless the one most likely to arise. Preliminary transformation may be desirable and one possibility is to consider transformations from (x, y) to

$$\left(\frac{x^{\lambda_1} - 1}{\lambda_1}, \quad \frac{y^{\lambda_2} - 1}{\lambda_2} \right)$$

and to estimate (λ_1, λ_2) by maximum likelihood (Box and Cox, 1964), assuming that on the transformed scale a bivariate normal distribution does apply. In some applications it may be reasonable to take $\lambda_1 = \lambda_2$.

If a bivariate normal distribution is fitted, estimates of five parameters are required and these might, for example, be the means (μ_x, μ_y), the variances (σ_x^2, σ_y^2) and the correlation coefficient ρ; see, however, Section 2, point (iv) for remarks on parametrization.

When there are k populations the problem will be to describe the set of populations in a concise way. There are many possibilities. Often separate descriptions will be attempted of (a) the means (μ_{xi}, μ_{yi}) $(i = 1, ..., k)$ and of (b) the parameters determining the covariance matrices. For (a) such questions will arise as whether the means lie on or around a line or curve and of whether their position can be linked with some other variable characterizing the populations. Ehrenberg's (1963) criticisms of regression applied to bivariate populations are partly directed at confusions of comparisons between populations with those within populations.

If the covariance matrices are not constant, it will be natural to look for aspects that are constant and these might include one or other regression coefficient, the ratio of the standard deviations, the correlation coefficient, etc. Any changes in covariance matrix may be linked with changes in mean.

Of course once a potentially reasonable representation is obtained, standard techniques, especially maximum likelihood, are available for fitting and for constructing significance tests. In many cases, however, the most challenging problem will be to discover the most fruitful concise representation among the many possibilities.

All the remarks of this section apply in principle to p variate problems.

5. MODELS WITH COMPONENTS OF VARIATION

In the mathematical theory of regression the most awkward problems are probably those in which the observations are split into components not directly observable, and the relationships between these components are to be explored. There is a very extensive theoretical literature on such situations; see, in particular, Lindley (1947), Madansky (1959), Tukey (1951), Sprent (1966), Fisk (1967), Kendall and Stuart (1967) and Nelder (1968).

In this section a few comments on such systems will be made, particularly on points which connect with the previous discussion.

The simplest situation is where only the dependent variable is split into components, a hypothetical true value and a measurement or sampling error. The main question, easily answered, is then to assess how much of the observed dispersion of Y about its regression on x is accounted for by the measurement or sampling error. For example Y might be the square root of a Poisson distributed variable, when the sampling error has variance nearly $1/4$. One would, in particular, want to know whether this accounted for all the random variation present. More difficult problems would arise if it were required to estimate the distributional form of the "hidden" component of random variation.

The more interesting cases are where both independent and regressor variables can be split into components:

$$X_i = \Phi_i + \xi_i, \quad Y_i = \Psi_i + \eta_i, \quad \Psi_i = \alpha + \beta \Phi_i + \epsilon_{\Psi\Phi,i}.$$

Here ξ_i, η_i are measurement or sampling errors of zero mean and $\epsilon_{\Psi\Phi,i}$ is a deviation from the regression line, again of zero mean. The simplest case is where Φ_i, the "true" value of the regressor variable, is a random variable. Random variables for different i are assumed independent and the triple $(\xi_i, \eta_i, \epsilon_{\Psi\Phi,i})$ is assumed independent of Φ_i. Various cases may arise for the covariance matrix of the triple, the simplest being that the three components are mutually independent. Fisk (1967) and Nelder (1968) have considered models in which the regression coefficient is a random variable.

Sometimes it is convenient to write $\beta_{\Psi\Phi}$ instead of β to distinguish it from β_{YX}, the population least squares regression of Y on X. In fact

$$\beta_{\Psi\Phi} = \beta_{YX} \frac{\mathrm{var}\,(X)}{\mathrm{var}\,(X) - \mathrm{var}\,(\xi)}.$$

If prediction of Y directly from X is the objective, β_{YX} is required, not $\beta_{\Psi\Phi}$, so long as X is a random variable; if, however, the future X's at which prediction is to be attempted are not random, or come from a different distribution, the presence of the components ξ does need consideration.

Much published discussion concentrates on the estimation of β and in particular on the circumstances under which β is consistently estimable; for some purposes it is enough to note that β is between β_{YX} and $1/\beta_{XY}$ (Moran, 1956). The simplest case is when $\mathrm{var}\,(\xi)$ can be estimated from separate data, for instance from within replicate variation, or theoretically. Quite often the correction factor $\beta_{\Psi\Phi}/\beta_{YX}$ is very near one.

Some further problems arise naturally and some, but not all, can be answered in a fairly direct way. In most cases separate estimates of at least part of the covariance matrix of (ξ, η) are required. If more than the minimum amount of information is available more searching test of the model is possible. The following illustrate the further problems:

(i) Estimate the three components of variance of Y, namely $\beta^2 \mathrm{var}\,(\Phi)$, $\mathrm{var}\,(\epsilon_{\Psi\Phi})$ and $\mathrm{var}\,(\eta)$.

(ii) In particular, are the data consistent with all the ϵ's being zero?

(iii) Is a discrepancy between the estimated regression coefficient of Y on X and a theoretical value explicable in terms of "errors" in the regressor variable?

(iv) Are apparent differences between groups in the regression coefficients of Y on X explicable in terms of "errors" in the regressor variable?

(v) In the context of Section 4, (X_i, Y_i) may refer to the sample means of the ith group, (Φ_i, Ψ_i) being the corresponding population means. The covariance matrix of (ξ_i, η_i) can be estimated: what can be said about the relation between Ψ_i and Φ_i?

(vi) How much more effectively could Y be predicted from X if X were measured more precisely, for example by additional replication?

(vii) Is non-linearity in the regression of Y on X explicable by errors in the regressor variable? (In non-normal cases, if Ψ has linear regression on Φ, Y will not have linear regression on X.)

When there is more than one regressor variable similar problems arise. The important techniques based on instrumental variables will not be considered here; see, however, Section 3, point (ii).

6. Miscellaneous Points

This final section deals with a number of miscellaneous topics not discussed earlier.

(i) *Graphical methods.* These are very important both for the direct plotting of scatter diagrams of pairs of variables, possibly distinguishing other variables by a coarse grouping, but also for the systematic plotting of residuals; see, for example, Anscombe (1961). Particularly with extensive data, the systematic plotting of residuals is likely to be the most searching way of testing and improving models. It is possible that developments in computer display devices will lead to valuable ways of inspecting relationships involving more than two variables.

(ii) *Outliers and robust estimation.* The screening of data for suspect observations will often be required. With limited data it will be usual to look at suspect values individually in order to decide whether to include them in any subsequent analysis; often analyses with and without suspect values will be needed. With p observations for each individual the best way of looking for outliers will depend on the type of effect expected. Thus

(a) if any extreme deviation is thought to be in a particular known variable, usually the dependent variable, residuals from its regression on the other variables should be examined. For further discussion, see Mickey *et al.* (1967);

(b) suppose that any extreme deviation is thought to be confined to one variable, but not necessarily the same variable for different individuals. This might be the case, for example, with occasional gross recording errors. One procedure is then to calculate p residuals for each individual, one for each variable regressed on all the others;

(c) if any individual may be subject to extreme deviations in one or more variables simultaneously, and the joint distribution is approximately p variate normal, it may be reasonable to calculate for the ith individual, with vector observation $\mathbf{Y}_i$, a standardized squared distance from the mean $\bar{Y}$, given by $D_i = (\mathbf{Y}_i - \bar{Y})' \mathbf{S}^{-1}(\mathbf{Y}_i - \bar{Y})$, where $\mathbf{S}$ is the estimated covariance matrix. Then the ordered D_i's can be plotted against the expected order statistics for samples from the chi-squared distribution with p degrees of freedom. Iteration of the procedure may be desirable. Wilk and Gnanadesikan (1964) have given a general discussion of graphical methods for multiresponse experiments.

With extensive data, however, it may be necessary to use methods of analysis that are insensitive to outliers, so-called methods of robust estimation; see, for example, Huber (1964).

(iii) *Missing values.* Afifi and Elashoff (1966, 1967) have reviewed the literature on missing values in multivariate data and have considered in some detail point estimation in simple linear regression. Univariate missing value theory concentrates on the computational aspects of exploiting the near-balance of a balanced design spoiled by a missing observation, but no information is contributed by the missing observations. In a multivariate case, however, information may be contributed by individuals for which some component observations are missing. In a multiple regression problem, there is usually no information from individuals in which a regressor variable is missing, unless that variable can be regarded as random. An exception is when there is, say, an individual with x_1 missing and analysis of the other individuals suggests the omission of x_1 from the regression equation. Suppose, however, that a regressor variable is random, and the individuals with that variable missing can be regarded as selected randomly, a quite severe assumption, which should be tested where possible. Then more can be done. In some applications nearly all individuals may have at least one missing component and then use of some missing value theory is essential. Roughly speaking, the covariance between any two random variables can be estimated from those individuals on which both variables are available; there seems scope for further work to settle just when it is wise to do this and when something more elaborate such as full maximum likelihood estimation is desirable.

(iv) *Non-normal variation.* The present paper is largely concerned with problems to which least squares methods are reasonably applicable, possibly after transformation. In regression-like problems in which particular non-normal distributions can be specified, we have usually to apply maximum likelihood methods. These are locally equivalent to least squares techniques and therefore a great deal of the above discussion, for example that on the choice of regressor variables, is immediately relevant. Anscombe (1967) considered in some detail the analysis of a linear model with non-normal distribution of error; Cox and Hinkley (1968) found the asymptotic efficiency of least squares estimates in such situations.

The justification of maximum likelihood methods is asymptotic but sometimes analogues of at least a few of the "exact" properties of normal-theory linear models can be obtained. The simplest case is when the ith observation on the dependent variable has a distribution in the exponential family (Lehmann, 1959, p. 50)

$$\exp\{A_i(y)\,B(\theta_i) + C_i(y) + D(\theta_i)\},$$

where θ_i is a single parameter and there is a linear model

$$B(\theta_i) = \sum x_{ir}\,\beta_r,$$

where the β's are unknown parameters and the x's known constants. Special cases are the binomial, Poisson and gamma distributions when the "linear" model applies to the logit transform, to the log of the Poisson mean and to the reciprocal of the mean of the gamma distribution. Sufficient statistics are obtained and in very fortunate cases useful "exact" significance tests for single regression coefficients emerge.

(v) *Experimental and observational data.* Many of the issues discussed in the paper apply less acutely to the analysis of controlled experiments than to the analysis of observational data and that is why the paper may seem overweighted towards the latter type of problem. In fact, in terms of the discussion in this paper, there are three rather different reasons why fewer difficulties arise in the analysis of experimental data, quite apart from the smaller random error to which such data are likely to be subject. These reasons are:

(1) the spacing of regressor variables is likely to be more suitable;

(2) substantial non-orthogonalities of estimation will be avoided;

(3) factors omitted from the treatments will be randomized and hence the worst difficulties associated with omitted variables (Section 2, point (ii)) will be avoided.

ACKNOWLEDGEMENT

I am grateful to Mrs E. J. Snell and to the referees for constructive comments.

REFERENCES

AFIFI, A. A. and ELASHOFF, R. M. (1966). Missing observations in multivariate statistics. I. Review of the literature. *J. Am. Statist. Ass.*, **61**, 595–604.
—— (1967). Missing observations in multivariate statistics. II. Point estimation in simple linear regression. *J. Am. Statist. Ass.*, **62**, 10–29.
ANSCOMBE, F. J. (1961). Examination of residuals. *Proc. 4th Berkeley Symp.*, **1**, 1–36.
—— (1967). Topics in the investigation of linear relations fitted by the method of least squares. *J. R. Statist. Soc.* B, **29**, 1–52.

BEALE, E. M. L., KENDALL, M. G., and MANN, D. W. (1967). The discarding of variables in multivariate analysis. *Biometrika*, **54**, 357–366.

Box, G. E. P. (1966). Use and abuse of regression. *Technometrics*, **8**, 625–630.

Box, G. E. P. and Cox, D. R. (1964). An analysis of transformations. *J. R. Statist. Soc.* B, **26**, 211–252.

Box, G. E. P. and TIDWELL, P. W. (1962). Transformation of the independent variables. *Technometrics*, **4**, 531–550.

Cox, D. R. (1960). Regression analysis when there is prior information about supplementary variables. *J. R. Statist. Soc.* B, **22**, 172–176.

—— (1961). Tests of separate families of hypotheses. *Proc. 4th Berkeley Symp.*, **1**, 105–123.

—— (1962). Further results on tests of separate families of hypotheses. *J. R. Statist. Soc.* B, **24**, 406–424.

Cox, D. R. and HINKLEY, D. V. (1968). A note on the efficiency of least squares estimates. *J. R. Statist. Soc.* B, **30**, 284–289.

DRAPER, N. R. and SMITH, H. (1966). *Applied Regression Analysis*. New York: Wiley.

EHRENBERG, A. S. C. (1963). Bivariate regression is useless. *Appl. Statistics*, **12**, 161–179.

—— (1968). The elements of law-like relationships. *J. R. Statist. Soc.* A, **131**, 280–302.

FISHER, R. A. (1956). *Statistical Methods and Scientific Inference*. Edinburgh: Oliver and Boyd.

FINR, B. (1967). Models of the second kind in regression analysis. *J. R. Statist. Soc.* B, **29**, 266 281.

GORMAN, J. W. and TOMAN, R. J. (1966). Selection of variables for fitting equations to data. *Technometrics*, **8**, 27–51.

HOCKING, R. R. and LESLIE, R. N. (1967). Selection of the best subset in regression analysis. *Technometrics*, **9**, 531–540.

HUBER, P. J. (1964). Robust estimation of location. *Ann. Math. Statist.*, **35**, 73–101.

JEFFERS, J. N. R. (1967). Two case studies in the application of principal component analysis. *Applied Statistics*, **16**, 225–236.

KEMPTHORNE, O. (1957). *An Introduction to Genetic Statistics*. New York: Wiley.

KENDALL, M. G. (1957). *A Course in Multivariate Analysis*. London: Griffin.

KENDALL, M. G. and STUART, A. (1967). *Advanced Theory of Statistics* (2nd ed.), Vol. 2. London: Griffin.

LEHMANN, E. L. (1959). *Testing Statistical Hypotheses*. New York: Wiley.

LINDLEY, D. V. (1947). Regression lines and linear functional relationships. *J. R. Statist. Soc.* B, **9**, 218–244.

—— (1968). The choice of variables in multiple regression. *J. R. Statist. Soc.* B, **30**, 31–66.

MADANSKY, A. (1959). The fitting of straight lines when both variables are subject to error. *J. Am. Statist. Ass.*, **54**, 173–205.

MICKEY, M. R., DUNN, O. J. and CLARK, V. (1967). Note on the use of stepwise regression in detecting outliers. *Comp. and Biomed. Res.*, **1**, 105–111.

MORAN, P. A. P. (1956). A test of significance for an unidentified relation. *J. R. Statist. Soc.* B, **18**, 61–64.

—— (1967). Testing for correlation between non-negative variates. *Biometrika*, **54**, 385–394.

NELDER, J. A. (1968). Regression, model-building and invariance. *J. R. Statist. Soc.* A, **131**, 303–315.

NEWTON, R. G. and SPURRELL, D. J. (1967a). A development of multiple regression for the analysis of routine data. *Applied Statistics*, **16**, 51–64.

—— (1967b). Examples of the use of elements for clarifying regression analysis. *Applied Statistics*, **16**, 165–172.

PLACKETT, R. L. (1960). *Regression Analysis*. Oxford: Clarendon Press.

—— (1965). A class of bivariate distributions. *J. Am. Statist. Ass.*, **60**, 516–522.

RAO, C. R. (1965). *Linear Statistical Inference and its Applications*. New York: Wiley.

SPRENT, P. (1966). A generalized least-squares approach to linear functional relationships. *J. R. Statist. Soc.* B, **28**, 278–297.

SPURRELL, D. J. (1963). Some metallurgical applications of principal components. *Applied Statistics*, **12**, 180–188.

TUKEY, J. W. (1951). Components in regression. *Biometrics*, **7**, 33–69.

—— (1954). Causation regression and path analysis. In *Statistics and Mathematics in Biology* (ed. O. Kempthorne). Iowa: Ames.

TURNER, M. E., MONROE, R. J. and LUCAS, H. L. (1961). Generalized asymptotic regression and non-linear path analysis. *Biometrics*, **17**, 120–143.

TURNER, M. E. and STEVENS, C. D. (1959). The regression analysis of causal paths. *Biometrics*, **15**, 236–258.
WILK, M. B. and GNANADESIKAN, R. (1964). Graphical methods for internal comparisons in multiresponse experiments. *Ann. Math. Statist.*, **35**, 613–631.
WILLIAMS, E. J. (1959). *Regression Analysis*. New York: Wiley.
YATES, F. (1939). Tests of significance of the differences between regression coefficients derived from two sets of correlated variates. *Proc. R. Soc. Edinb.*, **59**, 184–194.

Appl. Statist.,
(1974), **23**, *No.* 1, p. 51

The Choice of Variables in Observational Studies†

By D. R. Cox and E. J. Snell

Imperial College, London

SUMMARY

A review is given of considerations affecting the choice of explanatory variables in observational studies. Aspects of both design and analysis are considered. In particular the choice of explanatory variables in multiple regression is discussed and some recommendations made.

Keywords: MULTIPLE REGRESSION; ANALYTICAL SURVEYS; MEDICAL APPLICATION; SELECTION OF VARIABLES; DESIGN OF INVESTIGATIONS; OBSERVATIONAL STUDIES

1. INTRODUCTION

THIS paper reviews some general aspects of the choice of variables in observational studies. To keep the paper concise only outline examples have been included and to be specific these are medical, although the ideas apply widely.

Observational studies, where they are not purely descriptive, have as their objective the explanation or prediction of some response in terms of explanatory or predictor variables. It is useful to have two examples in mind.

Example 1. Consider an investigation into the incidence of a respiratory disease among a certain group of workers. The response variable may be severity of the disease, with possible explanatory variables being the worker's age, physical status, working conditions, previous employment, etc. Some variables may be more important than others in explaining the severity of the disease.

Example 2. A different situation is one of trying to predict the time to death among patients known to be suffering from a progressive and fatal disease. Possible predictive variables are type of treatment, treatment variables such as dose, clinical and biochemical measurements made on diagnosis, etc.

Although careful discussion of the most appropriate way to measure response is always important, and often several different measures will be called for, nevertheless what response variables to consider is frequently fairly clearcut. Thus in Example 1, severity may be assessed radiologically and graded according to standard levels. In Example 2, time to death is likely to be measured from time of diagnosis. In this paper we concentrate on the explanatory variables; how many such variables should be measured and, if many are observed, how should the analysis be handled to find the most relevant ones?

These are difficult issues. Many of the following points are rather trite when put in general terms and do not lend themselves very well to quantitative discussion. On the other hand, the decision as to what to do in any particular investigation can be hard.

† This paper is based on one prepared for the Division of Research in Epidemiology and Communications Science, World Health Organization.

2. Selection of Explanatory Variables for Measurement

The following general aspects of a study will influence the nature and number of explanatory variables measured:

(a) Whether the study is intended to investigate some rather specific hypothesis about the phenomenon or whether it is designed to screen out the most important variables from rather a large number of possibly relevant variables, the important variables to be examined in detail subsequently, possibly by experiments rather than by observational studies. In the former case it is important to try to anticipate the main explanations competing with the hypothesis under test and to measure relevant variables.

(b) Whether the response variables are observed quite quickly, so that the later parts of the study can be modified, if necessary, in the light of the earlier results. If this is not possible, it is more likely to be necessary to measure many variables on each individual.

(c) Questions of economy of time, ease of setting up instruments, difficulty of contacting individuals, loss of accuracy arising from increased work-load, availability of "good" official statistics, etc. will often be crucial in deciding how many variables can be measured.

(d) Variables may be included primarily to establish comparisons with previous related studies.

In many studies binary explanatory variables will be adequate in analysis, except for the most important variables, provided that the split between the two categories is appropriately made. On the other hand, poorly defined binary variables may be virtually useless and for this reason it will often be essential to record on a more than two-point scale. Usually it will be sensible to arrange that binary explanatory variables are constructed to have roughly a 50–50 split, in order for the effect on response to be as clear-cut as possible but if the response also is binary and its effect appreciable or if very non-linear effects are involved the position is more complicated (Cox, 1969).

Multiplicity in explanatory variables can arise in two rather different ways. We may measure a number of quite different properties, or we may have a number of ways of measuring what is essentially one property. For example, occurrence of particular symptoms may be elicited by a single question or probed by a battery of related questions.

Many possibilities exist for more sophisticated design, especially where the total number of explanatory variables is large and accuracy of measurement is likely to drop if all variables are measured on all subjects. One possibility is to measure only a subset of variables on each subject, the variables being chosen in a suitably balanced way. Another possibility, where some of the variables are arranged in batteries as indicated above, is not to measure each full battery on each subject, but to measure detailed variables only on subsets of individuals. Used with care these ideas should be fruitful, especially in very large-scale investigations. Of course simplicity in design remains a vitally important requirement.

3. Broad Problems of Analysis

Two main kinds of response variable commonly encountered are binary (e.g. occurrence, non-occurrence) and measurements that, possibly after transformation, lead to approximately normally distributed data. An important point is that while the precise techniques of analysis will be different for different types of response variable the broad strategy to be adopted and the difficulties of interpretation likely to be encountered are the same. We shall concentrate in Section 4 largely on the techniques for normal theory multiple regression, simply because this is the most thoroughly investigated case.

The following general points have to be considered:

(a) There is a working distinction between producing (i) a fit to the data useful for future prediction in the absence of major changes in the system and (ii) an "explanation" which will link with other studies, e.g. fundamental laboratory work, and will predict under quite different circumstances. For (i), two quite different models, involving different explanatory variables, are equally acceptable if they fit the data equally well. If a choice has to be made between them, it may be done on the basis of simplicity, e.g. in terms of the number of explanatory variables necessary or the ease with which the relevant variables can be measured; for a quantitative decision theoretic analysis, see Lindley (1968). In (ii), however, it is usually of central importance to find which explanatory variables have important effects.

(b) Even in the first case of prediction in the narrow sense it will not normally be wise to include all predictor variables. This is both for reasons of simplicity and because typically the mean square error of prediction will be raised by including too many variables.

(c) The main difficulties in dealing with observational studies stem from two rather different sources, the omission of relevant variables from those measured and the presence of fairly high dependencies among the explanatory variables.

As a simple illustration of the first situation consider the relation between time to death y and the level of some prescribed "dose" x. If the dose level is determined by the severity of disease, the dependence of y upon x as given by a simple linear regression cannot be interpreted as predicting the change in y for a particular individual given a change in x. Only if the omitted variable, severity of disease, is included as a further explanatory variable is a "causal" interpretation at all feasible. The second situation would, for example, arise in measuring the percentage of substances $A, B, \ldots$ in a compound, where there will be an exact linear relationship between the percentages in a compound. Although this is an extreme example, close dependencies are often unavoidable if a large number of explanatory variables are measured. Interpretation is difficult because many apparently different models may fit the data almost equally well. A principal component analysis of the explanatory variables may be tried in these circumstances (Jeffers, 1967).

In designed experiments, balance and randomization largely overcome these difficulties; the omission and hence randomization of an important variable leads to an increased error variance and to a seriously incomplete understanding, but not to a "biased" conclusion. It may sometimes be possible to take additional observations at values of the explanatory variables chosen so as to reduce non-orthogonalities in the data (Dykstra, 1966; Gaylor and Merrill, 1968; Silvey, 1969). In observational studies in which the objective is the comparison of, say, a treatment with a control, matching individuals to remove bias is likely to be useful (Cochran, 1965, 1972).

Of course in a very large-scale study the amount of data collected may be so great that quite apart from the difficulties of principle alluded to above there may be limitations imposed on computational grounds, or because of human limitations in what can be absorbed.

In deciding what variables to include it is important to take account of additional information, for example as to which variables are likely to be alternatives to one another and which it is almost certain to be necessary to include. It is equally important that any such "prior" knowledge inserted into the analysis should be tested for consistency with the data.

There are two approaches to the inclusion of general classification variables like age, sex, etc. One is to make separate analyses for men and women, combining the conclusions at a later stage if they seem compatible. The other is to fit a composite model in which say the sex difference is represented by a single parameter; this assumes that in some sense there is no interaction between the main explanatory variables and sex, an assumption that can be tested at least informally. With large sets of data it will, however, frequently be sensible to analyse in a series of sections merging the analyses in a second stage. In any case the examination of the consistency of conclusions from independent sets of data is an important and simple technique for assessing precision.

Note that there is a distinction implied here between genuine explanatory variables and classification variables that serve merely to define major subclasses of individuals. This is relevant when we are looking for proper "causal" relations. That is, it is not an "explanation" to say that a death rate for men is greater than that for women.

4. MORE DETAILED PROBLEMS OF ANALYSIS

We now consider in more detail situations where for each individual there is a continuous response variable y and a number of explanatory variables $x_1, ..., x_p$. Suppose that we work provisionally with the assumption that the expected value of y is a linear function of the explanatory variables. It is assumed that any preliminary transformation of the response variable, e.g. from response time to its logarithm has been made, also that the data have been edited to remove gross errors and to isolate suspect values.

We shall not describe the large body of statistical methods and theory associated with the linear model; for an introductory account, see Draper and Smith (1966) and for general comments Cox (1968).

Formal significance tests are a useful guide to the importance of different explanatory variables, but have not to be followed too rigidly. One reason for caution is that the tabulated significance levels of the F distribution refer to a single test carried out in isolation; in practice we are nearly always concerned with a chain of related tests and this makes the interpretation of the ordinary significance levels indirect (Draper et al., 1971; Pope and Webster, 1972; Spjøtvoll, 1972).

The difficulties of interpretation caused by non-orthogonality are less important if interest is purely in prediction over the range of explanatory variables covered by the data. Although several rather different looking equations will often have similar residual mean squares, it may be unimportant which equation is used. An equation with few explanatory variables will, however, give biased estimates if the omitted variables are at all relevant. The extent of bias, averaged over the observed distribution of the explanatory variables, in an equation with k variables is indicated by the statistic suggested by C. L. Mallows (Gorman and Toman, 1966)

$$C_k = (\text{residual sum of squares})/\hat{\sigma}^2 - (n - 2k),$$

where n denotes sample size and $\hat{\sigma}^2$ is a separate estimate of σ^2; in the absence of bias, $E(C_k) \simeq k$. Given several equations with similar residual mean squares, one with small bias is likely to be preferred. This does not necessarily mean one with many explanatory variables; increasing the number of variables may reduce the bias but at the expense of increasing the total error of prediction. If predicting outside the observed region of the explanatory variables, different equations will give vastly different predictions. Methods for selecting single well-fitting equations from a large set will be reviewed at the end of this section.

In most applications, however, the particular variables affecting response and the directions of their effects are of intrinsic interest and then the selection of just one well-fitting equation from among many is unsatisfactory and possibly very misleading.

In principle, the following procedure seems a sensibly cautious approach in such situations. All possible 2^p equations are fitted (Garside, 1965; Schatzoff *et al.*, 1968; Morgan and Tatar, 1972) and those clearly inconsistent with the data rejected; that is equations with a residual mean square significantly greater than the mean square residual from the full model are rejected. Typically if an equation involving a subset $\mathscr{S}$ of explanatory variables is consistent with the data, so is that based on a larger subset $\mathscr{S}'$, $\mathscr{S}' \supset \mathscr{S}$. (Any exceptions to this will be minor ones depending on the particular levels of significance used.) Such a subset we call primitive. A program to find the primitive models and associated information has been written at Imperial College by Mrs M. Ansell and is available for use on the CDC 6400 computer. If there is only one primitive model, the situation is fairly clear-cut; where there is more than one, a choice between them can be made only on the basis of additional information.

Unfortunately this procedure, even with sophisticated numerical analytic and programming techniques (Wampler, 1970; Mullet and Murray, 1971), does not seem feasible for more than 10–15 explanatory variables. If more explanatory variables are available, as will often be the case for example in large epidemiological studies, it follows that some reduction will be essential before the above method can be used. The use of several alternative reductions will usually be desirable.

The main methods for such reduction are as follows.

(a) We may examine sets of explanatory variables. If the data contain batteries of questions, some form of total score may be adequate. This can be tested for consistency with the data.

(b) Some variables may be specified for definite inclusion, for example if interest lies primarily in the supplementary effect of other variables.

(c) Classification variables (such as sex, age, etc.) may be used to split the data into sections for separate analysis.

(d) It may be thought on general grounds that a regression coefficient or regression coefficients, associated with a particular variable, even one not of primary importance, should be of a certain sign, e.g. should not be negative. Occasionally this may be helpful in clarifying the relationship. In any fitted model for which the regression coefficient is of the wrong sign, but not significantly different from zero, the coefficient is replaced by zero, i.e. the variable is in effect omitted. An estimate significantly different from zero and of the wrong sign implies that the prior assumption is wrong, or that the wrong form of relation is being fitted, or that an important variable has been omitted, or that by chance an extreme fluctuation has occurred.

(e) When special relationships can be postulated among the explanatory variables the methods of path analysis can be used; see, for example, Turner and Stevens (1959). These methods, originally developed in connection with genetics, have more recently been examined by sociologists, for example, by Blalock and Blalock (1968). The general idea is partly that the postulation or discovery of a series of special relationships between the variables will clarify the whole problem and partly that such relationships will increase the precision of estimates and hence help to resolve ambiguities.

The above approaches involve injecting some further external information. The remaining devices are essentially general computational devices; see Draper and Smith (1969) for a review up to that date.

(f) The most commonly used procedures for progressive selection of variables are forward selection, backward elimination (Hamaker, 1962; Oosterhoff, 1963; Abt, 1967; Mantel, 1970), stepwise regression (Efroymson, 1960; Breaux, 1968; Goodman, 1971) and "optimum" regression. These will not be described in detail. "Optimum" regression finds that equation which for a specified number of explanatory variables has the minimum residual sum of squares. An algorithm by Beale *et al.* (1967) makes it unnecessary to evaluate all regressions, the procedure is claimed to be manageable provided the number of variables is not much in excess of 20 (Beale, 1970).

(g) Newton and Spurrell (1967a, b) have proposed a method called element analysis for assessing the information provided by all 2^p fits; $2^p - 1$ elements, to be used in conjunction with certain rules, are calculated from the sums of squares attributable to regression.

(h) A suggestion of Gorman and Toman (1966) is to calculate a fractional factorial of the 2^p possible regressions and to select variables by a subjective inspection of the values of the residual mean square or the statistic C_k for the computed regressions. Further evidence is needed on the efficiency of this procedure; an example is given in Daniel and Wood (1971). Hocking and Leslie (1967) and La Motte and Hocking (1970) consider a technique to minimize C_k for given k, calculating a subset of the regressions; see also Rothman (1968).

(i) A procedure based on estimates which differ from the usual least squares estimates is that of ridge regression (Hoerl and Kennard, 1970a, b; Marquardt, 1970; Lindley and Smith, 1972). The least squares equations are modified to give estimates which are stable and which, although biased, give smaller mean square error of prediction. This is helpful when the main emphasis is on prediction and especially appropriate when the regression coefficients (or a subset of them) are generated by a random mechanism. It is not clear how useful the method is in the isolation of important variables.

We consider that the procedures (f) and (h) should be used, if at all, only where the particular variables selected are not of intrinsic interest or as a preliminary device in the reduction of variables, so that the recommended techniques for up to 10 variables can be followed; several different reductions should then normally be examined.

The possibility of interactions between the effects of different explanatory variables has usually to be borne in mind. They can be detected in essentially three ways, by graphical analysis of residuals, by fitting an extended model usually with fairly simple forms of interaction represented by cross products of primary explanatory variables or by analysing the data in sections. Of course in a problem with many explanatory variables the number of possible interaction terms, even of the simplest kind, is large. Then attention will often have to be restricted to those interactions thought particularly likely on general grounds and to interactions among variables with large "ordinary" effects.

With binary response variables essentially the same problems arise.

5. SOME MORE COMPLEX PROBLEMS

The difficulties discussed in Section 4 arise in the context even of the simplest multiple regression model. There are, of course, many other sources of difficulty of analysis. In addition to those arising from different kinds of response, e.g. binary, some further problems associated with normal theory regression that need caution are as follows:

(a) Missing values among the explanatory variables are a common source of difficulty (Buck, 1960; Afifi and Elashoff, 1966, 1967, 1969; Dagenais, 1971; Hartley and Hocking, 1971; Orchard and Woodbury, 1972). Current unpublished work by E. M. L. Beale and R. J. A. Little at Imperial College supports the method of Orchard and Woodbury.

(b) There may be a need for models non-linear in the parameters or variables.

(c) Major problems can arise when the individuals are arranged in groups. For example, the regressions between and within groups are likely to be different, and the errors of different individuals are unlikely to be mutually independent. The groups may be characterized by random variables and involve models with components of variance.

(d) There may be appreciable rounding or measurement errors in the explanatory variables (Swindel and Bower, 1972).

6. SOME RECOMMENDATIONS

It is difficult to give specific recommendations because of the widely differing situations that can arise in application. Some of the main points can be summarized as follows:

In design

(a) The nature of the study, and considerations of accuracy and economy determine how many variables are sensible.

(b) Divide the variables into batteries, where relevant and consider the possibility of a special design to omit some measurements.

In analysis, given that multiple regression techniques are applied,

(c) The distinction between predicting future observations and interpreting the data can influence the choice of variables.

(d) If interpretation is the objective, and p is not greater than 10–15, compute all 2^p regressions and examine those consistent with the data. Larger values of p should in some way be reduced to make the computations feasible.

(e) Automatic selection procedures, such as are commonly used in many generally available computer programs, should be used only as a preliminary device or if the particular variables selected are not of intrinsic interest.

(f) Use of supplementary information and assumptions may be crucial in clarifying relationships. Any such assumptions should, however, be tested for consistency with the data and the conclusions with and without the supplementary information should normally be compared.

(g) The possibility of interactions between the effects of different explanatory variables should be considered.

REFERENCES

ABT, K. (1967). On the identification of the significant independent variables in linear models.
I, II. *Metrika*, **12**, 1–15, 81–96.

AFIFI, A. A. and ELASHOFF, R. M. (1966, 1967, 1969). Missing observations in multivariate
statistics. I–IV. *J. Am. Statist. Assoc.*, **61**, 595–604; **62**, 10–29; **64**, 337–358, 359–365.

BEALE, E. M. L. (1970). Note on procedures for variable selection in multiple regression.
Technometrics, **12**, 909–914.

BEALE, E. M. L., KENDALL, M. G. and MANN, D. W. (1967). The discarding of variables in
multivariate analysis. *Biometrika*, **54**, 356–366.

BLALOCK, H. M. and BLALOCK, A. (editors) (1968). *Methodology in Social Research*. New York:
McGraw Hill.

BREAUX, H. J. (1968). A modification of Efroymson's technique for stepwise regression analysis.
Comm. ACM, **11**, 556–557.

BUCK, S. F. (1960). A method of estimation of missing values in multivariate data suitable for
use with an electronic computer. *J. R. Statist. Soc. B*, **22**, 302–306.

COCHRAN, W. G. (1965). The planning of observational studies of human populations. *J. R.
Statist. Soc. A*, **128**, 234–265.

—— (1972). Observational studies. In *Statistical Papers in Honor of George W. Snedecor*
(T. A. Bancroft, ed.), pp. 77–90. Iowa: Iowa State Press.

COX, D. R. (1968). Notes on some aspects of regression analysis. *J. R. Statist. Soc. A*, **131**,
265–279.

—— (1969). *Analysis of Binary Data*. London: Methuen.

DAGENAIS, M. G. (1971). Further suggestions concerning the utilization of incomplete observations
in regression analysis. *J. Am. Statist. Assoc.*, **66**, 93–98.

DANIEL, C. and WOOD, F. S. (1971). *Fitting Equations to Data*. New York: Wiley–Interscience.

DRAPER, N. R., GUTTMAN, I. and KANEMASU, H. (1971). The distribution of certain regression
statistics. *Biometrika*, **58**, 295–298.

DRAPER, N. and SMITH, H. (1966). *Applied Regression Analysis*. New York: Wiley.

—— (1969). Methods for selecting variables from a given set of variables for regression analysis.
Bull. Inst. Int. Statist., **43**, 7–15.

DYKSTRA, O. (1966). The orthogonalization of undesigned experiments. *Technometrics*, **6**,
279–290.

EFROYMSON, M. A. (1960). Multiple regression analysis. In *Mathematical Methods for Digital
Computers* (A. Ralston and H. S. Wilf, eds), Chapter 17. New York: Wiley.

GARSIDE, M. J. (1965). The best subset in multiple regression analysis. *Appl. Statist.*, **14**, 196–200.

GAYLOR, D. W. and MERRILL, J. A. (1968). Augmenting existing data in multiple regression.
Technometrics, **10**, 73–81.

GOODMAN, L. A. (1971). The analysis of multidimensional contingency tables: stepwise procedures
and direct estimation methods for building models for multiple classification. *Technometrics*,
13, 33–61.

GORMAN, J. W. and TOMAN, R. J. (1966). Selection of variables for fitting equations to data.
Technometrics, **8**, 27–51.

HAMAKER, H. C. (1962). On multiple regression analysis. *Statist. Neerlandica*, **16**, 31–56.

HARTLEY, H. O. and HOCKING, R. R. (1971). The analysis of incomplete data. *Biometrics*, **27**,
783–823.

HOCKING, R. R. and LESLIE, R. W. (1967). Selection of the best subset in regression analysis.
Technometrics, **9**, 531–540.

HOERL, A. E. and KENNARD, R. W. (1970a). Ridge regression: biased estimation for nonorthogonal
problems. *Technometrics*, **12**, 55–67.

—— (1970b). Ridge regression: applications to nonorthogonal problems. *Technometrics*, **12**,
69–82.

JEFFERS, J. N. R. (1967). Two case studies in the application of principal component analysis.
Appl. Statist., **16**, 225–236.

LA MOTTE, L. R. and HOCKING, R. R. (1970). Computational efficiency in the selection of regression variables. *Technometrics*, **12**, 83–93.

LINDLEY, D. V. (1968). The choice of variables in multiple regression. *J. R. Statist. Soc. B*,
30, 31–66.

LINDLEY, D. V. and SMITH, A. F. M. (1972). Bayes estimates for the linear model (with Discussion).
J. R. Statist. Soc. B, **34**, 1–41.

MANTEL, N. (1970). Why stepdown procedures in variable selection. *Technometrics*, **12**, 621–625.

MARQUARDT, D. W. (1970). Generalized inverses, ridge regression, biased linear estimation and non-linear estimation. *Technometrics*, **12**, 591–612.

MORGAN, J. A. and TATAR, J. F. (1972). Calculation of the residual sum of squares for all possible regressions. *Technometrics*, **14**, 317–325.

MULLET, G. M. and MURRAY, T. W. (1971). A new method for examining rounding error in least-squares regression computer programs. *J. Am. Statist. Assoc.*, **66**, 496–498.

NEWTON, R. G. and SPURRELL, D. J. (1967a). A development of multiple regression for the analysis of routine data. *Appl. Statist.*, **16**, 51–64.

—— (1967b). Examples of the use of elements for clarifying regression analysis. *Appl. Statist.*, **16**, 165–172.

OOSTERHOFF, J. (1963). On the selection of independent variables in a regression equation. Report 319. Math. Centre, Amsterdam.

ORCHARD, T. and WOODBURY, M. A. (1972). A missing information principle, theory and applications. *Proc. 6th Berkeley Symp.*, **1**, 697–715.

POPE, P. T. and WEBSTER, J. T. (1972). The use of an *F*-statistic in stepwise regression procedures. *Technometrics*, **14**, 327–339.

ROTHMAN, D. (1968). Comment on Hocking and Leslie's paper. *Technometrics*, **10**, 432.

SCHATZOFF, M., TSAO, R. and FIENBERG, S. (1968). Efficient calculation of all possible regressions. *Technometrics*, **10**, 769–779.

SILVEY, S. D. (1969). On choosing additional values of explanatory variables to counter multi-collinearity. *Bull. Inst. Int. Statist.*, **43**, 177–178.

SPJØTVOLL, E. (1972). Multiple comparison of regression functions. *Ann. Math. Statist.*, **43**, 1076–1088.

SWINDEL, B. F. and BOWER, D. R. (1972). Rounding errors in the independent variables in a general linear model. *Technometrics*, **14**, 215–218.

TURNER, M. E. and STEVENS, C. D. (1959). The regression analysis of causal paths. *Biometrics*, **15**, 236–258.

WAMPLER, R. H. (1970). A report on the accuracy of some widely used least squares computer programs. *J. Am. Statist. Assoc.*, **65**, 549–565.

A General Definition of Residuals

By D. R. Cox and E. J. Snell

Imperial College

[Read at a Research Methods Meeting of the Society, March 13th, 1968,
Professor R. L. Plackett in the Chair]

Summary

Residuals are usually defined in connection with linear models. Here a more general definition is given and some asymptotic properties found. Some illustrative examples are discussed, including a regression problem involving exponentially distributed errors and some problems concerning Poisson and binomially distributed observations.

1. Introduction

Residuals are now widely used to assess the adequacy of linear models; see Anscombe (1961) for a systematic discussion of significance tests based on residuals, and for references to earlier work. A second and closely related application of residuals is in time-series analysis, for example in examining the fit of an autoregressive model.

In the context of normal-theory linear models, the $n \times 1$ vector of random variables $\mathbf{Y}$ is assumed to have the form

$$\mathbf{Y} = \mathbf{X}\boldsymbol{\beta} + \boldsymbol{\epsilon}, \tag{1}$$

where $\mathbf{X}$ is a known matrix, $\boldsymbol{\beta}$ a vector of unknown parameters and $\boldsymbol{\epsilon}$ an $n \times 1$ vector of unobserved random variables of zero mean, independently normally distributed with constant variance. If $\hat{\boldsymbol{\beta}}$ is the vector of least-squares estimates of $\boldsymbol{\beta}$, the residuals $\mathbf{R}^*$ are defined by

$$\mathbf{Y} = \mathbf{X}\hat{\boldsymbol{\beta}} + \mathbf{R}^*. \tag{2}$$

Provided that the number of parameters is small compared with n, most of the properties of $\mathbf{R}^*$ are nearly those of $\boldsymbol{\epsilon}$, i.e. $\mathbf{R}^*$ should have approximately the properties of a random sample from a normal distribution. In fact, $\mathbf{R}^*$ being linear in $\mathbf{Y}$, the random variable $\mathbf{R}^*$ has, under (1), a singular normal distribution and hence the properties of significance tests can be studied in some detail (Anscombe, 1961).

The main types of departure from the model (1) likely to be of importance are:
(i) the presence of outliers;
(ii) the relevance of a further factor, omitted from (1), detected by plotting the residuals against the levels of that factor;
(iii) non-linear regression on a factor already included in (1), detected by plotting the residuals against the levels of that factor and obtaining a curved relationship;
(iv) correlation between different ϵ_i's, for example between ϵ_i's adjacent in time, detected from scatter diagrams of suitable pairs of R_i^*'s, or possibly from a periodogram analysis of residuals;
(v) non-constancy of variance, detected by plotting residuals or squared residuals against factors thought to affect the variance, or against fitted values;

(vi) non-normality of the distribution of the ϵ_i's, detected by plotting the ordered residuals against the expected order statistics from a standard normal distribution (Pearson and Hartley, 1966, Table 28).

Corresponding to the graphical analyses suggested in points (i)–(vi), statistics can be constructed for formal tests of significance. The idea of inspecting residuals is very old, but the systematic calculation of residuals, particularly from extensive data, has become practicable only recently; their thorough graphical analysis as a routine is feasible only with a suitable computer graphical output device.

The examination of the adequacy of a model by such analyses may be contrasted with a more formal approach in which there is fitted:

either (a) a more general model containing one or more additional parameters and reducing to (1) for particular values of the new parameters;

or   (b) a different family of models, adequacy of fit being assessed, say, by the maximum log likelihood achieved.

One example of (a) is the family of models considered by Box and Cox (1964), in which model (1) is considered as applying to an unknown power of the original observations. The advantages of the more formal techniques are that they have sensitive significance tests associated with them and that they are directly constructive in the sense that, if the initial model does not fit, a specific better-fitting model is obtained immediately from the analysis. On the other hand, analysis of residuals, especially by graphical techniques, does not require committal in advance to a particular family of alternative models. It will indicate the nature of a departure from the initial model, but not explicitly how to extend or replace the model. With very extensive data, significance testing is relatively unimportant and the types of departure that can be detected are more numerous than can be captured in advance in a few simple parametric models. It is in such applications that the analysis of residuals is likely to be most fruitful.

2. A More General Definition

The main object of the present paper is to give a more general definition of residuals and to illustrate some of its properties and applications. Consider a model expressing an observed vector random variable $\mathbf{Y}$ in terms of a vector $\boldsymbol{\beta}$ of unknown parameters and a vector $\boldsymbol{\epsilon}$ of independent and identically distributed unobserved random variables. More particularly we assume that each observation Y_i depends on only one of the ϵ's, so that we can write

$$Y_i = g_i(\boldsymbol{\beta}, \epsilon_i) \quad (i = 1, ..., n). \tag{3}$$

This assumption excludes applications to time series and also to component of variance problems in which several random variables enter into each observation. Models involving discrete distributions, such as the binomial and Poisson, are not in the first place included, because, for example, Poisson-distributed observations with different means cannot be expressed in terms of transformations of identically distributed observations. Later, however, in Sections 7–9, we extend the methods to deal with Poisson and binomial distributions.

To define residuals for (3), let $\hat{\boldsymbol{\beta}}$ be the maximum likelihood estimate of $\boldsymbol{\beta}$ from $\mathbf{Y}$. It would be possible to work with other asymptotically efficient estimates, or even with inefficient estimates, but the details of Section 4 would be different.

Now suppose that the equation

$$Y_i = g_i(\hat{\beta}, R_i) \tag{4}$$

has a unique solution for R_i, namely

$$R_i = h_i(Y_i, \hat{\beta}). \tag{5}$$

Note that

$$\epsilon_i = h_i(Y_i, \beta). \tag{6}$$

We take (5) as defining the residual corresponding to Y_i and the model (3); later we shall introduce a minor modification of (5) and then call R_i the crude residual.

Example 1. If (3) is the normal-theory linear model (1) with known variance, the residuals (5) are the same as those, R_i^*, of Section 1, equation (2). If the variance is an additional unknown parameter, then

$$R_i = R_i^* \{\sum R_j^{*2}/n\}^{-\frac{1}{2}},$$

and so R_i is essentially equivalent to R_i^*.

Example 2. Feigl and Zelen (1965) discussed some leukemia data in which for the ith individual Y_i is the time to death in weeks and x_i is the log of the initial white blood cell count. Feigl and Zelen considered primarily linear regression of Y_i on x_i with exponential errors, but here we work with the model, mentioned briefly by Feigl and Zelen,

$$Y_i = \beta_1 \exp\{\beta_2(x_i - \bar{x})\} \epsilon_i, \tag{7}$$

where $\epsilon_1, ..., \epsilon_n$ are independently exponentially distributed with unit mean and $\bar{x} = \sum x_i/n$. The advantage of (7) over a linear regression model is that for all $\beta_1 > 0$ and all β_2, x_i, the random variable on the left-hand side of (7) is non-negative.

For this model, if $\hat{\beta}_1, \hat{\beta}_2$ are maximum likelihood estimates of β_1, β_2, then

$$Y_i = \hat{\beta}_1 \exp\{\hat{\beta}_2(x_i - \bar{x})\} R_i,$$

i.e.

$$R_i = [\hat{\beta}_1 \exp\{\hat{\beta}_2(x_i - \bar{x})\}]^{-1} Y_i. \tag{8}$$

Example 3. For some purposes it is convenient for analysing a random sample $Y_1, ..., Y_n$ from a Weibull distribution to write the model in the form

$$Y_i = (\beta_1 \epsilon_i)^{\beta_2}, \tag{9}$$

where again $\epsilon_1, ..., \epsilon_n$ have an exponential distribution of unit mean.

Some further examples are given in Section 10.

Often the number of parameters is small compared with the number of observations and the configuration is such that all relevant combinations of parameters are estimated with small standard error of order $n^{-\frac{1}{2}}$. Then a residual R_i will differ from ϵ_i by an amount of order $n^{-\frac{1}{2}}$ in probability and most statistical properties of the R's will differ little from those of the ϵ's.

We examine the properties of the R's more carefully in Section 4. This will be done by expanding $R_i - \epsilon_i$ in a Taylor series in terms of $\hat{\beta}_s - \beta_s$. We need some of the properties of maximum likelihood estimates, in particular an expression for their bias, and these are developed briefly in Section 3.

3. Some Properties of Maximum Likelihood Estimation

Bartlett (1952), incidentally to his study of large-sample confidence intervals, gave a simple expression for the bias to order n^{-1} of the maximum likelihood estimate from a single random sample, there being one unknown parameter. Haldane (1953) and Haldane and Smith (1956) further discussed asymptotic expansions for the properties of a maximum likelihood estimate dealing with random samples and one or two unknown parameters; for further discussion and extensions see Shenton and Wallington (1962) and Shenton and Bowman (1963).

With a single parameter and observations that are independent, but not necessarily identically distributed, the log likelihood is

$$L(\beta) = \sum \log p_j(Y_j, \beta),$$

where $p_j(Y_j, \beta)$ is the p.d.f. of Y_j. For a regular problem the maximum likelihood equation $L'(\beta) = 0$ is to first order

$$L'(\beta) + (\hat{\beta} - \beta) L''(\beta) = 0. \tag{10}$$

Write

$$U^{(j)} = \frac{\partial \log p_j(Y_j, \beta)}{\partial \beta}, \quad V^{(j)} = \frac{\partial^2 \log p_j(Y_j, \beta)}{\partial \beta^2} \tag{11}$$

and replace $-L''(\beta)$ by its expectation

$$I = \sum E(-V^{(j)}),$$

where I is the total information in the sample.

We thus have the standard first-order expressions

$$\hat{\beta} - \beta = \frac{U^{(.)}}{I}, \quad \mathrm{var}(\hat{\beta}) = \frac{1}{I}, \tag{12}$$

where $U^{(.)} = \sum U^{(j)}$, the dot indicating a sum over the sample.

To obtain a more refined answer, we replace (10) by the second-order equation

$$L'(\beta) + (\hat{\beta} - \beta) L''(\beta) + \tfrac{1}{2}(\hat{\beta} - \beta)^2 L'''(\beta) = 0. \tag{13}$$

Take expectations in (13), thereby obtaining

$$E(\hat{\beta} - \beta) E\{L''(\beta)\} + \mathrm{cov}\{\hat{\beta} - \beta, L''(\beta)\} + \tfrac{1}{2} E(\hat{\beta} - \beta)^2 E\{L'''(\beta)\}$$
$$+ \mathrm{cov}\{\tfrac{1}{2}(\hat{\beta} - \beta)^2, L'''(\beta)\} = 0. \tag{14}$$

Now, approximately, by (12)

$$\mathrm{cov}\{\hat{\beta} - \beta, L''(\beta)\} = \frac{1}{I} \mathrm{cov}(U^{(.)}, V^{(.)}) = \frac{J}{I}, \tag{15}$$

where

$$J = \sum E(U^{(j)} V^{(j)}).$$

Also if

$$W^{(j)} = \frac{\partial^3 \log p_j(Y_j, \beta)}{\partial \beta^3}, \quad K = E(W^{(.)}),$$

then

$$E\{L'''(\beta)\} = K.$$

Note that I, J, K refer to a total over the sample and are of order n.

Finally, a calculation similar to (15) shows that the final term in (14) is $O(n^{-1})$, whence (14) gives (Bartlett, 1952), for the terms of order 1

$$-IE(\hat{\beta}-\beta)+\frac{J}{I}+\frac{K}{2I} = 0,$$

i.e.

$$b \equiv E(\hat{\beta}-\beta) = \frac{1}{2I^2}(K+2J), \qquad (16)$$

which is of order n^{-1}.

When there are parameters $\beta_1, ..., \beta_p$, we define

$$U_r^{(j)} = \frac{\partial \log p_j(Y_j, \boldsymbol{\beta})}{\partial \beta_r}, \quad V_{rs}^{(j)} = \frac{\partial^2 \log p_j(Y_j, \boldsymbol{\beta})}{\partial \beta_r \partial \beta_s},$$

$$W_{rst}^{(j)} = \frac{\partial^3 \log p_j(Y_j, \boldsymbol{\beta})}{\partial \beta_r \partial \beta_s \partial \beta_t},$$

$$I_{rs} = E(-V_{rs}^{(.)}), \quad J_{r,st} = E\{\sum U_r^{(j)} V_{st}^{(j)}\}, \quad K_{rst} = E(W_{rst}^{(.)}). \qquad (17)$$

Expansion of the equation

$$[\partial L/\partial \beta_r]_{\boldsymbol{\beta}=\hat{\boldsymbol{\beta}}} = 0$$

replaces the first-order equation (12) by

$$\hat{\beta}_r - \beta_r = I^{rs} U_s^{(.)}, \quad \text{cov}(\hat{\beta}_r, \hat{\beta}_s) = I^{rs}. \qquad (18)$$

The superscripts denote matrix inversion and the summation convention is applied to multiple suffices referring to parameter components.

The second-order equation (13) becomes

$$\frac{\partial L}{\partial \beta_r} + (\hat{\beta}_s - \beta_s)\frac{\partial^2 L}{\partial \beta_r \partial \beta_s} + \tfrac{1}{2}(\hat{\beta}_t - \beta_t)(\hat{\beta}_u - \beta_u)\frac{\partial^3 L}{\partial \beta_r \partial \beta_t \partial \beta_u} = 0.$$

On taking expectations, we have that

$$E(\hat{\beta}_s - \beta_s)I_{rs} = \tfrac{1}{2}I^{lu}(K_{rtu}+2J_{t,ru}), \qquad (19)$$

a set of simultaneous linear equations for the biases, with solution

$$b_s \equiv E(\hat{\beta}_s - \beta_s) = \tfrac{1}{2}I^{rs} I^{lu}(K_{rtu}+2J_{t,ru}). \qquad (20)$$

In the right-hand side of (20), which is of order n^{-1}, consistent estimates of parameters can be inserted.

4. Further Properties of Residuals

It would be useful to know the joint distribution of the R_i's defined by (5). The distribution of any suggested test statistic could then be found, and the properties of graphical procedures evaluated. It is, however, not feasible to determine this joint distribution in general and, as a first step, we consider the expectations and covariances of the R_i's.

For this, expand (5) in series, obtaining to order n^{-1}

$$R_i = \epsilon_i + (\hat{\beta}_r - \beta_r) H_r^{(i)} + \tfrac{1}{2}(\hat{\beta}_s - \beta_r)(\hat{\beta}_s - \beta_s) H_{rs}^{(i)}, \qquad (21)$$

where

$$H_r^{(i)} = \frac{\partial h_i(Y_i, \boldsymbol{\beta})}{\partial \beta_r}, \quad H_{rs}^{(i)} = \frac{\partial^2 h_i(Y_i, \boldsymbol{\beta})}{\partial \beta_r \, \partial \beta_s}. \qquad (22)$$

Thus

$$E(R_i) = E(\epsilon_i) + E(\hat{\beta}_r - \beta_r) E(H_r^{(i)}) + \mathrm{cov}\,(\hat{\beta}_r - \beta_r, H_r^{(i)})$$
$$+ \tfrac{1}{2}E\{(\hat{\beta}_r - \beta_r)(\hat{\beta}_s - \beta_s)\} E(H_{rs}^{(i)}), \qquad (23)$$

the neglected terms being $o(n^{-1})$.

In (23), the second term is given by (20). The fourth term is given to sufficient accuracy by the usual large-sample result (18) and to evaluate the third term, we have, again by (18),

$$\mathrm{cov}\,(\hat{\beta}_r - \beta_r, H_r^{(i)}) = E(I^{rs}\, U_s^{(.)}\, H_r^{(i)})$$
$$= I^{rs}\, E(U_s^{(i)}\, H_r^{(i)}); \qquad (24)$$

on the right-hand side the summation convention does not apply to the superscript i.

Thus, to order n^{-1},

$$E(R_i) = E(\epsilon_i) + b_r\, E(H_r^{(i)}) + I^{rs}\, E(H_r^{(i)}\, U_s^{(i)} + \tfrac{1}{2}H_{rs}^{(i)})$$
$$= E(\epsilon_i) + a_i, \qquad (25)$$

say. In the same way, squaring (21) and taking expectations, we have to the same order that

$$E(R_i^2) = E(\epsilon_i^2) + 2b_r\, E(\epsilon_i\, H_r^{(i)}) + 2I^{rs}\, E(\epsilon_i\, H_r^{(i)}\, U_s^{(i)} + \tfrac{1}{2}H_r^{(i)}\, H_s^{(i)} + \tfrac{1}{2}\epsilon_i\, H_{rs}^{(i)}) \quad (26)$$

and that for $i \neq j$

$$E(R_i\, R_j) = \{E(\epsilon_i)\}^2 + (a_i + a_j) E(\epsilon_i) + I^{rs}\, E(\epsilon_i\, H_r^{(j)}\, U_s^{(i)} + \epsilon_j\, H_r^{(i)}\, U_s^{(j)} + H_r^{(i)}\, H_s^{(j)}). \qquad (27)$$

We can summarize (25), (26), and (27) as follows:

$$\left.\begin{aligned}
E(R_i) &= E(\epsilon_i) + a_i, \\
\mathrm{var}\,(R_i) &= \mathrm{var}\,(\epsilon_i) + c_{ii}, \\
\mathrm{cov}\,(R_i, R_j) &= c_{ij},
\end{aligned}\right\} \qquad (28)$$

where a_i, c_{ii}, c_{ij} can be found in terms of the right-hand sides of (25), (26), and (27) and are of order n^{-1}. Note that the summation convention applies to the parameter suffices only and not to c_{ii}.

A simple example of these formulae is given in Section 6.

5. APPLICATION OF RESULTS OF SECTION 4

There are broadly three ways in which the above results can be used.

Firstly, if all the correction terms in (28) are numerically small this gives some assurance that treating the R_i's as having the same statistical properties as the ϵ_i's is reasonable. If, say, the correction terms are small for all residuals except one, we might look at that residual separately and then omit it from the rest of the analysis.

Secondly, for particular types of test statistic, the results can be used to approximate to its distribution. Thus if we consider

$$T = \sum R_i z_i,$$

where the z_i's are constants, then

$$E(T) = E(\epsilon_i) \sum z_i + \sum a_i z_i,$$
$$\text{var}(T) = \text{var}(\epsilon_i) \sum z_i^2 + \sum z_i z_j c_{ij}.$$

A statistic used for testing possible dependence on z_i of var (ϵ_i) is

$$T' = \sum R_i^2(z_i - \bar{z}), \tag{29}$$

where $\bar{z} = \sum z_i/n$. Here the results of Section 4 give only that

$$E(T') = \sum c_{ii}(z_i - \bar{z}) + 2E(\epsilon_i) \sum a_i(z_i - \bar{z}).$$

In principle, it is possible to extend the arguments to obtain $E(R_i^4)$ and $E(R_i^2 R_j^2)$ and hence to reach an approximation to var (T').

Thirdly, we may use (28) to define a modified residual R_i' having more nearly the properties of ϵ_i. How best to do this depends somewhat on the particular case, but one fairly general procedure is to write

$$R_i' = (1 + k_i) R_i + l_i, \tag{30}$$

where k_i, l_i are small constants. If we require that, to order n^{-1},

$$E(R_i') = E(\epsilon_i), \quad \text{var}(R_i') = \text{var}(\epsilon_i), \tag{31}$$

two equations determining k_i, l_i follow from (28).

A serious limitation to this discussion is that it applies only indirectly to the examination of distributional form, by plotting ordered residuals against the expected order statistics for the distributional form proposed for the ϵ_i's. For this we would like, in particular, to calculate $E(R_{(i)})$, where $R_{(i)}$ is the ith largest residual; alternatively, we would like to introduce modified residuals R_i'' such that

$$E(R_{(i)}'') = E(\epsilon_{(i)});$$

of course, it is easy to formulate even more ambitious aims. It is plausible, but not certain, that the modification (30), designed to produce residuals with approximately the same marginal mean and variance, is an advantage also from the point of view of plots to examine distributional form.

6. An Example

We consider further the data of Feigl and Zelen (1965). The model (7) is

$$Y_j = \beta_1 \exp\{\beta_2(x_j - \bar{x})\} \epsilon_j, \quad j = 1, 2, ..., n.$$

We first find the bias in $\hat{\beta}_1, \hat{\beta}_2$. Since ϵ_j is exponentially distributed with unit mean, we have, writing $(x_j - \bar{x}) = d_j$,

$$p_j(Y_j, \boldsymbol{\beta}) = [\exp\{- Y_j \exp(- \beta_2 d_j)/\beta_1\}]/\{\beta_1 \exp(\beta_2 d_j)\},$$
$$\log p_j(Y_j, \boldsymbol{\beta}) = -\{Y_j \exp(- \beta_2 d_j)/\beta_1\} - \log \beta_1 - \beta_2 d_j, \tag{32}$$

which, on differentiating and taking expectations, leads to

$$I_{rs} = \begin{cases} n/\beta_1^2, & r = s = 1, \\ \sum d_j^2, & r = s = 2, \\ 0, & r \neq s. \end{cases}$$

Equations (20) therefore give

$$b_1 = \tfrac{1}{2}I^{11}\{I^{11}(K_{111}+2J_{1,11})+I^{22}(K_{122}+2J_{2,12})\},$$

with a similar equation for b_2. From (17) and (32), we have, without the summation convention,

$$J_{t,1t} = E(\sum U_t^{(j)} V_{1t}^{(j)})$$

$$= \begin{cases} -2n/\beta_1^3, & t = 1, \\ -\sum d_j^4/\beta_1, & t = 2, \end{cases}$$

and

$$K_{1tt} = E(W_{1tt}^{(.)})$$

$$= \begin{cases} 4n/\beta_1^3, & t = 1, \\ \sum d_j^2/\beta_1, & t = 2. \end{cases}$$

Thus

$$b_1 = -\tfrac{1}{2}\beta_1/n.$$

A similar calculation gives

$$b_2 = -\tfrac{1}{2}\sum d_j^3/(\sum d_j^2)^2.$$

To find $E(R_i)$ and $E(R_i^2)$, given by (25), (26), we write

$$h_i(Y_i, \boldsymbol{\beta}) = Y_i \exp(-\beta_2 d_i)/\beta_1$$

from which we evaluate $H_r^{(i)}$, etc., and obtain

$$E(R_i) = 1 + \tfrac{1}{2}n^{-1} + \tfrac{1}{2}(d_i \sum d_j^3 - d_i^2 \sum d_j^2)/(\sum d_j^2)^2$$

$$= 1 + a_i, \tag{33}$$

$$E(R_i^2) = 2 + 2(d_i \sum d_j^3 - 2d_i^2 \sum d_j^2)/(\sum d_j^2)^2$$

$$= 2 + c_{ii}^\dagger, \tag{34}$$

where the connection with (28) is that $c_{ii}^\dagger = c_{ii} + 2a_i$.

Although Feigl and Zelen compared two groups of observations it is sufficient here to consider only one of the groups. The data are given in Table 1, together with the corresponding values $a_i, c_{ii}^\dagger$ calculated from (33), (34); values of the crude residuals R_i, defined by (5), are also given.

In order to calculate modified residuals R_i' we note, since ϵ_i has an exponential distribution, that we require a transformation which adjusts the mean and variance and yet restricts R_i' to be positive. Hence we take

$$R_i' = \{R_i/(1-l_i)\}^{1+k_i}, \tag{35}$$

where both l_i and k_i are small. Assuming this transforms R_i to an exponential distribution with unit mean, it can be shown that

$$E(R_i) = (1 - l_i)\,\Gamma\{1 + 1/(1 + k_i)\}$$

and

$$E(R_i^2) = (1 - l_i)^2\,\Gamma\{1 + 2/(1 + k_i)\}.$$

TABLE 1

Leukemia data (Feigl and Zelen, 1965). Log white blood cell count, x_i.
Survival time, weeks, Y_i. Crude and modified residuals, R_i and R_i'

x_i	Y_i	a_i	$c_{ii}^\dagger$	R_i	R_i'
3·36	65	−0·013	−0·340	0·56	0·49
2·88	156	−0·088	−0·946	0·79	0·70
3·63	100	0·013	−0·134	1·17	1·12
3·41	134	−0·007	−0·293	1·23	1·20
3·78	16	0·022	−0·063	0·22	0·19
4·02	108	0·029	−0·003	1·94	1·91
4·00	121	0·029	−0·005	2·13	2·11
4·23	4	0·028	−0·012	0·09	0·07
3·73	39	0·019	−0·082	0·51	0·46
3·85	143	0·025	−0·039	2·12	2·11
3·97	56	0·028	−0·009	0·96	0·90
4·51	26	0·015	−0·109	0·80	0·74
4·54	22	0·013	−0·131	0·71	0·65
5·00	1	−0·037	−0·528	0·05	0·03
5·00	1	−0·037	−0·528	0·05	0·03
4·72	5	−0·002	−0·249	0·19	0·15
5·00	65	−0·037	−0·528	3·47	4·17

If we equate these expressions to (33) and (34), take logarithms and expand, ignoring high-order terms of k_i and l_i, we find

$$a_i = -l_i - (1 - \gamma)\,k_i, \quad c_{ii}^\dagger = -4l_i - 2(3 - 2\gamma)\,k_i,$$

where $\gamma = \Gamma'(1)/\Gamma(1)$. Thus

$$k_i = \tfrac{1}{2}(4a_i - c_{ii}^\dagger), \quad l_i = 0{\cdot}21 c_{ii}^\dagger - 1{\cdot}85 a_i. \tag{36}$$

Values of R_i' are given in Table 1; the most noticeable difference between the crude and modified residuals occurs in the final entry, at $x = 5{\cdot}00$. A plot of R_i' against x_i shows no evidence that the mean or dispersion of the residuals vary systematically with x. The modified residuals R_i' are shown plotted in Fig. 1 against the expected values of exponential order statistics for a sample of size $n = 17$; the assumption of an exponential distribution is clearly confirmed.

A supplement to the graphical analysis is the calculation of a test statistic designed to examine consistency with the exponential distribution; see Cox and Lewis (1966, pp. 161–163) for a brief review of such tests applied to simple random samples.

One such test, although not normally the best, is based in effect on comparing the variance with the square of the mean. Now

$$T^* \equiv \sum R_i^2 = 31 \cdot 07$$

and the result (34) leads, with $n = 17$, to

$$E(T^*) = 2(n-2) = 30.$$

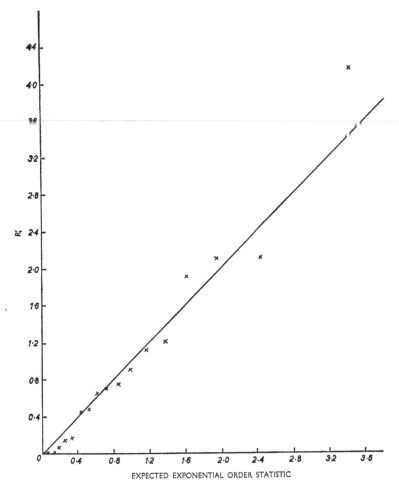

FIG. 1. Leukemia data. Modified residual, R_i', versus expected exponential order statistics. Straight line corresponds to unit exponential distribution.

Now if the residuals were obtained directly from a random sample of size n, i.e. $R_i = Y_i/\bar{Y}$, and the test statistic is

$$T^* = \sum (Y_i/\bar{Y})^2,$$

then it is easy to show that, to the order considered,

$$E(T^*) = 2(n-1).$$

This suggests that T^* should be regarded as derived from a random sample of size $n-1$, i.e. a "degree of freedom" subtracted for the extra parameter fitted. Again the close fit to the exponential distribution is confirmed.

The covariance between different residuals can be calculated from (25), (27). In fact

$$\mathrm{cov}(R_i, R_k) = -n^{-1} - d_i d_k / \sum d_j^2. \tag{37}$$

This is identical, to the order considered, with the corresponding formula for ordinary linear regression. Since the residuals here have approximately unit variance, (37) is also the correlation between residuals. In this example, the numerically greatest correlation is about 0·2. The presence of a substantial correlation between a particular pair of residuals would have been a warning of possible difficulties of interpretation, especially if both residuals had appeared to correspond to outliers.

In one way the fact that the R_i' and the R_i differ relatively little is an anticlimax. A more encouraging way of looking at the conclusions is, however, that they suggest that, even with the very small number of observations considered here, the distortion introduced by the fitting is small and that the unmodified residuals may in practice often be adequate.

7. POISSON DATA

In order to apply the methods of the preceding sections to Poisson-distributed observations, we must first consider how to define residuals so as to obtain nearly identically distributed variables. The difference from the earlier discussion is that, there, the model was defined directly in terms of independently distributed random variables. Here we proceed indirectly, defining R_i as

(a) $(Y_i - \mu_i)/\sqrt{\mu_i}$,

(b) $2(\sqrt{Y_i} - \sqrt{\mu_i})$,

or

(c) $\{\psi(Y_i) - \psi(\mu_i)\}/\{\psi'(\mu_i)\sqrt{\mu_i}\}$,

where $\mu_i = \mu_i(\boldsymbol{\beta})$ is the expected frequency, on some model dependent upon parameters $\beta_1, ..., \beta_p$, of the Poisson observation, Y_i. Each of these, asymptotically, defines a standard normal deviate; (c) is a generalization of (a) and (b), and in it $\psi(x)$ is an arbitrary function.

The choice of an appropriate transformation depends upon the requirements; the need for a direct interpretation might lead to (a) or, alternatively, for a multiplicative model, to (c) with $\psi(x) = \log x$. But since our immediate object is to detect departures, rather than to explain them, it is desirable to find a transformation to give a set of residuals with a distribution as near as possible to some known form. Anscombe (1953) suggests that an appropriate transformation for normalizing Poisson observations is

$$\{Y_i^{\frac{2}{3}} - (\mu_i - \tfrac{1}{6})^{\frac{2}{3}}\}/(\tfrac{2}{3}\mu_i^{\frac{1}{6}}). \tag{38}$$

This can be derived by considering the Taylor expansion for moments of a power Y_i^h and equating the coefficient of skewness to zero; see also Moore (1957).

We therefore define $h_i(Y_i, \boldsymbol{\beta})$ by (38); we have also

$$p_j(Y_j, \boldsymbol{\beta}) = \exp(-\mu_j)\mu_j^{Y_j}/Y_j!$$

and hence we can apply the results of Section 4 to obtain $E(R_i)$ and $E(R_i^2)$. To do so, however, involves certain approximations and it is necessary to distinguish between terms that become small when n, the number of distinct Poisson observations, is large and those that become small when μ, the expectation of a typical observation, is large. The general results of Section 5 refer to large n.

The biases are given by

$$b_s = -\tfrac{1}{2} I^{rs} I^{tu} \sum \frac{1}{\mu_j} \frac{\partial \mu_j}{\partial \beta_r} \frac{\partial^2 \mu_j}{\partial \beta_t \partial \beta_u} \tag{39}$$

with

$$I_{rs} = \sum \frac{1}{\mu_j} \frac{\partial \mu_j}{\partial \beta_r} \frac{\partial \mu_j}{\partial \beta_s}$$

and hence b_s is of order μ^{-1}.

In obtaining $E(R_i)$ and $E(R_i^2)$, there is appreciable simplification if we consider leading terms in expansions in powers of $1/\mu$; this leads to terms of order $\mu^{-\frac{1}{2}}$ for $E(R_i)$ and of order 1 and μ^{-1} for $E(R_i^2)$. Thus if μ_i is sufficiently large it will be adequate to take as an approximation only the terms of order 1 and this gives

$$E(R_i) = 0, \quad E(R_i^2) = 1 - I^{rs} \frac{\partial \mu_i}{\partial \beta_r} \frac{\partial \mu_i}{\partial \beta_s} \frac{1}{\mu_i}. \tag{40}$$

The expression for $E(R_i)$ to order $\mu^{-\frac{1}{2}}$ is, in fact,

$$E(R_i) = -\frac{b_r}{\mu_i^{\frac{1}{2}}} \frac{\partial \mu_i}{\partial \beta_r} + I^{rs} \left(\frac{1}{6\mu_i^{\frac{3}{2}}} \frac{\partial \mu_i}{\partial \beta_r} \frac{\partial \mu_i}{\partial \beta_s} - \frac{1}{2\mu_i^{\frac{1}{2}}} \frac{\partial^2 \mu_i}{\partial \beta_r \partial \beta_s} \right). \tag{41}$$

The transformations (a) and (b) also lead to the approximation (40), to the order considered.

8. Binomial Data

Following the arguments of Section 7, we define a transformation $\phi(Y_i/m_i)$ of the observation Y_i from a binomial distribution with parameters $\theta_i(\beta)$ and m_i. By considering the Taylor expansion and equating the skewness to zero, we obtain a differential equation of which a solution is

$$\phi(u) = \int_0^u t^{-\frac{1}{3}}(1-t)^{-\frac{1}{3}} dt, \quad 0 \leqslant u \leqslant 1. \tag{42}$$

Blom (1954) suggests equation (42) as a normalizing transformation but does not apply it. In order to simplify its application we have computed Table 2. This gives values of $\phi(u)/\phi(1)$, i.e. the incomplete beta function $I_u(\frac{2}{3}, \frac{2}{3})$, which is symmetrical about $u = 0.5$; multiplication by $B(\frac{2}{3}, \frac{2}{3}) = 2.0533$ gives the value of (42). For example, $\phi(0.2) = 2.0533 \times 0.257 = 0.528$, $\phi(0.8) = 2.0533(1 - 0.257) = 1.526$.

Introducing the mean and variance of the transformed binomial variate, we define

$$h_i(Y_i, \beta) = [\phi(Y_i/m_i) - \phi\{\theta_i - \tfrac{1}{6}(1 - 2\theta_i)/m_i\}]/\{\theta_i^{\frac{1}{3}}(1 - \theta_i)^{\frac{1}{3}}/\sqrt{m_i}\}; \tag{43}$$

this reduces to (38) for small θ_i. Plots on probability paper suggest that the transformation is very effective, even for values as small as $m_i = 5$, $\theta_i = 0.04$. Often the bias correction $-\tfrac{1}{6}(1 - 2\theta_i)/m_i$ can be omitted.

TABLE 2

Values of the incomplete beta function $I_u(\frac{2}{3}, \frac{2}{3})$

u	0·000	0·001	0·002	0·003	0·004	0·005	0·006	0·007	0·008	0·009
0·00	0	0·007	0·012	0·015	0·018	0·021	0·024	0·027	0·029	0·032
0·01	0·034	0·036	0·038	0·040	0·043	0·045	0·046	0·048	0·050	0·052
0·02	0·054	0·056	0·058	0·059	0·061	0·063	0·064	0·066	0·068	0·069
0·03	0·071	0·072	0·074	0·075	0·077	0·079	0·080	0·082	0·083	0·084
0·04	0·086	0·087	0·089	0·090	0·092	0·093	0·094	0·096	0·097	0·098
0·05	0·100	0·101	0·102	0·104	0·105	0·106	0·108	0·109	0·110	0·112
0·06	0·113	0·114	0·115	0·117	0·118	0·119	0·120	0·122	0·123	0·124
0·07	0·125	0·126	0·128	0·129	0·130	0·131	0·132	0·134	0·135	0·136
0·08	0·137	0·138	0·139	0·141	0·142	0·143	0·144	0·145	0·146	0·147
0·09	0·149	0·150	0·151	0·152	0·153	0·154	0·155	0·156	0·157	0·158
0·10	0·160	0·161	0·162	0·163	0·164	0·165	0·166	0·167	0·168	0·169
0·11	0·170	0·171	0·172	0·173	0·174	0·176	0·177	0·178	0·179	0·180
0·12	0·181	0·182	0·183	0·184	0·185	0·186	0·187	0·188	0·189	0·190
0·13	0·191	0·192	0·193	0·194	0·195	0·196	0·197	0·198	0·199	0·200
0·14	0·201	0·202	0·203	0·204	0·205	0·206	0·207	0·208	0·209	0·210
0·15	0·211	0·212	0·213	0·214	0·214	0·215	0·216	0·217	0·218	0·219
0·16	0·220	0·221	0·222	0·223	0·224	0·225	0·226	0·227	0·228	0·229
0·17	0·230	0·231	0·232	0·232	0·233	0·234	0·235	0·236	0·237	0·238
0·18	0·239	0·240	0·241	0·242	0·243	0·244	0·244	0·245	0·246	0·247
0·19	0·248	0·249	0·250	0·251	0·252	0·253	0·254	0·254	0·255	0·256
0·20	0·257	0·258	0·259	0·260	0·261	0·262	0·262	0·263	0·264	0·265
0·21	0·266	0·267	0·268	0·269	0·270	0·270	0·271	0·272	0·273	0·274
0·22	0·275	0·276	0·277	0·277	0·278	0·279	0·280	0·281	0·282	0·283
0·23	0·284	0·284	0·285	0·286	0·287	0·288	0·289	0·290	0·290	0·291
0·24	0·292	0·293	0·294	0·295	0·296	0·296	0·297	0·298	0·299	0·300
0·25	0·301	0·302	0·302	0·303	0·304	0·305	0·306	0·307	0·308	0·308
0·26	0·309	0·310	0·311	0·312	0·313	0·313	0·314	0·315	0·316	0·317
0·27	0·318	0·318	0·319	0·320	0·321	0·322	0·323	0·323	0·324	0·325
0·28	0·326	0·327	0·328	0·328	0·329	0·330	0·331	0·332	0·333	0·333
0·29	0·334	0·335	0·336	0·337	0·338	0·338	0·339	0·340	0·341	0·342
0·30	0·342	0·343	0·344	0·345	0·346	0·347	0·347	0·348	0·349	0·350
0·31	0·351	0·351	0·352	0·353	0·354	0·355	0·355	0·356	0·357	0·358
0·32	0·359	0·360	0·360	0·361	0·362	0·363	0·364	0·364	0·365	0·366
0·33	0·367	0·368	0·368	0·369	0·370	0·371	0·372	0·372	0·373	0·374
0·34	0·375	0·376	0·376	0·377	0·378	0·379	0·380	0·380	0·381	0·382
0·35	0·383	0·384	0·384	0·385	0·386	0·387	0·388	0·388	0·389	0·390
0·36	0·391	0·392	0·392	0·393	0·394	0·395	0·396	0·396	0·397	0·398
0·37	0·399	0·400	0·400	0·401	0·402	0·403	0·403	0·404	0·405	0·406
0·38	0·407	0·407	0·408	0·409	0·410	0·411	0·411	0·412	0·413	0·414
0·39	0·414	0·415	0·416	0·417	0·418	0·418	0·419	0·420	0·421	0·422
0·40	0·422	0·423	0·424	0·425	0·425	0·426	0·427	0·428	0·429	0·429
0·41	0·430	0·431	0·432	0·433	0·433	0·434	0·435	0·436	0·436	0·437
0·42	0·438	0·439	0·440	0·440	0·441	0·442	0·443	0·443	0·444	0·445
0·43	0·446	0·447	0·447	0·448	0·449	0·450	0·450	0·451	0·452	0·453
0·44	0·454	0·454	0·455	0·456	0·457	0·457	0·458	0·459	0·460	0·461
0·45	0·461	0·462	0·463	0·464	0·464	0·465	0·466	0·467	0·468	0·468
0·46	0·469	0·470	0·471	0·471	0·472	0·473	0·474	0·474	0·475	0·476
0·47	0·477	0·478	0·478	0·479	0·480	0·481	0·481	0·482	0·483	0·484
0·48	0·485	0·485	0·486	0·487	0·488	0·488	0·489	0·490	0·491	0·491
0·49	0·492	0·493	0·494	0·495	0·495	0·496	0·497	0·498	0·498	0·499

The p.d.f. of Y_j is

$$p_j(Y_j, \boldsymbol{\beta}) = \binom{m_j}{Y_j} \theta_j^{Y_j} (1-\theta_j)^{m_j - Y_j},$$

from which we obtain the biases

$$b_s = -\tfrac{1}{2} I^{rs} I^{tu} \sum \frac{m_j}{\theta_j(1-\theta_j)} \frac{\partial\theta_j}{\partial\beta_r} \frac{\partial^2\theta_j}{\partial\beta_t \partial\beta_u}, \tag{44}$$

To obtain $E(R_i)$ and $E(R_i^2)$, we consider expansions in powers of m^{-1} and get, analogous to (40), the approximation

$$E(R_i) = 0, \quad E(R_i^2) = 1 - I^{rs} \frac{\partial\theta_i}{\partial\beta_r} \frac{\partial\theta_i}{\partial\beta_s} \frac{m_i}{\theta_i(1-\theta_i)}. \tag{45}$$

In the numerical example which follows, we checked that the higher order terms neglected in (45) are indeed negligible.

9. A Further Example

Dyke and Patterson (1952) present the analysis for a 2^4 factorial design of the proportions of respondents who achieve good scores on cancer knowledge; some details of the data are given in columns (1)–(3) of Table 3. They assume a logit transformation of the proportions, the expected value of the transformed variate being a linear function of parameters representing main effects and interactions. Values of the parameters are estimated by maximum likelihood. We consider their solution and apply the methods of Section 8 to examine residuals from the fitted model.

Following Dyke and Patterson (with slight changes in notation) we write the model as

$$\theta_j(\boldsymbol{\beta}) = \{1 + \exp(-2z_j)\}^{-1},$$

where $z_j = l_{jr}\beta_r$, summed over the parameters; $l_{jr} = \pm 1$. Then

$$\frac{\partial\theta_j}{\partial\beta_r} = 2\theta_j(1-\theta_j) l_{jr},$$

$$\frac{\partial^2\theta_j}{\partial\beta_r \partial\beta_s} = 4\theta_j(1-\theta_j)(1-2\theta_j) l_{jr} l_{js}$$

and substitution into (44) gives

$$b_s = -4 I^{rs} I^{tu} \sum m_j \theta_j(1-\theta_j)(1-2\theta_j) l_{jr} l_{jt} l_{ju}.$$

From (45), we have

$$E(R_i) = 0$$

and

$$E(R_i^2) = 1 - 4 I^{rs} m_i \theta_i(1-\theta_i) l_{ir} l_{is}$$

$$= 1 + c_{ii}^\dagger.$$

Dyke and Patterson fit a model with five parameters, representing the overall mean and the four main effects. They quote the values of I^{rs} obtained in the course of their solution and we use these values to calculate b_s and $c_{ii}^\dagger$.

In order to calculate modified residuals, we use (30) writing $l_i = 0$ since $E(R_i) = 0$; solving for k_i, remembering k_i is small, we obtain

$$R_i' = (1 - \tfrac{1}{2}c_{ii}^\dagger) R_i.$$

Values of $c_{ii}^\dagger$, R_i and R_i' are given in Table 3. The biases b_s were all extremely small; none exceeded $2\frac{1}{2}$ per cent of the standard error of the estimate of the parameter.

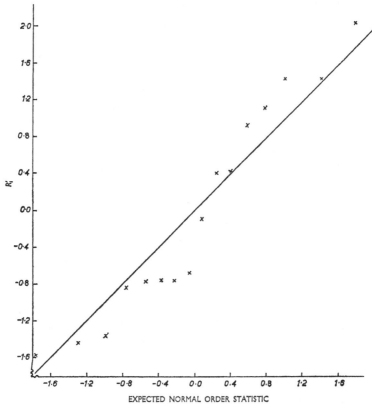

FIG. 2. 2^4 factorial design. Modified residual, R_i', versus expected normal order statistics. Straight line corresponds to unit normal distribution.

The modified residuals R_i' are plotted against the expected normal order statistics in Fig. 2 and show agreement with the assumption of a standard normal distribution.

Dyke and Patterson go on to fit a model with extra parameters to represent the interactions AD, BD and CD. Before proceeding to extend the model, we examine the residuals R_i' from the simple main effects model. If we regard the values of R_i' as observations in a 2^4 design, we can analyse them in the usual way and obtain the sums of squares given in Table 4. This is an unweighted analysis and, as such, can be used only for guidance; the existence of any apparent effect can be established only by fitting a model containing the appropriate parameters. Also, for the same reason, the sums of squares due to main effects in Table 4 are not zero. Nevertheless, the magnitude of the AC effect suggests it to be worth including in the model along with AD, BD and CD; the three-factor interaction ACD also is large but has not been included in the further analysis. We therefore fitted a model with nine parameters;

TABLE 3

2^4 *factorial design (Dyke and Patterson, 1952).* r_i *number of "good" responses out of* m_i. *Crude and modified residuals,* R_i *and* R_i'

Treatment	m_i	r_i	$-c_{ii}^\dagger$	R_i	R_i'
abcd	31	23	0·248	0·36	0·41
abc	169	102	0·540	−0·46	−0·68
abd	12	8	0·135	1·34	1·44
ab	94	35	0·372	−0·07	−0·09
acd	45	27	0·379	−0·59	−0·75
ac	378	201	0·711	−0·45	−0·84
ad	13	7	0·133	1·04	1·12
a	231	75	0·527	0·65	0·94
bcd	4	1	0·050	−1·33	−1·37
bc	32	16	0·190	0·38	0·43
bd	7	4	0·076	1·38	1·44
b	63	13	0·225	−0·68	−0·78
cd	11	3	0·123	−1·47	−1·57
c	150	67	0·505	1·45	2·06
d	12	2	0·100	−0·73	−0·77
1	477	84	0·705	−0·78	−1·43

TABLE 4

2^4 *factorial designs. Sums of squares of modified and crude residuals*

Effect	Sums of squares	
	R_i'	R_i
A	0·78	0·80
B	0·26	0·20
C	1·09	1·14
D	0·01	0·00
AB	0·04	0·04
AC	2·53	0·98
AD	1·85	1·57
BC	0·32	0·20
BD	2·06	1·69
CD	4·88	3·89
ABC	2·51	1·26
ABD	0·07	0·10
ACD	3·77	1·97
BCD	0·03	0·05
ABCD	0·03	0·03

the estimated values and standard errors are given in Table 5. Comparison of the estimates with their standard errors confirms that the AC interaction is at least as significant as the AD and BD interactions.

TABLE 5

2^4 *factorial design. Estimated parameters and their standard errors*

Parameter	Estimate	Standard error
Mean	−0·13	0·06
A	0·25	0·06
B	0·12	0·05
C	0·17	0·05
D	0·11	0·06
AC	−0·05	0·03
AD	0·10	0·06
BD	0·06	0·05
CD	−0·11	0·05

For comparison, the corresponding analysis on the crude residuals, R_i, is also given in Table 4; it is interesting that the AC interaction does not stand out in this case. Note, however, that 9 parameters are being fitted to 16 observations, so that the applicability of the asymptotic formulae is in doubt.

10. FURTHER WORK

Some possible extensions of the work fall under the following three broad headings.

(i) There are applications, for example to time series and components of variance problems, in which more than one random variable contributes to each observation. Durbin and Watson (1950, 1951) have considered a significance test for serial correlation based on residuals from a fitted regression.

(ii) There could be more refined distributional studies of the quantities discussed above. In particular, the more detailed study of order statistics of residuals would be useful both in connection with detecting outliers and for tests of distributional form. There is obvious scope for simulation.

(iii) Further special applications can be considered. Two general types concern (a) the examination of distributional form from random samples, for example by using the probability integral transformation to produce residuals having a theoretical rectangular distribution, and (b) more complex situations that are essentially generalizations of regression.

As an example of (a), consider the Weibull distribution (Section 2), where a transformation can be made to the unit exponential distribution, or the circular normal distribution, where the probability integral transformation can be applied. As further examples of (b), consider two problems closely associated with normal-theory regression, namely the calculation of residuals after fitting a transformation (Box and Cox, 1964) or after fitting a non-linear model by iterative least squares.

ACKNOWLEDGEMENT

We are grateful to Mrs E. A. Chambers and Mr B. G. F. Springer for programming the calculations. Their work was supported by the Science Research Council.

REFERENCES

ANSCOMBE, F. J. (1953). Contribution to the discussion of H. Hotelling's paper. *J. R. Statist. Soc.* B, **15**, 229–230.
—— (1961). Examination of residuals. *Proc. 4th Berkeley Symp.*, **1**, 1–36.
BARTLETT, M. S. (1952). Approximate confidence intervals. II. *Biometrika*, **40**, 306–317.
BLOM, G. (1954). Transformations of the binomial, negative binomial, Poisson and χ^2 distributions. *Biometrika*, **41**, 302–316.
BOX, G. E. P. and COX, D. R. (1964). An analysis of transformations. *J. R. Statist. Soc.* B, **26**, 211–252.
COX, D. R. and LEWIS, P. A. W. (1966). *The Statistical Analysis of Series of Events*. London: Methuen.
DURBIN, J. and WATSON, G. S. (1950). Testing for serial correlation in least squares regression. I. *Biometrika*, **37**, 409–428.
—— (1951). Testing for serial correlation in least squares regression. II. *Biometrika*, **38**, 159–178.
DYKE, G. V. and PATTERSON, H. D. (1952). Analysis of factorial arrangements when the data are proportions. *Biometrics*, **8**, 1–12.
FEIGL, P. and ZELEN, M. (1965). Estimation of exponential survival probabilities with concomitant observation. *Biometrics*, **21**, 826–838.
HALDANE, J. B. S. (1953). The estimation of two parameters from a sample. *Sankhyā*, **12**, 313–320.
HALDANE, J. B. S. and SMITH, S. M. (1956). The sampling distribution of a maximum likelihood estimate. *Biometrika*, **43**, 96–103.
MOORE, P. G. (1957). Transformations to normality using fractional powers of the variate. *J. Am. Statist. Ass.*, **52**, 237–246.
PEARSON, E. S. and HARTLEY, H. O. (1966). *Biometrika Tables for Statisticians*, 3rd ed. Cambridge University Press.
SHENTON, L. R. and BOWMAN, K. (1963). Higher moments of a maximum-likelihood estimate. *J. R. Statist. Soc.* B, **25**, 305–317.
SHENTON, L. R. and WALLINGTON, P. A. (1962). The bias of moment estimators with an application to the negative binomial distribution. *Biometrika*, **49**, 193–204.

Summary of discussion

When the paper was read to the Royal Statistical Society a vote of thanks was proposed by M. J. R. Healy, seconded by P. J. Harrison and continued by F. J. Anscombe, J. Durbin, V. Barnett, S. C. Pearce and A. M. Walker. There were written contributions by R. M. Loynes, C. L. Mallows, M. B. Priestley and J. W. Tukey.

Healy raises general issues and in particular suggests that simpler methods might often work. Harrison develops the role of residuals in time series analysis, especially their possible role in cusum analysis; Priestley raises further time series issues. Anscombe discusses computational issues, in particular the role of the APL programming language in statistical analysis. Durbin points out that test statistics based on residuals need careful discussion of the appropriate distribution for their assessment. Barnett discusses the choice of plotting positions for residuals when their distributional form is being analyzed. Walker and Tukey ask why the analysis of a model based on an exponential distribution with multiplicative errors had not been done on a log scale; the authors reply that they had wished to avoid strong dependence on small observations. Loynes and Mallows report developments in distributional theory relevant to the paper. Tukey questions the relevance of small adjustments in defining residuals and also raises a number of more detailed points.

Biometrika (1978), **65**, 2, pp. 263–72
Printed in Great Britain

Testing multivariate normality

BY D. R. COX AND N. J. H. SMALL

Department of Mathematics, Imperial College, London

SUMMARY

Previous work on testing multivariate normality is reviewed. Coordinate-dependent and invariant procedures are distinguished. The arguments for concentrating on tests of linearity of regression are indicated and such tests, both coordinate-dependent and invariant, are developed.

Some key words: Goodness of fit; Invariance; Multivariate normality; Nonlinearity; Probability plot; Transformation; Tukey's degree of freedom.

1. INTRODUCTION

There has been much recent work on testing univariate normality, stemming partly from work on weak convergence (Durbin, 1973) and partly from more empirical ideas. Unfortunately little of this work can be directly applied to testing multivariate normality. Even when v, the number of component variables, is only two immediate adaptation of univariate tests such as the chi-squared goodness of fit test is clumsy and if v is larger such tests are quite impracticable. Further, the absence of a simple yet general family of distributions extending the multivariate normal precludes the use of a likelihood ratio test; see, however, Barndorff-Nielsen (1977).

Just as in other applications of significance tests, the practical purpose of the test must be considered. It is a central theme of the present paper that a main objective of tests of multivariate normality is to see whether an estimated covariance matrix provides an adequate summary of the interrelationships among a set of variables; most practical applications of multivariate analysis depend either upon a direct interpretation of one or more covariance matrices or upon some further analysis of such matrices. While in particular applications very specific kinds of departure from multivariate normality might be of concern, the departure with the most serious consequences is often the occurrence of appreciable nonlinearity of dependence. In its simplest form, the covariance of two random variables is even qualitatively a poor indication of their association if appreciable curvature is present. Nonnormality of marginal distribution, as such, does not have this consequence. Therefore for the great majority of this paper we consider tests of linearity of regression rather than directly of normality.

There is a general distinction in multivariate analysis between procedures that are invariant under arbitrary nonsingular linear transformations of the v component variables and those that are dependent on the particular coordinate system used to record the data. Despite the great theoretical power and importance of invariance considerations in multivariate analysis, there are many practical situations where the particular choice of components is important, i.e. where effects are in some sense most usefully to be detected or interpreted in particular directions in the v-dimensional space of the variables. Therefore we give separate discussions of invariant and of coordinate-dependent techniques. The coordinate-dependent procedures are, however, all invariant under scale and location changes of the components.

An excellent broad review of the assessment of multivariate distributional properties has been given recently by Gnanadesikan (1977, pp. 150–95) so that only a brief outline of previous work on tests of multivariate normality need be given here. A quite powerful coordinate-dependent approach is to consider parametric transformations coordinate by coordinate, e.g. of y_s into

$$y_s^{(\lambda_s)} = \begin{cases} (y_s^{\lambda_s} - 1)/\lambda_s & (\lambda_s \neq 0), \\ \log y_s & (\lambda_s = 0), \end{cases}$$

for $s = 1, \ldots, v$. Then it may be reasonable to assume that for some unknown $\lambda = (\lambda_1, \ldots, \lambda_v)$ the transformed observations are multivariate normal with unknown mean and covariance matrix.

The required λ can be estimated by maximum likelihood and the null hypothesis $\lambda_s = 1$ ($s = 1, \ldots, v$) tested by a likelihood ratio test. In some applications in which the component variables are similar in kind it may be sensible to suppose that $\lambda_1 = \ldots = \lambda_v$. This generalization of the univariate technique of Box & Cox (1964) was probably first used in unpublished work by the late T. Burnaby; it was developed entirely independently and in more detail by Andrews, Gnanadesikan & Warner (1971, 1973). This approach is, of course, coordinate-dependent, although Andrews et al. (1973) have considered the possibility of a preliminary rotation of coordinates before the consideration of transformations. This general approach has the advantage over most of the others to be mentioned that it gives an explicit suggestion of the analysis to be adopted if clear evidence against the null hypothesis of multivariate normality is found. Indeed, the only other general procedure based on an explicit alternative model is the fitting of a mixture of normal components, usually with different means and the same covariance matrix (Day, 1969).

A widely useful invariant graphical procedure (Healy, 1968; Cox, 1968; Andrews et al., 1971) is based on the distribution of the ordered Mahalanobis distances of the individual points from their mean in the metric defined by the sample covariance matrix.

Thus if $Y_1, \ldots, Y_n$ are n independent observations of a v-dimensional vector, $\overline{Y}$ their sample mean and S their estimated covariance matrix, we compute

$$D_i^2 = (Y_i - \overline{Y})^{\mathrm{T}} S^{-1} (Y_i - \overline{Y})$$

and plot the ordered D_i^2 against the expected order statistics for samples of size n from the chi-squared distribution with v degrees of freedom. It would be useful to have a significance test based on this procedure. Often, too, it will be informative to supplement the information about the distances of the individuals from the mean by some consideration of angular position (Gnanadesikan, 1977, pp. 172–4).

Important tests of univariate normality are based on standardized third and fourth cumulants, these being of particular value because of their diagnostic power in indicating the qualitative nature of any departure from normality. One simple possibility for a co-ordinate-dependent multivariate procedure is to examine separately the marginal distribution of each component (Andrews et al., 1973). A conservative composite significance test can be obtained from the most significant of the individual component statistics by using a Bonferroni bound, i.e. by multiplying the most extreme significance level by v. Alternatively a more detailed analysis may be based on the estimated correlation matrix of the original variables; details are given in an unpublished paper by N. J. H. Small. An invariant procedure similar in spirit to that developed in §4 of the present paper was given by Malkovich

& Afifi (1973) who considered as a possible statistic the supremum of, for example, the standardized skewness over all linear combinations $a_1 y_1 + \ldots + a_v y_v$. They applied the same notion to other univariate statistics.

Mardia (1970) has obtained invariant combinations of the third- and fourth-order cumulants by examining those combinations that have maximum effect on the null hypothesis distribution of the Hotelling T^2 statistic. Estimates of these invariant combinations are suggested for use as test statistics. Subsequently (Mardia, 1975) the relation between these and the distances D_i^2 has been explored.

3. COORDINATE-DEPENDENT PROCEDURES

3·1. *General*

In this section we consider tests for linearity of regression relationships which are coordinate-dependent. We deal with situations in which the v component variables are to be treated symmetrically. Of course, if there are available both response and explanatory variables, we shall normally condition on the observed values of the explanatory variables and be directly interested in distributional properties only as they concern the conditional distribution of the response variable given the explanatory variables.

A complication in the discussion that follows is that when v is large a natural analysis leads to rather a large number of component statistics, and some way of simplifying this procedure will be necessary; see § 3·3.

3·2. *Two component variables*

While the bivariate case, $v = 2$, is not of great practical interest, it is worthwhile beginning with a discussion of it. Let the observations be (y_{i1}, y_{i2}) $(i = 1, \ldots, n)$, regarded as n independent observations. The null hypothesis is that these correspond to independent and identically distributed random variables (Y_1, Y_2) with a bivariate normal distribution.

A simple test of the linearity of the regression of Y_2 on Y_1 is provided by $Q_{2,1}$, the standard Student t statistic for the significance of the regression coefficient of Y_2 on Y_1^2 in a univariate linear model in which Y_2 is regressed on Y_1 and Y_1^2. In special circumstances nonlinear functions other than Y_1^2, for example $1/Y_1$, could be used, or an F statistic could be calculated for regression on a set of such functions. To treat the component variables symmetrically, we take with $Q_{2,1}$ the statistic $Q_{1,2}$.

The joint distribution of $(Q_{2,1}, Q_{1,2})$ is complicated, even though the individual distributions of $Q_{2,1}$ and $Q_{1,2}$ are simple, so that to work with a test depending on both components we consider the asymptotic distribution. In fact under the null hypothesis $(Q_{2,1}, Q_{1,2})$ is asymptotically bivariate normal with zero mean and unit variance. It remains therefore to find the asymptotic correlation coefficient of $Q_{2,1}$ and $Q_{1,2}$.

For this we may ignore the denominators of the Student t statistics and examine the correlation between the random variables

$$T_{21} = \Sigma Y_{i2}\{(Y_{i1} - \overline{Y}_{.1})^2 - (Y_{i1} - \overline{Y}_{.1})\, m_{30}/m_{20} - m_{20}\},$$

$$T_{12} = \Sigma Y_{i1}\{(Y_{i2} - \overline{Y}_{.2})^2 - (Y_{i2} - \overline{Y}_{.2})\, m_{03}/m_{02} - m_{02}\},$$

where, for example, $\overline{Y}_{.1} = \Sigma Y_{i1}/n$, $m_{r0} = \Sigma(Y_{i1} - \overline{Y}_{.1})^r/n$. We can without loss of generality take the random variables (Y_{i1}, Y_{i2}) as bivariate normal of zero mean, unit variances and correlation coefficient ρ. Then $Y_{i2} = \rho Y_{i1} + \tau Z_i$, where (Y_{i1}, Z_i) are independently standard normal, so that $\tau^2 = 1 - \rho^2$. It follows that

$$T_{21} = \Sigma Z_i(Y_{i1}^2 - 1) + O_p(1)$$

and that therefore,

$$\mathrm{cov}\,(T_{21}, T_{12}) \sim nE\{(Y_2 - \rho Y_1)(Y_1^2 - 1)(Y_1 - \rho Y_2)(Y_2^2 - 1)\} = 2n\rho(1-\rho^2)(2-3\rho^2),$$

on evaluating the relevant moments. Thus asymptotically

$$\mathrm{corr}\,(Q_{2,1}, Q_{1,2}) = \rho(2-3\rho^2), \tag{1}$$

where $\rho = \mathrm{corr}\,(Y_1, Y_2)$. This is consistently estimated on replacing ρ by r_{12}, the sample correlation coefficient of Y_1 and Y_2. Thus, if required, we can form a composite test statistic either as

$$\max\,(|Q_{2,1}|, |Q_{1,2}|) \tag{2}$$

or as the quadratic form

$$[Q_{2,1} \quad Q_{1,2}] \begin{bmatrix} 1 & r_{12}(2-3r_{12}^2) \\ r_{12}(2-3r_{12}^2) & 1 \end{bmatrix}^{-1} \begin{bmatrix} Q_{2,1} \\ Q_{1,2} \end{bmatrix}. \tag{3}$$

The statistic (2) can, for large samples, be tested for significance from tables of the bivariate normal distribution and the statistic (3) by the chi-squared distribution with two degrees of freedom.

Note from (1) that if ρ is small $Q_{2,1}$ and $Q_{1,2}$ have a correlation with the same sign as ρ, whereas if ρ is large the correlations have opposite signs; this last fact has a simple geometrical interpretation.

If information is available from several independent samples, all concerning the same two variables, a composite statistic can be formed in various ways.

3·3. *More than two component variables*

The ideas of § 3·2 can be generalized to v component variables in several ways. Among these methods are the following.

(a) We may regress each Y_r linearly on all other Y_s and on Y_t^2, and thus obtain a Student t statistic $Q_{r,t}^{(v)}$ $(t \neq r)$ for the quadratic contribution. There are $v(v-1)$ such statistics, and they can be regarded as forming a $v \times v$ array with empty main diagonal.

(b) The statistics considered in (a) can be supplemented by a further set of statistics $Q_{r,t,u}^{(v)}$ $(r \neq t \neq u)$, examining the regression of Y_r on $Y_t Y_u$, adjusting for the linear terms as before. This gives a further $\frac{1}{2}v(v-1)(v-2)$ statistics, and so in all $\frac{1}{2}v^2(v-1)$ statistics.

(c) Approaches (a) and (b) could be applied to marginal dependencies, regressing Y_r on Y_s^2 and Y_s, omitting all other variables, i.e. obtaining the statistics $Q_{r,s}^{(2)}$ of § 3·2. More generally suitable v'-dimensional definitions $(v' < v)$ could be examined.

(d) Instead of isolating single degrees of freedom, we may combine the contributions by forming in the standard way an F statistic for, in case (a), fitting all Y_s^2 $(s \neq r)$ in regressing Y_r on all Y_s and all Y_s^2.

(e) We may use Tukey's degree of freedom for nonadditivity (Tukey, 1949) to obtain one degree of freedom from each variable in turn. If Y_r is the dependent variable, let $\hat{Y}_r$ be the fitted value arising from linear regression on the remaining variables; then the degree of freedom gives the Student t statistic associated with including $\hat{Y}_r^2$ in the model, in addition to all linear terms. While this procedure has, especially for large v, the advantage of limiting the number of subsidiary statistics to be examined, empirical experience suggests that the dangers of overlooking major effects are too great for the procedure to be safely recommended, at least on its own.

In all these methods nonlinear functions such as reciprocals could be used instead of squares. The methods give rise to a set of test statistics, in general correlated. If there are an appreciable number of these they can be plotted against an appropriate probability scale, often the standard normal; it is known that moderate correlation between the values has little effect on the linearity of the plots. The plots can be replaced by or augmented by approximate significance tests and we discuss these briefly below.

The advantage of procedures that are based on single degrees of freedom is that they give more information for detailed diagnosis if evidence of a departure from the null hypothesis of multivariate normality is found. This suggests that for v not exceeding about 10, one of (a)–(c) should be used. For larger values of v, either the variates should be split into meaningful subsections, or (d), or, conceivably for very large v, (e), applied.

For the remaining discussion we concentrate on (a) and (c). In either approach, the natural graphical method is to plot the ordered Q's against the expected order statistics in samples of size m from the standard normal distribution, where m is the number of Q statistics to be plotted. It is assumed that the sample size from which the statistics are computed is such that the Student t distribution can be treated as effectively normal; if not a nonlinear transformation to marginal normality could be applied. For interpretation it is essential that at least the more extreme points in the plot should be labelled with the two defining suffixes. Note that the signs of the Q's are meaningful, provided that the signs of the original variables are, so that a normal plot is appropriate, rather than a half-normal plot of absolute values.

For more detailed numerical interpretation of the Q's, it is natural to consider them as a square array and to examine row and column sums $Q_{r,.}^{(v)}, Q_{.,s}^{(v)}$ or $Q_{r,.}^{(2)}, Q_{.,s}^{(2)}$, or sums of squares

$$S_{r,.}^{(v)} = \sum_{s \neq r} \{Q_{r,s}^{(v)}\}^2, \quad S_{.,s}^{(v)} = \sum_{r \neq s} \{Q_{r,s}^{(v)}\}^2. \tag{4}$$

It can be shown that approximately the statistics (4) have means $v-1$ and variances $2(v-1)\{1+2(v-2)/n\}$.

It is hard to give a firm discussion of the relative merits of the statistics $Q^{(v)}$ and $Q^{(2)}$. Computational simplicity to some extent favours $Q^{(2)}$ and this will also have advantages if v is comparable with n and the variables are almost independent, for then the fitting of linear regressions will effectively induce 'noise' masking the effects under study. On the other hand, if strong roughly linear relationships are known to be present, it seems sensible to eliminate them and hence to use the statistics $Q^{(v)}$. If n is large compared with v the simplest general procedure is to use the $Q^{(v)}$.

If information from several samples is combined it will usually be best to take the combined $Q_{i,j}^{(v)}$ as a weighted sum of the separate statistics, weighting by the sample size.

4. INVARIANT PROCEDURES

4·1. *General idea*

The procedures of §3 are coordinate-dependent. They in effect look for nonlinearities associated particularly with the variables $Y_1, ..., Y_v$; of course an initial transformation of the original data might be made. To obtain an invariant procedure examining nonlinearity the most direct approach is to find that pair of variables, linear combinations of the original variables, such that one has maximum curvature in its regression on the other. The amount of curvature so achieved is the test statistic, and the form of the two maximizing variables will, hopefully, be a useful diagnostic tool.

4·2. *Development of directions of maximum curvature: Population theory*

In the following discussion we can work either with samples and sample moments, or with random variables and corresponding population moments, which is what is done here.

Suppose that the variable $Y = (Y_1, ..., Y_v)^T$ is standardized so that its components have mean zero and let their covariance matrix be $\Sigma = ((\sigma_{ij}))$. For the higher moments, write for $r, s, t, u = 1, ..., v$,

$$E(Y_r Y_s Y_t) = \mu(r, s, t), \quad E(Y_r Y_s Y_t Y_u) = \mu(r, s, t, u).$$

Consider $X = a^T Y$ and $W = b^T Y$ with $a^T \Sigma a = b^T \Sigma b = 1$, so that X and W have zero mean and unit variance.

Let $\gamma = \gamma_{XW}$ denote the least squares regression coefficient of X on W^2, adjusting for linear regression on W. This is found most simply by considering the orthogonalized form,

$$X = \beta W + \gamma \{W^2 - W E(W^3) - 1\} + \varepsilon, \tag{5}$$

where ε is an error term uncorrelated with W and W^2, so that

$$\gamma_{XW} = \frac{E(X W^2) - E(W^3) E(X W)}{E(W^4) - 1 - \{E(W^3)\}^2}. \tag{6}$$

One population measure of the quadratic contribution to regression is

$$\eta_{XW} = \gamma_{XW} / [E(W^4) - 1 - \{E(W^3)\}^2]^{\frac{1}{2}}, \tag{7}$$

An interpretation of η_{XW}^2 is as the proportion of the total unit variance of X accounted for by the quadratic component in the least squares regression of X on W and W^2.

We can express γ_{XW} and η_{XW} in terms of a and b. For fixed b we wish to maximize the numerator of γ_{XW}, that is, to maximize

$$\zeta(a, b) = \Sigma a_r b_s b_t \mu(r, s, t) - \{\Sigma b_r b_s b_t \mu(r, s, t)\} (\Sigma a_r b_s \sigma_{rs}) \tag{8}$$

subject to $\Sigma a_r a_s \sigma_{rs} = \Sigma b_r b_s \sigma_{rs} = 1$. Consider $\zeta(a, b) - \frac{1}{2}\lambda \Sigma a_r a_s \sigma_{rs}$, where λ is a Lagrange multiplier, and differentiate with respect to a_u to give for $u = 1, ..., v$ at a stationary point,

$$\Sigma b_s b_t \mu(u, s, t) - (\Sigma b_t \sigma_{ud}) \{\Sigma b_r b_s b_t \mu(r, s, t)\} - \lambda \Sigma a_t \sigma_{ut} = 0. \tag{9}$$

Multiplication by b_u followed by summation over u gives $\lambda \Sigma a_t b_u \sigma_{ut} = 0$, and multiplication by a_u and summation gives $\zeta(a, b) - \lambda = 0$. Because it is clear that the maximized $\zeta(., b)$ is nonzero, unless all $\mu(r, s, t)$ are zero, it follows that $\Sigma a_t b_u \sigma_{ut} = 0$, that is that the associated X and W are uncorrelated. Further,

$$a_u = \{\Sigma b_r b_s \mu(r, s, t) \sigma^{tu} - b_u \Sigma b_r b_s b_t \mu(r, s, t)\} / \zeta(a, b), \tag{10}$$

where $((\sigma^{ij})) = \Sigma^{-1}$, which is assumed to exist.

Hence $\eta^2(b)$, the supremum of $\eta^2(a, b)$ over a for fixed b, is

$$\eta^2(b) = \frac{\Sigma b_s b_t b_u \mu(r, s, p) \mu(t, u, q) \sigma^{pq} - \{\Sigma b_r b_s b_t \mu(r, s, t)\}^2}{\Sigma b_s b_t b_u \mu(r, s, t, u) - 1 - \{\Sigma b_r b_s b_t \mu(r, s, t)\}^2}. \tag{11}$$

The required directions for maximum curvature are obtained by maximizing this expression subject to $\Sigma b_r b_s \sigma_{rs} = 1$; except possibly in extremely special cases, this maximization has to be done numerically. The value of b is obtained directly and that of a by substitution in (10).

4·3. Computation of the maximum curvature

We shall continue with the notation above, although thinking rather more of using sample moments and of calculating the maximum of $\eta^2(b)$ for use as a test statistic. To avoid computational instability arising from gross differences in scale, the variates should be standardized to have unit variance as well as zero mean.

Although the constraint $b^T \Sigma b = 1$ is important for the magnitude of the curvature, γ_{XW}, it is irrelevant for η_{XW}, and if no use is made in §4·2 of the relation $\Sigma b_r b_s \sigma_{rs} = 1$, then we ultimately obtain the form, homogeneous in b,

$$\hat{\eta}^2(b) = \frac{(\Sigma b_r b_s \hat{\sigma}_{rs})\{\Sigma b_r b_s b_t b_u \hat{\mu}(r,s,p)\hat{\mu}(t,u,q)\hat{\sigma}^{pq}\} - \{\Sigma b_r b_s b_t \hat{\mu}(r,s,t)\}^2}{(\Sigma b_r b_s \hat{\sigma}_{rs})\{\Sigma b_r b_s b_t b_u \hat{\mu}(r,s,t,u)\} - (\Sigma b_r b_s \hat{\sigma}_{rs})^3 - \{b_r b_s b_t \hat{\mu}(r,s,t)\}^2},$$

where a circumflex denotes a sample value. This expression is now to be maximized without constraint on b.

In nearly normal cases some simplification can be achieved by giving the denominator of $\hat{\eta}^2(b)$ its normal theory value of $2(\Sigma b_r b_s \hat{\sigma}_{rs})^3$ and then concentrating on the maximization of the numerator.

Given one or more starting values b_0 of b, the maximization of $\eta^2(b)$ can be carried out by the use of a 'hill-climbing' algorithm. Suitable b_0 may be selected from evaluations of $\hat{\eta}^2(b)$ for a sequence of b values defined by the intersections in a grid of lines of 'latitude and longitude' on a half-surface of a v-dimensional sphere, noting that $\hat{\eta}^2(b) = \hat{\eta}^2(-b)$. Such a grid may be formed by making uniform divisions of the angles in a system of spherical polar coordinates; the resulting points are spread fairly uniformly over the surface of the hypersphere. If each angular coordinate is divided into m parts $(m > 2)$ then there are $\{(m-1)^v - 1\}/(m-2)$ points; this increases very rapidly with v, even if a coarse grid, i.e. small m, is employed, in which case the probability that the selected b_0 lead to the global maximum, rather than merely to local maxima, is reduced. Also, the effort in evaluating $\eta^2(b)$ is roughly proportional to v^4, assuming that an array $\hat{\tau}(r,s,t,u) = \Sigma \hat{\mu}(r,s,p)\hat{\mu}(t,u,q)\hat{\sigma}^{pq}$ is used. These two facts combine to make $v = 6$ about the limit for computational feasibility. Larger numbers of variables could, for example, be dealt with by dividing them into subsets of size 6 or less.

For a test statistic we concentrate on the global maximum of $\hat{\eta}^2(b)$. For interpretation, however, in a nonnull case, it may well be useful to know the several roughly equal local maxima.

4·4. *The null hypothesis distribution*

Denote the maximum of $\hat{\eta}^2(b)$ by $\hat{\eta}^2_{\max}$. To apply a significance test based on $\hat{\eta}^2_{\max}$, we need to know at least approximately its distribution under the null hypothesis of multivariate normality. Clearly this distribution depends only on v and n. Analytical study of the distribution seems not to be feasible. Simulation shows that approximately for $n \geqslant 50$, $v \leqslant 6$, $\log \hat{\eta}^2_{\max}$ is normally distributed with mean $\log\{(5v^2)/(8n)\}$ and standard deviation $0·90$ $(v = 2)$, $0·59$ $(v = 3)$, $0·38$ $(v = 4)$, $0·31$ $(v = 5)$, $0·17$ $(v = 6)$.

It would be good to have some qualitative explanation both of the log normal shape and of the form of the mean and standard deviation. The nature of the dependence on n may be accounted for in general terms by the following argument. For fixed a and b, $\hat{\eta}^2(a,b)$ has a distribution with asymptotic mean and variance $3(v-1)/(2n)$ and $9(v-1)/(2n^2)$ respectively, under the null hypothesis. We now consider fitting this with the log normal distribution, corresponding to the distribution $N(\mu, \sigma^2)$, which has mean $e^{\mu+\frac{1}{2}\sigma^2}$ and variance $e^{2(\mu+\frac{1}{2}\sigma^2)}(e^{\sigma^2}-1)$. Upon equating moments, we obtain

$$\mu = \log\{\tfrac{3}{2}(v-1)^2(v+1)^{-1}n^{-1}\}, \quad \sigma^2 = \log\{(v+1)/(v-1)\}.$$

Of course $\hat{\eta}^2_{\max} \geqslant \hat{\eta}^2(a,b)$, so that in fitting a log normal distribution to $\hat{\eta}^2_{\max}$, the equalities above could only be maintained by the introduction of constants that were functions of n and p. However, the asymptotic dependence on n should remain unaltered, and hence in making the transition from $\hat{\eta}^2(a,b)$ to $\hat{\eta}^2_{\max}$ the similarity, as functions of n, between the μ and σ above and those obtained empirically, should be preserved.

Note that quite apart from its use as a test statistic $\hat{\eta}^2_{\max}$ has a direct numerical interpretation as a maximal proportion of variance accounted for by quadratic regression.

5. Interpretation

While the procedures of §§ 3 and 4 have been described in the first place as tests of significance, if evidence of nonlinearity is found some interpretation has always to be attempted. In the absence of a simple widely applicable alternative family of distributions, no general rules can be given, but the following comments may be helpful.

Inspection of scatter diagrams will always be required for interpretation. For the procedure of § 4 a first step will be to examine the plot for the derived variables for which η^2 is maximal. For the coordinate dependent approach of § 3, pairs of original variates may be plotted, or alternatively residuals of Y_r and Y_l^2 from their linear regression on the remaining variables, when $Q_{r,l}^{(v)}$ appears interestingly large. In clearcut cases the nonlinearity will arise either from a small number of aberrant points, which will then need special consideration, or from a consistent curvature. If the curvature arises in connection with only one or two component variables, it may be sensible to treat the remainder as multivariate normal and to describe separately the dependence of the anomalous variables on the remainder. Consistent patterns of signs in the curvature may indicate the general nature of appropriate transformations.

It would in principle be possible to develop techniques corresponding to those of §§ 3 and 4 but using some form of robust regression rather than least squares regression. We have not investigated this.

6. An example

To illustrate the way in which the results might be applied, we now give in brief outline an analysis of some data circulated some years ago by Dr P. D. P. Wood, Milk Marketing Board, to the Multivariate Study Group of the Royal Statistical Society. The data comprised 8 measurements on the pelvis of each of 90 Friesian cows.

The upper estimates in Table 1 give the first four estimated moments of the marginal distributions. For samples of size 90 from a univariate normal distribution, the lower and upper 5% points for g_1 are roughly -0.6 and 0.6, and for $g_2 - 0.85$ and 1.47, respectively. A frequency plot for variable 7 showed an observation at 292 mm, the range for the other animals being 150 to 223 mm. All observations on this extreme animal were omitted in the subsequent analyses. In particular this omission reduced g_1 and g_2 for variable 7 to -0.24 and 0.25. Study of the marginal distributions showed no other obvious outliers, although the first variable was markedly bimodal. Note that there is no evidence of systematic skewness; a log transformation was therefore not applied.

Table 1. *Pelvic measurements on cows. First four marginal moments.*
Upper values, all 90 cows. Lower values, selected 84 cows

	1	2	3	4	5	6	7	8
Mean (mm)	254·5	205·5	238·2	206·3	200·5	177·6	190·5	189·4
	251·1	203·6	240·6	209·2	202·7	180·6	189·0	189·2
St. dev. (mm)	25·9	12·4	20·8	20·2	17·5	21·2	16·7	15·6
	22·8	9·7	18·3	15·7	14·7	17·9	12·8	15·4
g_1	-0.01	1·01	-0.08	-1.13	-0.01	-0.60	2·26	-0.10
	-0.49	0·20	0·55	0·20	0·08	0·07	-0.25	-0.18
g_2	-0.22	1·63	1·14	3·60	2·82	1·52	13·33	1·13
	-1.03	-0.38	0·66	0·35	0·27	0·13	0·27	1·35

The next step was to use the coordinate-dependent methods of § 3·3. Method (e), using Tukey's degree of freedom for nonadditivity, was tried; the largest Student t statistic out of 8 is 2·58, for the regression of variable 2 on the square of the fitted value of its linear regression on the other variables, but such a value is not markedly extreme. The array of t statistics generated by method (a) contained a number of abnormally large values, mostly but not entirely connected with variable 2. Various scatter plots showed that there were 5 animals which in the space of the first 5 variables form an outlying group not on the main linear regression and therefore inducing curvature. They, too, were omitted for separate interpretation and method (a) of § 3·3 reapplied to the remaining 84 individuals. Table 2 shows the resulting Student t statistics and their marginal sums of squares: see (4). The ranked Student t statistics can be plotted against expected normal order statistics. There is nothing untoward.

The invariant procedure of § 4 was then applied to these remaining 84 individuals, taking 6 variables at a time for computational reasons. The largest value of η^2_{max} obtained was 0·341, which is not only well short of statistical significance but corresponds to only a modest degree of curvature.

Table 2. *Pelvic measurements. Curvature analysis for 84 selected cows*

Dependent variable	\multicolumn{8}{c	}{Squared variable}	Sum of squares for row
	1	2	3	4	5	6	7	8	
1	—	1·78	1·65	0·86	−0·63	−1·29	1·63	0·89	12·16
2	0·51	—	0·32	0·94	1·36	0·85	−0·48	−0·49	4·28
3	0·34	−2·40	—	−0·31	0·11	1·13	−1·07	−0·89	9·20
4	−0·09	2·08	0·34	—	−0·40	−1·76	−0·80	−1·03	9·41
5	−0·06	−0·51	−1·87	−1·21	—	2·35	−0·50	−0·82	11·66
6	−0·15	0·95	1·52	0·90	−0·10	—	1·99	1·23	9·53
7	0·33	−0·56	−0·92	−0·47	1·00	0·92	—	−1·33	5·11
8	−0·09	−0·57	−0·33	−0·74	−1·66	−0·33	0·39	—	4·00
Sum of squares for column	0·53	15·06	9·71	4·77	6·17	13·29	9·04	6·83	65·34

The marginal moments for the 84 individuals are recorded in the lower values of Table 1; there is some bimodality in variable 1.

To summarize, one outlier and five anomalous individuals have been detected. The remaining 84 individuals, while showing some evidence of marginal nonnormality, show no evidence of nonlinearity, and interpretation of the interrelationships among the 8 variables via their covariance matrix seems in order. No doubt these conclusions could be reached via other routes.

We are grateful to Dr P. D. P. Wood for permission to use the data analysed in § 6. N. J. H. Small's work was supported by the Science Research Council.

REFERENCES

ANDREWS, D. F., GNANADESIKAN, R. & WARNER, J. L. (1971). Transformations of multivariate data. *Biometrics* 27, 825–40.

ANDREWS, D. F., GNANADESIKAN, R. & WARNER, J. L. (1973). Methods for assessing multivariate normality. In *Multivariate Analysis*, Vol. 3, Ed. P. R. Krishnaiah, pp. 95–116. New York: Academic Press.

BARNDORFF-NIELSEN, O. (1977). Discussion of paper by D. R. Cox. *Scand. J. Statist.* 4, 67–9.

Box, G. E. P. & Cox, D. R. (1964). An analysis of transformations. *J. R. Statist. Soc.* B 26, 211–52.

Cox, D. R. (1968). Notes on some aspects of regression analysis. *J. R. Statist. Soc.* A 131, 265–79.

DAY, N. R. (1969). Divisive cluster analysis and a test for multivariate normality. *Bull. I.S.I.* 43, 2, 110–2.

Durbin, J. (1973). *Distribution Theory for Tests Based on the Sample Distribution Function*. Philadelphia: Society for Industrial and Applied Mathematics.

Gnanadesikan, R. (1977). *Methods for Statistical Data Analysis of Multivariate Observations*. New York: Wiley.

Healy, M. J. R. (1968). Multivariate normal plotting. *Appl. Statist.* 17, 157–61.

Malkovich, J. F. & Afifi, A. A. (1973). On tests for multivariate normality. *J. Am. Statist. Assoc.* 68, 176–9.

Mardia, K. V. (1970). Measures of multivariate skewness and kurtosis with applications. *Biometrika* 57, 519–30.

Mardia, K. V. (1975). Assessment of multinormality and the robustness of Hotelling's T^2 test. *Appl. Statist.* 24, 163–71.

Tukey, J. W. (1949). One degree of freedom for non-additivity. *Biometrics* 5, 232–42.

[*Received November* 1977. *Revised February* 1978]

Appl. Statist. (1978),
27, *No.* 1, *pp.* 4–9

Some Remarks on the
Role in Statistics of Graphical Methods

By D. R. Cox

Department of Mathematics, Imperial College, London

SUMMARY

Graphical methods are useful in statistics in exploratory analysis and in the presentation of conclusions, and to a more limited extent in very specific analyses. After some historical remarks, some general rules and principles are outlined. The role of smoothing is discussed and some techniques for introducing supplementary information outlined.

Keywords: DIAGRAMS; EXPLORATORY ANALYSIS; GRAPHS; HISTORY OF STATISTICS; PRESENTATION OF CONCLUSIONS

1. INTRODUCTION

THE object of this paper is to make a few general comments on the use in statistics of graphical methods. Throughout the discussion it is important to distinguish between their role:
(a) in exploratory analysis,
(b) in the specific analysis of data where relatively clearly formulated questions are to be answered, often in the light of an explicitly specified model, and
(c) in the presentation of conclusions.

For example, in the presentation of conclusions, especially to a fairly general audience, rather simple direct procedures will usually be desirable, whereas in explanatory analysis more specialized and complex techniques may be in order.

2. HISTORICAL BACKGROUND

It is trite to remark that history is loaded in favour of those who leave a written record of their work. Thus it is easier to find the history of graphical methods for the presentation of conclusions than of graphs as an everyday working tool of the scientist.

Table 1 is largely based on the paper of Funkhauser (1938) which gives a very detailed history of graphical methods, with some emphasis on their use in connection with official statistics. The table gives a very small and arbitrary selection of events and the following notes concern just a few of these.

Florence Nightingale's report on the British Army (Nightingale, 1857) contained much elaborate graphical analysis, including the use of colour. In the period 1860–80, which has been called the "golden age" of graphics, there was much work, especially among official statisticians in continental Europe, on elaborate techniques of graphical presentation, and there were a number of abortive attempts at international standardization. Note, however, that international standardization has been achieved to an appreciable extent in engineering (Arnell, 1963).

The French engineer L. Lalanne in 1843 studied in detail the idea of linearizing plots by non-linear transformation of the scales, what he called anamorphism, and so in particular introduced the log–log plot.

Of the great 18th- and 19th-century applied mathematicians, most of whom had some interest in statistics, d'Alembert and Fourier seem to have used graphical methods most

4

extensively. So far as I know, there is, for example, no record of graphical work by Gauss. The Belgian statistician Quetelet used graphical methods extensively in theoretical and applied work.

We take for granted the availability of reasonably accurately ruled graph paper. Early workers prepared their own. It seems that graph paper began to become fairly widely available

TABLE 1

The development of graphical methods*

(a) *Before 19th century*

1320–82	Oresme. Qualitative idea of graph of a function
1596–1650	Descartes. Co-ordinates
1623	Gunter scale
1690–	Halley *et al.* Plots and contour plots of empirical functions
1759–1823	W. Playfair. Systematic use of graphical methods for social and economic data

(b) *19th century*

1800	Empirical plot of physical time series with carefully prepared graph paper (*Phil. Mag*)
1821	Fourier. Plot of empirical cumultive distribution function
1843	Lalanne. Anamorphism. Contour plots. Stereogram
1857	Florence Nightingale's report
1860 (?)	Graph paper began to become common
1860–90	"Golden age" of graphical techniques. Attempts at international standardization
1874	Death of Quetelet
1875	W. Froude. Ruling machine
1883	Log–log paper patented
1884	D'Ocagne. Nomography
1899	Galton suggested probability paper

---

* Largely based on Funkhauser (1938).

commercially by about 1860, although it is not clear quite how generally it was used. Thus as late as 1875, Jevons described the construction of graph paper in a way suggesting that it was not familiar; at about that time the naval engineer W. Froude was sufficiently dissatisfied with the accuracy of commercial graph paper that he built a very accurate ruling machine solely for producing graph paper.

The more recent history of graphical methods is difficult to comment upon. My impression is that in the immediate postwar years graphical techniques were very widely used. Various special graph papers for probability plotting, including Professor J. W. Tukey's extremely ingenious "binomial paper", were readily available commercially. In the period 1955–70, a central emphasis among statisticians was on the computer implementation of "standard" statistical calculations in a form for relatively painless use. With this largely accomplished, and with the increasing availability of graphical output from the computer, there has been a recent upsurge of interest in graphical methods, well shown by the large attendance at the Sheffield Conference.

The role of the computer is, of course, partly to enable previously available methods to be used easily and partly to allow new developments; to begin with, at least, the first role is likely to be predominant.

3. SOME GENERAL ASPECTS

There is a major need for a theory of graphical methods, although it is not at all clear what form such a theory should take. Of course, theory is not to be taken as meaning mathematical theory! What are required are ideas that will bring together in a coherent way things that previously appeared unrelated and which also will provide a basis for dealing systematically with new situations.

At a simpler level, some elementary but important suggestions for the clarity of graphs are as follows:
(i) the axes should be clearly labelled with the names of the variables and the units of measurement;
(ii) scale breaks should be used for false origins;
(iii) comparison of related diagrams should be made easy, for example by using identical scales of measurement and placing diagrams side by side;
(iv) scales should be arranged so that systematic and approximately linear relations are plotted at roughly 45° to the x-axis;
(v) legends should make diagrams as nearly self-explanatory, i.e. independent of the text, as is feasible;
(vi) interpretation should not be prejudiced by the technique of presentation, for example by superimposing thick smooth curves on scatter diagrams of points faintly reproduced.

Some published diagrams, and many submitted for publication, break one or more of these points.

The following suggestions are concerned with the planning of what is to be plotted and hence are rather more fundamental. So far as is feasible, diagrams should be planned so that
(a) departures from "standard" conditions should be revealed as departures from linearity, or departures from totally random scatter, or as departures of contours from circular form;
(b) different points should have approximately independent errors;
(c) points should have approximately equal errors, preferably known and indicated, or, if equal errors cannot be achieved, major differences in the precision of individual points should be indicated, at least roughly;
(d) individual points should have clearcut interpretation;
(e) variables plotted should have clearcut physical interpretation;
(f) any non-linear transformations applied should not accentuate uninteresting ranges;
(g) any reasonable invariance should be exploited.

The snag with these requirements is, as is fairly obvious, that they tend to be mutually contradictory. A common dilemma is typified by the choices between a cumulative sum chart versus an "ordinary" control chart, a cumulative distributional plot versus a histogram and a cumulative periodogram versus an estimated spectral density. In each case, the first method gives smooth results, and correspondingly lends itself to the detection of certain kinds of smooth change, and moreover often can be arranged to be linear in simple situations, whereas the second allows an easier appreciation of random scatter. Thus if visual assessment of random error is especially important, points (b) and (c) will get considerable weight.

The embarrassing variety of choices that may be available for very simple problems is shown rather cryptically in Table 2 which summarizes some properties of ten different graphical methods for assessing the fit of values to an exponential distribution.

4. SMOOTHING

As indicated above, an important choice in a number of contexts is of the severity of smoothing appropriate. This has been much discussed in the theoretical literature both in the context of regression and of density estimation. The work concentrates on minimization of mean squared error. This may be reasonable when the objective is the final presentation of, say, a calibration curve for direct practical use. If, however, the graph is to be used for exploratory purposes, or for the final presentation of conclusions in a context where the reader may wish to assess error for himself, the mean squared error criterion seems largely irrelevant; for one thing, it is likely to be changes in or unusual features of the graph that are of most interest, rather than individual ordinates as such.

There are two consequences. One is that in this second set of circumstances use of relatively light smoothing will often be wise. Secondly, in most cases it will be good to use apparently crude methods in which non-overlapping blocks of points are averaged in a simple way to produce "smoothed" points with independent errors, and if possible points with equal and known standard errors. Thus in spectral estimation this suggests simple Daniell smoothing

TABLE 2

Some graphical methods for assessing consistency with an exponential distribution

	Technique	*Criteria by which technique is*	
		Satisfactory	*Unsatisfactory*
1.	Ordered observations versus expected exponential order statistics	a, d, e	b, c
2.	Ordered log observations versus expected extreme value order statistics	a, d, e	b, c, f
3.	Log survivor function versus y	a, e	b, c
4.	Histogram with fitted exponential curve	b, d, e	a
4a.	Residual of histogram from fitted curve	a, b	d, e
5.	Rootogram (square root of histogram) with fitted curve	b, c, d	e
5a.	Residual of rootogram from fitted curve	a, b, c	d, e
6.	"Time on test" plot	a	b, c, d
7.	Hazard estimated from graphical data	a, b, d	
8.	Log hazard estimated from groups with equal numbers of observations	a, b, c, d	

rather than use of one of the more elaborate windows. For an ingenious special method of spline-like smoothing based on reference points with independent errors, see Jeffreys (1968, Section 4.5).

5. DISPLAY OF SUPPLEMENTARY INFORMATION

Most graphs used in the analysis of data consist of points arising in effect from distinct individuals, although there are certainly other possibilities, such as the use of lines dual to points. In many cases of exploratory analysis, however, the display of supplementary information attached to some or all of the points will be crucial for successful interpretation. The primary co-ordinate axes should, of course, be chosen to express the main dependence explicitly, if not initially certainly in the final presentation of conclusions.

There are two general decisions to be made when displaying supplementary information, the first concerning the amount of such information and the second the precise technique to be used.

The amount of supplementary information that it is sensible to show depends on the number of points. The possibility of showing such information only for relatively extreme points and possibly for a sample of the more central points should be considered when the number of points is large; thus in a probability plot of contrasts from a large factorial experiment it may be enough to label only the more extreme values.

The main methods for showing supplementary information are by
(a) naming a point;
(b) the shape used to mark a point;
(c) colour;
(d) the use of lines of various lengths emanating from a point in various directions (Anderson, 1957);
(e) isometric plots;

(f) stereographic techniques;
(g) cinematographic techniques.

When the supplementary information is of further quantitative variables, additional to those used in defining the co-ordinate axes, and when the number of points is fairly small, (d) is an especially effective technique.

6. Interpretation of Complex Shapes

There is an interesting interplay between vectors of numbers and curves or shapes. A recurring theme in pattern recognition and stereology is the description of a complex shape by a vector of numbers, which can then be the basis of quantitative analysis. Yet a central idea of the graphical analysis of multivariate data is, in a sense, the opposite one. Numerical formulation is, of course, very desirable for any form of automatic decision-making and data recording and, if the dimensionality can be made small, for incisive summarization.

Two contrasting techniques for replacing data vectors by shapes are the orthogonal function method of Andrews (1972) and the use of faces (Chernoff, 1973). In the first the data components are used as coefficients in an expansion in orthogonal functions, usually a Fourier series, each individual thus giving a curve. In the second the data components determine various aspects of a diagram of a face, each individual thus giving a face. In the first procedure, but not the second, the different component variables can be assigned in a rational way and, despite the undoubted ingenuity of the second technique, it seems relatively limited as a systematic aid to rational understanding.

Oxnard (1973) reviews a wide variety of techniques for the description of shape.

7. Special Statistical Fields

In a very detailed account of graphical methods, it would be possible to discuss separately techniques for the main special statistical topics, e.g. single sample methods, regression, factorial experiments, time series and multivariate analysis; for the last, see Gnanadesikan (1977). As emphasized throughout the paper, rather different techniques may be suitable for exploratory analysis, including the checking of assumptions, from those for the presentation of conclusions. Appropriate graphical analysis may make the conclusions so clearcut that detailed specific analysis is unnecessary; simple sign tests based directly on the graphs may then be enough to reassure the sceptical or fainthearted. Inspection of the form of a likelihood function or surface, in particular to prevent inappropriate summarization by point estimates and standard errors, can be important in some contexts. On the whole, however, the role of graphical methods in specific analysis is limited.

Graphical methods can be valuable for presenting the results of theoretical work in concise form. For instance, a probit transformation will often transform power curves at different significance levels into approximately parallel curves or even lines. For a particularly ingenious method of comparing the powers of tests via the contours of a stylized sensitivity function, see Prescott (1976).

For decision-making, graphical methods of presentation may be particularly appropriate for displaying data obtained sequentially, whether or not a formal sequential stopping rule is operated.

The specialized techniques of nomography for calculating graphically functions of several variables are made largely but not entirely obsolete by the pocket calculator. The techniques of construction involve some ingenuity with "old-fashioned" projective geometry, and there are some deep problems connected with existence.

8. Discussion

This final section is devoted to a few miscellaneous comments.

First, the role of graphical methods in teaching is of great importance, ranging from the use of simple idealized diagrams to illustrate theoretical points in concrete form, to the direct teaching of graphical methods as such, especially important in inculcating a critical attitude to assumptions and model formulation.

The role of graphs in the publication of conclusions, especially formally in journals, deserves comment. The cost per page of diagrams tends to be lower than that of verbal text, which in turn is lower than that of tables. In terms of amount of information presented, however, especially where detailed quantitative conclusions are involved, graphs are relatively expensive. There is at present intense pressure on journals for economies in production and this will usually preclude double presentation of information by both tables and graphs. In some ways these pressures are dangerous in that the money saved by omitting information on publication may be very small compared with the cost of obtaining that information. On the other hand, the relation between conciseness and lucidity is non-monotonic; up to a point conciseness in presenting conclusions is highly desirable and pressures towards such conciseness are to be welcomed by all except the immediate recipient. There is some scope for economy by the use of cheaper (and nastier) methods of production; I feel, however, that journals should aim to maintain present standards of quality, as part of a system of incentives for careful presentation.

Finally, there has been no attempt in the paper to discuss explicitly the relative merits of tables and graphs for the presentation of conclusions. Both call for considerable care in design and in presentation. Bachhi (1968), by very ingenious use of symbols of various shapes and by colour coding, has given methods for graphical display of material that would normally be given in tabular form and this is in the spirit of the 19th-century work mentioned in Section 2. While there are situations where the choice between graphical and tabular presentation is a matter of taste, the arguments seem persuasive for tabular presentation of detailed numerical information, especially when it is of some permanent importance (Ehrenberg, 1977).

ACKNOWLEDGEMENTS

This paper is based on an introductory talk given at a Conference on Graphical Methods in Statistics, Sheffield, March 1977, and I am grateful to the Organizing Committee for their invitation. I had very valuable help with the historical section of the paper from my colleagues Ms Brenda Sowan and Dr Eduardo Ortiz and from Ms Judit Torok, Science Library.

REFERENCES

ANDERSON, E. (1957). A semi-graphical method for the analysis of complex problems. *Proc. Nat. Acad. Sci.*, **43**, 923–927.
ANDREWS, D. F. (1972). Plots of high-dimensional data. *Biometrics*, **28**, 125–136.
ARNELL, A. (1963). *Standard Graphical Symbols*. New York: McGraw-Hill.
BACHHI, R. (1968). *Graphical Rational Patterns*. Jerusalem: Israel University Press.
CHERNOFF, H. (1973). The use of faces to represent points in k-dimensional space graphically. *J. Am. Statis. Assoc.*, **68**, 361–368.
EHRENBERG, A. S. C. (1977). Rudiments of numeracy. *J. R Statist. Soc.* A, **140**, 277-297.
FUNKHAUSER, H. G. (1938). Historical development of the graphical representation of statistical data. *Osiris*, **3**, 269–404.
GNANADESIKAN, R. (1977). *Methods for Statistical Data Analysis of Multivariate Observations*. New York: Wiley.
JEFFREYS, H. (1968). *Theory of Probability*, 3rd ed. Oxford University Press.
NIGHTINGALE, F. (1857). Notes on matters affecting the health, efficiency and hospital administration of the British Army. London. Private publication.
OXNARD, C. (1973). *Form and Pattern in Human Evolution*. Chicago University Press.
PRESCOTT, P. (1976). Comparison of tests for normality using stylized sensitivity surfaces. *Biometrika*, **63**, 285–289.

Biometrika (1986), **73**, 3, *pp.* 543–54
Printed in Great Britain

Analysis of variability with large numbers of small samples

By D. R. COX

Department of Mathematics, Imperial College, London SW7 2BZ, U.K.

AND P. J. SOLOMON

Department of Statistics, University of Adelaide, Adelaide, S.A., Australia 5001

SUMMARY

Procedures are discussed for the detailed analysis of distributional form, based on many samples of size r, where especially $r = 2, 3, 4$. The possibility of discriminating between different kinds of departure from the standard normal assumptions is discussed. Both graphical and more formal procedures are developed and illustrated by some data on pulse rates.

Some key words: Graphical methods; Kurtosis; Nonnormality; Order statistics; Overdispersion; Pulse rate; Skewness.

1. INTRODUCTION

It is common both in some kinds of balanced experimental design and in schemes of routine testing to have quite large numbers of small groups of observations, each group obtained under the same conditions. For example, in a large study, blood pressure measurements might be taken in duplicate on each patient visit, or in certain routine chemical analyses triplicate samples from the same source might be dealt with independently.

There can be major problems in ensuring the independence of such replicate observations necessary to ensure that the relevant source of variability is not underestimated. Assuming that this independence is achieved, preferably by appropriate 'blinding', we use the replicate observations primarily to improve the precision of the mean, or other location estimate, but also to estimate a component of variance within samples, usually measuring sampling or measurement error. A further use is to check for gross errors affecting a single observation.

In the present paper, however, we examine what further information can be extracted from such data. Throughout we suppose that, possibly after transformation, the 'standard' assumption of a normal distribution with constant variance is at least a reasonable starting point for an analysis. Of course more detailed analysis is likely to be worthwhile only if the variation within samples is of intrinsic interest. We assume throughout that rounding errors, digit preferences and the like are relatively unimportant and that the variability studied is not an artefact.

2. SOME GENERAL IDEAS

2·1. *Formulation of models*

Suppose that we have m independent samples each of size r, the observations in the ith sample being $y_{i1}, \ldots, y_{ir}$ $(i = 1, \ldots, m)$. As a basis for the analysis we consider a number of possible models for the corresponding random variables $\{Y_{ip}\}$, of which the simplest is the standard, normal theory one.

Normal theory model, M_N. The Y_{ip} are independently normally distributed with constant variance σ^2 and with $E(Y_{ip}) = \mu_i$.

As explained in § 1 the focus of interest in the present paper is not the $\{\mu_i\}$, which may indeed have some additional structure specified, for instance, by a regression or factorial model. We shall typically suppose that there are substantial differences in mean present and that no useful information about the internal variability can be recovered from the between-sample variation.

There are numerous ways in which interesting departures from M_N may occur and we shall consider just three.

Systematic changes in variance, M_{SY}. Here the assumptions of M_N hold except that var $(Y_{ip}) = \sigma_i^2$, which is not constant but is a function either of an explanatory variable z characterizing the ith population or of μ_i. Such variation can, if necessary, be represented in various ways (Cook & Weisberg, 1983), e.g.

$$\sigma_i^2 = e^{\beta z_i} \sigma_0^2, \quad \sigma_i^2 = e^{\beta \mu_i} \sigma_0^2, \quad \sigma_i^2 = e^{\beta \log \mu_i} \sigma_0^2 = \mu_i^\beta \sigma_0^2,$$

where σ_0^2 is a 'baseline' variance and β captures the systematic dependency present. An important extension allows for more complex multidimensional dependence: for example, if the samples are arranged in a row × column array, there may be systematic differences in variance between rows, between columns, or both.

Complementary to M_{SY} is the possibility that, while each population has a different variance, the changes in variance are random, unrelated to any observed feature. This is a model of overdispersion relative to M_N.

Overdispersion model, M_{OD}. Again the normal theory assumptions of M_N are modified only by allowing each population to have a different variance, var $(Y_{is}) = \sigma_i^2$, but now $\tau_i = \sigma_i^2$ $(i = 1, \dots, m)$ are independent unobserved values of a random variable T having a probability density function $h(t)$. In the type of application we have in mind, it will often not be feasible to estimate the form of $h(t)$ with any precision, and it may then be adequate to assume that T has an inverse gamma distribution, with density

$$(\tfrac{1}{2}f_0\tau_0')^{\frac{1}{2}f_0} t^{-\frac{1}{2}f_0-1} \exp\left(-\tfrac{1}{2}f_0\tau_0'/t\right)\{\Gamma(\tfrac{1}{2}f_0)\}^{-1}, \tag{1}$$

where f_0 plays the role of an 'effective degrees of freedom' and

$$E(T) = \tau_0' f_0 (f_0 - 2)^{-1} = \tau_0,$$

say.

For our final model, we suppose that, except for location, all populations have the same distribution, which is, however, nonnormal.

Nonnormal model, M_{NN}. All $\{Y_{is}\}$ are independent and the density of Y_{is} is $g(y - \mu_i)$, where $g(x)$ is a nonnormal density of zero mean. We write κ_r for the rth cumulant of Y_{is} and $\rho_r = \kappa_r/\sigma^r$ $(r \geq 3)$.

Of course, there are other possibilities and, in particular, we could combine M_{SY}, M_{OD} and M_{NN} in various ways, and also introduce models involving serial correlation.

We are interested in methods for detecting departures from M_N, in the estimation of relevant parameters in M_{SY}, M_{OD} and M_{NN}, and in studying the feasibility of discriminating between these three kinds of departure, in particular between M_{OD} and M_{NN}. While for very small values of r, separation of M_{OD} and M_{NN} may often not be possible, note that if there is substantial underdispersion in the sample estimates of variance, M_{OD} can at once be eliminated from consideration.

No special model for gross errors has been included. Both M_{OD} and M_{NN} can simulate such errors, in the first case via an occasional very large 'true' variance and in the second via a very long-tailed error distribution. Note that with $r > 2$ if gross errors were detected to be predominantly in one direction, M_{NN} rather than M_{OD} would be required.

To some extent further analysis and interpretation may be quite strongly influenced by what kind of departure from M_N is most appropriate. Thus, as between M_{OD} and M_{NN}, the former prompts the question 'why are some groups more variable than others?', whereas under the latter the errors may more reasonably be presumed to have a totally homogeneous structure.

2·2. *Some simple properties of M_{OD} and M_{NN}*

For the ith sample, we write

$$\bar{y}_{i.} = \Sigma\, y_{ip}/r, \quad S_i = \Sigma\, (y_{ip} - \bar{y}_{i.})^2$$

for the sample mean and sum of squares. These are sufficient under both M_N and M_{OD}. It is convenient also to write $\tilde{\sigma}_i = \{S_i/(r-1)\}^{\frac{1}{2}}$ and $\tilde{\tau}_i = \tilde{\sigma}_i^2$ for the usual estimates of standard deviation and variance.

For the overdispersion model M_{OD} with the inverse gamma compounding density (1), the likelihood of the ith sample is

$$\frac{(f_0\tau_0')^{\frac{1}{2}f_0}\Gamma(\frac{1}{2}f_0+\frac{1}{2}r)}{\Gamma(\frac{1}{2}f_0)\pi^{ir}\{f_0\tau_0'+r(\bar{y}_{i.}-\mu_i)^2+S_i\}^{\frac{1}{2}f_0+\frac{1}{2}r}}, \tag{2}$$

which is of the Student's t form. If the $\{\mu_i\}$ are regarded as nuisance parameters, we examine the marginal likelihood based on S_i which is

$$\frac{S_i^{\frac{1}{2}r-3/2}\{1+S_i/(f_0\tau_0')\}^{-\frac{1}{2}f_0-\frac{1}{2}r+\frac{1}{2}}}{(f_0\tau_0')^{\frac{1}{2}r-\frac{1}{2}}B(\frac{1}{2}f_0,\frac{1}{2}r-\frac{1}{2})}. \tag{3}$$

For the general compounding density $h(\tau)$, (3) is replaced by

$$\int_0^\infty \tau^{-1}q_{r-1}(S_i\tau^{-1})h(\tau)\, d\tau, \tag{4}$$

where $q_{r-1}(x)$ is the probability density of the chi-squared distribution with $r-1$ degrees of freedom.

Estimation under the model M_{OD} can be based on (3). Alternatively, and also for discriminating between M_{OD} and M_{NN}, it is useful to record the cumulants $\{\kappa_j(\tilde{\tau}_i)\}$ of $\tilde{\tau}_i$ in terms of the cumulants $\{\lambda_j\}$ of the compounding density $h(\tau)$. These are, with $\sigma^2 = \lambda_1 = E(\tau)$,

$$E(\tilde{\tau}_i) = \sigma^2, \quad \text{var}\,(\tilde{\tau}_i) = \sigma^4\left(\frac{2}{r-1}+\frac{r+1}{r-1}\frac{\lambda_2}{\sigma^4}\right) = \sigma^4\left(\frac{2}{r-1}+\psi_2\right), \tag{5}$$

say, and

$$\kappa_3(\tilde{\tau}_i) = \sigma^6 \left\{ \frac{8}{(r-1)^2} + 12 \frac{r+1}{(r-1)^2} \frac{\lambda_2}{\sigma^4} + \frac{(r+1)(r+3)}{(r-1)^2} \frac{\lambda_3}{\sigma^6} \right\}.$$

Because λ_2/σ^4, λ_3/σ^6, ... are the cumulants of a nonnegative random variable of unit mean, $\lambda_3/\sigma^6 \geq (\lambda_2/\sigma^4)(\lambda_2/\sigma^4 - 1)$, with equality attained by a two-point distribution with one atom at zero. Therefore

$$\kappa_3(\tilde{\tau}_i) \geq \sigma^6 \left\{ \frac{8}{(r-1)^2} + \frac{(r+1)(9-r)}{(r-1)^2} \frac{\lambda_2}{\sigma^4} - \frac{(r+1)(r+3)}{(r-1)^2} \left(\frac{\lambda_2}{\sigma^4} \right)^2 \right\} \tag{6}$$

$$= \sigma^6 \left\{ \frac{8}{(r-1)^2} + \frac{(9-r)}{(r-1)} \psi_2 - \frac{(r+3)}{(r+1)} \psi_2^2 \right\}.$$

Note that negative skewness in the distribution of $\tilde{\tau}_i$ is possible, although for small values of i, very large values of ψ_2 would be necessary to bring this about.

By comparison we have under the nonnormal model M_{NN} (Kendall & Stuart, 1969, Ch. 12)

$$E(\tilde{\tau}_i) = \sigma^2, \quad \text{var}(\tilde{\tau}_i) = \sigma^4 \left(\frac{2}{r-1} + \frac{\rho_4}{r} \right),$$

$$\kappa_3(\tilde{\tau}_i) = \sigma^6 \left\{ \frac{8}{(r-1)^2} + \frac{4(r-2)}{r(r-1)^2} \rho_3^2 + \frac{12\rho_4}{r(r-1)} + \frac{\rho_6}{r^2} \right\}. \tag{7}$$

From (5) and (7) we can compute fairly directly via the first two moments of the $\{\tilde{\tau}_i\}$ estimates of either λ_2/σ^4 in M_{OD} or ρ_4 in M_{NN}. Both are dimensionless measures of the departure from the standard conditions M_N.

For $r > 2$ we can compute for each sample scale and location invariant measures of 'shape', e.g. the standardized third cumulant. Under M_N and M_{OD} these are distributed independently of $\bar{y}_i$ and S_i, but this independence is in general lost under M_{NN}.

We shall in the subsequent discussion suppose that the population means $\{\mu_i\}$ are such that no useful information about error properties can be gleaned from the sample means $\bar{y}_i$. Note, however, that under M_{NN}

$$\text{cov}(\bar{y}_i, S_i) = (r-1)\kappa_3/r.$$

If the $\{\mu_i\}$ are distributed with variance σ_μ^2, this covering both systematic and random variation, we have

$$\text{var}(\bar{y}_i) = \sigma^2/r + \sigma_\mu^2, \quad \text{var}(S_i) = 2(r-1)\sigma^4 + (r-1)^2\kappa_4/r,$$

so that

$$\text{corr}(\bar{y}_i, S_i) = \rho_3(1 - 1/r)^{\frac{1}{2}} \{1 + \tfrac{1}{2}(r-1)\rho_4/r\}^{-\frac{1}{2}} (1 + r\sigma_\mu^2/\sigma^2)^{-\frac{1}{2}}. \tag{8}$$

Under M_{NN} there is thus some possibility that a plot of S_i versus $\bar{y}_i$ might show correlation which could be misinterpreted as evidence of a systematic relation between σ_i^2 and μ_i. The correlation (8) is, however, likely to be negligible as soon as $r\sigma_\mu^2/\sigma^2 \gg 1$ and this we assume.

2·3. *Procedures based on the distribution of S_i*

In § 2·2 some of the properties of the sample sums of squares have been outlined, especially under M_{OD} and M_{NN}. We now consider some corresponding statistical procedures.

For preliminary analysis a probability plot of the ordered $\{S_i\}$ against the expected order statistics of the chi-squared distribution with $r-1$ degrees of freedom (Pearson & Hartley, 1972, Table 20) is natural.

When r is not too small, say $r \geqslant 5$, and a number of S_i are available, a powerful general method of analysis (Bartlett & Kendall, 1946) under M_N is to employ linear methods for $\log S_i$ using the fact that under sampling a normal distribution

$$\text{var} (\log S_i) = \psi(\tfrac{1}{2}r - \tfrac{1}{2}), \tag{9}$$

where $\psi(z)$ is the digamma function and the values for $r = 2, 3, 4$ are $4\cdot93$, $1\cdot64$, $0\cdot935$.

We shall not make extensive use of this in the present work, partly because of the severe loss of efficiency when $r = 2, 3$ and partly because of the undue sensitivity to small values of S_i and the failure of the method without ad hoc modification if $S_i = 0$; such values could quite easily arise from rounding errors. As one example of its use, however, one could test for departure from M_N in the direction of M_{OD} or M_{NN} by comparing

$$(m-1)^{-1}\{\Sigma (\log S_i)^2 - (\Sigma \log S_i)^2/m\}$$

with (9).

For M_{OD} expansion of (3) or (4) for small $1/f_0$ or small dispersion of $h(x)$ shows that the locally most powerful test is based on the distribution of ΣS_i^2 given ΣS_i, or equivalently the marginal distribution of the dispersion index

$$I = \frac{\{\Sigma S_i^2 - (\Sigma S_i)^2/m\}m^{-1}}{(\Sigma S_i/m)^2\{2/(r-1)\}}. \tag{10}$$

The divisor $2/(r-1)$ ensures that, under M_N, $I \to 1$ in probability as $m \to \infty$. Note that detailed specification of $h(x)$ is unnecessary for the local optimality property. For large m, I is asymptotically normal: because of the skewness of I, a rather better approximation is obtained by taking $\log I$ to be normal with mean $-(r+1)/\{m(r-1)\}$ and variance $2(r+1)/\{m(r-1)\}$.

2·4. *Detection of systematic changes*

We now consider the examination of possible systematic relations between variance and an explanatory variable or between variance and mean. Again graphical analysis will usually be a natural first step supplemented where appropriate by a test statistic which we take in the form

$$T = \Sigma a_i S_i / \Sigma S_i, \tag{11}$$

where we can without loss of generality scale the explanatory variable so that $\Sigma a_i = 0$, $\Sigma a_i^2 = m$. For instance if the sample means $\bar{y}_i$ are taken as the explanatory variable

$$a_i = (\bar{y}_{i.} - \bar{y}_{..})\{\Sigma (\bar{y}_{j.} - \bar{y}_{..})^2/m\}^{-\frac{1}{2}},$$

where $\bar{y}_{..} = \Sigma \bar{y}_{i.}/m$.

There are two ways of obtaining a null hypothesis distribution of T. The first is to note that under M_N all permutations of $\{S_1, \ldots, S_m\}$ are equally likely; hence, with $\bar{S}_. = \Sigma S_i/m$,

$$E(T) = 0, \quad \text{var} (T) = \{\Sigma (S_i - \bar{S}_.)^2/m\}/\bar{S}_.^2. \tag{12}$$

Under weak assumptions, asymptotic normality will hold as $n \to \infty$: for higher permutation moments, see Cox & Hinkley (1974, p. 185). An 'exact' permutation test is in principle possible.

Under M_N a more sensitive analysis is possible, by using the normal theory distribution of the ratio of quadratic forms rather than the permutation distribution. A reasonably simple approximate test is derived by writing

$$\text{pr}\,(T' \leq x) = \text{pr}\,\{\Sigma\,(a_i - x/\sqrt{m})S_i \leq 0\} \simeq \Phi\left[\frac{x}{\{2(r-1)(1+x^2/m)\}^{\frac{1}{2}}}\right].$$

Thus for a two-sided equi-tailed test at level 2α we need critical limits x_α^* defined by

$$\frac{x_\alpha^*}{\{2(r-1)(1+x_\alpha^{*2}/m)\}^{\frac{1}{2}}} = \pm k_\alpha^*,$$

where $\Phi(-k_\alpha^*) = \alpha$. That is,

$$x_\alpha^* = \pm\{2(r-1)\}^{\frac{1}{2}} k_\alpha^* \{1 - 2(r-1)k_\alpha^{*2}/m]^{-\frac{1}{2}}. \tag{13}$$

For a more refined calculation it is possible to introduce a correction based on the standardized skewness and kurtosis of the random variable $\Sigma\,(a_i - x/\sqrt{m})S_i$, namely

$$\rho_3^{(1)} = \frac{2\sqrt{2}}{\sqrt{(r-1)}\sqrt{m}}\left\{\rho_{3a} - \frac{3x_\alpha^*}{\sqrt{m}} + O\left(\frac{1}{m}\right)\right\},$$

$$\rho_4^{(1)} = \frac{12}{(r-1)m}\left\{\rho_{4a} - \frac{4x_\alpha^*}{\sqrt{m}}\rho_{3a} + O\left(\frac{1}{m}\right)\right\},$$

where $\rho_{3a} = \Sigma\,a_i^3/m$, $\rho_{4a} = \Sigma\,a_i^4/m - 3$.

Numerical work in the special case $r = 2$ suggests that for the purpose of the present paper, where m is likely to be quite large, use of (13) is entirely adequate. Note that the simpler approximation (13) with the $1/m$ term omitted is equivalent to the permutation test (12) with a normal theory value for var (T), as is clear on general grounds.

3. Samples of size two

We now consider in more detail what is probably the most common case in applications, namely $r = 2$, when a large number of duplicate observations are available. Provided that the numbering of observations within a pair is uninformative, we may replace the ith pair of observations (y_{i1}, y_{i2}) by

$$\bar{y}_i = \tfrac{1}{2}(y_{i1} + y_{i2}), \quad S_i = \tfrac{1}{2}(y_{i1} - y_{i2})^2.$$

An initial analysis for detecting systematic dependencies is to plot S_i against $\bar{y}_i$ or some other suitable explanatory variable z_i. For a formal test we use § 2·4.

If there is no systematic relation, attention is focused on the marginal distribution of the $\{S_i\}$. Under the standard model M_N, S_i/σ^2 has the chi-squared distribution with one degree of freedom. Probably the simplest graphical analysis is a seminormal plot of the ordered $\sqrt{S_i}$; see Pearson & Hartley (1972, Table 21). Departure in the direction of underdispersion would indicate M_{NN} whereas overdispersion could indicate either M_{OD} or M_{NN}.

Under M_{NN}, it is clear that even with a very large value of m only even-order cumulants of the distribution of the y's can be determined. Thus there is no way of studying the skewness of y, at least so long as there is no relevant information in the variation between the $\{\bar{y}_{i.}\}$.

To supplement the plot of S_i versus $\bar{y}_i$, the index of dispersion can be calculated, and by (5) and (7) regarded as estimating $1+\frac{3}{2}(\lambda_2/\sigma^4)$ under M_{OD}, and $1+\frac{1}{4}\rho_4$ under M_{NN}.

4. SAMPLES OF SIZE THREE

4·1. *General discussion*

When we have samples of size three rather than samples of size two, there is appreciable extra sensitivity in having estimates of variance with two degrees of freedom rather than one, but in some ways the more interesting difference lies in the richer possibilities for examining distributional shape and in particular of detecting skewness under the non-normal model, M_{NN}.

Thus an initial analysis will often consist of:
(i) a plot of S_i versus $\bar{y}_i$ or a similar explanatory variable z_i, supplemented where necessary by the test statistic (11);
(ii) a plot of the ordered S_i versus the expected order statistics from an exponential distribution, supplemented where necessary by the test statistic (10). Note that e_{pm}, the expected value of the pth smallest observation in samples of size m from the unit exponential distribution, is

$$e_{pm} = \frac{1}{m} + \ldots + \frac{1}{m-p+1} \simeq \begin{cases} \log\{(m+\frac{1}{2})/(m-p+\frac{1}{2})\} & (p \neq m), \\ \log(m+\frac{1}{2}) + \gamma & (p = m), \end{cases}$$

where γ is Euler's constant.

We now concentrate on the more detailed analysis of the variation and this is most naturally done in terms of the order statistics.

4·2. *Order statistics in samples of three*

It is convenient to drop temporarily the suffix indicating the particular sample and to write $y_{(1)} \leq y_{(2)} \leq y_{(3)}$ for the order statistics. The central results were given by Fisher (1930) in connection with a study of the exact distribution of sample estimates of skewness, in which induction on sample size was used, samples of three being the 'starting point'.

If $\tilde{\sigma}^2 = \frac{1}{2}S$ is the usual estimate of variance, the one 'degree of freedom' that describes the skewness of the sample can be taken in dimensionless form as

$$w = (y_{(3)} - 2y_{(2)} + y_{(1)})/\tilde{\sigma}.$$

Under the standard normal model, M_{N}, and also under M_{OD}, (w, S) are independent, so that clear dependence between (w, S) is an indication that M_{NN} should be considered.

To obtain the joint distribution of $(\tilde{\sigma}, w)$ under M_{N} and M_{NN} it is convenient, following Fisher (1930), to write

$$y_{(3)} = y_{(2)} + \tilde{\sigma}\cos v + \sqrt{3}\tilde{\sigma}\sin v, \quad y_{(1)} = y_{(2)} - \tilde{\sigma}\cos v + \sqrt{3}\tilde{\sigma}\sin v,$$

where $-\pi/6 \leq v \leq \pi/6$, the extreme angles corresponding to samples with $y_{(2)} = y_{(3)}$ or

$y_{(1)}$. Note that $w = 2\sqrt{3} \sin v$ and that

$$t = \frac{6}{\pi} \sin^{-1}\left(\frac{w}{2\sqrt{3}}\right)$$

takes values in $(-1, 1)$.

Now the joint density of $\{y_{(1)}, y_{(2)}, y_{(3)}\}$ is $6g(y_{(1)})g(y_{(2)})g(y_{(3)})$ and, on transforming to new variables $(\tilde{a}, y_{(2)}, v)$, the density of $(\tilde{\sigma}, v)$ is

$$12\sqrt{\tilde{\sigma}} \int_{-\infty}^{\infty} g(x - \tilde{\sigma} \cos v + \sqrt{3}\tilde{\sigma} \sin v)g(x)g(x + \tilde{\sigma} \cos v + \sqrt{3}\tilde{\sigma} \sin v)\, dx. \qquad (14)$$

It follows immediately that for M_N the statistics $\tilde{\sigma}$ and v are indeed independent and that the marginal density of v is uniform; i.e. that of t is uniform on $(-1, 1)$. Incidentally the usual unbiased estimate of the third cumulant $\frac{3}{2}\Sigma(y_p - \bar{y})^3 = 3\tilde{\sigma}^3 \sin(3v)$, showing explicitly that consideration of w or v is essentially equivalent to that of the standardized third cumulant ratio. Note that the individually standardized third cumulant lies in $(-\sqrt{3}, \sqrt{3})$.

There are the following open questions connected with (11).

(i) Does knowledge of the joint distribution of $(\tilde{\sigma}, v)$ determine $g(x)$ uniquely, except for a translation?

(ii) Does independence of $\tilde{\sigma}$ and v imply that $g(x)$ is normal?

(iii) Is there a simple necessary and sufficient condition for a given distribution of $(\tilde{\sigma}, v)$ to be representable in the form (14)?

(iv) What statistic or statistics are theoretically most sensitive for detecting departures from M_N?

Note that under M_{OD}, $\tilde{\sigma}$ and v are independent, v is uniform and $\tilde{\sigma}$ or S has the overdispersed distribution discussed in § 2·2. Also if it is required to estimate skewness under the model M_{NN} averaging of individually standardized estimates is inappropriate, in particular in the light of the constraint mentioned above: a consistent estimate of κ_3/σ^3 is

$$\frac{3}{2m}\Sigma_{i,p}(y_{ip} - \bar{y}_i)^3/(m^{-1}\Sigma\tilde{\sigma}_i^2)^{3/2}.$$

A reasonable analysis in practice is:

(a) the examination of the marginal distribution of the S_i via the probability plot and test statistic mentioned in § 4·1;

(b) inspection of a scatter plot of t_i versus $\tilde{\sigma}_i$, clear dependence showing evidence against M_N and M_{OD};

(c) inspection of the marginal distribution of t_i supplemented by calculation of the mean of t_i as a statistic nonzero values of which would indicate skewness in the model M_{NN}.

5. Samples of size more than three

We discuss only briefly corresponding procedures for samples of size four and more. The natural extension of the statitic w of § 4 is provided via linear functions of order statistics with a simple interpretation and zero expectation under normality. Thus for samples of size four we start with measures of skewness and kurtosis

$$(y_{(4)} - y_{(3)}) - (y_{(2)} - y_{(1)}), \quad (y_{(4)} - y_{(3)}) + (y_{(2)} - y_{(1)}) - k(y_{(3)} - y_{(2)}),$$

choosing k so that the second of these has zero expectation under M_N: in fact $k = 2.465$. This leads us to define

$$w' = \{y_{(4)} - y_{(3)} - y_{(2)} + y_{(1)}\}/\tilde{\sigma}, \quad w'' = \{y_{(4)} - 3.465y_{(3)} + 3.465y_{(2)} - y_{(1)}\}/\tilde{\sigma}.$$

It is easily shown that, under M_N and M_{OD}, w' and w'' are independent of $\tilde{\sigma}$,

$$\text{var}(w') = 0.771, \quad \text{var}(w'') = 2.550.$$

They are also uncorrelated but far from independent and far from normally distributed.

It is likely, therefore, that the most effective procedure is to compute w' and w'' from each set of data and to examine the sample means and variances for consistency with the normal theory values. Departures would have a fairly clear diagnostic value. The sample estimate of cov (w', w'') could also be calculated, although it is unclear what interpretation is to be put on a nonzero value.

6. SOME EXAMPLES

Samples of size three. As an illustration, we analyse some pulse rates from the International Prospective Primary Prevention Study in Hypertension, a large scale clinical trial. Before entry and randomization to treatments, patients typically attended three qualifying visits at least one day apart. Pulse rate, the number of beats per minute, was one of several variables measured at each pre-entry visit.

The data analysed here consist of three pulse rates, y, for a sample of one hundred men and one hundred women. The analysis goes in three broad steps.

First, a plot, not given here, of the within-patient sum of squares S_i versus the mean pulse rate $\bar{y}_i$ shows clear evidence that S_i increases with $\bar{y}_i$, thus indicating either M_{SY} or, just conceivably, sampling correlation between S_i and $\bar{y}_i$ under M_{NN}. Clearly M_N is inappropriate.

However, use of the reciprocal pulse rate removes the systematic relation between sum of squares and mean. The reciprocal pulse rate has a direct interpretation as the mean time between beats. The changes induced by reciprocal transformation are supported by the statistic T; for women, $T = 0.897$ on the original scale, but this changes to -0.096 for reciprocals; correspondingly for men, T changes from 0.631 to 0.094 under the reciprocal transformation.

The second step is a probability plot of the ordered S_i versus the expected exponential order statistics. Both the pulse rates and reciprocal pulse rates show overdispersion, but less for reciprocals; see Fig. 1. The very large values of S_i correspond to women with large variation in pulse rate which variation, however, may be clinically meaningful. Therefore it is on the whole sensible to regard such values as part of the population under study rather than as aberrant values to be rejected.

Supplementing the plot by the index of overdispersion I of (10) verifies the improved distributional properties of the reciprocals. Under M_{NN}, from (7), I leads to an estimate of the dimensionless fourth cumulant, namely 5.556 and 8.238 for pulse rates for women and men, and on the reciprocal scale, 2.283 and 4.602, respectively.

The third step in the analysis plots S_i versus the angular measure of skewness t_i. We found it useful to examine for comparison simulated data from other distributions such as the normal, exponential and Student's t distributions. Normal data produced a random scatter as expected whereas exponential data gave a plot which was clearly positively skew.

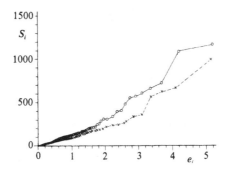

Fig. 1. Entry pulse rate for 100 women. Triples. Crosses, S_i versus exponential score, e_i; circles, $10S_i$ versus exponential score, e_i, after reciprocal transformation.

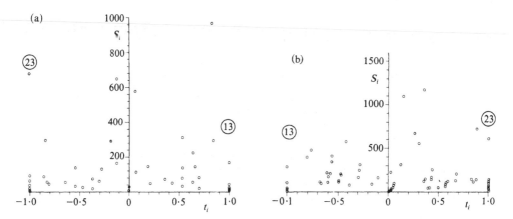

Fig. 2. Entry pulse rate for 100 women. Triples. (a) S_i versus angular statistic, t_i. (b) $10S_i$ versus angular statistic, t_i, after reciprocal transformation.

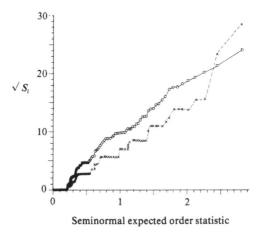

Seminormal expected order statistic

Fig. 3. Entry pulse rate for 100 women. Pairs. Crosses, seminormal plot of ordered $\sqrt{S_i}$; circles, seminormal plot of ordered $\sqrt{S_i}$ after reciprocal transformation.

Figure 2(a) shows that S_i and t_i for pulse rates are not independent. Reciprocals improve the symmetry; see Fig. 2(b). The skewness measure t_i is characterized by concentrations of points at -1, 1 and 0, the first two values a consequence of grouping; its distribution is certainly nonuniform, both on the original and reciprocal scales, so that M_{OD} is not appropriate. The women exhibit slightly negatively skew pulse rates, giving mean $t_i -0\cdot078$ and standard deviation $0\cdot719$. On the reciprocal scale, t_i has mean $0\cdot114$ and standard deviation $0\cdot705$: the mean is exaggerated by one very large S_i. By contrast, the men exhibit slight positive skewness in pulse rate with mean t_i $0\cdot134$ and standard deviation $0\cdot739$; for reciprocals, mean t_i is $0\cdot097$ with standard deviation $0\cdot747$. The assumption of symmetry in the distribution of the reciprocals is further supported by the usual standardized measure of skewness which takes values of $0\cdot315$ and $-0\cdot362$ for women and for men.

Exactly unit values of t_i are accounted for by rounding. Simulations suggest, however, that rounding cannot entirely explain the anomalous distribution of t_i. The grouping could have been aggravated by some temporal correlation. In particular, when two out of three rates agree, do they occur together as consecutive values? Pooling men and women, of 81 individuals with two rates the same, 19 had a different rate at the second visit, whereas 34 had a different rate at the first, and 28 at the third visit. This tends to suggest that some doctors may recall the previous pulse rate at the subsequent visit.

In summary, the reciprocal pulse rates appear to have a symmetric, nonnormal, long-tailed distribution, suggesting that Student's t distribution with a small number of degrees of freedom may be appropriate. This provides a suitably flexible family of symmetrical distributions if parametric representation is required.

Samples of size two. Partly in order to compare results, we have analysed a series of pairs of observations formed by taking just the first and last pulse rate from the data discussed above.

Reciprocals certainly reduce the size of the kurtosis, being typically positive. However, reciprocals for women exhibit slight negative kurtosis as shown in Fig. 3, although testing with I, which under M_{NN} leads to an estimate of $\hat{\rho}_4$ as $-0\cdot774$ for reciprocals, this is by no means statistically significant. Also, the other values of I vary little. In particular, for women, pulse rates give $\hat{\rho}_4$ as $3\cdot335$, and for men $6\cdot054$, which changes to $4\cdot493$ for reciprocals.

Samples of size four. Data on strengths of yarn (Cox & Snell, 1981, p. 131) in the form of 12 samples each of size four show excellent agreement with the normal theory values for w' and w''. Namely, the observed mean w' for bobbins is $-0\cdot0178$ with observed standard deviation $0\cdot710$, and for w'', the observed mean is $-0\cdot466$ with standard deviation $1\cdot561$; the theoretical standard deviations are respectively $0\cdot878$ and $1\cdot597$ and the standard errors of the means thus $0\cdot253$ and $0\cdot461$.

ACKNOWLEDGEMENTS

We are grateful to the Science and Engineering Research Council and to Ciba-Geigy Ltd (Basle) for support of this work.

REFERENCES

BARTLETT, M. S. & KENDALL, D. G. (1946). The statistical analysis of variance-heterogeneity and the logarithmic transformation. *Suppl. J. R. Statist. Soc.* **8**, 128–38.

Cook, R. D. & Weisberg, S. (1983). Diagnostics for heteroscedasticity in regression. *Biometrika* **70**, 1–10.
Cox, D. R. & Hinkley, D. V. (1974). *Theoretical Statistics*. London: Chapman & Hall.
Cox, D. R. & Snell, E. J. (1981). *Applied Statistics*. London: Chapman & Hall.
Fisher, R. A. (1930). The moments of the distribution for normal samples of measures of departure from normality. *Proc. R. Soc.* A **130**, 16–28.
Kendall, M. G. & Stuart A. (1969). *Advanced Theory of Statistics*, 1, 3rd ed. High Wycombe: Griffin.
Pearson, E. S. & Hartley, H. O. (1972). *Biometrika Tables for Statisticians*, 2. High Wycombe: Griffin.

[*Received December* 1985. *Revised April* 1986]

Biometrika (1992), **79**, 1, *pp.* 1-11
Printed in Great Britain

Nonlinear component of variance models

By P. J. SOLOMON

Department of Statistics, University of Adelaide, GPO Box 498, Adelaide SA 5001, Australia

AND D. R. COX

Nuffield College, Oxford OX1 1NF, U.K.

SUMMARY

General aspects of nonlinearity in the context of component of variance models are discussed, and two special topics are examined in detail. Firstly, simple procedures, both formal and informal, are proposed for describing departures from normal-theory linear models. Transformation models are shown to be a special case of a more general formulation, and data on blood pressure are analyzed in illustration. Secondly, an approximate likelihood is proposed and its accurate performance is examined numerically using examples of exponential regression and the analysis of several related 2×2 tables. In the latter example, the approximate score test has improved power over the Mantel-Haenszel test.

Some key words: Cumulant; Likelihood; Logit; Nonlinearity; Nonnormality; Score test; Transformation; Variance component.

1. INTRODUCTION

Nonlinearity is an important theme underlying many current developments in statistics and in applied mathematics. In this paper we discuss a few general aspects of nonlinearity in connexion with components of variance and then examine two special topics in more detail.

Our emphasis is on nonlinear models rather than on nonlinear methods of estimation, the latter being needed for efficient estimation even in unbalanced linear normal theory problems. Nonlinear models may be considered because they are of intrinsic interest or because they may be a convenient technical device for testing the adequacy of standard linear models. There is also a distinction between situations in which the variance components are of primary interest and those in which the components are required only for the assessment of the precision of contrasts.

If the amount of nonlinearity is small, local linearization will produce a model of linear form, typically unbalanced.

2. TYPES OF NONLINEARITY

We shall not attempt a careful classification of types of nonlinearity. In linear theory, the simplest and most general models are respectively

$$Y_{is} = \mu + A_i + B_{is} \quad (i = 1, \dots, m; s = 1, \dots, r), \tag{1}$$

$$Y = x\beta + aU. \tag{2}$$

In (1), $A_1, A_2, \ldots, A_m$ are random variables of zero mean and variance σ_A^2, independently normally distributed or, in second-order theory, uncorrelated and the B_{is} have analogous properties and variance σ_B^2. In (2), Y is the full $n \times 1$ vector of observed response variables, the $n \times q$ matrix x and the $q \times 1$ vector β of unknown parameters define the systematic part of the variation, whereas the matrix a and the vector U of random variables define the random structure; typically U contains components of k different types with variances $\sigma_1^2, \sigma_2^2, \ldots, \sigma_k^2$, say.

Nonlinearity can arise from (1) and (2) in the following ways.

(i) The systematic part $x\beta$ is replaced by a nonlinear form; for example, Rudemo, Ruppert & Streibig (1989) consider such models in application to bioassay data.

(ii) The random components, for example A_i and B_{is} in (1), combine nonlinearly; e.g. the nonlinearity may be modelled approximately as

$$Y_{is} = \mu + A_i + B_{is} + \alpha_{20}A_i^2 + \alpha_{11}A_iB_{is} + \alpha_{02}B_{is}^2.$$

(iii) The random and systematic parts in (2) combine nonlinearly, for example in an exponential growth model with random doubling time; i.e.

$$Y_{is} = \exp\{\mu + (\beta + A_i)x_{is}\} + B_{is}. \qquad (3)$$

Linear models combining random and systematic components, so-called regression models of the second kind, are presented in a broad context by Nelder (1977). See Racine-Poon (1985) for a Bayesian approach to estimation in such nonlinear models.

(iv) The essentially normal-theory based structure of (1) and (2) can be replaced by an analogous form for the exponential family. For the Poisson and binomial distributions with one overlaid component of variation this has a long history in the context of the negative binomial (Greenwood & Yule, 1920) and beta binomial (Skellam, 1948) distributions respectively. Much recent work has focused on the conceptual and numerical analytical problems of models for the canonical parameters of exponential family models of essentially linear structure, i.e. applying representations analogous to (2) not to observations but to canonical parameters.

In § 3 we examine the simplest model of type (ii) in which nonlinearity enters a one-way balanced structure. In § 4 we examine a technique for approximating the likelihoods for nonlinear variance component models.

3. BALANCED STRUCTURES

3·1. *Formulation of model*

Consider first a one-way arrangement in which r independent repeat continuous observations are obtained on each of m groups, the groups being regarded as a random sample from a population of groups, so that model (1) is potentially applicable. Now if μ is defined as the overall population mean, and the mean for the ith group is defined to be $\mu + A_i$, then the properties $E(A_i) = 0$, $E(B_{is}|A_i = a) = 0$ are matters of definition; all other properties of $\{A_i\}$, $\{B_{is}\}$ are to some extent capable of empirical test. Note that, in the absence of additional information about the groups, it is not meaningful to make, say, the properties of $\{B_{is}\}$ depend on the label i as such. To represent local departures from the normal-theory version of (1) we shall investigate the representation

$$Y_{is} = \mu + A_i + B_{is} + \alpha_{20}A_i^2 + \alpha_{11}A_iB_{is} + \alpha_{02}B_{is}^2, \qquad (4)$$

where $\{A_i\}$ and $\{B_{is}\}$ are zero mean, independently normally distributed random variables with variances σ_A^2, σ_B^2. We assume $(\alpha_{20}, \alpha_{11}, \alpha_{02})$ to be small and retain only first order terms; a formal asymptotic treatment would require them to be of order $1/\sqrt{m}$ as $m \to \infty$.

The second-order representation (4) models a quite general form of nonlinearity involving both skewness of the random effects and heterogeneity of the within-group variation. To interpret the parameters α_{20}, α_{11} and α_{02} which capture these features, note first that we can write

$$\mu + A_i + \alpha_{20} A_i^2 = \mu' + A_i', \tag{5}$$

where $\mu' = \mu + \alpha_{20}\sigma_A^2$ and A_i' is a nonnormal random variable of zero mean, approximate variance σ_A^2 and third moment $9\alpha_{20}\sigma_A^4$, so that the standardized third cumulant of A' is

$$\rho_3^A = 9\alpha_{20}\sigma_A, \tag{6}$$

with a similar interpretation for ρ_3^B. Because ρ_3^A, ρ_3^B are dimensionless, they are more helpful for qualitative interpretation than α_{20}, α_{02}.

Equation (5) can be regarded as two terms of the Fisher-Cornish inversion of the Edgeworth expansion of a nearly normal random variable A_i' and suffers from the usual disadvantage of such expansions of nonmonotonicity in A_i. The importance of this depends on the context. Note that the nonmonotonicity with respect to A_i, or to B_{is}, but so far as we know not both simultaneously, can be removed by rewriting (5) in the form

$$\mu + (2\alpha_{20})^{-1}(e^{2\alpha_{20}A_i} - 1), \tag{7}$$

equivalent to the first order in α_{20}; we shall not do this here.

In some ways, however, the most interesting parameter is α_{11}. Note that given $A_i = a\sigma_A$, the random variable attached to observation (i, s) is $B_{is}(1 + \alpha_{11}a\sigma_A) + \alpha_{02}B_{is}^2$, which has approximate standard deviation $(1 + \alpha_{11}a\sigma_A)\sigma_B$; thus

$$\rho_{11} = \alpha_{11}\sigma_A \tag{8}$$

is a dimensionless measure of the rate of change of the conditional standard deviation of within group variation with the group mean, thus completing the interpretation of the parameters in (4).

An important special case of (4) is the family of transformation models which is discussed in § 3·2 and exemplified in § 3·5.

3·2. *A transformation model*

The model (4) with three additional parameters can be contrasted with a transformation model (Solomon, 1985) with one additional parameter, namely that for some κ

$$Y_{is}^{1/\kappa} = \mu + A_i^* + B_{is}^*, \tag{9}$$

the right-hand side being a normal-theory representation. Provided that μ is very much greater than $\sqrt{(\sigma_A^2 + \sigma_B^2)}$ we have an expansion that, to quadratic terms,

$$Y_{is} = \mu^\kappa \{1 + (A_i^* + B_{is}^*)/\mu\}^\kappa$$

$$= \mu^\kappa + \kappa\mu^{\kappa-1}A_i^* + \kappa\mu^{\kappa-1}B_{is}^* + \tfrac{1}{2}\kappa(\kappa - 1)\mu^{\kappa-2}(A_i^{*2} + 2A_i^* B_{is}^* + B_{is}^{*2}), \tag{10}$$

which is of the form (4) with, however, the special feature that $\alpha_{20} = \alpha_{02} = \tfrac{1}{2}\alpha_{11}$.

The transformation model (9) is clearly much more restrictive than the general model (4); however transformation models are widely used in practice and often provide a simple basis for statistical analysis and interpretation. Choice between the models will depend on the context, as well as on practical considerations. Estimation based on (4) separates three distinct ways in which the statistical model may fail, and thus, in particular, throws light on what is achieved by transformation.

By way of example, and comparison, § 3·5 discusses an analysis of data on blood pressure using both models.

3·3. *Statistical analysis*

One possible approach to statistical analysis based on (4), especially for testing the null hypothesis $\alpha_{20} = \alpha_{11} = \alpha_{02} = 0$, is to compute log likelihood derivatives. We shall follow the equivalent but less formal route of considering, together with the second-order statistics, the cubic statistics

$$S_{30} = \sum_{s,i} (\bar{Y}_{i.} - \bar{Y}_{..})^3, \quad S_{21} = \sum_{s,i} (\bar{Y}_{i.} - \bar{Y}_{..})^2 (\bar{Y}_{is} - \bar{Y}_{i.}),$$

$$S_{12} = \sum_{\eta,i} (\bar{Y}_{i.} - \bar{Y}_{..})(Y_{is} - \bar{Y}_{i.})^2, \quad S_{03} = \sum_{s,i} (Y_{is} - \bar{Y}_{i.})^3, \tag{11}$$

of which the second is identically zero. Direct evaluation under (4) gives

$$E(S_{30}) = \alpha_{20}\sigma_A^4 6 \frac{r}{m}(m-1)(m-2) - \alpha_{11}\sigma_A^2\sigma_B^2 \frac{6}{m}(m-1) + \alpha_{02}\sigma_B^4 \frac{6}{rm}(m-1)(m-2),$$

$$E(S_{12}) = \alpha_{11}\sigma_A^2\sigma_B^2 2(m-1)(r-1) + \alpha_{02}\sigma_B^4 \frac{6}{r}(m-1)(r-1),$$

$$E(S_{03}) = \alpha_{02}\sigma_B^4 6 \frac{m}{r}(r-1)(r-2).$$

Thus unbiased estimates of α_{20}, α_{11}, α_{02} can be found if σ_A^2, σ_B^2 are known. If, as would typically be the case, σ_A^2, σ_B^2 are unknown and replaced by the usual estimates, we obtain as consistent estimates of the α's

$$\tilde{\alpha}_{20} = \frac{m}{\tilde{\sigma}_A^4 6r(m-1)(m-2)} \left\{ S_{30} + \frac{3}{m(r-1)} S_{12} - \frac{(m-1)(m+1)}{m^2(r-1)(r-2)} S_{03} \right\},$$

$$\tilde{\alpha}_{11} = \frac{1}{\tilde{\sigma}_A^2\tilde{\sigma}_B^2 2(m-1)(r-1)} \left\{ S_{12} - \frac{(m-1)}{m(r-2)} S_{03} \right\}, \quad \tilde{\alpha}_{02} = \frac{r}{\tilde{\sigma}_B^4 6m(r-1)(r-2)} S_{03}.$$

These are converted into estimates $\tilde{\rho}_3^A$, $\tilde{\rho}_{11}$, $\tilde{\rho}_3^B$ of the dimensionless parameters via the standard estimates of the components of variance σ_A^2, σ_B^2.

In this preliminary study we shall not develop in detail the sampling properties of the resulting estimates. Under the null hypothesis of normality, for large r and m, the standard error of $\tilde{\rho}_3^B$ is $\sqrt{\{6/(rm)\}}$; the corresponding standard error for $\tilde{\rho}_3^A$ is $\sqrt{(6/m)}$ if the 'correction terms' are small, i.e. if the estimate is essentially the standardized third cumulant of the m sample means.

Note that a very general mathematical development for calculating cumulants in component of variance models is given by Speed in a series of papers that appeared in the 1980's; see, e.g. Speed & Silcock (1988).

There are numerous ways of testing whether $\alpha_{11} = 0$. A graphical method closely corresponding to the estimation of α_{11} is to plot the within group sum of squares for the ith group against the group mean. Indeed because under the null hypothesis $\alpha_{20} = \alpha_{11} = \alpha_{02} = 0$ the quantities plotted are independent, an 'exact' permutation test of regression is available. This could take various forms depending on the choice of test statistic. The most effective practical procedure is likely to be to regress the log mean square within groups on the sample mean applying standard least squares methods. In doing this the residual mean square has a known theoretical value under normality and a permutation test could be applied in those cases where careful, cautious assessment of significance is important.

3·4. *Generalizations*

While the above discussion has been for the one-way arrangement, essentially similar arguments apply to general balanced cross-classified and nested arrangements. For example, the hierarchical model with three nested effects has a representation analogous to (4),

$$Y_{ist} = \mu + A_i + B_{is} + C_{ist} + \alpha_{200}A_i^2 + \alpha_{020}B_{is}^2 + \alpha_{002}C_{ist}^2$$
$$+ \alpha_{110}A_iB_{is} + \alpha_{101}A_iC_{ist} + \alpha_{011}B_{is}C_{ist},$$

and straightforward calculation shows that the cubic statistics, again defined analogously to (11), have expectations

$$E(S_{003}) = \alpha_{002}\sigma_C^4 6m \frac{r_1}{r_2}(r_2-1)(r_2-2),$$

$$E(S_{030}) = \alpha_{020}\sigma_B^4 6m \frac{r_2}{r_1}(r_1-1)(r_1-2) - \alpha_{011}\sigma_B^2\sigma_C^2 6\frac{m}{r_1}(r_1-1) + \alpha_{002}\sigma_C^4 6\frac{m}{r_2}(r_1-1),$$

and so on.

More importantly, under the transformation model,

$$\alpha_{200} = \alpha_{020} = \alpha_{002} = \tfrac{1}{2}\alpha_{110} = \tfrac{1}{2}\alpha_{101} = \tfrac{1}{2}\alpha_{011}$$

which is simply a generalization of the special feature of (4), and similarly for general balanced cross-classified arrangements.

3·5. *An example*

To illustrate the above discussion, we analyze data on the blood pressure of hypertensive males obtained in the International Prospective Primary Prevention Study in Hypertension (Solomon, 1985). The groups are $m = 25$ individual patients and the repeat observations refer to the $r = 16$ distinct measurements per patient; of course we are here ignoring the time series structure of the data. Table 1 gives the main statistics computed from diastolic and systolic blood pressures (mmHg) and various transforms of these.

Table 1. *Estimates of the α parameters and their dimensionless equivalents for data on blood pressure for 25 hypertensive males from the International Prospective Primary Prevention Study in Hypertension*

Blood pressure	$\tilde{\alpha}_{20}$	$\tilde{\alpha}_{02}$	$\tilde{\alpha}_{11}$	$\tilde{\rho}_3^A$	$\tilde{\rho}_3^B$	$\tilde{\rho}_{11}$
diastolic	0·0166	0·0114	0·0248	0·7565	0·7150	0·1254
√ diastolic	0·2851	0·0924	0·2908	0·6686	0·3113	0·0758
log diastolic	1·1656	0·0410	0·7568	0·5703	0·0276	0·0411
systolic	0·0092	0·0043	0·0121	1·1428	0·5026	0·1669
√ systolic	0·1915	0·0455	0·1773	0·9587	0·2251	0·0986
log systolic	0·9409	0·0151	0·5308	0·7710	0·0117	0·0479

Under the transformation model, we expect $\alpha_{20} = \alpha_{02} = \frac{1}{2}\alpha_{11}$. The corresponding esti-mates of the parameters for both diastolic (0·0166, 0·0114, 0·0248) and systolic (0·0092, 0·0043, 0·0121) pressures reasonably approximate this expected relationship. The estimates suggest therefore that the transformation model, which we have noted as being really quite restrictive, is appropriate. Indeed, previous analysis (Solomon, 1985) showed that, assuming a common transformation in a bivariate model, log transformation for both diastolic and systolic blood pressures is close to maximizing the likelihood.

The moment calculations presented in Table 1 do more than validate the application of the transformation model. The parameter estimates provide greater insight into the underlying relationships in the data, as well as into what the transformation model is doing. The parameters ρ_3^A and ρ_3^B represent the degree of skewness of the between- and within-patient effects. For diastolic blood pressure, $\tilde{\rho}_3^A = 0·7565$ and $\tilde{\rho}_3^B = 0·7150$ are of moderate and comparable skewness on the original scale, but the distribution of the between-patient effects for systolic pressure has a relatively longer tail ($\tilde{\rho}_3^A = 1·1428$). This difference between the blood pressures is likely to be a reflection of the fact that diastolic, but not systolic, pressure was subject to a treatment target level of 95 mmHg. Note, however, that log transformation virtually eliminates the within-patient skewness, and reduces the between-patient skewness.

The estimated rate of change of the conditional standard deviation of the within-patient variation with the patient mean, $\tilde{\rho}_{11}$, is 16·7% for systolic, and 12·5% for diastolic pressure. Thus a moderate proportion of the within-patient variation is attributable to the fractional change of the standard deviation within-patients. In practice, such variability may well be important biologically. For instance, published standards of treatment dosage for hypertensives typically assume that the relationship between the 'true' patient mean and the 'true' within-patient variation is known, and that the conditional standard deviation of the within-patient variation does not change with changes in the mean; clearly such assumptions may not be appropriate and the example illustrates that superficial under-standing of the underlying relationships in the data on blood pressure may well lead to inappropriate treatment. Again, note that both square root and log transformation reduce this change in the dependency between the patient mean and within-patient variation.

4. LIKELIHOODS FOR NONLINEAR VARIANCE COMPONENT MODELS

4·1. *An approximate likelihood*

We now turn to a quite different issue, namely the approximate evaluation of likelihoods for models containing at least one additional level of random variation entering non-linearly into a likelihood of given form. An example is the likelihood for the exponential regression model (3) with random doubling times. We propose one such approximation.

Let $A_1, \ldots, A_m$ be independently, identically normally distributed with mean μ_A and variance σ_A^2, the variance being in some sense relatively small. Suppose that for fixed $A_1, \ldots, A_m$ the likelihood has the form $\Pi\, L_j(\theta, A_j)$, where θ is a parameter which may or may not be of primary interest. Then the required likelihood is $\Pi\, E\{L_j(\theta, A_j)\}$, where the expectation is over the distribution of A_j. Thus we consider approximations to $E\{L(\theta, A)\}$, dropping the suffix j temporarily. Write

$$l(\theta, A) = \log L(\theta, A), \quad l^{(r)} = [\partial^r l(\theta, A)/\partial A^r]_{A=\mu_A},$$

and define $A^* = (A - \mu_A)/\sigma_A$, which has a standard normal distribution.

Then

$$L(\theta, A) = L(\theta, \mu_A) \exp(\sigma_A l^{(1)} A^* + \tfrac{1}{2}\sigma_A^2 l^{(2)} A^{*2})(1 + \tfrac{1}{6}\sigma_A^3 l^{(3)} A^{*3} + \tfrac{1}{24}\sigma_A^4 l^{(4)} A^{*4}). \quad (12)$$

We now take expectations over the distribution of A^*. The first term gives

$$\frac{1}{\sqrt{(1 - \sigma_A^2 l^{(2)})}} \exp\left\{\frac{\sigma_A^2 l^{(1)2}}{2(1 - \sigma_A^2 l^{(2)})}\right\}, \quad (13)$$

the leading term of a Laplace expansion. It is exact for a one-way normal theory arrangement. Inclusion of further terms in (12) multiplies (13) by

$$1 + (\tfrac{1}{2}l^{(3)} l^{(1)} + \tfrac{1}{8}l^{(4)})\sigma_A^4 + o(\sigma_A^4),$$

which to ensure positivity we rewrite in exponentiated form as

$$\exp\{(\tfrac{1}{2}l^{(3)} l^{(1)} + \tfrac{1}{8}l^{(4)})\sigma_A^4\}\{1 + o(\sigma_A^4)\}. \quad (14)$$

Thus the final approximation to the full likelihood is

$$\prod L_j(\theta, \mu_A)\{(1 - \sigma_A^2 l_j^{(2)})\}^{-\frac{1}{2}} \exp\left\{\frac{\sigma_A^2 l_j^{(1)2}}{2(1 - \sigma_A^2 l_j^{(2)})} + \tfrac{1}{2}(l_j^{(3)} l_j^{(1)} + \tfrac{1}{4}l_j^{(4)})\sigma_A^4\right\}\{1 + o(\sigma_A^4)\}. \quad (15)$$

Of course, were the final term in (14) and (15) appreciable, this would be a warning of the probable failure of the expansions employed. We point out that large values of the α's defined in § 3 take us outside the range of the approximation, but point qualitatively in the right direction.

For example, if the random variables Y_{js} $(j = 1, \dots, m; s = 1, \dots, r)$ have the structure

$$Y_{js} = \exp\{(\theta + A_j)x_{js}\} + B_{js}, \quad (16)$$

where $\{B_{js}\}$, $\{A_j\}$ are independently normal with zero mean and variances σ_B^2, σ_A^2, and the $\{x_{js}\}$ are fixed constants, we may apply the above results taking L_j to refer to the set $\{Y_{j1}, \dots, Y_{jr}\}$. There is a slight gain in simplicity by considering the special case in which $x_{js} = x_s$ $(j = 1, \dots, m)$.

Let $T_j(\theta) = \Sigma Y_{js} x_s^r e^{\theta x_s}$, $m_r(\theta) = 2^{(r-1)} \Sigma x_s^r e^{2\theta x_s}$. Then

$$l_j^{(r)} = \{T_j(\theta) - m_r(\theta)\}/\sigma_B^2, \quad (17)$$

so that the contribution to the likelihood from the jth sample is, from (15), approximately

$$[1 - \sigma_A^2\{T_2(\theta) - m_2(\theta)\}/\sigma_B^2]^{-\frac{1}{2}}(2\pi\sigma_B^2)^{-\frac{1}{2}r}$$

$$\times \exp\left(-\frac{\Sigma(Y_{js} - e^{\theta x_s})^2}{2\sigma_B^2} + \frac{\sigma_A^2\{T_1(\theta) - m_1(\theta)\}^2}{\sigma_B^4 2[1 - \sigma_A^2\{T_2(\theta) - m_2(\theta)\}/\sigma_B^2]}\right.$$

$$\left. + \tfrac{1}{2}\sigma_A^4\left[\frac{\{T_3(\theta) - m_3(\theta)\}\{T_1(\theta) - m_1(\theta)\}}{\sigma_B^4} + \frac{T_4(\theta) - m_4(\theta)}{4\sigma_B^2}\right]\right). \quad (18)$$

We check the accuracy of the approximation numerically, divorced from stochastic aspects, by comparing it with the exact answer

$$\int_{-\infty}^{\infty} \frac{1}{(2\pi)^{\frac{1}{2}r}\sigma_B^r} \exp\left\{-\frac{\Sigma(y_{js} - e^{(\theta + A_j)x_s})^2}{2\sigma_B^2}\right\} \frac{1}{\sqrt{(2\pi)}\sigma_A} \exp\left(-\frac{A_j^2}{2\sigma_A^2}\right) dA_j, \quad (19)$$

based on simulated rather than empirical data for a particular choice of parameter values.

For illustration we take $m = 10$ samples with $r = 5$ observations in each sample, and values of the parameters $\sigma_B = 0{\cdot}25$, $\theta = 0{\cdot}25$, $\{x_s\} = \{1, 2, 3, 4, 5\}$ and $\sigma_A = 0{\cdot}01$. Figure 1

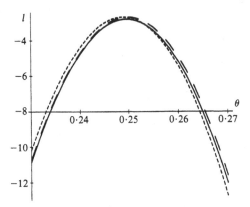

Fig. 1. Exact and approximate log likelihood functions
l for the exponential regression example (16) over
reasonable range of θ. Exact log likelihood, shown by
solid line; approximate log likelihood with leading
term only, dotted line; approximate log likelihood with
correction term, dashed line.

shows the log likelihood functions, over a reasonable range of θ, corresponding to the
exact likelihood, the approximation with only the leading term, and the approximation
with the correction term, evaluated at the true parameter values. The plot demonstrates
clearly that, even with a moderate standard deviation of A, the approximation performs
well, the log likelihood functions being almost indistinguishable at this level. Note that
for a single sample, the range of values of θ of interest will be wider than that depicted.

For the same m, r, σ_B, θ and $\{x_s\}$ as above, and with σ_A increasing from 0·0025 to
0·02, Table 2 sets out the exact and approximate likelihoods evaluated at the true
parameter values, together with the correction factor. For small to moderate σ_A^2 the
agreement between the exact and approximate likelihoods is extremely good. It is worth
noting that the variation in slope is detectable visually even at very small standard
deviations of A. At these small levels, the correction term typically takes close to unit
values.

Table 2. *Exact and approximate likelihoods for*
θ, σ_A, σ_B in exponential model (16), and correc-
tion factor, evaluated at the true parameter
values, for increasing σ_A

σ_A	Exact likelihood	Approximate likelihood (with correction)	Correction factor
0·0025	2·70416	2·70419	0·99991
0·0075	0·16998	0·16807	0·94724
0·01	0·06164	0·06900	1·02115
0·0125	0·10198	0·09835	0·68910
0·015	0·01643	0·01015	0·32259
0·0175	4×10^{-9}	3×10^{-9}	0·38876
0·02	0·000201	0·00000852	142·868

However, Table 2 illustrates that as soon as the correction factor adjusts the leading term by 20 to 30%, the method may fail badly. For instance, for values of σ_A nearer 0·02, the correction term can make much more than a minor difference. We recommend using the approximation provided the correction factor lies between 0·75 and 1·25. Although Table 2 shows that in particular cases the approximation may be reasonable for values of the correction factor lying outside this range, we caution that care is needed in application.

When the standard deviation of A is known to be large relative to σ_B, an alternative approach is required.

4·2. *Several 2×2 tables*

We consider briefly the combination of information from several 2×2 contingency tables assumed to have a constant logistic difference Δ between the two 'treatments' under comparison via a binary response. Again, for simplicity, suppose that there are in the jth table n individuals receiving treatment 0, n receiving treatment 1, that the numbers of successes are (Y_{j0}, Y_{j1}) and that the logit probabilities of successes are $(\alpha_j, \alpha_j + \Delta)$, so that the contribution to the likelihood is

$$\frac{\exp\{\alpha_j(y_{j0}+y_{j1})+\Delta y_{j1}\}}{(1+e^{\alpha_j})^n(1+e^{\alpha_j+\Delta})^n}.$$

If the α_j are totally arbitrary nuisance parameters, inference about Δ is based on the likelihood for Δ conditional on $y_{j0}+y_{j1}$ $(j=1,\ldots,m)$, that is on a product of generalized hypergeometric forms. For testing $\Delta=0$, the Mantel–Haenszel test results. If the α_j are assumed to be all the same, we pool the tables. An intermediate situation is to suppose the α_j to be random variables A_j independently normally distributed with mean μ_A and variance σ_A^2.

The approximation (15) can now be used for the ensuing likelihood. In this outline investigation we concentrate on testing $\Delta=0$ via the score statistic, i.e. the gradient of log likelihood with respect to Δ evaluated at $\Delta=0$. Further we shall use a pre-specified value of σ_A^2, in particular because reasonably precise estimation of σ_A^2 is unlikely to be feasible unless the amount of data is very large. The unknown parameter μ_A will be replaced by the logit of the overall proportion of successes, which is the maximum likelihood estimate under $\Delta=\sigma_A^2=0$ and very close to the maximum likelihood estimate under $\Delta=0$ and the values of σ_A^2 employed.

For our numerical work we have considered a one-sided test of $\Delta=0$ with alternative $\Delta>0$, taking $m=10$ and studying power over the range $0<\Delta\leqslant 1$. Reasonable values of σ_A^2 are often 0·05, 0·10; if the overall success rate is $\frac{1}{2}$, probabilities corresponding to α_j would have standard deviations approximately 0·056, 0·079, respectively.

The first derivative of the log likelihood $(\partial l/\partial\Delta)_{\Delta=0}$, depends on

$$t_2=\sum(y_{j2}-n\pi), \quad t_.=\sum(y_{j1}+y_{j2}-2n\pi),$$

where $\pi=e^{\mu_A}/(1+e^{\mu_A})$. Now t_2 is closely related to the Mantel–Haenszel statistic, in the derivation of which $t_.$ is held constant as a conditioning variable. In detail

$$(\partial l/\partial\Delta)_{\Delta=0}=t_2+\sigma_A^2(a_1 t_.+b_1)+\sigma_A^4(a_2 t_.^2+b_2 t_.+c_2),$$

where

$$a_1 = -n\pi(1-\pi)\{1+2\sigma_A^2 n\pi(1-\pi)\}^{-1}, \quad b_1 = n\pi(1-\pi)(\pi-\tfrac{1}{2})\{1+2\sigma_A^2 n\pi(1-\pi)\}^{-1},$$

$$a_2 = n\pi(1-\pi)(\pi-\tfrac{1}{2})\{1+2\sigma_A^2 n\pi(1-\pi)\}^{-2}, \quad b_2 = -\tfrac{1}{2}n\pi(1-\pi)(6\pi^2-6\pi+1),$$

$$c_2 = n^2\pi^2(1-\pi)^2(\pi-\tfrac{1}{2}) - \tfrac{1}{8}n\pi(1-\pi)(12\pi^2-10\pi+1) + \tfrac{1}{2}n\pi^2(1-\pi)(6\pi^2-6\pi+1).$$

The second derivative $(\partial^2 l/\partial\Delta^2)_{\Delta=0}$ has an elementary but lengthy form which will not be reproduced here; it is available from the authors on request.

In the first instance we examine the distribution of the score test statistic under the null hypothesis of no logistic difference; the mean and variance of the statistic should be 0 and 1 respectively. Based on 200 simulations with $n = 10$ individuals receiving each treatment, and assuming the true values $\mu_A = 0$ and $\sigma_A^2 = 0.05$, the estimated mean of the statistic is 0.097 with standard deviation 1.012. Similarly, assuming a true value of $\sigma_A^2 = 0.10$, the corresponding estimates are -0.021 and 0.953. We found that the distribution of the score test statistic is not much changed when the wrong σ_A^2 is substituted.

Table 3 sets out the results from analogous simulations, but here we substitute the maximum likelihood estimate of μ_A for its true value $\mu_A = 0$ in calculating the mean and variance of the test statistic. We also take $n = 25$ individuals, $m = 25$ and assume the true value of σ_A^2 over a range. Again, the aim is to see whether the estimated mean and variance of the score test statistic are near 0 and 1. The observed negative values of the means are all very small, although the standard deviations are consistently less than one. This may suggest that using the wrong value of the mean μ_A is potentially more serious than using the wrong variance. Increasing both m and n would be expected to improve the asymptotic behaviour of the test statistic, which is nonetheless reasonable even under the present assumptions.

We were interested to compare the power of the score test to that of the Mantel–Haenszel test with continuity correction. Whichever test has the larger mean is likely to be more

Table 3. *Mean and standard deviation of score test statistic under the null hypothesis $\Delta = 0$ with true value of σ_A^2 and maximum likelihood estimate of μ_A. Results from 200 simulations with $m = 25$, $n = 25$, $\mu_A = 0$*

σ_A^2	Mean	Standard deviation
0	-0.029	0.929
0.05	-0.009	0.923
0.1	-0.076	0.829

Table 4. *Mean and standard deviation of score test statistic and Mantel–Haenszel test statistic, with continuity correction, for ranges of true values of Δ and σ_A^2; $n = m = 10$, 30 simulations. Standard deviation in parentheses*

Δ	$\sigma_A^2 = 0.05$		$\sigma_A^2 = 0.1$	
	Score	Mantel–Haenszel	Score	Mantel–Haenszel
0	0.132 (0.972)	0.184 (0.919)	0.049 (0.977)	0.234 (0.961)
0.1	0.232 (0.933)	0.009 (1.081)	0.467 (0.805)	0.231 (1.113)
0.5	2.241 (0.818)	1.421 (0.830)	2.288 (0.823)	1.841 (0.984)
1	3.721 (1.053)	2.747 (1.065)	3.754 (0.847)	3.139 (0.956)

sensitive to departures from the null hypothesis, as we are looking to recognize values beyond the 5% tail. Table 4 presents selected distributions for the test statistics evaluated over a range of true parameter values. In each case the mean and standard deviation are based on 30 paired simulations, the standard errors of the differences being approximately $0 \cdot 12$. The results from simulations illustrate that under departures from the null hypothesis, the score test is the more sensitive.

The score test thus has increased power under certain conditions. Whether it is wise to use the score test in practice is doubtful. As compared with the Mantel–Haenszel procedure, the new test assumes more, is more difficult to apply and lacks the simple direct appeal of the Mantel–Haenszel test. Nevertheless the possibility of improved power in problems with many nuisance parameters of similar interpretation illustrates a general phenomenon of broad theoretical interest. We stress that there are a number of aspects of this problem for which detailed exploration is desirable but will not be attempted here.

ACKNOWLEDGEMENTS

P. J. Solomon thanks SERC and the 1990/91 Australian Bicentennial Fellowship for support. The work was completed while P. J. Solomon was at the National Centre for Epidemiology and Population Health, Canberra.

REFERENCES

GREENWOOD, M. & YULE, G. U. (1920). An enquiry into the nature of frequency distributions representative of multiple happenings with particular reference to the occurrence of multiple attacks of disease or of repeated accidents. *J.R. Statist. Soc.* **83**, 255-75.

NELDER, J. A. (1977). A reformulation of linear models. *J.R. Statist. Soc.* A **140**, 48-76.

RACINE-POON, A. (1985). A Bayesian approach to nonlinear random effects models. *Biometrics* **41**, 1015-23.

RUDEMO, M., RUPPERT, D. & STREIBIG, J. C. (1989). Random-effect models in nonlinear regression with applications to bioassay. *Biometrics* **45**, 349-62.

SKELLAM, J. G. (1948). A probability distribution derived from the binomial distribution by regarding the probability of success as variable between two sets of trials. *J.R. Statist. Soc.* B **10**, 257-61.

SOLOMON, P. J. (1985). Transformations for components of variance and covariance. *Biometrika* **72**, 233-9.

SPEED, T. P. & SILCOCK, H. L. (1988). Cumulants and partition lattices V: calculating generalized *k*-statistics. *J. Aust. Math. Soc.* A **40**, 171-96.

[*Received December* 1990. *Revised May* 1991]

Scand J Statist 9: 147–152, 1982

On Partitioning Means into Groups

D. R. COX and E. SPJØTVOLL

Imperial College, London and Agricultural University of Norway, Aas

Received February 1981, in final form June 1981

ABSTRACT. A simple method based directly on standard *F* tests is given for partitioning means into groups. Several illustrative examples are discussed.

Key words: analysis of variance, contrasts, grouping of means, multiple comparisons

1. Introduction

In comparing treatment means, for example in analysis of variance, it is nearly always necessary and may sometimes be sufficient to give the sample means together with an estimated standard error for simple differences, so that confidence limits can be calculated for differences, or for more complicated contrasts. In other cases, interest may be concentrated on particular contrasts. In the present paper, however, we suppose that it is reasonable to describe any variation in the treatment means by partitioning the treatments into groups with hopefully the same mean for all treatments in the same group: this makes particular sense if on general grounds it is likely that the treatments fall into a fairly small number of such groups.

The object of this paper is to point out that this partitioning can be achieved without any complicated probability calculations. Related papers are by Spjøtvoll (1977) & Aitkin (1974).

The central idea is that all sufficiently simple partitions consistent with the data are considered. If a standard significance test is used a correct grouping, if one exists, will be 'rejected' only with probability the nominal level of the test, provided that partitioning into the appropriate number of groups is included. This is essentially the same argument as used to produce confidence regions as sets of values not rejected in a significance test. The probabilistic properties of the procedure are discussed more fully in Section 3.

We shall not study in detail the relation with the numerous other procedures for grouping or clustering means that have been suggested in the literature; see, however, Section 5 for some brief comparative remarks. The features of the procedure of the present paper that are distinctive, and which we feel made it attractive for applications, are as follows:

(i) the procedure may produce several different groupings consistent with the data and does not force an essentially arbitrary choice among several more-or-less equally well fitting configurations;

(ii) the groupings are not restricted to those in the observed order of the sample means;

(iii) the procedure is based solely on standard significance tests and complex probability calculations including sequences of interrelated choices are avoided, at the cost, however, of some approximations.

2. Method

Suppose that the population means are $\mu_1, ..., \mu_a$ with estimates $\hat{\mu}_1, ..., \hat{\mu}_a$. The estimates are assumed to be independent $N(\mu_i, \sigma^2/b_i^2)$, for $i=1, ..., a$, where $b_1, ..., b_a$ are known constants. Available also is an independent estimate s^2 of σ^2 with f degrees of freedom.

The simplest model is the one where all μ_i are equal. This is our first hypothesis. If this hypothesis is rejected we try a more complicated model. The next step is a model where the means are equal in two groups. There is a total of $2^{a-1}-1$ partitions in two groups, but there are usually a small number, if any, that are consistent with the data. To measure the consistency (or fit) with the data we use the significance level of the test that the means have a given grouping. If there is no satisfactory partitioning of the means in subsets of two groups of equal means, we proceed by trying three groups and so on. If at a certain stage we have clearly satisfactory consistency with the data, we normally stop, since a further splitting up of the means will, except for minor fluctuations, give a better fit but at the expense of a more complicated model. It is in any case implicit that any group consistent with the data can reasonably be 'refined' by splitting into subgroups.

When using this technique we need to evaluate consistency with hypotheses of the form

$$H_G: \mu_i = \theta_k, \quad i \in G_k, \quad k = 1, \ldots, q$$

where $G_1, \ldots, G_q$ are disjoint subsets of $\{1, \ldots, a\} = \bigcup_{k=1}^{q} G_k$, and the means $\theta_1, \ldots, \theta_q$ are unknown. Let g_k be the number of elements of G_k $(k = 1, \ldots, q)$. This hypothesis says that there are q groups where the μ_i's are equal and where group number i consists of the μ_i's with indices equal to the elements of G_k.

To judge consistency with the hypothesis H_G, we use the standard statistic F, namely

$$F_G = \frac{1}{a-q} \frac{\sum_{k=1}^{q} \sum_{i \in G_k} b_i (\bar{y}_i - \bar{y}_{G_k})^2}{s^2}$$

where

$$\bar{y}_{G_k} = \left(\sum_{i \in G_k} b_i \bar{y}_i \right) \bigg/ \left(\sum_{i \in G_k} b_i \right)$$

is the weighted average over the g_k means in group G_k. Large values of F_G indicate that the hypothesis H_G is false. We use the usual significance probability

$$P_G = P(\mathcal{F}_{a-q,f} \geq F_G),$$

where $\mathcal{F}_{a-q,f}$ denotes an F distributed random variable with $a-q$ and f degrees of freedom.

Commonly there will be several different partitions into q, the minimum number of groups, that are consistent with the data. Presentation of conclusions in simple form is of great importance and often a single statement can be synthesized by specifying those means that always fall in the same group and those that always fall in different groups. For instance, if the consistent partitions were

(abc,	def)	(ab,	cdef)
(abd,	cef)	(abcd,	ef)

then the conclusion would be in terms of the sets ab and ef. Treatments in the same set always occur in the same group, whereas treatments in different sets always lie in different groups; the assignment of c and d is not established by the data, at the level in question. Of particular use in simple summarization in general are those sets of treatments with the properties just mentioned.

The conclusions to be drawn via the proposed method are as follows:

(i) we obtain a smallest number q such that the data are consistent with q groups of differing means. We conclude that at least q groups are needed to describe the data satisfactorily;

(ii) we list the partitions into q groups that are consistent with the data. In some cases we may go further and list also partitions into $q+1$ or more groups;

(iii) we may assemble the information from the different groupings of size q in the way outlined above.

It will frequently be wise to examine two levels of significance, e.g. 0.05 and 0.01 or 0.1 and 0.01, and in any case the significance points should not be interpreted very rigidly.

3. Probability statements

We now consider the probability statements associated with this procedure. Throughout α denotes the significance level used in the F tests.

First suppose that in fact the true means are partitioned into ω groups. The true grouping, if tested, will be 'rejected' only with probability α and hence the value of q given by our procedure is a conservative lower $1-\alpha$ confidence limit for ω, the 'true' number of groups. The limit is conservative because of the possibility that, whereas the 'true' model would be rejected as inconsistent with the data, some other partition with the same or smaller number of groups is judged consistent with the data.

A rather similar approximate property holds for the condensation of the consistent partitions in which we assemble sets of means such that two means in different sets are never together in a consistent partition. Under the rather crude approximation that the correct value of q is observed, again a $1-\alpha$ conservative confidence statement is achieved.

A final stronger probabilistic property applies to

Table 1. *Duncan's barley data: yields in bushels/acre*

(a) *Mean yields*

a	b	c	d	e	f	g
49.6	58.1	61.0	61.5	67.6	71.2	71.3

(b) *Some partitions*

No. of groups	Partition	F	P
1		4.61	0.0020
2	*a, bcdefg*	2.42	0.059
	ab, cdefg	2.14	0.088
	abc, defg	2.01	0.11
	abcd, efg	1.50	0.22
	abcde, fg	2.59	0.046
	abcdef, g	4.29	0.0046
	abd, cefg	2.19	0.082
	ac, bdefg	3.09	0.023
	b, acdefg	5.13	0.0016
	acd, befg	3.11	0.023
	ad, bcefg	3.25	0.018
3	*a, bcd, efg*	0.29	0.88
	ab, cd, efg	0.85	0.50
	a, bcde, fg	0.90	0.48
	a, bc, defg	1.28	0.30
	a, bd, cefg	1.43	0.25
	abc, d, efg	1.49	0.23
	ac, bd, efg	1.50	0.23
	abcd, e, fg	1.71	0.17
	a, bcde, fg	1.97	0.12
	a, b, cdefg	1.99	0.12
	a, bce, dfg	2.09	0.11
	a, bcdef, g	2.16	0.098
	a, bcdeg, f	2.19	0.094

an extension of the procedure recommended here which is desirable in principle even if often too complex for practice. Suppose that instead of stopping the procedure when consistency with one or more partitions into say q groups is achieved, we examine all possible partitions. Then the correct partition, whatever it is, will be included with probability $1-\alpha$, i.e. under the assumptions made, we have a confidence set of partitions consistent with the data. Normally this gives a very cumbersome way of representing the data and therefore we make an approximate summarization of the confidence set as follows:

(i) We assume that if a partition into q groups is in the confidence set, so is any 'refinement' of that partition into $q+1$ or more groups. This, while often not precisely true if the significance points are used rigidly, is usually very nearly the case;

(ii) we assume that there is no important consistent partition into more than q groups different from those obtained by 'refining' the partitions listed in our recommended technique.

Then the procedure we have suggested is a concise approximate representation of the exact $1-\alpha$ confidence set of partitions.

4. Some examples

To illustrate the method we look at three simple sets of data that have been analyzed in the literature by various methods.

Example 1. The following data used by Duncan (1955) concern seven varieties of barley grown in six blocks. Table 1*a* gives mean yields in bushels/acre. For convenience the varieties have been named $a, \ldots, g$ in order of increasing mean yield; the same has been done for the later examples. The estimate of variance is 79.64 with 30 degrees of freedom, so that the estimated standard error of a mean is 3.64 and of the difference between two means is 5.15.

We now try partitioning into groups, and the conclusions are summarized in Table 1*b*. Note that we consider both 'natural' partitions, in which the

Table 2. *Fat absorption (g) in mixtures of doughnuts*

(a) *Mean absorptions*

a	b	c	d	e	f	g	h
161	162	165	172	175	178	181	185

(b) *Some partitions*

No. of groups	Partition		F	P
1			3.56	0.0043
2	a,	bcdefgh	3.06	0.015
	ab,	cdefgh	1.82	0.12
	abc,	defgh	0.79	0.58
	abcd,	efgh	0.87	0.53
	abcde,	fgh	1.38	0.25
	abcdef,	gh	1.92	0.10
	abcdefg,	h	2.92	0.019
	ac,	bdefgh	2.41	0.044
	bc,	adefgh	2.55	0.035
	ad,	bcefgh	3.45	0.0077
	abd,	cefgh	2.18	0.065
	abe,	cdfgh	2.81	0.022
	abce,	dfgh	1.67	0.15
	abcf,	degh	2.03	0.084
	abcg,	defh	2.67	0.028
	abde,	cfgh	2.81	0.022
	abch,	defg	3.07	0.015
	abcdf,	egh	1.77	0.13
	abcdg,	efh	2.47	0.040
	abcdeh,	fg	3.13	0.013
	abcdeg,	fh	2.67	0.028

groups correspond with the rank ordering of the sample means, and some other partitions. The more useful partitions into two groups are given, with the 'natural' ones first. Some partitions into three groups are given partly because none of the partitions into two groups gives an excellent fit and partly to show the effect of refining the acceptable two-group partitions.

There are models with the means in just two groups that give a satisfactory explanation of the observations, in the sense that the corresponding hypotheses are not rejected by an ordinary significance test. If we draw the line for acceptable models at $P = .05$, we have only the five possible models

a, bcdefg
ab, cdefg
abc, defg
abcd, efg
abd, cefg

Thus a and efg must belong to different groups while b, c and d could belong to any of these two

groups, although not all assignments are acceptable at 0.05 level.

In a final presentation of conclusions it would normally be sensible to include the individual treatment means and the means of the more interesting groups.

Example 2. Table 2 concerns an experiment on fat absorption of different mixtures of doughnuts. The means in Table 2a are based on 6 observations each and the estimate of variance is 141.6 with 40 degrees of freedom. Table 2b summarizes some partitions.

Two groups are enough to give a satisfactory explanation for the observations. Using the 5% level as the border between satisfactory and unsatisfactory models it is seen that all satisfactory partitions are such that *ab* belongs to one group and that *gh* belongs to the other. The intermediate means *cdef* could belong to any of the two groups. For instance, the partition *ab cdefg* corresponds to $P = 0.12$ and *abcdef gh* to $P = 0.10$. Although these P's are on the low side, the corresponding partitions cannot be rejected.

Scand J Statist 9

Table 3. *Loaf volumes (ml) of varieties of wheat*

(a) Mean volumes

a	b	c	d	e	f	g	h	i
654	729	755	801	828	829	846	853	861

j	k	l	m	n	o	p	q
903	908	922	933	951	977	987	1 030

(b) Some partitions

No. of groups	Partition	F	P
1		28.35	10^{-11}
2	a–i, j–q	10.11	10^{-11}
3	a–c, d–i, j–q	4.48	1.3×10^{-6}
4	a–c, d–i, j–m, n–q	2.48	0.0084
5	a, bc, d–i, j–n, o–q	1.43	0.18
	a, bc, d–i, j–m, n–q	1.60	0.11
	a, b–d, e–i, j–n, o–q	1.61	0.11
	a, b–d, e–i, j–m, n–q	1.78	0.071
	a, bc, d–i, j–o, pq	1.85	0.058
	ab, cd, e–i, j–n, o–q	1.90	0.050
	a, b–d, e–i, j–n, o–q	2.03	0.036
	ab, c, d–i, j–n, o–q	2.03	0.036
	a, bc, d–i, j–l, m–q	2.07	0.032
	ab, cd, e–i, j–m, n–q	2.07	0.032
	a–c, d–i, j–m, n–p, q	2.24	0.020
	a, b–d, e–i, j–m, n–q	2.24	0.020
	a, bc, d–i, j–p, q	2.25	0.019
	a–c, d, e–i, j–n, o–q	2.30	0.017
	a–c, d–i, j–n, op, q	2.31	0.016

Example 3. As a final example we take one involving a larger number of means. It concerns an experiment (Duncan, 1965) with the loaf volumes after baking of 17 varieties of wheat. The observed means are each the average of 5 observations. The estimate of variance is 1713 with 64 degrees of freedom.

There need to be at least 5 groups of means. If we require $P \geq 0.10$ for acceptable consistency with the data the only possible partitions are

a, bc, defghi, jklmn, opq
a, bc, defghi, jklm, nopq
a, bcd, efghi, jklmn, opq

which show that a constitutes one group, bc belong to an other, efghi to a third, jklm to a fourth and opq to a fifth group. The mean d could belong to group two or to group three while n could belong to groups four or five. If we require only $P \geq 0.05$, the following partitions also are acceptable

a, bcd, efghi, jklm, nopq
a, bc, defghi, jklmno, pq
ab, cd, efghi, jklm, opq

Then the five groups will consist of a, c, efghi, jklm and pq, respectively, with b, d, n, o undecided. This would also be the conclusion even if we included all partitions with $P \geq 0.025$.

5. Discussion

The procedure outlined above can be used in any problem where a set of parameters is to be partitioned into groups. The F test would be replaced by an appropriate test of homogeneity, such as the likelihood ratio test. Rather similar arguments can be applied to the selection of variables in multiple regression and allied problems (Spjøtvoll, 1977).

Scott & Knott (1974) used a cluster analysis approach for grouping means. Their clustering is based on the sum of squares within groups, and in that sense their method is similar to ours. The kind of conclusions they obtain, however, are different. The clustering algorithm points only to one partition for each number q of groups. For the data in Example 1 they obtain the partition abcd efg for $q=2$, but none of the others which also may be considered as acceptable as measured by a standard F test.

Gabriel (1964) is concerned with simultaneous tests of all hypotheses

$$H: \mu_{i_l} = \ldots = \mu_{i_k},$$

where $\{i_l, \ldots, i_k\}$ is some subset of $\{1, \ldots, a\}$. This is a different set of hypotheses than the ones we consider in Section 2. The results obtained may, of course, be different. For the data in Example 2 the hypothesis that the group means for a, b, c, d, e and g are equal, is not rejected. Our conclusion, however, was that ab and gh belonged to different groups.

When rather a large number of groups are necessary to fit the data, some doubt is thrown on the suitability of description in terms of groups rather than of an essentially continuous distribution of individual means. In Example 3, however, although the number of groups is quite large, the notion of a number of groups of almost identical genotypes is plausible and it could be of interest to try to identify such groups.

References

Aitkin, M. A. (1974). Simultaneous inference and the choice of variable subsets in multiple regression. *Technometrics* **16**, 221–227.

Duncan, D. B. (1955). Multiple range and multiple F tests. *Biometrics* **11**, 1–42.

Duncan, D. B. (1965). A Bayesian approach to multiple comparisons. *Technometrics* **7**, 171–222.

Gabriel, K. R. (1964). A procedure for testing the homogeneity of all sets of means in analysis of variance. *Biometrics* **20**, 459–477.

Scott, A. J. & Knott, M. (1974). A cluster analysis method for grouping means in the analysis of variance. *Biometrics* **30**, 507–512.

Spjøtvoll, E. (1977). Alternatives to plotting C_p in multiple regression. *Biometrika*, **64**, 1–8.

Emil Spjøtvoll
Department of Mathematics and Statistics
Agricultural University of Norway
N-1432 Aas
Norway

[*From* Biometrika, *Vol.* 42. *Parts* 1 *and* 2, *June* 1955.]

SOME QUICK SIGN TESTS FOR TREND IN LOCATION AND DISPERSION

BY D. R. COX

Statistical Laboratory, University of Cambridge

AND A. STUART

Division of Research Techniques, London School of Economics

1. INTRODUCTION AND SUMMARY

Many distribution-free tests have been devised to test the hypothesis of randomness of a series of N observations, i.e. the hypothesis that N independent random variables have the same continuous distribution function. Of these, the rank correlation tests are the most efficient tests against normal trend alternatives, but others are of some use in situations where speed and simplicity of computation are important.

In this paper, we discuss a class of simple sign tests, considered first as tests against trend in location. Optimum tests are found from the standpoint of asymptotic relative efficiency (a measure of local power in large samples), and it appears that the best of these tests may be preferred to the other simple tests considered in the literature, although they are, of course, less efficient than the rank correlation tests.

Similar tests are available for trend in dispersion, and the efficiency of these, in the normal situation, is investigated and compared with the test based on the maximum-likelihood estimator. Finally, we add a few remarks on sequential sign tests.

Readers not interested in the theory should look at §§ 10, 11 and 14, where there are brief statements of the tests and numerical examples.

2. THE SIGN TESTS FOR TREND IN LOCATION

We consider a series of N independent observations from a standardized normal regression model with an upward trend, i.e.

$$H_1: \quad y_i = \alpha + \Delta i + \epsilon_i \quad (i = 1, 2, ..., N),$$

where $\Delta \geqslant 0$ and the ϵ_i are independent standardized normal variates. We wish to test the null hypothesis

$$H_0: \quad \Delta = 0,$$

using a distribution-free test statistic, so that our test will remain valid whatever the continuous distribution of the ϵ terms in the model, although naturally its efficiency will vary with the form of the distribution.

The most efficient known distribution-free tests of H_0 are those based on the rank correlation coefficients (Stuart, 1954), but our object here is specifically to find tests which are quick and simple to compute. We define for $i < j$ the score

$$h_{ij} = \begin{cases} +1 & \text{if } y_i > y_j, \\ 0 & \text{if } y_i < y_j. \end{cases}$$

h_{ij} is thus based on a comparison of the ith and jth in the series of observations. The distribution of the observations will be assumed continuous so that the possibility of ties can be ignored (see, however, § 16).

We confine ourselves throughout to comparisons of *independent* pairs of observations, i.e. no observation is compared with more than one other observation. (This is in contrast to the procedure used in calculating the rank correlation coefficients, where every observation is compared with every other observation in the series.) Since there are N observations, there can be no more than $\frac{1}{2}N$ such independent comparisons. We now assume N to be even and always take $i < j$. Our problem is to find the set of comparisons and the appropriate weights w_{ij} which will make the statistic

$$S = \Sigma w_{ij} h_{ij} \tag{1}$$

as efficient a test of H_0 as possible. The summation in (1) contains $\frac{1}{2}N$ terms, no suffix being repeated. All tests of the form (1) are distribution-free, since on the null hypothesis any h_{ij} is a $0-1$ variate with probabilities $(\frac{1}{2}, \frac{1}{2})$, whatever the distribution of the ϵ_i.

3. Asymptotic relative efficiency

We shall use as our criterion of efficiency the asymptotic relative efficiency (A.R.E.) of a test. If there are two consistent tests, s and t, of a hypothesis $H_0: \Delta = 0$, the A.R.E. is the reciprocal of the ratio of sample sizes required to attain the same power against the same alternative hypothesis H_1, taking the limit as the sample size N tends to infinity and as H_1 tends to H_0. (This second limiting process is necessary to keep the power of consistent tests bounded away from 1.) Pitman (1948) and Noether (1955) have shown that, if s and t both have normal limiting distributions on H_0 and H_1, the A.R.E. of s compared to t is given by

$$\text{A.R.E. } (s, t) = \lim_{N \to \infty} \left(\frac{R^2(s)}{R^2(t)} \right)^{1/r}, \tag{2}$$

where

$$R^2(X) = \left\{ \left[\frac{\partial}{\partial \Delta} E(X \mid \Delta) \right]_{\Delta=0} \right\}^2 \Big/ D^2(X \mid \Delta = 0), \tag{3}$$

provided that r satisfies the equations

$$\lim_{N \to \infty} R^2(s) N^{-r} = R_1, \quad \lim_{N \to \infty} R^2(t) N^{-r} = R_2. \tag{4}$$

Here E and D^2 denote mean and variance as usual, while R_1 and R_2 are constants independent of N. The interpretation of the A.R.E. is discussed critically in § 9 below.

4. The best sign test

Since h_{ij} is a 0–1 variate,
$$E(h_{ij}) = \text{prob } (y_i > y_j),$$

and as $(y_j - y_i)$ is a normal variate with mean $(j-i)\Delta$ and variance 2 this is

$$E(h_{ij}) = G\left\{ -\frac{(j-i)\Delta}{\sqrt{2}} \right\}, \tag{5}$$

where

$$G(x) = \int_{-\infty}^{x} \frac{1}{\sqrt{(2\pi)}} e^{-\frac{1}{2}t^2} dt.$$

Now

$$\left[\frac{\partial}{\partial x} G(x) \right]_{x=0} = \frac{1}{\sqrt{(2\pi)}}, \tag{6}$$

so that, by (5) and (6),

$$E'(h_{ij}) = \left[\frac{\partial}{\partial\Delta}E(h_{ij})\right]_{\Delta=0} = \frac{(i-j)}{2\sqrt{\pi}}.\tag{7}$$

We now write $(j-i) = r_{ij}$. Using (7) in (1), we obtain

$$E'(S) = \Sigma w_{ij}E'(h_{ij}) = -\frac{1}{2\sqrt{\pi}}\Sigma w_{ij}r_{ij}.\tag{8}$$

We also have

$$D^2(S\,|\,\Delta=0) = \Sigma w_{ij}^2\,V(h_{ij}\,|\,H_0) = \tfrac{1}{4}\Sigma w_{ij}^2.\tag{9}$$

Equations (3), (8) and (9) give

$$R^2(S) = \frac{1}{\pi}\frac{(\Sigma w_{ij}r_{ij})^2}{\Sigma w_{ij}^2},\tag{10}$$

and we now wish to maximize (10) to obtain the highest possible A.R.E. We do this in two stages. First we maximize (10) with respect to the w_{ij}, regarding the r_{ij} as fixed, and then we choose the supremum of these maxima for variations in the r_{ij}.

To maximize (10) for fixed r_{ij} and variation in the w_{ij}, we must maximize $\Sigma w_{ij}r_{ij}$ subject to Σw_{ij}^2 being held constant, i.e. we must unconditionally maximize

$$F = \Sigma w_{ij}r_{ij} - \lambda\Sigma w_{ij}^2.$$

It is clear from the conditions of the problem that each w_{ij} will be a function of the corresponding r_{ij}, so that on differentiating F for a stationary value we get

$$r_{ij} + w_{ij}\frac{\partial r_{ij}}{\partial w_{ij}} - 2\lambda w_{ij} = 0,$$

i.e.

$$\frac{r_{ij}}{w_{ij}} + \frac{\partial r_{ij}}{\partial w_{ij}} = 2\lambda.$$

This is satisfied by

$$w_{ij} = \lambda r_{ij},\tag{11}$$

so that the required set of weights are proportional to the distances apart of the observations compared. The stationary value is a maximum. Substituting (11) into (10), we have

$$R^2(S) = \frac{1}{\pi}\Sigma r_{ij}^2.\tag{12}$$

This is the maximum value of $R^2(S)$ for a fixed set of r_{ij}. The r_{ij} are a set of $\tfrac{1}{2}N$ differences between pairs of integers chosen from the integers $1,2,...,N$. It is easily seen that Σr_{ij}^2 is largest when the pairs are $(1,N),(2,N-1),(3,N-2)$ and so on. In general

$$r_{ij} = (N-k+1)-k = N-2k+1 \quad (k=1,2,...,\tfrac{1}{2}N)\tag{13}$$

so that

$$\Sigma r_{ij}^2 = \sum_{k=1}^{\frac{1}{2}N}(N-2k+1)^2 = \tfrac{1}{6}N(N^2-1),\tag{14}$$

and the supremum value of (12) is therefore

$$R^2(S_1) = \frac{N(N^2-1)}{6\pi} \sim \frac{N^3}{6\pi}.\tag{15}$$

We have denoted by S_1 the optimum S statistic

$$S_1 = \sum_{k=1}^{\frac{1}{2}N}(N-2k+1)h_{k,N-k+1},$$

83   D. R. Cox and A. Stuart

for which

$$E(S_1 \mid \Delta = 0) = \tfrac{1}{2}\Sigma(N - 2k + 1) = \tfrac{1}{8}N^2,$$
$$D^2(S_1 \mid \Delta = 0) = \tfrac{1}{4}\Sigma(N - 2k + 1)^2 = \tfrac{1}{24}N(N^2 - 1).$$

(16)

The test based on S_1 is essentially a simplified version of Spearman's rank correlation test, which is in effect defined by

$$V = \sum_{i<j} (j - i) h_{ij},$$

(17)

where the summation extends over all possible $\tfrac{1}{2}N(N-1)$ pairs of observations. Stuart (1954) has shown that

$$R^2(V) \sim \frac{N^3}{4\pi},$$

(18)

so that using (15) and (18) in (2) and (4) with $r = 3$, we obtain for the A.R.E. of S_1 compared to V

$$\text{A.R.E.} (S_1, V) = (\tfrac{2}{3})^{\frac{1}{3}} = 0.87.$$

(19)

The loss of A.R.E. involved in reducing the number of comparisons from $\tfrac{1}{2}N(N-1)$ to $\tfrac{1}{2}N$ is as little as 13%.

These values of the A.R.E. depend on the assumption of normality, but the calculation of the form of the optimum statistic, S_1, and also of the statistic, S_3, of § 5, does not. For (7) remains true, with a changed numerical factor, for general continuous distributions.

5. An unweighted sign test

The relatively high efficiency of S_1 compared to V leads us to construct, by analogy, a simplified version of Kendall's rank correlation test, which may be defined by

$$Q = \sum_{i<j} h_{ij},$$

(20)

and gives equal weight to all $\tfrac{1}{2}N(N-1)$ comparisons. Q has the same A.R.E. as V (Stuart, 1954). The analogous sign test, based on $\tfrac{1}{2}N$ equally weighted independent comparisons, is

$$S_2 = \Sigma h_{ij},$$

(21)

and using (10), we obtain, with all $w_{ij} = 1$,

$$R^2(S_2) = \frac{2}{N\pi} (\Sigma r_{ij})^2.$$

(22)

We now require to choose $\tfrac{1}{2}N$ pairs from the first N integers so that (22) or, equivalently, $\Sigma(j - i) = \Sigma r_{ij}$, takes its maximum value. This occurs whenever every i is chosen from the first $\tfrac{1}{2}N$ integers and every j from the last $\tfrac{1}{2}N$ integers. In particular, it occurs when every $r_{ij} = \tfrac{1}{2}N$ exactly, so that

$$S_2 = \sum_{k=1}^{\frac{1}{2}N} h_{k, \frac{1}{2}N+k},$$

(23)

and (22) becomes

$$R^2(S_2) = \frac{N^3}{8\pi}.$$

(24)

Using (24) and (18), we obtain

$$\text{A.R.E.} (S_2, V) = (\tfrac{1}{2})^{\frac{1}{3}} = 0.79,$$

(25)

while from (15)

$$\text{A.R.E.} (S_2, S_1) = (\tfrac{1}{2})^{\frac{1}{3}} = 0.91.$$

(26)

Thus the simplified version of Kendall's rank correlation test is 21 % less efficient than Q or V, and 9 % less efficient than the simplified Spearman coefficient S_1. The use of S_2 is equivalent to a test considered by Theil (1950).

6. THE BEST UNWEIGHTED SIGN TEST

However, we can improve on the efficiency of S_2, and in fact get very nearly as high an efficiency as that of S_1, by 'throwing away' some of the $\frac{1}{2}N$ comparisons and retaining equal weights for the others. This was suggested by one of the present authors in the discussion of Foster & Stuart (1954); it leads to an increase in efficiency because, by comparing observations further apart, individual comparisons are made more sensitive.

In (1), let every w_{ij} be either 0 or 1, and let m ($\leqslant \frac{1}{2}N$) be the number of non-zero w_{ij}. For our new statistic S_3 we have, as in (8),

$$E'(S_3) = -\frac{1}{2\sqrt{\pi}}\Sigma w_{ij}r_{ij} \quad (w_{ij} = 0 \text{ or } 1), \tag{27}$$

and from (9)
$$D^2(S_3 \mid \Delta = 0) = \tfrac{1}{4}m, \tag{28}$$
so that (3), (27) and (28) give

$$R^2(S_3) = \frac{1}{m\pi}(\Sigma w_{ij}r_{ij})^2 \quad (w_{ij} = 0 \text{ or } 1). \tag{29}$$

To maximize this by choice of m and r_{ij}, we again work in two stages. For fixed m, (29) will take its largest value when the comparisons given zero weights are based on the middle $(N-2m)$ observations, while every i is chosen from the first m observations and every j is chosen from the last m observations. In particular, this will be so when every $r_{ij} = (N-m)$ exactly, so that

$$S_3 = \sum_{k=1}^{m} h_{k,N-m+k} \tag{30}$$

and (29) becomes
$$R^2(S_3) = \frac{m(N-m)^2}{\pi}. \tag{31}$$

(31) is the largest possible value of $R^2(S_3)$ for fixed m. (S_2 is the special case of S_3 when $m = \frac{1}{2}N$.) We now choose m to maximize (31). Differentiating, we get

$$m = \tfrac{1}{3}N \tag{32}$$

for a maximum, so that finally
$$S_3 = \sum_{k=1}^{\frac{1}{3}N} h_{k,\frac{2}{3}N+k},$$

for which
$$\left.\begin{aligned} E(S_3) &= \tfrac{1}{6}N, \\ V(S_3) &= \tfrac{1}{12}N, \end{aligned}\right\} \tag{33}$$

and from (31),
$$R^2(S_3) = \frac{4N^3}{27\pi}. \tag{34}$$

From (34) and (18), we have
$$\text{A.R.E.}\,(S_3, V) = (\tfrac{16}{27})^{\frac{1}{2}} = 0\cdot84, \tag{35}$$

while from (15)
$$\text{A.R.E.}\,(S_3, S_1) = (\tfrac{8}{9})^{\frac{1}{2}} = 0\cdot96. \tag{36}$$

Compared with either V or S_1, S_3 has about 5 % higher efficiency than S_2, and in fact its efficiency is 96 % of that of S_1, so that for practical purposes it may be recommended instead of S_1 because it requires no weighting of the comparisons.

7. Comparison of the sign tests

In Table 1, the A.R.E. of the sign tests are tabulated, compared to each other, to the rank correlation tests, and to the best (parametric) test against normal regression, based on the sample regression coefficient b, which has a value of (3) given by

$$R^2(b) \sim \frac{N^3}{12},\tag{37}$$

as follows immediately from the fact that b is an unbiased estimator of Δ with variance $12/\{N(N^2-1)\}$.

Table 1. *Asymptotic relative efficiencies of sign tests*

Test statistic	Asymptotic relative efficiency		
	Compared to S_1	Compared to rank correlation tests	Compared to best parametric test
$S_1 = \overset{\frac{1}{2}N}{\underset{k=1}{\Sigma}} (N-2k+1)\, h_{k,\,N-k+1}$	1·00	0·87	0·86
$S_2 = \overset{\frac{1}{2}N}{\underset{k=1}{\Sigma}} h_{k,\,\frac{1}{2}N+k}$	0·91	0·79	0·78
$S_3 = \overset{\frac{1}{2}N}{\underset{k=1}{\Sigma}} h_{k,\,\frac{3}{2}N+k}$	0·96	0·84	0·83

From (2), (18) and (37), it follows that the A.R.E. of either rank correlation coefficient compared to b is

$$\text{A.R.E.}\,(V,b) = \left(\frac{3}{\pi}\right)^{\frac{1}{2}} = 0\cdot98,\tag{38}$$

and not $3/\pi = 0\cdot95$ as given by Stuart (1954).

8. Comparison with A.R.E. of other tests

Apart from the two rank correlation tests already discussed, Stuart (1954) investigated the A.R.E. of three other distribution-free tests for trend in location. Two of these, the rank serial correlation test and the turning point test, were found to have zero values of R as defined by (3); the third, the difference-sign test, was found to have a value of r equal to 1 in (4), as against $r = 3$ for all the tests considered in this paper. It followed that the three tests mentioned all have A.R.E. zero compared to the rank correlation tests (and hence to all the tests discussed here). Noether (1955) gives general results which rigorize these conclusions.

A well-known and simple test which has not, as far as we know, previously been discussed from the point of view of A.R.E. is the median test, due to Brown & Mood (1951). The N (even) observations are divided into two sets of $\frac{1}{2}N$ consecutive observations. The test statistic is simply the number of observations in the first set which exceed the sample median y_m, and it is therefore defined by

$$B = \sum_{i=1}^{\frac{1}{4}N} b_{im}, \tag{39}$$

where
$$b_{im} = \begin{cases} 1 & \text{if} \quad y_i > y_m, \\ 0 & \text{if} \quad y_i < y_m. \end{cases}$$

The A.R.E. of B is easily obtained. We know that y_i is a normal variate with mean $(\alpha + i\Delta)$ and unit variance. It follows that the sample median y_m is asymptotically a normal variate with mean $(\alpha + \frac{1}{2}(N+1)\Delta)$ and variance of order N^{-1}. Since y_i and y_m are asymptotically independent, $(y_i - y_m)$ is asymptotically normal with mean $\Delta[i - \frac{1}{2}(N+1)]$ and unit variance, so that for $i < \frac{1}{2}(N+1)$

$$E(b_{im}) = \text{prob}\,(y_i > y_m) \sim 1 - G\{\Delta[\tfrac{1}{2}(N+1)-i]\}. \tag{40}$$

Using (6) in (40), we obtain
$$E'(b_{im}) \sim -\frac{1}{\surd(2\pi)} [\tfrac{1}{2}(N+1)-i], \tag{41}$$

so that from (39) and (41),

$$E'(B) = \sum_{i=1}^{\frac{1}{4}N} E'(b_{im}) \sim -\frac{1}{\surd(2\pi)} \sum_{i=1}^{\frac{1}{4}N} [\tfrac{1}{2}(N+1)-i] \sim -\frac{N^2}{8\,\surd(2\pi)}. \tag{42}$$

Also (Brown & Mood, 1951) $\qquad D^2(B \mid \Delta = 0) \sim \dfrac{N}{16}. \tag{43}$

(42) and (43) give, in (3),

$$R^2(B) \sim \frac{N^3}{8\pi}. \tag{44}$$

Comparison of (44) with (24) shows that B has precisely the same A.R.E. as S_2, and is therefore slightly less efficient than S_1 and S_3. If the observations are available in serial order, S_3 is simpler to compute than B, which involves ranking all the observations to find the median, and then making $\frac{1}{2}N$ comparisons, as against $\frac{1}{3}N$ for S_3. There is therefore no reason to prefer B to S_3 in this case. If, however, the data were available graphically, B would be considerably easier to compute, and this would outweigh the slight loss of efficiency compared to S_3.

9. COMPARISON OF THE POWERS OF TESTS

So far we have compared tests by the A.R.E. in the usual way. Before considering the power of the test S_3 in small samples it is convenient to examine the meaning of the A.R.E. more carefully. If the A.R.E. of a quick test relative to an efficient test is A, then asymptotically A^{-1} as many observations have to be made for the quick test to give the same local power as the efficient test. This is directly relevant if in designing an experiment a choice has to be made between, on the one hand, using an efficient method of analysis and on the other taking more observations and using a quick method of analysis. But it is not directly relevant to the choice of a method of analysis for a given body of data, because it depends in part on r, defined by (4), measuring the rate at which power increases with increasing N. For a given problem r is fixed and so the A.R.E. can be reinterpreted in terms of the power attained at a fixed sample size, but it seems preferable to compare tests directly in terms of power.

Consider a test based on a statistic S normally distributed with mean $E(S \mid \Delta)$ and standard deviation $D(S \mid \Delta)$, where the null hypothesis is $\Delta = 0$. The null hypothesis is rejected at the significance level α if

$$S > E(S \mid 0) + \lambda_\alpha D(S \mid 0). \tag{45}$$

where

$$G(-\lambda_\alpha) = \alpha. \tag{46}$$

The power of the test is $G[p(\Delta)]$, where

$$p(\Delta) = \frac{E(S \mid \Delta) - E(S \mid 0) - \lambda_\alpha D(S \mid 0)}{D(S \mid \Delta)}. \tag{47}$$

Now

$$p'(0) = \left(\frac{\partial p(\Delta)}{\partial \Delta}\right)_{\Delta=0} = \frac{E'(S \mid 0) + \lambda_\alpha D'(S \mid 0)}{D(S \mid 0)}. \tag{48}$$

In all the applications in this paper $D'(S \mid 0) = 0$, so that

$$p'(0) = \frac{E'(S \mid 0)}{D(S \mid 0)} = R(S). \tag{49}$$

Near $\Delta = 0$,

$$p(\Delta) = \Delta R(S) - \lambda_\alpha + O(\Delta^2), \tag{50}$$

and in applications the first two terms give, asymptotically in N, the whole of the power curve. Moreover, $R(S) \sim RN^{-\frac{1}{2}r}$ as $N \to \infty$ and comparable tests of a given hypothesis will have the same r; hence we usually need to consider just R. We call $p(\Delta)$ the *power deviate* and $p'(0)$ the *power derivative*. Asymptotically the graph of $p(\Delta)$ against Δ is linear, and tests at different significance levels are given by parallel straight lines. Or to put the same fact another way, the power curves are asymptotically linear when plotted on arithmetical probability paper.

Now consider the small sample theory with S possibly not normally distributed. Then if the power curves are plotted on probability paper they can be expected to form an approximately parallel set of curves approaching a set of parallel lines as the sample size increases and the distribution of S tends to normality. This is of course only a method of presenting the results of power calculations, but we shall find it very convenient both in assessing the small-sample behaviour and in comparing different tests.

Consider now two tests for which the asymptotic values of $R(S)$ are R_1, R_2. Then asymptotically in N the power curves for a given α are two lines on probability paper, the ratio of their slopes being R_1/R_2 independent of α; both lines intersect the probability axis at α.

A first consequence is that there is no simple general relation between the difference in the power of the two tests and the ratio R_1/R_2. If $R_1 \neq R_2$ we can, by taking α sufficiently small, make the difference in power between the two tests arbitrarily near unity. In practice we are probably only interested in $0 \cdot 20 \geqslant \alpha \geqslant 0 \cdot 001$, but the general conclusion remains that the difference in power between a quick test and an efficient test will be greatest for small α. Table 2 expresses this quantitatively; it shows for given R_1/R_2 the powers of the two tests at the point at which the difference in powers is greatest. The values in Table 2 are independent of N, but the values of Δ at which these powers are attained do depend on N. This is the restriction on the alternative hypothesis referred to in § 3. Thus if $R_1/R_2 = 0 \cdot 7$ and $\alpha = 0 \cdot 05$, the difference in power is greatest for the value of Δ at which the power of the efficient test is 77 % and of the quick test 51 %.

Now consider the power of S_3 in small samples. Two methods can be used. The first is to take the expansion (50) to higher powers of Δ and to introduce a correction for the non-normality of S based on an Edgeworth expansion. This may be shown to give good results even for very small N, and is a general method which could be used where direct numerical calculation is difficult. However, for S_3 it is much easier to calculate the power directly from the National Bureau of Standards tables of the binomial distribution (1950).

Table 2. *Asymptotic theory. Powers (per cent) of quick and efficient tests at points at which difference in power is greatest**

R_1/R_2 $\quad\quad\alpha$	0·10	0·05	0·01	0·001
0·9	67, 73	63, 71	49, 60	54, 67
0·8	61, 74	56, 72	49, 71	43, 72
0·7	59, 80	51, 77	42, 77	39, 83
0·6	54, 84	47, 84	39, 86	29, 87
0·5	48, 88	41, 89	30, 90	20, 93
0·3	35, 96	27, 96	14, 97	7, 99

The power was computed in this way for $N = 15\,(15)\,135$, the significance level being the largest value $\leqslant 0\cdot05$. Under the null hypothesis the test statistic is distributed as $(\frac{1}{2}+\frac{1}{2})^{\frac{1}{3}N}$ and under the alternative hypothesis as $(p+q)^{\frac{1}{3}N}$, where

$$p = G\left(-\frac{\sqrt{2\,N}\Delta}{3}\right). \tag{51}$$

The power corresponding to given p, $\frac{1}{3}N$ can be read off directly from the tables and (51) solved for Δ. The results are given in Table 3. For comparative purposes the exact power of the parametric test based on the regression coefficient, b, has been computed for the same values of N and Δ. When the standard deviation about the regression line is known, the power is exactly $G[p(\Delta)]$, where

$$p(\Delta) = \{\tfrac{1}{12}N(N^2-1)\}^{\frac{1}{2}}\,\Delta - \lambda_\alpha. \tag{52}$$

To avoid rewriting the values of Δ in Table 4 the rows of both tables have been lettered, and each entry in Table 4 relates to the value of Δ shown above the corresponding entry in Table 3.

Asymptotically, the ratio of the R values of the two tests is, by (34) and (37), $4/(3\sqrt{\pi}) = 0\cdot75$; the interpretation of this in terms of power can be obtained from Table 2. The full curve in Fig. 1 and the full curves in Fig. 2 for $k = 0$ show the power curves for $N = 15, 30$, and the dotted lines are the corresponding asymptotic power curves. The small-sample power is lower than the value given by the asymptotic theory, the difference being quite appreciable in the region of 80–90 % power. The power curves of the most efficient test are exactly linear and differ from their asymptotic form only because of the very small difference between $\{N(N^2-1)\}^{\frac{1}{3}}$ and $N^{\frac{3}{2}}$. Hence the test S_3 is less efficient relative to b than the asymptotic

* These values were obtained graphically by drawing on probability paper lines whose ratio of slopes is R_1/R_2 and reading off the maximum difference in probability between them. The differences in power are determined accurately, but it is rather difficult to find the precise point of maximum difference. The values in Table 2 involve R_1, R_2 only through their ratio R_1/R_2.

Table 3. *Exact power of S_3 test against normal regression alternatives*

Values of the standardized regression coefficient, Δ, are given in parentheses, and the corresponding power appears immediately below.

Sample size (N) ...	15	30	45	60	75	90	105	120	135
Significance level α ...	0·031	0·011	0·018	0·021	0·022	0·049	0·045	0·040	0·036
a	(0·0035)	(0·0018)	(0·0012)	(0·0009)	(0·0007)	(0·0006)	(0·0005)	(0·0004)	(0·0004)
	0·035	0·013	0·021	0·026	0·027	0·062	0·057	0·053	0·048
b	(0·0178)	(0·0089)	(0·0059)	(0·0044)	(0·0036)	(0·0030)	(0·0025)	(0·0022)	(0·0020)
	0·050	0·023	0·042	0·055	0·064	0·135	0·134	0·133	0·130
c	(0·0358)	(0·0179)	(0·0119)	(0·0090)	(0·0072)	(0·0060)	(0·0051)	(0·0045)	(0·0040)
	0·078	0·046	0·091	0·126	0·154	0·291	0·306	0·317	0·327
d	(0·0545)	(0·0272)	(0·0182)	(0·0136)	(0·0109)	(0·0091)	(0·0078)	(0·0068)	(0·0060)
	0·116	0·086	0·173	0·245	0·306	0·508	0·542	0·572	0·598
e	(0·0742)	(0·0371)	(0·0247)	(0·0185)	(0·0148)	(0·0124)	(0·0106)	(0·0093)	(0·0082)
	0·168	0·149	0·297	0·416	0·512	0·730	0·773	0·807	0·836
f	(0·0954)	(0·0477)	(0·0318)	(0·0238)	(0·0191)	(0·0159)	(0·0136)	(0·0119)	(0·0106)
	0·237	0·244	0·461	0·617	0·727	0·894	0·924	0·946	0·961
g	(0·1190)	(0·0595)	(0·0397)	(0·0298)	(0·0238)	(0·0198)	(0·0170)	(0·0149)	(0·0132)
	0·328	0·376	0·648	0·804	0·891	0·974	0·986	0·992	0·996
h	(0·1466)	(0·0733)	(0·0489)	(0·0366)	(0·0293)	(0·0244)	(0·0209)	(0·0183)	(0·0163)
	0·444	0·544	0·823	0·933	0·975	0·997	0·999	1·000	1·000
i	(0·1812)	(0·0906)	(0·0604)	(0·0453)	(0·0362)	(0·0302)	(0·0259)	(0·0227)	(0·0201)
	0·590	0·736	0·944	0·989	0·998	1·000	1·000	1·000	1·000
j	(0·2326)	(0·1163)	(0·0775)	(0·0582)	(0·0465)	(0·0388)	(0·0332)	(0·0291)	(0·0258)
	0·774	0·914	0·995	1·000	1·000	1·000	1·000	1·000	1·000

Table 4. *Exact power of b test against normal regression alternatives*

Values of Δ are given in parentheses above the corresponding entry in Table 3.

Sample size (N) ...	15	30	45	60	75	90	105	120	135
Significance level α ...	0·031	0·011	0·018	0·021	0·022	0·049	0·045	0·040	0·036
a	0·036	0·013	0·023	0·027	0·030	0·066	0·062	0·057	0·054
b	0·059	0·030	0·056	0·074	0·088	0·179	0·182	0·183	0·183
c	0·103	0·074	0·143	0·201	0·249	0·429	0·457	0·481	0·502
d	0·171	0·157	0·300	0·416	0·509	0·722	0·764	0·799	0·827
e	0·267	0·294	0·519	0·673	0·775	0·919	0·945	0·962	0·973
f	0·395	0·485	0·746	0·877	0·941	0·988	0·994	0·997	0·999
g	0·551	0·699	0·911	0·975	0·993	0·999	1·000	1·000	1·000
h	0·772	0·880	0·984	0·998	1·000	1·000	1·000	1·000	1·000
i	0·879	0·977	0·999	1·000	1·000	1·000	1·000	1·000	1·000
j	0·979	0·999	1·000	1·000	1·000	1·000	1·000	1·000	1·000

theory suggests. The difference is greater for smaller α. The corresponding graphs to Figs. 1 and 2 for higher N show that for α in the range 0·01–0·05, the asymptotic theory applies well for $N \geqslant 60$.

The next thing is to investigate whether the form of S_3, involving the rejection of the middle third of the set of observations, can profitably be modified in small samples. Suppose that ($\frac{1}{3}N - 2k$) observations are rejected so that the number of comparisons is ($\frac{1}{3}N + k$); the exact power function can be worked out from the binomial tables as before, but an immediate comparison is not possible because the significance levels for different values of α cannot be made equal, except by the artificial device of randomized tests. However, if the curves are plotted on probability paper they are almost parallel for different α and an

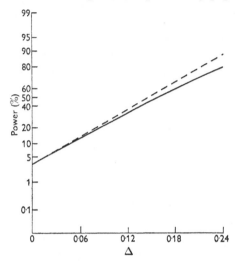

Fig. 1. Power of S_3 for $N = 15$. —— Exact power. ---- Value from asymptotic theory.

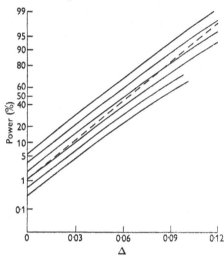

Fig. 2. Power of S_3 for $N = 30$. —— Exact power. ---- Value from asymptotic theory. Full curves are in descending order $k = 0, 1, 2,$ $0, 1, 2$, the second set having lower values of α than the first. Broken curve is $k = 0$.

increase in power with change in k would be shown by decreasing curvature. As would be expected, a negative k leads to a loss of power. Fig. 2 shows for $N = 30$ the curves for $k = 0, 1, 2$. There is a tendency for the curvature to increase as α decreases, but there does not seem to be any systematic change with k. Therefore, although an increase in k increases the number of available significance levels, it does not appreciably increase power. Hence there seems to be little value in modifying the $\frac{1}{3}$ rule in small samples; similar calculations for $N = 15$ confirm this.

We have not made the corresponding investigations for S_1.

10. Examples of use of the sign tests against trend in location

To illustrate the S_1 and S_3 tests, we use the figures of annual rainfall at Oxford for the years 1858–1952, quoted by Foster & Stuart (1954, Table 9).

For S_1 we compare the kth observation with the $(N - k + 1)$th, scoring 1 when the former is the larger and 0 when it is the smaller. The unit scores are then weighted by the distance apart of the observations compared, i.e. by $(N - 2k + 1)$. In this case $N = 95$ and is odd, so that we must ignore the middle observation and proceed with $N = 94$. The unit scores are those with weights as follows:

89, 83, 79, 77, 75, 71, 65, 59, 57, 55, 51, 49, 47, 45, 33, 31, 27, 15, 3.

The value of S_1 is the sum of these weights, 1011. From (16), with $N = 94$, we have

$$E(S_1) = 1104 \cdot 5, \quad D^2(S_1) = 34603 \cdot 75, \quad D(S_1) = 186 \cdot 0.$$

The observed value of S_1 thus represents a deviation from expectation of almost exactly one-half its standard error and is therefore in good agreement with the null hypothesis of zero trend.

If, alternatively, we were to use the simpler S_3 test we compare the kth observation with the $(\frac{2}{3}N + k)$th, scoring 1 or 0 as before, but no weighting is necessary. Since $N = 95$ and is not a multiple of three, we retain the extra observations (in accordance with our findings at the end of §9 above) and compare each of the first thirty-two observations with the corresponding observation in the last thirty-two. For our sequence of scores we obtain

1 0 1 0 1 0 0 1 1 0 0 0 0 0 0 0 0 1 1 1 1 1 1 1 1 0 0 0 1 0 0 0,

the total score S_3 being 14. This clearly agrees well with the expected value of $\frac{1}{2} \times 32 = 16$. The standard error of S_3 is, from (33), $\sqrt{(\frac{1}{4} \times 32)} = 2 \cdot 83$, so that the deviation from expectation, corrected for continuity, is just over one-half of a standard error.

11. SIGN TESTS FOR TREND IN DISPERSION

We consider now tests for a trend not in location but in the dispersion about a fixed location. For example, in a regression problem we may want to test quickly whether the scatter about the regression curve increases as the independent variable increases.

Divide the series $x_1, \ldots, x_N$ into sets $x_1, \ldots, x_k; x_{k+1}, \ldots, x_{2k}; \ldots$, rejecting a few observations in the centre of the original series if N is not exactly divisible by k. The best choice of k is discussed below. For each set of k observations find the range, w, thus getting a series of ranges $w_1, \ldots, w_r$, where r is the integral part of N/k. The ranges are then tested for trend by one or other of the tests S_1 and S_3.

If the null hypothesis is that $x_1, \ldots, x_N$ are independently distributed with constant dispersion about a regression line, $w_1, \ldots, w_r$ are independent and identically distributed. If the regression is not linear the w's will be approximately identically distributed unless the trend within sets of observations varies appreciably.

In the next section, a valid test is obtained for any k, and the best value of k for detecting certain special forms of trend is found for large samples. The behaviour for small samples has not been investigated. The following provisional rules are suggested:

$$
\begin{aligned}
&\text{if} &N \geqslant 90 &\quad \text{take} \quad k = 5, \\
&\text{if} \quad 90 > N \geqslant 64 &&\quad \text{take} \quad k = 4, \\
&\text{if} \quad 64 > N \geqslant 48 &&\quad \text{take} \quad k = 3, \\
&\text{if} \quad 48 > N &&\quad \text{take} \quad k = 2.
\end{aligned}
$$

Except when N is very large it is probably advisable to use the weighted, rather than the unweighted, sign test.

12. Choice of k for dispersion tests

To investigate the theory of the test for trend in dispersion we take a special form for the null and alternative hypotheses. Suppose that $x_1, \ldots, x_N$ are independently normally distributed with constant mean and with standard deviations $\sigma(1), \ldots, \sigma(N)$, where $\sigma(n)$ varies at most slowly with n. Then the ranges $w_1, \ldots, w_r$ defined in §11 have very nearly the distribution of ranges of k observations drawn from normal populations of standard deviations $\sigma_1 = \sigma(\tfrac{1}{2}k)$, $\sigma_2 = \sigma(\tfrac{3}{2}k)$, ... Patnaik (1950) has shown that a range of k observations can be represented to a close approximation as a multiple of a χ-variate with suitably chosen degrees of freedom, ν_k. Therefore $w_i^2 \sigma_j^2/(w_j^2 \sigma_i^2)$ is approximately an F variate with (ν_k, ν_k) degrees of freedom.

Hence

$$\text{prob}\,(w_i/w_j > 1) \simeq \int_0^{(\sigma_i/\sigma_j)^2} \frac{\Gamma(\nu_k)}{[\Gamma(\tfrac{1}{2}\nu_k)]^2} \frac{x^{\frac{1}{2}\nu_k - 1}}{(1+x)^{\nu_k}}\, dx$$

$$\simeq \frac{1}{2} + \left\{ \left(\frac{\sigma_i}{\sigma_j}\right)^2 - 1 \right\} A_k$$

where

$$A_k = \frac{\Gamma(\nu_k)}{[\Gamma(\tfrac{1}{2}\nu_k)]^2} (\tfrac{1}{2})^{\nu_k},$$

provided that $(\sigma_i/\sigma_j)^2 - 1$ is small.

If we assume that the trend in standard deviation is such that $\sigma(n) = \sigma_0 e^{\gamma n} \sim \sigma_0(1 + \gamma n)$, where $\gamma N \ll 1$, so that γ is the fractional increase in standard deviation per observation, we have

$$(\sigma_i/\sigma_j)^2 - 1 \sim 2k\gamma(i-j)$$

and

$$\text{prob}\,(w_i/w_j > 1) \simeq \tfrac{1}{2} + 2k\gamma(i-j)\,A_k.$$

Consider first the application to the ranges of the unweighted sign test, S_3. From the $r \simeq N/k$ ranges we make approximately $\tfrac{1}{3}N/k$ independent comparisons in each of which $i - j \simeq \tfrac{2}{3}r$. Therefore if S is the total score, its mean and standard deviation are given by

$$E(S \mid \gamma) \simeq \frac{1}{3}\frac{N}{k}\,4k\gamma\,\frac{2rA_k}{3} = \frac{8N^2 A_k \gamma}{9k},$$

$$D(S \mid \gamma) = \left(\frac{N}{3k}\right)^{\frac{1}{2}} + O(\gamma^2).$$

Therefore the power derivative, $p_3'(0)$, of the test is

$$p_3'(0) = \frac{E'(S \mid 0)}{D(S \mid 0)} \simeq \frac{8\sqrt{3}\,N^{\frac{3}{2}}A_k}{9\sqrt{k}}. \tag{53}$$

An exactly analogous calculation for the weighted sign test S_1 gives

$$p_1'(0) \simeq \frac{2\sqrt{6}\,N^{\frac{3}{2}}A_k}{3\sqrt{k}}. \tag{54}$$

Thus in both cases the asymptotically best value of k is the one that maximizes $A_k/\sqrt{k}$. From Patnaik's table of ν_k the values in Table 5 have been computed.

Now the number of ranges is N/k and the number of comparisons is one-half or one-third of this and is therefore small even when N is, by usual standards, quite large. Therefore it is advisable to use smaller values of k than the theoretical optimum in large samples. In the absence of an investigation of the small-sample properties of the test, the rule of §11

Table 5. *Determination of efficiencies of different set sizes, k, for testing trend in dispersion*

k	$A_k/\sqrt{k}$	k	$A_k/\sqrt{k}$
2	0·112	6	0·167
3	0·141	7	0·169
4	0·158	8	0·170
5	0·164	9	0·169

is suggested. This is based on the considerations that there is little gain in taking $k > 5$ and that it is advisable, whenever possible, to have at least sixteen ranges.

If we substitute, in (53) and (54), the value $A_k/\sqrt{k} \simeq 0\cdot 16$, we have

$$\left.\begin{array}{l} n_1'(0) \sim 0\cdot 246 N^{\frac{3}{2}}, \\ p_1'(0) \simeq 0\cdot 261 N^{\frac{3}{2}}. \end{array}\right\} \tag{55}$$

It remains to compare (55) with the corresponding quantity for the maximum-likelihood test of the corresponding parametric hypotheses.

13. A.R.E. OF DISPERSION TESTS

For simplicity assume that $x_1, \ldots, x_N$ are normally and independently distributed with zero mean and that the standard deviation of x_n is $\sigma_0 e^{\gamma n}$, where γN is small. The log likelihood is

$$L = -\tfrac{1}{2}N \log(2\pi) - N \log \sigma_0 - \gamma \sum_1^N n - \frac{1}{2\sigma_0^2} \Sigma x_n^2 e^{-2\gamma n}.$$

If we differentiate and take expectations, retaining only the terms independent of γ, and letting N tend to infinity, we get

$$E\left(\frac{\partial^2 L}{\partial \sigma_0^2}\right) \sim -\frac{2N}{\sigma_0^2}, \quad E\left(\frac{\partial^2 L}{\partial \sigma_0 \partial \gamma}\right) \sim -\frac{N^2}{\sigma_0}, \quad E\left(\frac{\partial^2 L}{\partial \gamma^2}\right) \sim -\tfrac{2}{3}N^3. \tag{56}$$

The large-sample variance of $\hat{\gamma}$, the maximum-likelihood estimate of γ, is given by inverting the Hessian matrix with elements (56). We get when γ is small and N is large

$$V(\hat{\gamma}) \sim 6/N^3. \tag{57}$$

Thus the power derivative of the test based on the maximum-likelihood estimate is

$$p_m'(0) = \frac{N^{\frac{3}{2}}}{\sqrt{6}} = 0\cdot 408 N^{\frac{3}{2}}. \tag{58}$$

(58) still applies if the mean in unknown or if a linear trend in mean has to be estimated.

From the formulae (55) and (58), and the fact that $p_x'(0) = R(x)$, it follows, on using (2) with $r = 3$, that the A.R.E.'s of the tests S_3, S_1 compared with the maximum-likelihood test are about 71 and 74 % respectively.

A test entirely analogous to the above tests can be found by calculating the variances within each set of k instead of the range. This is slightly more efficient in the parametric case but much of the simplicity of the test is lost and the increase in power may be shown to be trivial.

14. EXAMPLES OF THE USE OF SIGN TESTS AGAINST TREND IN DISPERSION

We again use for illustrative purposes the rainfall data quoted by Foster & Stuart (1954, Table 9). Using the provisional rule given in § 11 above, we take ranges of sets of five observations. Since $N = 95$, this gives us exactly nineteen sets, no rejection of observations being necessary. The nineteen ranges are:

9·64,   12·30,   12·01,   11·45,   5·43,   13·05,   9·86,   10·89,   6·95,   15·03,

11·34,   6·63,   12·19,   8·55,   4·80,   11·00,   7·76,   7·03,   10·98.

If we apply the test S_1 we drop the middle value and take the signs of 10·98–9·64, 7·03–12·30, ... down to 11·34–6·95, thus obtaining

score:	0	1	1	1	1	1	0	1	0
weight:	17	15	13	11	9	7	5	3	1

The total score is therefore 58, and from (16) with $N = 18$ the mean score is 40·5 and the variance is 242·25, so that the standard error is 15·6. The deviation from expectation is about 1·12 standard errors, and so the two-sided normal significance level is about 27 %. The exact significance level is, by enumeration, $73/256 \simeq 28\frac{1}{2}$ %.

If we use the test S_3 we reject the middle five of the nineteen ranges and take the signs of 12·19–9·64, etc. This gives scores

0   1   1   1   0   1   0

There is clearly good agreement with an equal probability for zeros and ones; significance would be tested in the binomial distribution $(\frac{1}{2} + \frac{1}{2})^7$. The test S_3 is not to be recommended in the present instance because with only seven comparisons the loss of sensitivity compared to the S_1 test would be considerable.

Thus although there is a slight indication that the dispersion decreases with time, both tests suggest that this could easily be a sampling fluctuation.

15. SEQUENTIAL TESTS

Finally, we point out the possibility of constructing sequential tests for trend related to the tests considered above. While this paper was in preparation an abstract (Noether, 1954) appeared describing briefly a test rather similar to the one we had developed. Hence a full discussion will not be attempted here. However, some calculations in a special case suggest that the average sample size under the null and alternative hypotheses are, for the sequential sign tests, only a little greater than the corresponding parametric fixed sample size.

Sequential sign tests for trend are only likely to be of practical value under rather exceptional circumstances. For they require that observations are sufficiently easy to obtain for it to be worth while to use inefficient methods of analysis, and yet sufficiently difficult to obtain for the saving from the use of a sequential method to be important. A possible application is to the marking of a large number of examination scripts. If they are marked in alphabetical order it may be useful to test, as the marking proceeds, for a trend in the marks, which would indicate a changing standard of marking. A sequential method is appropriate and yet elaborate calculations would be out of place.

16. General comments

The calculation of the efficiency of the above tests and the determination of optimum weightings, etc., has been based on a particular type of alternative hypothesis. It is clear in a general way that the tests will remain effective for detecting monotone trends. Positive serial correlation among the observations would increase the chance of a significant answer even in the absence of a trend.

The occurrence of ties has been ignored in the above work. A small number of ties can be dealt with by counting one-half a comparison in each direction, i.e. if $y_i = y_j$ we calculate as if one-half a comparison has $y_i > y_j$ and one-half has $y_i < y_j$. If a substantial proportion of the comparisons are ties a special investigation is necessary or the comparisons should be randomized.

Estimates for the trend could be constructed from the test statistics S_1, S_3. It is very doubtful if such estimates would be of value; in any case, in much work with quick tests, if the trend is shown to be significant it can be estimated graphically.

REFERENCES

Brown, G. W. & Mood, A. M. (1951). On median tests for linear hypotheses. *Proceedings of the Second Berkeley Symposium*, pp. 159–66. University of California Press.

Foster, F. G. & Stuart, A. (1954). Distribution-free tests in time-series based on the breaking of records. *J. R. Statist. Soc.* Series B (Methodological), **16**, 1–22.

National Bureau of Standards (1950). *Tables of the Binomial Probability Distribution*, Applied Mathematics Series, no. 6. Washington.

Noether, Gottfried, E. (1954). Abstract of 'A sequential test of randomness against linear trend'. *Ann. Math. Statist.* **25**, 176.

Noether, Gottfried E. (1955). 'The asymptotic relative efficiencies of tests of hypotheses'. *Ann. Math. Statist.* (to be published).

Patnaik, P. B. (1950). The use of mean range as an estimator of variance in statistical tests. *Biometrika*, **37**, 78–87.

Pitman, E. J. G. (1948). Lecture notes on 'Non-parametric statistical inference', University of North Carolina Institute of Statistics (mimeographed).

Stuart, Alan (1954). The asymptotic relative efficiencies of distribution-free tests of randomness against normal alternatives. *J. Amer. Statist. Ass.* **49**, 147–57.

Theil, H. (1950). A rank-invariant method of linear and polynomial regression analysis. *Indag. math.* **12**, 85–91, 173–7.

3

Applications

M OST IF NOT ALL of the papers in the earlier part of this volume arose out of an application, or more commonly a group of statistically related applications not necessarily in the same field of study. This final set of papers concerns very specific applications, although I hope that the ideas involved are of some broader interest.

The statistician's involvement in applications can be of two broad types. One is what is commonly called consulting and may involve making suggestions or even doing analyses without a very deep involvement in the subject-matter aspects of the work. The other is collaborative work in which, usually although not necessarily as part of a group of workers, a substantial effort is made to integrate the subject-matter and statistical aspects of the investigation. Experience suggests that the latter is much more likely to be fruitful than the former, although, especially in a university environment, some time has to be spent on essentially short-term consulting work.

Most of the papers that follow arose out of collaborative work. Some of the applied fields are not represented here. Hydrology, in which I have had a long interest, is covered in Volume II. Epidemiology, especially veterinary epidemiology, is a more recent concern.

A1 The relation between strength and diameter of wool fibres. *J. Textile Inst.* **41** (1950), T481–T491 (with S. L. Anderson).

Careful study of the relation between the breaking load and breaking extension of textile fibres and their diameter is complicated by the strong correlation between fibre diameter and fibre length. Other things being equal, the so-called weakest link theory of strengths of materials implies that the longer the test specimen the lower the strength; the strength of a chain is the strength of its weakest link as the slogan goes, although the parallel between a textile fibre and a chain of discrete links is only approximate.

Mr Anderson had the ingenious idea of cutting fibres of various lengths into 1.5 cm lengths before testing. The longer fibres thus supply a number of measurements and the dependence on diameter can be examined for fixed length rather than indirectly by some form of multiple regression. The primary data are reproduced in the paper. The more statistical parts of the data analysis emphasize the importance of distinguishing very clearly between partial and total regression coefficients. There is also given an elementary, but I think important, formula linking partial regression coefficients in one random system with total regression coefficients in a different random system. A small detail of presentation is that, because the primary readers of the paper were likely to be physicists, a parallel was drawn between total and partial regression coefficients and total and partial derivatives.

A2   Some statistical aspects of mixing and blending. *J. Textile Inst.* **45** (1954), T113–T122.

A theoretical analysis is made of the problem of mixing white and coloured fibres to produce a uniform mixture. This mixing is achieved in the textile industry by repeated application of a process of attenuation and doubling together. That is, a thick long web of material (*sliver*) is thinned by, say, a factor of six and then six such thinned sections merged together and the process repeated until satisfactory uniformity and mixing is achieved and a final thin product (*yarn*) then produced. Somewhat similar procedures are probably used in other industries to produce relatively uniform material from very variable input.

On the basis of a very simplified model, a calculation is made of the probability that a white fibre will end up next to a coloured fibre and this is used as an index of mixing. The number of stages of processing needed to produce adequate mixing as given by the model is many fewer than the number conventionally used in the industry and the reasons for the discrepancy are discussed.

A3   On a discriminatory problem connected with the works of Plato. *J. R. Statist. Soc.* B **21** (1959), 195–200 (with L. Brandwood).

Paper A3 concerns a curious and interesting problem. The essence is that two of Plato's major works, *Republic* and *Laws*, in effect define an origin and the unit point on a line and it is required to place a number of smaller works on that line. The data for each work consists of a multinomial distribution with 32 cells giving the distribution of stress in the ends of sentences.

The method advocated was to use the reference works to generate a single-parameter exponential family defined by a single unknown parameter for each of the minor works. Estimation of this places each work on the defining line.

In fact the line is interpreted as defining temporal order, but of course that is an additional assumption. Also the exponential family form was chosen to give a simple answer. A conceptually interesting point is that the data were complete enumerations; no sampling was involved. Also it is hardly sensible to think of Plato writing at random. Why then should a probability model be at all appropriate?

While this is not reported in the paper, an attempt was made to address the issue of the meaning of the probabilistic model by dividing each work into quarters and comparing the variation between quarters within works with that expected under a multinomial model. The agreement was gratifyingly close. That is, Plato for some purposes wrote mimicking totally random variation. This might be regarded as a primitive form of cross-validation, but the object is quite different.

The method has been used on other linguistic problems. So far as I understand it, classical scholars were divided between those who regarded the conclusions as ridiculous and those who claimed that they were what sensible classical scholars had always known. In 2004 I met Dr Brandwood again after very many years and he, in particular, confirmed the last statement!

A4   The distribution of the incubation period for the acquired immunodeficiency syndrome (AIDS). *Proc. R. Soc. Lond.* B **233** (1988), 367–377 (with G. F. Medley, L. Billard and R. M. Anderson).

Paper A4 was written fairly early in the period of scientific interest in AIDS when data were very limited. It examined the experience of a U.S. cohort of haemophiliacs known to have received a transfusion with contaminated blood. A parametric model was fitted to the distribution of incubation times, many of which were censored. The method of fitting may have been insufficiently conditional and therefore unnecessarily complicated. The conclusion of an incubation period of roughly gamma form with a mean of about 11 years has stood the test of time remarkably well; it was essentially the first such estimate.

Critics pointed out that, because the largest observed incubation period was about 10 years, only the conditional distribution given that the incubation period is less than 10 years can be estimated nonparametrically. This is, of course, correct and a valuable warning about the need for sensitivity analysis and of the dependence of the conclusions on assumptions. All the same that conditional distribution is of little or no scientific interest and extrapolation is essential to get worthwhile conclusions.

A5   Quality and reliability: some recent developments and a historical perspective. *J. Operational Res. Soc.* **41** (1990), 95–101.

This invited lecture, a Blackett Memorial Lecture to the Operational Research Society, reviewed the history of industrial statistics and associated operational research issues. It emphasizes the implicit interest in such matters of some of the great industrial pioneers. In the part of the paper concerned with contemporary issues, a statement is made which, if regarded as a prediction, was spectacularly incorrect, namely that the techniques of industrial management are inapplicable in an educational setting. As an indication of whether they would be applied, I failed to see the quite large-scale taking-over, at least in the United Kingdom, of vice-chancellorships by those with largely a business background and bringing with them, naturally enough, the attitudes and jargon of their previous trade. I hope that before too long there will be a number of careful studies of the effect on the school and university systems of such managerial approaches.

A6   Quality-of-life assessment: can we keep it simple? (with discussion). *J. R. Statist. Soc.* A **155** (1992), 353–393 (with R. Fitzpatrick, A. E. Fletcher, S. M. Gore, D. J. Spiegelhalter and D. R. Jones).

Paper A6 was the outcome of a long and happy sequence of meetings which began as an informal discussion group dealing with the analysis of data becoming of increasing interest in the mid to late 1980s dealing with patients' self-assessed quality of life. A better term is 'health-related quality of life'. Among the points emphasized in the paper are the desirability, so far as possible, of keeping various dimensions of what is a very complicated idea separate and also, as far as feasible, using simple sum scores in dimensions assessed via multiple questions. The less technical parts of the paper were rewritten in a number of sections and published serially in the *British Medical Journal*.

Additional publication

Applied Statistics: Principles and Examples. London: Chapman and Hall, 1981 (with E. J. Snell).

This book aims first to set out some general principles underlying applied statistical work and then to illustrate them with simple examples used in courses that Joyce Snell and I had given at Imperial College, London. The examples are all in a sense real but, as is inevitable in a book of this kind, only the barest outline of background to the problems could be given.

32—THE RELATION BETWEEN STRENGTH AND DIAMETER OF WOOL FIBRES

By S. L. ANDERSON and D. R. COX

1. SUMMARY

Fibres selected from the fore body-region of a one-year-old Romney lamb were measured for length, cut into 2 cm. pieces, measured for diameter, and the breaking load and extension of the pieces determined. The statistical analysis of the results gives the relations between breaking load, breaking extension, diameter and fibre length.

One of the main conclusions is that fibre breaking load per unit cross-section increases with diameter when the length tested is constant. Possible applications of the results to studies of drafting are briefly indicated.

2. INTRODUCTION

The main aim of this work is to find whether the breaking load per unit cross-section of wool fibres depends on the diameter.

Some unpublished work by the late W. L. Semple on the strength of wool tops suggested that for fibres within a top, breaking load per unit cross-section increases with fibre diameter. Now tops contain fibres grown in different times of the year, from different flocks, sheep and body positions and N. Galpin[1] has shown that these factors greatly influence fibre properties. In particular she found that fibre cross-section falls in winter by an amount depending on the severity of the winter, the skin expansion ratio and the number of fibres per unit skin area.

To avoid these complications the present investigation was confined to fibres from a single region of one sheep, grown during a short, summer, good-growing period. At the same time the work gave information on other points related to breaking load and other fibre properties.

Previous work both on bundles and single fibres has been reported but none restricted to samples of the above kind. Reimers and Swarts[2] measured the breaking load on 50 separate fibres taken from the shoulder region of eight different rams. They concluded that the breaking load per unit cross-section of fibres of all diameters was about the same with a slight tendency to be less in very thick fibres. These authors give a summary of work up to 1928, and of the eight references quoted, three report a negative correlation between breaking load per unit cross-section and diameter. Bosman, Waterston and Van Wyk[3] measured the breaking load of bundles of 100 fibres. Although most of their samples were taken from clips and bales a part of their work relates to fibres within the same staple of Merino wool. In this part they report that "the fine fibres within the same staple thus tend to have a higher tensile strength than the coarser ones".

3. GENERAL DISCUSSION OF METHOD

There are two possible effects of fibre length and fibre diameter on fibre breaking load. Firstly, assuming that the breaking load of a fibre is the breaking load of its weakest section, a decrease in breaking load with increase in length is to be expected. This effect was discussed thoroughly by Peirce.[4] Secondly, the fibre breaking load per unit cross-section may depend on the fibre diameter. This is the main effect in which we are interested in the present paper. The separation of the two effects is made more complicated by the high correlation between length and diameter for the fibres.

The method adopted was to choose fibres of all possible lengths, but each fibre was cut into 2 cm. pieces, all of which were measured for diameter and breaking load. The only artificial restriction was thus the exclusion of fibres shorter than 2 cm. A fixed jaw setting of $1 \cdot 5$ cm. was used, approximately $0 \cdot 25$ cm. at each end of the fibre being gripped in the jaws. In this way the main effect can be investigated directly while the magnitude of the weakest link effect can be investigated as follows. Assuming the weakest link theory to hold, the breaking load of a complete fibre $1 \cdot 5$ n cm. long would be the smallest of the n breaking loads of the separate sections which are actually tested.

4. MATERIALS AND PREPARATION OF SAMPLE

The fibres were all taken from a few locks of wool shorn from the forebody region of a one-year-old lamb. This lamb was one of the Romney flock used by N. Galpin for studies of wool growth. The fibres were grown in the three months summer period 2nd June to 30th September, 1946, during which a small change only in skin area took place. Little sign of medullation was found in the fibres so that estimates of solid cross-section could be made from measurements of their optical diameter.

The fibres were rinsed in petrol ether to remove grease and given a light washing treatment in a weak solution of sodium oleate after which they were dried at room temperature. Previous experience of this washing and separating process showed that it sometimes led to breakage of the extreme ends of the fibres. This was reflected in the present instance in a correlation between length and diameter ($r = 0 \cdot 87$) slightly lower than is obtained when more refined methods are used. The fibres were then separated, measured for length to the nearest mm. and left on a velvet board to await testing.

The strength tester used was similar to the Cambridge Textile Extensometer, a complete trace of the load-extension curve of each fibre piece being obtained. Throughout the tests one spring was used giving a constant rate of loading of $0 \cdot 1205$ g. per sec. which led to breaking times between 20 and 130 sec.

The procedure was as follows. A fibre length group was selected at random and one fibre extracted from it. This was cut into 2 cm. pieces, the identity of the ends of each piece with respect to fibre tip being noted. These fibre pieces were then placed on a glass slide and the diameter of each was measured at 5 places in the central $1 \cdot 5$ cm. region. A projection microscope of standard pattern was used for this giving a magnification of 500 X, measurements being made to the nearest micron. A piece was then mounted in the jaws of the extensometer which were exactly $1 \cdot 5$ cm. apart. A fan was directed on to the piece for one minute to ensure uniform conditions and then the extensometer was started and the fibre piece loaded to rupture. The position of the break was noted, being classified as root, middle or tip. Results from pieces which broke in the jaws were disregarded. A room at constant temperature and relative humidity was not available, but the random order of testing made spurious correlations unlikely.

4.1 The Rate of Loading Effect

Since the extensometer worked at constant rate of loading the time that a fibre piece was under load was proportional to its breaking load (in the present instance this varied from 1 to 20 g.). However, it is better to consider breaking loads for a *constant time under load* and this is now the general practice. One reason for using this concept is that since the load-extension relation depends on the time effect, the internal states of stressed fibres are more likely to be the same at a given constant time from the start of stressing than under other conditions. The observed values of breaking load F were therefore converted to 10 *second values, i.e.,* to the breaking load F_{10} which would have been realised if the fibre piece had been loaded to rupture in a time of 10 sec. The equation deduced from that given by Midgley and Peirce[1] was used,

$$F = F_{10}(1 - 0 \cdot 1 \log_{10} 0 \cdot 1 t)\dots\dots\dots\dots\dots\dots(1)$$

relating F and F_{10} to the time t in sec. during which the fibre piece was loaded.* With the $0 \cdot 1205$ gm. per sec. rate of loading $F = 0 \cdot 1205 t$ and thus from (1),

$$F_{10} = F(1 \cdot 008 - 0 \cdot 1 \log_{10} F)^{-1}\dots\dots\dots\dots\dots(2)$$

As is explained later the qualitative conclusions are unaffected if this correction for rate of loading is not applied to the results.

5. DESCRIPTION OF RESULTS AND METHODS OF ANALYSIS

For each of the $1 \cdot 5$ cm. fibres pieces we have,

(i) breaking load converted to 10 sec. breaking time;
(ii) breaking extension;
(iii) approximate point of break;
(iv) diameter at five points;
(v) length of parent fibre.

Fig. 1 we have plotted (breaking load)/(estimated cross-sectional area) against diameter to show that the breaking load per unit cross-section increases with fibre diameter (Section 6.2). In the remainder of the detailed analysis, the main technique used is the multiple regression analysis of the logarithms of the observed values. It is not essential to use logs but we have preferred to do so because: —

(i) their use leads immediately to linear relations:
(ii) the variances about the regression lines are constant:
(iii) it seems more reasonable to expect relations of the form:
breaking load = const. $\times$ *(diam.)*$^{const.}$ *+ const. + random error,*
which corresponds under certain assumptions to:
log breaking load = cont. $\times$ *log. diameter + const. + random error,*
than relations:
breaking load = const. $\times$ *diam. + const. + random error,* or
breaking load/(diam.)^2 *= const.* $\times$ *diam. + const. + random error.*

In discussing the results we usually quote the relevant regression coefficient and its standard error. Correlation coefficients are given in some cases to show the degree of dependence in the sample but since the fibres were not selected at random with respect to diameter the correlation coefficients are *not* estimates of the correlation coefficients in the bulk.

6. DISCUSSION OF RESULTS

6.1 Relation between Breaking Point and Diameter

For each $1 \cdot 5$ cm. piece of fibre we have measurements of diameter at five points and in considering the relation between breaking load and diameter we have to decide which value of diameter to use. As would be expected the fibres tended to break at thin points.

The most natural value to correlate with the breaking load is the diameter at the point of break. To test whether this gave the best results we did the multiple regression analysis of log breaking load on log of (1) mean diameter,

---

* A subsidiary experiment showed that this equation applies reasonably well to wool fibres.

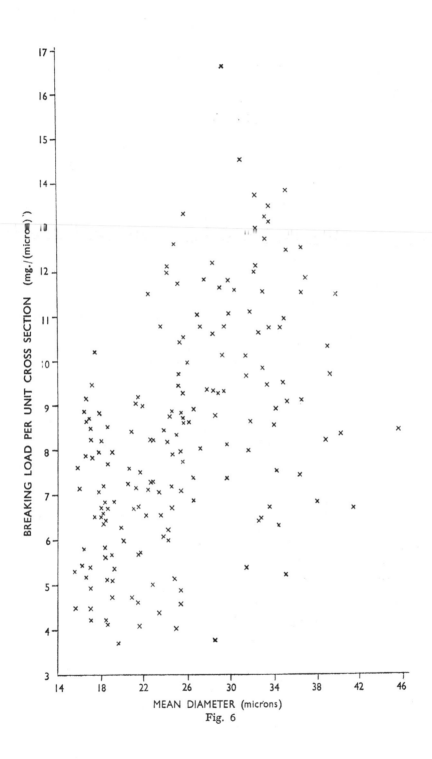

MEAN DIAMETER (microns)

Fig. 6

(2) the observed diameter nearest to the point of break, (3) minimum diameter. A significantly higher correlation was found using mean diameter (1) than for (2) and (3). This fact implies that the mean diameter contains all the relevant information about diameter and so it has been used in the remainder of this work. The general conclusions are however unaffected if the diameter at the point of break is used instead.

The probable reasons why breaking load is better correlated with estimated mean diameter than with diameter near to point of break are as follows. Firstly the nearest measurement of diameter to the point of break may be up to about 5mm. from the actual point of break. Secondly due to the experimental errors in the determination of diameter the mean of 5 values may give a better estimate of the cross-sectional area at the breaking point than the diameter measured there.

6.2 Breaking Load per Unit Cross-section

Fig. 1 shows the scatter diagram of breaking load per unit cross-section and diameter for the 185 fibre pieces each 1·5 cm. long. The breaking load is that for a 10 sec. breaking time and the cross-sectional area was estimated as $\frac{1}{4}\pi$ (diam.)^2. Despite a large random scatter the diagram shows that the breaking load per unit cross-section increases with diameter. The increase is highly significant statistically.

Table I
Mean Values for 1·5 cm. Fibre Pieces

Mean breaking load	8·180 gm.
Mean diameter	25·66 microns
Mean extension	39·4 %
Mean breaking load per unit cross-section	8·37 mg./(micron)^2
Number of pieces tested	185

6.3 Breaking Load, Breaking Extension and Diameter

Table II gives some of the regression coefficients * and correlation coefficients in the analysis of the relation between log breaking load, log extension and log diameter.

The regression coefficient of log breaking load on log diameter is significantly greater than 2. This confirms the conclusion of 6·2 that the breaking load per unit cross-section increases with diameter. This conclusion also follows if we use the diameter at the point of break or the unstandardised breaking load.

Table II
Breaking Load, Breaking Extension and Diameter of 1·5 cm. Fibre Pieces

	Regression coefficient	Standard error	Correlation coefficient
Log breaking load on log diam. (total regression)	+2·595	0·077	0·928
Log breaking load on log extension for fixed log diam. (partial regression) ...	+0·562	0·038	0·720
Log extension on log diam. (total regression)	+0·326	0·105	0·225

* If y, x_1 and x_2 are three variables, the *total* regression coefficient of y on x_1 is found as follows. Let x_1 change by one unit. Then x_2 also changes since it is in general correlated with x_1. The total regression coefficient is the average increase in y due to the combined increases of x_1 and x_2. The *partial* regression coefficient of y on x_1 for fixed x_2 is the average increase in y for unit change in x_1 with x_2 fixed. Partial and total regression coefficients are closely analogous to partial and total differential coefficients.

The mean value of breaking load for a fibre of 25 microns diameter is about 5 g. and the regression coefficient of log breaking load on log diameter is approximately 2·5. Thus we can summarise the results in the simple approximate rule:

Fibre breaking load (10 sec. breaking time) $= 5(d/25)^{2·5}$ g. where the fibre diameter d is expressed in microns.

For fibres of given diameter, breaking extension increases on the average with breaking load. The regression coefficient of log extension on log diameter is $0·326 \pm 0·105$ so there is a slight tendency for the thicker fibres to break at higher extensions.

6.4 Effect of Fibre Length

We explained in section 3 that the breaking load of a whole fibre may depend on the length of the fibre. To investigate this we have to make some estimate of the breaking load of a whole fibre. We assume that the smallest of the breaking loads of the n separate constituent sections is the breaking load of a fibre $1·5$ n cm. long. Thus for example the third fibre in the table given in the Appendix has two sections. The breaking loads are $1·44$ g. and $1·12$ g. with mean diameter $(19·0 + 18·4)/2 = 18·7$ microns. Therefore we take $1·12$ g. as the breaking load of a fibre 3 cm. long and diameter $18·7$ microns. On doing the multiple regression analysis of the logs of these values we find that the partial regression coefficient of log breaking load on log fibre length for fixed diameter is $0·049 \pm 0·130$. The regression on fibre length is thus not significant. The reason for this is probably that a jaw break on one or more sections of any fibre precluded the use of that fibre for calculating the above partial regression which is thus based on only 55 fibres. Of these 80 per cent. are formed from one or two pieces.

The partial regression coefficient of log breaking load on log diameter for fixed fibre length is $2·388 \pm 0·260$ and is an estimate derived indirectly from fibres of all lengths. We can use this to examine the internal consistency of the results since the value $2·595 \pm 0·077$ of Section 6.3 is a direct estimate of the same quantity taken from fibre pieces all of the same length, $1·5$ cm. The two values agree reasonably well. An exact test of the difference between them has not been made as the correlation between the two estimates is unknown.

It is interesting to make a rough estimate from these results of the relation between fibre breaking load, diameter and length to be expected in tops. Our last result suggests that the relation between fibre breaking load and fibre diameter is not appreciably affected by changes in the length of the fibre tested. Now a small amount of breakage occurred in preparing the sample and very much more would have taken place in making the fibres into tops. Suppose that after this further breakage had taken place, the fibres had been measured for breaking load diameter and length. What would the new relation have been? Our results suggest that provided the breaking load of the fibre *pieces* is unchanged after processing, then:

   (i) *there should be little relation between fibre breaking load and fibre length for fixed fibre diameter*:

   (ii) *fibre breaking load per unit cross-section should increase with fibre diameter*.

These two conclusions are valid for the lock of fibres sampled for the present experiment. Now tops contain fibres grown in different times of the year from different flocks, sheep and body positions. To apply the results we thus have to make the big assumption that the above conclusions (i) and (ii) apply also to the variation between flocks, sheep, *etc.* If this assumption is valid these conclusions will apply to fibres in tops. We now wish to estimate from these conclusions the relation between fibre breaking load and fibre length in tops. The total regression coefficient of log breaking load on log fibre length for our data cannot be used because it depends on the length-diameter correlation for our sample and this sample was not chosen randomly with respect to length and diameter. We need an equation to tell us the relation between breaking load and length for any given relation between fibre length and fibre diameter. In fact

$$b_{12} = b_{12 \cdot 3} + b_{13 \cdot 2} b_{23} \dots \dots \dots (3)$$

where x_1, x_2, x_3 are three variables, b_{12} is the total regression coefficient of x_1 on x_2 for fixed x_3, etc. In equation (3) take $x_1 = $ log breaking load, $x_2 = $ log length and $x_3 = $ log diameter. Then (3) enables us to deduce the relation between breaking load and length for any given relation between length and diameter (b_{23}). In wool tops the correlation between length and diameter is usually small and positive. Thus with $b_{12 \cdot 3} = -0 \cdot 049 \pm 0 \cdot 130$, $b_{13 \cdot 2} = 2 \cdot 388 \pm 0 \cdot 260$ and b_{23} small and positive, our data suggest that

(iii) *fibre breaking load should increase slowly with fibre length in tops.* The second conclusion is in agreement with the unpublished work of the late W. L. Semple referred to in the Introduction. We have no direct evidence on the first and third conclusions.

6.5 Variation of Diameter

On each $1 \cdot 5$ cm. piece of fibre, measurements of diameter at five points are available. These provide an estimate of the coefficient of variation of measured diameter *within* a $1 \cdot 5$ cm. length of fibre. We found the percentage coefficient of variation to be $10 \cdot 3 \pm 4 \cdot 25$.

There is no evidence that the coefficient of variation of the five values of fibre diameter is related to the mean fibre diameter.

7. PRACTICAL APPLICATIONS OF THE RESULTS

In this section we discuss very briefly two practical situations in which the above results might be relevant.

Firstly certain types of irregularity in drafting—the so-called drafting waves—produce groups of short fibres in the yarn. Thus the thin places in the yarn contain more than the average proportion of long fibres and these if they have large diameters are intrinsically stronger. This means that the strength of the yarn should be rather greater than would be deduced from the cross-sectional area of the thinnest point.

Secondly the results can be applied to show that long fibres are more likely to break in drafting than short fibres. To prove this we have to show that fibre tension in drafting increases more rapidly with fibre length than does fibre breaking load. We first prove that fibre tension is approximately proportional to fibre length in most drafting systems. Now the tension in a fibre gripped by the front rollers is the sum of the frictional forces acting on the fibre due to contact with fibres moving at back roller speed. If the frictional force per unit length due to this cause were constant, the fibre tension would be exacly proportional to fibre length. Actually the frictional force on a short section of the fibre depends on (1) the proportion of fibres in the cross-section with the speed of the back rollers, and (2) the pressure in the sliver at the section. (1) increases towards the back roller while (2) depends on the design of the drafting zone. The variation of (1) tends to make fibre tension increase more rapidly than fibre length and so unless (2) increases rapidly towards the front rollers, fibre tension will increase more rapidly than proportionally to the fibre length. A fibre experiences its greatest tension near its leading end and so the point of break should be near the leading end. Thus fibre breakage in drafting should lead to an increase in the number of *very* short fibres. Our results (Section 6.4) showed that fibre breaking load increased only slowly with fibre length. If this applied also to fibres in tops, the conclusion that long fibres are more likely to break than short fibres is proved. This is of course what would be expected but the conclusion would not follow if fibre breaking load increased rapidly with fibre length.

8. CONCLUSIONS

A summary of the main conclusions is as follows.

  (i) Fibre breaking load per unit cross-section increased with fibre diameter.

 (ii) Fibre breaking load was positively correlated (for fixed diameter) with fibre breaking extension.

(iii) There was a small positive correlation between fibre breaking extension and fibre diameter.

(iv) The percentage coefficient of variation of fibre diameter within 1.5 cm. lengths was 10.3 ± 4.25.

It should be emphasised that the above conclusions apply to a sample of fibres that were grown during a good growing period of three summer months.

ACKNOWLEDGEMENTS

We wish to thank Mr. R. C. Palmer for suggesting the use of fibres from a good growing period and for detailed criticism of the draft of the paper. Thanks are due also to Dr. N. Galpin for supplying us with fibres from one of her Romney flock. The strength tests were done by Miss Brearley and the computations by Messrs. G. D. Whiteside and F. Cousins. Our thanks are due to the Council of the Wool Industries Research Association for permission to publish this account.

REFERENCES

[1] Galpin. *J. Agric. Science*, 1948, **38**, 303-313.
[2] Reimers and Swarts. *Annals Univ. Stellenbosch*, 1930, **8**, 1-33.
[3] Bosman, Waterston and Van Wyk. *Onderstepoort J.*, 1940, **15**, 313-324.
[4] Peirce. *J. Textile Institute*, 1926, **17**, T355-368.
[5] Midgley and Peirce. *J. Textile Institute*, 1926, **17**, T337.

APPENDIX

The main experimental results are given below. They are arranged in groups corresponding to whole fibres. Thus, for example, the third group gives the results from the two pieces into which a fibre 4.5 cm. long was cut. All results from fibres which broke at the jaws have been omitted. Asterisks have been used to mark those fibres for which jaw breaks occurred, the number of asterisks denoting the number of jaw breaks. Thus the double asterisk against the second result means that this fibre was cut into three pieces two of which gave jaw breaks. (These fibres were omitted from all calculations involving estimates of the behaviour of whole fibres.) The results for the different pieces of the same fibre have been recorded in order beginning with the root end.

Fibre length (cm.)	10 sec. breaking load of piece (g.)	Mean diameter of piece (microns)	Breaking extension of piece (%)	Fibre length (cm.)	10 sec. breaking loan of piece (g.)	Mean diameter of piece (microns)	Breaking extension of piece (%)
2·4	1·49	15·8	47	9·2*	8·85	32·6	50
					7·71	36·4	37
6·7**	3·76	24·2	50		4·15	31·4	20
4·5	1·44	19·0	40	7·8**	12·71	37·0	60
	1·12	18·4	20		5·35	32·6	17
6·1**	5·96	33·6	27	4·8	2·37	20·4	57
					2·86	20·8	50
3·9	3·30	21·4	53				
	3·36	21·8	60	6·0*	2·04	22·8	23
					1·82	18·6	30
2·8	1·55	17·4	57				
				3·4	1·93	19·8	33
3·4*	2·80	22·4	13				
				6·1**	6·95	25·8	53
5·9*	2·86	24·2	37				
	2·53	20·6	50	4·5	1·49	18·4	27
					1·87	18·2	40
4·9*	2·25	19·0	50				
				6·1	4·71	23·6	57
7·2*	11·43	33·2	67		4·09	26·6	43
	12·11	36·6	60		3·30	22·6	37
5·2	2·09	18·0	50	6·4*	3·76	23·8	47
	1·55	19·2	43		3·19	24·6	40
8·4	7·76	38·0	33	9·8	13·15	36·6	50
	10·09	37·6	57		9·02	41·4	33
	11·27	32·4	60		13·77	45·6	40
	7·71	29·8	37		10·57	40·2	27
8·7*	12·22	39·0	57	4·5*	3·98	25·6	33
	9·51	33·6	40				
	9·51	33·6	37	3·8	4·44	25·6	47
					1·98	25·0	13
7·2*	2·48	24·8	37				
	6·08	24·8	47	2·8	1·93	17·0	37
4·4*	6·02	28·8	47	4·9	2·09	18·6	37
					1·71	18·0	33
2·0	1·01	15·6	37				
				8·1*	8·40	33·0	40
2·4	1·71	16·6	40		12·13	35·2	56
					5·00	35·0	13
5·3	1·82	17·2	50				
	1·87	16·6	53	6·6	5·57	24·2	43
					5·41	25·6	50
5·9*	4·76	25·6	37		5·85	25·2	43
	5·51	24·2	47				
				3·4	1·60	19·0	47
6·7*	4·21	24·6	43				
	5·28	25·4	60	6·5**	2·48	25·4	17
3·6*	2·31	18·6	43	5·6*	1·71	18·4	23
					1·60	20·8	20
4·9*	2·75	21·6	40				
				7·7*	11·27	29·4	60
8·4*	9·98	32·4	67		10·94	31·0	60
	5·46	32·8	30		6·71	28·4	33
2·7	1·98	17·8	30	2·2	1·33	19·0	17

Fibre length (cm.)	10 sec. breaking load of piece (g.)	Mean diameter of piece (microns)	Breaking extension of piece (%)	Fibre length (cm.)	10 sec. breaking load of piece (g.)	Mean diameter of piece (microns)	Breaking extension of piece (%)
7·3	9·75	32·2	30	8·1*	9·02	34·8	53
	10·99	33·2	50		8·39	30·4	47
	6·90	34·2	43		5·57	29·6	40
3·8*	1·76	17·8	40	7·4*	13·43	35·2	63
8·4*	11·64	39·2	67		14·25	39·8	57
	11·91	33·6	57	7·8***	8·81	35·2	50
2·7	1·87	16·4	43	5·4	4·66	27·2	47
7·6	5·90	28·4	57		4·54	22·4	37
	7·71	28·4	50	3·3	2·20	17·2	40
	7·06	27·6	63				
	5·62	28·6	50	6·8*	2·42	28·6	17
7·1*	5·68	27·8	33		4·09	24·4	53
	8·22	29·8	43	4·2	2·31	21·0	40
2·4	1·01	17·0	13		1·82	18·4	23
5·1*	1·12	19·6	20	3·8*	1·12	18·6	37
7·2	6·31	29·4	23	5·0	2·53	22·2	30
	7·31	29·4	47		2·09	21·6	20
	4·94	26·6	40	5·7	2·53	21·2	33
8·3**	8·81	31·8	47		3·36	22·8	30
	11·61	33·6	60		2·31	25·4	13
2·6	1·12	16·2	23	7·6**	5·85	34·4	23
4·2*	1·12	16·6	23		4·48	25·4	30
7·8**	9·86	33·0	47	2·8	1·71	18·2	63
	10·52	35·0	33	7·9*	8·17	34·2	40
2·7	1·93	16·8	60		10·09	31·6	60
2·3	2·42	17·4	60		7·46	31·4	50
6·8**	3·59	25·4	43	6·6*	2·75	24·2	17
5·9*	6·25	27·2	57		3·03	23·4	40
	7·69	29·0	47	2·3	1·12	17·0	30
4·9	1·66	18·2	30	3·8*	1·39	18·6	27
	1·98	16·6	33	8·1*	6·84	31·8	27
1·8	1·22	16·4	23		7·82	31·4	40
6·4	4·83	25·2	37		5·06	29·6	30
	6·78	29·2	40	7·4	10·09	34·6	40
	5·28	26·0	37		12·76	35·6	47
5·1*	1·44	16·0	47		9·68	38·8	33
3·1	1·22	17·0	20	4·8	2·91	22·6	30
4·0	1·55	18·4	27		2·42	21·4	27
	0·96	17·0	27	3·1	1·98	19·2	40
				5·3	1·66	18·0	30
					1·87	17·0	33

Fibre Length (cm.)	10 sec. Breaking Load of Piece (g.)	Mean Diameter of Piece (microns)	Breaking Extension of Piece (%)	Fibre Length (cm.)	10 sec. Breaking Load of Piece (g.)	Mean Diameter of Piece (microns)	Breaking Extension of Piece (%)
5·6*	1·66	21·4	13	5·9*	1·87	20·0	27
	3·81	26·6	30		1·49	21·6	13
6·3*	6·31	27·0	40	7·5**	8·28	33·4	43
	6·25	31·6	33	3·7	1·98	17·8	50
6·1*	2·04	21·4	20		2·20	17·8	33
	2·70	23·8	30	6·4*	2·86	23·6	27
4·8	1·87	23·4	30		3·41	24·6	33
	3·19	21·2	50	4·4	4·09	25·0	40
5·2	4·04	25·4	33		3·76	24·6	37
	4·71	25·2	33	2·0	0·86	15·6	40
8·4**	7·76	34·0	40				
	10·69	32·4	57				
6·6	4·49	25·6	50				
	4·66	26·2	37				
	2·97	22·8	23				

Wool Industries Research Association,
Torridon, Headingley, Leeds.

Received 2/5/50

9—SOME STATISTICAL ASPECTS OF MIXING AND BLENDING

By D. R. Cox

A theoretical approach to the study of irregularities in slivers and yarns arising from the uneven mixing of fibres of different types. The paper is mainly concerned with mixing by doubling in the drawing operation. Statistical details for those who are interested are given in separate sections.

1. INTRODUCTION

The majority of work on the irregularity of yarns and slivers has dealt with variations either in the number of fibres per cross-section, or in closely related properties such as weight per unit length, yarn diameter or twist. The present paper outlines a possible approach to the study of irregularities arising from the uneven mixing of fibres of different types, for example fibres of different colours. It is mainly concerned with mixing by doubling in the drawing operation although similar methods could no doubt be applied to mixing taking place in other processes.

The treatment is entirely theoretical, although emphasis has been placed on the physical assumptions and conclusions rather than on details of statistical analysis. Readers not interested in statistical details should omit §§ 3, 4.2 and 5.2.

2. TYPES OF IRREGULARITY

It will be assumed for simplicity that there are just two types of fibre, which to be explicit will be called coloured and white. We may distinguish three sorts of yarn irregularity arising from inadequate mixing :

(i) irregularities due to an uneven arrangement of coloured and white fibres in the cross-section of the yarn ;

(ii) irregularities due to variations in the relative numbers of coloured and white fibres in the cross-section of the yarn ;

(iii) very-long-term irregularities.

The distinction between the first two types of irregularity requires careful explanation. Even with a very large number of doublings the best that can be expected is a random dispersion of coloured and white fibres in the cross-section of the yarn, so that from time to time the surface fibres will be predominantly of one colour. However, with a small number of doublings, greater aggregations of coloured or white fibres may be expected to occur, with a correspondingly increased chance of a predominantly white or coloured group of surface fibres. Irregularity of the first type may be said to arise whenever a randomization of the fibre arrangement in the cross–section, without a change of the relative numbers of coloured and white fibres per cross-section, would improve the mixing. A second consequence of the reduction in the number of doublings is an increase in the variation of the relative numbers of coloured and white fibres per cross-section. Irregularities arising from this constitute the second type.

In considering the first two sorts of irregularity it is supposed that all fibres of one type are equivalent so that there is no need to consider the mixing of fibres of one type among themselves.

The three types of irregularity will be discussed separately in §§ 4, 5, 6 but first it is convenient to derive some general formulae for yarn irregularity.

3. SOME GENERAL FORMULAE CONNECTED WITH YARN IRREGULARITY

Consider first a yarn formed from fibres of one type. If at any stage there are two ends up, the fibres in any cross-section of the resulting sliver can be divided into two groups according as to which of the initial slivers they belong. In the same way, if Ω is the total number of doublings, the fibres in any yarn cross-section can be divided into Ω groups, some possibly empty, corresponding to the Ω original slivers from which the yarn is formed. In the yarn cross-section x cm from some fixed point on the yarn, let the number of fibres in the Ω groups be $y_1(x), \ldots, y_\Omega(x)$. The total number of fibres at x is $f(x)$, where

$$f(x) = y_1(x) + \ldots + y_\Omega(x).$$

If $\bar{f}$ is the mean number of fibres per yarn cross-section, and if all the initial slivers have the same mean number of fibres per cross-section, the average, m, of each $y_i(x)$ is given by

$$\bar{f} = m\Omega, \qquad \qquad \ldots\ldots(1)$$

Now it has been found experimentally that even for large Ω, var $(f) > \bar{f}$, and so we introduce, in the usual way, an irregularity index, k, defined by

$$k = \frac{\text{var }(f)}{\bar{f}} . \qquad \qquad \ldots\ldots(2)$$

There are two extreme ways in which an irregularity index greater than unity can arise :

(a) the $y_i(x)$ may be independent, each with a variance k times its mean ;

(b) each $y_i(x)$ may have a variance equal to its mean, but $y_i(x)$'s with different suffixes may be correlated.

The first possibility could be true only when m is very small (i.e., when the number of doublings is very large), if the frequency with which two or more fibres from the same original sliver occur together in the yarn cross-section remained of the same order as the frequency with which single fibres occur. This would mean that, even with a large number of doublings, the system would be incapable of breaking down the original groupings of fibres into single fibres. While a detailed investigation of this possibility may be of value in studying 'coarse' systems of mixing, it seems unlikely that (a) will arise in worsted drawing. In this preliminary study, it is natural, therefore, to concentrate on possibility (b).

It will therefore be assumed that when m is small,

$$\text{var }(y_i) = m, \qquad \qquad \ldots\ldots(3)$$

and that each y_i has a correlation coefficient ρ with r of the other quantities y and is uncorrelated with the rest. Equation (3) is roughly equivalent to the assumption that prob $(y_i \geqslant 2) = O(m^2)$. The correlation is supposed to arise in drafting from the movement together of sets of fibres. We have

$$\text{var }(f) = \Omega m + \Omega m \, r \, \rho,$$

so that

$$k = \text{var }(f)/\bar{f} = 1 + r \, \rho. \qquad \qquad \ldots\ldots(4)$$

The quantities r, ρ enter only through their product $r\rho$. There is strong experimental evidence[1] that in producing yarn of a given count k is independent of the number of doublings, Ω, unless Ω is small.

These ideas can now be applied to a mixture of white and coloured fibres. Let the (length–biased) proportion of white fibres be p and suppose that $y_1, \ldots, y_{p\Omega}$ refer to white fibres and $y_{p\Omega+1}, \ldots, y_\Omega$ to coloured fibres. If the distribution across the yarn cross-section of white and coloured *groups* is random and if $\Omega \gg 1$, each 'white' y will be correlated, on the average, with rp 'white' y's and with $r(1-p)$ 'coloured' y's. Therefore if $f_w(x)$, $f_c(x)$ denote the numbers of white and coloured fibres at x

$$f_w(x) = y_1(x) + \ldots + y_{p\Omega}(x)$$
$$f_c(x) = y_{p\Omega+1}(x) + \ldots + y_\Omega(x),$$

and
$$\bar{f}_w = p\bar{f}, \quad \bar{f}_c = (1-p)\bar{f}.$$

Also
$$\mathrm{var}\,(f_w) = p\Omega m + p\Omega . rp . \varrho m$$

$$= p\bar{f}[1 + (k-1)p]. \qquad \ldots\ldots(5)$$

Thus if k_w is the irregularity index for white fibres,

$$k_w = \mathrm{var}\,(f_w)/\bar{f}_w = 1 + (k-1)\,p$$

Similarly
$$k_c = 1 + (k-1)(1-p).$$

Further
$$\mathrm{cov}\,(f_w, f_c) = p\Omega\, r(1-p)\, m\varrho$$
$$= p(1-p)\bar{f}\,(k-1). \qquad \ldots\ldots(6)$$

Thus
$$\mathrm{corr}\,(f_w, f_c) = (k-1)\left(\frac{p\,(1-p)}{k_w\,k_c}\right)^{\frac{1}{2}}.$$

In particular if $p = \frac{1}{2}$

$$\mathrm{corr}\,(f_w, f_c) = \frac{k-1}{k+1}.$$

If p is small $k_w \simeq 1$. This means that if white fibres form a small proportion of the whole, the number of white fibres per cross-section should follow a Poisson distribution, even if the distribution of the total number of fibres per cross-section departs appreciably from the Poisson form.

A typical value of k is 1·3 so that when $p = \frac{1}{2}$, the predicted correlation between f_w, f_c is not large.

The above formulae apply for a large number of doublings. With a small number of doublings irregularity depends not only on drafting irregularities but also on that part of the initial irregularity of the tops that has not been eliminated by doubling. Foster and, from different assumptions, Cox and Ingham[1] have proposed a formula to represent this. The formula takes no account of very-long-term variations (§6) and, even with this proviso, there is some evidence that it underestimates the contribution of top irregularity to yarn irregularity. However the formula will be assumed here to be approximately valid. It is

$$CV^2(f) = \frac{k}{\bar{f}} + \frac{A}{\Omega},$$

where A measures, in a sense defined precisely elsewhere[1], the coefficient of variation squared of thickness of the initial top. Hence we may expect

$$CV^2(f_w) = \frac{k_w}{p\bar{f}} + \frac{A_w}{\Omega_w},$$

$$\qquad \ldots\ldots(7)$$

$$CV^2(f_c) = \frac{k_c}{(1-p)\bar{f}} + \frac{A_c}{\Omega_c},$$

where k_w, k_c are defined above and A_w is the initial irregularity of the white tops and Ω_w is the number of doublings experienced by the white fibres considered separately. The top irregularities of the white and coloured fibres should be statistically independent and the drafting irregularities correlated ; accordingly we expect, as in (6),

$$\text{cov}\,(f_w, f_c) = p(1-p)\,\bar{f}\,(k-1) \qquad\qquad \ldots\ldots(8)$$

These formulae are used in §5.

4. IRREGULAR ARRANGEMENT OF COLOURS IN THE YARN CROSS-SECTION

4.1 General Explanation

An extreme form of irregularity of the first type would arise if coloured and white rovings were drawn separately and finally doubled together in spinning, giving an appearance similar to that of a marl yarn. Less extreme irregularities may arise when the number of doublings in the drawing operation is inadequate, resulting in a patchy appearance, even though the variations in the relative numbers of white and coloured fibres per cross-section may not be excessive.

It is possible to make a rough theoretical estimate of the number of doublings that should be necessary to attain adequate fibre arrangement.

Suppose that there have been Ω total doublings in which coloured and white fibres have been mixed together, and that the initial coloured and white slivers contain the same mean number of fibres per cross-section. Then, as explained in §2, the fibres in any yarn cross-section may be separated into Ω groups, some possibly empty, corresponding to the Ω original slivers from which the yarn has been formed. If the mean number of fibres per yarn cross-section is $\bar{f}$, the mean number of fibres per group is $m = \bar{f}/\Omega$.

The next step is to define some quantity measuring the degree of mixing. To obtain a direct correspondence with what is likely to be observed in the piece, it would be desirable to calculate the frequency with which the surface fibres are predominantly of one colour over some appreciable length of yarn. However, in the present introductory stage of the work, it is advisable to adopt a much simpler and more empirical approach. Imagine the yarn opened out into a plane web, so that the fibres in any one cross-section are arranged in order along a line. Suppose that the fibres in each group remain together and that the order of groups along the line is random. Denote by π the proportion of white fibres that have white right-hand neighbours. For very bad mixing π will be nearly one, while for perfect random mixing π will be equal to the proportion of white fibres in the yarn.

The assumption that the fibres in one group remain together means that there is no lateral mixing of the slivers. One further assumption is necessary before π can be calculated in terms of the number of doublings. This concerns the variation in the numbers of fibres in the different groups ; they will be assumed to vary in Poisson distributions, not necessarily independently.

4.2 Theory

Let the average proportion of white fibres be p. In each cross-section there are Ω groups of fibres, the number of fibres per group varying in Poisson distributions of mean m, so that the probability that a group contains r fibres is

$$q_r = \frac{e^{-m}\,m^r}{r!}\,. \qquad\qquad \ldots\ldots(9)$$

Assuming that the white and coloured groups are arranged randomly, we have that the remaining fibre, the right-hand one, has a probability $(p\Omega-1)/(\Omega-1) \simeq p$ of having a white neighbour, since $p\Omega-1$ of the remaining $\Omega-1$ groups are white. Thus counting all fibres in the group, the probability that a white fibre has a white right-hand neighbour is approximately $(r-1+p)/r$. The probability that a white fibre selected at random lies in a group of size r is proportional to rq_r, so that

$$\pi = \frac{\sum\limits_{r=1}^{\infty} \left(\frac{r-1+p}{r}\right) rq_r}{\sum\limits_{r=1}^{\infty} rq_r}$$

$$= \frac{\bar{r} - (1-p)(1-q_o)}{\bar{r}},$$

where $\bar{r}$ is the average of r including zero values. Introducing the particular expression (1), we have that

$$\pi = 1 - \frac{(1-p)(1-e^{-m})}{m}. \qquad \dots\dots(10)$$

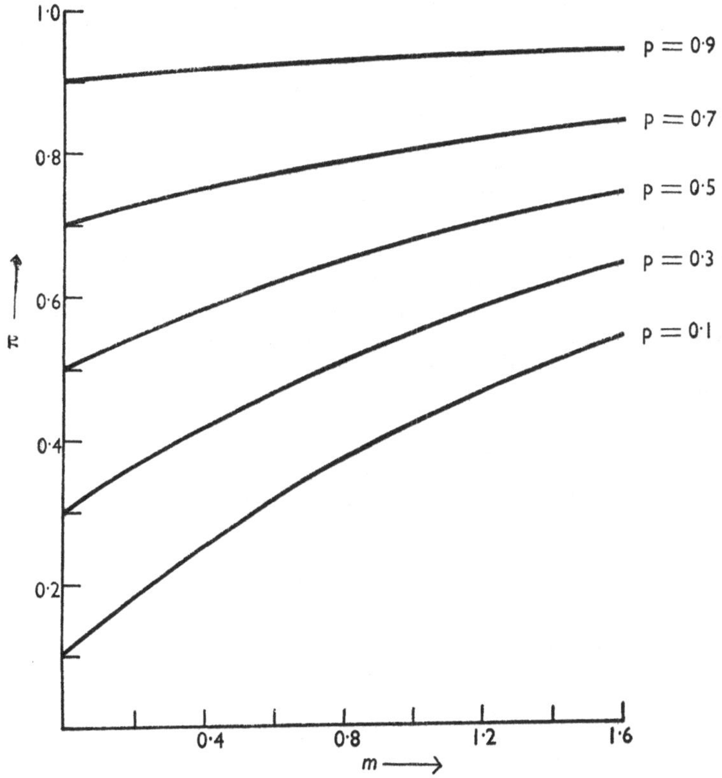

Fig. 1.
Index of mixing, π, as a function of m. Mixture of p parts white with $1-p$ parts coloured.

4.3 Discussion

Fig. 1 gives the index of mixing π plotted against $m =$ (mean number of fibres per yarn cross-section)/number of doublings. The quantity π is difficult to interpret physically because we do not know what deviation of π from its limiting value p represents an observable additional irregularity in the cloth. As a rough estimate suppose that π must be within 10 per cent of p for the mixing to be satisfactory. Then for a 50–50 mixture, $p = \frac{1}{2}$, we must have

$$m \quad < \quad 1/5, \qquad \qquad \ldots \ldots (11)$$
$$\text{No. of doublings} \quad > 5\,\bar{f}.$$

The number of doublings needs to be more for unequal mixes ; thus with $p = 0.1$, we need $m < \frac{1}{45}$ to achieve $\pi = 0.11$, i.e.,

$$\text{no. of doublings} > 45\,\bar{f},$$

expressing the result that it is more difficult to mix unequal proportions than equal proportions. (The comparison of $\pi = 0.55$, $p = 0.50$ with $\pi = 0.11$, $p = 0.10$ may however be a misleading one ; it is not obvious how one should define equal departures from perfect mixing in such cases.)

It will be noted that the two mixtures in which the proportions of white fibres are p and $1-p$ are represented by separate curves. The physical meaning of this is that in a mixture of say one part white and nine parts coloured, $p=0.1$, large aggregations of coloured fibres are inevitable and a reduction in the number of doublings has only a slight effect on the degree of aggregation. This is shown by the slowly rising curve for $p = 0.9$. For the white fibres, however, a reduction in the number of doublings may have a large effect on the amount of aggregation so that the curve for $p = 0.1$ rises steeply. In judging the theoretical adequacy of mixing, the more critical curve should normally be used, i.e., the curve for $p \leqslant \frac{1}{2}$.

We now consider briefly the effect of failure of some of the assumptions. Failure of the Poisson formula (9), which enters the expression for π only through q_o, should not have a serious effect on the conclusions, provided that the departure from (9) is not caused by the repeated occurrence together of substantial groups of fibres from the same original sliver (see §3).

It has been assumed that there is no lateral mixing of the constituent slivers. This will be untrue when the slivers are doubled in the form of thin webs with a large number of surface fibres. Fewer doublings will then be required. Better mixing will also be produced by any tendency towards a systematic alternation of white and coloured groups.

Even when the condition (11) is satisfied there will be an irregular distribution of colours across the fibre cross-section, but this will arise from the hazards of random sampling and will not be improved by further doubling.

VARIATIONS IN RELATIVE NUMBERS OF FIBRES PER YARN CROSS-SECTION

5.1 General Explanation

In the previous section we considered irregularities arising from an uneven arrangement of fibres in the yarn cross-section. We now suppose that this arrangement is effectively random and that the only irregularity we need consider is that due to variation in the relative numbers of white and coloured fibres per cross-section.

The yarn is described by two irregularity curves $\{f_w(x)\}$, $\{f_c(x)\}$ giving respectively the numbers of white and coloured fibres in the cross-section x cm from some fixed point on the yarn. It will be assumed that, after a large number of doublings, the variances and correlation of $f_w(x)$ and $f_c(x)$ are given by formulae (5) and (6) of §3. The precise correlation between $f_w(x)$ and $f_c(x)$ is, for the present purpose, comparatively unim although a substantial negative correlation would seriously affect the conclusions. It is difficult to see how such a correlation could arise with conventional machinery, unless differences between the physical properties of the two types of fibre produced some complicated effect on the relative fibre motion in drafting.

We can describe the efficiency of mixing by the ratio $R(x) = f_w(x)/f_c(x)$ and in particular by the coefficient of variation of $R(x)$. In the production of yarn of a given count some variation in $f_w(x)$ and $f_c(x)$, and hence also in $R(x)$, always occurs. In §5·2 this minimum variation in $R(x)$ is calculated and a rough estimate is then made of the number of doublings that should, in theory, be necessary to attain this minimum variation.

5.2 Theory

To calculate the minimum variation of $R(x)$ we assume that the variances and covariances of $f_w(x)$ and $f_c(x)$ are given by (5) and (6). Now for any random variables Y, Z with small coefficients of variation

$$CV(Y) \simeq \text{st. dev. } (\log_e Y)$$
$$\text{corr } (Y, Z) \simeq \text{corr } (\log_e Y, \log_e Z).$$

Thus

$$\text{var}(\log_e R) \simeq CV^2(f_w) + CV^2(f_c) - 2CV(f_w) \, CV(f_c) \, \text{corr } (f_w, f_c)$$

$$= \frac{1}{\bar{f} p \, (1-p)}, \qquad \ldots \ldots (12)$$

where $\bar{f}$ is the mean number of fibres per cross-section and p the proportion of white fibres. If we assume that $\log_e R$ is approximately normally distributed, it follows that, for roughly 99 per cent of the yarn, R lies between the limits

$$\left(\frac{p}{1-p} \right) \exp \left\{ \pm 2 \cdot 58 \left(\frac{1}{\bar{f} p \, (1-p)} \right)^{\frac{1}{2}} \right\}. \qquad \ldots \ldots (13)$$

Some numerical values derived from (13) are given in §5.3.

To attempt a calculation of the number of doublings we use formulae (7), (8) of §3, which give

$$\text{var } (\log_e R) \simeq \frac{1}{\bar{f} p \, (1 - p)} + \frac{A_w}{\Omega_w} + \frac{A_c}{\Omega_c},$$

where A_w, A_c are the initial irregularities of the white and coloured slivers and Ω_w, Ω_c the numbers of doublings experienced by the white and coloured slivers separately. In the particular case of 50–50 mixes with $p = \frac{1}{2}$, $A_w = A_c = A$, $\Omega_w = \Omega_c = \Omega/2$, we have that

$$\text{var } (\log_e R) \simeq 4 \left(\frac{1}{\bar{f}} + \frac{A}{\Omega} \right). \qquad \ldots \ldots (14)$$

5.3 Discussion

Equation (13) gives the limits within which 99 per cent of the values of R should lie in a yarn formed with a very large number of doublings. Table I gives some values computed from this equation. The table has been constructed in terms of the proportion of white fibres in the cross-section, which is equal to $R/(1 + R)$. For example the first entry in the third column means that in a 1 : 4 mixture with 30 fibres per cross-section, the proportion of white fibres is between 0·071 and 0·45 for 99 per cent of the yarn.

The variations are large. The effect on the appearance of the cloth will probably depend not only on the amount of variation but also on the lengths of sections of yarn predominantly of one colour. These lengths should be approximately proportional to the mean fibre length, suggesting that, other things being equal, short fibres should give the appearance of better mixing than long fibres. Further work is required before a direct correspondence can be established between the variation of R and what is observed in the piece.

Table I

99 per cent Limits for Proportion of White Fibres

Mean Number of Fibres per Cross-section	50–50 Mixture $p = \frac{1}{2}$		1 : 4 Mixture $p = \frac{1}{5}$	
30	0·28	0·72	0·071	0·45
50	0·32	0·68	0·091	0·38
100	0·38	0·62	0·12	0·32
200	0·41	0·59	0·14	0·28

The formula (14) gives, for a 50–50 mixture, the variance of $\log_e R$ in terms of the number of doublings, Ω, the initial irregularity, A, and the mean number of fibres per cross-section, $\bar{f}$. To make the second term negligible we require, say,

$$\frac{A}{\Omega} < \frac{1}{50}\frac{1}{\bar{f}},$$

i.e.,

$$m = \frac{\bar{f}}{\Omega} < \frac{1}{50A}. \qquad \dots\dots(15)$$

Now a practical value of A is roughly 0·01, corresponding to a coefficient of variation of 10 per cent, so that (15) gives

$$m < 2.$$

This may be compared with the condition (11) for adequate fibre arrangement

$$m < \tfrac{1}{8}.$$

Equation (11) is thus the more difficult to satisfy, i.e., it takes more doublings to ensure a random arrangement of fibres in the yarn cross-section than it does to control the total variation of the relative numbers of white and coloured fibres. The condition for adequate mixing has been formulated in terms of the total variation of $R(x)$. If, however, long-term variations in $R(x)$ are of importance in themselves, the doublings would need to be many more, in fact as many as are required to minimize long-term variations in yarn count. This follows because the second term in (14) represents the contribution of long-term variations.

To apply the formulae to a special case, consider the production of a 50–50 mixture, $p = \frac{1}{2}$, with 100 fibres per cross-section, $\bar{f} = 100$. The theory predicts that, if the number of doublings is below about 50, i.e., $\frac{1}{2}\bar{f}$, there will be excessive variations in R resulting, presumably, in bad appearance in the piece. As the number of doublings is increased from 50 to 500, i.e., $5\bar{f}$, no appreciable decrease in the variation of R should be observed but better mixing across the cross-section should be obtained. Under ideal conditions there should be no improvement by increasing the number of doublings above 500. It will be clear from the simplified and approximate methods that have been used that the figures 50 and 500 are only crude approximations (see §7 for further discussion).

6. VERY-LONG-TERM IRREGULARITIES

In the previous sections it has been assumed that all fibres of one colour are equivalent; Miss Hannah[1] has pointed out the importance of considering very-long-term variations in colour such as systematic differences between bags of tops. These differences are reduced only slowly, if at all, by doubling, although the reduction can be accelerated by a suitably planned arrangement of doubling. Practical recommendations for this have been put forward by Miss Hannah[2] and independently by Cox and Raper[3]. Very-long-term irregularities will not be discussed further here.

7. DISCUSSION AND CONCLUSIONS

The number of doublings predicted theoretically by equations (5) and (13) is of course very many fewer than is normally used in the industry. Some discrepancy is to be expected from the necessity of applying a 'safety factor' to guard against occasional very irregular sections, or such mishaps as the simultaneous breakage of several white ends at the back of the 1st can gill. The difference between theory and practice is, however, too large to be accounted for in this way and some other explanation must be sought.

At least four possibilities present themselves. One is that the function of the majority of doublings is the reduction of very-long-term irregularities of colour. The second is that the need for many doublings arises because the initial white and coloured tops will not be of exactly the same lengths. When they are first doubled together there will, therefore, be thin ends entirely of one colour, unless production is stopped as soon as the shorter tops run out. Many doublings will be needed to ensure that these thin ends are well mixed up with the remainder of the production. The third possibility is that there is something seriously wrong with the theory; for example, fibres adjacent in the top may have a strong tendency to remain together despite the relative motion that should occur in drafting. The fourth possibility is that the theory is substantially correct and that adequate mixing is attainable with a comparatively small number of doublings, the large number normally used in the industry being either unnecessary or required for the reduction of count variation.

The main conclusions can be summarised as follows :

(i) Three types of irregularity are distinguished arising from uneven mixing.

(ii) The theoretical number of doublings necessary to control the first two types of irregularity is much smaller than the number normally used in the industry.

(iii) The theoretical number of doublings necessary to obtain adequate mixing within the yarn cross-section is greater than the number to minimize variations in the relative numbers of white and coloured fibres per cross-section.

(iv) For a 50–50 mixture, the minimum number of doublings to control the mixing within the yarn cross-section is, in theory, approximately five times the mean number of fibres per yarn cross-section. With the number of doublings used in the industry the arrangement in the cross-section should be effectively random.

(v) Large variations in the relative numbers of white and coloured fibres are to be expected no matter how great the number of doublings.

(vi) The number of doublings to control the variation in the relative numbers of white and coloured fibres is not more than, and may be very much less than, the number to control yarn count variation.

(vii) After a large number of doublings a yarn with a small mean fibre length should, other things being equal, give the appearance of better mixing than a yarn with a longer mean fibre length.

ACKNOWLEDGEMENT

Some of this work was done at the Wool Industries Research Association and I am grateful to the Director of Research for permission to publish the paper. I also wish to thank Mr. J. Ingham, Mr. G. F. Raper and Mr. B. H. Wilsdon for some very helpful comments.

REFERENCES

1 D.R. Cox and J. Ingham. *J. Text. Inst.*, 1950, **41**, P376, including discussion.
2 M. Hannah. *ibid.* 1953, **44**, T436.
3 D. R. Cox and G. F. Raper. *Wool Industries Research Association Bulletin*, 1950, **13**, 57 (restricted publication).

Received 3.6.1953.
Accepted for publication 1.9.1953.

Statistical Laboratory,
St. Andrew's Hill,
Cambridge.

On a Discriminatory Problem Connected with the Works of Plato

By D. R. Cox and L. Brandwood

Birkbeck College, University of London

[Received November, 1958]

SUMMARY

A FORM of discriminant analysis for qualitative data is developed and used to place in order some of the works of Plato.

1. Introduction

Between writing the *Republic* (*Rep.*) and the *Laws*, Plato wrote the *Critias* (*Crit.*), *Philebus* (*Phil.*), *Politicus* (*Pol.*), *Sophist* (*Soph.*) and *Timaeus* (*Tim.*), but it is not known in what order. Kaluscha (1904) was the first to use statistical methods in an attempt to assign an order to these works. Billig (1920) continued this line of investigation, but the conclusions from it have not been generally accepted by classical scholars, partly, perhaps, because of the very subjective methods used to interpret the statistical tables. Brandwood (1958) has applied more objective statistical methods to the problem and the aim of the present note is to explain briefly the technique used.

The stylistic property on which the statistical analysis is based is the distribution of quantity over the sentence ending (clausula). The last five syllables only of each sentence are considered, each syllable being classed as long or short. The sentence endings are thus divided into $2^5 = 32$ types. For each work a frequency distribution is obtained showing the number of endings of each type (Table 1). There is a marked difference between the distributions for *Rep.* and *Laws*; the problem is in effect to order the other works in decreasing order of affinity with, say, *Rep.*, and then, provided that Plato's change in literary style was monotone in time, we will have estimated the order in which the works were written.

The technique used is essentially discriminant analysis. The optimum discriminator between *Rep.* and *Laws* is obtained, giving therefore a method of scoring each type of sentence ending. The mean score for all sentences in *Rep.* is substantially negative, that for *Laws* is substantially positive; the mean scores for the other works then give the required ordering. The calculation is completed by obtaining an approximate significance test for the difference between two mean scores.

The work may be compared with that of Barnard (1935), who used discriminant analysis to date Egyptian skulls. Her observations, however, were on continuous and approximately normally distributed variates, and ordinary linear discriminant analysis was used. Here, however, the observations are on five qualitative variables (long, short). It would be possible to replace "long" by a 1 and "short" by a 0 and then to use a linear discriminant function as an approximation; this procedure is not likely to take proper account of the importance of particular special patterns of quantity among several syllables, and it is better, and a lot simpler, to regard the 32 different types of ending as qualitatively different. Problems of discrimination with qualitative variables arise also in other fields, for example in medical diagnosis.

2. Theory

Let there be two populations Π_0, Π_1 of individuals and let each individual fall into one of k types, the probabilities being θ_{01}, . . . , θ_{0k}; θ_{11}, . . . , θ_{1k}, assumed for the moment to be known. Assume that different individuals are statistically independent. In the application that we have in mind Π_0 is *Rep.*, Π_1 is *Laws*, an individual is a sentence ending and $k = 32$.

Suppose first that we have N individuals from one or other population, n_i being of the ith type, and that we require to decide which population has in fact been sampled. Welch (1939) showed that optimum discrimination between the populations is based on the likelihood ratio. The probability of the observations in sampling from Π_0 is of the multinomial form

$$\frac{N!}{\Pi\, n_i!}\, \Pi\, \theta_{0i}^{n_i} \tag{1}$$

so that the log-likelihood ratio is

$$\Sigma\, n_i \log \left(\frac{\theta_{1i}}{\theta_{0i}}\right), \tag{2}$$

where any system of logarithms can be used, but in fact natural logarithms were used in the calculations. That is, each sentence is given a score

$$s_i = \log \frac{\theta_{1i}}{\theta_{0i}} \tag{3}$$

and the optimum discriminator is the total score.

Now suppose that there exist not just populations Π_0 and Π_1, but a series of populations representing a gradual change from the distribution in Π_0 to that in Π_1. A natural way of representing such a series is to define populations Π_λ, $0 \leqslant \lambda \leqslant 1$, the probability $\theta_{\lambda i}$ of the ith type in Π_λ being

$$\theta_{\lambda i} = \frac{\theta_{0i}^{1-\lambda}\, \theta_{1i}^{\lambda}}{\sum\limits_{j=1}^{k} \theta_{0j}^{1-\lambda}\, \theta_{1j}^{\lambda}}. \tag{4}$$

For a random sample of N individuals, n_i of the ith type, the log-likelihood is, except for a constant,

$$\sum_{i=1}^{k} [n_i(1 - \lambda) \log \theta_{0i} + n_i \lambda \log \theta_{1i}] - N \log \left[\sum_{i=1}^{k} \theta_0^{1-\lambda}\, \theta_{1i}^{\lambda} \right],$$

so that a sufficient statistic for λ is the total score (2), or more conveniently, the mean score

$$\bar{s} = \frac{1}{N}\, \Sigma\, n_i s_i = \frac{\Sigma\, n_i \log (\theta_{1i}/\theta_{0i})}{N}. \tag{5}$$

The random variable $\bar{s}$ converges in probability, as N increases, to $\phi(\lambda) = \Sigma\, \theta_{\lambda i} s_i$, which is a monotone function of λ; it would be possible to convert $\bar{s}$ into an estimate of

λ, but this seems unnecessary, and we shall instead take $\phi(\lambda)$ as the parameter of interest. The variance of $\bar{s}$, considered as an estimate of $\phi(\lambda)$, is

$$V(\bar{s}) = \frac{1}{N} \{\Sigma \, \theta_{\lambda i} s_i^2 - (\Sigma \, \theta_{\lambda i} s_i)^2\} \tag{6}$$

and an unbiased estimate of this is

$$V_e(\bar{s}) = \frac{1}{N(N-1)} \left\{ \Sigma \, n_i s_i^2 - \frac{(\Sigma \, n_i s_i)^2}{N} \right\}. \tag{7}$$

Now suppose that we have samples of sizes N', and N'', from two populations (two of the "doubtful" works in the application that we are describing). Calculate the corresponding mean scores $\bar{s}'$ and $\bar{s}''$ and the estimated variance of the difference,

$$V_e(\bar{s}' - \bar{s}'') = V_e(\bar{s}') + V_e(\bar{s}''). \tag{8}$$

In large samples $(\bar{s}' - \bar{s}'')$ will be nearly normally distributed and a significance test can be made in the usual way. If a significant difference is obtained, the populations can confidently be placed in order on the λ-scale, provided that (4) is a reasonable representation of the populations. The adequacy of (4) could in principle be examined by a χ^2 test, although this was not done in the present application. The correctness of (4) matters only in the derivation of (5) as an optimum statistic and not in the calculation of the standard error; (5) has in any case a certain intuitive appeal.

If the population probabilities are known for Π_0 and Π_1, this completes the theory. In the present application the numbers of sentences in *Rep.* and in *Laws* are appreciably greater than those in the other books, so that it is reasonable to replace θ_{0i} and θ_{1i} by observed frequencies and to neglect the consequent error in s_i. In general the large-sample variance of $\bar{s}' - \bar{s}''$ is given by (8) plus

$$\Sigma \, (\theta_{\lambda i} - \theta_{\mu i})^2 \, V(\hat{s}_i) + 2 \sum_{i > j} (\theta_{\lambda i} - \theta_{\mu i})(\theta_{\lambda j} - \theta_{\mu j}) \, C(\hat{s}_i, \hat{s}_j), \tag{9}$$

where $\theta_{\lambda i}$, $\theta_{\mu i}$ refer to the two populations under comparison, and where $\hat{s}_i$ is a sample estimate of $\log (\theta_{1i}/\theta_{0i})$ obtained by replacing probabilities by sample frequencies. If M_0, M_1 denote the numbers of observations on the reference populations Π_0, Π_1, then a simple asymptotic calculation gives

$$V(\hat{s}_i) \sim \Sigma \left\{ \frac{1 - \theta_{0i}}{M_0 \theta_{0i}} + \frac{1 - \theta_{1i}}{M_1 \theta_{1i}} \right\} (\log e)^2, \tag{10}$$

$$C(\hat{s}_i, \hat{s}_j) \sim - k \left(\frac{1}{M_0} + \frac{1}{M_1} \right) (\log e)^2. \tag{11}$$

Note however that under the null hypothesis that the two populations under comparison are identical, i.e. $\lambda = \mu$, the additional terms (9) vanish, so that (7) and (8) may be used for a significance test, with s_i replaced by $\hat{s}_i$.

3. Application and Discussion

The first two columns of Table 1 give the percentage distribution among the 32 types for the reference populations Π_0, *Rep.*, and Π_1, *Laws*. The third column gives the estimated scores, $\hat{s}_i$, obtained on replacing probabilities in (3) by sample frequencies. Thus the first entry is $\log_e (2 \cdot 4/1 \cdot 1) = 0 \cdot 779$. The remaining columns give the percentage frequency distributions for the doubtful works *Crit.*, . . . , *Tim.*

Table 2 gives the mean scores for the doubtful works; for example that for *Crit.* is

$$\frac{1}{100} (3 \cdot 3 \times 0 \cdot 779 + 2 \cdot 0 \times 0 \cdot 867 + \ldots + 4 \cdot 0 \times 0 \cdot 215).$$

Mean scores have been calculated also for *Rep.* and *Laws*; these are the expected values that would apply to any work for which the distribution of quantities differs only randomly, from that in *Rep.* or *Laws*.

TABLE 1

Percentage Distribution of Sentence Endings

Type of Ending	Π_0 Rep.	Π_1 Laws	$\hat{s}_i$=Natural Log of Ratio	Crit.	Phil.	Pol.	Soph.	Tim
U U U U U .	1·1	2·4	0·779	3·3	2·5	1·7	2·8	2·4
– U U U U .	1·6	3·8	0·867	2·0	2·8	2·5	3·6	3·9
U – U U U .	1·7	1·9	0·113	2·0	2·1	3·1	3·4	6·0
U U – U U .	1·9	2·6	0·315	1·3	2·6	2·6	2·6	1·8
U U U – U .	2·1	3·0	0·358	6·7	4·0	3·3	2·4	3·4
U U U U – .	2·0	3·8	0·642	4·0	4·8	2·9	2·5	3·5
– – U U U .	2·1	2·7	0·255	3·3	4·3	3·3	3·3	3·4
– U – U U .	2·2	1·8	−0·199	2·0	1·5	2·3	4·0	3·4
– U U – U .	2·8	0·6	−1·541	1·3	0·7	0·4	2·1	1·7
– U U U – .	4·6	8·8	0·647	6·0	6·5	4·0	2·3	3·3
U – – U U .	3·3	3·4	0·030	2·7	6·7	5·3	3·3	3·4
U – U – U .	2·6	1·0	−0·956	2·7	0·6	0·9	1·6	2·2
U – U U – .	4·6	1·1	−1·430	2·0	0·7	1·0	3·0	2·7
U U – – U .	2·6	1·5	−0·548	2·7	3·1	3·1	3·0	3·0
U U – U – .	4·4	3·0	−0·385	3·3	1·9	3·0	3·0	2·2
U U U – – .	2·5	5·7	0·824	6·7	5·4	4·4	5·1	3·9
– – – U U .	2·9	4·2	0·372	2·7	5·5	6·9	5·2	3·0
– – U – U .	3·0	1·4	−0·761	2·0	0·7	2·7	2·6	3·3
– – U U – .	3·4	1·0	−1·224	0·7	0·4	0·7	2·3	3·3
– U – – U .	2·0	2·3	0·140	2·0	1·2	3·4	3·7	3·3
– U – U – .	6·4	2·4	−0·982	1·3	2·8	1·8	2·1	3·0
– U U – – .	4·2	0·6	−1·946	4·7	0·7	0·8	3·0	2·8
U U – – – .	2·8	2·9	0·039	1·3	2·6	4·6	3·4	3·0
U – U – – .	4·2	1·2	−1·253	2·7	1·3	1·0	1·3	3·3
U – – U – .	4·8	8·2	0·536	5·3	5·3	4·5	4·6	3·0
U – – – U .	2·4	1·9	−0·231	3·3	3·3	2·5	2·5	2·2
U – – – – .	3·5	4·1	0·157	2·0	3·3	3·8	2·9	2·4
– U – – – .	4·0	3·7	−0·077	4·7	3·3	4·9	3·5	3·0
– – U – – .	4·1	2·1	−0·668	6·0	2·3	2·1	4·1	6·4
– – – U – .	4·1	8·8	0·765	2·0	9·0	6·8	4·7	3·8
– – – – U .	2·0	3·0	0·405	3·3	2·9	2·9	2·6	2·2
– – – – – .	4·2	5·2	0·215	4·0	4·9	7·3	3·4	1·8
Number of sentences .	3,778	3,783	—	150	958	770	919	762

There are minor errors in the third decimal place of the scores $\hat{s}_i$.

The estimated variances of the mean scores, also given in Table 2, are calculated from (7), again replacing s_i by $\hat{s}_i$. Thus for *Crit.* the estimated variance is

$$\frac{1}{149}\left\{\frac{(3\cdot3\times0\cdot779^2+\ \ldots\ +4\cdot0\times0\cdot215^2)}{100} -\frac{(3\cdot3\times0\cdot779+\ \ldots\ +4\cdot0\times0\cdot215)^2}{100^2}\right\},$$

the divisors 100 being inserted because the frequencies in Table 1 are percentages.

TABLE 2

Mean Scores and Their Standard Errors

	Crit.	Phil.	Pol.	Soph.	Tim.	Rep.	Laws
Mean score	−0·0346	0·1996	0·1303	−0·0407	−0·1170	−0·2652	0·2176
Estimated variance	0·003799	0·0003342	0·0003973	0·0005719	0·0007218		
Estimated st. error	0·0616	0·0183	0·01993	0·0239	0·0269		

The variance, and hence the standard error, of the difference between any two mean scores can now be found, and the order determined. The more critical comparisons are shown in Table 3; those not shown in the table are very highly significant statistically.

TABLE 3

Some Comparisons of Mean Scores

	Difference	Estimated St. Error	Level of Significance Attained (%)
Tim. v. Soph.	0·0763	0·0360	5
Tim. v. Crit.	0·0824	0·0672	20
Soph. v. Crit.	0·0061	0·0661	>90
Crit. v. Pol.	0·1649	0·0648	1
Pol. v. Phil.	0·0693	0·0270	1

The final ordering is *Rep., Tim., Soph., Crit., Pol., Phil., Laws*: there is reasonably strong evidence that *Tim.* is correctly placed before *Soph.*, but the position of *Crit.* could be anywhere between somewhat before *Tim.* to before *Pol.* This conclusion agrees broadly with the one arrived at in the earlier work mentioned in §1. It is not in accord with the views held by the majority of classical scholars, although there is a minority group who have reached a similar ordering by apparently independent arguments.

Brandwood (1958) has described further aspects of the analysis and given a detailed discussion of the conclusions. One further point that may be noted briefly is that *Rep.* and *Laws* are each divided into a number of books; these have been treated as homogeneous in the above analysis. This can be checked by calculating mean scores separately for each book and comparing the dispersion of these means with that to be expected from (7). The agreement is good. Further, it was expected on general grounds that there would be no systematic change in style with serial number; this was confirmed.

A final general remark is that the frequency distributions in Table 1 are based not on samples, but on complete enumeration. In applying probabilistic arguments we are in effect assuming that certain aspects of Plato's writings are adequately described by the laws of probability for independent events; this assumption receives some support from the fact just noted about the dispersion between books within works.

REFERENCES

BARNARD, M. M. (1935), "The secular variations of skull characters in four series of Egyptian skulls", *Ann. Eug. London*, **6**, 352–362.

BILLIG, L. (1920), "Clausulae and Platonic chronology", *J. Philol.*, **35**, 225–256.

BRANDWOOD, L. (1958), *The dating of Plato's works by the stylistic method—a historical and critical survey.* Ph.D. thesis, University of London.

KALUSCHA, W. (1904), "Zur Chronologie der platonischen Dialoge", *Wiener Studien*, **26**, 190–204.

WELCH, B. L. (1939), "Note on discriminant functions", *Biometrika*, **31**, 218–219.

Proc. R. Soc. Lond. B **233**, 367–377 (1988)
Printed in Great Britain

The distribution of the incubation period for the acquired immunodeficiency syndrome (AIDS)

By G. F. Medley[1], L. Billard[3], D. R. Cox[2], F.R.S.,
and R. M. Anderson[1], F.R.S.

[1] *Department of Biology, Imperial College, University of London, London SW7 2BB, U.K.*
[2] *Department of Mathematics, Imperial College, University of London, London SW7 2AZ, U.K.*
[3] *Department of Statistics, University of Georgia, Athens, Georgia 30602, U.S.A.*

(*Received* 29 *December* 1987)

This paper contains details of methods and full results of estimates of the incubation period of acquired immunodeficiency syndrome (AIDS) that were outlined in a previous paper by Medley *et al.* (*Nature, Lond.* **328**, 719 (1987)). The original model is modified to assess the influence of age and sex on the incubation period: age, but not sex, is statistically significant. The difficulties associated with interpretation of these data, and the additional information that would be required to resolve these difficulties are discussed.

Introduction

One of the striking features of acquired immunodeficiency syndrome (AIDS) is that the incubation period appears to be both long and very variable. The incubation period is defined as the time between infection and diagnosis of disease, the latter usually involving the onset of severe illness. Data on the incubation period are, for obvious reasons, difficult to acquire. One substantial set of such data kindly provided by Dr T. Peterman, Centers for Disease Control, Atlanta, Georgia, is that on patients who have no other risk of acquiring infection other than having received a transfusion of infected blood or blood products.

Previous attempts at measuring the incubation period of AIDS vary from more direct, non-parametric methods (Goedert *et al.* 1986; Barton 1987; Taylor *et al.* 1986; Peto 1988) to more indirect estimation of distribution parameters (Lui *et al.* 1986; Rees 1987a; Jensen 1988; Wilkie 1988). The data used have either been a summarized form of essentially the same data set as used here, or data from the large cohort studies in the U.S.A. The estimated mean incubation periods for the current data vary from 4.5 years (Lui *et al.* 1986) to 15 years (Rees 1987a), although the methods for obtaining the latter figure have been questioned (Barton 1987; Lui *et al.* 1987; Beal 1987; Costagliola & Downs 1987). The distributions used to describe the incubation period have included the gamma, Gompertz, lognormal, normal and Weibull.

A summary of our analyses has been reported previously (Medley *et al.* 1987), and the object of the present paper is to provide the details of the method of

[367]

analysis which had to be omitted from the earlier note. However, in the present paper we have extended the analysis by developing the models to include subgroups explicitly.

The paper is arranged as follows. The data are described in more detail before the basic model and the fitting procedure are defined. A selection of particular models is shown in more explicit detail, and the results are given for all the models fitted, and our conclusions outlined.

DESCRIPTION OF DATA

In January 1987, there were 494 transfusion associated cases of AIDS known in the U.S.A. diagnosed before July 1986. For each patient a time of transfusion and a time of diagnosis are given, the former being determined retrospectively for each case. The times were recorded to the nearest month, and were treated as continuous for the present purpose. All cases where any doubt existed over the accuracy of either time were excluded, reducing the number of cases to 297. Table 1 summarizes the data with times grouped into biannual intervals. The data were divided into different age classes chosen to correspond to concepts of immunological development, while simultaneously attempting to provide reasonable sample sizes: younger than 5 years ($n = 36$), older than 4 years and younger than 60 years ($n = 126$) and older than 59 years ($n = 135$). There are 185 male and 112 female patients. Important *caveats* concerning the interpretation of these data were listed in our previous note and will not be repeated. The present data give no information about the individuals who have not developed AIDS before June 1986, which precludes the use of more conventional analyses. The data contains no information about the proportion of infected patients ultimately developing AIDS.

PRIMARY MODEL

We begin by treating the patients as a homogeneous group, although later they are split into subgroups. We suppose that in the population of individuals under study, infection leading ultimately to AIDS occurs as a Poisson process with rate $a(t; \alpha)$, where time t is measured from the first known infection. The probability density function of the incubation period is assumed to have a parametric form, say $p(x; \eta)$. It is possible that a case of AIDS is not diagnosed as such, or that an AIDS case does not fulfill exactly the definition of AIDS (T. A. Peterman, personal communication). We denote the probability of positive diagnosis at time t by $q(t; \theta)$. The α, η and θ are vectors of unknown parameters. We shall concentrate on a number of special cases, namely,

$$a(t; \alpha) = \alpha_0 \exp(\alpha_1 t) \quad \text{or} \quad \alpha_0 + \alpha_1 t;$$

$$p(x; \eta) = \eta_0 \eta_1^{\eta_0} x^{\eta_0 - 1} \exp\{-(\eta_1 x)^{\eta_0}\} \quad \text{or} \quad \eta_1^{\eta_0} x^{\eta_0 - 1} \exp\{-x\eta_1\}/\Gamma(\eta_0);$$

$$q(t; \theta) = \Phi\{(t - \theta_0)/\theta_1\} \quad \text{or} \quad 1,$$

where $\Phi(\cdot)$ is the standard normal integral. The choices for these functions are discussed more fully below.

TABLE 1. EXPECTED AND OBSERVED NUMBERS OF ALL PATIENTS BY
HALF-YEAR OF TRANSFUSION AND DIAGNOSIS

(The observed numbers of patients are grouped into half-year intervals of transfusion and diagnosis, although they are treated by month in the analysis. The totals in each year are given, as are the expected values calculated assuming exponential incidence and Weibull incubation.)

year and half-year of diagnosis		year and half-year of transfusion																
		1978		1979		1980		1981		1982		1983		1984		1985		1986
		1	2	1	2	1	2	1	2	1	2	1	2	1	2	1	2	1
1986	1	0	0	2	1	2	1	11	5	8	5	14	9	9	5	7	0	1
		0		3		3		16		13		23		14		7		1
		(0.72)		(2.19)		(4.62)		(8.80)		(14.86)		(21.49)		(24.24)		(15.04)		(0.83)
1985	2	0	1	1	0	3	3	6	2	1	11	4	2	8	6	4	1	
	1	0	0	1	3	5	3	2	6	6	7	9	8	8	4	0		
		1		5		14		16		25		23		26		5		
		(1.58)		(4.54)		(8.88)		(15.49)		(23.44)		(28.66)		(22.61)		(4.73)		
1984	2	0	0	1	0	0	4	3	5	2	5	4	11	2	0			
	1	0	0	2	2	1	4	3	4	4	7	4	1	0				
		0		5		9		15		18		20		2				
		(1.67)		(4.38)		(7.64)		(11.56)		(14.13)		(11.15)		(2.33)				
1983	2	0	0	1	2	1	0	1	2	2	1	1	2					
	1	1	0	0	0	1	0	2	2	4	0	0						
		1		3		2		7		7		3						
		(1.58)		(3.77)		(5.70)		(6.97)		(5.50)		(1.15)						
1982	2	0	0	0	1	2	0	1	2	1	0							
	1	0	0	0	1	0	0	2	0	0								
		0		2		2		5		1								
		(1.34)		(2.81)		(3.43)		(2.71)		(0.57)								

A person infected at time t and diagnosed at time u defines a point in the region $\mathscr{D} = \{0 \leqslant t < u \leqslant t_0\}$ in the plane, where t_0 is the time at which data entry is closed, June 1986. Under the above assumptions points occur in $\mathscr{D}$ in a non-homogeneous Poisson process of rate $a(t;\alpha)\, p(u-t;\eta)\, q(u;\theta)$. Thus with n cases at points (t_i, u_i) the log likelihood is

$$\sum_i \{\log a(t_i;\alpha) + \log p(u_i - t_i; \eta) + \log q(u_i; \theta)\}$$

$$-\int_{\mathscr{D}} a(t;\alpha)\, p(u-t;\eta)\, q(u;\theta)\, du\, dt \quad (1)$$

by generalization of the theory for one dimensional nonhomogeneous Poisson processes and where the sum is over all patients. This is maximized numerically by using a standard general algorithm (Nelder & Mead 1965) with a convergence testing parameter of 10^{-6}, and standard results of maximum likelihood estimates applied. Where necessary, integrals were evaluated numerically to an accuracy of 10^{-8} with standard routines.

More explicit forms of (1) have been found analytically for some special cases. We give here only the important case of exponential growth, Weibull incubation and unit diagnosis function, i.e.

$$a(t;\alpha) = \alpha_0 \exp(\alpha_1 t),$$

$$p(t;\eta) = \eta_0 \eta_1^{\eta_0} t^{\eta_0-1} \exp\{-(\eta_1 t)^{\eta_0}\},$$

$$q(t;\theta) = 1.$$

Then equation (1) becomes

$$n(\log \alpha_0 + \eta_0 \log \eta_1 + \log \eta_0) + \alpha_1 \sum_i t_i + (\eta_0-1) \sum_i \log(u_i - t_i)$$

$$- \eta_1^{\eta_0} \sum_i (u_i - t_i)^{\eta_0} - \alpha_0 (e^{\alpha_1 t_0}-1)/\alpha_1 - \int_0^{t_0} \exp\{\alpha_1 t - \eta_1^{\eta_0}(t_0-t)^{\eta_0}\}\, dt, \quad (2)$$

where n is the total number of patients. In particular, if $\eta_0 = 1, 2$ the integral in (2) takes the simpler forms:

$$\eta_0 = 1: \frac{\alpha_0}{\alpha_1 + \eta_1} \{\exp(\alpha_1 t_0) - \exp(-\eta_1 t_0)\};$$

$$\eta_0 = 2: \frac{\alpha_0}{\eta_1} \sqrt{\pi} \exp\left\{\alpha_1 t_0 + \left(\frac{\alpha_1}{2\eta_1}\right)^2\right\} \left\{\Phi\left(t_0 \eta_1 \sqrt{2} + \frac{\alpha_1}{\eta_1 \sqrt{2}}\right) - \Phi\left(\frac{\alpha_1}{\eta_1 \sqrt{2}}\right)\right\}.$$

Subdividing the data and fitting the models to each set of data separately reveals interesting differences, but it is not possible to test directly from (2) for the significance of those differences because the models are not suitably nested. Generally, with a diagnosis function of unity, the log likelihood (1) may be rewritten to consider w subgroups where the rth subgroup contains s_r individuals $(r = 1, \ldots, w)$:

$$\sum_r \left\{\sum_i \log\{\pi_r a(t_{ri};\alpha)\} + \sum_i \log p(u_{ri} - t_{ri};\eta_r) - \pi_r \int_{\mathscr{D}} a(t;\alpha)\, p(u-t;\eta_r)\, du\, dt\right\},$$

$$(3)$$

where t_{ri} and u_{ri} are respectively the times of infection and diagnosis in case i in subgroup r, and η_r are the parameters to be estimated for subgroup r. Here π_r is the proportion of the total population within subgroup r, and is treated as an estimable parameter. We consider a separate incubation period for each group, and suppose that the incidence function is the same for all groups. We first fit the model with the incubation function identical for each subgroup, and then with separate incubation functions so as to compare the maximized likelihoods.

The results are given in the form of tables in which the parameter estimates, log likelihood (l), expected total number transfused and mean, median and standard deviation of the fitted incubation distribution are given. The time units used throughout are years. The total number given infective transfusions is estimated via

$$N* = \int_0^{t_0} \alpha_0 \exp{(\alpha_1 t)}\, dt = \alpha_0(e^{\alpha_1 t_0} - 1)/\alpha_1$$

and

$$N* = \int_0^{t_0} (\alpha_0 + \alpha_1 t)\, dt = t_0(\alpha_0 + \alpha_1 t_0/2)$$

for the exponential and linear incidence functions respectively. The mean, variance and median of the Weibull are

$$\mu = (1 + 1/\eta_0)/\eta_1,$$

$$\sigma^2 = (1 + 2/\eta_0)/\eta_1^2 - \mu^2,$$

$$M = (0.693\,147)^{1/\eta_0}/\eta_1.$$

The respective forms for the gamma distribution are

$$\mu = \eta_0/\eta_1, \quad \sigma^2 = \eta_0/\eta_1^2$$

TABLE 2. PARAMETER ESTIMATES OF MODELS WITH NO DIFFERENTIAL DIAGNOSIS

(The functional forms for the different incidence functions and incubation density functions are defined in the text. The estimated mean (μ), median (M) and standard deviation (σ) of the incubation period are given in years. The log-likelihood value is denoted by l, and $N*$ is the number of people estimated to develop AIDS as a result of receiving an infective transfusion.)

	total	age < 5	4 < age < 60	age > 59	female	male
(a) exponential incidence and Weibull incubation functions						
l	526.06	−6.18	102.24	157.11	85:21	245.45
μ	6.33	2.33	8.06	5.57	8.86	5.55
M	6.14	2.18	7.80	5.51	8.44	5.43
σ	2.80	1.25	3.58	2.08	4.32	2.29
$N*$	6705	102	3938	4146	3684	3583
α_0	14.48	0.72	13.61	3.41	9.11	7.23
α_1	0.71	0.53	0.64	0.85	0.69	0.72
η_0	2.41	1.94	2.40	2.91	2.16	2.60
η_1	0.14	0.38	0.11	0.16	0.10	0.16
(b) exponential incidence and gamma incubation functions						
l	526.21	−5.72	102.11	157.69	85.34	245.35
μ	12.09	2.43	27.44	8.56	17.69	9.83
M	10.61	2.13	23.84	7.76	15.21	8.74
σ	7.41	1.49	17.46	4.57	11.67	5.73
$N*$	14197	107	24269	6640	8231	7587
α_0	32.81	0.76	83.88	6.28	21.77	15.31
α_1	0.70	0.53	0.64	0.83	0.68	0.72
η_0	2.66	2.67	2.47	3.51	2.30	2.95
η_1	0.22	1.10	0.09	0.41	0.13	0.30
(c) linear incidence and Weibull incubation functions						
l	499.98	13.69	94.61	137.83	72.32	233.08
(d) linear incidence and gamma incubation functions						
l	501.51	−13.74	94.77	139.56	72.89	234.06

G. F. Medley and others

TABLE 3. PARAMETER ESTIMATES OF MODELS INCLUDING DIFFERENTIAL DIAGNOSIS

	total	4 < age < 60	age > 59	female	male
(a) exponential incidence and Weibull incubation functions					
l	534.18	108.49	160.51	91.37	249.15
μ	5.21	5.54	5.55	5.22	5.22
M	4.99	5.29	5.44	4.88	5.08
σ	2.46	2.67	2.26	2.78	2.24
N^*	2873	1151	2692	851	2375
α_0	16.78	11.90	4.74	8.27	8.21
α_1	0.56	0.47	0.74	0.48	0.64
η_0	2.24	2.19	2.64	1.96	2.49
η_1	0.17	0.16	0.16	0.17	0.17
θ_0	4.22	4.17	4.25	4.52	3.78
θ_1	0.49	0.00	0.00	0.27	0.19
(b) exponential incidence and gamma incubation functions					
l	534.86	108.40	162.90	91.85	249.25
μ	6.95	9.00	4.86	6.54	7.68
M	6.09	7.84	4.41	5.62	6.83
σ	4.28	5.67	2.60	4.32	4.50
N^*	3648	1946	1193	957	3631
α_0	24.24	18.91	8.45	10.53	13.41
α_1	0.54	0.48	0.53	0.46	0.63
η_0	2.64	2.52	3.50	2.29	2.92
η_1	0.38	0.28	0.72	0.35	0.38
θ_0	4.27	4.17	5.13	4.55	3.80
θ_1	0.52	0.00	0.67	0.28	0.20
(c) linear incidence and Weibull incubation functions					
l	535.24	106.37	163.26	89.98	252.88
μ	4.04	5.92	3.69	3.57	4.03
M	3.73	5.46	3.51	3.22	3.78
σ	2.25	3.36	1.83	2.17	2.12
N^*	1342	682	599	396	1089
α_0	7.98	4.39	0.01	12.30	0.00
α_1	38.30	19.38	17.96	8.86	32.66
η_0	1.86	1.83	2.12	1.69	1.99
η_1	0.22	0.15	0.24	0.25	0.22
θ_0	5.53	4.49	5.84	5.30	6.31
θ_1	1.41	0.39	1.01	0.83	2.15
(d) linear incidence and gamma incubation functions					
l	537.23	106.63	165.32	90.86	254.11
μ	4.61	8.28	3.95	3.86	4.67
M	3.97	6.99	3.53	3.28	4.08
σ	3.01	5.75	2.25	2.62	2.91
N^*	1472	894	635	410	1241
α_0	8.88	5.85	0.00	14.43	0.00
α_1	41.96	25.38	19.03	8.77	37.22
η_0	2.35	2.07	3.08	2.16	2.57
η_1	0.51	0.25	0.78	0.56	0.55
θ_0	5.53	4.49	5.84	5.33	6.28
θ_1	1.40	0.40	1.03	0.84	2.12

and M which satisfies

$$\frac{1}{2} = \int_0^{2M\eta_1} t^{\eta_0-1}\, e^{-\frac{1}{2}t}\, dt \Big/ \int_0^{\infty} t^{\eta_0-1}\, e^{-\frac{1}{2}t}\, dt.$$

Table 2 contains results for the four possible combinations of incidence and incubation functions with no differential diagnosis, and table 3 likewise for a diagnosis function of probit form. The full results are not given for the linear incidence functions in table 2 because of the much-reduced log likelihood values, and the youngest age class has been omitted from table 3 because the small number of cases made the numerical fitting difficult and unreliable. Note that in both tables the gamma and Weibull fits are indistinguishable for all subdivisions of data. In table 2 the linear incidence is a much worse fit than the exponential form: compare log likelihood values of 526.06 with 499.98 and 526.21 with 501.51. However, the introduction of the probit diagnosis function leads to the linear incidence being a rather better fit, although not significantly so ($\chi^2 = 2.12, p > 0.3$ and $\chi^2 = 4.74$, $p > 0.075$). In conjunction with exponential incidence, the probability of correct diagnosis is almost a step function moving from 0 to 1 during 1982. This contrasts with the more elongated diagnosis function obtained with linear incidence, which has a mean during 1983, and a value of about 0.75 in June 1986.

TABLE 4. PARAMETER ESTIMATES OF MODELS WITH EXPLICIT SUBGROUPS

(Results of fitting model adapted for explicit groupings (equation 3 in the text). The functional forms for the incidence function, $a(t;\alpha)$, and incubation density function, $p(x;\eta)$, were exponential and Weibull respectively.)

(a) sex

	same incubation distribution		separate incubation distribution	
	male	female	male	female
l	334.63		336.03	
μ	6.82		9.84	5.55
M	6.62		9.41	5.43
σ	2.98		4.72	2.28
$N*$	18869		23108	
α_0	35.57		43.56	
α_1	0.73		0.73	
η_0	2.44		2.20	2.62
η_1	0.13		0.09	0.16
π	0.38	0.62	0.57	0.43

(b) age

	same incubation distribution			separate incubation distribution		
	age < 5	4 < age < 60	age > 59	age < 5	4 < age < 60	age > 59
l		246.14			336.03	
μ		6.82		2.68	6.87	4.95
M		6.63		2.53	9.63	4.87
σ		2.95		1.37	4.12	1.91
$N*$		22272			35987	
α_0		36.62			59.17	
α_1		0.75			0.75	
η_0		2.47		2.05	2.57	2.80
η_1		0.13		0.33	0.09	0.18
π	0.12	0.42	0.46	0.01	0.79	0.20

Table 4 shows the results of including age and sex explicitly, with exponential growth in incidence, a Weibull incubation period and no differential diagnosis. The fits with the same incubation periods give results very similar to those of table 2; note that the estimated proportions within each group agree exactly with the data. When separate incubation periods are considered, the estimated proportions within each group alter to reflect that a longer incubation period causes a smaller proportion of those infected to be observed over a given time interval. Although the introduction of a sex difference is not statistically significant ($\chi^2 = 2.8$, $p > 0.2$), age is highly statistically significant ($\chi^2 = 14.97$, $p < 0.005$).

We do not know the precise date of introduction of screening for the presence of human immunodeficiency virus (HIV) antibodies to HIV antigens in blood for transfusion in the U.S.A., but it is likely that it varied in different regions in the U.S.A. during 1985. Consequently, we artificially truncated the data to exclude those patients transfused (1) after December 1985, and (2) after June 1985 to look for alteration in the estimated incidence function. The calculations for table 2 were repeated for both data sets, and the only qualitative difference was the alteration of the exponential parameters in the youngest age class as shown in table 5. As noted in our previous paper (Medley *et al.* 1987), the change in estimated doubling time to that of the remainder of the population is probably caused by the introduction of screening for cytomeglovirus in blood intended for children, a test which coincidentally detects HIV antibodies (T. A. Peterman, personal communication).

TABLE 5. PARAMETER ESTIMATES FROM TRUNCATED DATA

(Results of fitting exponential incidence and Weibull incubation density to the youngest age class (less than 5 years). The data have been artificially truncated, and should be compared with the table 2 (first row, second column).)

	Dec. 1985	Jun. 1985
l	0.86	-2.69
μ	2.16	1.97
M	2.03	1.89
σ	1.12	0.92
N^*	160	182
α_0	0.37	0.32
α_1	0.70	0.74
η_0	2.02	2.27
η_1	0.41	0.45

DISCUSSION AND CONCLUSIONS

Before summarizing our conclusions, we emphasize three major difficulties associated with interpretation of these data. The first is the reporting lag (time between the diagnosis of a case and its notification). The data are collected up to January 1987 (6 months after the last diagnosis) so that only the longer lags are excluding the later diagnoses. The second difficulty of interpretation arises from the introduction of screening of blood for HIV antibodies. Future work with more recent information will have to consider this. The third aspect concerns the death of patients from causes unrelated to AIDS who would have contracted AIDS had they lived long enough. This will bias our data against the longer incubation times, especially in the older patients. This may account for the decreased estimated incubation period in patients older than 59 years. It is not possible to adjust for this because we do not know the age distribution of patients transfused. The shorter incubation period in younger age classes may be influenced by the causes that necessitated the blood transfusion and that are not related to HIV. Deaths that occur soon (with 3 months or so) after transfusion are less relevant, as these will influence the incidence function rather than the incubation distribution. It may be possible to consider age-specific survival rates in future analyses.

Our fitting procedure is an attempt to account for missing data: those patients who have been given infective blood and who will develop AIDS at some time in the future, but who have not yet been identified. This is accomplished by the inclusion of the incidence function. We prefer the exponential to the linear form for two reasons. Firstly, the rise in seropositivity to HIV is very well described by the exponential curve in the U.S.A., and the fitted value of the doubling time agrees well with that derived from the serological data taken from a large cohort of homosexual males in San Francisco during the early stages of the epidemic (Centers for Disease Control 1985). Secondly, the likelihood values when the diagnosis function is unity indicate a much better fit assuming an exponential incidence. The probit diagnosis function combines with the linear incidence to introduce some nonlinearity into the incidence function. The estimated diagnosis function when incidence is linear indicates that about 75 % of all AIDS cases are not diagnosed as such, and is much less likely than the estimate obtained in conjunction with the exponential incidence.

Assuming constant diagnosis, the combination of incidence and incubation functions attempts to reproduce the observed data (table 1). Obviously, a longer incubation period can be offset by a higher incidence to a great extent over the short period of data available, and vice versa. The exponential incidence is the same 'shape' (= doubling time) for both gamma and Weibull incubation periods, and as they both model the observations equally well, the incubation functions must also have the same 'shape'. However, to obtain that shape over the period of the data, the incubation functions are of completely different scales, which explains the difference in estimated mean incubation periods (gamma ≫ Weibull) and difference in estimated number infected, N^* (gamma ≫ Weibull). Without any additional constraints, we are unable to distinguish between these results and no doubt other fits. In particular, some estimate, no matter how crude, of the number of infective transfusions given would have been invaluable. We are wary of any method that compares directly data from patients infected by any other route and those from transfusion-associated patients, purely because the incubation periods may be very different (Jensen 1988; Rees 1987 a, b).

The possibility of the incubation period being influenced by explanatory variables of epidemiological significance such as age and sex of the patient is clearly important. Children have a much shorter incubation period than older patients. The introduction of a sex difference is not statistically significant; however, the estimated means are so different (by a factor of 2) and the epidemiological consequences so important that the possibility of different incubation periods for different sexes, and indeed that of differences between other groups (e.g. intravenous drug abusers versus homosexuals versus heterosexuals), should be borne in mind when interpreting data and developing models. Data on the relative rates of transfusion (of all blood) to the different age and sex groups would be invaluable in investigating this aspect further.

The estimated mean value depends strongly on the parametric form chosen to describe the incubation period. Other methods of estimating the incubation period will always be seriously affected by this problem of unobserved cases. Nonparametric methods can only estimate relative survival rates over a restricted period, and are of rather limited interest as it is not possible to extrapolate beyond the data or assign absolute values without the additional information implicit in the parametric form. Recent work (Blythe & Anderson 1988) suggests that the exact form of the incubation period does not have a great impact on the general shape of the epidemic, which is predicted to be most influenced by the first two moments of the distribution.

In conclusion, although we regard these attempts to characterize the incubation period of AIDS as accurate, they must be tentative until more of the incubation period is observed, and the processes that govern the incubation period within individual patients are better understood.

R. M. A. gratefully acknowledges financial support from the Medical Research Council, and D. R. C. holds a Science and Engineering Research Council Senior Research Fellowship. L. B. was partly supported by the National Institutes of Health, U.S.A., and the Office of Naval Research, U.S.A. This work has benefited from discussions with Dr Stephen Blythe, Dr Anne Johnson, Dr Angela McLean and Professor Julian Peto.

REFERENCES

Barton, D. E. 1987 *Nature, Lond.* **326**, 734.

Beal, S. 1987 On the sombre view of AIDS. *Nature, Lond.* **328**, 673.

Blythe, S. P. & Anderson, R. M. 1988 Distributed incubation and infectious periods in the transmission dynamics of the human immunodeficiency virus (HIV). *IMA J. Math. appl. Med. Biol.* **5**. (In the press.)

Costagliola, D. & Downs, A. M. 1987 Incubation time for AIDS. *Nature, Lond.* **328**, 582.

Centers for Disease Control 1985 *Morbidity Mortality wkly Rep.* **34**, 573–582.

Goedert, J. J., Biggar, R. J., Weiss, S. H., Eyster, M. E., Melbye, M., Wilson, S., Ginzburg, H. M., Grossman, R. J., DiGioia, R. A., Sanchez, W. C., Giron, J. A., Ebbesen, P., Gallo, R. C. & Blattner, W. A. 1986 Three-year incidence of AIDS in five cohorts of HTLV-III-infected risk group members. *Science, Wash.* **231**, 992–995.

Jensen, J. L. 1988 On the incubation time distribution and the Danish AIDS data: discussion contribution at the RSS seminar on AIDS, 25th November 1987. *J. R. statist. Soc.* (In the press.)

Lui, K.-J., Lawrence, D. N., Morgan, W. M., Peterman, T. A., Haverkos, H. W. & Bregman, D. J. 1986 A model-based approach for estimating the mean incubation period of transfusion-associated acquired immunodeficiency syndrome. *Proc. natn. Acad. Sci. U.S.A.* **83**, 3051–3055.

Lui, K.-J., Peterman, T. A. & Lawrence, D. N. 1987 Comments on the sombre view of AIDS. *Nature, Lond.* **329**, 207.

Medley, G. F., Anderson, R. M., Cox, D. R. & Billard, L. 1987 Incubation period of AIDS in patients infected via blood transfusion. *Nature, Lond.* **328**, 719–721.

Nelder, J. A. & Mead, R. 1965 A simplex method of function minimisation. *Computer J.* **7**, 308–313.

Peto, J. 1988 Discussion contribution at the RSS seminar on AIDS, 25th November 1987. *J. R. statist. Soc.* (In the press.)

Rees, M. 1987*a* The sombre view of AIDS. *Nature, Lond.* **326**, 343–345.

Rees, M. 1987*b* Incubation period for AIDS. *Nature, Lond.* **330**, 427–428.

Taylor, J. M. G., Schwartz, K. & Detels, R. 1986 The time from infection with human immunodeficiency virus (HIV) to the onset of AIDS. *J. infect. Dis.* **154**, 694–697.

Wilkie, A. D. 1988 An actuarial model for AIDS: discussion contribution at the RSS seminar on AIDS, 25th November 1987. *J. R. statist. Soc.* (In the press.)

J. Opl Res. Soc. Vol. 41, No. 2, pp. 95–101, 1990
Printed in Great Britain. All rights reserved

0160-5682/90 $3.50 + 0.00

The Blackett Memorial Lecture
20 November 1989

Quality and Reliability: Some Recent Developments and a Historical Perspective

D. R. COX

Nuffield College, Oxford

A general review is given of the development of ideas on quality and reliability, starting with Josiah Wedgwood in the late 18th century, through the intense activity of the 1930s, to the present, including a brief account and assessment of ideas associated with the names of Deming and Taguchi. While there is some mention of statistical aspects, no technical details are given. The second part of the paper contains comments on a number of general issues, ranging from the implications for methodological research to the potential special role of operational research and to the implications for education.

INTRODUCTION

I greatly appreciate the invitation to give this lecture. The subject is a very topical one of broad importance, but, because my own involvement nowadays is as an interested observer rather than as an active participant, I shall concentrate on some brief historical remarks and on a critique of current themes rather than on technical detail or on specific applications.

The historical remarks fall conveniently into a number of periods. The objective is not to give a complete scholarly account of what is, after all, a very large subject, but rather to pick out a few themes which link strongly with modern ideas.

HISTORICAL REVIEW

18th and 19th centuries

A convenient, if somewhat arbitrarily chosen, starting figure is Josiah Wedgwood. An exhibition in the Science Museum a few years ago including some of his notebooks brought home the almost laboratory-like nature of the careful experimentation that went into the development of Wedgwood's products. It seems that, much of the time, Wedgwood china was in such demand and the Company's profits so high that careful control of costs was not important, but in 1772 the Company was in severe difficulties, and in response Wedgwood undertook a very thorough study of the costs of all stages of his work. We thus see Wedgwood as a pioneer in two key themes in industrial quality control: careful technological experimentation and the analysis of economic aspects.

Wedgwood was not unique. In the case of Boulton and Watt's Soho engineering works, the founding fathers were primarily technologists, their sons keenly concerned about studies of costs. By the late 18th century the library of the Soho works contained 'many volumes of statistics', and accounts of the time indeed stress the detailed statistical analysis employed, i.e. the careful study of empirical data.

Wedgwood and the younger Boulton and Watt's attitude to this aspect of industry was nevertheless atypical of the period to the extent that when similar methods were promulgated some 50 years later, they aroused a lot of interest. Charles Babbage's book, *On the Economy of Machinery and Manufacture*, which ran to four editions in four years and sold 7,000 copies in its first year (1833), emphasized cost control, after, in effect, dismissing the technological side with the strikingly modern comment that 'The first objective of every person who attempts to make any article of consumption is, or ought to be, to produce it in perfect form'. It is incidentally an interesting sidelight on the intellectual climate of the time that a book by a Cambridge Professor of Mathematics, an observer of rather than a participant in the industrial world, should have been so widely distributed.

95

Despite the popularity of Babbage's book, the systematic foundations for control of industrial production were laid not in the UK but 50 years later in North America. By this time the scientific and technological underpinning of UK industry had to some extent been forgotten. Thus, in the textile industry, machines functioned satisfactorily but it seems likely that the subtle understanding of physics and engineering that explained why the machines worked had largely been lost and was, in a sense, no longer crucially needed. It has been suggested that part of the reason why North American industry adopted a more 'scientific' approach lay in having to redevelop to some extent from first principles. Scientific management was identified with, in particular, the ideas of Taylor. Note that the adjective 'scientific' refers not to the underlying physics, chemistry or engineering of the production process but rather to the systematic analysis of the costs and other such aspects of the work. Taylor appreciated that the base of much loss of quality and efficiency was unnecessary variability—another very modern theme—and he saw the answer as lying in rigid standardization of all phases of work. This led eventually to a socially unacceptable position, to excesses of time and motion study, and in a sense to the emergence of a third theme to add to the technological and economic aspects of industrial quality, namely the sociological.

Early 20th century

In 1908, Student (W. S. Gosset) developed the statistical test and distribution named after him. He did this, and other highly original statistical work, in order to help analyse the data on quality obtained at Guinness's brewery. It has often been remarked, but is worth repeating, that Student had extremely limited mathematics. Many mathematicians before him could have derived the distribution with little effort, but none saw the fruitful formulation. At about the same time, the Copenhagen Telephone Company appointed a mathematician, A. K. Erlang, to study telephone traffic problems. These are concerned essentially with the provision of a reliable service in the face of the randomness inherent in the nature of the traffic. It seems likely that Erlang was the first mathematician employed specifically on industrial problems of this nature, and his work is the beginning of applied probability in an industrial setting, leading in particular to modern queueing theory and reliability theory.

Of great importance but of less immediately obvious relevance is the development in the 1920s by R. A. Fisher of principles of experimental design, initially largely in an agricultural context. In particular, the advantages of studying the effects of several factors simultaneously as compared with a one-at-a-time approach are particularly striking in an agricultural context, where investigations usually take a long time to complete. A wider appreciation of the role of multifactor experiments in industrial design, research and production is one major modern theme.

The 1930s

Based largely on Bell Labs, major developments in what are now regarded as traditional notions of statistical quality control came to fruition in the early 1930s. Shewhart developed the theoretical concepts of processes in and out of control, of assignable causes and so on, and Dodge and Romig provided parallel notions of acceptance sampling. These ideas spread rapidly, being taken up in the UK at GEC by Dudding and Jennett and by Professor E. S. Pearson, leading in 1935 to a major British Standards Institution publication, BS600.

The formation in the early 1930s of the Royal Statistical Society Agricultural and Industrial Research Section represented a wider interest in statistical methods in industry. In particular, Tippett at the British Cotton Industry Research Association and H. E. Daniels, then at the Wool Industries Research Association, made major studies of variability, control of variability being a key theme in achieving quality control in the traditional textile industries. They and others also made effective use of the ideas on experimental design referred to above. Statistical ideas, especially of components of variance, developed largely in industrial connections, have been important in many fields, ranging from human genetics to educational testing.

The Second World War

Statistical methods of quality control and acceptance sampling were widely used in the UK and North America in the Second World War. BS600 was issued in a simplified and shortened form as BS600R, and was a main source of information. A major emphasis in this period was on making

96

methods appealing to the 'shop floor'; management had little choice in that government departments, especially the Ministry of Supply, required the adoption of suitable procedures. At the same time, centring particularly in the UK on the group SR17 at the Ministry of Supply under the intellectual leadership of G.A. Barnard, there were major investigations of technique. In particular, R.L. Plackett and J. P. Burman gave an elegant theory of designs for investigating a large number of factors in a small number of runs, showing that in certain cases, k two-level factors could be investigated in $k + 1$ runs using designs based on Hadamard matrices; the underlying mathematical theory arises also in connection with error-correcting and error-detecting codes. This design notion connected with the idea of so-called fractional replication has since been extensively developed, largely, although not entirely, with industrial applications in mind.

A simple illustration showing the investigation of (up to) seven factors in eight runs is summarized in Table 1. Note that, provided there are no interactions, i.e. that the effects of the separate factors are additive, the balance of the design allows the simple separate estimation of the effects of the different factors by comparison of simple averages. Extensions include the idea, popular for a time, of so-called supersaturated designs in which the number of factors could exceed the number of runs; these designs are effective only when virtually all the factors have null effects and the primary issue is to screen out the one or two factors that do have an effect on response. While the existence of such designs is of academic interest, it seems unlikely that it would often be wise to use them.

TABLE 1. *Key to design for investigating up to seven factors in eight runs*

Run	X_1	X_2	X_3	X_4	X_5	X_6	X_7
1	-1	-1	$+1$	-1	$+1$	$+1$	-1
2	$+1$	-1	-1	-1	-1	$+1$	$+1$
3	-1	$+1$	-1	-1	$+1$	-1	$+1$
4	$+1$	$+1$	$+1$	-1	-1	-1	-1
5	-1	-1	$+1$	$+1$	-1	-1	$+1$
6	$+1$	-1	-1	$+1$	$+1$	-1	-1
7	-1	$+1$	-1	$+1$	-1	$+1$	-1
8	$+1$	$+1$	$+1$	$+1$	$+1$	$+1$	$+1$

The seven columns represent seven factors, each of which can occur at one of two levels, 1 and -1. Thus, in run 3, X_2 occurs at the upper level, 1. The balance of the arrangement ensures that, provided the effects of the different factors are additive, simple separate estimation of the effect of each factor is possible by a direct comparison of means at upper and lower levels. With fewer than seven factors, any choice of the required number of columns will give a design with the same 'main effect' balance; it may, however, be possible to make a special choice of design that will in addition allow the study of more complicated structures—in particular, the examination of some kinds of interaction.

Post-war period

While active interest and work in statistical quality control continued after the end of the War, to some extent the enthusiasm waned, in part perhaps because the methods were taken for granted, and in part because customers were less explicit in their demands. Cumulative sum charts were an important development from simple control charts. The most important developments were, however, in the process industries, very particularly the work of G.E.P. Box, then at ICI, on design of experiments, with an emphasis on factorial experiments with quantitative factor levels and on designs for finding optimal operating conditions. Combined with this was his development of the idea of management by evolutionary operation (EVOP), i.e. that plant should be subject to continuous experimentation to achieve continuing improvements. This idea, which never received the widespread recognition and adoption that it deserved, is somewhat related to the recent notion of management by continuous improvement.

As viewed by an academic statistician, however, UK interest in the problems of industrial statistics in general and quality control in particular was at a peak in the early 1950s and has been in decline since, at least until a very recent resurgence in the light of the developments to be described in a moment. Of course, a distinction has to be drawn between nominal adherence to

97

quality-control ideas and their effective implementation. In North America, interest has been at a higher level, in particular as represented by the activities of ASQC (American Society for Quality Control).

Recent developments and the impact via Japan

In 1950, two US scientists, the statistician W. E. Deming and the engineer J. M. Juran, visited Japan as advisers over the reconstruction of Japanese industry. Deming had been an associate of Shewhart and had earlier had major success in applying quality-control ideas to the operations of the US Bureau of the Census. (He was at that stage also well known for a number of innovative technical statistical ideas.) These visits had a major impact, leading in particular to:

—an emphasis on training in methodological issues, in particular via the Japanese Union of Scientists and Engineers;
—an emphasis on careful experimentation;
—stress on the achievement of high quality, preferably on defect-free quality, rather than on acceptance sampling for the detection of bad quality;
—the development of attitudes and working practices throughout an organization that will encourage constant search for high quality.

These last points, encapsulated in Deming's 14 principles, emphasize the sociological component of quality achievement as of comparable weight with the technological and economic. The 14 points represent a civilized approach to these issues; in his remarks to the recent 150th Anniversary celebrations of the American Statistical Association, Deming emphasized particularly the importance of collaboration between individuals, and was severely critical of what he regarded as the over-competitive nature of American society, at least at an individual level.

Deming worked in the US as a consultant over a long period, but it was with the general awareness, perhaps 10 years or so ago, in the US of the impact of Japanese competition that he became a major public figure and that he persuaded top management of some major US companies of the importance of quality as a key goal. Much of the current concern with quality worldwide can be traced to his influence, directly by his lectures and personal advice, via the work of Deming Associations and by diktat to the European associates of US companies.

Another individual whose work has had a major impact in the US, and to some extent outside, is the Japanese engineer G. M. Taguchi. His work is not easy to follow, partly because of language difficulties, but among the points he stresses are the following:

—that quality and reliability are achieved by minimizing variability;
—that techniques of multifactor experimental design should be used in the design phases of a project as well as in industrial research and operation (the emphasis on the design phase accounts for the term 'off-line quality control');
—that it will often be fruitful to distinguish factors that affect means and not variances, those that affect variances and not means, and those that affect both; a position can then be reached where variance is minimized, and then the mean can be adjusted to target level via the first kind of factor;
—that factors that represent outside uncontrollable noise, such as unusual conditions encountered in use, should be investigated to find conditions where their impact is minimized.

He also places some emphasis on the formulation of loss functions as a basis for determining target levels of key properties.

Note that the emphasis in Box's work on process optimization is on getting the mean of some property (perhaps cost per unit yield) to an optimum level. In Taguchi's context, the mean can often be relatively easily brought to any desired level and the objective is the minimization of variability. As noted above, careful study of variability was a major concern of industrial research in the 1930s, and indeed, except for the use of fractionated experiments, there are fairly close parallels with, in particular, some of the textile work of that period. In assessing Taguchi's contributions, it is important to distinguish the innovative and fruitful concepts he has emphasized from some of the particular statistical techniques and rather rigid procedures he has advocated, which at a detailed level are open to criticism from the standpoint of both design and analysis. The Institute of Statisticians has set up an organization concerned with industrial quality achieved via Taguchi's ideas.

98

The current position

The preceding, very incomplete review is meant primarily to illustrate that some of the key ideas have quite a long history. To see this is in no way to undervalue the recent contributions. It is hard to assess the present position, even if one confines attention to the UK. It seems likely, however, that, as so often in such comparisons, the best British practice is at a very high level, but that there are large areas of industry where there is major scope for improvement.

SOME GENERAL CONSIDERATIONS

There now follow brief general comments, some of them expressions of personal opinion, on some of the broad issues connected with quality and reliability. These issues are treated in order of broadly decreasing technical interest and broadly increasing importance and difficulty.

Development of new methods

Consider the role of new statistical or OR techniques in the present context. Now it seems likely that very rarely will a major specific improvement in quality hinge crucially on a new methodological development—for example, in the statistical design of experiments. Nevertheless, the healthy empirical tradition that pervades British OR, statistics and, no doubt, other topics can, I believe, lead to an undervaluing of theory and of new development more generally. Theoretical discussion of methods, and more generally of concepts, is one intellectual driving-force for these subjects, which, without theoretical underpinning, are in danger of becoming sterile. Further, one implicit message of my historical review is the obvious one of the unifying role of theory, and of mathematical theory in particular. Finally, it takes time before what begins perhaps as a rather academic exercise gets translated, perhaps not by its innovator, into something immediately useful, blessed, for example, by the ultimate seal of approval: simple software, not too user-hostile. I believe that there is a serious danger, in particular, in the present academic climate in the UK, of a neglect of longer-term theory. Also, the intellectual discipline of meticulous attention to technique can be an antidote to reliance on vague exhortation and fashionable slogans.

Application of current advanced methodological knowledge

An easier and, in a sense, more immediately important aspect is the need to make widely available current knowledge of methodological notions of a relatively advanced kind. The Japanese experience in this respect is obviously relevant. One key issue here concerns the kind of knowledge called for. It seems to me that it needs to be based on genuine understanding not of technical details but of broad concepts, and yet to be sufficiently specific as to be clearly relevant. Thus an account of design of experiments tied to analysis of variance is not appropriate, nor is an account that presents the subject as a mysterious list of trick combinatorial arrangements. One needs an account which, for example, explains in compelling general terms why factorial-type arrangements are a good idea and what the limitations are, supplemented by advice on how to find designs for the simpler, relatively standard situations.

Application of simple quantitative notions

Even more important is likely to be the very wide dissemination of a critical and constructive attitude to quantitative information. This includes the use of control charts and other graphical techniques, and some understanding of random variability and its measurement—the simplest notion of statistical distribution, in fact.

Role of computers

The actual and potential role of computers in the various aspects of quality control ranges from the storage of information, the painless and hopefully flexible provision of the results of simple analyses to the use of relatively sophisticated algorithms for the choice of experimental plans and for the analysis of the resulting data, including the display of the conclusions in widely understandable form. In the latter respect, the increasingly wide acceptance of computer-based methods can lead to the strange and potentially dangerous situation where suggestions on experimental design may be accepted 'because the computer says so', when advice from an individual scientist might be regarded with suspicion. Or again, the analysis of some data may be thought to depend on no special assumptions about the process under investigation, because some standard package is being used, whereas in fact some wholly inappropriate assumptions may be involved. It is certainly possible to exaggerate these fears; they point to the importance stressed above of a genuine understanding of key underlying concepts.

Educational aspects

The points discussed above have major educational implications, ranging from the teaching of quantitative ideas in primary and secondary schools and the course content of the training of technologists in tertiary education to the provision of continuing education in the workplace and outside. Societies such as the Operational Research Society have a major role to play, especially, but not only, in the last mentioned. The role of the Japanese Union of Scientists and Engineers in providing extensive courses and handbooks on technique is particularly noteworthy.

Role of operational research

While I am, of course, aware of a substantial interest in quality and reliability in the operational research world, I hope I am not being unfair in suggesting that the contribution so far has not been as great as might have been expected. I know that the content of current journals is not always a sound guide as to where the balance of present interests lies, but there do seem, for example, rather few papers on the quality theme in the Society's journal. It has been suggested in the earlier parts of the paper that consideration of and balance between technological, economic and social aspects of the issue is called for, and it is precisely this balance that the operational research approach might be expected to provide. Has there been too much emphasis on short-term narrowly economically formulated objectives? *Post hoc* explanations and even justifications for any such misemphasis would be easy to construct.

Definition of quality and its appropriate level

A different kind of lecture on this topic might well have started by considering two key questions; how is quality to be measured, and what is the appropriate level at which to aim? In some very specific industrial contexts, especially in the role of a subcontractor, one answer is easy: quality consists in meeting a (usually multidimensionally formulated) specification. In other contexts, the answer 'keeping the customer happy' may be a quite useful first formulation. It is, however, often open to the objections of being insufficiently analytical, of being insufficiently concerned with what the customer might want tomorrow rather than today, and with not putting sufficient weight on the differing needs of different potential customers.

When a particular quantitative definition of quality is available, it is tempting to suggest that decision analysis should indicate the appropriate level of quality at which to aim, balancing the high costs of high quality against the losses resulting from lower quality. While this may be difficult to fault in principle, there are major snags. The multidimensional character of quality raises the 'satisficing'-'optimizing' issue. Again, in costing loss of quality it may be easy to overlook the long-term consequences of loss of goodwill. Finally, consider a very specific context of a clearly defined criterion and well-established costs, so that the conditions for optimal operation, say, of a production process will be computable and perhaps even made automatic via some control-theoretic technique. Even if the effect of uncontrolled perturbations is minimized in the spirit of Taguchi's suggestions, such perturbations are likely to occur, sometimes in unexpected form. Such perturbations are, on evolutionary grounds, likely on the whole to lower quality, perhaps rather seriously. These last two points suggest that a target quality higher than the nominal optimum will often be wise, reinforcing the general spirit of Deming's recommendations.

Public service implications

In the previous part of the paper I have deliberately refrained from defining precisely the nature of the operations involved, but much of the emphasis has implicitly been on industrial processes of some kind or other. A very important issue, however, concerns the analysis and improvement of

100

quality in public service contexts, of which health and education serve as important examples. Rational discussion is not helped by the organizational conservatism of both fields and by the massive skill of the former and roughly equal failure of the latter in mobilizing public support for defence of the *status quo*. There is not space, nor am I equipped, to discuss these issues in depth. A key common difficulty, however, is clearly that of defining and measuring quality.

In medical contexts there is at the moment intense interest in measuring what is rather grandiosely called 'quality of life', usually as assessed directly by the patient. When this is used as an aspect of the comparison of alternative treatment regimes or as a basis for choice by the patient between alternative treatments, and the multidimensional character of quality is explicitly recognized, these developments seem important and relatively uncontroversial. Much more difficult, and highly controversial, however, is the proposed use of essentially unidimensional quality-of-life measures for resource allocation by the use of qualys (quality adjusted life years). Here the notion is not only to score quality-of-life via a one-dimensional score, with the unit qualy corresponding to a year of life in a state of unimpaired health, survival at various impaired levels receiving less than unit score, but also to allocate resources to maximize the long-run average number of qualys achieved. While such calculations, combined with careful sensitivity analysis, could well be enlightening, their relatively mechanical use is fraught with major hazards.

In educational matters, even in the relatively well-defined objective of measuring the quality of teaching, there are major difficulties, not the least being the need to distinguish content from style. It is interesting that, some years ago, Deming entered a strong plea against the involvement of students in assessing teaching, essentially on the grounds that students are not in a position to judge importance of content and that content, not style, is the key issue. Of course, the explicit use of student comment formally and informally to monitor style of teaching is now widespread in the UK and North American University systems. There are further difficult issues connected with assessing the quality of research.

In both these fields, health and education, any notion that the 'market' could bypass the difficult issues of assessment and choice involved seems simplistic in the extreme.

CONCLUSION

This lecture has concentrated on generalities and not on specific technical issues. References to the extensive literature have therefore not been given. One starting-point is the Royal Society publication[1] based on a joint meeting with the Royal Statistical Society. This contains 14 papers on various fairly general topics, plus a glossary of technical terms. Deming's ideas can be studied via his very informal book.[2] For an introduction to and comment on some of Taguchi's ideas, see Kackar,[3] and for more detailed statistical discussion, see Box *et al.*[4]

Acknowledgements—I am very grateful to G. E. P. Box, T. P. Davis, G. H. Freeman, A. A. Greenfield and S. Lewis for helpful advice on recent developments, to E. Newell for historical information and to the President of the Operational Research Society, Dr J. C. Ranyard, for suggesting the topic as one suitable for the Blackett Lecture.

REFERENCES

1. ROYAL SOCIETY (1989) Discussion meeting: Industrial quality and reliability. *Phil. Trans. R. Soc. Lond.* **A327**, 479–638. Also available from Royal Society as book.
2. W. E. DEMING (1986) *Out of the Crisis*. British Deming Association (2 Castle St, Salisbury).
3. R. N. KACKAR (1985) Off-line quality control, parameter design and the Taguchi method (with discussion). *J. Qual. Techn.* **17**, 176–188.
4. G. E. P. BOX, S. BISGAARD and C. A. FUNG (1989) An explanation and critique of Taguchi's contributions to quality engineering. In *Taguchi Methods: Applications in World Industry* (A. BENDELL, J. DISNEY and W, A. PRIDMORE, Eds). IFS Publications.

101

A6

J. R. Statist. Soc. A (1992)
155, Part 3, *pp.* 353–393

Quality-of-life Assessment: Can We Keep It Simple?

By D. R. COX and R. FITZPATRICK,   A. E. FLETCHER,

Nuffield College, Oxford, UK   *Royal Postgraduate Medical School, London, UK*

S. M. GORE and D. J. SPIEGELHALTER   and   D. R. JONES†

Medical Research Council Biostatistics Unit, Cambridge, UK   *University of Leicester, UK*

[*Read before* The Royal Statistical Society *on Wednesday, March 11th, 1992,*
the President, Professor T. M. F. Smith, *in the Chair*]

SUMMARY
The importance of general statistical principles of study design and analysis to quality-of-life assessment in clinical trials is emphasized. Basic methods are reviewed briefly, with reference to three examples. Careful use of standard tools supplemented with context-specific scales is recommended. Problems of weighting and aggregation are discussed; the use of simple weighting schemes supplemented by sensitivity analysis is suggested. Some technical issues are explored, including factorial question structure, components of variance to distinguish mean treatment and patient-specific treatment effects and informative loss to follow-up. Simplicity of design, analysis and presentation are stressed.

Keywords: ANALYSIS OF VARIANCE; CLINICAL TRIALS; FACTORIAL DESIGNS; INFORMATIVE CENSORING; QUALITY OF LIFE; SENSITIVITY ANALYSIS; SURVIVAL ANALYSIS; WEIGHTING

1. INTRODUCTION

In the last decade interest has increased in quality-of-life (QOL) measures in four broad health contexts: measuring the health of populations, assessing the benefit of alternative uses of resources, comparing two or more interventions in a clinical trial and making a decision on treatment for an individual patient (Katz, 1987; Spitzer, 1987; Aaronson and Beckmann, 1987; Walker and Rosser, 1988; McDowell and Newell, 1987). Each context requires an assessment of the impact of ill health on aspects of the everyday life of the individual.

Simple measures of health status—such as the Karnofsky scale in cancer (Karnofsky and Burchenal, 1949) and the American Rheumatism Association (ARA) scale (Steinbrocker *et al.*, 1949)—began to appear within clinical medicine in the 1940s in response to inadequacies of traditional measures of mortality and morbidity as descriptions of outcome in many chronic diseases. In the USA in the late 1960s and early 1970s, concerns about the effectiveness and economic costs of health services led the National Center for Health Services Research to support the development of measures of health status. Similar considerations in the UK have motivated the recent emphasis on an audit of clinical practice (Secretaries of State for Health, 1989).

Increasing and sometimes indiscriminate use of QOL measures, including quality-adjusted life years (QALYs), has provoked concern about these methods in the four

† *Address for correspondence*: Department of Epidemiology and Public Health, Clinical Sciences Building, Leicester Royal Infirmary, PO Box 65, Leicester, LE2 7LX, UK.

0035-9238/92/155353 $2.00

contexts above, especially when important consequences, such as treatment decisions or resource allocation, depend on them (Smith, 1987; Loomes and McKenzie, 1989; Carr-Hill, 1989; Fletcher, 1991). Instruments developed in one context are sometimes applied in studies with different objectives. QOL assessment in clinical trials comparing treatments differs from the resource allocation and individual treatment decision contexts. In making individual treatment decisions, for example, any QOL assessment has often been collapsed to a single scale to allow a direct comparison between options. Clinical trial results, in contrast, are best left multidimensional and reported so as to make it straightforward for future users to weigh up the alternatives with their own, currently unspecified, values.

In this paper we comment on issues of weighting and validation, make some general comments about trial design, analysis and reporting and address some more technical points such as the use of factorial designs in questionnaires, analysis of components of variance to distinguish mean treatment and patient-specific treatment effects, and the analysis of QOL in conjunction with length of survival, with particular reference to problems of informative loss to follow up. We concentrate primarily on applications to clinical trials, although the discussion may be more widely applicable.

The current practice in the use of standard instruments in trials is reviewed briefly in the next section. Three specific examples are then introduced, to be used in Section 4 to illustrate design issues. Sections 5–7 deal with aggregation, analysis and reporting issues respectively, which, although placed in the order in which they are met in a given study, are nevertheless interdependent. Practical recommendations in each section are intended not as 'tablets of stone' but to emphasize a commonsense approach to design and analysis issues. In our conclusions in Section 8, we recommend the careful use of standard tools supplemented with simple context-specific scales used without elaborate weighting schemes. Analysis of variance can be used to investigate heterogeneity both among response measures and between individuals. Finally, the importance of clinical interpretability is emphasized.

2.   BRIEF APPRAISAL OF USE AND PROPERTIES OF STANDARD INSTRUMENTS

Aaronson (1989) reviewed some recent examples of the use of QOL measures in clinical trials. Such measures are not universally accepted as legitimate. Although QOL could be considered as a global holistic judgment, a more reductionist approach is common. The nebulous concept is broken into components or *dimensions*; these dimensions may be further decomposed into a number of questions or *items*, which in turn may be answered on a *scale*. In describing or comparing groups we may want to recompose these detailed responses back into dimensions or even into a single global assessment. This process of recomposition involves the use of weighting schemes; see Section 5.

Most clinical trials have made use of standard instruments that have been to some extent evaluated by psychometric criteria. We describe these procedures only briefly, without making recommendations on the use of specific instruments. Several systematic reviews of some standard scales are available (Maguire and Selby, 1989; Spilker, 1990; Fallowfield, 1990; Bowling, 1991). Other reviews of QOL assessment in trials concentrate on specific disease or patient groups, such as cancer (see, for example, Fayers and Jones (1983)), cardiovascular disease (Fletcher, 1988) and the elderly (Kane and Kane, 1981).

TABLE 1
Extract from the NHP

Question	Weight given to positive response
I am waking up in the early hours of the morning	12.57
It takes me a long time to get to sleep	16.10
I sleep badly at night	21.70
I take tablets to help me to sleep	22.37
I lie awake for most of the night	27.26

2.1. *Use of Standard Instruments*

Many 'standard' instruments are currently available although the extent of experience of their use in clinical trials varies widely. The first and probably most widely used scale, the Karnofsky index (Karnofsky and Burchenal, 1949), measures the physical performance of patients with cancer on a scale of 0–100. Other multidimensional instruments were developed for general use on patient populations: two of the best known are the sickness impact profile (SIP) which comprises 136 questions scoring on 12 dimensions (Bergner *et al.*, 1981) and the Nottingham health profile (NHP) which asks 38 questions on six dimensions: pain, physical mobility, energy, emotions, sleep and social contacts, and seven more general questions (Hunt *et al.*, 1986). Many such instruments employ weighted items; the questions concerning sleep in the NHP are shown in Table 1, with the weights contributing to an overall score on that dimension.

2.2. *Evaluation of Instruments*

The criteria for evaluation of instruments have been the concern mainly of psychometricians (van Knippenberg and de Haes, 1988). Briefly, scales should be

(a) valid,
(b) reliable and
(c) responsive or sensitive.

2.2.1. *Validity of instruments*

Criterion validity involves assessing an instrument against an accepted absolute standard, which is unavailable for QOL instruments since they measure phenomena which are experiential and subjective. Other methods of evaluating a QOL instrument entail varying degrees of formality.

An assessment of *face (or content) validity* of an instrument involves checking whether items in an instrument appear to cover its intended topics clearly and unambiguously. It has been suggested that face validity may be maximized by including individuals of diverse backgrounds (patients, doctors, nurses, social scientists, etc.) among the assessors (Aaronson, 1989).

The *construct validity* of an instrument may be assessed by a more formal inspection of the overall pattern of relationships between the instrument and other measures. Thus instruments such as the NHP and the SIP have been assessed by comparing responses in groups with apparently different health statuses. There should also be some agreement between the instrument and other measures of overlapping constructs, such as correlations with various biochemical, radiological or clinical measures of severity of disease. For example, in a study of anxious patients, higher levels of distress measured by the symptom rating test (Kellner and Sheffield, 1973) were associated with lower concentrations of benzodiazepam in the blood (Robin *et al.*, 1974). At its most formal level this may be approached by the multitrait–multimethod analysis developed in psychometrics (Campbell and Fiske, 1959) or by factor analytic methods (Brown *et al.*, 1984; Mason *et al.*, 1988).

2.2.2. *Reliability*

The reliability of an instrument is a measure of its ability to yield the same results on repeated trials under the same conditions. Although reliability is generally considered easier than validity to test using standard methods (Nunnally, 1978; Fleiss, 1975), inappropriate methods, such as the use of correlation coefficients as measures of agreement, should be avoided (Bland and Altman, 1986; Chinn, 1990). A basic approach to the assessment of the reliability of an instrument is to examine its internal reliability at a single administration (Cronbach, 1951). Split-test reliability is examined by dividing instruments into equivalent halves for all dimensions and inspecting the degree of agreement between halves. The alternative approach of test–retest measures (Chinn and Burney, 1987) may, however, prove to be practically difficult in the context of a clinical trial. Further, if patients undergo significant clinical changes in health between the test and retest, random error and true change may be impossible to distinguish.

Reliability between interviewers or raters should also be assessed if instruments are not self-administered. High interrater reliability reported for some instruments may be the result of intensive training or special expertise not necessarily available in subsequent usage.

2.2.3. *Responsiveness*

Any QOL instrument used in a clinical trial should be able to detect clinically significant changes over time (Kirshner and Guyatt, 1985). Such responsiveness has been studied far less than has validity. Responsiveness of QOL instruments may be examined through associations with other changes in health status (Fitzpatrick *et al.*, 1987) or in physiological measures (Meenan *et al.*, 1984), or the sensitivity and specificity of instruments compared against an established criterion of change if it exists, e.g. by examination of receiver operator curves (Mackenzie *et al.*, 1986). However, there are difficulties in deciding the criterion: should assessments by the patient, the doctor, a consensus of the two or physiological measures be used as the standard against which to judge QOL instruments (Deyo and Centor, 1986; Fitzpatrick *et al.*, 1989)? Discrimination of treatment responses from placebo or trial participation effects (e.g. through learning effects) may be problematical, since patients will report improvements due to both.

Finally, the degree of generalizability of an instrument is important. When using an instrument in a new context, its measurement properties should be tested before the main study.

3. THREE EXAMPLES

QOL assessments arise in a range of applications; reference will be made in subsequent sections to some of the issues and the examples introduced here:

(a) a life threatening disease in children in which QOL forms a secondary outcome measure;
(b) treatment to reduce the risk from an asymptomatic condition in which delayed potential benefits are traded off against immediate but possibly long-term side-effects;
(c) conditions of chronic disability with no implications for survival.

3.1. *Randomized Control Trial of Treatments for Advanced Neuroblastoma*
Advanced forms of neuroblastoma (stages III and IV), a childhood solid tumour, are aggressive cancers (Evans *et al.*, 1971), with generally very poor prognosis. They are usually treated with a cocktail of cytotoxic drugs; severe side-effects are associated with these treatment regimens. A multicentre trial of a randomized comparison of a single high dose of melphalan (requiring barrier nursing techniques) *versus* no further treatment following the induction therapy is reported elsewhere (Pinkerton *et al.*, 1988). The length of survival was the primary outcome measure in this trial.

3.2. *Trials of Anti-hypertensive Therapy*
Hypertension is a risk factor for cardiovascular disease. Although very high levels of blood pressure may be associated with headache and dizziness, most patients are asymptomatic. The benefits from the drug treatment of hypertension, primarily in reducing strokes, have been well established, but controversy remains over the risk–benefit ratio for mild hypertension where treatment benefits may be offset by adverse effects. In general, the main issue is which treatment produces minimal interference with a patient's QOL.

3.3. *Evaluation of Treatments for Arthritis and Rheumatism*
Rheumatoid arthritis (RA) is a chronic disease which is rarely fatal; its main health impacts are pain, stiffness and functional limitations. Several laboratory, radiological and clinical measures are used to assess the course of the disease, but there is by no means complete agreement on which measure or combination of measures best represents the short-term or long-term outcome. Perhaps as a result, health status measures have received particular attention in rheumatology.

4. DESIGN ISSUES IN USE OF
QUALITY-OF-LIFE MEASURES IN CLINICAL TRIALS

4.1. *Standard Design Considerations*
QOL measurements are particularly susceptible to *systematic errors* associated with observer effects and the conditions under which the measurement is made. Consequently, standard precautions for avoiding bias should be adopted whenever feasible. These precautions include randomization, blindness of the person administering the questionnaire and standardization of recording procedures.

4.2. *What to Measure?: Choice of Quality-of-life Dimensions*

What to measure in a clinical trial depends on the nature of the disease, the expected benefits and adverse effects of treatment and the length of the observation period. For example, in severe illness, patients are unlikely to be working or leading physically active lives and questions on these dimensions would not be included. An evaluation of QOL in a clinical trial can usefully be preceded by a descriptive study in the same patient population group, both to identify the QOL dimensions impaired by the disease and to suggest those on which treatment might have an effect.

For a few health problems, there are disease-specific instruments which concentrate on particularly relevant aspects of QOL. In many studies, however, the range of possible adverse effects cannot reliably be specified in advance and a wide battery of QOL scales or wider ranging generic scales may be needed (Croog *et al.*, 1986). This may lead to patients being overburdened with inappropriate questions, and to an approach that is insensitive to small or specific but important changes.

4.2.1. *Neuroblastoma trial*

The QOL questionnaire for the neuroblastoma trial addressed both physical and psychological aspects of the child's QOL. Topics included functional status (restriction of physical activity), symptoms (including pain), side-effects (nausea and vomiting, loss of appetite, difficulties in hearing), worry about side-effects (such as hair loss) and about future treatment, and overall assessment of enjoyment of life. An inclusion of additional items on social functioning of the child, and on the impact of the disease and treatment on the family and friends of the child, would have been desirable.

In some trials patients are allowed to select their own QOL objectives at the beginning of the trial, e.g. by selecting from the QOL instrument items that they consider to be of particular concern. This involves a loss of standardized information and of the ability to generalize, although it may provide a more accurate indicator of responsiveness (Tugwell *et al.*, 1987). The '$N=1$' trial design performed on an individual patient to decide their future treatment (Guyatt *et al.*, 1986) represents an extreme example.

4.3. *How to Measure?: Choice of Instruments*

A key choice is that between the use of a standard instrument and an *ad hoc* or specially developed questionnaire. Standard instruments should ideally have a history of successful use, but they may not be entirely appropriate to the intended application; in particular, they may lack sensitivity compared with an especially developed instrument. Their validity and reliability should not be assumed without question (Hutchinson *et al.*, 1979; Maguire and Selby, 1989).

Our general recommendation is to use a standard, validated core instrument, with customized additions for the particular application. The European Organization for Research on Treatment of Cancer's modular questionnaire (Aaronson *et al.*, 1988), for example, provides a suitable starting point for many cancer and chronic disease trials.

Standardization of technique is widely accepted for measurements such as blood pressure but is even more necessary in the QOL context because of the potentially greater variety and impact of subjective measurement errors.

The issue of the burden on the patient is also crucial; lengthy questionnaires are an imposition on sick patients, and even on healthy ones. If it is felt that supplementary questions are appropriate, general principles of questionnaire design should be followed (Sudman and Bradburn, 1982). Wording should avoid ambiguity or bias, and attention should be given to a pleasing layout and ease of completion to encourage a good response, and to clear coding specification to allow rapid and accurate data processing. It is also vital to pilot both the questionnaires and their processing.

4.3.1. *Rheumatoid arthritis studies*

Several RA-specific instruments have been developed, with the health assessment questionnaire (HAQ) (Fries *et al.*, 1982) and the arthritis impact measurement scales (Meenan *et al.*, 1982) the most widely used. However, generic health status instruments have also been quite widely used as outcome measures in RA. The SIP has been compared with the earlier and simpler four-point ARA functional scale and found to have higher test–retest reliability and to distinguish clinically meaningful differences within a single ARA class (Deyo and Inui, 1984). This is important because the kinds of change produced by interventions are generally too small to be detected by the gross categories of the ARA functional scale.

In terms of construct validity, these instruments correlate with measures of severity of disease such as grip strength, indices of joint damage and laboratory tests (Fitzpatrick *et al.*, 1989), although there are limitations of correlation analyses in this context. Health status instruments have detected changes from the use of non-steroidal anti-inflammatory drugs within quite short time periods such as 4–12 weeks, and that changes occur on scales such as pain and mobility rather than on scales such as social activities that might be expected to respond over longer time periods (Anderson *et al.*, 1989; Parr *et al.*, 1988). In patients experiencing less dramatic clinical changes, a shorter instrument (the HAQ) was as responsive to change as the SIP (Fitzpatrick *et al.*, 1989) and an abbreviated instrument with selected items was more sensitive than the full length version of the SIP (Deyo and Centor, 1986).

4.3.2. *Construction of specific scale*

There are several ways of recording the required information:

(a) as a binary response, e.g. condition present or absent;
(b) as a response on a k-point ordinal scale representing increasing (or decreasing) severity, often $k = 3, 5, 7$;
(c) on a visual analogue scale, by marking a point on a line, often 10 cm in length, on which positions to the left or right represent increasing or decreasing severity.

For methods (b) and (c) it is desirable that, so far as is feasible, scale points should be anchored by clear verbal descriptions, preferably between categories (rather than at their centres) to facilitate assignment of borderline cases.

An advantage of the visual analogue and ordinal scales with larger values of k is the protection against loss of information resulting from a bunching of responses into one or two cells. The simplicity of a scale with a limited number of points, however, will often be very appealing in implementation (McCormack *et al.*, 1988). Firm empirical evidence of the superiority of visual analogue scales over categorical scales is difficult to find (Remington *et al.*, 1979).

In terms of reliability, there is evidence (Lissitz and Green, 1975) that there is little to be gained by using scales comprising more than five points, but no clear advantage to the use of even or odd numbers of points (Remington *et al.*, 1979). When typical responses are distributed around the centre of the scale, theoretical studies of grouping based on a latent normally distributed response give efficiencies of 80% and 90% for optimally placed groupings of three and five cells respectively (Cox, 1957). These results further support the widely held, if ill-documented, belief that there is no gain in going beyond 5–7 categories in such cases.

Visual analogue scales involve more labour in analysis and will be less familiar to many patients. Their main advantage lies not in the apparent gain in precision of measurement but in the possibility that they offer for appropriate *post hoc* specification of groupings for analysis. Loss of information due to inappropriate prior choice of categories may thus be avoided.

The discussion is different in detail if responses cluster around one end of the scale, as they may, for example, if the absence of a side-effect to which the question relates is common; the broad conclusions are similar, however. For simple methods for the analysis of visual analogue scale data of this kind see Solomon (1989).

4.3.3. *Asking long series of questions: technical solution*

There are some design issues that are more specific to answering a long series of questions. Here we point out some possibilities. Suppose that there are k dimensions, each comprising a number of questions. The questions in each dimension are divided into two equal sections, any absolutely key question being placed in both sections. If the questionnaire is administered just once to each patient, the patient receives one section of each dimension. If, say, each patient receives the questionnaire twice, they receive one section of a particular dimension the first time and the complementary set the second time. There are now various designs that can be used associated with the fractional replication and confounding patterns in the 2^k factorial system, regarding the possible questions to be asked of an individual as forming the 2^k possible treatments in the associated factorial scheme.

There are many variants on this theme. With two questioning times per patient and a modest value of k we could divide the patients into homogeneous sets of 2^{k-1} for allocation to a treatment and its complement. For a larger value of k and one application per patient we could use a suitable fraction of the 2^k system as the basis.

4.4. *Who and Whom to Measure?*

The essence of the QOL approach is the expression of a subjective viewpoint and therefore the main respondent should always be the patient. Inability of a patient, e.g. some mentally disabled, demented elderly or terminally ill patients, to respond adequately may necessitate proxy assessment by a relative or professional. The use of standard health questionnaires such as behavioural rating scales in such circumstances involves changing from a subjective individual assessment of QOL to a social valuation assuming, for example, that deviant behaviour reflects a poor QOL.

4.4.1. *Neuroblastoma trial*

As the median age of children in the trial was approximately 3 years at diagnosis, proxy assessment and reporting of QOL was generally necessary. The emphasis was on parental assessments, although a few (simple!) questions were included in the clinician's follow-up assessment forms. The agreement between parental and clinician assessments was, where evaluable, found to be rather poor. An attempt to separate parental perception of the child's QOL from a parental report of the child's perception of the child's QOL was made via both the form and the wording of questions.

4.5. *Where and When to Measure?*

The response to a QOL assessment may be changed by the physical context in which it takes place and the respondent's perception of the degree to which it is under the control of, say, the treating clinician. In some (but not all) societies it may be acceptable as well as convenient to administer QOL surveys by telephone, but it remains possible that this will lead to loss of sensitivity (Tandon *et al.*, 1989).

When to measure QOL is largely dictated by the objectives of the trial. General statistical principles apply: the need for base-line observations, a final assessment of QOL in patients who withdraw from the follow-up, avoidance of unnecessary assessments to the detriment of doctor or patient compliance and targeting assessments

(a) to distinguish early from late treatment effects,
(b) to reflect the pattern of treatment administration, e.g. cycles of chemotherapy, and
(c) to concentrate measurements when maximum treatment response is expected.

The assessment should be specific to some explicit time period (e.g. the last few days or last month). Some QOL changes may not be apparent if the follow-up period is too short since patients may take time to modify a life style adapted to a chronic disability.

5. ISSUES IN WEIGHTING AND AGGREGATION

In Section 2 we noted that the concept of QOL is generally decomposed into dimensions and items, but that even in clinical trials there is often a need for some recombination. This requires sets of weights, implicitly quantifying different states of health. In this section we raise some of the difficulties of obtaining and using weights, particularly when attempting to combine QOL and length of life in a single response measure.

5.1. *How May Weights be Assigned?*

Two main approaches have been used to assign weights. First, data analytic techniques such as factor analysis are intended to identify distinct 'constructs' that underlie the responses given by individuals to a battery of questions, and in so doing can also provide weights for questions within identified factors (Olschewski and Schumacher, 1990). Alternatively, there have been attempts to scale the states according to implicit or explicit personal valuations. Sometimes arbitrary values have been assigned to ordered sets of health states: for example, the Karnofsky index places the states at equal intervals (100, 90, 80, . . .), whereas Fanshel (1972) adopts an equally arbitrary power law $(1, 1-2^{-8}, 1-2^{-7}, . . .)$.

The effect of the differential weights used in some standard instruments can be severe. For example, in the quality of well-being (Kaplan and Anderson, 1988), some individual symptoms are weighted much more highly than items on the three subscales describing physical and social activity and mobility. The weight associated with confinement to a wheelchair is 0.06, while that for cough or wheezing is 0.257, which could lead to extraordinary conclusions. In most questionnaires, however, the range of weights is not so extreme and the scores are aggregated from several items.

Torrance (1987) describes some techniques for eliciting personal valuations for states of health. The most basic is *direct rating* on a scale from 0 to 100, while the most sophisticated requires a *simulated gamble* in which a utility, in the full decision theoretic sense, is assessed. Other techniques include the *time trade-off* method, in which the relative values of alternative health states A and B are implicitly derived by varying the periods of time spent in these states until the subject judges them equivalent. *Ratio scaling* similarly derives implicit valuations by requiring the subject to weigh the equivalence of differing numbers of people being in those states. Equivalence of weights obtained by different methods in a given study is far from guaranteed (Froberg and Kane, 1989).

5.2. *Critique of Weighting Methods*

Consider, for example, the four states: no angina, moderate angina, severe angina and death. Suppose being symptom free were given the value 1 and being dead given value 0. Where should the other two states lie on this scale? One possibility is direct judgment of the value with no operational definition: Torrance (1987) suggests 'utilities' of 0.7 and 0.5 respectively for moderate and severe angina. Ignoring for the moment the interpersonal variation that may be encountered, we may consider how such values might be used in a clinical trial on patients with severe angina, where 'partial relief' indicates a change from severe to moderate angina.

The stated values would imply that treatments A and B (see Table 2) were equivalent, since each would yield an increase in total utility of 0.2 per patient. This 'patient-equivalent' interpretation is used as the basis of an assessment procedure by Rosser and Kind (1978). For A and B to be considered equivalent requires the assumption that between-patient variation in response is unimportant.

If serial measurements are made on the patients, we might declare treatment C also equivalent to treatments A and B. Time trade-off has been extensively used as an assessment technique by Torrance and others and forms the basis of the use of QALYs as an outcome measure combining both quality and duration of life.

TABLE 2

Potentially equivalent (hypothetical) treatments in severe angina

---

Treatment A gave — partial relief for all patients
Treatment B gave — no relief to 60% of patients but total relief to 40% of patients
Treatment C gave — all patients total relief for 40% of the time, and no relief for the rest of the time
Treatment D gave — no pain relief, but extended life expectancy by 9.6 months
Treatment E gave — 50% of patients no relief *and* 1 year reduction in life expectancy, while remaining 50% experience full pain relief and life expectancy increased by 3.6 months
Treatment F gave — total relief and no change in life expectancy to 70%, but 30% die immediately

---

Suppose further that all the patients had a life expectancy of 2 years, giving each patient a mean prospect of 1 QALY (since they have severe angina with value 0.5). Then we would also consider treatments D, E and F equivalent to treatments A, B and C, since each on average provides an additional 0.4 QALYs per patient.

In practice, however, many people may not consider treatments A–F as equally desirable, even though a particular pair may be considered equivalent. This is *not* simply a consequence of people assigning different valuations, although this is undoubtedly also a problem. The primary issue is that there is no strong reason why values assigned on the basis of one type of aggregation, say over patients or over time, should be valid when used differently. Also, any use of clinical trial results to treat individual patients equates aggregation over patients with aggregation over personal uncertainty, in which true 'utilities' are relevant. Thus, for example, a future patient would need to feel that certain partial relief was equivalent to a 60% chance of no relief and a 40% chance of total relief.

The use of QALYs in resource allocation can lead to even more extreme equivalences, since different numbers of patients are balanced. Hence equally expensive interventions on the patients with severe angina described above would be considered equivalent (each giving 10 additional QALYs) if, for instance, they

(a) gave one person 20 years of extra life,
(b) gave 20 people 1 year of extra life,
(c) gave 2000 people 3.65 days of extra life each,
(d) gave 10 people total pain relief (with 2-year life expectancy) or
(e) gave 20 people total pain relief (life expectancy 2 years), but cut 1 year off life expectancy of 20 others without relief of symptoms.

It seems unlikely that any valuations could make these investments equally desirable. We conclude that it is generally inappropriate to aggregate QOL and length of life in reporting clinical trial results; some alternative approaches are discussed in Section 6.4. Moreover, since no valuations can be expected to be appropriate under the set of aggregations used in clinical trials, it is pointless to conduct elaborate experiments to derive sophisticated weighting schemes. Instead, the crucial need is for simple scales that appear reasonable (i.e. have face validity) and for sensitivity analysis of results to different valuations. Fortunately, the accepted robustness of linear scoring schemes should aid this approach (Dawes, 1979).

5.3.   *What Pragmatic Advice on Weighting Can be Offered?*

The scoring of different items on a k-point scale can to a limited extent be regarded as a technical statistical problem in that one can examine the circumstances under which elaborate systems of weighting give heightened sensitivity over straight integer scores. More elaborate weights might be based on

(a) discriminant analysis,
(b) normal scores, for responses which are approximately symmetrically distributed around a central value, or
(c) exponential scores for responses concentrated around one end, representing the absence of a particular symptom.

In most cases, sensible design demands that the end points of the scale are not extravagantly different from neighbouring values, and provided that this is ensured simple integer scoring is likely to be enough for many purposes.

Given the general lack of a pre-eminent weighting scheme, a prudent course includes checks on the robustness of conclusions to alternative, but still arbitrary, choices of weighting schemes. We suggest that reliance on an explicit weighting scheme to yield a global QOL index should be avoided where possible. An exception may arise when a limited number of states is involved and a sensitivity analysis can be explicitly presented (Gore, 1988; Glasziou *et al.*, 1990). Instead, we recommend the use of simple weighting schemes within dimensions, followed by the application of the analytical techniques outlined in the next section to investigate patterns of variation, together with sensitivity analyses. This approach can be extended to allow a presentation of the profile of weights representing the boundaries between preferences for the alternative treatments under comparison (Glasziou *et al.*, 1990). More detailed recommendations for simple approaches to the analysis of QOL data are made in the following section.

6. ANALYSIS

6.1. *General Approach*

Special aspects of the analysis of QOL data in a clinical trial setting stem partly from the essentially multidimensional character of the concept and partly from the substantial components of variation not only between patients within treatment groups but also across time within patients. There are often the further complications of patient withdrawal and a need to consider QOL data alongside information on survival. Initially, we ignore problems associated with censoring; see, however, Section 6.4.

For instrument development, formal psychometric multivariate methods may be helpful, but for interpretation of particular trials we favour a much simpler approach. Essential features of complex serial data, for example, may be highlighted by use of the straightforward methods described by Matthews *et al.* (1990).

For an instrument with a fairly small number of clearly identified dimensions, each with a number of items, we suggest that cross-sectional and summary profiles through time be dealt with as follows:

(a) a simple, largely uniform, scoring scheme is adopted for the possible answers to a particular item, e.g. scores 0, 1, 2, . . .;
(b) for a particular patient and dimension, the unweighted average over item answers is taken and expressed in standardized form, e.g. as a percentage of the maximum achievable dimension score;
(c) for an overall summary, the unweighted average of (b) over the separate dimensions is taken.

In the absence of specific *a priori* considerations, the separate dimension means, dimension-by-treatment means and the overall sum of the treatment means are the key values for interpretation. There are two broad strategies to the combination across dimensions. First, it may on general grounds be quite possible that the treatment effects under study are qualitatively different in the different dimensions; for example, a drug may improve anxiety but increase depression. If a summary score is required then the separate dimensions should be merged only subject to the absence of treatment-by-dimension qualitative interaction. If it is necessary to test the significance of departures from a global null hypothesis of treatment equivalence, the multivariate character of the analysis should be recognized, e.g. by a Bonferroni-type adjustment to the most significant p-value (Miller, 1981).

Secondly, if it is likely that the treatments affect all dimensions similarly, and especially in a small trial, it may be sensible to put more emphasis on an overall mean across dimensions, checking, however, for treatment-by-dimension interaction, despite the likely lack of power of such checks. Such comparisons need, however, to be supplemented by checks for anomalous items and, usually more importantly, for interaction of treatment and prognostic features. Although it may be useful to look for patient characteristics at entry that affect the preferred choice of treatment, care is necessary to avoid the familiar trap of 'data dredging' and subgroup analysis.

For some items, e.g. the presence of and severity of particular adverse reactions, it may be wise to deal separately with

(a) the proportion of patients for whom the effect is present at all and
(b) some measure of its average importance when present, perhaps even distinguishing frequency from severity.

As already noted, the multitude of issues open to study in an investigation of typical complexity raises familiar and major issues of multiplicity. The restriction of analyses to a few key issues nominated *a priori* avoids the difficulties of anarchic search for 'significant' effects, yet may lead to important matters being overlooked. There is no simple approach without snags; we do not see the formal techniques of multiple comparisons as particularly helpful or relevant. A structured application of Bonferroni-type adjustments in the analysis of a limited number of key issues nominated in advance, with other effects thrown up in further analysis being regarded as primarily hypothesis generating, at least offers simplicity of approach.

6.2. *Additional Role of Components of Variance*

To consider analysis in a little more detail, suppose for simplicity that there is a single medically homogeneous group of patients and that on each patient a single score is considered. For the moment it is immaterial whether the score is a dimension total or an overall total, and whether or not it is weighted. Then in principle we can determine components of variance, σ_p^2 between patients, σ_o^2 between observers and $\sigma_{t,po}^2$ between times within patients within observers over a short time span. More than three variance components are involved in more complex situations.

From these variance components, correlation coefficients can be determined if required. Of these probably the most relevant is $\sigma_{t,po}^2/(\sigma_p^2 + \sigma_{t,po}^2)$, the inferred correlation between repeat measurements by the same observer on the same patient over a short time span. This is indeed a useful dimensionless summary statistic for a particular study but is a poor measure for comparisons because it is strongly dependent on the true variability of the patient group involved. Thus for broad comparisons and planning it is important to record separately the 'error' component of variance, which is likely to be relatively stable between different studies, and the component of variance between patients, which is likely to be relatively study specific.

It is important to distinguish between average treatment effects over populations of patients and the treatment effect encountered by an individual patient, particularly whenever a treatment-by-patient interaction is appreciable in a comparative trial. For two treatments, one crude measure is the proportion π of individuals showing a long run difference favouring treatment A over treatment B. If there is no treatment-by-patient interaction, π is either 0 or 1. All patients are then assumed to show the same treatment effect or more generally any changes in treatment effect are assumed to be captured via suitable concomitant variables. The estimation of π is difficult and requires assumptions that are often largely uncheckable.

The simplest case is of a two-period two-treatment crossover design. Here the second period results can be corrected for any overall period effect and the proportion of patients for whom the second treatment observation exceeds the first treatment observation found. Alternatively, if $\hat{\tau}$ denotes an estimated treatment difference and $\hat{\sigma}_T^2$ the estimated variance between times within treatment groups, the normal theory estimate $\Phi(\hat{\tau}/\hat{\sigma}_T)$ can be used. This will tend to be nearer to $\frac{1}{2}$ than it should be because a part of σ_T^2 will be 'noise'. In particular there will be a contribution from variation between times within patients within treatments. This suggests that we should correct σ_T^2 by subtracting an estimate σ_e^2 of the noise component and therefore use the estimate $\Phi\{\hat{\tau}/\sqrt{(\hat{\sigma}_T^2 - \hat{\sigma}_e^2)}\}$. There is also the design consideration that data should be collected on some if not all patients to permit the estimation of σ_T^2.

From a study that is not of crossover form the estimation of π is even more speculative. We can estimate a component of variance between patients within treatments, preferably eliminating a variation between times within patients. There are now two extreme possibilities. One is that when two different treatments are used this source of variability is sampled independently, when essentially the same arguments as for crossover designs are available. The other is that the same random component is involved for both treatments when π is either 0 or 1. The real situation will be intermediate but π cannot be explored without some information about the correlation coefficient between the two associated random components or some equivalent quantity. The only simple way of estimating this would be via crossover data obtained under broadly comparable conditions.

To estimate the proportion π, the most satisfactory parametric route is now via a balanced analysis of variance which includes the following degrees of freedom, mean squares and expected mean squares:

treatments (T)	1		
patient (P)	—		
$T \times P$	d_{tp}	MS_{tp}	$\sigma_e^2 + r\sigma_{tp}^2$
error	d_e	MS_e	σ_e^2.

If τ is the overall population difference of treatment means and $\hat{\tau}$ is the sample mean difference, the proportion of individuals with a positive true difference is $\Phi(\xi)$, say, where ξ is estimated by $\hat{\tau}/\sqrt{\{(MS_{tp} - MS_e)/r\}}$.

If σ_e^2 is negligible compared with σ_{tp}^2, confidence limits for ξ can be obtained via the non-central t-distribution. More generally, the best way to obtain confidence limits for ξ is to reduce by sufficiency to the consideration of the three independent random variables $\hat{\tau}$, MS_{tp} and MS_e, to write down their likelihood, to reparameterize in terms of ξ, σ_{tp} and σ_e and thence to derive the profile likelihood for ξ. With a large amount of high quality data it would be possible in principle to obtain a nonparametric estimate. In practice the main difficulty is likely to be in separating the real treatment-by-individual interaction from noise.

6.3. *Complex Weighting and Sensitivity Analysis*

We have argued in favour of simple weighting schemes and methods of analysis. The arguments against the use of formal multivariate methods and associated complex weightings are essentially that an interpretation is made less direct and, most importantly, that the basis for determining weights, however well defined on formal statistical grounds, is in no way guaranteed to be meaningful either clinically or to individual patients. Further, 'standard' weights determined from a previous study may not be appropriate for the investigation under analysis, yet the use of non-standard weights destroys comparability between investigations. The simple procedures that we have suggested make it relatively easy to assess the effect of changes of emphasis, especially between dimensions.

If, nevertheless, 'complex' weights are used, sensitivity analysis is highly desirable. For this, one systematic approach is as follows. If weights $w_1, \ldots, w_m$ are initially used for m items or dimensions, choose reasonable practical perturbations $d_1, \ldots, d_m$. If m is reasonably small, the primary analysis can be repeated with all 2^m possible systems of weights $w_1(1 \pm d_1), \ldots, w_m(1 \pm d_m)$.

Otherwise, a fractional replicate should be examined to allow estimation of at least the main effects of the separate perturbations. A refinement of this would recognize that the system has the structure of a mixture experiment in that equal proportional changes in all weights would have no effect on the conclusions drawn.

6.4. *Combination with Other Types of Data*

For a full interpretation, it may be important to combine QOL data with other kinds of information including objective measures such as the results of exercise tests, as well as physiological and biochemical measurements. Here substantive research hypotheses (Wermuth and Lauritzen, 1990), often concerning conditional independences, may arise on subject-matter grounds. A particularly important case arises when QOL data have to be combined with survival data.

6.4.1. *Quality of life and survival*

Patients with incomplete follow-up raise particular problems for the analysis and reporting of serial QOL data. If QOL can be summarized as a short list of health states, then it may be possible to adopt the techniques of multistate survival analysis with death as an 'absorbing state'; transition intensities between states may, for example, be structured as a proportional hazard model including a covariate representing treatment (Kay, 1982; Andersen, 1988). However, this requires extensive data, which are rarely available, and quite strong assumptions about the nature of the dependences.

Other approaches have assumed that QOL has been placed on a single dimension between 0 and 1, and so a patient's progress can be plotted as a profile ('Carlen's vitagram') which reduces to 0 at death. The standard QALY approach takes the area under this profile as the summary response measure (Olschewski and Schumacher, 1990). Glasziou *et al.* (1990) emphasize that this assumes that the value of a health state is independent of time and previous states of health, and that QOL is additive over time, thus relating to the problems of aggregation raised in Section 5.

If only a limited number of states are considered, sensitivity analyses can be used to guide the choice of treatment for a particular patient with their own personal values (Simes, 1986; Hilden, 1987; Gore, 1988). Since an individual may be averse to risks in the immediate future, the assumption of additivity over time can be dropped by 'discounting' time, making a year in the distant future less valuable. Glasziou *et al.* (1990) give plots that show how the choice of treatment in breast cancer may depend on the valuations given to health states and to the discount rate.

A major problem arises when the follow-ups are censored—what QOL should be assigned to the patients' unknown futures? Certain logical bounds could be assumed (Gelber *et al.*, 1989), or it may be naïvely assumed that techniques of censored survival analysis could be adopted, simply by using quality-adjusted survival times as the response measure. Unfortunately, this may introduce substantial bias since the censoring mechanism is now informative. For example, in two groups with identical censoring mechanisms, the group with poorer QOL is accumulating QALYs slowly, and hence will tend to be censored earlier on the QALY scale, thus systematically underestimating the hazard in that group. This phenomenon is illustrated in detail by Glasziou *et al.* (1990).

If patients always progress monotonically through a limited sequence of states, then, even with censored data, estimates of the mean time spent in each state are obtainable and hence treatments may be compared. This approach has been illustrated (Goldhirsch *et al.*, 1989; Glasziou *et al.*, 1990) in a study of operable breast cancer, in which patients progress through stages of treatment toxicity (TOX), remission (TWIST—time without symptoms or toxicity) and relapse (REL) to death. For a particular treatment the survival curves representing the time to reach each stage, restricted to an upper time limit, can be plotted as shown in Fig. 1 for a limit of 7 years post-operative follow-up. The shaded areas then estimate the mean time in each state. Zero weight may, for example, by given to TOX and REL, and only TWIST used as a response measure (Gelber *et al.*, 1989), or sensitivity analyses provided for a range of weights, with confidence intervals being given for the division between preferred treatments (Goldhirsch *et al.*, 1989; Glasziou *et al.*, 1990).

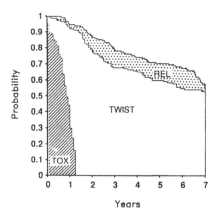

Fig. 1. Restricted survival by stage, from Glasziou *et al.* (1990) (reproduced by permission of John Wiley and Sons, Ltd)

This approach, although quite attractive, has clear limitations. More generally, it is preferable to avoid the explicit combination of QOL and length of life and instead to present them as multiple end points of a trial. These end points will have to be formally or informally combined for making treatment decisions, but responsibility for this is left to the clinician and the patient. This places a special onus on clarity of presentation of results.

In some simple situations it may be possible to provide an approximate description of the distribution of QOL *versus* time, conditional on survival time. Another possibility is to provide the survival curve, but juxtaposed with a description of the QOL of those patients still in the risk set. The analysis can then take the form of epoch-specific comparison of QOL. Thus treatments are compared with respect to survival and with respect to QOL of survivors: if one treatment is superior on both counts then reporting is straightforward. If, however, there is conflict then the relationship between QOL and survival should be investigated. The existence of conflict between measures of QOL and length of life exactly characterizes circumstances in which they should *not* be combined into a single measure.

We illustrate some problems associated with serial QOL assessment in the presence of censoring by death by reference to data on patients in the year following a heart transplant at Papworth Hospital. Here we consider data from the NHP (Hunt *et al.*, 1986) collected at 3, 6, 12, 18 and 24 months post transplant; previous analyses have illustrated pre- to post-transplant improvements (Caine *et al.*, 1990). Problems that arise include sporadic missing observations, censoring of follow-up at April 1990 and loss to follow-up due to either death or a persistent refusal to return questionnaires. Approximately 7% of the cohort were lost between each follow-up, apart from those censored by time. Our discussion is preliminary and informal.

Table 3 displays summary statistics for the sleep section of the NHP, details of which were given in Section 2.1. Such a table gives no idea of the progress of individual patients and does not indicate the extent of selective withdrawal of the more severe patients. However, it is apparent that a reduction in sleep scores would be attributable to an increasing proportion with zero sleep scores (i.e. no sleep problems) rather than a reduced severity among those who experience sleep problems. A fuller version of the display in Table 3 would give a histogram of the ordered scores at each follow-up. Profiles of individual patients chosen at random from the cohort could be superimposed on the population graph to illustrate the variation over time in individual patients.

The issue of selective withdrawal may be investigated by subdivision of responders at any point into those who do and those who do not withdraw before the next follow-up. Table 4 shows a breakdown of the total NHP score (obtained by adding the scores over the six main sections) according to status at next follow-up, to give a simple illustration of an approach to exploration of the data.

Without formal analysis, this suggests that those about to die or to be lost to follow-up tend to have worse current scores, whereas those who miss their next follow-up tend to have somewhat better current scores. This accords with the idea that current health status may be predictive of future clinical outcome, and that those experiencing fewer problems may not be so diligent in returning questionnaires.

TABLE 3
QOL following heart transplant: summary measures of NHP sections—proportion zero and mean and standard deviation of non-zero scores

Sleep score measurement time	3 months	6 months	12 months	18 months	24 months
Proportion zero	0.50	0.52	0.59	0.59	0.61
(total)	(151)	(165)	(135)	(128)	(109)
Mean (standard deviation) of	26.5	30.6	31.2	30.4	26.2
non-zeros	(19.8)	(21.0)	(20.2)	(20.0)	(20.8)

TABLE 4
Mean (and standard deviation) of total NHP score and numbers at risk in the heart transplant study, broken down by status at next follow-up

| | Results after the following times: | | | | |
	3 months	6 months	12 months	18 months	24 months
Total	52.2 (6.1)	52.0 (6.0)	38.4 (4.8)	38.6 (5.5)	36.3 (6.2)
	151	165	134	128	108
Status at next follow-up					
Recorded	53.7 (6.5)	40.2 (5.1)	31.8 (4.6)	37.6 (6.6)	36.3 (6.2)
	127	115	102	94	108
Missing	19.8 (10.8)	29.2 (12.8)	44.5 (18.6)	27.6 (12.0)	
	13	21	13	9	
Censored		122.9 (31.6)	57.6 (22.9)	33.4 (18.3)	
	0	17	9	14	
Lost	73.5 (34.5)	95.8 (40.2)	53.7 (26.3)	65.6 (21.1)	
	11	9	7	9	
Dead		127.7 (82.9)	145.7 (67.4)	52.0 (20.4)	
	0	3	3	2	

7. REPORTING AND INTERPRETATION OF CONCLUSIONS

Simple, accurate and lucid reporting of conclusions is the goal, to facilitate the combination of results across similar studies and the use of results for individual decision-making by clinicians and patients.

QOL assessments are likely to be reported in medical journals in which there are strong restrictions on the number of tables and figures. A reporting strategy is set out below: it concentrates on the content of tables and figures and focuses on results pertaining to QOL *dimensions*. In general, detailed reporting of items within dimensions is likely to be possible only exceptionally.

Reports of QOL measures in clinical trials should address

(a) summary measures (over individuals at each assessment time or over time for each individual) against the backdrop of individual profiles (see Table 3),

(b) important differences of, say, *d* units between treatments or over time, by displaying typical profiles which exhibit such differences (see below),

(c) censoring of QOL assessment by death or other cause, to indicate whether censoring is informative and whether informativeness is related to treatment or time (see Section 6.4 and Table 4),

(d) conflict between dimensions, or between survival and measures of quality (see Sections 6.1 and 6.4),

(e) components of variance (assuming assessments are missing at random) (see Section 6.2) and

(f) sensitivity analysis to quantify the dependence of the results obtained on the weighting scheme employed (see Section 6.3).

Implicit in requirement (b) is an important difficulty associated with an interpretation of systematic differences between patient groups, namely the subject-matter importance of a difference of, say, d units between treatments. What magnitude of difference in score can be judged substantively significant either from a clinical viewpoint or from the perspective of an individual patient? A systematic approach to elicitation of, say, clinicians' views on the scale of important changes can sometimes be adopted (Freedman and Spiegelhalter, 1983). Alternatively, gross changes can sometimes be inferred when using questionnaires with defined cut points to indicate severity; for example, a high score on the Goldberg health questionnaire (Goldberg and Williams, 1988) suggests that further psychiatric evaluation may be necessary. More often, however, we are concerned with more subtle changes, such as those of mood.

One approach is to examine typical score profiles which exhibit the difference in question. If a component of variance between times within patients is approximately known, it is plausible that a systematic difference of more than one or two corresponding standard deviations would be perceptible by individual patients. The latter (two standard deviations) is effectively the recovery of mean QOL to the patient's previous maximum: assessment times may be chosen to estimate this component of variance by incorporating a dual base-line (see Section 4.6), or replication at some time after randomization. Overall it can be argued that values of d small compared with the standard deviation between patients within times are unlikely to have much importance.

The use of QOL results from trials in individual decision-making requires further comment. Methods of eliciting patients' preferences in a clinical rather than research context have begun to emerge (McNeill et al., 1978). However, it is less clear how the aggregated values and preferences incorporated into QOL instruments play a role in this context, other than in generally sensitizing clinicians to the issue. Indeed it has been suggested that, for the patient, data regarding probabilities of various outcomes are of greater importance than the values that others have attached to outcomes (Drummond, 1989).

Clinical trial results may be used to estimate prognostic probabilities, but it is important to adopt the patient's own values into any decision in which competing objectives, such as QOL and length of life, are being traded off (Simes, 1986). A general analysis allows recommendations to be expressed conditional on any individual's values (Simes, 1986), these values to be inserted later in specific cases. Since we are concerned with decisions in the face of uncertainty, values placed on states of health should be interpretable as utilities; in other words, any mean QOL values used should be considered as personal expectations.

8. CONCLUSIONS

Both their subjectivity and their multiplicity increase difficulty and complexity of QOL assessments compared with the use of a single objective measure. It is thus essential that basic tenets and principles of design, analysis and presentation be adhered to; it cannot be overemphasized that they *do* apply here. In particular, we note the following.

(a) Simplicity should be the keynote wherever possible (whether with respect to design, analysis or presentation): hence our advocacy of simple weighting schemes, supplemented by sensitivity analyses.

(b) Distributions of responses to QOL assessments are at least as important as mean (etc.) values, in view of the subjectivity of the measures and the need to allow some personalization of scales, scoring systems and interpretation.

(c) Despite the desirability of allowing (some) personalization or special choices for particular applications, the use of standard (*evaluated*) scales is advocated as a core with more detailed customized extensions.

(d) A presentation of results of QOL assessments should, as well as being as simple as possible, facilitate their use in treatment decisions for individual patients, and interpretation of the clinical significance of effects.

(e) Sensitivity analyses should play an important role in the presentation and interpretation of the results of QOL assessments.

Despite practical and analytical difficulties, satisfactory assessment of QOL in clinical trials is worth pursuing, since the relevance of traditional approaches is limited and their objectivity to some extent illusory. To our original question 'Quality-of-life assessment: can we keep it simple?' our answer is a qualified 'yes'.

ACKNOWLEDGEMENTS

We acknowledge with thanks the experience and access to data afforded by collaboration with several groups performing clinical trials and studies, in particular the European Neuroblastoma Study Group and Dr Noreen Caine of Papworth Hospital.

REFERENCES

Aaronson, N. K. (1989) Quality of life assessment in clinical trials: methodologic issues. *Contr. Clin. Trials*, **10**, suppl., 195S–208S.

Aaronson, N. K. and Beckmann, J. (eds) (1987) The quality of life of cancer patients. *European Organisation for Research on Treatment of Cancer Monograph Series*, vol. 17. New York: Raven.

Aaronson, N. K., Bullinger, M. and Ahmedzai, S. (1988) A modular approach to quality of life assessment in cancer clinical trials. *Rec. Reslts Cancer Res.*, **111**, 231–249.

Andersen, P. K. (1988) Multi-state models in survival analysis: a study of nephropathy and mortality in diabetes. *Statist. Med.*, **7**, 661–670.

Anderson, J., Firschein, H. and Meenan, R. (1989) Sensitivity of a health status measure to short-term clinical changes in arthritis. *Arth. Rheum.*, **32**, 844–890.

Bergner, M., Bobbitt, R., Carter, W. and Gilson, B. (1981) The Sickness Impact Profile: development and final revision of a health status measure. *Med. Care*, **19**, 787–805.

Bland, J. M. and Altman, D. G. (1986) Statistical methods for assessing agreement between two methods of clinical measurement. *Lancet*, i, 307–310.

Bowling, A. (1991) *Measuring Health: a Review of Quality of Life Measurement Scales*. Milton Keynes: Open University Press.

Brown, J., Kazis, L., Spitz, P., Gertman, P., Fries, J. and Meenan, R. (1984) The dimensions of health outcomes: a cross-validated examination of health status measures. *Am. J. Publ. Hlth*, **74**, 159–161.

Caine, N., Sharples, L. D., English, T. A. H. and Wallwork, J. (1990) Prospective study comparing quality of life before and after heart transplantation. *Transplnt Proc.*, **22**, 73–75.

Campbell, D. and Fiske, D. (1959) Convergent and discriminant validation by the multitrait-multimethod matrix. *Psychol. Bull.*, **56**, 81–105.

Carr-Hill, R. (1989) Assumptions of the QALY procedure. *Socl Sci. Med.*, **29**, 469–477.

Chinn, S. (1990) The assessment of methods of measurement. *Statist. Med.*, **9**, 351–362.

Chinn, S. and Burney, P. G. (1987) On measuring repeatability of data from self administered questionnaires. *Int. J. Epidem.*, **16**, 121–127.

Cox, D. R. (1957) A note on grouping. *J. Am. Statist. Ass.*, **52**, 543–547.

Cronbach, L. J. (1951) Coefficient alpha and the internal structure of tests. *Psychometrika*, **16**, 297–334.

Croog, S., Levine, S., Testa, M., Brown, B., Bulpitt, C., Jenkins, C., Klerman, G. and William, G. (1986) The effects of antihypertensive therapy on the quality of life. *New Engl. J. Med.*, **314**, 1657–1663.

Dawes, R. M. (1979) The robust beauty of improper linear models. *Am. Psychol.*, **34**, 571–582.

Deyo, R. and Centor, R. (1986) Assessing the responsiveness of functional scales to clinical change: an analogy to diagnostic test performance. *J. Chron. Dis.*, **39**, 897–906.

Deyo, R. and Inui, T. (1984) Towards clinical applications of health status measures: sensitivity of scales to clinically important changes. *Hlth Serv. Res.*, **19**, 276–289.

Drummond, M. (1989) Output measurements for resource allocation decisions in health care. *Oxf. Rev. Econ. Poly*, **5**, 59–74.

Evans, A. E., D'Angio, G. J. and Randolph, J. (1971) A proposed staging for children with neuroblastoma. *Cancer*, **27**, 374–378.

Fallowfield, L. (1990) *The Quality of Life*. London: Souvenir.

Fanshel, S. (1972) A meaningful measure of health for epidemiology. *Int. J. Epidem.*, **1**, 319–337.

Fayers, P. M. and Jones, D. R. (1983) Measuring and analysing quality of life in cancer clinical trials: a review. *Statist. Med.*, **2**, 429–446.

Fitzpatrick, R., Bury, M., Frank, A. and Donnelly, T. (1987) Problems in the assessment of outcome in a back pain clinic. *Int. Disabil. Stud.*, **9**, 161–165.

Fitzpatrick, R., Newman, S., Lamb, R. and Shipley, M. (1989) A comparison of measures of health status in rheumatoid arthritis. *Br. J. Rheumatol.*, **28**, 201–206.

Fleiss, J. L. (1975) Measuring agreement between two judges on the presence or absence of a trait. *Biometrics*, **31**, 651–659.

Fletcher, A. E. (1988) Measurement of quality of life in clinical trials of therapy. *Rec. Reslts Cancer Res.*, **111**, 216–230.

——(1991) Pressure to treat and pressure to cost: a review of cost effectiveness analysis. *J. Hypertens.*, **9**, 193–198.

Freedman, L. S. and Spiegelhalter, D. J. (1983) The assessment of subjective opinion and its use in relation to stopping rules for clinical trials. *Statistician*, **32**, 153–160.

Fries, J., Spitz, P. and Young, D. (1982) The dimensions of health outcome: the Health Assessment Questionnaire: disability and pain scales. *J. Rheumatol.*, **9**, 789–793.

Froberg, D. and Kane, R. (1989) Methodology for measuring health-state preferences: II, Scaling methods. *J. Clin. Epidem.*, **42**, 459–471.

Gelber, R. D., Gelman, R. S. and Goldhirsch, A. (1989) A quality-of-life-oriented endpoint for comparing therapies. *Biometrics*, **45**, 781–795.

Glasziou, P. P., Simes, R. J. and Gelber, R. D. (1990) Quality adjusted survival analysis. *Statist. Med.*, **9**, 1259–1276.

Goldberg, D. P. and Williams, P. (1988) *A User's Guide to the General Health Questionnaire*. Windsor: NFER–Nelson.

Goldhirsch, A., Gelber, R. D., Simes, R. J., Glasziou, P. P. and Coates, A. (1989) Costs and benefits of adjuvant therapy in breast cancer: a quality-adjusted survival analysis. *J. Clin. Oncol.*, **7**, 36–44.

Gore, S. M. (1988) Integrated reporting of quality and length of life—a statistician's perspective. *Eur. Heart J.*, **9**, 228–234.

Guyatt, G., Sackett, D., Taylor, D. W., Chong, J., Roberts, R. and Pugsley, S. (1986) Determining optimal therapy—randomised trials in individual patients. *New Engl. J. Med.*, **314**, 889–892.

Hilden, J. (1987) Reporting clinical trials from the viewpoint of a patient's choice of treatment. *Statist. Med.*, **6**, 745–752.

Hunt, S. M., McEwen, J. and McKenna, S. P. (1986) *Measuring Health Status*. London: Croom Helm.

Hutchinson, T. A., Boyd, N. F. and Feinstein, A. R. (1979) Scientific problems in clinical scales as demonstrated in the Karnofsky index of performance status. *J. Chron. Dis.*, **32**, 661-666.

Kane, R. A. and Kane, R. L. (1981) *Assessing the Elderly: a Practical Guide to Measurement.* Lexington: Heath.

Kaplan, R. M. and Anderson, J. P. (1988) The Quality of Well-Being Scale: rationale for a single quality of life index. In *Quality of Life: Assessment and Application* (eds S. R. Walker and R. Rosser), pp. 51-77. Dordrecht: Kluwer.

Karnofsky, D. and Burchenal, J. (1949) The clinical evaluation of chemotherapeutic agents in cancer. In *Evaluation of Chemotherapeutic Agents* (ed. C. Macleod), pp. 191-205. New York: Columbia University Press.

Katz, S. (1987) The science of quality of life. *J. Chron. Dis.*, **40**, 459-465.

Kay, R. (1982) The analysis of transition times in multi-state stochastic processes using proportional hazards regression models. *Communs Statist. Theory Meth.*, **11**, 1743-1756.

Kellner, R. and Sheffield, B. F. (1973) A self rating scale of distress. *Psychol. Med.*, **3**, 88-100.

Kirshner, B. and Guyatt, G. (1985) A methodologic framework for assessing health indices. *J. Chron. Dis.*, **38**, 27-36.

van Knippenberg, F. C. E. and de Haes, J. C. J. M. (1988) Measuring quality of life of cancer patients: psychometric properties of instruments. *J. Clin. Epidem.*, **41**, 1043-1055.

Lissitz, R. W. and Green, S. B. (1975) Effect of the number of scale points on reliability: a Monte Carlo approach. *J. Appl. Psychol.*, **60**, 10-13.

Loomes, G. and McKenzie, L. (1989) The use of QALYs in health care decision making. *Socl Sci. Med.*, **28**, 299-308.

Mackenzie, R., Charlson, M., DiGioia, D. and Kelley, K. (1986) Can the Sickness Impact Profile measure change?: an example of scale assessment. *J. Chron. Dis.*, **39**, 429-438.

Maguire, P. and Selby, P. (1989) Assessing quality of life in cancer patients. *Br. J. Cancer*, **60**, 437-440.

Mason, J., Anderson, J. and Meenan, R. (1988) A model of health status for rheumatoid arthritis: a factor analysis of the Arthritis Impact Measurement Scales. *Arth. Rheum.*, **31**, 714-720.

Matthews, J. N. S., Altman, D. G., Campbell, M. J. and Royston, P. (1990) Analysis of serial measurements in medical research. *Br. Med. J.*, **300**, 230-235.

McCormack, E. M., Horne, D. J. de L. and Sheather, S. (1988) Clinical applications of visual analogue scales: a critical review. *Psychol. Med.*, **18**, 1007-1019.

McDowell, I. and Newell, C. (1987) *Measuring Health: a Guide to Rating Scales and Questionnaires.* Oxford: Oxford University Press.

McNeill, B., Weichselbaum, R. and Pauker, G. (1978) Fallacy of the five year survival in cancer. *New Engl. J. Med.*, **299**, 1397-1401.

Meenan, R., Anderson, J., Kazis, L., Egger, M., Altz-Smith, M., Samuelson, C., Willkens, R., Solsky, M., Hayes, S., Blocka, K., Weinstein, A., Guttadauria, M., Kaplan, S. and Klippel, J. (1984) Outcome assessment in clinical trials; evidence for the sensitivity of a health status measure. *Arth. Rheum.*, **27**, 1344-1352.

Meenan, R., Gertman, P., Mason, J. and Dunaif, R. (1982) The Arthritis Impact Measurement Scales: further investigation of a health status instrument. *Arth. Rheum.*, **25**, 1048-1053.

Miller, R. G. (1981) *Simultaneous Statistical Inference*, 2nd edn. New York: Springer.

Nunnally, J. (1978) *Psychometric Theory*, 2nd edn. New York: McGraw Hill.

Olschewski, M. and Schumacher, M. (1990) Statistical analysis of quality of life data. *Statist. Med.*, **9**, 749-763.

Parr, G., Darekar, B., Fletcher, A. and Bulpitt, C. (1988) Joint pain and quality of life: results of a randomised trial. *Br. J. Clin. Pharm.*, **27**, 235-242.

Pinkerton, R., Pritchard, J., de Kraker, J., Jones, D. R., Germond, S. and Love, S. (1988) ENSG1—randomised study of high-dose melphalan in neuroblastoma. In *Autologous Bone Marrow Transplantation* (eds K. A. Dicke, G. Spitzer and S. Jagannath). Houston: University of Texas Press.

Remington, M., Tyrer, P. J., Newson-Smith, J. and Cicchetti, D. V. (1979) Comparative reliability of categorical and analogue rating scales in the assessment of psychiatric symptomatology. *Psychol. Med.*, **9**, 765-770.

Robin, A., Curry, S. H. and Whelpton, R. (1974) Clinical and biochemical comparisons of clorazepate and diazepam. *Psychol. Med.*, **4**, 388-392.

Rosser, R. and Kind, P. (1978) A scale of valuations of states of illness: is there a social consensus? *Int. J. Epidem.*, **7**, 347-358.

Secretaries of State for Health (1989) Working for patients: medical audit. *Working Paper 6*. London: Her Majesty's Stationery Office.

Simes, R. J. (1986) Application of statistical decision theory to treatment choices: implications for the design and analysis of clinical trials. *Statist. Med.*, **5**, 411–420.

Smith, A. (1987) Qualms about QALYs. *Lancet*, i, 1134–1136.

Solomon, P. J. (1989) Analysis of patient self-assessment visual analogue scales. *Appl. Stoch. Models Data Anal.*, **5**, 153–164.

Spilker, B. (1990) *Quality of Life Assessments in Clinical Trials*. New York: Raven.

Spitzer, W. O. (1987) State of science 1986: quality of life and functional status as target variables for research. *J. Chron. Dis.*, **40**, 465–471.

Steinbrocker, O., Traeger, C. and Battman, R. (1949) Therapeutic criteria in rheumatoid arthritis. *J. Am. Med. Ass.*, **140**, 659–662.

Sudman, S. and Bradburn, N. M. (1982) *Asking Questions: a Practical Guide to Questionnaire Design*. San Francisco: Jossey-Bass.

Tandon, P. K., Stander, H. and Schwarz, R. P. (1989) Analysis of quality of life data from a randomised, placebo-controlled heart-failure trial. *J. Clin. Epidem.*, **42**, 955–962.

Torrance, G. W. (1987) Utility approach to measuring health-related quality-of-life. *J. Chron. Dis.*, **40**, 593–600.

Tugwell, P., Bombardier, C., Buchanan, W., Goldsmith, C., Grace, E. and Hanna, B. (1987) The MACTAR patient preference disability questionnaire: an individualised functional priority approach for assessing improvement in physical disability in clinical trials in rheumatoid arthritis. *J. Rheumatol.*, **14**, 446–451.

Walker, S. R. and Rosser, R. M. (1988) *Quality of Life: Assessment and Application*. Dordrecht: Kluwer.

Wermuth, N. and Lauritzen, S. L. (1990) On substantive research hypotheses, conditional independence graphs and graphical chain models. *J. R. Statist. Soc.* B, **52**, 21–50.

Summary of discussion

When the paper was read to the Royal Statistical Society a vote of thanks was proposed by D. Ashby and seconded by A. Hopkins and the discussion continued by J. L. Hutton, C. J. Bulpitt, H. M. Goodare, R. A. Carr-Hill, T. Sheldon, J. Hibbert, K. Abrams, P. Fayers, and R. G. Newcombe. Written contributions were sent by M. Farquhar, R. D. Gelber and A. Goldhirsch, D. J. Girling, S. Kreiner, M. Olschewski, G. Schulgen and M. Schumacher, G. Parr, S. Senn, P. J. Solomon, R. J. Stephens, S. P. Stenning, M. K. B. Parmar and D. Machin, G. W. Torrance, N. Wermuth and by A. Williams. A collective reply by the authors concluded the paper.

The emphasis in the paper on the multidimensional character of health-related quality of life as contrasted with any need to force a one-dimensional scale on the matter is welcomed by most contributors, especially in a moving contribution from a patient (Goodare); the multidimensional character seems essential to clarify options for individual patient choice in difficult contexts. A counterview is put by some (Torrance, Williams) leading to a defence of such indices as QALYs, these comments coming from the viewpoint of those more concerned with resource allocation than with detailed analyses of clinical trials and with individual choice. A particular aspect of this concerns the circumstances under which survival time should be combined with other measures into a single index such as TWIST, time without symptoms or toxicity (Gelber and Goldhirsch) rather than studied by their joint distribution.

Among other topics outlined in the discussion are the role of multistate models (Abrams, Schulgen et al.) and the use of Rasch models as a base for interpretation (Kreiner).

Acknowledgements

The editors are grateful to copyright holders for permission to reproduce the papers in this volume.

American Statistical Association

[D5] Some systematic supersaturated designs. *Technometrics* **4** (1962), 489–495.

The Biometrika Trustees

[D1] Some systematic experimental designs. *Biometrika,* 1951, **38**, 312–323, reproduced by permission of the Biometrika Trustees.
[D2] The design of an experiment in which certain treatment arrangements are inadmissible. *Biometrika,* 1954, **41,** 287–295, reproduced by permission of the Biometrika Trustees.
[D3] The use of a concomitant variable in selecting an experimental design. *Biometrika,* 1957, **44**, 150–158, reproduced by permission of the Biometrika Trustees.
[D7] A note on design when response has an exponential family distribution. *Biometrika,* 1988, **75**, 161–164, reproduced by permission of the Biometrika Trustees.
[D10] On sampling and the estimation of rare errors. *Biometrika,* 1979, **66**, 125–132, reproduced by permission of the Biometrika Trustees.
[SM5] Two further applications of a model for binary regression. *Biometrika,* 1958, **45**, 562–565, reproduced by permission of the Biometrika Trustees.
[SM12] Response models for mixed binary and quantitative variables. *Biometrika,* 1992, **79**, 441–461, reproduced by permission of the Biometrika Trustees.
[SM20] Testing multivariate normality. *Biometrika,* 1978, **65**, 263–272, reproduced by permission of the Biometrika Trustees.
[SM22] Analysis of variability with large numbers of small samples. *Biometrika,* 1986, **73**, 543–554, reproduced by permission of the Biometrika Trustees.
[SM23] Nonlinear component of variance models. *Biometrika,* 1992, **79**, 1–11, reproduced by permission of the Biometrika Trustees.
[SM25] Some quick sign tests for trend in location and dispersion. *Biometrika,* 1955, **42**, 80–95, reproduced by permission of the Biometrika Trustees.

Blackwell Publishing Ltd

[D4] The use of control observations as an alternative to incomplete block designs. *J. R. Statist. Soc.* B **24** (1962), 464–471.

[D6] Present position and potential developments: some personal views. Design of experiments and regression. *J. R. Statist. Soc.* A **147** (1984), 306–315.

[SM1] Some statistical methods connected with series of events. *J. R. Statist. Soc.* B **17** (1955), 129–164.

[SM3] On the estimation of the intensity function of a stationary point process. *J. R. Statist. Soc.* B **27** (1965), 332–337.

[SM4] The regression analysis of binary sequences. *J. R. Statist. Soc.* B **20** (1958), 215–242.

[SM6] The analysis of multivariate binary data. *J. R. Statist. Soc.* C **21** (1972), 113–120.

[SM7] The analysis of exponentially distributed life-times with two types of failure. *J. R. Statist. Soc.* B **21** (1959), 411–421.

[SM8] Regression models and life-tables. *J. R. Statist. Soc.* B **34** (1972), 187–220.

[SM9] A note on multiple time scales in life testing. *J. R. Statist. Soc.* C **28** (1979), 73–75.

[SM13] An analysis of transformations. *J. R. Statist. Soc.* B **26** (1964), 211–252.

[SM17] Notes on some aspects of regression analysis. *J. R. Statist. Soc.* A **131** (1968), 265–279.

[SM18] The choice of variables in observational studies. *J. R. Statist. Soc.* C **23** (1974), 51–59.

[SM19] A general definition of residuals. *J. R. Statist. Soc.* B **30** (1968), 248–275.

[SM21] Some remarks on the role in statistics of graphical models. *J. R. Statist. Soc.* C **27** (1978), 4–9.

[A3] On a discriminatory problem connected with the works of Plato. *J. R. Statist. Soc.* B **21** (1959), 195–200.

[A6] Quality-of-life assessment: can we keep it simple? *J. R. Statist. Soc.* A **155** (1992), 353–393.

The Board of the Foundation of the Scandinavian Journal of Statistics

[SM24] On partitioning means into groups. *Scand. J. Statist.* **9** (1982), 147–152.

Sir David Cox

[D8] Randomization and concomitant variables in the design of experiments. In *Statistics and Probability: Essays in Honor of C. R. Rao*, edited by G. Kallianpur, P. R. Krishnaiah and J. K. Ghosh, pp. 197–202. Amsterdam: North-Holland, 1982.

[D9] Some sampling problems in technology. In *New Developments in Survey Sampling*, edited by N. L. Johnson and H. Smith, Jr., pp. 506–527. New York: Wiley, 1969.

[SM14] The role of statistical methods in science and technology. Inaugural Lecture, Birkbeck College, London, 1961.

Elsevier

[SM11] On the calculation of derived variables in the analysis of multivariate responses. *J. Multivar. Anal.* **36** (1992), 162–170. Reprinted with permission from Elsevier.

Institute of Mathematical Statistics

[SM10] Linear dependencies represented by chain graphs. *Statistical Science* **8** (1993), 204–283. Reprinted with the permission of the Institute of Mathematical Statistics <www.imstat.org>.

International Statistical Institute

[SM16] Interaction. *Inter. Statist. Rev.* **52** (1984), 1–31.

Palgrave Macmillan

[A5] Quality and reliability: some recent developments and a historical perspective. *J. Operational Res. Soc.* **41** (1990), 95–101. Reproduced with permission of Palgrave Macmillan.

Regents of the University of California

[SM2] Multivariate point processes. In *Proceedings of the Sixth Berkeley Symposium on Mathematical Statistics and Probability Theory* **3**, edited by L. M. Le Cam, J. Neyman and E. L. Scott, pp. 401–448. Berkeley: University of California Press, 1972.

The Royal Society

[A4] The distribution of the incubation period for the acquired immunodeficiency syndrome (AIDS). *Proc. R. Soc. Lond.* B **233** (1988), 367–377.

The Textile Institute

[A1] The relation between strength and diameter of wool fibres. *J. Textile Inst.* **41** (1950), T481–T491.
[A2] Some statistical aspects of mixing and blending. *J. Textile Inst.* **45** (1954), T113–T122.

John Wiley & Sons, Inc.

[SM15] Combination of data. In *Encyclopedia of Statistical Sciences* **2**, edited by S. Kotz and N. L. Johnson, pp. 45–53. © 1982 John Wiley & Sons. Reprinted with permission of John Wiley & Sons, Inc.

Printed in the United States
By Bookmasters